26th Annual Conference of the International Society of Exposure Science (ISES 2016)

Interdisciplinary Approaches for Health and the Environment

Book of Abstracts

Utrecht, The Netherlands
9 - 13 October 2016

ISBN: 978-1-5108-4724-8

Printed from e-media with permission by:

Curran Associates, Inc.
57 Morehouse Lane
Red Hook, NY 12571

CURRAN ASSOCIATES INC.
proceedings
.com

Some format issues inherent in the e-media version may also appear in this print version.

Printed by Curran Associates, Inc. (2017)

For permission requests, please contact International Society of Exposure Science (ISES) at the address below.

International Society of Exposure Science (ISES)
c/o Infinity Conference Group
1035 Sterling Road, Suite 202
Herndon VA 20170
USA

Phone: (703) 925-9620
Fax: (703) 925-9453

info@intlexposurescience.org

Additional copies of this publication are available from:

Curran Associates, Inc.
57 Morehouse Lane
Red Hook, NY 12571 USA
Phone: 845-758-0400
Fax: 845-758-2633
Email: curran@proceedings.com
Web: www.proceedings.com

UTRECHT
THE NETHERLANDS
OCTOBER 9-13

ISES 2016

INTERDISCIPLINARY APPROACHES for HEALTH and the ENVIRONMENT

October 9-13, 2016

Abstract book

Content

Presenting author Index 8

Sunday, October 9, 2016 15

Plenary Address 1: Water as integrator of uses, stakes and exposures 15

Monday, October 10, 2016 16

Plenary Address 2: Respiratory health effects and livestock farming related to microbial and dust exposure 16

Mo-SY-A1: Harmonization, access, transparency: improving environmental epidemiology for public health decision-making - I 18

Mo-SY-B1: Targeted activities for improving workplace exposure assessments 23

Mo-SY-C1: Intermittent Exposure in Risk Assessment 30

Mo-SY-D1: From external to internal exposure: the necessity of toxicokinetic information 34

Mo-SY-E1: Exposure to atmospherically dispersed hazards: assessment, public information and perspectives – I 39

Mo-SY-F1: Extending participatory sensing to personal exposure and policy support - I 47

Mo-SY-G1: Wastewater-based epidemiology (WBE) - from measuring illicit drug use towards understanding population health status - I 53

Mo-SY-H1: Advances in consumer exposure assessment - I 58

Mo-PL-I1: Transportation Related Air Pollution - I 63

Mo-SY-A2: Harmonization, access, transparency: improving environmental epidemiology for public health decision-making - II 70

Mo-SY-B2: Firefighters and Chemical Exposures: Protection Under Fire 73

Mo-SY-C2: Quantitative in vitro to in vivo extrapolation (QIVIVE): Advances in tools to quantify exposure-response relationships for risk assessment – I 77

Mo-SY-D2: 15 years of Human Biomonitoring in Flanders: surveillance feeding policy and research - I 82

Mo-SY-E2: Exposure to atmospherically dispersed hazards: assessment, public information and perspectives – II 86

Mo-SY-F2: Extending participatory sensing to personal exposure and policy support – II 91

Mo-SY-G2: Wastewater-based epidemiology (WBE) - from measuring illicit drug use towards understanding population health status – II 95

Mo-SY-H2: Advances in consumer exposure assessment - II 98

Mo-PL-I2: Transportation-Related Air Pollution – II 101

Mo-SY-A3: Multiple route exposure to multiple chemicals, the cocktail effect 105

Mo-SY-B3: From occupational to environmental biomonitoring: lessons to be learned. 109

Mo-SY-C3: Quantitative in vitro to in vivo extrapolation (QIVIVE): Advances in tools to quantify exposure-response relationships for risk assessment - II 114

Mo-SY-D3: 15 years of Human Biomonitoring in Flanders: surveillance feeding policy and research – II 119

Mo-SY-E3: Assessing exposure to SVOCs in dust 124

Mo-SY-F3: Thresholds of Toxicological Concern: an exposure-driven approach to risk assessment 132

Mo-SY-G3: The role of analytical chemistry within exposure science. 138

Mo-SY-H3: Aggregate exposure assessment of contact allergens in consumer products 145

Mo-PL-I3: Risk Assessment 151

Mo-SY-A4: The Children's Health Exposure Analysis Resource 161

Mo-SY-B4: New Biomarkers for Human Biological Monitoring in Occupational Health 167

Mo-SY-C4: Quantitative in vitro to in vivo extrapolation (QIVIVE): Advances in tools to quantify exposure-response relationships for risk assessment - III 172

Mo-SY-D4: Evidence-Based Research on Interventions to Reduce Personal Exposures to Environmental Pollutants 178

Mo-SY-E4: Toward an Understanding of Indoor exposures 184

Mo-SY-F4: Exposure-Based Toxicity Testing 189

Mo-SY-G4: Advanced mass spectrometric techniques for the analysis of environmental organic contaminants 194

Mo-SY-H4: Analysis of Patterns of Co-Exposure: Methodologies and Applications 201

Mo-PL-I4: VOCs and SVOCs 206

Poster sessions Monday October 10, 2016 210

Tuesday, October 11, 2016 ..... N/A

Plenary Address 3: The exposome: moving from concept to reality ..... N/A

Tu-SY-A1: The Exposome: From concept to practice - I 278

Tu-SY-B1: Uncertainty in scientific assessments: Recent efforts by governmental bodies to develop guidance for assessors 282

Tu-PL-C1: Aggregate and Cumulative Exposure Evaluations 287

Tu-PL-D1: Land Use Regression Modeling – I 294

Tu-SY-E1: Real-time measurements and integrated models to estimate traffic exposures in complex urban environments. 299

Tu-SY-F1: OECD Task Force on Exposure Assessment - Better exposure science for better lives – I 304

Tu-SY-G1: Environmental Justice: Developing the Scientific Foundation Supporting Cumulative Exposures/Risks/Impacts and Disparate Impacts Research - I 309

Tu-PL-H1: Analytical Methods – I 315

Tu-PL-I1: Quantitative Methods 321

Tu-SY-A2: The Exposome: From concept to practice – II 329

Tu-SY-B2: Uncertainty in scientific assessments: Recent efforts by governmental bodies to develop guidance for assessors 331

Tu-SY-C2: Use of Agent Based Models in Exposure Assessment 335

Tu-PL-D2: Land Use Regression Modeling – II 338

Tu-SY-E2: Exposure science meets social science: Tools for the effective communication of the health risks associated with air pollution exposure and implications for policy development 341

Tu-SY-F2: OECD Task Force on Exposure Assessment - Better exposure science for better lives - II 344

Tu-SY-G2: Environmental Justice: Developing the Scientific Foundation Supporting Cumulative Exposures/Risks/Impacts and Disparate Impacts Research – II 347

Tu-PL-H2: Analytical Methods - II 350

Tu-PL-I2: Close Contact: Contaminants in Clothing 353

Tu-SY-A3: The Exposome: From concept to practice – III 359

Tu-PL-B3: Urinary Biomarkers 364

Tu-SY-C3: Health effects of air pollutant 372

Tu-SY-D3: How can knowledge of toxicokinetics, mode of action and biomonitoring help you in human exposure risk assessment of chemicals? 378

Tu-SY-E3: The Effects of Climate Change on Human Exposures to Air Pollution 383

Tu-SY-F3: Current opportunities and challenges in exposure surveillance to implement prevention strategies at the national and European scale 389

Tu-SY-G3: Advancing human exposure metrics in Life Cycle Assessment (LCA) and Chemical Alternatives Assessment (CAA) – I 394

Tu-PL-H3: Pesticides 400

Tu-PL-I3: Understanding Exposure Measurement Error 406

Tu-SY-A4: The Exposome: From concept to practice – IV 412

Tu-PL-B4: Occupational Exposures 415

Tu-SY-C4: Advanced methods for characterizing air pollution exposures at community scale 420

Tu-SY-D4: Human Biological Monitoring Following Chemical Incidents 425

Tu-SY-E4: Air pollution exposure assessment getting personal: a European perspective 432

Tu-SY-F4: Advancing Exposure Science to Address Complex Environmental Issues 441

Tu-SY-G4: Advancing human exposure metrics in Life Cycle Assessment (LCA) and Chemical Alternatives Assessment (CAA) - II 447

Tu-PL-H4: Kinetics 453

Tu-PL-I4: Neurotoxicants 458

Poster sessions Tuesday October 11, 2016 464

Wednesday, October 12, 2016 533

We-SY-A1: Detection of new and emerging risks of chemicals (NERCs); the need for interdisciplinary cooperation 533

We-SY-B1: The Worker Health and Efficiency (WE) Program: Understanding and mitigating the risks of Chronic Kidney Disease in El Salvadorian Sugarcane Cutters. 538

We-SY-C1: What are the requirements for nanomaterial exposure models? – I 543

We-SY-D1: Biomonitoring: The Genie is out of the Bottle: Challenges in Data Quality and Interpretation 550

We-SY-E1: Exposure to SVOCs in the Indoor Environment - Products, Emissions, Exposure, Pharmacokinetics and Biomarkers – I 555

We-SY-F1: Exposure science informing policy decision-making - I 560

We-SY-G1: Pesticide Exposure: Developing Monitoring, Methods and Modeling in Human Health Risk Assessments (Consumer and Worker Risk) - I 565

We-SY-H1: Tool and methods for an exposure driven safe by design approach for nanomaterials - I 571

We-PL-I1: Waterborne Contaminants 577

We-SY-A2: New Data Streams for 21st Century Exposure Science 583

We-PL-B2: Exposure Factors 590

We-SY-C2: What are the requirements for nanomaterial exposure models? – II 594

We-SY-D2: Biomonitoring: Uses in Policy and Regulations and Enhancements as a Result of Collaborative Efforts 600

We-SY-E2: Exposure to SVOCs in the Indoor Environment - Products, Emissions, Exposure, Pharmacokinetics and Biomarkers – II 607

We-SY-F2: Exposure science informing policy decision-making – II 613

We-SY-G2: Pesticide Exposure: Developing Monitoring, Methods and Modeling in Human Health Risk Assessments (Consumer and Worker Risk) – II 620

We-SY-H2: Tool and methods for an exposure driven safe by design approach for nanomaterials - II 625

We-PL-I2: Indoor Environment 631

We-SY-A3: New Frontiers in Toxicology Create New Challenges for Risk Assessment: What must Exposure Scientists do to Meet the Challenge? 637

We-SY-B3: Occupational Exposure Models - Development and/or Evaluation of REACH and other European and US models and tools (including tool for nanomaterials) - I 641

We-SY-C3: Wristband Samplers Advancing Chemical Exposure Science – I 648

We-SY-D3: UBA HBM Colloquium I - Human Biomonitoring in International Population Studies Improving our Knowledge of Environmental Public Health 654

We-SY-E3: Exposure to SVOCs in the Indoor Environment - Products, Emissions, Exposure, Pharmacokinetics and Biomarkers – III 661

We-SY-F3: Measuring marijuana exposure in a changing legal landscape 668

We-SY-G3: Exposure Science and 21st century oil and gas development – I 673

We-PL-H3: Spatio-Temporal Measures – I 679

We-PL-I3: Sensor Technology 685

We-SY-A4: Exposure Sciences with Stakeholders in Contested Societal Debates About the Risk of Toxic Substances 692

We-SY-B4: Occupational Exposure Models - Development and/or Evaluation of REACH and other European and US models and tools (including tool for nanomaterials) - II 696

We-SY-C4: Wristband Samplers Advancing Chemical Exposure Science – II 701

We-SY-D4: UBA HBM Colloquium II - Human Biomonitoring in Europe Harmonising Instruments and Data for Science and Policy-making 707

We-SY-E4: Exposure to SVOCs in the Indoor Environment - Products, Emissions, Exposure, Pharmacokinetics and Biomarkers – IV 712

We-SY-F4: E-Cigarettes, Exposures, and (Health) Effects 717

We-SY-G4: Exposure Science and 21st century oil and gas development – II 721

We-PL-H4: Spatio-Temporal Measures – II 726

We-PL-I4: Continuous/Real Time Measures 732

Poster sessions Wednesday, October 12, 2016 739

Thursday, October 13, 2016 807

Plenary Address 4: Potential of metabolomics in chemical risk analysis 807

Th-SY-A1: The exposome: a transdisciplinary paradigm for improved environment and health associations - I 808

Th-SY-B1: Tooth-matrix biomarkers to reconstruct the early life exposome 813

Th-SY-C1: Development of personal sampling devices and chemical screening methods for large-scale epidemiology and human biomonitoring studies - I 819

Th-PL-D1: Exposure Modeling 824

Th-SY-E1: Methodologies in finding new and/or emerging risks of chemicals (NERCs) - I 830

Th-SY-A2: The exposome: a transdisciplinary paradigm for improved environment and health associations - II 836

Th-SY-B2: Aspects to consider for Fungi and Mycotoxins occupational exposure and risk assessment 839

Th-SY-C2: Development of personal sampling devices and chemical screening methods for large-scale epidemiology and human biomonitoring studies – II 842

Th-SY-D2: Environmental Exposure Monitoring in Birth & Early Life Cohort Studies 846

Th-SY-E2: Methodologies in finding new and/or emerging risks of chemicals (NERCs) - II 850

Presenting author Index

Abdallah, Mohamed	Tu-Po-32
Adetona, Anna	Tu-PL-B4.1
Adetona, Olorunfemi	Mo-Po-41
Adgate, John	We-SY-G3.4, We-SY-G4.2
Aggarwal, Manoj	Mo-SY-F3.5, Tu-SY-D3.1, Tu-SY-D3.3
Aggerwal, Manoj	Mo-Po-60
Aherrera, Angela	Tu-LBA-16
Ahrens, Andreas	We-SY-F1.2
Alam, Mohammed	Mo-SY-G4.4
Albergamo, Vittorio	Tu-PL-H1.4
Amouei Torkmahalleh, Mehdi	Mo-Po-14
Ananyeva, Oksana	Th-PL-D1.5
Andersen, Mikael Skou	Tu-SY-C3.3
Anderson, Kim A.	We-SY-C3.1
Andra, Syam	Th-SY-B1.3
Appenzeller, Brice M.R.	Tu-PL-H3.4, We-Po-03
Armitage, James	Mo-SY-C2.2
Arnone, Mario	Tu-Po-31
Arnot, Jon	Mo-SY-C3.2, Tu-SY-F4.3, We-SY-A2.1, We-SY-E4.3
Arora, Manish	Th-SY-B1.1
Arunachalam, Saravanan	Tu-SY-C4.3, Tu-SY-C4.4
Aubin, Daniel	Mo-LBA-04
Austin, Christine	Th-SY-B1.2, Mo-Po-16
Bader, Michael	Mo-PL-I4.4, Tu-SY-D4.4
Baduel, Christine	Tu-PL-H1.3
Baier-Anderson, Cal	Tu-SY-G3.5
Bakker, Joost	Th-SY-E1.4
Bakker, Martine	Mo-SY-H3.1
Ban, Hyunkyung	Mo-Po-49
Barouki, Robert	Th-SY-A2.1
Barratt, Benjamin	Tu-SY-E4.5, Tu-SY-E2.3
Bartha, Bernadett	We-PL-I1.5
Bartonova, Alena	Tu-SY-E4.6
Bartzis, John	Th-SY-A1.2
Barzyk, Timothy	Tu-SY-G2.1, Tu-SY-G2.2
Baxter, Lisa	Tu-SY-E3.1, We-PL-B2.5
Béchaux, Camille	Tu-SY-F3.4
Beckmann, Gwendolyn	Tu-Po-24
Beelen, Rob	We-SY-A4.2
Been, Frederic	Mo-SY-G1.3
Beevers, Sean	Tu-SY-C4.1
Bekö, Gabriel	Tu-PL-I2.3
Bergmann, Alan	We-SY-C4.5
Besseling, Ellen	We-SY-C1.5
Biesterbos, Jacqueline	Mo-SY-B4.4
Bijlsma, Lubertus	Mo-SY-G3.2
Blount, Benjamin	We-SY-F3.2
Boerleider, Romilda	We-Po-29
Boije af Gennäs, Urban	We-SY-F2.5
Bolte, John	We-PL-I3.4, We-Po-23
Boogaard, Peter J.	Mo-SY-B3.1, Mo-SY-B4.5
Bos, Peter	Mo-SY-D1.1
Bovee, Toine	Th-SY-E2.3
Brandsma, Sicco	Mo-Po-53
Breen, Michael	Tu-SY-C4.2
Breysse, Patrick	We-SY-F4.1
Brochot, Céline	Tu-PL-H4.2, Mo-LBA-09
Brokamp, Cole	Tu-PL-D1.1
Brook, Jeffrey	Th-SY-D2.2
Brown, Richard	Th-SY-E2.2
Brown, Kenneth	We-PL-I4.1
Bruinen de Bruin, Yuri	Th-SY-E1.5
Buckley, Timothy	Mo-PL-I1.4
Bukalasa, Joseph	Tu-PL-C1.5
Burgess, Jeff	Mo-SY-B2.1
Bury, Daniel	We-Po-02
Butler, David	We-PL-I2.5
Calafat, Antonia	Th-SY-C1.1, We-SY-D2.1
Calamandrei, Gemma	Th-SY-A2.2
Calkins, Miriam	Tu-PL-B4.2
Campo, Laura	Mo-SY-B4.2
Casey, Joanna	We-SY-G3.3
Castiglioni, Sara	Mo-SY-G1.1, Mo-SY-G1.4
Chadeau-Hyam, Marc	Tu-SY-A2.1
Chang, Howard	Tu-SY-E3.4
Chang, Irene	Th-SY-D2.3
Chapizanis, Dimitris	Th-SY-A1.4

Chartier, Ryan	Tu-Po-51
Chata, Caroline	Tu-PL-H3.1
Chen, Rui	We-LBA-19
Cherrie, John	We-SY-B3.1
Ching-Chun, Lin	Tu-Po-38
Choi, Judy	Mo-PL-I3.2
Chovolou, Yvonni	Tu-PL-B3.5
Ciffroy, Philippe	Th-PL-D1.4
Clarke, James	We-SY-G2.5
Claus Henn, Birgit	Th-SY-B1.5
Clavaguera, Simon	We-SY-H2.2
Clewell, Harvey	Mo-SY-F3.3, We-SY-A3.2
Clougherty, Jane	Tu-SY-E1.3, We-PL-H3.2
Colles, Ann	Tu-Po-49
Connolly, Alison	We-SY-G1.2
Corea, Namali	Mo-SY-H2.3
Corlin, Laura	Mo-Po-43
Correia-Sá, Luísa	Mo-Po-34, Tu-Po-03
Covaci, Adrian	Mo-SY-D3.3, Mo-SY-G2.3, We-SY-E4.5
Crépet, Amélie	Mo-SY-A3.4
croes, kim	Mo-SY-D2.3
Curl, Cynthia	Mo-SY-D4.3
Curto, Ariadna	We-PL-I3.5
Curwin, Brian	We-SY-B1.2
Cyrys, Josef	Tu-PL-D1.5, Mo-SY-D4.6
Davis, Benjamin	We-Po-34
De Brouwere, Katleen	We-SY-F1.3, We-SY-F2.4
De Craemer, Sam	Mo-SY-D3.1
De Donno, Antonella	We-Po-22
de Hoogh, Kees	Tu-PL-D2.2
de Kluizenaar, Yvonne	Mo-PL-I2.1
de Nazelle, Audrey	Mo-SY-F1.5, Tu-SY-E4.2
de Rooij, Myrna	Tu-PL-D1.3
de Weger, Letty A	Mo-SY-E1.4
deCastro, Rey	Tu-PL-I1.2
Dekoninck, Luc	Mo-SY-F1.4, Mo-SY-F2.2
Delmaar, Christiaan	Mo-SY-H1.1, Mo-SY-E3.2
Dereumeaux, Clémentine	Tu-SY-F3.3
Devilee, Jeroen	We-SY-A4.1

Deziel, Nicole	We-SY-G4.3, Nicole
Deziel,	Mo-Po-33
Dionisio, Kathie	Tu-SY-E3.2, Tu-SY-G4.4
Dixon, Holly	We-SY-C4.1
Dodson, Robin	Mo-SY-E3.1, We-PL-I2.1, We-Po-57
Donald, Carey	We-SY-C4.4
Dons, Evi	Tu-SY-E4.4
Dorne, Jean Lou	Mo-SY-D1.3
Dornic, Nicolas	Mo-SY-H3.5, Tu-Po-21, We-Po-19
Doutsi, Artemis	Th-SY-D2.1
Douziech, Mélanie	Mo-Po-25
Dudzina, Tatsiana	We-SY-F1.4
Duncan, Sara	Mo-PL-I4.3
Durant, John	We-PL-H4.4
Eeftens, Marloes	We-PL-I4.5
Eichler, Clara	We-SY-E2.4
El Yamani, Mounia	Tu-PL-H3.2
Embry, Michelle	Mo-SY-F3.2
Emke, Erik	Tu-PL-H1.5
Enkhbat, Undarmaa	Mo-Po-46
Ernstoff, Alexi	Tu-SY-G3.3, Tu-Po-09
Evans, Greg	We-PL-H3.4
Ezani, Eliani	Mo-Po-48
Fabian, M. Patricia	Tu-PL-C1.1, Tu-SY-G1.4, Mo-Po-18
Fan, Zhi-Hua (Tina)	Tu-Po-07a
Fantke, Peter	Mo-SY-E4.2, Tu-SY-G3.6, Tu-SY-G4.3, Tu-SY-G4.6
Fecht, Daniela	Tu-PL-C1.2
Fehrenbacher, Cathy	Mo-SY-H1.2, Tu-SY-F2.2
Fent, Kenneth	Mo-SY-B2.2
Ficheux, Anne Sophie	Mo-Po-07, Tu-Po-29
FILLOL, Clémence	Tu-SY-F3.2
Fisher, Jeffrey	Tu-PL-H4.4
Fito, Carlos	We-SY-H2.3
Florentin, Arnaud	Tu-Po-27
Focant, Jef	Th-SY-C2.2
Forouzanfar, Mohammad	Tu-SY-C3.1
Foss Hansen, Steffen	Tu-SY-G3.4

Franken, Remy	We-PL-I2.3
Fransman, Wouter	We-SY-H2.1
Fustinoni, Silvia	Mo-SY-B3.3
Gaborek, Bonnie	Tu-PL-H4.1
Ganci, Aristide	Mo-SY-G4.1, Mo-SY-G4.2
Garcia Hidalgo, Elena	Mo-SY-H3.2
Gennings, Chris	Th-SY-B1.4
Geraets, Liesbeth	Mo-SY-D1.2
Germini, Andrea	Tu-SY-B1.2
Gilbert, Dorothea	Mo-SY-C4.3
Gillespie, Andrew	Mo-PL-I3.6
Glaser, Jason	We-SY-B1.5
Glorennec, Philippe	Mo-SY-E3.5
Göen, Thomas	Mo-SY-B3.2, Tu-PL-H2.4, Mo-Po-22
Gombojav, Enkhjargal	Mo-SY-D4.5
Goodman, Michael	Mo-SY-A1.3
Govarts, Eva	Mo-SY-D3.4
Gras, Ronda	Th-SY-C2.1
Greggs, Bill	Tu-SY-G4.1
Greiner, Matthias	Tu-PL-I1.4
Guiseppi-Elie, Annette	Tu-SY-F4.5
Gulliver, John	Tu-SY-E1.1, Tu-SY-E4.3
Gundert-Remy, Ursula	Mo-SY-D1.4
Gurung, Anobha	Tu-PL-D2.3
Habre, Rima	Mo-PL-I2.3, We-PL-H3.3
Hall, Eric	Tu-SY-G2.3
Hamm, Nicholas	We-PL-I4.3
Hammel, Stephanie	We-SY-C3.5, We-SY-E2.1
Handakas, Evangelos	Mo-Po-05
Hänninen, Otto	Tu-SY-C3.5
Harnpicharnchai, Kallaya	We-Po-13
Hart, Andy	Tu-SY-B1.1
Hasegawa, Takahiro	Tu-SY-F1.1, Tu-SY-F2.1
Haynes, Erin	We-SY-G3.1
Hays, Sean	Mo-SY-C3.3, Mo-SY-F4.5, Tu-SY-D3.2
Heiland, Astrid	Tu-Po-33, Tu-Po-36
Heinemeyer, Gerhard	Tu-SY-B1.3
Henneberger, Luise	Mo-SY-C2.1
Hernandez, Felix	Mo-SY-G3.1
Hertoghs, Kirsten	Tu-Po-61
Hesse, Susanne	Mo-SY-B1.2, We-SY-B3.3
Hickey, Christina	We-SY-A3.4
Hilpert, M.	We-Po-16
Hjerpe, Andrea	Mo-SY-A4.3
Ho, Kin Fai	We-PL-H4.5
Hoek, Gerard	Tu-SY-E1.4
Hollnagel, Heli M	Mo-SY-F3.4, Mo-SY-F4.1, Mo-SY-F4.4, Mo-Po-09
Hopf, Nancy B.	Tu-LBA-11
Hopke, Philip	Tu-PL-B4.3
Hoppe-Jones, Christiane	Mo-SY-B2.3
Horton, Megan	Tu-PL-I4.5, We-Po-08
Horvat, Milena	Th-SY-A2.3
Houba, Remko	We-SY-A1.3
Huang, Yuli	Mo-Po-15, Tu-Po-16
Hudda, Neelakshi	Mo-PL-I1.1
Huizer, Daan	We-SY-F2.1
Ingenbleek, Luc	Mo-Po-59
Int Panis, Luc	Mo-SY-F2.3
Isaacs, Kristin	Mo-SY-H2.2, We-SY-A2.2, We-SY-A2.3, We-SY-E1.3
Ito, Yuki	Tu-Po-26
Jacobs, Cor	Mo-SY-E2.1
Janssen, Stijn	Mo-SY-E2.2
Jantunen, Matti	Mo-SY-E4.3, Mo-Po-27
Jarmul, Stephanie	Tu-Po-60
Jerrett, Michael	Tu-SY-A3.2
Jesen, Keld Alstrup	We-SY-C1.1, We-SY-H1.3, We-Po-59
Johnson, Markey	Tu-PL-C1.3, Tu-PL-D1.4
Johnson, Jona	We-SY-F4.5
Johnston, Jason	We-SY-G2.4
Jolliet, Olivier	Tu-PL-I1.5, Tu-SY-G4.2
Jones, Dean	Tu-SY-A3.3
Jones, Kate	We-SY-G1.1
Jyethi, Darpa	We-PL-I2.4
Kadiri, Shamusideen	We-Po-05
Kaibaldiyeva, Ulmeken	We-Po-47
Kamijima, Michihiro	Th-SY-C2.3

Kang, Habyeong	We-LBA-26, We-LBA-25
Karakitsios, Spyros	Th-SY-A1.5
Karr, Catherine	Mo-SY-D4.2
Karrer, Cecile	Mo-SY-A3.3
Kaufman, John	We-PL-I1.2, We-PL-I1.3
Keith, John	Mo-PL-I3.3
Kennedy, Marc	Mo-SY-A3.2
Kephalopoulos, Stylianos	We-Po-44
Kilaru, Vasu	Tu-SY-E1.5
Kile, Molly	We-SY-C3.4
Kim, Chloe	Tu-Po-43
Kim, Daeseon	We-Po-15
Kissel, John	We-SY-E4.1
Klenow, Stefanie	Mo-SY-E3.4
Klimowska, Anna	Tu-Po-06, Tu-Po-58
Klompmaker, Jochem	Tu-SY-E1.2
Knudsen, Lisbeth E.	Mo-SY-D1.5, Tu-Po-30
Koch, Holger	We-SY-D1.2
Koelmans, Albert	We-SY-C1.4
Koivisto, Antti Joonas	We-SY-C1.3
Koivisto, Joonas	Mo-Po-37
Kolbaum, Anna Elena	We-Po-26
Kolkman, Annemieke	Mo-SY-G3.5
Kolossa-Gehring, Marike	We-SY-D2.3, We-SY-D2.5
Koopal, Cees	Tu-Po-04
Koppisch, Dorothea	We-SY-F2.3
Kosemund, Kirstin	Mo-SY-F3.1, Tu-Po-23
Kovatch, Patricia	Mo-SY-A4.1
kovtunenko, irina	Tu-Po-48
Kraft, Martin	Tu-PL-B3.3
Kramer, Nynke	Mo-SY-C2.4
Krause, Sophia	Mo-SY-C2.3
Kruszewski, Francis	Tu-Po-59
Kuester, Christian	We-SY-G1.5
Kuijpers, Eelco	Th-SY-A1.3
Kwong, Laura	Tu-SY-G1.1
Lai, Chane-Yu	We-Po-30, We-Po-31, We-Po-32
LAI, Foon Yin	Mo-SY-G2.2
Laird, Brian	Tu-Po-02
LaKind, Judy	Mo-SY-A1.1, Mo-SY-A2.3, We-SY-D1.1
Lal, Raj	We-Po-17
Lamoree, Marja	Mo-SY-G3.6, Mo-LBA-03
Lapierre, Florian	Th-SY-C1.3
Laumbach, Robert	Mo-PL-I1.3
Lautenschalaeger, Daniele	We-SY-G1.3
Laverge, Jelle	Mo-SY-E4.1
Lebot, Barbara	Mo-PL-I4.5
Lee, Kiyoung	Th-SY-E1.1, Mo-Po-47
Lee, Eun Gyung (Emily)	We-SY-B3.6
Lee, Boram	Mo-Po-35
Lehtomäki, Heli	Tu-SY-C3.2
Lemazurier, Emmanuel	Mo-SY-C4.2
Lenderink, Annet	We-SY-A1.1
Leonards, Pim	Tu-PL-I4.2, We-Po-28
Lessmann, Frederik	Tu-Po-05
Levin, Hal	Mo-SY-E4.5
Liang, Donghai	Tu-PL-I3.1, Tu-SY-E3.5, We-PL-I3.2, Mo-Po-04
Lim, Miyoung	We-PL-H3.5
Lim, Chris	We-Po-58
Lin, Yu-Wen	We-Po-01
Lin, Wei-Cheng	We-LBA-27
Lind, Lars	Mo-SY-H4.2
Little, John	We-SY-A2.4, We-SY-E1.1
Liu, Shi	Mo-Po-28
Lofts, Stephen	We-SY-C2.1
Loh, Miranda	Mo-SY-F1.3, Tu-SY-A3.1
Longley, Ian	We-PL-I3.1
Lorber, Matthew	Tu-PL-H4.3
Lucas, Rebekah	We-SY-B1.3
Lunchick, Curt	We-SY-G1.4
Luo, Zhiwen	We-Po-39
Lye, Ellen	We-SY-D2.2
Ma, Jin	Tu-Po-47
Maiheu, Bino	Mo-SY-E1.2
Makris, Konstantinos	We-PL-I1.1
Malmqvist, Ebba	Tu-SY-C3.4
Mampaey, Maja	Mo-SY-D2.1
Masion, Armand	We-SY-H1.4
Mason, Ann	Tu-SY-G3.2
Mattison, Donald	Mo-SY-A1.2
Mayer, Philipp	Mo-SY-C3.1
Mayer, Stefan	Th-SY-B2.2
McCormick, Lindsay	We-SY-C3.3

McKenzie, Lisa	We-SY-G3.2, We-SY-G4.4
McKernan, Lauralynn	We-SY-B4.2
McKone, Tom	Tu-SY-G3.1, We-SY-E4.2
Meek, Bette	Mo-SY-H1.4, Tu-SY-B1.5
Meesters, Johannes	We-SY-C2.3
Megson, David	Mo-SY-B3.5
Melnyk, Lisa	Tu-SY-G1.2, Mo-Po-42
Merten, Caroline Gabrielle	Tu-Po-28
Messier, Kyle P	We-PL-H4.2
Meuwissen, Iris	Tu-PL-C1.4
Mielke, Howard	Mo-SY-D4.1
Miller, Aubrey	Mo-SY-A1.5, Mo-SY-A2.2, Tu-SY-F4.4, We-SY-C3.2
Minet, Laura	Mo-SY-F1.2
Mol, Hans	Mo-SY-B4.1, Mo-SY-G3.3
Money, Chris	Mo-SY-B1.5, Mo-LBA-05
Montforts, Mark	We-SY-A4.4
Morgan, Neil	We-SY-G2.5
Morley, David	Mo-PL-I1.2
Morrens, Bert	Mo-SY-D3.2
Morrison, Glenn	We-SY-E3.5
Mortensen, Mary	We-SY-D1.5, Mo-Po-02
Mortensen, Ninell	Tu-LBA-14
Morton, Jackie	Mo-SY-B3.4
Mueller, Michael	Tu-SY-D4.1
Mullen, Kathleen	We-SY-C4.3
Naito, Wataru	We-PL-H3.1
Nakai, Satoshi	Tu-Po-34
Nakayama, Shoji	Th-SY-C2.4, Tu-SY-A1.3
Neisel, Friederike	Mo-SY-C1.1
Newton, Ashley	Mo-Po-39
Nguyen, Vy	Th-PL-D1.2, Mo-Po-44
Nguyen, Nhan	Tu-SY-F1.2, We-SY-B4.5
Nguyen, Khanh Hoang	Mo-SY-G4.3
Noij, Dook	Mo-SY-B1.1
North, Michelle	We-PL-I2.2, Tu-Po-53

Nylander-French, Leena A	Tu-PL-B4.6, Tu-Po-44, Tu-Po-45, We-Po-04
O' Mahony, Cian	Mo-SY-H3.3, We-SY-A3.3
O'Brien, Jake	Mo-SY-G1.5
Odabasi, Danyal	We-Po-53
Oerlemans, Arné	Mo-SY-B4.3
Oiamo, Tor	We-Po-60
Okeme, Joseph	We-SY-E3.6, Mo-Po-55, Tu-Po-55, We-Po-33
Olmedo-Palma, Pablo	We-SY-D1.3, Tu-LBA-18, Tu-LBA-17
Orme-Zavaleta, Jennifer	Tu-SY-F4.1
O'Rourke, Mary Kay	Mo-Po-19
Ospina, Maria	Mo-Po-20
Ottenbros, Ilse	Tu-Po-56
Oya, Naoko	Tu-Po-25
Paini, Alicia	We-Po-35, We-Po-36
Palmen, Nicole	Th-SY-E1.3, We-SY-A1.5
Pardo, Larissa	Tu-LBA-15
Park, Jeongim	We-LBA-24
Parsons, Patrick	Mo-SY-A4.2
Patel, Chirag	Mo-SY-H4.3, Tu-SY-A4.2
Patton, Allison	Th-PL-D1.1
Paulik, L. Blair	We-SY-C4.2, We-SY-G3.5
Pawly, Jessica	Tu-PL-H2.3
Pearce, John	Tu-SY-E3.3
pelletier, Maud	Tu-Po-37
Persijn, Stefan	Tu-PL-H2.1
Peters, Erica	We-SY-F3.5
Peters, Leon	Mo-SY-G4.5
Petrosian, Arina	Tu-Po-13
Phillips, Kirk	We-PL-B2.2
Pilorget, Corinne	Tu-SY-F3.1
Pirard, Philippe	Tu-SY-D4.3
Plichta, Veronika	Mo-Po-36
Portengen, Lützen	Tu-SY-A4.1
Price, Paul	Tu-SY-C2.1, Tu-SY-C2.3
Price, Oliver	Tu-SY-F2.3
Pritchard, Helen	Tu-SY-E2.2
Qian, Hua	Mo-SY-H1.3
Quik, Joris	We-SY-C2.4

UTRECHT THE NETHERLANDS OCTOBER 9-13 ISES 2016

QUINDROIT, Paul	We-Po-46
Radomyski, Artur	Tu-PL-H4.5
Ramirez, Noelia	Tu-PL-H1.2, Mo-LBA-07
Ramirez-Santana, Muriel	We-SY-G2.1
Ramos-Bonilla, Juan	Tu-PL-I1.3
Rashid, Audil	We-PL-B2.4, Tu-Po-12
Rasmussen, Sara	We-SY-G4.1
Reif, David	Tu-Po-17
Reiner, Jessica	Th-SY-C1.4
Reis, Stefan	Mo-SY-F1.1
Rinkevich, Joseph	Tu-SY-G4.5
Romanoff, Lovisa	Mo-Po-06
Rossner, Alan	We-PL-I4.2, Mo-Po-40
Rudzok, Susanne	Tu-PL-H3.5
Ruprecht, Ario Alberto	Mo-PL-I1.5
Russell, Armistead	Tu-PL-D1.2, Tu-PL-I3.4
Rustemeyer, Thomas	We-SY-A1.2, We-SY-A4.3
Safruk, Adam	Tu-PL-I4.4
SAKAI, Haruya	We-Po-06
Salmon, Maëlle	We-PL-I3.3, Mo-Po-26
Salthammer, Tunga	Tu-PL-I2.1, We-SY-E2.3
Sanaa, Moez	Tu-SY-B1.4
Sanchez, Araceli	We-SY-B4.4, We-SY-H2.4
Sarigiannis, Dimosthenis	Th-SY-A1.1, Tu-SY-A2.2, Tu-SY-C2.2
Savic, Nenad	We-SY-B4.1
Scheepers, Paul	Tu-SY-D4.2, Mo-Po-12
Schinkel, Jody	Mo-SY-B1.3, We-SY-B4.3
Schlueter, Urs	Mo-SY-B1.4
Schoeters, Greet	Mo-SY-D2.2, We-SY-D2.4
Schuur, Gerlienke	Tu-SY-F1.3
Seitz, Teresa	We-SY-A1.4
Seo, SungChul	Tu-Po-19
Serio, Francesca	We-LBA-28
Shairsingh, Kerolyn	Mo-SY-F2.1
Sharma, Raju	Tu-PL-I4.3, Tu-Po-18a
Shen, Beverly	Tu-Po-54

Shen, Heqing	Mo-Po-21
Shendell, Derek	Mo-LBA-02
Shin, Hyeong-Moo	Mo-PL-I4.1
Silva, Diana	Tu-SY-E2.1
Simon, Matthew	We-Po-56
Simpson, Christopher	Mo-Po-03
Sioutas, Constantinos	Mo-PL-I2.2
Slautenko, Eugene	Tu-Po-35
slob, wout	Tu-SY-B2.2
Smink, Beitske	We-SY-F3.1
Smith, Anthony	Tu-SY-B2.3
Soeteman-Hernandez, Lya	Th-SY-E1.2
Sommer, Yasmin	We-Po-27
Soneja, Sutyajeet	We-PL-H4.1, We-Po-37, We-Po-38
Spinelle, Laurent	Tu-LBA-12
Spooner, Neil	Th-SY-C1.2
Sprong, Corinne	Mo-SY-A3.1
Stadnicka-Michalak, Julita	Mo-SY-C4.4
Stahlmecke, Burkhard	We-SY-H1.2
Stapleton, Heather	We-SY-A2.5, Tu-LBA-13
Steinle, Susanne	Tu-SY-E4.1, Mo-Po-13
Stephenson, Dale	Mo-Po-38
Strak, Maciej	We-Po-40
Stylianou, Katerina	Th-PL-D1.3
Sukiene, Vilma	Mo-SY-E3.3
Sumner, Susan	Mo-SY-A4.5
Svendsen, Claus	We-SY-H2.5
Svyatets, Ekaterina	We-LBA-21
Sy, Mouhamadou	We-Po-48
Taylor, Kyla	We-Po-18
Teeguarden, Justin	Mo-SY-C4.5
Teeguraden, Justin	Mo-PL-I3.1
Teitelbaum, Susan	Tu-PL-B3.1, We-SY-D1.4
ter Burg, Wouter	Mo-SY-C1.2, Mo-SY-C1.3
ter Laak, Thomas	Mo-SY-G3.4
Thomas, Kevin	Mo-SY-G1.2
Thornburg, Jonathan	We-SY-F4.2
Tischer, Martin	We-SY-B3.2, We-SY-B3.5
Tornero-Velez, Rogelio	Mo-SY-H4.4, Mo-SY-H4.5
Tozer, Sarah	Mo-SY-H2.1, Mo-SY-H3.4
Tranfo, Giovanna	Tu-Po-01

Tripathy, Sheila	Tu-PL-I4.1
Troutman, John	We-SY-A3.5
Tsai, Meng-Shan	Tu-Po-39
Tsai, Shih-Wei	Mo-LBA-06
Tulve, Nicolle	We-PL-B2.1, We-Po-51
Turczynowicz, Leonid	We-PL-I1.4
Tyrrel, Sean	We-PL-I4.4
Underhill, Lindsay	We-Po-54
Urbanus, Jan	We-SY-F1.1
Vallero, Daniel	We-SY-E1.4, We-SY-F1.5
van den Dries, Michiel	We-LBA-29
van den Elshout, Sef	Mo-SY-E1.3
Van der Pol, Peggy	We-SY-F3.4
van der Veen, Ike	Tu-PL-I2.2
van Klaveren, Jacob	Mo-PL-I3.5
Van Leuken, Jeroen	Mo-SY-E1.5
van Loon, Cornelis	We-Po-24
Van Nieuwenhuyse, An	Tu-SY-D4.5
van Nuijs, Alexander	Mo-SY-G2.1
van Tongeren, Martie	We-SY-B3.4, We-SY-H1.1
Varshavsky, Julia	Mo-Po-01
Vázquez-Campos, Socorro	We-SY-H1.5
Verbist, Koen	We-SY-F2.2
Vermeire, Theo	Mo-SY-F4.3
Vermeulen, Roel	Tu-SY-A4.3
Verner, Marc-André	Tu-PL-I3.2, Mo-Po-29
Vesper, Hubert	Mo-SY-A2.1
Viegas, Carla	Th-SY-B2.1, Tu-PL-B4.4, Mo-Po-32, Tu-Po-07b
Viegas, Susana	Th-SY-B2.3, We-Po-20, Tu-Po-20
Viet, Susan	Tu-PL-H1.1
Vila, Javier	We-Po-45
Villeneuve, Paul	We-Po-11
Vilone, Giulia	We-SY-G2.3
Visser, Wouter	We-SY-F4.4
Vlaanderen, Jelle	Tu-SY-A3.4
Volten, Hester	Mo-SY-E2.4
von Goetz, Natalie	Tu-SY-B2.1, We-SY-E3.3
Vongmany, Natalie	Th-SY-E2.1
Vrijheid, Martine	Tu-SY-A1.2
Walendzik, Gudrun	Tu-SY-F1.4
Watkins, Timothy	Tu-SY-F4.2
Webster, Thomas	Mo-SY-H4.1, We-SY-E1.5
Weiss, Ilana	We-SY-B1.1
Weschler, Charles	Mo-SY-E4.4, We-SY-E3.1, We-SY-E3.2
Wesseling, Joost	Mo-SY-E1.1, Mo-SY-E2.3
Wesseling, Catharina	We-SY-B1.4
Westerhout, Joost	Tu-PL-I1.1
Wetmore, Barbara	Mo-SY-C3.5, Mo-SY-C4.1, Mo-SY-F4.2
Wiesner, Mark	We-SY-C2.2
Wijnhoven, Susan	We-SY-C1.2
Wilson, Karen	We-SY-F3.3
Wolfe, Christopher	Mo-Po-50
Wormuth, Matthias	We-LBA-22
Wright, Robert	Mo-SY-A4.4, Tu-SY-A1.4
Wu, Hao	Tu-PL-D2.1
Xu, Ying	We-SY-E2.2
Xue, Jianping	Tu-SY-G1.3, Tu-SY-G1.5
Yang, Guiling	We-Po-25
Ye, Sherry (Xiaoyun)	We-PL-H4.3
Young, Bruce	We-SY-G2.2
Yu, Sol	We-LBA-20
Zaleski, Rosemary	Mo-SY-H1.5
Zare Jeddi, Maryam	Mo-PL-I3.4, Mo-Po-10
Zarn, Jürg	Mo-SY-C1.4
Zauli Sajani, Stefano	Tu-PL-I3.3
Zeman, Florence	Tu-PL-B3.4
Zhang, Yinping	Tu-PL-I2.4, We-SY-E1.2
Zhang, Ting	We-LBA-23
Zhou, Jiaqi	Tu-PL-I3.5
Zhou, Ying	Tu-PL-B3.2

Sunday, October 9, 2016

Plenary Address 1: Water as integrator of uses, stakes and exposures

Annemarie van Wezel, Copernicus Institute of Sustainable Development, Utrecht, The Netherlands

This plenary focuses on exposures to (emerging) chemicals via the water cycle, the various sources relating to societies use of water and chemicals, pathways, possible health and economic effects and mitigation options. The use of advanced analytical techniques is highlighted, i.e. suspect screening using high resolution mass spectrometry to analyze a broad set of industrial chemicals, pharmaceuticals and pesticides including their transformation products, and the subsequent prioritization techniques. Also attention will be given to the analysis of nanoparticles such as nanoplastics, including field flow fractionation techniques. In addition to these analytical techniques, modelling techniques using water and land uses to predict water quality are highlighted and examples of predictions are given based on communal and industrial emissions. The interpretation of health risks using in vitro and in vivo toxicity data, including generalizations such as the use of thresholds of toxicological concern (TTC) to assess risks when scarce data are available, is discussed with a focus on relatively polar substances that are hard to remove using water treatment technologies. Also possibilities posed by epidemiological techniques using big data health statistics combined with spatiotemporal information on water quality are reviewed. Some examples are provided from the way data water quality data can feed epidemiological analyses, using sewer epidemiological approaches. As in water because of the integration typically complex environmental mixtures occur, some attention is paid to the used of effect-directed triggers compared with triggers based on concentrations of individual substances. The risks of new and emerging technologies and ways to manage and prevent these risks will be described, using examples form nanotechnology and new technologies for oil and gas. The awareness of these risks by technology developers can be enlarged by risk and technology assessment, ultimately in support of the technology development and its acceptance by society. Finally mitigation science in combination with exposure science is elaborated, connecting the water cycle and the chemical life cycle. Mitigation options can be assessed by coupling emissions, exposure and effects in hydrological models. A solution-focused and systems-oriented perspective combined with a mitigation database can offer a common perspective amongst relevant actors, inform policy making and stimulate cross-sectoral learning. Finally, the plenary will conclude seeking parallels between aqueous exposure science and other environmental exposures.

Monday, October 10, 2016

Plenary Address 2: Respiratory health effects and livestock farming related to microbial and dust exposure

Dick Heederik, Utrecht University, Institute for Risk Assessment Sciences, Utrecht, The Netherlands

Several studies have investigated the effect of livestock farm emissions on respiratory health of local residents but results are inconsistent. The most comprehensive study thus far measured lung function of adults living in a rural German area (Radon et al. Epidemiology 2007). Results indicated a decreased FEV1 in adults with a high number of animal houses in the proximity of their home address. The current study aims to explore associations between proximity to livestock farms and lung function in a general, non-farming population of adults in the Netherlands. This study made use of GP records from approximately 100,000 patients, questionnaire survey data from approximately 14,000 individuals, and a medical survey among approximately 2,500 residents living near livestock farms (IgE serology, lung function) in the south east of the Netherlands (figure). Exposure assessment comprised of PM10 sampling over a one year period on 62 locations with different livestock densities around the sampling sites. Two week average PM10 samples were taken and analyzed for some specific pathogenic bacteria by qPCR (Coxiella burnetii) and endotoxin by the LAL assay. Several associations between livestock density and proximity from livestock facilities and health effects were observed:
- an increased pneumonia risk around poultry farms. No indications exist that this elevated risk has a zoonotic nature. Impaired host defense mechanisms resulting from high dust and endotoxin exposure seems a more likely explanation;
- individuals with COPD more frequently reported respiratory symptoms and increased use of medication when they lived in close proximity to a livestock farm;
- people living in areas with a high number of livestock farms in a radius of 1 kilometer around the house have a reduced lung function (FEV1, MMEF) compared to individuals with a low livestock farm density around the home. Individuals who were measured on days with high NH3 levels had a lower lung function adjusted for age, standing height, gender, smoking;
- people living in areas with high number of livestock farms were less often sensitized to common allergens (HDM, cat, dog, pollen), adjusted for being raised on a farm, in comparison with people with a low livestock density around the home.

Livestock density and distance from a livestock farm were associated with elevated PM10 and endotoxin levels. First attempts were undertaken to develop Land Use Regression models for PM10 and in particular endotoxin exposure. Endotoxin levels were more clearly explained by spatial variables such as livestock farm density, in particular by the presence of pig and poultry farms.

Results from these studies indicate that spatial and temporal livestock related respiratory effects exists associated with most likely primary and secondary particle emissions. Endotoxin emissions very likely contribute to the environmental exposure from livestock farms and the resulting respiratory health effects. This study contributes to the evidence that livestock production contributes to ambient particulate exposure and the burden of disease resulting from this exposure.

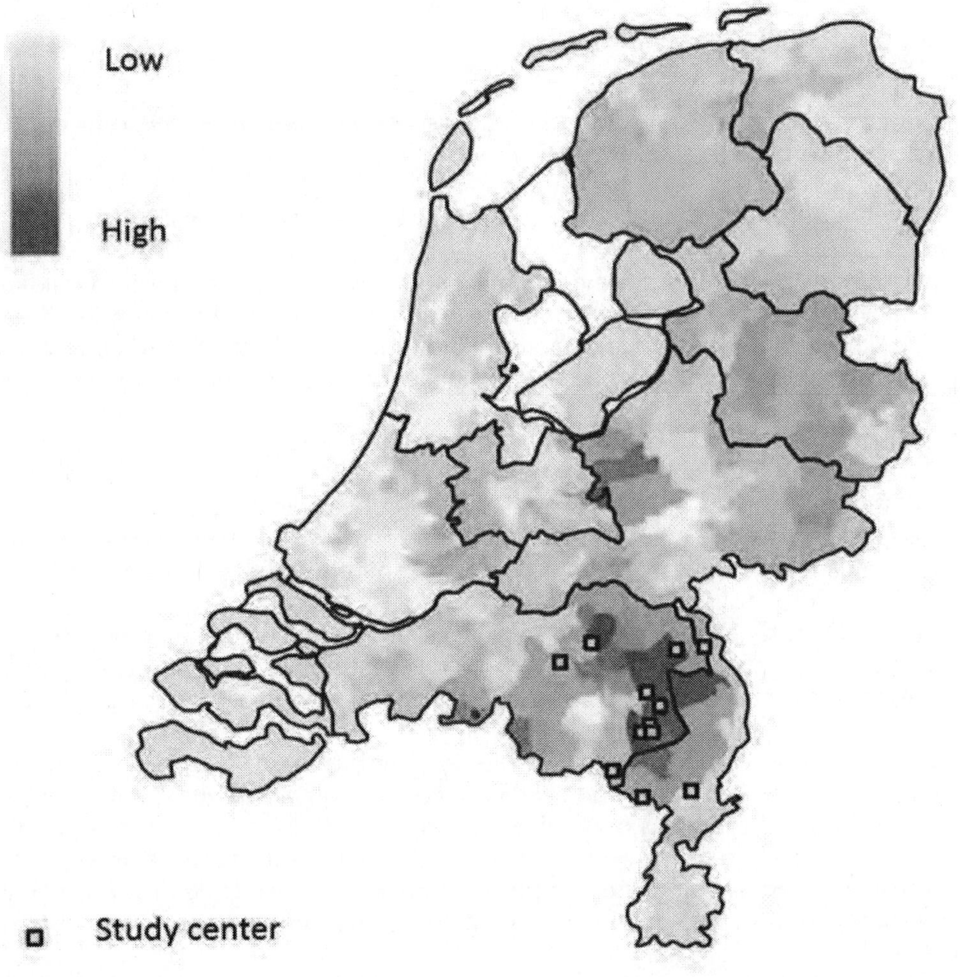

Livestock output in the Netherlands in standardized kg animal (blue shading) and location of the study centers (yellow)

Mo-SY-A1: Harmonization, access, transparency: improving environmental epidemiology for public health decision-making - I

Mo-SY-A1.1

Concordance, transparency, and access: Why do we need these in exposure science and health outcome research?

Judy LaKind, *LaKind Associates, LLC, Catonsville, MD, United States*

In observational research, evidence is usually derived from multiple studies, and any single result is rarely considered sufficient for public health decision making. Despite more than five decades of research and thousands of studies published, the ability to draw robust conclusions regarding the presence or absence of causal links between specific environmental exposures and human health remains limited. To develop policies that are protective of public health and can withstand scrutiny, agencies need to rely on investigations of satisfactory quality that follow sufficiently concordant protocols in terms of exposure assessment, outcome ascertainment, data analysis, and reporting of results. Absent such concordance, the ability of environmental epidemiology studies to inform decision making is greatly diminished.

Systems and tools are proposed here to improve concordance among environmental epidemiology studies. Specifically, working systems in place in other fields of research are critically examined and used as guidelines to develop analogous policies and procedures for environmental epidemiology. A three-part path forward (Figure 1) toward more concordant, transparent, and readily accessible environmental epidemiology evidence that parallels ongoing efforts in medical research is described. The three parts address methods for improving quality and accessibility of systematic reviews, access to information on ongoing and completed studies, and principles for reporting. The goals are to increase the value of epidemiological research in public health decision making.

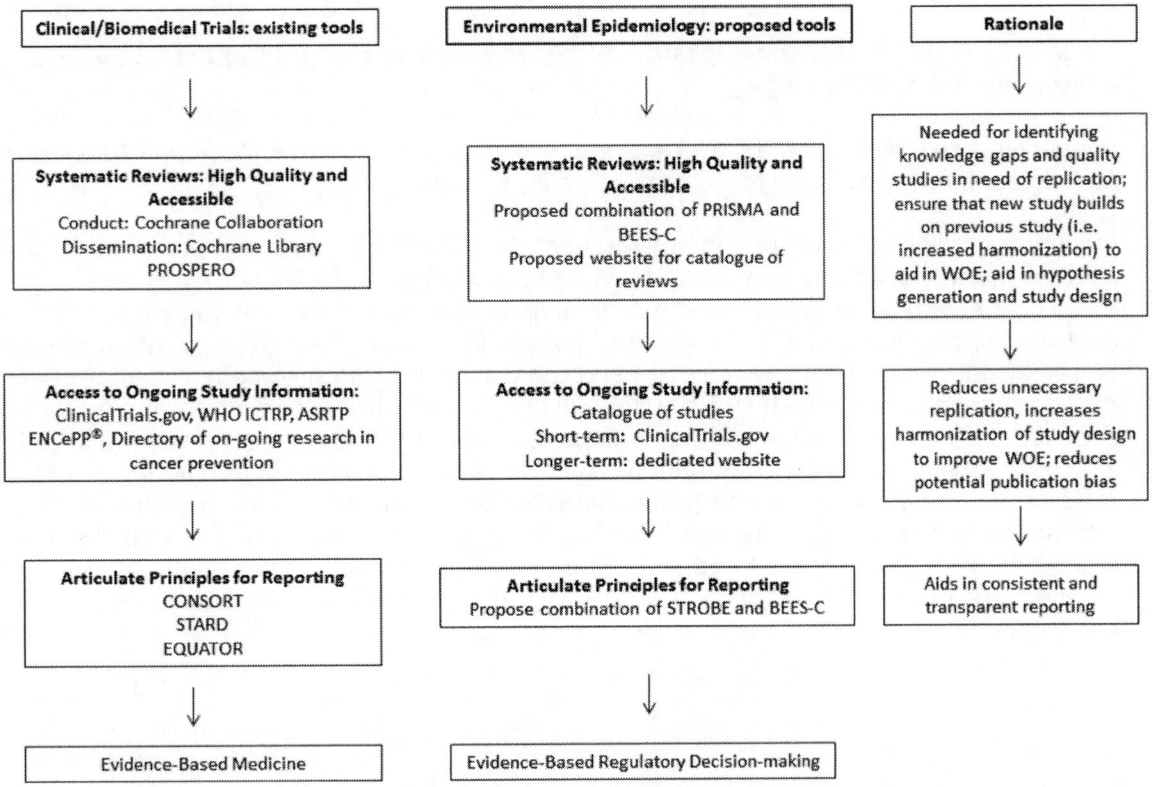

FIGURE 1. Process for concordance in clinical and biomedical research and proposed parallel process for environmental epidemiology research to improve public health decision making.

Figure 1

Mo-SY-A1.2

Combining large datasets on exposure and health outcome - evolution of environmental epidemiology.

Donald Mattison, *Risk Sciences International and McLaughlin Centre for Population Health Risk Assessment, University of Ottawa, Ottawa, Canada*

Challenges facing health care and environmental epidemiology can be approached through analysis of large observational datasets. This presentation will review the use of electronic health records (EHR) and geocoded pollutant concentrations to assess environmental exposures in the context of population health. One example of an EHR will be reviewed (Cerner HealthFacts, CHF) and the use of this HIPAA compliant deidentified data for understanding the beneficial and adverse health effects of drug treatment will be described. The presentation will then discuss the use of CHF data combined with 3-digit zip codes as well as the zip code tabulation areas to characterize environmental exposures. These approaches allow characterization of changes in health status following alterations in exposures, while also including the prior health status of the individuals being followed. While these large and complex health and environmental exposure datasets present challenges, they also offer useful approaches for characterizing population health.

Mo-SY-A1.3

Lessons learned from registration of clinical studies, past to the present

Michael Goodman, *Emory University Rollins School of Public Health, Atlanta,, Georgia, United States*

Large-scale research studies, particularly intervention trials, are time- and resource intensive and may take years to complete; however, only some studies result in peer-reviewed publications. The availability of a full picture of both past and current studies facilitates a better understanding of knowledge gaps, identifies studies that warrant replication, helps avoid redundancies, and allows development of ideas for future research. Further, in the absence of concordant protocols for various aspects of study design, clear and transparent information on study design of ongoing studies would assist researchers in ensuring that their proposed study protocols will build on past research. One method for improving access to information on ongoing and completed clinical research is early registration of studies. Study registration has been termed a "scientific, ethical and moral responsibility" because informed decision making is not possible when publication bias and selective reporting are present; in addition, the availability of information in study registries assists researchers and funders in avoiding unnecessary duplication, identifying gaps, and encouraging collaboration. Several opportunities for clinical trial registration are now available, including ClinicalTrials.gov, the World Health Organization (WHO) International Clinical Trials Registry Platform (ICTRP), and the International Standard Randomized Controlled Trial Number (ISTRP) register. ClinicalTrials.gov provides a publicly-available site for information on both clinical trials and observational studies of investigational drugs. For each study, the site contains information on its title and design, disease or condition of interest, intervention, eligibility criteria for - and description of - participants, location(s), analytic methods, and outcome. In recent years, availability of data from clinicaltrials.gov has allowed a number of meaningful analyses. Examples of such analyses include an evaluation of terminated trials, assessment of discrepancies between registered results and peer-reviewed publications, and analyses of time to publication. These analyses would be difficult, if not impossible, without study registration.

Mo-SY-A1.5

Inclusion Of Citizen Science And Atypical Environmental Data In Support Of Environmental Health Decision-Making

Aubrey Miller, *National Institutes of Health, Bethesda, Maryland, United States*
Liam O'Fallon, *NIEHS, Durhan, North Carolina, United States*
Joseph Hughes, *NIEHS, Durham, North Carolina, United States*
April Bennett, *NIE, Bethesda, Maryland, United States*
Richard Kwok, *N, Du, N, United States*

Aim
Discuss the rapidly evolving areas of citizen science (CS) and community based participatory research including highlights of both the opportunities and challenges for furthering our understanding of environmental exposures and health effects. Such information, in addition to traditional epidemiology approaches, can be used to help further site specific exposure and risk assessments, as well as providing additional weight-of-evidence to help underpin risk management decisions.
Methods
Disaster responses and other investigations, such as the Gulf Oil Spill and hydraulic fracturing, incorporating community engaged research will be discussed to provide additional context and understanding of data acquired through such non-traditional forums. Challenges include the use of non-validated data, harmonization of disparate data sets, and the interpretation and communication of findings. Case studies will highlight the benefits, as well as the challenges, in using CS approaches to promote environmental public health. Focus will include user-friendly tools and technologies for assessing environmental exposures, data quality & management, and interpretation of information and risk communications.
Results
Attendees will gain insight into: 1) the important need and role of CS in supporting responses to emerging environmental threats and disasters; 2) the use and limitations of CS exposure assessment tools and data; 3) the generation of useful data through effective planning and partnerships, 4) interpretation of data, ethical considerations, and communicating results; and 5) the need to foster CS to help measure and understand environmental exposures.
Conclusions
While complex and often challenging, CS approaches have the positive benefits of bringing local expertise and knowledge to the table, improving community involvement, understanding, and acceptance of the environmental information used for decision-making. Lastly, such efforts further educate and strengthen communities to take an active role in understanding and monitoring their environment to promote environmental health and to build infrastructure and resiliency in responding to environmental disasters and other challenges.

Mo-SY-B1: Targeted activities for improving workplace exposure assessments

Mo-SY-B1.1

ECETOC TRA FOR WORKERS: Lessons learned from its use under reach

Dook Noij, *Dow Benelux, Terneuzen, Netherlands*
Andreas Ahrens, *European Chemicals Agency, Helsinki, Finland*

Introduction:
The ECETOC Targeted Risk Assessment (TRA) model for workers is intended to evaluate the risks for workers arising from the manufacture and use of chemicals. The versions 2 and 3 of the model have been extensively used for worker chemical safety assessments during the first and second REACH registration phases (2007-2013). The model is also incorporated in the ECHA CHESAR tool (Chemical Safety and Reporting tool). The version 3 model was developed based on the solicited feedback from TRA users on their experiences with the tool in the 2009-2011 period, as well as the feedback by ECHA (integration into CHESAR). Since 2014 ECETOC is working closely with ECHA to improve the guidance on the application (domain) of the tool and to investigate the need for future modifications. This was prompted by ECHA and industry experiences with the application of the tool in Chemical Safety Assessments, and the updates of the ECHA guidance documents on use description and on occupational exposure assessment.
Results:
For version 3, the suggestions received from users have resulted in the implementation of improvements in the core of the model (exposure estimates), as well as in related exposure determinants (operational conditions and risk management measures).
The cooperation between ECETOC and ECHA has led, among others, to more precise description of PROC's (Process Categories),better guidance on PROC-assignment for modeled activities, improved clarifications of the applicability domain (e.g. PROC's for closed systems, and substances of very high concern), as well as a more transparent approach to address elevated operation temperature. These improvements are partly built into the new CHESAR version (V3.0) and are reflected in the updates of the ECHA guidance documents. The decision to develop an updated version of the model is pending the analysis of the results of the ETEAM study (Evaluation of Tier 1 Exposure Assessment Models) and the results of a limited dermal exposure assessment validation study.
Discussion:
The discussion will address the experiences and challenges with implementation of proposed modifications of the model and further definition of its application (domain). Examples are the balancing of user requests/expectations for enhancements versus principles to maintain its Tier 1 characteristics and scientific integrity, the pros/cons of adaptation of the model during its application within the REACH implementation timelines, the level of training/expertise required for users to adequately apply the model, and securing consistency of model application/outcomes in different settings (stand-alone tool versus integration in CHESAR).

Mo-SY-B1.2

Describing the Effectiveness of Sector Specific Exposure Controls

Susanne Hesse, *Fraunhofer Institute for Toxicology and Experimental Medicine, Hannover, Germany*
Katharina Blümlein, *Fraunhofer Institute for Toxicology and Experimental Medicine, Hannover, Germany*
Stefan Hahn, *Fraunhofer Institute for Toxicology and Experimental Medicine, Hannover, Germany*

Aims and background:
A number of relevant risk management measures (RMMS) were identified by ESIG (European Solvents Industry Group) for use in their chemical safety assessments. These include various levels of containment in combination with ventilation, use of drum pumps for filling procedures as well as draining and flushing of equipment before maintenance operations. Provisional efficiency suggestions for these RMMs have been provided by ESIG. But up to now they could not be supported by quantitative experimental data. Fraunhofer ITEM evaluated the influence of sector specific risk management measures on inhalation exposure by 1. Reviewing publicly available information and 2. Conducting experimental studies to determine the impact on air concentrations.
Methods:
Publicly available information was gathered by searching literature and by contacting relevant industry representatives.
Inhalation exposure was experimentally assessed for three representative solvent scenarios: Gravity transfer, drum pump transfer as well as draining and flushing. The lab-based simulations considered different levels of containment and ventilation. Each simulation scenario was assigned to a specific phrase code. For the measurements a standardised laboratory set up including reservoir and collection containers, a drum pump for transfer, a fume hood and a portable IR spectrometer as a detector were used.
Results and Conclusion:
In conclusion, quantitative data could not be identified by reviewing publicly available information and experimental studies had to be conducted. The experimental data seem to support the original suggestions made by ESIG in regards to the efficiencies of the individual RMMs. As an example, full containment (closed fume cupboard with ventilation switched on) leads to an efficiency of >99% while the transfer via drum pump results in a reduction of 93.5%. All measures show efficiencies above 90% when compared to the worst case scenario (no RMM in place).

Mo-SY-B1.3

Can worker dermal exposure predictions be improved based on current data sets?

Jody Schinkel, *TNO, Zeist, Netherlands*
Henk Goede, *TNO, Zeist, Netherlands*
Wouter Fransman, *Zeist, Zeist, Netherlands*
Anjoeka Pronk, *TNO, Zeist, Netherlands*
Suzanne Spaan, *TNO, Zeist, Netherlands*
Rianda Gerritsen, *TNO, Zeist, Netherlands*

Aims and background
Occupational exposure models have been developed to estimate dermal exposure concentrations (or ranges) to hazardous chemicals. In spite of the acceptance of these tools, however, these models have not been comprehensively validated. In this part of the symposium the validation of the dermal of ECETOC TRA will be presented, as an example of the possibilities to improve dermal models based on currently available data.
Methods
In a project sponsored by CEFIC-LRI, the dermal model of ECETOC TRA is validated using published dermal exposure measurement. TRA exposure estimates are compared with exposure levels presented in public available literature, creating an overview of available exposure measurement data. When improvements of more resolution in dermal exposure models is needed, insight in dermal exposure determinants and dermal exposure processes is needed. Availability of information on these elements is investigated in two recent initiatives to improve dermal exposure assessment, namely the BROWSE project (www.browseproject.eu) and during the development of a dermal module of the Advanced REACH Tool (dART).
Results and conclusions
Although we know quite a lot about dermal exposure process, good quality information on exposure to underpin our exposure models is scares and fragmented. An uniform format to collect and store exposure data is missing. Furthermore, agreement on standard measuring methods is lacking. In addition, the dermal exposure process is rather complex, with various processes like dermal loading, dermal transfer, resuspension and cleaning are important. The effect of different determinants on these processes has not been systematically investigated yet. If we want to improve the current assessment methods for dermal exposure in the workplace, a clear and coordinated research program is needed.

Mo-SY-B1.4

How can worker exposure models be comprehensively validated? Implications of the ETEAM project results

Urs Schlueter, *BAuA - Federal Institute for Occupational Safety and Health, Dortmund, Germany*
Martin Tischer, *BAuA - Federal Institute for Occupational Safety and Health, Dortmund, Germany*

The German Federal Institute for Occupational Safety and Health (BAuA) has initiated a comprehensive Evaluation of the Tier 1 Exposure Assessment Models (ETEAM) on the performance of REACH Tier 1 worker exposure models. Carried out by the Institute of Occupational Medicine (IOM Edinburgh) and the Fraunhofer Institute for Toxicology and Experimental Medicine (ITEM Hannover) ETEAM was intended to compare different REACH Tier 1 exposure models using an integrated approach including

- conceptual evaluation,
- external validation and
- between user reliability (BURE) / user-friendliness.

Reviewing the findings of ETEAM within the context of the constraints presented by available exposure data and exposure models that are intended to be broadly applicable (wide range of substances, use situations) results in the following conclusions:

- Overall comparison of model estimates with measurements suggests that the tools tend to be conservative.
- They may not be sufficiently conservative in all situations.
- Statistical analyses of the BURE study suggest that user variation in the choice of input parameters lead to very different results.

Both underestimation of exposure and the impact of user variation can have serious impacts. As a consequence more confidence in the level of conservatism and accuracy of models is necessary. Industry can help to reduce uncertainty within exposure assessment by comparing the estimates from different sources, including other tools and measured data. For authorities REACH offers a regulatory basis to request independent measurement data.

Further model developments and improvements should consider user friendliness implications and the ability of users to choose the correct input parameters. The BURE has shown that there are some parameters which induce a high level of variability due to their vague definition. In particular these are:

- use categorisation for all tools,
- intrinsic dustiness (only qualitatively defined),
- type of setting (professional/ industrial) and
- definition of risk management measures.

Obviously, the definition of the corresponding parameters should be as precise as possible. However, the user's knowledge about the tools is also of high relevance. Therefore to decrease the total level of uncertainty, it is crucial that users are well informed about both the models and the assessed situations.

The results ETEAM will assist in choosing the most appropriate model for a given exposure situation. In addition, they will help authorities to assess whether an exposure scenario is safe and to estimate how conservative the estimates are. Finally they will support model developers in identifying areas where models can be improved.

This page intentionally left blank.

Mo-SY-B1.5

Re-thinking strategies for improving exposure assessments

Chris Money, *Cynara Consulting, Brockenhurst, United Kingdom*

Background
In order that chemicals risk assessments can be seen to be relevant and reliable, they must be able to effectively undertake both hazard and exposure assessment. However, the fact that the use of many chemicals is widespread can be associated with a high degree of uncertainty in the ability to accurately predict exposure. This uncertainty can undermine steps being taken to improve hazard assessment. The introduction of tiered approaches to exposure assessment provides opportunities for improving the relevance and reliability of exposure assessments through the efficient and targeted acquisition and application of key information on exposure parameters.
Strategy Development
Ensuring reliable and relevant exposure assessments not only demands that the exposure model is accurate but also requires that the determinants used as the basis of predictions reflect real world conditions. This consideration is particularly important when the determinants are factors that enable users of chemicals to determine how best to reduce exposures, such as those which may be communicated to workers and consumers within safety data sheets and labels.
Within Europe, following the introduction of REACH, suppliers and users of chemicals have begun to more widely share such information, but its provision could become piecemeal unless suitably targeted and coordinated. The presentation will examine the considerations necessary for implementing effective strategies for collecting exposure information which are summarised in the figure . Particular attention will be paid to recent initiatives aimed at securing refined exposure information for worker and consumer assessments.
Conclusion
Recent developments aimed at structuring how information on exposure determinants might be collected by suppliers and users of chemicals provide opportunities for ensuring that exposure assessments are relevant and accurate. Implementing suitably coordinated approaches across these groups should avoid any tendency for the information to become piecemeal or deficient.

	Need	Benefit	Requirements
Tier 1	Simple and common basis for developing Exposure Scenarios *Use Maps, Use Descriptors*	• Consistent • Understandable • Relevant • Communicable	• Dialogue within/between supply chains • Generic descriptions of use • Universal set of Use Descriptors • Associated standard phrases
Tier 1+	When and what further information helps refine the exposure assessment? *SWEDs, SCEDs, SpERCs*	• Targeted • Structured / optimised • Enables coordination • Rewards engaged groups	• Dialogue within supply chains • Published data on key determinants • Ability of Use Descriptors to differentiate certain uses (e.g. sub PCs) • Extended set of standard phrases
Tier 1++	Further refinement of scenarios of likely particular interest *Limited current activity*	• Targeted at uses and substances of concern • Provides added assurance • Allow for 'unique' controls to be addressed	• Activity for concerned end user groups • Openly available supporting data • Use Descriptors define domain of reliability • Extended set of standard phrases

Considerations for effective exposure data collections strategies

Mo-SY-C1: Intermittent Exposure in Risk Assessment

Mo-SY-C1.1

Introduction to intermittent exposure and examples from regulatory risk assessment of consumer exposure situations

Friederike Neisel, *Federal Institute for Risk Assessment, Berlin, Germany*
Ulrike Bernauer, *Federal Institute for Risk Assessment, Berlin, Germany*
Astrid Heiland, *Federal Institute for Risk Assessment, Berlin, Germany*
Matthias Herzler, *Federal Institute for Risk Assessment, Berlin, Germany*
Gerhard Heinemeyer, *Federal Institute for Risk Assessment, Berlin, Germany*

Background:
Risk assessment of consumer exposure is generally based on chronic health-based limit values that shall exclude adverse health effects in case of long-term exposure up to exposure over life-time. However, in reality, exposure of the general population will often fluctuate between higher and lower levels. In addition, many consumer products are used infrequently, with an on-off exposure profile called intermittent exposure. In these cases, questions arise whether a refinement of the risk assessment is possible, if exposure during the use event is higher than the respective chronic health-based limit value.
Objectives:
The presentation shall introduce to determinants of risks from intermittent consumer exposure.
Methods:
Examples of current risk assessments for the general population are discussed, e.g. exposure to outdoor air contaminants or to do-it yourself products like paints. This includes different approaches for risk assessment of long-term and short-term effects. An accent is given to different exposure time patterns resulting from one consumer use and to the quality of input data on consumer exposure frequency and time, including the variability of consumer behavior.
Results:
At present, there is still no common approach in dealing with fluctuating and intermittent exposure of the general population. Refinements of risk assessments for intermittent consumer exposure should consider toxicological information as well as the quality of the underlying exposure data.

Mo-SY-C1.2

Setting typical intermittent or peak exposure profiles, focusing on consumer non-food and dietary exposure

Wouter Ter Burg, *RIVM, Bilthoven, Netherlands*

In risk assessment of chemicals normally a chronic health based limit value is compared to an averaged chronic or lifetime exposure. Traditionally, the default approach is to derive a chronic exposure estimate, whereas in real-life, the exposure is in many cases not chronic or constant over time. In most situations, the exposure fluctuates over time, is intermittent, showing peaks, or only occurs at elevated levels for certain periods of time. Averaging these exposures to a lifetime exposure can lead to false conclusions when compared to a chronic based health limit value. A shift in the risk assessment approach is needed, where starting point should be the actual exposure as a given fact and making sure the toxicological reference value is appropriate to assess the risks for that exposure. As an important start in the approach set up by Geraets et al (submitted), the many types of exposures have been categorized into certain exposure profiles. An exposure profile involves the contact of a substance with a subject over a period of time. The exposure profiles describe the external exposure resulting from use of dietary and non-food products or residential exposure, exemplified by daily life products. Thus far the focus has been on the use of products, but substance properties on absorption, kinetics and mode of action may influence the categorization of the exposure profiles. It shows that in risk assessment the exposure assessment and hazard characterization are closely linked and dependent on each other.

Mo-SY-C1.3

Application of Haber's rule, mode of action and role of toxicokinetics

Wouter Ter Burg, *RIVM, Bilthoven, Netherlands*

The exposure assessment should describe the actual exposure as accurate as possible. The way the exposure is looked at in detail then depends heavily on the hazard characterization of a chemical. The toxicokinetics and mode(s) of action play an important role, both for the exposure and toxicity alike. Information on these aspects has impact on what elements to consider in risk assessment, e.g. do you focus on the concentration in a certain media or on the dose expressed per body weight. Further, how does duration change the focus of the assessment.
Haber's rule, describing the relationship between concentration/dose, time and an effect, has been used in risk assessment extensively. Basically, he assumed a constant relationship between concentration/dose and time eliciting a predefined effect, C x t = k. More recently, a modification was applied to the rule indicating that C^n x t = k with a substance specific value for n provided better predictions. The rule has been determined for acute lethal effects, but nevertheless has been applied to many other endpoints. The presentation will focus on different endpoints, with their mode of actions and whether or not the modified Haber's rule may apply.

Mo-SY-C1.4

Less than lifetime exposures to pesticide residues might need a risk assessment

Jürg Zarn, *Federal Food Safety and Veterinary Office (FSVO), Bern, Switzerland*

The ARfD and ADI concepts, related to acute or chronic exposures, are used to assess the dietary risk of pesticide residues in crops. Chronic exposure estimation utilizes per capita lifetime averaged food consumption values and is therefore often significantly lower than acute exposure estimates because these are based on high percentiles of food consumption surveys.

In current dietary risk assessment, all pesticides are assessed with regard to chronic exposure, but only those that have shown to be acutely toxic and therefore have an ARfD allocated, also with regard to acute exposure. For exposures longer than one day but shorter than chronic there are no exposure scenarios available. To evaluate whether such scenarios covering intermittent exposures are necessary, the effect of extending the exposure duration on the toxic potency of pesticides was investigated.

Ratios of subacute/subchronic, subchronic/chronic and subacute/chronic no observed adverse effect levels (NOAEL) from rodent pesticide feeding studies were calculated. Ratio distributions differing significantly from 1 would indicate an effect of exposure duration on NOAELs.

No statistical significant effect of exposure duration on NOAEL ratio distributions was found. On the contrary, the number of animals and the dose spacing in the studies were identified as factors pretending an effect of exposure duration on NOAELs as well as the dose decrement over time observed in feeding studies applying the test substances in constant feed concentrations.

The results indicate that toxic potencies of pesticides are similar over a wide exposure duration range. However, for pesticides not allocated an ARfD, the dietary risk assessment is based only on the ADI and related low-level exposure estimates. As no short-term exposure scenarios are available, the short-term potency of pesticides residues in crops remain unassessed.

Zarn, J. A., B. E. Engeli and J. R. Schlatter (2011) Regul. Toxicol. Pharmacol 61(2): 243-250.

Zarn, J. A., B. E. Engeli and J. R. Schlatter (2013) Regul. Toxicol. Pharmacol 67(2): 215-220.

Mo-SY-D1: From external to internal exposure: the necessity of toxicokinetic information

Mo-SY-D1.1

Introduction to kinetics as connector of external and internal dose: importance for risk assessment

Peter Bos, *RIVM, Bilthoven, Netherlands*

An introduction on the importance of insight in kinetics for the estimation of internal exposure will be given. Human risk assessment is generally based on external dose metrics. There is an increasing awareness that this process can be improved if exposure assessment focuses on internal rather than external exposure estimates. One of the main advantages hereof is that exposure from different sources can be combined, especially if exposure to a substance occurs via multiple routes, i.e., oral, inhalation and/or dermal exposure. At present, the most common way is to add up the doses taken up via the respective exposure routes, at best adjusted for the fraction absorbed per route of exposure. The internal exposure then is often expressed as internal dose per kg body weight.
However, each route of exposure is characterized by kinetic processes that determine the fate of a substance in the body and thus the internal exposure. Four steps can be distinguished: absorption, distribution, metabolism and excretion. Insight in these processes is essential for an adequate estimation of the internal exposure, i.e., the internal exposure at the site (organs, tissues) where toxicity is expected.
The first question then raised is what the appropriate internal dose metric will be. Some toxic effects are related to tissue concentration while for other effects tissue dose is the best descriptor. This understanding underlines the necessity of taking into account the exposure rate, i.e., the impact on the internal exposure of the exposure dose in combination with the exposure duration. A short exposure to a high dose will lead to a different internal exposure pattern (and thus a different outcome of toxicity) than exposure to the same dose spread over a longer period of time. The more so, since at high doses saturation of kinetic processes may occur.
A second point is knowledge about whether the toxicity is caused by the parent compound one is exposed to or by one of its metabolites. For instance, if a metabolite is the toxic agent, oral exposure may lead to different toxicity than dermal exposure to the same dose since following oral exposure all substance first passes the liver (the main organ for metabolism) before being distributed throughout the body.
Internal exposure estimates that can be related to toxic effects can reduce uncertainties involved in human risk assessment but an adequate insight in route-specific kinetics is a prerequisite.

Mo-SY-D1.2

The importance of information on toxicokinetics for human health risk assessment: specific issues for inhalation exposure

Liesbeth Geraets, *RIVM - National Institute for Public Health and the Environment, Utrecht, Netherlands*
Marco Zeilmaker, *RIVM - National Institute for Public Health and the Environment, Bilthoven, Netherlands*
Peter Bos, *RIVM - National Institute for Public Health and the Environment, Bilthoven, Netherlands*

Background: Human health risk assessment includes in general a high-to-low dose extrapolation. Experimental animals are usually exposed in toxicity studies using dose levels which vary over one order of magnitude. Moreover, the experimental animal dose levels are in most cases significantly higher than the human exposure levels. Although high-to-low dose extrapolation is a common step in human risk assessment, it introduces various uncertainties. One of these uncertainties is related to the toxicokinetics of the specific chemical. Many kinetic processes such as absorption, metabolism or excretion can be subject to saturation at high dose levels, resulting in disproportionate increases in internal blood or tissue concentration relative to the external dose administered.
Objectives: The aim is to emphasize the importance of kinetic information for the determination of a safe exposure in human risk assessment of inhalation exposures assessed by conversion from a high animal exposure to a low exposure in man.
Methods: For two selected compounds, i.e. methyl tert-butyl ether and 1,2-dichloroethane, Physiologically Based Toxicokinetic (PBTK) modeling was used to follow the extrapolation and conversion steps as performed in existing risk assessments for these compounds. The modeled human exposure scenarios include both continuous long-term exposure for the general population and occupational exposure for 8h/d for 5 days per week. Both blood Cmax and AUC are considered as internal dose metrics. It is presented how internal exposure might change due to high-to-low dose extrapolation applied on the external exposure. Also, the impact of conversion of an intermittent exposure to continuous exposure on these internal dose metrics is included.
Results: In order to reduce the uncertainties related to high-to-low dose extrapolation, information on toxicokinetics is considered essential in order to create an adequate margin of safety at the appropriate dose metric. Human health-based limit values based on an external dose metric without sufficient knowledge on kinetics might be too high to be sufficiently protective. Without insight in the actual internal exposure, the toxic agent, the appropriate dose metric, and whether an effect is related to internal concentration or dose, application of assessment factors on an external dose metric and the conversion to continuous exposure results in an uncertain human health risk assessment of inhalation exposures.

Mo-SY-D1.3

Toxicokinetics for the oral route: human risk assessment of chemicals and food safety

Jean Lou Dorne, *EFSA, Parma, Emilia Romagna, Italy*

Human risk assessment of chemicals in the food safety area involves the classic steps bringing the hazard and exposure dimension together for risk characterisations. Sound hazard identification and hazard characterisation requires an understanding of both toxicokinetic (TK) and toxicodynamic (TD) processes for compounds entering the human body via the oral route. This enables the translation of external dose (exposure) into internal dose (TK) to integrate absorption, distribution, metabolism, excretion processe (ADME) and toxicity (TD) for sound dose response modelling and risk characterisation.
A number of historical developments have enabled risk assessors to integrate interspecies differences and human variability in TK (ADME) processes for chemicals entering the body via the oral route. In this context, the Mode of Action (MoA) and Adverse outcome pathway (AOP) frameworks provide science-based starting points to integrate TK and TD taking into account the biological relevance of test species compared with humans. Of utmost importance for the oral route is the understanding of the role of the intestine and the liver in ADME processes particularly regarding absorption, bioavailability, intestinal versus liver metabolism and transport and their consequences (e.g. bioactivation, detoxification, entero-hepatic re-circulation). Tools available to address these issues include in vitro cell systems in humans and test species and physiologically-based (PB) models such as PB-TK and PB-TK -TD models as part of integrated testing strategies (ITS). Integrating TK knowledge requires understanding the practical needs of risk assessors and decision makers from problem formulation, level of ADME knowledge available for the chemical to be assessed (known knowns, known unknowns and unknowns unknowns) as well as the resources and time available for the assessment. Three examples of TK integration in the food safety area are illustrated namely an emerging mycotoxin in food with scarce TK data, a regulated product with TK data from a dossier and a well-known contaminant with full PB-TK or PB-TK-TD models.
Further developments in this area include further integration of ITS as mechanistic alternatives to animal testing and involve the integration of human in vitro models to depict ADME processes with quantitative knowledge of human variability for in vitro to in vivo (IVIVE) extrapolation and the development of in silico tools including generic QSAR tools and PB models. In this context, international cooperation between national and international scientific advisory bodies is critical and will enable sharing data and experience, developing harmonised databases, tools and models and training of future generations of risk assessors.

Mo-SY-D1.4

Dermal Exposure - Special Considerations

Ursula Gundert-Remy, *Berlin, Germany*
Hans Mielke, *Federal Institute for Risk Assessment, Berlin, Germany*

For the risk assessment route-to-route extrapolation is often done by correcting the non-oral route exposure by the route specific absorption and comparing the result with the (oral) health based guidance value (HBGV). This procedure gives incorrect results if the absorption on the oral route is less than 100%. With this contribution we give two examples for estimating the risk by dermal exposure.
Example 1: Bisphenol A. The target organs for toxicity on the oral route are liver and kidney. BPA has a high presystemic first pass, it is anticipated that the route of exposure is an important determinant for the concentration in blood and organs. A PBTK analysis was done to tackle the problem.
For the parametrisation of the human physiologically based toxicokinetic (PBTK) model the following data were used: in silico data for the partitioning in the tissues, human in vitro data for Vmax and Km, in vitro data and in vivo human data for the rate and extent of dermal and oral absorption.
The modeling results are in line with the physiology of dermal versus oral absorption. After absorption through the skin the blood in the portal vein has a lower concentration as compared to the oral route. Cmax in the liver is several fold lower after dermal as compared to the oral administration.
Example 2: Coumarin is a liver toxicant with a high first pass. For the parametrisation of the PBTK model we used in silico data for the partitioning in the tissues, in vitro data from human hepatozytes for the metabolic constants Vmax and Km, in vitro data and in vivo data for the rate and extent of dermal and of oral absorption. At identical doses compared to oral exposure, the dermal exposure resulted in a lower simulated liver peak concentration. Are Cmax or AUC values in the liver relevant for the risk assessment? For doses and study durations in 11 oral rat studies we derived 31 Cmax and AUC values in a rat PBTK model. We graded 31 hepatotoxic responses in the 11 studies on a five-point grading scale (0= no effect, 4= massive liver toxicity). In a graphical analysis the severity grade of hepatotoxicity increases systematically with increasing Cmax-hep, which was not observed for AUC-hep indicating that Cmax-hep is the relevant toxicokinetic metric.
PBPK modelling is a helpful tool to perform risk assessment for dermal route of exposure.

Mo-SY-D1.5

Kinetic Issues when Performing Biomonitoring

Lisbeth E. Knudsen, *University of Copenhagen, Copenhagen, Denmark*

Background
Human biomonitoring represent the integrated exposure in humans and reflects the uptake, metabolism and excretion, considered affected by genetic and other factors. In the DEMOCOPHES project we were able to monitor exposures of school children and their mothers to a number of environmental toxicants.
Objectives
The current presentation stresses the importance of taking kinetic issues into account when interpretatting results.
Method
Danish participation in the large European Human biomonitoring (HBM) pilot project DEMOCOPHES Demonstration of a study to Coordinate and Perform Human biomonitoring on a European Scale investigated the exposure to 65 different biomarkers in a group of Danish school children aged 6-11 years and their mothers from rural and urban areas in 2011. The environmental exposure to chemicals from a variety of different sources including foods, plastics, electronics, cosmetics, and housing was investigated. Nine polychlorinated biphenyls (PCBs), 4 dichlorodiphenyltrichloro-ethane (DDT) metabolites, hexachlorobenzene (HCB) and beta-hexachlorocyclohexane (ß-HCH), 6 polyfluoroalkyl substances (PFASs) and 7 polybrominated diphenyl ethers (PBDEs) were analysed in plasma samples. The blood samples were also analysed for dioxin-like activity and the biomarker of effect, micronuclei frequency. Hair samples were analysed for mercury. Urine samples were analysed fifteen phthalate metabolites, 7 parabens, and 9 phenols as well as cadmium, paracetamol and cotinine.
Some of the substances are short-lived, while others have long half-life. This influenced how the biomonitoring was performed, i.e. which type of samples were taken and at what time.
Results
There was a significant association between the intake of fish and mercury hair concentrations. The exposure to most chemicals analysed seem to follow a family related pattern and the concentrations of chemicals within the same groups are significantly correlated.
As some of the compounds were measured in higher levels in children compared to mothers, increased focus on the exposure in young children in the future is recommended. Further investigations of the correlation between biomarkers of exposure, as well as follow-up studies of the participants with repeated biomarker measurements and register-based investigations is suggested. For more detailed investigation of specific exposure sources more studies with increased power and detailed questionnaire is needed. Toxicokinetic considerations will be included in the planned European Human Biomonitoring Initiative.
Mørck TA, et al. The Danish contribution to the European DEMOCOPHES project: A description of cadmium, cotinine and mercury levels in Danish mother-child pairs and the perspectives of supplementary sampling and measurements. Environ Res. 2015 141:96-105 and PhD thesis of TA Mørck.

Mo-SY-E1: Exposure to atmospherically dispersed hazards: assessment, public information and perspectives - I

Mo-SY-E1.1

Exposure maps for NO2 and Particulate Matter in the Netherlands; spatial resolution versus temporal resolution

Joost Wesseling, *National Institute for Public Health and the Environment, Bilthoven, Netherlands*
Ronald Hoogerbrugge, National Institute for Public Health and the Environment, Bilthoven, Netherlands
Peter Zandveld, *TNO, Utrecht, Netherlands*

Yearly average concentrations of NO2, PM10, PM2.5 and EC in the Netherlands are prepared every year. These maps have a high spatial resolution of 25 by 25 meter and serve to provide high resolution concentration distributions for the concentrations. For this, all important roads are taken into account in a detailed way. The effects of emissions from other sources are taken into account by way of background maps that are calibrated using measurements from the National Air Quality monitoring Network. Although concentrations are calculated for every adres in the Netherlands, only average values on a 25 by 25 meter grid are shown on the maps.
Apart from the yearly average maps, also hourly maps for NO2, O3 and PM10 are calculated on a scale of 125 by 125 meter. The results of these calculations are provided to the public using a website and apps. For these hourly maps, concentrations measured at (urban) background locations are used to derive background concentration maps. The concentration contributions from traffic on the major roads are calculated and subsequently calibrated (scaled) using measurements at locations near large roads. For this calibration all available relevant measurements in the Netherlands are employed. The hourly maps are also used to provide an hourly air quality index for all of the Netherlands. Furthermore, they serve as input for several other projects.
Hourly prognoses are calculated using the HIRLAM and LOTOS-EUROS models and a model for contributions from traffic on highways. The results of estimations for the background concentrations and the traffic contributions are combined to yield total concentration maps for the next 48 hours.
The construction of both sets of maps, high spatial resolution combined with low time resolution and vice versa, will be discussed and examples wil be presented.

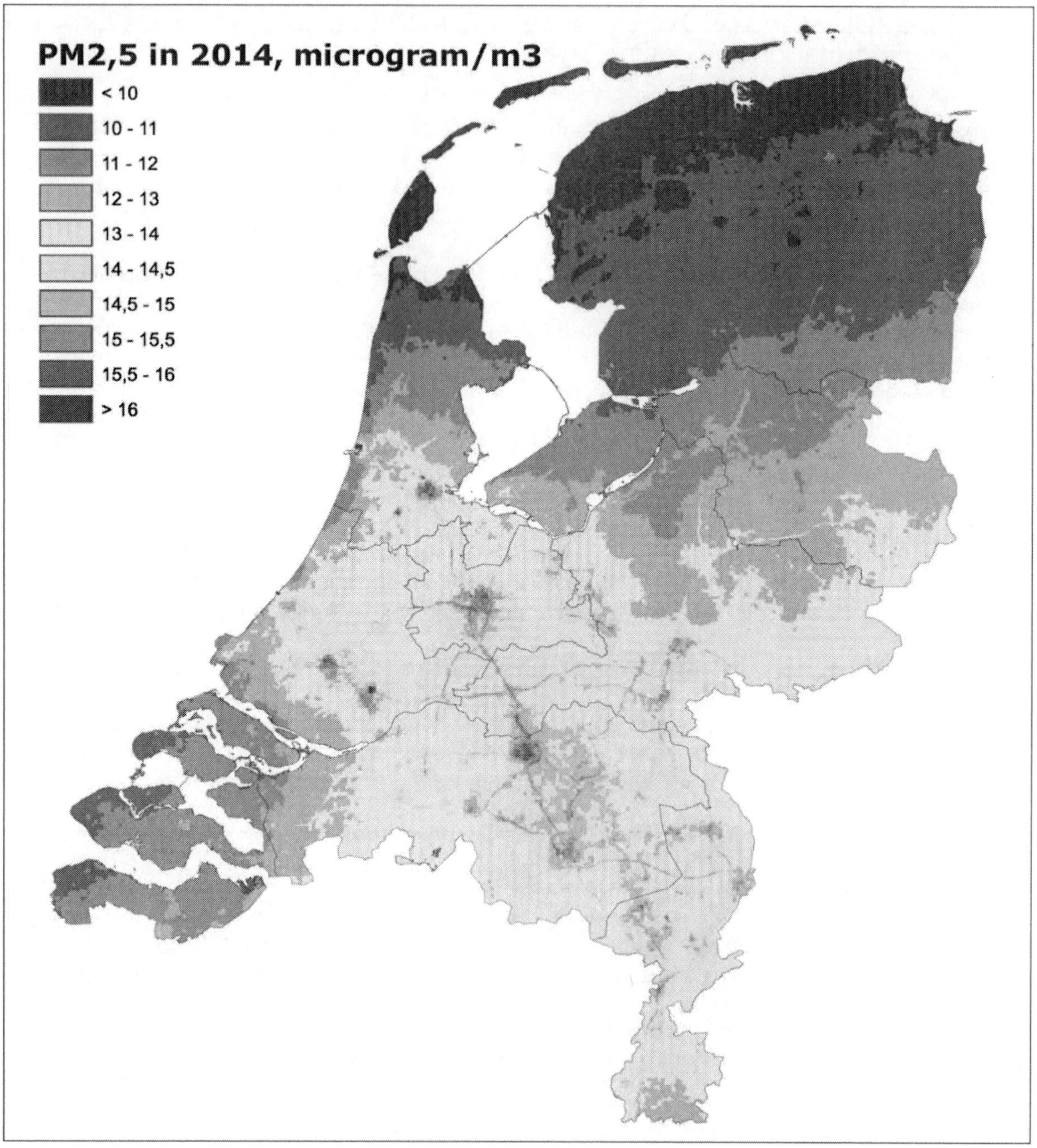

PM2,5 concentrations in The Netherlands in 2014 in microgram/m3.

Mo-SY-E1.2

Air Quality Impact Assessement for Ringland

Bino Maiheu, *Flemish Institute for Technological Research, VITO nv, Mol, Belgium*
Stijn Vranckx, *Flemish Institute for Technological Research, VITO nv, Mol, Belgium*
Wouter Lefebvre, *Flemish Institute for Technological Research, VITO nv, Mol, Belgium*
Stijn Janssen, *Flemish Institute for Technological Research, VITO nv, Mol, Belgium*

The Antwerp ring road (R1) has been at the centre of a public/political debate in Flanders for more than a decade, involving different action groups. The Ringland initiative crow-sourced over a 100 kEUR from enthusiasts to fund environmental, health and economic impact assessment studies. This contribution will report on the air quality impact of their solution to the ring road problem.
The tunnel complex which the Ringland initiative proposes, separates the R1 into a local traffic tunnel and a tunnel for through traffic, with several entrances and exits and number of surface-level interchanges along the way. A good estimation of the redistribution of traffic emissions from the tunnel portals is therefore essential for any air quality impact assessment.
To redistribute the traffic emissions to the different tunnel portals, the Hardy-Cross method, typically used for hydraulic network calculations, was used. In this method a network of pipes (traffic tunnels in our case) can be solved iteratively for total pressure in its nodes and flow speed in its pipe segments when the pressure losses (or gains) are known in each segment. Next to friction losses near the walls, a pressure gain is induced by the vehicle piston effect, allowing us to compute the flow in the complete tunnel system depending on tunnel geometry and traffic intensity/composition.
The model estimated respectively 14.6 % of the total R1 traffic emissions are released near the western exit/entrance of the tunnel complex, 21 % to the north and 10.1 % to the south and 15.7 % to the eastern main exit/entrance. The remainder of the emissions are released along the several exits along the trajectory (with a maximum of 7 % of the total near the busiest section of the R1).
The IFDM dispersion model was used to compute the redistributed emissions on the ambient annual averaged concentrations for the whole agglomeration. Sizeable reductions were found, not only near the original location of the R1, but also in the urban background. The tunnel portals remain hotspots, with significant increases, however limited to a few 100 m from the portal. Overall, 97 % of the population see a reduction in concentrations levels, whereas only 3 % see an increase. The results were subsequently used by Belgium's leading epidemiologists concluding that the project would result in an annual decrease of 21 deaths.
This work was supported by the Ringland initiative and the contributing citizens of Antwerp.

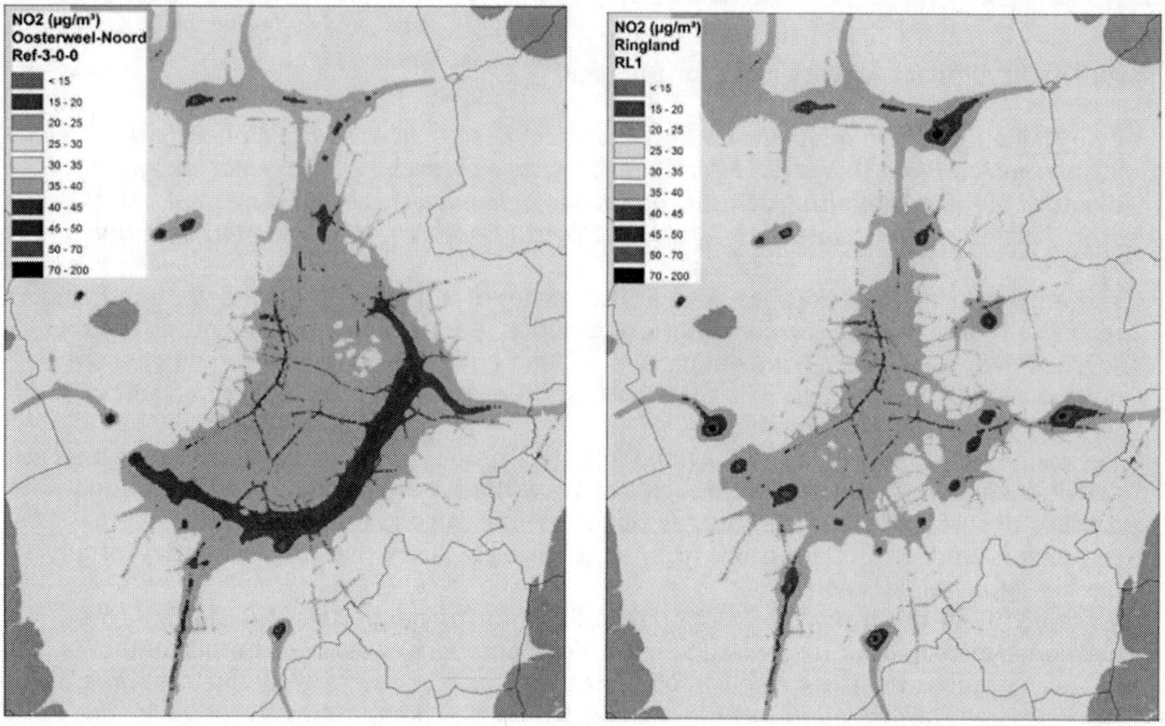

NO2 concentrations in 2020 in one of the likely future scenario's for Antwerp (left), right: Ringland NO2 map.

Mo-SY-E1.3

China's revised air quality index (AQI) compared to other AQI-s. Exposure and communication aspects of AQI design.

Sef Van den Elshout, *DCMR Environmental Protection Agency Rijnmond, Schiedam, Netherlands*
Lisa Blyth, *VITO Flemish Institute for Technology Research, Mol, Belgium*
Yu Lei, *Chinese Academy for Environmental Planning, Beijing, China, People's Republic of*
Wei Liu, *Chinese Academy for Environmental Planning, Beijing, China, People's Republic of*
Nele Veldeman, *VITO Flemish Institute for Technology Research, Mol, Belgium*

Background
An AQI (air quality index) is a communication tool. China revised its AQI in 2012 (HJ633-2012 - on trial). The revised AQI takes into account recent insights in relevant pollutants and the health effects associated to exposure levels. PM2.5 was included in the AQI based on international consensus on its relevance, and following public demand. It is published in real-time with hourly updates. In 2014 experiments with hourly PM index readings (previously 24h moving average) started to improve the consistence between perceived and reported air quality. The practical performance and the appropriateness of the revised AQI is assessed.
Methods
First the design and concept of the Chinese AQI is compared to other AQIs (US, UK, Europe, HongKong). AQI scale and breakpoints and health messages at different exposure levels are compared. Secondly, the selected AQI-s are applied to Chinese monitoring data (unvalidated online data) to verify their practical usability of the AQIs under Chinese conditions.
Results and discussion
Almost all AQIs differ in design and interpretation of the air quality. The Chinese authorities designed their AQI taking into account both recent exposure science as wells as the prevailing air quality conditions. In fact, the results show that applying AQI breakpoints currently in use in Europe is not feasible as the AQI would be in or above the highest class for most of the time. This is an undesirable communication property.
Most AQI-s address short-term health effects of air pollution and provide behavioral advice but air quality has health effects both for short- and long-term exposure. This is a communication challenge anywhere and for every AQI. The revised Chinese AQI is used to assess exposure and provide behavioral advice. The results show that given the current structural level of air pollution, AQIs aimed at communication on the short-term exposure face difficulties in China. Frequently occurring pollution levels are judged as harmful. Though this might represent an accurate exposure risk, it hardly makes sense to communicate (day after day) that exposure is too high: there is no perspective to act for the individual, the pollution cannot be escaped. The communication relevance is reduced and message-fatigue might occur.
The revised Chinese AQI is a pragmatic compromise between recent thinking on exposure and practical feasibility as a communication tool, given the current Chinese air quality. Several recently updated AQI-s use short PM-averaging times, just like the Chinese AQI.

Mo-SY-E1.4

Spatiotemporal correlations between air pollutants and eye-, nose- and lung symptoms of individuals collected by a citizen science platform

Letty A De Weger, *Leiden University Medical Center, Leiden, Netherlands*
Joost Wesseling, *National Institute for Public Health and the Environment, Bilthoven, Netherlands*
Ronald Hoogerbrugge, *National Institute for Public Health and the Environment, Bilthoven, Netherlands*
Arnold JH van Vliet, *Wageningen University, Wageningen, Netherlands*
Pieter S Hiemstra, *Leiden University Medical Center, Leiden, Netherlands*
Jacob K Sont, *Leiden University Medical Center, Leiden, Netherlands*

Background: According to the last report of the European Environmental Agency (1), up to 93% of the urban population in the European Union is exposed to levels of PM2.5 above World Health Organization guidelines and up to 98% to levels of ozone above these guidelines. Health risks resulting from air pollution are substantial, and include heart disease, stroke, respiratory diseases and premature deaths. However, the proportion of the population affected by less severe disease is much larger than the proportion suffering from severe diseases. In spite of this, most epidemiological studies focus on the severe outcomes, because these data are usually better available. In this study we focus on nose-, eye- and lung symptoms, which can have a strong impact on public health and on economy (e.g. health care costs, lost work and school days).
Objective: The aim of this study was to determine whether symptom data collected by the citizen science platform Allergieradar.nl (2) can be used to identify individuals that experience adverse health effects following exposure to the most relevant pollutants: PM10, O3 and NO2 and pollen.
Methods: Local pollutant concentrations were calculated for every hour using adaptations of the Dutch standard calculation models for air quality(3). We assessed to what extent these data and daily pollen counts (Leiden University Medical Center) were related to eye, nose and lung symptom scores in individuals participating in the citizen science platform Allergieradar.nl (2) between 1 Jan-31 Aug 2014 and 1 Jan-26 Oct 2015. The data were analysed using the software package Intercooled STATA 11.0 (StataCorp, USA). Data from participants with less than 6 entries were not used in this study.
Results: In the analyses the correlation coefficients between daily symptoms scores of 237 participants (5174 valid entries) and pollen or local pollutant concentrations were determined. Correlations coefficients >0.7 were found for either nose, eye or lung symptoms of 30, 9, 18 and 13 individual participants with pollen, NO2, PM10 and O3, respectively.
Conclusions Our study indicates that marked interindividual differences exist regarding the strength of the relationship between symptoms and one or more component of air pollution or pollen. In the future this may enable us to provide personalized information to individuals regarding potential risk of exposure to air pollution or pollen.
1.Air quality in Europe (2015) European Environmental Agency, Copenhagen, Denmark
2.de Weger LA, et al. (2014) Allergy 69:1085-1091
3.http://wetten.overheid.nl/BWBR0022817/

Mo-SY-E1.5

Atmospheric dispersion modelling of the large Q fever outbreak in the Netherlands in 2009

Jeroen Van Leuken, *RIVM National Institute for Public Health and the Environment, Bilthoven, Netherlands*
Jan Van de Kassteele, *RIVM National Institute for Public Health and the Environment, Bilthoven, Netherlands*
Ferd Sauter, *RIVM National Institute for Public Health and the Environment, Bilthoven, Netherlands*
Wim Van der Hoek, *RIVM National Institute for Public Health and the Environment, Bilthoven, Netherlands*
Dick Heederik, *Utrecht University, Utrecht, Netherlands*
Arie Havelaar, *University of Florida, Gainesville, Florida, United States*
Arno Swart, *RIVM National Institute for Public Health and the Environment, Bilthoven, Netherlands*

Aim
From 2007 through 2010, the Netherlands experienced the largest human and veterinary Q fever epidemic ever described. Over 4,000 human cases were notified and approximately a twelve-fold higher number was probably infected by Coxiella burnetii, the causative agent of Q fever. Dairy goat farms, and to a lesser extent dairy sheep farms, were identified as the major source of these human infections with high Coxiella burnetii shedding rates during parturition of the animals. This project aimed at simulating the airborne dispersion of Coxiella burnetii through the atmosphere and to verify the correlation between exposure and Q fever observed in humans. Thus, we could verify whether meteorological conditions might have played a role during the epidemic.
Methods
We used an atmospheric dispersion model (Operational Priority Substances Short Term model) to simulate the dispersion of Coxiella burnetii. We correlated the modelled exposure levels to the observed human Q fever incidence data. As a comparison, we also defined two simple models with no meteorological information: a spatial uniform concentration distribution ('null model') and a distance-dependent concentration distribution ('distance model'). Since Coxiella burnetii emission rate data were lacking we defined three simple emission profiles as an input for the atmospheric dispersion model.
Results
Exposure levels modelled with the atmospheric dispersion model explained the spatial distribution of human cases better than the distance and null model. Mainly wind (both speed and direction) and convection were important. The wind conditions determined the size of the exposed area; atmospheric stability influenced the exposure levels at surface level.
Conclusions
We concluded that atmospheric dispersion models are suitable for dispersion modelling of airborne pathogenic micro-organisms. Modelled airborne Coxiella burnetii concentrations were a better predictor for Q fever incidence than distance alone. Although additional information is needed - especially regarding emission data - these results provide a basis for the use of atmospheric dispersion models to predict and to visualise the spread of airborne pathogens during livestock, industry or bio-terroristic related outbreaks to a surrounding human population. Follow-up research now focuses on parameterisation of

emission profiles, dispersion modelling, and translating exposure to disease for other pathogenic micro-organisms from intensive livestock farming.

Mo-SY-F1: Extending participatory sensing to personal exposure and policy support - I

Mo-SY-F1.1

Integrating low-cost sensor and model data to improve the assessment of personal exposure to air pollution in the urban-rural nexus.

Stefan Reis, *Centre for Ecology & Hydrology, Edinburgh, United Kingdom*
Massimo Vieno, *Centre for Ecology & Hydrology, Edinburgh, United Kingdom*
Susanne Steinle, *Institute of Occupational Medicine, Edinburgh, United Kingdom*
Rachel Beck, *Centre for Ecology & Hydrology, Edinburgh, United Kingdom*
Edward Carnell, *Centre for Ecology & Hydrology, Edinburgh, United Kingdom*

Individual mobility includes commuting to places of work and education, shopping, leisure activities or spending time outdoors for recreational activity. Despite a growing understanding of the substantial variety of individual lifestyles and how they affect the time spent and activities undertaken in a wide range of micro-environments on an individual level, the assessment of public health effects of air pollution often relies on quantification methods which assume a static population at their residential locations. With the emergence of low-cost personal sensors capable of measuring individual level exposure, as well as providing much better spatio-temporally resolved data on pollution fields for key air pollutants, the assessment of health effects from exposure to air pollution can make step changes towards a more comprehensive picture of exposure for individuals, population sub-groups and the general public. To achieve this, the integration of both sensor networks at different levels and scales, and of models from local to urban to regional scale is an essential building block. In this paper, we will demonstrate the relevance of taking into account individual mobility and population level spatial variability, as well as the benefits of personal sensing to inform and validate models. The implications of these findings are relevant to policy makers as well as local authorities, regulators and urban planners alike.
In our paper, we present the results from a model analysis of population-level exposure to fine particulate matter (PM2.5) and nitrogen dioxides (NOx) in the UK, using new Census 2011 population data, which includes both residential population distribution, and population during working hours (see Figure). Applying the atmospheric chemistry transport model EMEP4UK, we will simulate population-weighted exposures to PM2.5 and NOx for a whole year, in order to quantitatively assess the influence of location and behavioural patterns on potential exposure for population sub-groups. In addition, we will compare results from a personal exposure monitoring study (using GPS and portable particle monitors) with model results for the same period and location.
The results of this analysis will inform the development of more appropriate air pollution control policies by identifying the relative contribution of urban vs. non-urban pollution levels, and is builds on a recent paper by Vieno et al. (2016).
Reference: Vieno, M., Heal, M. R., Williams, M. L., Carnell, E. J., Nemitz, E., Stedman, J. R., and Reis, S.: The sensitivities of emissions reductions for the mitigation of UK PM2.5, Atmos. Chem. Phys., 16, 265-276, doi:10.5194/acp-16-265-2016, 2016.

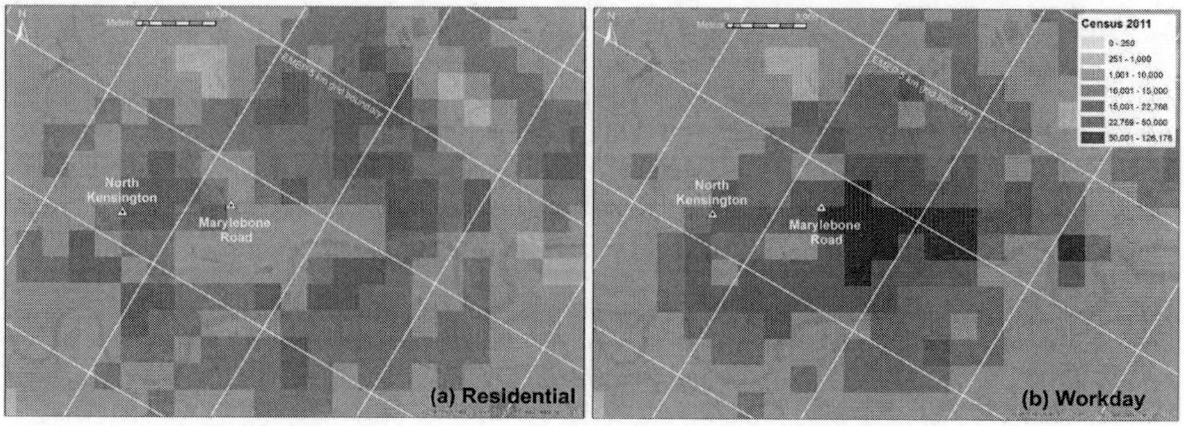

Illustration of variability of urban population density comparing (a) residential and (b) workday populations based on UK Census 2011 and CEH Land Cover Map 2007 datasets

Mo-SY-F1.2

The challenges of developing reliable air pollution exposure surfaces using ad-hoc participatory sensing data

Laura Minet, *McGill University, Montreal, Canada*
Ronald Gehr, *McGill University, Montreal, Canada*
Jeffrey Brook, *Environment Canada, Toronto, Canada*
Ilan Levy, *Hebrew University of Jerusalem, Jerusalem, Israel*
Marianne Hatzopoulou, *University of Toronto, Toronto, Ontario, Canada*

The objective of this study is to compare exposure surfaces developed based on static and dynamic measurements of ambient NO2 and O3. For this purpose, a mobile monitoring campaign was conducted in summer 2015 in Montreal. Volunteers carried micro sensors as they walked and cycled around the city. Pollutants levels of 1411 road segments were measured. Using data from repeated visits at each segment, various land-use regression (LUR) models were developed. One year earlier, during summer 2014, a campaign with fixed monitor locations had taken place at 76 sites using the same micro sensors. LUR models as well as associated exposure surfaces were developed based on participatory sensing in 2015, and compared with summer 2014 results. The exposure surfaces resulting from both campaigns were highly dissimilar, and several possible explanations can be put forward. LUR models based on road segments with a small number of repeated observations are associated with poor coefficients of determination (R^2) and the exposure surfaces derived from them are poorly correlated with the summer 2014 exposure surface. On the other hand, restricting the LUR models to road segments with a high number of observations leads to poor predictive capability out of sample. This study highlights the sensitivity of LUR models based on mobile monitoring campaigns to the number of visits per road segment and to the location of the roads. This stresses the importance of careful design of mobile data collection campaigns.

UTRECHT
THE NETHERLANDS
OCTOBER 9-13
ISES 2016

Mo-SY-F1.3

Participant Use of Wireless Sensing Technologies in an Exposure Study

Miranda Loh, *Institute of Occupational Medicine, Edinburgh, United Kingdom*
Susanne Steinle, *Institute of Occupational Medicine, Edinburgh, United Kingdom*
Miranda Loh, *Institute of Occupational Medicine, Edinburgh, United Kingdom*
Denis Sarigiannis, *Aristotle University of Thessaloniki, Thessaloniki, Greece*
Dimitris Chapizanis, *Aristotle University of Thessaloniki, Thessaloniki, Greece*
Anjoeka Pronk, *TNO, Utrecht, Netherlands*
Eelco Kuijpers, *TNO, Utrecht, Netherlands*
Thomas Maggos, *NCSR Demokritos, Athens, Greece*
Mina Stamatelopoulou, *NCSR Demokritos, Athens, Greece*
John Cherrie, *Institute of Occupational Medicine, Edinburgh, United Kingdom*

The public now has access to numerous wireless technologies and Smartphone apps for monitoring various aspects of people's lives. Many aspects of this "quantified self" are also factors which are important to understand individuals' environmental exposures. Consumer wireless technologies can provide a convenient and useful instrument for data collection for researchers, while also providing participants with an opportunity for greater engagement with their data. For the HEALS (EU FP7) study, such devices have been used to measure physical activity (Fitbit), location (Moves app), diet (Fatsecret app), and indoor climate (Netatmo) variables in a pilot project in cities in 3 countries (UK, Greece, and Netherlands) over a 6-month period. Participants were parents of children aged less than 3 years. Data collected indicated that mean (SD) levels of temperature and relative humidity in homes around Edinburgh (UK) were 19 (2.0) C and 63 (8.2) % respectively, while mean (SD) carbon dioxide levels were 879 (457) ppm. Further analyses are planned to determine room air change rates and to examine longitudinal patterns in time use data for the study population. From a participant perspective, apps and devices that did not require a lot of effort but provided feedback on their situation were received well. Participants expressed interest in being able to track their physical activity and see information about carbon dioxide, temperature, humidity, and noise in their homes, although apps (e.g. Fatsecret) which require more active participation were of mixed benefit.

Mo-SY-F1.4

High resolution modeling of instantaneous personal exposure to traffic related Black Carbon using noise exposure as a proxy for traffic

Luc Dekoninck, *University of Ghent, Ghent, OVL, Belgium*
Dick Botteldooren, *University of Ghent, Ghent, OVL, Belgium*
Luc Int Panis, *University of Hasselt, Hasselt, Belgium*

Traffic simultaneously emits noise and air pollution. A strong correlation between the two environmental exposures can be expected in all micro-environments. In the specific case of particulate matter emission, an actual physical relation exists with the engine noise emission. Capturing the spectral content in the noise exposure provides instantaneous information on the traffic and the traffic dynamics. This relation is used to improve the predictive models for traffic related PM exposure.
Bicyclists are exposed in a direct way to traffic related noise and particulate matter (BC and UFP). Combining the instantaneous traffic information sensitive to the acceleration of the nearby vehicle flow with meteorological data and background exposure in a resolution of 10 seconds enables the disentanglement of the local traffic variation and meteorological disturbances. This results in an additive model predicting the local and background exposure as two independent components. Noise maps are used to extrapolate the noise measurements to the dwellings in the validation dataset. The indoor exposure is modeled as two independent components: the local traffic exposure and background. An I/O ratio, function of the daily average temperature, combined with the extrapolated bicycle model resolves the indoor exposure for the at home indoor micro-environment.
In the in-vehicle micro-environment the background concentration is not the dominant component and the additive approach fails. Increasing the spatiotemporal resolution of the model and including spatial traffic information results in a valid model despite the missing instantaneous traffic data. The complex and non-linear interaction of the diurnal patterns of background concentration, diurnal traffic pattern and traffic dynamics is resolved by fitting a diurnal pattern to the large participatory sensing dataset.
The combination of the three activity specific models results in a daily BC exposure model route sensitive to the personal time-activity pattern and instantaneous meteorology and background exposure. An external participatory sensing campaign acts as an independent validation set and reaches a correlation of 0.65.

Mo-SY-F1.5

On the use of smart phones to promote healthy and sustainable behaviors

Audrey De Nazelle, *Imperial College London, London, United Kingdom*
Roseline Polle, *Imperial College, London, United Kingdom*

Background
Going beyond the purely tailpipe and engine-based technologies, new solutions are sought to address air pollution in cities. Growth in information and communication technology (ICT) innovations and use offers new approaches to tackle this challenge. Digital health and app-based behavioural interventions are booming research and business areas. However, despite increasing public interest in air pollution, the unique opportunities offered by digital technologies for a 2-way process of collecting data, reshaping behaviour, and influencing public policies hasn't yet been explored. We propose a novel ICT-based method, embedded within a transtheoretical model of behavioural change framework, to raise awareness and promote behavioural changes to reduce exposures in urban populations.
Method
An experimental pilot study was conducted in London, UK, summer 2014. Thirty volunteers were tracked using the Moves app during 5 days on two occasions. Data collected by Moves was then processed and combined with local air quality maps to estimate personal exposures to air pollution. Half of the volunteers were provided with personalized feedback on their exposures and contributions to air pollution. All were interviewed three times: before, in-between, and after the tracking sessions. Their stage of change in their consideration of minimizing air pollution exposure was assessed before and after the experiment to estimate the potential for moving individuals towards healthier and more sustainable behaviours through the ICT-derived personalized feedback approach. To provide guidance on how to best enhance such approaches in the future, existing apps and literature on digital technology to promote healthy and sustainable behaviour were reviewed.
Results
Most participants (90%) were in pre-contemplation, meaning they had never considered a change in travel behaviour to decrease their exposures to air pollution. Following personalized feedback, 30% of the participants progressed towards contemplating a behaviour change in the future, while none of the participants in the control group changed. Lessons can be drawn from successes in mhealth to further develop air pollution feedback apps- such as incorporating key features that will develop a sense of community, show progress and enable a sense of autonomy.
Conclusion
There is clear potential to change people's awareness and guide them towards healthier and more sustainable behaviours using novel ICT methods and personalized feedback. Further explorations are needed on precise personalized messages and app features that might produce the greatest effect for stage of change transitions towards actual changes in behaviour, and also to create support for air pollution reduction policies.

Mo-SY-G1: Wastewater-based epidemiology (WBE) - from measuring illicit drug use towards understanding population health status - I

Mo-SY-G1.1

Introduction to wastewater-based epidemiology: novelty and advancements of the approach

Sara Castiglioni, *Mario Negri Institute, Milan, Italy, Italy*

Wastewater-based epidemiology (WBE) is a novel potent approach for monitoring habits and lifestyle of an entire community. It is based on the chemical analysis of urban wastewater for the excreted biomarkers of endogenous human metabolism. Mass spectrometry is the analytical technique used for this approach since it is sensitive and specific enough to detect analytes at trace levels even in a complex matrix such as raw urban wastewater. The general concept was proposed as a New Non-Intrusive Tool to evaluate the use of illicit and misused drugs in a community. In 2005 it was implemented in Italy to estimate cocaine use, and later it was extended to other illicit drugs such as heroin, cannabis, and amphetamine-like stimulants. A rapidly developing scientific discipline was born with the exciting potential for monitoring spatial and temporal patterns of use of illicit drugs on local and national scale, track changes of drug use over time, and identifying new drug use patterns and with the unique ability to provide objective and updated estimates. In 2010, a Europe-wide network (Sewage analysis CORE group - SCORE) was initiated to standardize the WBE approach and to coordinate international studies through the establishment of a common protocol of action. SCORE group assessed the overall uncertainty related to the WBE estimates critically and established a best practice protocol to quantify and reduce the uncertainty and to improve the reliability of estimates. Successive studies were conducted in the following years (2011-2014) involving up to 50 European cities in 16 different countries and contemporaneous studies were conducted in Australia, USA, Canada and Asia. The potential of WBE to complement and extend the existing epidemiologically based approaches was recognized and explored by the European Monitoring Centre for Drugs and Drug Addiction (EMCDDA). They supported multidisciplinary conferences and exploratory international studies. Current research aims to bring together wastewater analysis and drug epidemiology by sharing knowledge from different disciplines.
Recently, the same approach was successfully employed to evaluate the use of new psychoactive substances, counterfeit medicines, alcohol, tobacco and biomarkers of human health related not only to lifestyle of the population, but also to health status and exposure to environmental or food contaminants. Some oxidative stress and pesticides biomarkers have been already tested as well as novel DNA biosensors. Future perspectives of WBE are therefore very promising and novel fields of investigation could be succesfully explored.

Mo-SY-G1.2

5 years of flushing out drug use with sewage-based epidemiology

Kevin Thomas, *Norwegian Institute for Water Research (NIVA), Oslo, Norway*

A collaborative study on the analysis of wastewater for estimating the use of illicit drugs has been performed between 2011 and 2015 allowing spatial differences and temporal changes in illicit drug use to be performed in a large population (i.e. between 11.5-24.7 million people in up to 42 cities; www.score-cost.eu). The study was performed according to best practice where the sewer networks and sampling systems were characterized by the use of a questionnaire and analytical performance by means of repeated interlaboratory test studies. The questionnaire highlighted where differences in sewer design and sampling protocols may influence the data, while the interlaboratory tests showed that the analytical data could be safely compared. In 2011 the approach was simultaneously applied in 19 European cities, making it possible to directly compare illicit drug loads over a 1-week period. The main findings from 2011 were distinct spatial patterns in drug use across Europe. Cocaine use was higher in Western and Central Europe and lower in Northern and Eastern Europe. The total consumption for Europe as a whole is extrapolated to 356 kg daily, which would account for approximately 10 - 15 % of the global supply of cocaine (as estimated by the United Nations Office on Drugs and Crime). High per capita ecstasy loads were measured in Dutch cities, as well as in Antwerp and London. Cocaine and ecstasy loads were significantly elevated during the weekend compared to weekdays. As the collaborative network developed over 70 different cities, including a number from North America and Australia, many of these have been tested within this collaboration over a number of years. Increased spatial coverage has allowed a more robust comparison with surveillance data, which generally were in good agreement. Temporal differences show relatively stable loads overall for all of the investigated drugs across Europe. Wastewater analysis on the above scale has allowed the large spatial and temporal coverage of illicit drug use. This has been achieved quickly and cheaply and highlights the unique data that wastewater provides on illicit drug use that can be used for comparisons with surveillance data or used to plug gaps in available data. Wastewater analysis offers clear benefits when used alongside existing surveillance methods and we recommend its implementation on a formal basis.

Mo-SY-G1.3

Integration of wastewater-based epidemiology in the national drug monitoring system of various countries

Frederic Been, *University of Antwerp, Antwerp, Belgium*
Foon Yin Lai, *University of Antwerp, Antwerp, Belgium*
Juliet Kinyua, *University of Antwerp, Antwerp, Belgium*
Lisa Benaglia, *University of Lausanne, Lausanne, Switzerland*
Olivier Delémont, *University of Lausanne, Lausanne, Switzerland*
Pierre Esseiva, *University of Lausanne, Lausanne, Switzerland*
Alexander L. N. van Nuijs, *University of Antwerp, Antwerp, Belgium*
Adrian Covaci, *University of Antwerp, Antwerp, Belgium*

Aim: Illicit drug residues, in particular their urinary metabolites, are now being analysed in wastewater in various countries to assess and monitor the consumption of illicit drugs at the community level. The objective of this work is to present examples of how results form wastewater analysis can be used, together with conventional surveillance methods, to evaluate current drug policies, identify new threats to public health, develop new strategies and guide law enforcement, both at the regional and national level.
Methods: Results from wastewater sampling campaigns conducted in Switzerland and Belgium are scrutinised to understand their implications from the perspective of policy makers, social workers, addiction researchers and law enforcement. Current knowledge about the epidemiology of drug use in the investigated areas as well as police intelligence about the functioning and structure of local drug markets are included in the investigation. Focus is set on data about cocaine, heroin and amphetamine-type stimulants.
Results: Results from wastewater analysis provided an indirect estimate of the size of the drug market in the investigated areas, as well as a mean to monitor its evolution over time. This information is of particular interest for law enforcement trying to assess the share held by specific criminal organisations on local drug markets. This information can also help understand the structure of criminal organisations, as well as identify their supply routes. From an epidemiological perspective, the information derived from wastewater analysis facilitates the identification of potential changes in drug use (e.g. consumption habits, availability of new substances) compared to conventional surveillance methods. Furthermore, when data derived from the latter is combined to wastewater results, current drug policies and harm reduction measures can be better evaluated.
Conclusion: Wastewater-based epidemiology provides additional and highly relevant information to evaluate illicit drug use both from its epidemiological and criminal dimensions. Yet, to unfold its full potential and allow understanding the implications of wastewater results from a public health perspective, the approach needs to be integrated into the existing surveillance methods.

Mo-SY-G1.4

Bridging the fields of wastewater-based epidemiology with classical epidemiology

Sara Castiglioni, *Mario Negri Institute, Milan, Italy, Italy*
Liesbeth Vandam, *European Monitoring Centre for Drugs and Drug Addiction, Lisbon, Portugal*
Paul Griffiths, *European Monitoring Centre for Drugs and Drug Addiction, Lisbon, Portugal*

Wastewater-based epidemiology (WBE) is a novel approach able to provide estimates of illicit drugs use in a population through the chemical analysis of specific urinary metabolites in urban wastewater. It was demonstrated over the last years that WBE results can complement the conventional approaches used in drug epidemiology to help facing a complex phenomenon such as drug abuse. More recently WBE has been considered as a promising tool to monitor the use of new psycoactive substances (NPS). The potential of using these two approached together seems enormous. Wastewater analysis is able to produce objective, near-real-time estimates of drug use in a defined population, and provides an excellent means of quantifying changes in drug-use over time. It can give timely information and is extremely adaptable since experiments can be designed to study drug use in a specific area or to compare the use between different areas during defined periods of the year. On the other side, traditional epidemiological indicators can provide much more detailed information about patterns of use (e.g. the frequency and mode of use) of a substance and characteristics of the user populations, including the potential harms (e.g. health risks) on individuals.
Few attempts have been made to date to compare drug use estimates obtained through WBE and traditional epidemiological data, but they confirmed the high complementary characteristics of these different approaches. In view of these promising results, WBE will be tested to provide and complement other information in the field of drug epidemiology such as estimation of market size, evaluation of the presence of dumping in a defined area and outcome measurements.
WBE is a new, multidisciplinary field of investigation which requires a high degree of collaboration with drug epidemiologists and stakeholders from public health sector, addiction and prevention institutions in order to improve the comparability of approaches and optimize their complementary characteristics.

Mo-SY-G1.5

A modelling approach to estimate the number of people contributing to a wastewater sample using population biomarkers

Jake O'Brien, *University of Queensland, Brisbane, Queensland, Australia*
Phong Thai, *Queensland University of Technology, Brisbane, Queensland, Australia*
Christoph Ort, *EAWAG, Dübendorf, Switzerland*
Kevin Thomas, *Norwegian Institute for Water Research (NIVA), Oslo, Norway*
Malcolm Reid, *Norwegian Institute for Water Research (NIVA), Oslo, Norway*
José Antonio Baz Lomba, *Norwegian Institute for Water Research (NIVA), Oslo, Norway*
Yeonsuk Ryu, *Norwegian Institute for Water Research (NIVA), Oslo, Norway*
Foon Yin Lai, *University of Antwerp, Antwerp, Belgium*
Jochen Mueller, *University of Queensland, Brisbane, Queensland, Australia*

An important uncertainty of wastewater-based epidemiology (WBE) is the size and variability of the de facto population in the catchment of interest. In the absence of a day-specific direct population count, an indirect surrogate model to estimate population size is required. Such a model requires the use of population markers which should be 1) specific to human consumption; 2) quantifiable within wastewater samples; 3) representative of the population and 4) resistant to in-sewer degradation. In an earlier study we proposed and preliminarily showed that a suite of pharmaceuticals and personal care products (PPCPs) including caffeine and an artificial sweetener satisfy the first 3 of the 4 criteria and could be used as population markers. The de facto population was then estimated through Bayesian inference by updating the population size provided by WWTP staff (prior knowledge) with measured chemical mass loads of the PPCPs. Cross validation showed that large populations can be estimated fairly accurately with a few chemical mass loads quantified from 24-h composite samples. In contrast, the prior knowledge for small population sizes cannot be improved substantially despite the information of multiple chemical mass loads. A follow up study was conducted on the in-sewer stability of these 14 PPCPs to assess their applicability as population markers against the fourth criteria and their applicability for future population models. Our data analysis will also provide some new quantitative parameters of what makes a good population marker as well as their impact to the model. Recommendations on using these markers to better estimate the population size and potential sources of variability in the data set will also be provided.

Mo-SY-H1: Advances in consumer exposure assessment - I

Mo-SY-H1.1

ConsExpo Web - Introduction of Updated Tools for Consumer Exposure Assessment

Christiaan Delmaar, *RIVM, bilthoven, Netherlands*
susan wijnhoven, *RIVM, Bilthoven, Netherlands*
benjamin makkes van der deijl, *RIVM, Bilthoven, Netherlands*
roel schreurs, *RIVM, Bilthoven, Netherlands*
martin kosterman, *RIVM, Bilthoven, Netherlands*
gerlienke schuur, *RIVM, Bilthoven, Netherlands*

The consumer exposure modelling tool ConsExpo is widely used in product safety assessments for REACH, biocides and cosmetics. The previous version dates from 2005 and was available as a stand-alone (downloadable) software tool. In 2016, a new version has been developed as a freely accessible web application, the ConsExpo Web tool.
The tool includes a variety of exposure models to estimate chemical exposure via the inhalation, dermal and oral routes. Models include simple, typical low-tier exposure models but also more advanced, physics based models suitable for more refined exposure analysis. ConsExpo can be used to perform deterministic exposure analyses, but contains also options for probabilistic uncertainty and variability assessment.
Next to an updated version of the ConsExpo software, a separate tool has been developed for the exposure assessment of nano materials in consumer spray products, the ConsExpo Nano tool. This tool is also freely accessible. The ConsExpo Nano tool can be used to simulate the alveolar dose of nano materials resulting from single or repeated use of consumer spray products. To this end, ConsExpo Nano combines external exposure models with a deposition and clearance model. Doses can be expressed in different dose metrics. The ConsExpo Web and ConsExpo Nano tools will be presented. An overview of their features will be given and future developments will be discussed.

Mo-SY-H1.2

Overview of Consumer Exposure Model (CEM) Updates

Cathy Fehrenbacher, *U.S. Environmental Protection Agency, Washington, District of Columbia, United States*
Charles Bevington, *U.S. Environmental Protection Agency, Washington, District of Columbia, United States*
Heidi Hubbard, *ICF International, Durham, North Carolina, United States*

This presentation will describe the updated Consumer Exposure Model (CEM) that was developed based on the EPA's Exposure and Fate Assessment Tool (E-FAST). The updated version now includes six inhalation models, five ingestion models, and four dermal models. All CEM models are used to estimate chemical concentrations in exposure media, including indoor air, airborne particles, settled dust, and soil. The models also evaluate dermal flux of a chemical through the skin and the migration of a chemical from an article to saliva. These are combined with media contact rates and exposure factors to determine the daily dose and chronic average daily dose of chemical resulting from product and article use scenarios CEM is parameterized for a variety of indoor use environments, including residences and specific rooms within residences, offices, schools, automobiles, and outdoor scenarios. Notably, models to estimate exposure to semi-volatile organic compounds (SVOCs) from consumer articles has been incorporated, including a mass balanced model for estimating emissions and indoor fate and transport of SVOCs. CEM requires that the chemical molecular weight, vapor pressure, Kow, and Koa be provided. All other input variables, including mass transfer, partition, and diffusion coefficients can either be estimated within CEM from these baseline physical-chemical parameters and model defaults or, if data are available, can be supplied by the user. The latest version of CEM also has the option to model higher exposure associated with product use near the breathing zone. CEM has been developed as a flexible tool for assessing both data-rich and data-poor chemicals. This presentation will include case studies using the CEM model and an overview of updates to CEM as a result of beta testing and peer review.
Disclaimer: The views expressed in this abstract are those of the authors and do not necessarily reflect Agency policy or endorsement.

Mo-SY-H1.3

Effect of Varying Vapor Pressure Bands on Inhalation Exposure in REACH Consumer Exposure Modelling Tools

Hua Qian, *EMBSI, Annandale, United States*
Rosemary Zaleski, *EMBSI, Annandale, United States*
Chris Money, *Cynara Consulting, Brockenhurst, United Kingdom*

Vapor pressure (VP) bands have been applied in several screening level exposure tools to estimate human inhalation exposure. The VP bands use conservative assumptions relating to how much of a substance in a product/article is likely to be released into air based on its vapor pressure during 'standard' applications. For example, the European Centre for Ecotoxicology and Toxicology of Chemicals (ECETOC) Target Risk Assessment (TRA) consumer module assumes that 100% of a substance will be released to air if its VP is ≥10Pa. A substance with VP<10Pa will only release a fraction of the substance into the air. To better understand the most appropriate use of consumer exposure predictive tools that use this banding approach (e.g. ECETOC TRA and The European Solvents Industry Group (ESIG) Generic Exposure Scenario (GES) Risk and Exposure Tool (EGRET)), it is necessary to describe the level of conservatism of the current VP bands and their specificity (whether they provide sufficient discrimination for different exposure scenarios). The ECETOC TRA and EGRET consumer exposure models have been compared with the predictions of a higher tier consumer model (ConsExpo) for multiple exposure scenarios and multiple VP bands. The comparisons have been extended to include measured data from a range of consumer scenarios where solvents are used. This analysis confirms the inherent conservatism of the banding approach also suggests a potential refinement for substances at medium high VP range by introducing an additional VP band.

Mo-SY-H1.4

Framework for Human Health Risk Assessment of Noncancer Effects Resulting from Short-duration and Intermittent Exposures to Chemicals

Bette Meek, *McLaughlin Centre for Risk Science, Ottawa, Ontario, Canada*
Lynne Haber, *Toxicology Excellence for Risk Assessment (TERA) Center/University of Cincinnati, Cincinnati, OH, United States*

Durations of exposure to chemicals for single, repeated or intermittent periods, often vary from those upon which most guidance values are based. Since it is not feasible to conduct toxicity studies or develop Toxicity Reference Values (TRVs) specific to each scenario of interest, a framework was developed to improve the scientific basis for the evaluation of short-duration and intermittent exposures in a variety of settings, drawing as much as possible on existing TRVs. The framework incorporates the use of TRVs based on exposure periods as similar as possible to the 'actual' exposure periods and an integrated, tiered approach that addresses the potential for non-cancer effects resulting from continuous short-duration and intermittent exposures.
Higher tiers entail more effort and consideration of additional information. The framework outlines considerations relevant to identifying when existing TRVs can be applied directly (in lower tiers), or appropriately adapted to assess the acceptability of short-duration or intermittent exposure scenarios, taking into account toxicokinetic and toxicodynamic considerations.
The framework is illustrated by a number of case studies and based on the outlined approach, recommends application of dose averaging only under limited, specified conditions.

Mo-SY-H1.5

Exploring Online Global Resources Useful for Consumer Exposure Assessments

Pertti Hakkinen, *National Institutes of Health, Bethesda, Maryland, United States*
Rosemary Zaleski, ExxonMobil Biomedical Sciences, Annandale, NJ, United States

Many consumer exposure assessment-related databases and other resources are available online globally at no cost. For example, PubMed from the (U.S.) National Library of Medicine (NLM) in the National Institutes of Health (NIH) provides access to more than 25 million citations from the biomedical literature. When available, the PubMed citations include links to full-text content from PubMed Central and publisher web sites. NLM also offers the TOXNET® suite of databases, with consumer exposure-relevant examples being the Hazardous Substances Data Bank (HSDB®) and the Household Products Database (HPD). NLM has been enhancing the HSDB and HPD in recent years. This includes new product categories in the HPD. The HSDB enhancements include new materials (e.g., nanomaterials), and state-of-the-science toxicology, exposure, and risk assessment information. In addition, NLM offers the Enviro-Health Links (EHLs) bibliographies of links to authoritative and trustworthy online resources in toxicology, environmental health, exposures, risk assessment, and risk management. Examples are the EHLs pages on "Indoor Air" and "Toxicology Web Links." Selected non-NLM databases and other resources will be noted, along with examples of how their product-specific exposure factors and other information can be used for developing consumer-related exposure scenarios and assessments. Finally, a look into the future for databases and other online resources relevant for consumer exposure assessments will be provided.

Mo-PL-I1: Transportation Related Air Pollution - I

Mo-PL-I1.1

Identifying and Quantifying Airport-related Impacts on Ambient Particle Number Concentrations in Boston Area

Neelakshi Hudda, *Tufts University, Medford, MA, United States*
Matthew Simon, *Tufts University, Medford, MA, United States*
Wig Zamore, *Somerville Transportation Equity Partnership, Somerville, MA, United States*
Douglas Brugge, *Tufts University, Boston, MA, United States*
John Durant, *Tufts University, Medford, MA, United States*

Aircrafts emit ultrafine particles (UFP) at high rates (1) and recent studies (2,3) suggest that aviation emissions may be a significant, underestimated UFP source; its impacts extending tens of kilometers downwind in contrast to traffic emissions impacts. Our aim was to examine newly-available (3 month to 3.67 years) particle number concentration (PNC) datasets (4,5) from three stationary sites within 7.3 km of the airport in Boston (~1000 flights/day) for the evidence of airport-related emissions impact on ambient PNC. Our hypothesis that flight activity was associated with PNC when prevailing winds positioned these sites downwind of the airport was supported by the results. Controlling for meteorology (temperature, wind speed, and solar radiation), temporal variation (hour of the day and weekday or weekend differentiation) and concurrent or modeled local traffic, the correlation between PNC and flight activity was positive and significant ($r=0.22$ [1.67 year dataset at 4 km downwind site] and 0.29 [3.67 year dataset at 7.3 km downwind site], $p<6.5\times10^{-9}$). Long-term monitoring sites were downwind of the airport for 3.6% and 5.3% of the time in 2014 but weighted contribution of PNC under these winds to the annual average was 4.7% (at 7.3 km site) and 10% (at 4 km site), respectively. Further, we analyzed the dependence of PNC on wind speed. When sites were downwind of the airport, PNC were higher at higher wind speeds (maximum PNC occurred in the 25-30 km/h range) indicating that the impact was likely from buoyant aircraft plumes. We also analyzed co-located measurements of other pollutants (CO, BC, NO, NO_2, NO_X, SO_2, $PM_{2.5}$). Other than PNC, only oxides of nitrogen (NO_x and NO_2) were significantly ($r=0.18$ and 0.2, respectively, and $p<2.2\times10^{-6}$) correlated with flight activity taking meteorology, temporal variation and traffic into account. However, their concentrations decreased with increasing wind speed when winds were from the airport, indicating that the dominant source was likely roadway traffic emissions. Our analysis suggests that there is a need to take UFP concentrations into account in epidemiological studies of airport-related health effects and exposure prediction models in urban areas may be improved by inclusion of airport-related emissions.
References: (1) Lobo, P. Atmos. Environ. 2015. (2) Hudda, N. Environ. Sci. Technol. 2014. (3) Keuken, M. P. Atmos. Environ. 2015. (4) Fuller, C. H. Rev. Environ. Health 2013. (5) Tucker, K. L. BMC Public Health 2010.

Mo-PL-I1.2

Methods to Improve Traffic Flow and Noise Exposure Estimation on Minor Roads

David Morley, *Imperial College London, London, United Kingdom*
John Gulliver, *Imperial College London, London, United Kingdom*

Background
Address-level estimates of exposure to road traffic noise for epidemiological studies are dependent on obtaining data that is both accurate and with good geographical coverage on annual average daily traffic (AADT) flows as input to empirical noise models. National agencies often have reliable traffic count data for major roads, but for residential areas served by minor roads, such information is often not available. For national scale noise exposure assessment traffic flow data on minor roads is often assigned a constant average value. With no consideration of how traffic flow varies over the minor road network, estimated noise exposures will also show little variability.
Objectives
Here we present a method to predict AADT at the national scale for minor roads using a routing algorithm within a geographical information system (GIS) to rank roads by importance based on simulated journeys through the road network. The aim is to obtain more accurate results from standard noise models, however this dataset will be of use in any exposure models concerning traffic levels such as air pollution.
Methods
Routing importance was derived for each minor road segment in the UK using routing enabled OpenStreetMap data. From every point where a minor road intersects the major road network, the shortest driving route to every other minor road accessible without crossing another major road is calculated. A count of how many times each road segment is transversed gives an indication of road importance. From a training set of known minor road AADT, routing importance is used to predict AADT on all UK minor roads in a regression model along with the road class, urban or rural location and AADT on the nearest major road.
Results
Validation with both independent traffic counts and noise measurements (242 sites) show that this method gives a considerable improvement in noise prediction capability when compared to models that do not give adequate consideration to minor road variability (Spearman's rho increases 0.46 to 0.72, RMSE decreases 6.28 to 4.80 Lday). This has significance for epidemiological cohort studies attempting to link noise exposure to adverse health outcomes.

Mo-PL-I1.3

Commuting Patterns and Estimated Air Pollutant Exposures in the Rutgers Commuter Community Cohort (RC3) Study

Robert Laumbach, *Rutgers School of Public Health, Piscataway, NJ, United States*
Kathleen Black, *Rutgers School of Public Health, Piscataway, NJ, United States*
Judith Graber, *Rutgers School of Public Health, Piscataway, NJ, United States*
Zhongyuan Mi, *Rutgers School of Public Health, Piscataway, NJ, United States*
Panos Georgopoulos, *Rutgers School of Public Health, Piscataway, NJ, United States*
Howard Kipen, *Rutgers School of Public Health, Piscataway, NJ, United States*

AIM: Among automobile commuters in urban areas, in-vehicle microenvironments can account for a substantial proportion of daily exposures to traffic-related air pollutants (TRAP), which have been associated with adverse pulmonary, cardiovascular, neurological, and reproductive health outcomes. The high degree of spatial and temporal variability of these exposures presents a challenge for studies of potential adverse health effects of exposures to TRAP among daily commuters. METHODS: To assemble a cohort of commuters for studies of short- and long-term exposure to TRAP and other physical and psychosocial stressors during commuting, we administered an on-line survey to characterize commuting patterns among 18,196 Rutgers University faculty and staff at the New Brunswick/Piscataway NJ campus. Using survey data on commute routes and times, we are estimating individual-level exposures to diesel particulate matter and other TRAP during daily commutes using atmospheric dispersion modeling. Survey data and estimated exposures will be used to recruit subjects for more-intensive studies of measured exposures and changes in health-related biomarkers. RESULTS: 5,008 (28% of those invited) employees provided survey data, including 2,788 (61%) full-time staff and 1,132 (25%) full-time faculty. 60% were female. 4,145 (91%) commuted by private automobile. 2,004 (44%) of respondents had usual commutes to work of ≥ 40 minutes, with the majority of respondents spending time on major highways. The geographic distribution of the locations of the origin of automobile commutes to the central campus location was plotted (figure). 2,464 respondents consented to be contacted for future studies. CONCLUSIONS: The survey results provide the basis for planning studies of impacts of commuting-related air pollution and other stressors on health. The diversity of observed commuting patterns will enable selection of commuters with contrasting commuting characteristics for studies of TRAP exposures, acute biomarker responses, and acute and chronic health outcomes.

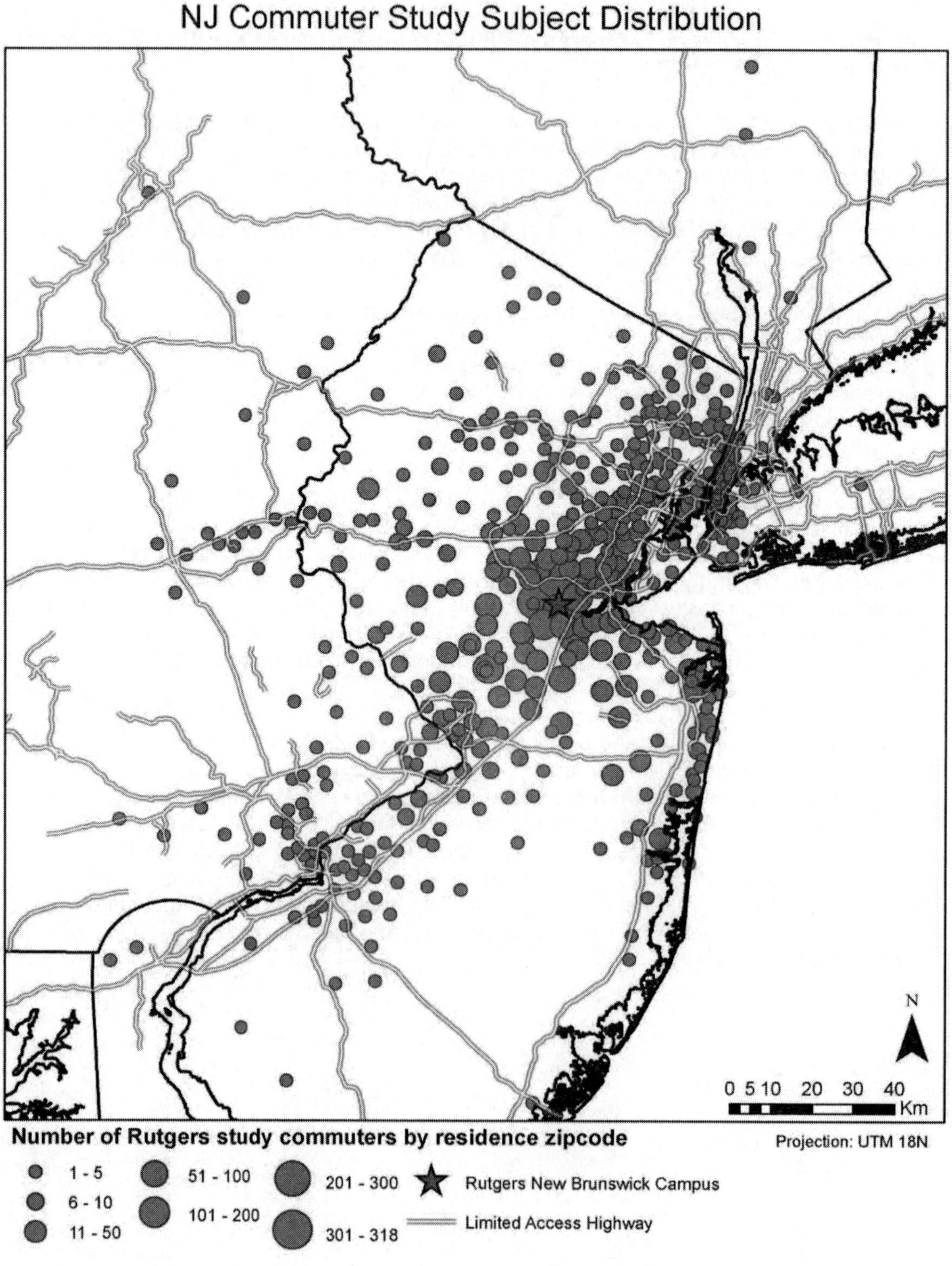

Distribution of the residential locations of RC3 survey respondents.

UTRECHT
THE NETHERLANDS
OCTOBER 9-13
ISES 2016

Mo-PL-I1.4

The Influence of Gas Price on Near-Road Air Quality

Timothy Buckley, *U.S. Environmental Protection Agency Office of Research & Development, Research Triangle Park, North Carolina, United States*
Bryan Hubbell, *U.S. EPA Office of Research & Development, Research Triangle Park, North Carolina, United States*
Lindsay Stanek, *U.S. EPA Office of Research & Development, Research Triangle Park, North Carolina, United States*

With an average of 1.8 vehicles per household (2013) across 117 million U.S. households, mobile sources have a strong influence on public health as a major source of air pollution that is highly integrated within communities. There is growing interest in the social and economic determinants of environmental health. We investigate gasoline prices as one such factor that may influence vehicle use (and emissions from gasoline combustion) and near-road air quality. Gas prices are one part of the cost of utilizing an auto, with other factors including vehicle fuel efficiency, value of operator's time, and maintenance costs. Daily (or weekly) gas prices may be the element of cost which would be best reflected in short term fluctuations in vehicle use. The transportation economics literature has found that there is some impact of gas prices on volume of driving, but it is highly variable and depends on factors such as availability of substitute modes of transportation. Studies have tended to show that demand for gas tends to be inelastic, especially over the short run, and that VMT is even more inelastic, because of the other non-gas price factors that drive VMT demand. However, most of these studies evaluated demand over longer periods of time or large spatial areas. To evaluate how short-term fluctuations in gas prices affect short-term traffic density and near-road air quality, we obtained time-resolved near-road air quality data (NO_2 and $PM_{2.5}$) from a subset of the 70 EPA national near-road monitoring stations where time-resolved meteorology and traffic counts were simultaneously recorded. Over the relevant period of monitoring, gas price has varied tremendously. Locality-specific gas price information was obtained from GasBuddy.com. Using this data, we will explore the relationship between gas prices and traffic counts, and between predicted traffic counts and air quality to determine the effect of gas prices on air quality. We will construct statistical models that account for potential confounders in both the gas price to traffic count relationship and the traffic count to air quality relationships. We will consider alternative lags between gas price changes and changes in traffic counts, and various averaging times for gas prices and traffic counts. Improved understanding of local traffic density and air quality responses to local gas price variability will provide valuable insights into how economic factors can impact air quality and ultimately public health.

Mo-PL-I1.5

Have urban traffic limitations influence on air quality?

Ario Alberto Ruprecht, *Fondazione IRCCS Istituto Nazionale dei Tumori and Italian College GPs SIMG, Milan, Italy*
Cinzia De Marco, *Fondazione IRCCS Istituto Nazionale dei Tumori, Milan, Italy*
Roberto Mazza, *Fondazione IRCCS Istituto Nazionale dei Tumori, Milano, Italy*
Paolo Contiero, *Fondazione IRCCS Istituto Nazionale dei Tumori, Milan, Italy*
Andrea Tittarelli, *Fondazione IRCCS Istituto Nazionale dei Tumori, Milan, Italy*
Bruno Villavecchia, *Agenzia Mobilità Ambiente e Territorio, Milan, Italy*
Roberto Boffi, *Fondazione IRCCS Istituto Nazionale dei Tumori, Milan, Italy*
Silvia Moroni, *Agenzia Mobilità Ambiente Territorio, Milan, Italy*

Background: In the City of Milan, the local Authorities, in order to reduce the traffic-related pollution, introduced a Limited Traffic Zone (LTZ) by charging a fee to all vehicles entering into the city center since 2008. This initiative, started as a 'pollution charge' called ECOPASS was re-formulated in 2012 as a 'congestion charge' called AREA C. In addition, during high pollution episodes due to persistent stable atmosphere events, typical of the Po Valley winter, they may decide to adopt 'Car-free days' or traffic stops as done on several Sundays in 2011, 2013 and on working days on 2015.
Objective: To compare the Black Carbon (BC) content in Particular Matter (PM) - the BC/PM10 ratio - an empirical indicator of presence of Polycyclic Aromatic Hydrocarbons, inside/outside the ECOPASS/AREA C zone and during/after the time period interested in the mitigation actions.
Method: BC was measured using the aethalometers model AE31 (Magee Scientific) and AE51 (Aethlab) and PM using pre-calibrated Optical Particle Counter (OPC) Aerocet 531 (Metone Instruments).
Procedure: Instruments operated at kerbside and third floor levels in several measurements campaigns, in Summer, Autumn, Winter, and during/after the Car-free days or traffic stops.
Results: No statistically significant difference resulted in the PM10 concentrations in all measurements campaigns, meanwhile, on average BC/PM10 ratio decreased in Summer by 50%, and in Autumn and Winter by 32%, despite the interference of the residential heating power plants within the Ecopass/AREA C zone as compared to no-restriction zone. Improvement of the BC/PM10 ratio was measured not only at the kerbside but also at the third floor level: reduction of about 25% as compared to third floor levels in no-restriction zone. On Sunday days traffic stops reductions in BC/PM10 was ranging from 41 to 53%. During working days traffic stops the reduction was 51% at kerbside and 26% at third floor level.
Conclusion: Measured reductions are corresponding to a difference of one to three BC change units between in/out LTZ and during/after Car-free days or traffic stops with a corresponding personal exposure reduction to traffic generated toxic and carcinogenic pollutants and consequent important health benefits for the residents, commuters and 'city users' with particular benefits for most sensitive ones such as children, pregnancy women, elder or sick people, reducing the burden of diseases and mortality due to traffic-proximity pollutants exposure.

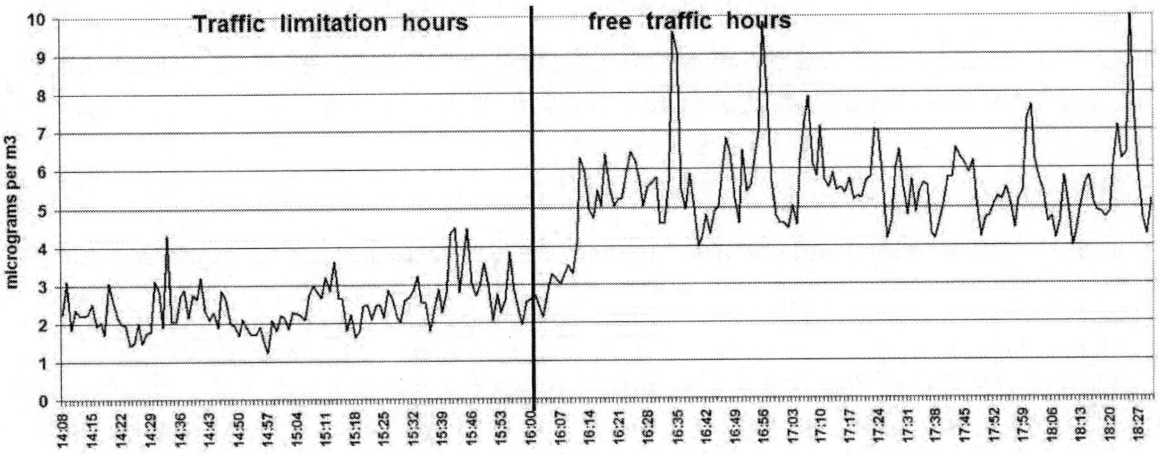

Typical Black Carbon sharp increase immediately after traffic increase from 300/400 to 1,800/2,000 vehicles/hr measured at one first floor balcony facing a high traffic street before and after traffic restrictions

Mo-SY-A2: Harmonization, access, transparency: improving environmental epidemiology for public health decision-making - II

Mo-SY-A2.1

Ensuring harmonized and comparable laboratory measurements to improve public health programs

Hubert Vesper, *Centers for Disease Control and Prevention/National Center for Environmental Health, Atlanta, GA, United States*

Aim
Measurements of exposure biomarkers are important for the assessment of health risks, the development of clinical and public health decision points, and for monitoring exposures over time. The lack in comparability of biomarker measurements performed in different studies and at different times prevents the effective assessment, implementation, and monitoring of public health activities.
Results
Harmonization programs which create measurement results that are traceable to one accuracy basis and thus are comparable across methods, locations, and over time were successfully implemented for clinical analytes such as cholesterol. These programs focus on the accuracy and the analytical reliability of the measurement result and do not require laboratories to use the same analytical method. This is accomplished by providing laboratories with panels of specimens with reference values for the relevant analyte. These panels are used for assessment of measurement accuracy and for identifying potential sources that cause inaccurate and unreliable results. Common sources causing inaccurate measurements identified through harmonization programs are method calibration and analytical specificity. By continuously providing panels of specimens and evaluating analytical performance using consistent protocols and criteria, harmonized and thereby comparable laboratory measurements are achieved. Because these harmonization programs focus on analytical performance using panels of specimens, they can easily be adopted to improve the comparability of exposure biomarker measurements.
Conclusions
Harmonization programs to determine, improve and track the analytical performance of biomarker measurements are successfully used for clinical biomarkers and can be applied to biomarkers used in exposure assessments.

Mo-SY-A2.2

Opportunities from (U.S.) National Library of Medicine (NLM) to Identify, Access, Share, and Discuss information

Pertti Hakkinen, *National Institutes of Health, Bethesda, Maryland, United States*
Aubrey Miller, *National Institutes of Health, Bethesda, MD, United States*

The (U.S.) National Library of Medicine (NLM), a part of the National Institutes of Health (NIH), is the world's largest biomedical library. NLM's online databases and other resources such as smartphone apps and web pages optimized for mobile devices are available globally at no cost. This presentation will describe NLM's current approaches for users to identify, access, share, and discuss information related to environmental epidemiology, exposure science, toxicology, public health, and other topics. For example, NLM's PubMed provides access to more than 25 million citations from the biomedical literature. When available, the PubMed citations include links to full-text content from PubMed Central and publisher web sites. Recently, PubMed Commons was developed to allow authors to share opinions and information about publications in PubMed. All authors of publications in PubMed are eligible to become members. PubMed Commons authors can comment on any publication in PubMed, rate the helpfulness of comments, and invite other eligible authors to join. PubMed Commons Journal Clubs is another recent initiative to offer access to discussions of publications and connect them to the relevant PubMed citations. It is open to journal clubs holding regular discussions for research, graduate and postgraduate education, or for continuing education. NLM also offers PubMed Health, specializing in reviews of clinical effectiveness research, and providing easy-to-read summaries and access to full technical reports. NLM's ClinicalTrials.gov provides researchers, patients, and others with easy access to information on publicly and privately supported clinical studies on a wide range of diseases and conditions from around the world. Noteworthy is that ClinicalTrials.gov also includes many studies associated with environmental factors and exposures, collection of environmental specimens, and environmental controls, interventions, and outcomes. Finally, NLM offers the TOXNET® suite of databases such as the Hazardous Substances Data Bank (HSDB®) and resources such as the Enviro-Health Links (EHLs) bibliographies of links to authoritative and trustworthy online resources in toxicology, environmental health, exposures, and other topics.

Mo-SY-A2.3

Panel Discussion

Judy LaKind, *LaKind Associates, LLC, Catonsville, MD, United States*

The Symposium will conclude with a panel discussion. Symposium participants will share concluding thoughts, offer insights for paths forward and answer audience questions.

Mo-SY-B2: Firefighters and Chemical Exposures: Protection Under Fire

Mo-SY-B2.1

Cancer Prevention in the Fire Service: Exposure Assessment, Toxic Effects and Risk Management

Jeff Burgess, *University of Arizona, Tucson, Arizona, United States*
Stephanie Griffin, *University of Arizona, Tucson, Arizona, United States*
Shane Snyder, *University of Arizona, Tucson, Arizona, United States*
Christiane Hoppe-Jones, *University of Arizona, Tucson, Arizona, United States*
Jin Zhou, *University of Arizona, Tucson, Arizona, United States*
Kenneth Fent, *Centers for Disease Control and Prevention, Cincinnati, Ohio, United States*

Cancer is a leading cause of fire service morbidity and mortality. Exposure to carcinogens occurs through skin contamination, through the lungs when respiratory protection is not worn during all phases of fire suppression and overhaul, and through inhalation during standby, operation of apparatus and off-gassing of equipment. In addition to fire smoke, diesel exhaust exposure can occur from operation of apparatus at the fire ground and in the station. Since cancer has a long latency period between exposure and disease onset, measurements are needed that can determine the effectiveness of new interventions on a much shorter time interval. The purpose of the current research is to identify effective methods of reducing firefighter exposure to carcinogens and associated toxic effects through completion of the following specific aims: 1) Evaluate exposure to carcinogens throughout the work shift; 2) Measure biomarkers of carcinogenic effect in relation to workplace exposures; and 3) Within a risk management framework, test the effectiveness of interventions to reduce fire service carcinogen exposure and effects. In this ongoing study collaborative study with the Tucson Fire Department (TFD), exposure to particulates and volatile chemicals is being measured at the fireground and in-transit, as well as diesel particulate matter during responses and in the fire station. Urine collected during annual medical surveillance evaluations and post-fireground activities is being analyzed for chemical contaminants. Epigenetic biomarkers of carcinogenic effect will be analyzed at baseline in incumbent firefighters and new recruits, and repeated after 1-2 years in the same new recruits. Based on initial exposure monitoring, exposure reduction interventions will be put in place by TFD, and the extent to which firefighter chemical exposures and biomarkers of effect can be reduced by following these risk management steps will be determined. We anticipate that the proposed research will identify carcinogenic exposures throughout the fire shift and measure the effectiveness of interventions designed to reduce cancer risks.

Mo-SY-B2.2

Firefighters' Unique Exposure Profiles

Kenneth Fent, *National Institute for Occupational Safety and Health, Cincinnati, Ohio, United States*
Gavin Horn, *University of Illinois-Urbana/Champaign, Urbana-Champaign, Illinois, United States*
Joachim Pleil, *U.S. Environmental Protection Agency, Research Triangle Park, North Carolina, United States*

Firefighters have an increased risk for several types of cancer compared to the general population. It has been hypothesized that chemical exposures encountered during firefighting work contributes to this elevated risk. Firefighters' have unique exposure profiles and can be exposed to hundreds of combustion byproducts through both the inhalation and dermal routes, including single-ring and polycyclic aromatic hydrocarbons, acid gases, flame retardants, and dioxins. This presentation will discuss recent exposure assessment findings related to firefighters' external and systemic exposures and workplace factors that may influence those exposures, including use of respiratory protection, job function, firefighting tactics, contamination and decontamination of protective ensembles, and skin cleaning. We have assessed firefighters' exposures during responses to vehicle fires, training involving simulated smoke, and most recently, structure fires with modern furnishings. Although firefighters typically wear self-contained breathing apparatus (SCBA) when entering burning structures, they do not always wear SCBA for exterior operations on the fire ground, vehicle fires, or simulated smoke training. Our studies have shown that hazardous levels of chemicals can be generated in these scenarios. We have also found that—even when wearing protective ensembles and SCBA—firefighters can have polycyclic aromatic hydrocarbon contamination on their neck and can absorbed single-ring and polycyclic aromatic hydrocarbons into their bodies (as measured in urine and breath). We are currently studying firefighters' exposures and physiologic strain during responses to controlled residential fires, where foams, carpet padding, and other materials containing flame retardants were present. Our preliminary data show that single-ring and polycyclic aromatic hydrocarbons and flame retardants are released into the air during these fires and deposit onto protective ensembles. The volatile and semi-volatile substances will evaporate over time and could pose an inhalation risk, whereas the non-volatile substances can be transferred to firefighters' skin with repeated donning and doffing of protective gear. Our preliminary data indicate that gross decontamination with soapy water and scrubbing is able to remove the majority of this contamination. However, most fire departments do not perform gross decontamination after fire responses and may not launder protective clothing on a regular basis. As a result of these and other studies, fire departments have begun revising and instituting new policies and procedures to lessen firefighters' exposures and health risk.

Firefighters at controlled residential fire

Mo-SY-B2.3

Analysis of Combustion Byproducts in Fire Fighter urine using Mass Spectrometry and Bioassays

Christiane Hoppe-Jones, *University of Arizona, Tucson, Arizona, United States*
Shawn Beitel, *University of Arizona, Tucson, Arizona, United States*
Byron Hempel, *University of Arizona, Tucson, Arizona, United States*
Jeff Burgess, *University of Arizona, Tucson, Arizona, United States*
Shane Snyder, *University of Arizona, Tucson, Arizona, United States*

Elevated rates of lung, gastrointestinal and kidney cancer occurrences have been reported in firefighters. During a fire, firefighters are exposed to smoke, diesel particulates and elevated levels of organic chemicals, such as flame retardants and others originating from furniture, carpets, etc. as well as their combustion by-products. In this study, biomarkers of exposure were chosen based on previous studies. Polycyclic aromatic hydrocarbons (PAH) metabolites and methoxy-phenols are two classes of compounds commonly found in urine of people exposed to smoke and fire. While many combustion byproducts have been identified, many are still unknown. The analysis and potential identification of unknown compounds is being conducted using high-resolution mass spectrometry instruments. Analysis of unknowns combines the tools that are used to compare the mass spectra of baseline urine and urine after exposure as well as the statistical data mining that follows. After extraction of the urine, volatile and semi-volatile compounds are being analyzed on a GC-qTOF while non-volatile compounds are being analyzed on a LC-qTOF. Many flame retardants, such as PBDEs contain bromine. The comparison of organobromine levels is being conducted using an ICP-MS instrument coupled with a GC or LC. Preliminary results show that PAH metabolites, such as naphthol, fluorenol and phenanthrols, in the urine increase after 2-4 hours of exposure to smoke. Statistical analysis of the baseline urines prior to exposure compared to post-exposure urine is being conducted for each analyte. This presentation will discuss the analysis and results of known combustion byproducts like PAHs, as well as unknown byproducts, using time-of-flight mass spectrometry in conjunction with statistical software can help identify unknown compounds in complex matrices like urine. In addition to the identification of chemical contaminants, the use of in vitro bioassays, including an AhR activation assay and a p53 reporter gene activity assay will be evaluated for potential use in exposure determination.

Mo-SY-C2: Quantitative in vitro to in vivo extrapolation (QIVIVE): Advances in tools to quantify exposure-response relationships for risk assessment - I

Mo-SY-C2.1

Utilizing mass balance modeling for the assessment of internal exposure in cell-based bioassays

Fabian Fischer, *UFZ - Helmholtz Centre for Environmental Research, Leipzig, Germany*
Luise Henneberger, *UFZ - Helmholtz Centre for Environmental Research, Leipzig, Germany*
Lukas Linden, *UFZ - Helmholtz Centre for Environmental Research, Leipzig, Germany*
Kai Bittermann, *UFZ - Helmholtz Centre for Environmental Research, Leipzig, Germany*
Kai-Uwe Goss, *UFZ - Helmholtz Centre for Environmental Research, Leipzig, Germany*
Beate I. Escher, *UFZ - Helmholtz Centre for Environmental Research, Leipzig, Germany*

The application of in vitro bioassays for chemical risk assessment requires tools that allow extrapolation from the effects observed in the in vitro bioassays to the whole organ or organism. A key parameter that needs to be understood is the intracellular concentration of the chemicals in the bioassays that triggers an effect. The main assumption of in vitro to in vivo extrapolations (IVIVE) is that the effective target concentrations are in principle the same across equal cell types and independent whether tested in vivo or in vitro. Target concentrations can often be approximated by cellular or tissue concentrations, which are accessible in vivo via physiologically based toxicokinetic modeling. However, in vitro bioassays typically report effects in nominal concentrations only. Cell concentrations cannot be directly measured in miniaturized systems with only few microliters of volume in 384 or 1154 well plates, but can be predicted by equilibrium partition models. Recently, Armitage et al. have developed a simple mass balance model for the calculation of the mass distribution of neutral organic chemicals in in vitro test systems. In this work, we present an extended mass balance modeling approach for the assessment of internal exposure in cell-based bioassays for neutral and ionogenic organic chemicals (IOCs). As reliable partition coefficients for the test medium compartments and cells are essential for the mass balance model, the use of experimentally determined partition coefficients was preferred. If no experimental values were available, the partition coefficients for neutral chemicals were predicted using polyparameter linear free energy relationships (PP-LFERs). For organic ions partition coefficients to serum albumin were estimated using 3D quantitative structure activity relationships (3D-QSARs). Liposome-water partition coefficients of IOCs were calculated using the quantum chemically based software COSMOtherm. The mass balance model was applied to effect data from the US EPA ToxCast database on the example of GeneBLAzer bioassays, which are reporter gene assays for various hormone receptors and transcription factors. We experimentally determined the lipid and protein content of assay media and cell lines used in the Tox21 program, because first modeling results suggested that the mass distribution of the test chemicals is highly depended on the medium composition. The cellular concentrations of 100 chemicals (49 neutral chemicals and 51 IOCs) that cover a wide range of hydrophobicity and are non-volatile were predicted from reported nominal effect concentrations. This exercise is a first step towards a meaningful dose metric in in vitro assays suitable for IVIVE.

Mo-SY-C2.2

Simulating the distribution and kinetics of neutral organic chemicals in in vitro test systems

James Armitage, *University of Toronto Scarborough, Toronto, Ontario, Canada*
Jon Arnot, *ARC Arnot Research & Consulting, Toronto, Ontario, Canada*

Background
Data from in vitro toxicity tests are being used to assess the potential hazards and risks related to exposure to organic chemicals and will be increasingly relied upon in the future. Such data may be used directly (e.g., comparing EC50s of different chemicals) or as part of in vitro to in vivo extrapolation exercises (Q-IVIVE). Concerns over the use of nominal (total) concentrations rather than measured and/or freely-dissolved concentrations in in vitro testing have long been discussed in the literature. Nevertheless, nominal concentrations are frequently the only metrics reported for in vitro tests, especially for high-throughput screening (HTS) applications where conducting measurements is technically challenging.
Objectives
The main objectives of this study are:
i) To develop a generic equilibrium partitioning-based model to simulate the distribution of neutral organic chemicals in in vitro toxicity test systems
ii) To develop a generic time-variant (dynamic) model to simulate the toxicokinetics of neutral organic chemicals in in vitro toxicity test systems
iii) To provide guidance on test conditions and partitioning property combinations for which the use of nominal concentrations may be particularly problematic
Methods
The models were developed following established approaches for describing equilibrium partitioning and time-variant (dynamic) distribution of neutral organic chemicals. The equilibrium partitioning-based model requires partitioning data and information on the volumes of the various components present in the test system (e.g., serum albumin, cell seeding density). Inputs characterizing diffusivity (water), permeability (membrane) and susceptibility to degradation are additionally required for the time-variant version. A preliminary approach for estimating sorption to the vessel wall was also implemented. The models were parameterized to represent different test systems and conditions (e.g., well plate size, volume fraction of serum, cell-free vs. cell-based assays) and applied to neutral organic chemicals spanning a wide range of property values.
Results
A key factor influencing the discrepancy between nominal (total) concentrations and more relevant metrics (i.e., freely-dissolved concentration) in cell-free assays is sorption to the vessel wall. For the more hydrophobic chemicals considered, the nominal (total) concentration can easily exceed the freely-dissolved concentration by an order of magnitude or more. In cell-based assays, the presence of serum albumin and lipids is a key consideration as the sorption capacity of these constituents is greater than the vessel wall (given the approaches implemented). Degradation in medium and cells can be influential but the influence of these processes is dependent on other test conditions and model assumptions.

Mo-SY-C2.3

How the desorption kinetics of organic chemicals from albumin may impact QIVIVE

Sophia Krause, *Helmholtz Centre for Environmental Research, Leipzig, Germany*
Nadin Ulrich, *Helmholtz Centre for Environmental Research, Leipzig, Germany*
Kai-Uwe Goss, *Helmholtz Centre for Environmental Research, Leipzig, Germany*

As the liver is considered a particularly important organ for transformation of chemicals, in vitro experiments with liver S9 fractions, hepatocytes or liver tissue are often used to predict in vivo metabolic conversion. The translation of in vitro results into in vivo effects is done with the aid of pharmacokinetic models. Most of these models rely on the assumption that only unbound chemicals can be taken up in cells. As most substances are not present freely dissolved in blood plasma, they consequently have to desorb from sorbing blood components first. Among these components the blood protein albumin, which is the most important transport protein in blood, might be an especially important compartment. The desorption process has been generally assumed to be rapid enough not to cause kinetic limitations, although desorption kinetics from albumin has not been investigated systematically yet. If desorption occurs rather slowly, it could be a limiting factor for the hepatic uptake of chemicals, because the blood is passing the liver within seconds [1].
We determined desorption rate constants for various chemicals from albumin by a modified method introduced by Kopinke et al [2]. First experiments with PAHs indicate a clearly incomplete desorption from BSA after 19 s, which is about the residence time of blood in the liver. Consequently, for these substances desorption can become a limiting factor for hepatic uptake or, in case the subsequent metabolic conversion occurs much faster, it can be even limiting for the entire hepatic transformation process. In that case desorption rates would be important parameters for modelling.
Our aim is to investigate the desorption kinetics from albumin (bovine serum albumin and human serum albumin) for a broad range of chemicals. Based on the measured data set, principles for predicting desorption kinetics for further substances will be examined. For instance, the first experiments revealed desorption rate constants showing a correlation to BSA-water partition coefficients. And as there are also other blood components apart from albumin, which are potentially important compartments for binding, we might extend the measurements towards other blood components or even whole blood to develop a broad understanding of all kinetic processes, which could be important for the rate of hepatic uptake.

1. Berezhkovskiy, L.M., Determination of hepatic clearance with the account of drug-protein binding kinetics. J Pharm Sci, 2012. 101(10).
2. Kopinke, F.D., Kinetics of desorption of organic compounds from dissolved organic matter. Environ Sci Technol, 2011. 45(23).

Mo-SY-C2.4

Dose Metrics in Repeatedly Dosed In Vitro Toxicity Assays

Nynke Kramer, *Utrecht University, Utrecht, Netherlands*
Emma Di Consiglio, *Istituto Superiore di Sanità, Rome, Italy*
Bas Blaauboer, *Utrecht University, Utrecht, Netherlands*
Emanuela Testai, *Istituto Superiore di Sanità, Rome, Italy*

Aim: The emphasis in toxicological risk assessment is moving away from animal testing towards integrating in vitro cell-based assays with computational modeling to extrapolate in vitro derived effective concentrations to human-relevant doses. To improve in vitro assays to predict repeat-dose systemic toxicity, testing strategies are developed where batteries of complex in vitro assays with highly differentiated cells of human origin and representing multiple organs are repeatedly dosed for extended periods of time. One such strategy was developed during the EU FP7 Predict-IV project, where RPTEC/TERT1 kidney cells, primary rat and human hepatocytes, HepaRG liver cells and 2D and 3D primary brain cultures were dosed daily or every other day for 14 days to a selection of drugs varying in their mechanism of pharmacological action. Molecular perturbations induced by the drugs were assessed by integrating a suite of omics technologies. Dose response analyses and physiologically based pharmacokinetic models were developed to relate daily oral exposure to in vitro derived points of departures. To address the increasingly acknowledged problem that traditionally used in vitro nominal concentrations are not necessarily proportional to the concentration at the target site, which is directly related to the initial molecular changes caused by the drugs, one aim of the project was to assess whether and how knowledge of the kinetics of drugs in in vitro assays helps explain the variations in observed effect between drugs, cell types and assay setup.
Methods: The concentration of a selection of drugs in cells, labware, cell attachment matrices, and exposure medium was measured over time.
Results: Results indicate that lipophilic drugs like chlorpromazine bind significantly to serum constituents in the exposure medium and plastic labware. A few drugs, including less lipophilic drugs like ibuprofen, bind to cell-attachment matrices. Chemicals that reach high concentrations is cells, including cyclosporin A and amiodarone, significantly accumulate over time after repeated dosing, partly explaining their increased toxicity after repeated dosing, compared to a single dose.
Conclusions: The results of these studies clearly suggest that integrating knowledge of the differences in concentration over time of drugs in cells in vitro between dosing regimens, as well as knowledge of transporter and biotransformation enzyme function, allows for the development of a mechanistic understanding of the observed in vitro toxicity.

Mo-SY-D2: 15 years of Human Biomonitoring in Flanders: surveillance feeding policy and research - I

Mo-SY-D2.1

Human biomonitoring studies in Flanders to support policy action

Ann Colles, *Flemish Institute for Technological Research, Mol, Belgium*
Dries Coertjens, *University of Antwerp, Antwerp, Belgium*
Bert Morrens, *University of Antwerp, Antwerp, Belgium*
Gudrun Koppen, *Flemish Institute for Technological Research, Mol, Belgium*
Greet Schoeters, *Flemish Institute for Technological Research, Mol, Belgium*
Ilse Loots, *University of Antwerp, Antwerp, Belgium*
Maja Mampaey, *Flemish government, Brussels, Belgium*
Karen Van Campenhout, *Flemish government, Brussels, Belgium*

Aim
Human biomonitoring (HBM) is part of the Flemish Environment and Health Surveys (FLEHS). In FLEHS I (2002-2006) 4458 participants of 3 different age groups were recruited in 8 areas with different environmental pressures. Results show significant geographical differences. In rural areas, expected to represent background levels, higher persistent organic pollutants (POPs) levels were observed. New questions emerged: what are the factors influencing exposure? How do we manage knowledge gaps and uncertainty? Which policy measures can be taken? The aim of the phased action plan is to translate human biomonitoring results into policy action.
Methods
In cooperation with the Flemish government a structured and participatory procedure was developed to translate FLEHS data into policy action. This procedure is based on an analytic-deliberative and iterative approach involving experts, policy makers and stakeholders. After prioritizing the importance of HBM results on the basis of health, social and policy criteria, the top priorities are subjected to further interpretation and policy formulation.
Results
The higher POPs blood levels in rural areas were selected as one of the priorities of FLEHS I. More detailed data analysis identified consumption of locally grown food and combustion habits as contributing factors. This evidence, together with expert consultations and stakeholder debate, led to the formulation of policy measures by the risk managers, such as changes in legislation on residential combustion and open fires, expansion of the monitoring network for dioxins and polychlorinated bifenyls (PCBs) in ambient air to include residential and agricultural sites, investments in new research on chemical contamination of home-grown food resulting in tools for custom-made citizens' advice, collection of old chlorinated pesticides and promoting healthy gardening. More recent FLEHS HBM data showed declining time trends for POPs, suggesting the success of these actions.
Conclusions
Interpreting HBM results by using a combination of analysis and deliberation in a structured and participatory procedure in which different stakehoders such as participants, policy advisors, industry eg. are involved, yields a broadened spectrum of

policy options to address environmental health issues and leads to well-informed and socially robust policy.

Mo-SY-D2.2

Human exposure trends in Flanders: 15 years of human biomonitoring

Greet Schoeters, *VITO, Mol, Belgium*
Eva Govarts, *VITO, Mol, Belgium*
Liesbeth Bruckers, *Hasselt University, Hasselt, Belgium*
Nathalie Lambrechts, *VITO, Mol, Belgium*
Ann Colles, *VITO, Mol, Belgium*
Elly Den Hond, *VITO, Mol, Belgium*
Vera Nelen, *PIH, Antwerp, Belgium*
Tim Nawrot, *Hasselt University, Hasselt, Belgium*
Ilse Loots, *University of Antwerp, Antwerp, Belgium*
Isabelle Sioen, *Ghent University, Ghent, Belgium*
Bert Morrens, *University of Antwerp, Antwerp, Belgium*
Nick Van Larebeke*Free University Brussels, Brussels, Belgium*
Willy Baeyens, *Free University Brussels, Brussels, Belgium*

Background:
Since 2002 a human biomonitoring program has been established in Flanders (Belgium) for environmental health surveillance. Three multi-annual cycles of the programme have been finished by now. This allows to evaluate time trends of internal exposure over the last decade.
Methods:
In each HBM cycle, exposure biomarkers have been analysed in cord samples of newborn-mother child pairs, in urine and blood samples of 14 and 15 years old adolescents and of adults between 50 and 65 years of age. The current biomonitoring program in Flanders includes more than 50 exposure biomarkers.
Results:
Exposure levels of traditional pollutants which have been banned since the 70's such as polychlorinated biphenyls (PCB's), dichlorodiphenyldichloroethylene (p,p'-DDE), hexachlorobenzene (HCB) have decreased significantly, the decrease is most pronounced for HCB. Heavy metals such as cadmium and lead are decreasing with time, a trend which is not confirmed for arsenic. Tt-muconic acid levels decrease. Tt-muconic acid is measured in urine as an indicator of benzene exposure. Urinary 1-hydroxypyrene, an indicator for exposure to polyaromatic hydrocarbons (PAHs), remains at the same level. The metabolite levels of pollutants that were more recently introduced in consumer products and in the environment and that have been recently restricted such as di(2-ethylhexyl) phthalate (DEHP) and mono-n-butyl phthalate (MnBP) have also decreased over the last 5 years in the adolescent population. The levels of perfluorooctanoic acid (PFOA) and perfluorooctanesulfonic acid (PFOS) were measured in cord blood samples and showed also a decreasing trend over the last 5 years.
Conclusion
The exposure profiles indicate the potential of environmental policies to protect environmental health and support poilcies to prioritise actions.
Acknowledgements:

The FLEHS studies were commissioned, financed, and steered by the Ministry of the Flemish Community.

Mo-SY-D2.3

Determinants of exposure to POPs and pesticides in the Flemish population

Kim Croes, *Vrije Universiteit Brussel, Brussels, Belgium*
Nathalie Lambrechts, *VITO, Mol, Belgium*
Ann Colles, *VITO, Mol, Belgium*
Liesbeth Bruckers, *Hasselt University, Hasselt, Belgium*
Ilse Loots, *University of Antwerp, Antwerp, Belgium*
Vera Nelen, *PIH, Antwerp, Belgium*
Greet Schoeters, *vito, Mol, Belgium*
Stefaan Dehenauw, *University of Ghent, Ghent, Belgium*
Adrian Covaci, *University of Antwerp, Antwerp, Belgium*
Tim Nawrot, *Hasselt University, Hasselt, Belgium*
Willy Baeyens, *Vrije Universiteit Brussel, Brussels, Belgium*

BACKGROUND: In 2001, the first cycle of the Flemish Environment and Health Study (FLEHS) started. During the past 15 years, multiple pollutants were measured in the blood and urine of newborns, adolescents and adults, residing in whole Flanders. Furthermore, information on lifestyle, food consumption, socio-economic status, occupation, living conditions, tobacco smoke, diseases, and medication intake was obtained through self-administered questionnaires.
OBJECTIVES: The objective of this study is to give an overview of the determinants that significantly influence the body burden of classical pollutants, like PCDD/Fs and dl-PCBs, HCB and p,p'-DDE and more recent pesticides, like metabolites from organophosphate pesticides and glyphosate.
METHODS: The dioxin-like activity of PCDD/Fs and dl-PCBs in the serum was obtained with the CALUX bioassay, while the marker PCBs and pesticides were measured with GC-MS. Confounders and possible covariates were selected and tested with univariate regression analysis. In the adult and newborn campaigns, selected determinants with a p-value below 0.20 in univariate analysis were used in the multiple regression model, but only stayed in the model when significant ($p<0.05$).
RESULTS: Throughout the three biomonitoring campaigns, blood fat content, BMI and intake of fat-rich food were important predictors for the lipid-dependent pollutants, like dioxins, HCB, PCBs, and p,p'-DDE, as was the case for local egg consumption and being breastfed as a newborn. For the currently used pesticides, the most important determinants were season of sampling and residing close to professional vegetable cultivation.
CONCLUSIONS: This study showed that several lifestyle factors significantly influence the body burden of both persistent, accumulative and less persistent pollutants, measured in the Flemish population.

This page intentionally left blank.

Mo-SY-E2: Exposure to atmospherically dispersed hazards: assessment, public information and perspectives - II

Mo-SY-E2.1

Assessment of exposure to vapour from plant protection products around treated fields

Cor Jacobs, *Alterra Wageningen UR, Wageningen, Netherlands*
Erik Van Den Berg, *Alterra Wageningen UR, Wageningen, Netherlands*

A significant fraction of the dosage of plant protection products applied in agriculture may volatilize after application. Bystanders and residents may be directly exposed to the vapour from these products. In addition, the vapour may deposit onto non-target surfaces such as ditches and private garden, sometimes leading to indirect exposure.
The goal of this study is to assess exposure of humans and non-target surfaces to vapour of plant protection products after application to crops.
The emission model PEARL has been coupled to the atmospheric dispersion model OPS. The special version of OPS applied here allows assessment of dispersion of gaseous substances nearby surface sources. The volatilisation of plant protection products from a crop fully covering the surface is first calculated using PEARL, which provides the emission strength for OPS. The coupling ensures a consistent use of meteorological conditions that drive the models. Two hypothetical products with different vapour pressure are considered. Five-year time series of weather conditions in The Netherlands are used. Exposure assessment is based on a series of weekly applications during the growing season (April-October). Thus, a wide range of meteorological conditions is covered. Exposure is evaluated at various distances from the treated fields. Timing of the application is varied to demonstrate the possibility to evaluate effects of differing application scenarios on exposure.
Like expected exposure is sensitive to the vapour pressure of the plant protection product. Volatilization generally reaches a maximum within 24 hours after the application and shows a fast decline thereafter. However, depending on the weather conditions and the competing processes such as penetration in plant tissue, emission peaks may be postponed and volatilization may continue for several days after the application, up to a week. In spite of the fact that application in the evening hours results in lower emissions, atmospheric concentrations are calculated to be higher than when the plant protection product is applied in the morning. Stable atmospheric conditions in the evening strongly reduce atmospheric mixing. PEARL-OPS can also be used to assess exposure at a regional scale, for multiple applications varying in space and time. The gaseous deposition pattern of plant protection products around treated fields, evaluated for an example region in the North of the Netherlands, is strongly influenced by the surface characteristics. However, the definition of these surface characteristics is highly uncertain and needs further study.

Mo-SY-E2.2

Dynamic Assessment of Exposure to Air Pollution Using Mobile Phone Data

Bart Dewulf, *Ghent University, Ghent, Belgium*
Tijs Neutens, *Ghent University, Ghent, Belgium*
Wouter Lefebvre, *VITO, Mol, Belgium*
Gerdy Seynaeve, *Proximus, Brussels, Belgium*
Charlotte Vanpoucke, *Irceline, Brussels, Belgium*
Carolien Beckx, *VITO, Mol, Belgium*
Nico Van de Weghe, *Ghent University, Ghent, Belgium*
Stijn Janssen, *Ghent University, Ghent, Belgium*

Aim Exposure to air pollution can have major health impacts, such as respiratory and cardiovascular diseases. Traditionally, only the air pollution concentration at the home location is taken into account in health impact assessments and epidemiological studies. Our aim is to incorporate individual travel patterns to limit the bias in air pollution exposure assessments. Methods In this work, we present a novel approach to calculate the daily exposure to air pollution using mobile phone data of approximately 5 million mobile phone users living in Belgium. At present, this data is collected and stored by telecom operators mainly for management of the mobile network. Yet it represents a major source of information in the study of human mobility. We calculate the exposure to NO2 using two approaches: assuming people stay at home the entire day (traditional static approach), and incorporating individual travel patterns using their location inferred from their use of the mobile phone network (dynamic approach). Results The mean exposure to NO2 increases with 1.27 µg/m3 (4.3%) during the week and with 0.12 µg/m3 (0.4%) during the weekend when incorporating individual travel patterns. During the week, mostly people living in municipalities surrounding larger cities experience the highest increase in NO2 exposure when incorporating their travel patterns, probably because most of them work in these larger cities with higher NO2 concentrations. Conclusions It is important for health impact assessments and epidemiological studies to incorporate individual travel patterns in estimating air pollution exposure. Mobile phone data is a promising data source to determine individual travel patterns, because of the advantages (e.g. low costs, large sample size, passive data collection) compared to travel surveys, GPS, and smartphone data (i.e. data captured by applications on smartphones).

Mo-SY-E2.3

The value of citizen air pollution measurements for participants, NGOs and science: the 'Together for Healthy Air' campaign in the Netherlands

Anne Knol, *Friends of the Earth, Amsterdam, Netherlands*
Joost Wesseling, *National Institute for Public Health and the Environment, Bilthoven, Netherlands*
Lan Nguyen, *National Institute for Public Health and the Environment, Bilthoven, Netherlands*

Air pollution is one of the most important environmental health problems in the Netherlands. Friends of the Earth (FoE) Netherlands set up the 'Together for Healthy Air' project to engage citizens in air quality measurements and to put air quality on the political agenda. RIVM, the Dutch national institute for public health and the environment, works on air quality monitoring and analyses. The two institutes have a shared goal of providing the public with information on air quality and have cooperated in this citizen science project. Citizens were engaged by FoE and together measurement locations were selected. NO2 measurements were carried out using Palmes diffusion tubes for one year (13 periods). RIVM calculated air quality concentrations using standard models for air quality. Measurement and model results were compared. Participants provided metadata on the measurements and evaluated the campaign. On average, an annual concentration of 31 ug/m3 NO2 was measured. At twelve locations the EU limit value of 40,5 ug/m3 was exceeded. Comparability of measurements and model results was generally well, although a few large differences were found. Differences between measured and modelled values can be caused by errors in measurement, model parameters or model input data. Participants were generally satisfied with the campaign. Air pollution still exceeded limit values at several locations. The value of the citizen air quality measurement campaign was considerable for all parties. In addition to providing potentially valuable measurement data, the campaign engaged and informed citizens, brought science and society closer together and raised political attention.

Mo-SY-E2.4

Atmospheric measurements: do it yourself, do it together!

Hester Volten, *RIVM, Bilthoven, Netherlands*
Joost Wesseling, *RIVM, Bilthoven, Netherlands*
Edith van Puttten, *RIVM, Bilthoven, Netherlands*
Annemarie van Alphen, *RIVM, Bilthoven, Netherlands*
Erik Tielemans, *RIVM, Bilthoven, Netherlands*
Jeroen Devilee, *RIVM, Bilthoven, Netherlands*
Frank Kresin, *Waag Society, Amsterdam, Netherlands*
Christine van den Horn, *Waag Society, Amsterdam, Netherlands*
Linda Carton, *Radboud University Nijmegen, Nijmegen, Netherlands*
Michel Grothe, *Geonovum, Amersfoort, Netherlands*

As sensor and communication technology is developing rapidly, the use of low-cost sensors to measure air quality is becoming more and more attractive to citizen communities and city governments. Alternative monitoring networks complementing official monitoring networks are within reach. Not only offers this the exciting possibility of more data and more detailed spatial resolution in monitoring, it also offers new possibilities to inform the public, create awareness about atmospheric hazards, and to empower citizens by enabling them to do their own measurements. The Dutch Environmental Protection Agency, RIVM has the ambition to support these activities. Theoretically, we would have a role in assuring data quality, providing calibration facilities, knowledge and a (political) context for interpreting the results. But what do citizens or cities need from us in practice?
In order to explore this question, we joined several projects such as the Amsterdam Smart Citizen Lab (see Figure, https://waag.org/en/news/amsterdam-smart-citizens-lab-publication) and the project Smart Emission in Nijmegen (http://www.ru.nl/gpm/onderzoek/research-projects/smart-emission). We learned that taking citizen science seriously is a self-fulfilling prophecy. By joining a citizen project governmental institutes raise trust and enthusiasm in the participants, and the chance of success is increased. We experienced that citizens are quite prepared to take responsibility, but expertise and help from the government is often appreciated. We conclude that citizen science is something to do together: citizens, sensor makers, experts from universities, and environmental protection agencies. We will present what we learned about our role in citizen science as an environmental protection agency and what we plan to contribute to the goal of citizen science becoming an integral part of a healthy society.

Citizens and experts working together to build a low-cost air quality sensor in the Amsterdam Smart Citizen Lab of the Waag Society in Amsterdam.

Mo-SY-F2: Extending participatory sensing to personal exposure and policy support - II

Mo-SY-F2.1

High-resolution Characterization of the Spatial Variability of Traffic Related Air Pollution Exposure at the Neighbourhood Scale

Weitao CHE, *The Chinese University of Hong Kong, Hong Kong, Hong Kong*
Mingyuan HU, *The Chinese University of Hong Kong, Hong Hong, Hong Kong*
Kerolyn Shairsingh, *University of Toronto, Toronto, Canada*

As a type of environmental-centric and human-based Mobile Sensing System(MSS), mobile device-based participatory sensing has been gradually applied for collection of environmental data with multiple granularities in space and time to improve the update frequency and reduce the cost of environmental data collection. However, participatory data are generally contributed by volunteer participants at arbitrary locations and times in most cases. Consequently, collected samples are randomly distributed in space and time, thus making processing and management of these data difficult, especially for the purpose of supporting simulations and calculating personal exposure. Therefore, effective organization and management of these participatory data are needed to further support the data input for regional simulations and personal exposure.
This topic presents a new platform for bridging the gap between participatory sensing and regional simulations personal exposure, which is called Virtual Geographic Environment (VGE). VGE is proposed as a new generation of geographic language that is characterized by 'feeling it in person, knowing it beyond reality'. In general, VGE has the features of handling data, models and human behavior. The aim of this topic is to effectively organize the participatory data by VGE, and then to use the models managed by VGE to support regional simulations, and to calculate personal exposure with the help of human behavior simulation of VGE.
In this topic, we first discuss the potential limitations of the current participatory sensing. Next, we introduce the VGE platform and its advantages of not only data handling but also model management. In which, participatory noise and air quality data are taken as examples. Spatial and temporal characteristics of participatory data are systematically investigated and spatio-temporal patterns are analyzed. Then these data are reorganized by using data reconstruction methods. After that, regional simulations based on the reconstructed data is processed, and to a relative "full dataset" is created. Finally, incorporated with the human behavior simulation of VGE, automatic data extraction from the the "full data set" is conducted by meeting the demand of personal exposure applications.

Mo-SY-F2.2

Extending Participatory Sensing to Personal Exposure Assessments using Microscopic Land Use Regression Models

Luc Dekoninck, *University of Ghent, Ghent, OVL, Belgium*
Dick Botteldooren, *University of Ghent, Ghent, Belgium*
Luc Int Panis, *Hasselt University, Diepenbeek, Belgium*

The exposome is commonly understood as the combination of all environmental exposures across multiple disciplines. It is sensitive to the personal features and behavior of the individual and is projected as a potential improvement for epidemiological investigations. Participatory sensing measurements assess the spatial and temporal variability of environmental indicators and are used to quantify the interpersonal variability. Transferring the acquired information from the participatory sensing campaigns to a study population is a basic requirement to include personal exposure into the epidemiological results.
Instantaneous microscopic land-use regression modeling (μLUR) is proposed as an innovative solution to reduce the void between participatory sensing and epidemiology. It is based on data science techniques to model the participatory sensing data in detail. The μLUR models are activity specific and result in indicator functions which capture and predict the indicator variability in a high spatiotemporal resolution. The potential of adding complex dose corrections function in the temporal resolution of the μLUR is illustrated.
Personal exposure is the combination of two spatiotemporal components: the personal behavior and the spatiotemporal variation of the indicator. The personal behavior is a set of activities performed for a specific purpose and in a specific micro-environment. The indicator is also sensitive to the properties of the activity. The activity is the common spatiotemporal object. The μLURs are resolving the spatiotemporal variability for each type of activity. A data workflow combining the personal behavior and the activity specific models can build and use μLUR models with any spatiotemporal resolution. The combined functionality of the μLUR and the data workflow transfers the measured variability in the participatory sensing campaigns to any mobile population. Multiple indicators can be calculated on the same population. The data workflow is compatible with the requirements of the exposome. A validated indicator can be used in policy applications. The policy scenarios are implemented by providing variants of the personal behavior for the population under investigation. The policy scenarios can be sensitive to origin-destination matrices, activity pattern, modal choice, route choice and changing traffic networks. The sensitivity of the policy scenarios is identical to the spatiotemporal resolution of the underlying μLURs. The μLUR technique and data workflow are illustrated with existing models for the case of traffic related exposure to Black Carbon.

Mo-SY-F2.3

What can health policy learn from personal exposure measurements?

Luc Int Panis, *VITO, Mol, Flanders, Belgium*
Michelle Laeremans, *Hasselt University, Diepenbeek, Flanders, Belgium*
Ione Avila-Palencia, *CREAL, Barcelona, Catalonia, Spain*
Juan Pablo Orjuela-Mendoza, *Imperial College London, London, England, United Kingdom*
Natalie Müller, *CREAL, Barcelona, Catalonia, Spain*
Arnout Standaert, *VITO, Mol, Flanders, Belgium*
Evi Dons, *Hasselt University, Diepenbeek, Flanders, Belgium*
Tom Cole-Hunter, *CREAL, Barcelona, Catalonia, Spain*
Esther Anaya, *Imperial College London, London, United Kingdom*
Thomas Götschi, *University of Zuerich, Zurich, Switzerland*
Audrey de Nazelle, *Imperial College London, London, England, United Kingdom*
Sonja Kahlmeier*University of Zurich, Zuerich, Zurich, Switzerland*
David Rojas-Rueda, *CREAL, Barcelona, Catalonia, Spain*
Mark Nieuwenhuijsen, *CREAL, Barcelona, Catalonia, Spain*
Christian Brand, *Oxford University, Oxford, England, United Kingdom*
Elisabeth Raser,*BOKU, Vienna, Wien, Austria*

Aim:
The objective of our participatory measurement campaigns is to collect longitudinal data on physical activity levels and personal air pollution exposure, with a focus on active transport. In addition we want to evaluate the combined health effect of personal air pollution exposure and physical activity.
Methods:
We recruit volunteers for participatory sensing campaigns through classic and social media. After completing an online "baseline" questionnaire, people are selected for experimental measurements based on age, gender and physical activity level. 120 people living in Antwerp, Barcelona and London measured levels of Black Carbon by wearing an aethalometer for 1 week during all of their normal activities. At the same time their level of physical activity was measured with a Sensewear armband and a Zephyr heart-rate monitor. The exposure measurements were complemented with measurements of cardiovascular and respiratory health. Cardiovascular health was evaluated using fundus photography and image analysis of the retinal blood vessels. Respiratory health was studied with spirometry to evaluate lung function. All measurements were repeated in three different seasons. Prior consent was obtained from all participants and all experiments were approved by the appropriate ethics committees in each of the countries.
Results:
Between November 2014 and March 2016 more than 8000 people were recruited throughout Europe. Recruitment through workplaces and social media proved to be especially efficient. Each participant filled in an extended "baseline" questionnaire, providing information on their health, physical activity level and modes of transport. 120 people (40 in each city) completed a total of 3 weeks of measurements during their normal activity.
Personal exposure to Black Carbon is very variable and depends on city, location, local traffic and physical activity level.

At each visit prior to and following the measurement week they performed a lung function test and had their retinal vessels photographed. This constitutes one of the largest environmental health related participatory campaigns and data cleaning will be the first hurdle. Further analysis of the health data is planned during the summer of 2016.
Conclusions:
Collection of large amounts of longitudinal data through an on-line application is possible but language issues are major hurdle in European multi-center studies. Despite the current hype, participatory measurement campaigns and experiments remain cumbersome and expensive. Available air quality sensors are not yet very reliable. Measuring only concentrations is a poor proxy for inhalation of black carbon which also depends on physical activity level.

Mo-SY-G2: Wastewater-based epidemiology (WBE) - from measuring illicit drug use towards understanding population health status - II

Mo-SY-G2.1

Wastewater-based Epidemiology to Track Down the Actual Use of New Psychoactive Substances : Challenges and Recommendations

Alexander Van Nuijs, *University of Antwerp, Antwerp, Belgium*
Juliet Kinyua, *University of Antwerp, Antwerp, Belgium*
Adrian Covaci, *University of Antwerp, Antwerp, Belgium*

Background
New Psychoactive Substances (NPS) are substances that mimic psychoactive effects of illicit drugs like cocaine, cannabis and amphetamine and are produced to evade national and international drug control legislations by introducing slight modifications to chemical structures of controlled substances. NPS are easily acquired legally through online vendors and smart shops where they are sold under false labels with misleading information about their effects and safety. They are considered a growing problem in many communities and are responsible for numerous fatal intoxications. Detection of NPS is a challenge due to their rapid appearance in and out of the drug scene and due to the constantly increasing amount of new substances that appears on the drug market.
Objectives
In order to perform an evidence-based risk assessement of NPS, it is necessary to gather detaled information on the types and amounts of NPS that are used in the general population. Wastewater-based epidemiology, which analyses wastewater samples for the presence of biomarkers of NPS, is a promising approach to gain knowledge on the actual use of NPS. However, there exist several challenges that hampers the routine application of wastewater-based epidemiology for detection of NPS use:
1) Very little scientific information on the metabolic fate of NPS is available. Therefore, it is not always clear which biomarker (parent compound or metabolite) needs to be targeted in wastewater-based epidemiology studies.
2) If the use of NPS is limited to only a few individuals within a community, concentrations of the biomarker in wastewater will be too low to detect with the existing analytical methods.
This presentation gives an overview on the work that has been carried out so far regarding the detection of the actual use of NPS in the general population based on wastewater analysis, discusses the challenges and issues that still exist in this research field, and will provide some recommendations for future research directions.

Mo-SY-G2.2

Using wastewater-based epidemiology to monitor population alcohol and tobacco use

Foon Yin LAI, *University of Antwerp, Antwerp, Belgium*
Wayne Hall, *The University of Queensland, Brisbane, Australia*
Coral Gartner, *The University of Queensland, Brisbane, Australia*
Alexander van Nuijs, *University of Antwerp, Antwerp, Belgium*
Adrian Covaci, *University of Antwerp, Antwerp, Belgium*
Jochen Mueller, *The University of Queensland, Brisbane, Australia*

Over the last decade, wastewater-based epidemiology has become a solid approach to back-estimate illicit drug use in a population. The potential of using wastewater-based epidemiology to assess other population indicators of lifestyle-related health risk has been discussed in the literature. Recently, wastewater-based epidemiology has been used to monitor community-wide alcohol and tobacco use. This presentation aims to (a) introduce how alcohol and tobacco use can be estimated using wastewater-based epidemiology; (b) provide an overview of previous studies that have used this methodology to estimate use of these substances in different countries; and (c) discuss the potential limitations of the method.
Similar to illicit drugs, human excreted biomarkers of alcohol and tobacco use in wastewater samples are analysed using state-of-the-art analytical instruments. These biomarkers include ethyl sulfate for alcohol use and cotinine and trans-3'-hydroxycotinine for nicotine use (a proxy for tobacco use). Concentrations of these biomarkers are measured using liquid chromatography coupled with tandem mass spectrometry. Consumption of these substances is back-calculated through multiplying the biomarker concentration measured by the total daily wastewater flow rate and the molar excretion factor. The approach was first used in Norway for monitoring alcohol use and in Italy to assess tobacco use. Similar studies have been conducted in Australia, Belgium, China, New Zealand, and Spain. These studies generally showed elevated alcohol drinking on the weekends compared to weekdays, whereas tobacco smoking was relatively steady throughout the week. The level and spatial profile of tobacco use identified by the approach in the northern and southern Italy was consistent with that described in the national population survey. Different spatial patterns in alcohol consumption were observed among different countries e.g. higher levels in large cities compared to small villages in Belgium and greater levels in rural towns than urban areas in Australia. Also, temporal variations in alcohol use were observed in Belgian cities. Overall, wastewater-based epidemiology can provide objective information on alcohol and tobacco use in the population of different communities within and between countries and over time. This method will be useful for authorities in identifying regions with high priority, and planning and evaluating interventions for reducing alcohol and tobacco use and their related harms to the society.

Mo-SY-G2.3

Future perspectives for wastewater-based epidemiology: Testing urban water for community-wide public health assessment

Adrian Covaci, *University of Antwerp, Wilrijk-Antwerp, Belgium*
Frederic Been, *University of Antwerp, Wilrijk-Antwerp, Belgium*
Foon Yin Lai, *University of Antwerp, Wilrijk-Antwerp, Belgium*
Alexander van Nuijs, *University of Antwerp, Wilrijk-Antwerp, Belgium*

Over the last decade, wastewater-based epidemiology (WBE) has grown as a solid methodology to monitor and back-estimate illicit drug use in a population. The potential of using WBE to assess other population indicators of lifestyle-related health risks has been discussed in the literature. Very promising results have been obtained, which support the potential of this approach to obtain valuable information about hidden or hardly measurable phenomena. Yet, the information from wastewater that can be gathered is not limited to illicit drugs. If seen as a pooled sample of human excretion (mainly urine and faeces), wastewater potentially bears valuable information relevant to the population's lifestyle and health. Numerous urinary biomarkers have been reported in the literature as potential indicators for diagnosis and prognosis of diseases. If present and stable in wastewater, these biomarkers could potentially be used to obtain valuable information about the health status of large populations.
With regard to the potential of this approach, we need to i) identify biomarkers specific to various diseases, such as diabetes, cancer, stress, etc, which could potentially be analysed in wastewater; ii) develop reliable and robust analytical methods; iii) monitor the occurrence of the selected biomarkers at different time points and sampling locations and iv) evaluate the findings in perspective of relevant epidemiological data, providing an innovative strategy to monitor and assess public health directly at population level.
Only a limited amount of research has been done in this area, with the most notable findings related to the investigation of the cumulative oxidative stress biomarker 8-iso-prostaglandin F2α by LC-MS/MS and measurement of DNA by amperometric sensors. Yet, there are numerous other promising biomarkers which could provide useful information about the health status of the population, such as tobacco specific nitrosamines or markers of alcohol-induced liver disease to name just a few.
Using this approach, disease prevalence could thus be noninvasively monitored over longer periods of time and at different spatial resolutions (local and (inter)national), potentially allowing to setup early-warning systems. Moreover, it could be used to evaluate public health policies and prevention campaigns.

Mo-SY-H2: Advances in consumer exposure assessment - II

Mo-SY-H2.1

Effective Use of Human Exposure Data for Aggregate Consumer Exposure Assessment

Sarah Tozer, *Procter & Gamble, Egham, Surrey, United Kingdom*
Cian O'Mahony, *Creme Global, Dublin, Ireland*

Background: A tiered approach is recommended for exposure assessment, including aggregate exposure assessment in the consumer product domain. However, guidance is needed on the best use of data and tools used to make exposure estimates at various tiers. With these challenges in mind, the European Center for and Ecotoxicology and Toxicology of Chemicals (ECETOC) has created a task force to address this question.
Objectives: The objectives of this task force are twofold: 1) to provide an overview of the current exposure landscape for consumer products, detailing the main exposure input data sources, models and tools that are available for exposure assessment in the food, cosmetics, household, and consumer products domains; 2) to provide guidance on the best use of data and tools for conducting aggregate exposure assessments for chemicals in consumer products, using examples based on the preservatives triclosan and phenoxyethanol.
Methods: For the landscaping work exposure input data for key consumer product categories was collected from the literature and internet sites by task force members, and categorised into exposure algorithms, habits and practices data, co-use data, product composition data (chemical occurrence data and presence probability). The data was collated together with information on available exposure tools, and their usefulness was commented upon. Data were categorised in accordance with a tiered exposure strategy, indicating at which tier they would be most suitably applied. Aggregate exposure case studies were carried out for the preservatives triclosan and phenoxyethanol using the tiered approach, with subject-oriented tools utilised at the high tier.
Results: The landscaping exercise demonstrated that tools and data exist to estimate exposures from consumer products. The tabulated information is made available that provides a useful resource for individuals seeking to perform consumer exposure estimation. Some key gaps were identified, such as the need for refined models in the household product domain. This project also provides much needed guidance on conducting aggregate exposure assessments for consumer products, showing triclosan and phenoxyethanol assessed at low and high tiers. At the high tier, the greatest exposure refinements were observed through use of better habits and practices data with product co-use data, and the use of presence probabilities. The refinement of concentration values has less impact on predicted exposure, because chemical concentration in product is directly proportional to the resulting exposure, whereas it is seen that refining product co-use and the frequency of chemical occurrence can impact the estimated exposure by several orders of magnitude.

Mo-SY-H2.2

Development of a Harmonized Database of Reported and Predicted Consumer Product Ingredient Information

Kristin Isaacs, *US EPA, Research Triangle Park, North Carolina, United States*
Katherine Phillips, *Oak Ridge Institute for Science and Education, Research Triangle Park, North Carolina, United States*
Kathie Dionisio, *US EPA, Research Triangle Park, North Carolina, United States*
Derya Biryol, *Oak Ridge Institute for Science and Education, Research Triangle Park, North Carolina, United States*
Paul Price, *US EPA, Research Triangle Park, North Carolina, United States*

Near-field exposure to chemicals in consumer products has been identified as a significant source of exposure for many chemicals. Quantitative data on product chemical composition and weight fraction is a key parameter for characterizing this exposure. While data on product composition are scarce, recent efforts have obtained such information from publically-available Material Safety Data Sheets (MSDS). To supplement these data, techniques have been developed to predict associated weight fractions based on ordered product ingredient lists (e.g., for personal care products). In addition, quantitative structure property relationship (QSPR) methods have been used to develop models which the functions that an arbitrary chemical might perform in consumer products based on its chemical structure. Here, we describe a new harmonized database to provide reported and predicted weight fractions and chemical functions of chemicals in consumer products. The database contains 1) MSDS-based product ingredient data (new and existing) for over 10,000 products, 2) data collected from publically-available ingredient lists and corresponding weight-fraction predictions for 5,907 products, 3) reported chemical function data for 14,000 chemicals, and 4) QSPR-predicted chemical functional use information for over 8,000 chemicals. The products in the database were mapped to a harmonized set of over 300 consumer product categories suitable for use in human exposure modeling. This product composition database integrates information on the chemical use-category linkages, product composition data, and predictive approaches necessary in estimating near-field human exposures to thousands of chemicals in products in support of risk-based decision-making. This abstract does not necessarily reflect U.S. EPA policy.

Mo-SY-H2.3

The Application of Specific Consumer Exposure Determinants (SCEDs) to Refine Consumer Exposure Estimates

Namali Corea, *SC Johnson, Frimley Green, Surrey, United Kingdom*
Carlos Rodriguez, *Proctor & Gamble, Brussels, Belgium*
Francesca Angiulli, *AISE, Brussels, Belgium*

An assessment of the potential risks to consumer health is required under the EU REACH Regulation when a registered substance is classified as hazardous. All known consumer uses of the substance must be assessed and consequently the ECETOC Targeted Risk Assessment tool was developed as a high throughput, lower tier risk assessment tool for REACH. The tool provides conservative estimates of consumer exposure which can be further refined by the application of SCEDS, if required. SCEDs have been developed for a number of product categories by sectors such as AISE (International Association for Soaps, Detergents and Maintenance Products). The SCEDs contain habits and practices data that can be incorporated directly into the ECETOC TRA algorithms to provide more realistic exposure estimates. They also serve to provide a harmonised source of generalised, consumer exposure information in a transparent and user-friendly manner. Application of SCEDS results in refined exposure estimates which are still exaggerated when compared to the exposure estimates from higher tier models. An example is provided for non-aerosol, continuous action Air Care products. Adult inhalation exposure to these products was approximately 20-fold lower using the SCEDS when compared to the ECETOC TRA. However, the inherent conservatism of the SCEDs was demonstrated when exposure was compared to estimates from the higher tier, RIFM 2-Box Air Dispersion Model. Inhalation exposure using the 2-box model was a further 13-fold lower than the value calculated with the SCEDs. The use of SCEDS provides a convenient, easy to use, scientific-based refinement for the ECETOC TRA model.

Mo-PL-I2: Transportation-Related Air Pollution - II

Mo-PL-I2.1

Personal exposure monitoring of UFP in different micro-environments

Yvonne De Kluizenaar, *TNO, Delft, Netherlands*
Eelco Kuijpers, *TNO, zeist, Netherlands*
Iris Eekhout, *TNO, Leiden, Netherlands*
Roel Vermeulen, *Utrecht University, Utrecht, Netherlands*
Gerard Hoek, *Utrecht University, Utrecht, Netherlands*
Marita Voogt, *RIVM, de Bilt, Netherlands*
Reinier Sterkenburg, *TNO, Delft, Netherlands*
Frank Pierik, *TNO, Delft, Netherlands*
Anjoeka Pronk, *TNO, Zeist, Utrecht, Netherlands*

Introduction: Particulate matter air pollution has been linked to adverse health effects in a large number of epidemiological and experimental studies. It has been hypothesized that the ultrafine fraction may be particularly relevant to health. However epidemiological studies on ultra-fine particles (UFP) are limited. The aim of this study is to gain insight in personal real time UFP exposure patterns by different micro environments, by activities and by time of the day.
Methods: For 11 healthy volunteers (Eindhoven region within the AiREAS network, The Netherlands), personal UFP exposure assessment (10-500 nm) with a sample rate of 10 seconds was performed continuously for 5 days with a personal monitoring device (DiSCmini, MatterAerosol). Individual GPS tracks were simultaneously recorded (Qstarz BT 1000XT) and classified into micro-environments (indoors-home, indoors-other, outdoors: motorized transportation, cycling, walking and stationary) based on cluster detection and speed. Peak UFP exposures were determined by identifying all instances where UFP levels were above the 90th, 95th or 99th percentile for at least 5 minutes. Activity diaries were kept by a selection of the volunteers.
Results: Overall the UFP levels varied over a wide range (P5-P95: 103 and 107 particles/cm3), with highly skewed distributions. Median and quartile (P25 and P75) outdoor UFP levels were slightly higher than indoor UFP levels. However P95 indoor levels were higher compared to outdoor P95 levels. This was more pronounced in the home compared to other indoor locations. When comparing the different modes of transportation, the highest levels (P50, P75, P95) were observed during motorized transport > cycling > walking. An exploration of the occurrence of UFP peaks by time of the day demonstrated an increased number of peaks between 7 am and 9 pm, both indoors and outdoors. Outdoors, the number of peaks increased between 3-7 pm. Indoors, more peaks seemed to occur in the morning (7-9 am), afternoon (12am-3pm) and evening (5-8pm). Activity diaries will be used to obtain more insight in the indoor sources.
Conclusions: Although exposure to UFP is ubiquitous, this study demonstrated differences in real time exposure levels and the occurrence of exposure peaks between different micro environments, activities and times of the day. This information may contribute to further refinement of exposure characterization in epidemiological studies, thereby contributing to better quantification of exposure-effect relationships.

Mo-PL-I2.2

Increased oxidative potential of fine particulate matter (PM2.5) in major freeways of Los Angeles, CA

Constantinos Sioutas, *University of Southern California, Los Angeles, California, United States*
Farimah Shirmohammadi, *University of Southern California, Los Angeles, California, United States*
Dongbin Wang, *University of Southern California, Los Angeles, CA, United States*
Sina Hasheminassab, *University of Southern California, Los Angeles, CA, United States*
Vishal Verma, *University of Illinois at Urbana-Champaign, Champaign, Illinois, United States*
Martin Shafer, *University of Wisconsin-Madison, Madison, Wisconsin, United States*
James Schauer, *University of Wisconsin-Madison, Madison, Wisconsin, United States*

In this study an on-road sampling campaign was conducted using a mobile instrumentation platform to assess the chemical composition and oxidative potential of fine particulate matter (PM2.5) on three major representative roadways environments for daily commuters in Los Angeles, including: 1) I-110, a freeway consisting of almost exclusively light-duty vehicles (LDVs); 2) I-710, a freeway with high fraction of heavy-duty vehicles (HDVs), and; 3) Wilshire/Sunset (WS) boulevards, which are two of the busiest surface streets in Los Angeles. Sampling was also conducted at University of Southern California (USC) in an urban background site. The PM samples were analyzed for elemental carbon (EC), organic carbon (OC), polycyclic aromatic hydrocarbons (PAHs) and elemental compositions. PM2.5 oxidative potential was quantified using two different assays: the macrophage ROS assay and the dithiothreitol (DTT) assay. Overall, higher mass fractions of the PAHs, EC, OC as well as major transition metals (i.e. Al, Ba, Cu, Fe, Mn, Ni, Sb, Ti, Pb and Zn) were observed in both freeways compared to surface street and urban background site. More importantly, the highest per PM mass OP levels were observed in freeways: the per PM mass ROS activity was highest at I-110 (µg Zymosan/ mg PM), while the per PM mass DTT activity was greatest at I-710 and (32.35±13.26 nmol/min mg PM). The higher PM redox activity observed on freeways indicates the increased intrinsic toxicity of PM in freeways comparing to other micro environments. The DTT activity levels measured at the studied freeways were compared with the DTT activity levels reported from previous dynamometer studies, which capture only tailpipe emissions. The higher freeway DTT activity levels of measured PM in our study compared to those in dynamometer facilities illustrate the important contribution of non-tailpipe emissions (e.g. re-suspended road dust and vehicular abrasions of brake and tire wear) on the PM2.5 oxidative potential. Finding from this study will provide significant insight on PM-induced toxicity exposure to daily commuters driving in different roadway environments.

Mo-PL-I2.3

Sources of Quasi-Ultrafine, Fine and Coarse Particulate Matter in the Southern California Children's Health Study Communities

Rima Habre, *University of Southern California, Los Angeles, CA, United States*
Robert Urman, *University of Southern California, Los Angeles, CA, United States*
Scott Fruin, *University of Southern California, Los Angeles, CA, United States*
Theresa Bastain, *University of Southern California, Los Angeles, CA, United States*
Fred Lurmann, *Sonoma Technology Inc, Petaluma, CA, United States*
Martin Shafer, *University of Wisconsin-Madison, Madison, WI, United States*
Patrick Gorski, *University of Wisconsin-Madison, Madison, WI, United States*
Rob McConnell, *University of Southern California, Los Angeles, CA, United States*
Edward Avol, *University of Southern California, Los Angeles, CA, United States*
Frank Gilliland, *University of Southern California, Los Angeles, CA, United States*

Aim: To resolve and quantify the main sources contributing to particulate matter (PM) air pollution in the quasi-ultrafine, fine and coarse size fractions, in eight southern California Children's Health Study (CHS) communities
Methods: The CHS Intra-Community Variation PM sampling campaign was conducted in 2008-9 in the communities of Anaheim, Glendora, Long Beach, Mira Loma, Riverside, San Dimas, Santa Barbara and Upland, California. Month-long integrated PM samples were collected in the quasi-ultrafine (PM0.2), fine (PM2.5) and coarse (PM2.5-10) size fractions (n= 476, 265 and 298, respectively) using specially-designed Harvard Cascade Impactors. In addition to gravimetric mass, concentrations of total and water-soluble metals, elemental and organic carbon, water-soluble organic carbon and major ions were obtained. Enrichment factors (EF) relative to the PM2.5 fraction were calculated for each chemical species in the PM0.2 and PM10 size fractions. The Positive Matrix Factorization (EPA PMF v5.0) model was used to resolve and estimate the contributions of major factors in each of the three size fractions.
Results: The elements Al and Ca had among the highest median EF in PM10 relative to PM2.5, while EC, OC, and B had the highest median EF in PM0.2 relative to PM2.5. Six, seven and five sources (tracers in parentheses) were resolved in the quasi-ultrafine, fine and coarse size fractions, respectively. Traffic (EC, water-soluble and water-insoluble OC), fuel oil (Ni, V), and brake and tire wear (Sb, Zn, Ba) contributed 57.8%, 14.6% and 10.3% to total PM0.2 mass, respectively. Gasoline vehicles (OC, B), ammonium sulfate (NH4+, S), ammonium nitrate (NH4+, NO3-), fuel oil/diesel (Ni, V, EC), and non-tailpipe abrasive brake and tire wear (Sb, Zn, Ba) contributed 25.8%, 21.3%, 12.4%, 11.1%, and 3.5%, to PM2.5 mass, respectively. Finally, secondary formation (NH4+), crustal (Al, Ca, Fe), sea salt (Na, Mg, Cl-), and non-tailpipe abrasive wear (Ba, Cu, Sb, Zn) contributed 27.8, 27.1%, 23.5%, and 7.6% to PM2.5-10 mass, respectively. The largest traffic impacts were seen in Anaheim, Glendora, Upland and Riverside, while the largest fuel oil impacts related to ports activity were seen in Long Beach. We were also able to distinguish tailpipe from non-tailpipe traffic sources in the quasi-ultrafine and fine size fractions.
Conclusions: PMF-resolved source factors displayed significant seasonal and within- and between-community variability in all three size fractions, reflecting the heterogeneity of air pollution exposures in the Los Angeles basin. Ongoing spatiotemporal modeling will allow us to examine associations between these exposures and respiratory outcomes in the CHS.

UTRECHT
THE NETHERLANDS
OCTOBER 9-13
ISES 2016

Mo-SY-A3: Multiple route exposure to multiple chemicals, the cocktail effect

Mo-SY-A3.1

Dietary exposure assessment to mixtures of pesticides and other substances

Corinne Sprong, *RIVM, Bilthoven, Netherlands*
Hilko van der Voet, *Wageningen University and Research Centre, Wageningen, Navassa Island*
Amélie Crépet, *French Agency for Food, Environmental and Occupational Health & Safety, Maisons-Alfort, France*
Jacob van Klaveren, *National Institute for Public Health and the Environment, Bilthoven, Nepal*

Every day we are exposed to multiple possibly harmful chemicals in our diet. Traditionally, risk assessment was performed for single (groups of) chemicals. However, there is a growing awareness of assessing the risk of complex mixtures of chemicals having the same toxicological effect.
In 2012, the European Food Safety Authority (EFSA) published guidance on the use of the probabilistic methodology for modelling dietary exposure to pesticide residues for single and multiple compounds (1). EFSA included two model runs, the optimistic model run and pessimistic model run in which major uncertainties are treated liberal and conservative, respectively. The outcome of these model runs can be used to assess whether refinement is useful. Currently, further refinements are under discussion at the European level (2,3). The EFSA methodology for cumulative risk assessment is implemented in the Monte Carlo Risk Assessment (MCRA) program, a web-based exposure assessment tool. Within the software, the settings of both model runs are clearly indicated, making it easy to select the right parameters. The use of this tool will be demonstrated.
For cumulative exposure assessment of pesticides, cumulative assessment groups (CAGs) are defined or still under development. These CAGs can consist of over 100 pesticides at level 1 (affect the same target organ) and dozens at level 2 (same effect in target organ). Other chemicals, such as pollutants or toxins, are also known to affect the toxicological endpoints defined for pesticide CAGs. Probably, these substances should also be considered in cumulative risk assessment, resulting in even larger amount of chemicals belonging to the same CAG. Furthermore, the assumptions when grouping pesticides into CAGs are often based on the precautionary principle due to a lack of information on mixture effects in current assessments. Therefore, refinement is needed to obtain realistic exposure assessments allowing better decision-making. Better identification of mixtures using user-friendly but powerful software to quickly identify key mixtures of potential risk to human health is crucial in this process. Within the EuroMix project, a tool is developed and implemented within MCRA for mixture identification. This mixture selection tool, to be further explained and exemplified in the presentation of Amélie Crepet, will be demonstrated.

References:

1. EFSA, 2012. Guidance on the Use of Probabilistic Methodology for Modelling Dietary Exposure to Pesticide Residues. EFSA J. 10(10), 2839

2. http://ec.europa.eu/food/plant/pesticides/max_residue_levels/cumulative_risk/index_en.htm
3. http://www.efsa.europa.eu/en/press/news/160127

Mo-SY-A3.2

Aggregate exposure to pesticides from dietary and non-dietary exposure: A UK case study for residents, bystanders and spray operators

Marc Kennedy, *Fera, York, United Kingdom*
David Garthwaite, *Fera, York, United Kingdom*

Exposure to pesticides can occur from multiple sources and distinct sub-populations can be affected very differently. In the EU Horizon 2020 project Euromix, new tools are being added to the Monte Carlo Risk Assesment (MCRA) software which allow calculation of aggregate exposure in a very flexible way. An external model, designed to calculate a particular source of exposure, can be run to generate simulated non-dietary exposures. The results can then be integrated with the dietary calculations of MCRA.
We present a particular example based on the BROWSE model (Bystander, Resident, Operator and WorkerS Exposure), which has various built-in scenarios for different populations, spray application types and crops. BROWSE generates probabilistic output to represent variation in exposure conditions. By default, many assumptions such as scenario definitions and input parameters have conservative default values. However, by repeated runs of the model with realisations of actual spray amounts we aim to produce a more realistic exposure distribution.
The annual UK pesticide usage survey (PUS) provides detailed field level information about real combinations of pesticides applied to crops at field level. It includes the main crops grown in the UK including arable, orchard, outdoor vegetable, soft fruit and covered crops.
The examples presented will illustrate how the new probabilistic aggregate model implemented in MCRA, as part of the Acropolis and Euromix projects, can generate a detailed distribution of exposures to multiple pesticides.
BROWSE is used to generate a matrix of exposure simulations representing population variation for defined sub-populations. This matrix is then linked to MCRA and aggregated with the relevant dietary exposure for the UK population to produce total exposures. The resulting MCRA outputs highlight the relative sources of exposure, and the most significant pesticides. An important potential use of the results is to prioritise testing of chemical mixtures to those that occur in real exposures.

Mo-SY-A3.3

Linking probabilistic exposure models for non-food and food sources to calculate aggregate consumer exposure: Case study on Bisphenol A

Cecile Karrer, *ETH Zurich, Zurich, Switzerland*
Natalie von Goetz, *ETH Zurich, Zurich, Switzerland*
Christiaan Delmaar, *National Institute for Public Health and the Environment (RIVM), Bilthoven, Netherlands*
Konrad Hungerbühler, *ETH Zurich, Zurich, Switzerland*

Background
There is a rising demand for more holistic risk assessments to better ensure consumers' safety. In the estimation of chemical risks, the quantification of the overall exposure is a crucial step. For this purpose, an exposure assessment needs to account for multiple routes and multiple chemicals that have the same effect. Realistic exposure estimates can be obtained by using probabilistic models, which account for variability in human behavior.
Objectives
Within the EuroMix EU project, different food and non-food exposure models are linked to facilitate the exposure assessment from different sources for different chemicals. In this work, we in particular link the results of the probabilistic models MCRA and PACEM to assess aggregate exposure from food and consumer products. As a test case, the exposure to the endocrine disrupting substance Bisphenol A (BPA) is assessed. BPA is present in both food and non-food sources, the latter being polycarbonate plastics, epoxy resins and thermal paper. Since the metabolization of BPA differs depending on the intake routes, aggregation can only be performed on the level of internal exposure.
Methods
MCRA is a web-based program for Monte Carlo risk assessment with emphasis on exposure from food sources. PACEM is a probabilistic aggregate consumer exposure model, which was first used for exposure assessments for ingredients in cosmetics and personal care products. For the case study on BPA, PACEM is adapted to account for all non-food sources containing BPA and all age groups, including toddlers and children. The internal exposure to BPA is calculated by feeding the external aggregate exposure from each exposure source into a PBPK model. Human biomonitoring studies are used for validation.
Results
The case study on BPA results in individual-based probabilistic aggregate consumer exposure estimates. All relevant routes and sources are considered, and internal exposure is calculated for all routes. To our best knowledge, this is the first probabilistic assessment of internal aggregate exposure of BPA across all possible pathways. These results are important because they will help to evaluate every-day exposure to and risk from BPA in an integrated and realistic way.

Mo-SY-A3.4

Mixtures selection from combined exposure and PBPK modelling to aggregate exposure

Amélie Crépet, *ANSES, French agency for health and safety, MAisons-Alfort, France*

Due to the large number of chemicals found in the environment, individuals are daily exposed from various sources to complex mixtures of chemicals which can interact and cause health diseases. The risk related to chemical mixtures is difficult to characterize. One reason lies in the multitude of possible combinations of chemicals for which it is unrealistic to test toxicological combined effects. For this reason, risk assessment is usually performed for chemicals belonging to a same chemical family and having same mode of action. However, those mixtures could not reflect the reality of exposures.
As an introduction, an integrated approach will be proposed to help to overcome new challenges related to exposure assessment to mixtures coming from various sources. Methods recently developed to define mixtures from combined exposures will be explained. The first approach is based on the decomposition of the co-exposure matrix into two matrices to extract the main mixtures which are relevant to study will be explained. A second approach includes a clustering of individuals with similar food patterns to determine food vectors of the mixtures. The last method combines exposure levels and data on the toxicity of the substances to characterize mixtures. Finally, physiologically-based pharmacokinetic models will be presented as tools to aggregate exposure from various sources and to estimate internal exposures.
The methods will be illustrated by examples on pesticides residues and various substances in food, and on metals from food, air, dust and cigarettes.

Mo-SY-B3: From occupational to environmental biomonitoring: lessons to be learned.

Mo-SY-B3.1

Performing and understanding biological monitoring: how the experience in occupational toxicology can help.

Peter J. Boogaard, *Shell International bv, The Hague, Netherlands*

Human biomonitoring, the measurement of the levels of chemicals and associated biological or biochemical effects in human body fluids and tissues, has been routinely applied in occupational settings for many decades. The increased availability of analytical technologies with ever decreasing detection limits makes human biomonitoring not only more accessible but also more sensitive. These developments have allowed a shift from occupational to environmental human biomonitoring.
Human biomonitoring has many benefits, both in its ability to integrate human exposures to chemical substances via all routes of exposure and in its potential to improve the efficiency and accuracy of related health risk assessments. Any valid health risk assessment requires a basic knowledge and understanding of the underlying dose-effect relationships, however, this information is often lacking, especially at low exposure levels. Even when it is available, a variety of other factors need to be taken into account for a valid interpretation of human biomonitoring data.
In this presentation the considerations will be set out that need to be applied in order that available biomonitoring data can be reliably interpreted within the context of their associated uncertainties.
To allow a reliable interpretation of human biomonitoring results, several considerations have to be taken into account. If human biomonitoring information is to be used to evaluate and describe health risks, information is required on 4 key elements: (1) the analytical integrity, (2) the extent to which biokinetic considerations have been taken into account, (3) the relevance of the available data for health effects, and (4) how the data align with other available information.
In this presentation, using specific examples, the level of understanding that is required for each of these four key elements is described. In addition, it will be discussed how the application of data varies according to the level of understanding, including the relative importance of each element. If some information pertaining to one or more of the 4 key elements is incomplete or lacking, a valid human health risk assessment may be impossible. However, the human biomonitoring data may still be useful for other purposes, such as policy support or checking the efficiency of mitigation measures. A framework will be presented that incorporates the 4 key elements, based upon established scientific criteria. The framework may help to evaluate any human biomonitoring data with respect to the portion of the risk assessment process in which it can be reliably applied.

Mo-SY-B3.2

Reference values: tools used to interpret biomonitoring

Hans Drexler, *University Erlangen-Nuremberg, Erlangen, Germany*
Thomas Göen, *University of Erlangen-Nuremberg, Erlangen, Germany*

In Germany biological monitoring is part of the medical surveillance program. Biological limit values are health-based (BAT-value) or risk-based (for carcinogenic substances). In general, health based biological exposure limits are established for the use of biological monitoring in the prevention of occupational diseases. For substances for which the concept of health-based or risk-based threshold values is not applicable, the Working Group Setting of Threshold Limit Values in Biological Materials of the DFG Commission for the Investigation of Health Hazards of Chemical Compounds in the Work Area has established "Biologische Arbeitsstoff-Referenzwerte" (BARs, Biological Reference Values for Chemical Compounds in the Work Area) as an approach for evaluating biomonitoring data. The BAR represents the upper reference concentration of a biomarker in the general adult population without occupational exposure to the agent. It is derived from biomonitoring data of a sample of a defined population group. In general, a BAR corresponds to the 95th percentile of the sample distribution. Ideally, national environmental surveys including human biomonitoring results are used as basis for deriving BARs. The influence of age, sex, social status, residential area and life style factors on background exposure is considered in the evaluation of these values. Because tobacco smoking is the most frequent influencing factor, several BARs have been determined for non-smokers only. To date, BARs for 17 substances or substance groups are listed in the List of MAK and BAT Values 2011. BARs for another five substances have been discussed, but have not been established because of the insufficient scientific database. Establishing the BARs aims to facilitate the evaluation of human exposure to chemical compounds for which no health-based threshold values can be derived but an adequate assessment of exposure is required due to their toxicity. The application of BARs does not permit a toxicological evaluation, but does allow the occurrence and the extent of occupational exposure to hazardous substances to be proved.

Mo-SY-B3.3

Biomonitoring for benzene exposure: from occupational exposure to environmental pollutant

Silvia Fustinoni, *University of Milan, Milan, Italy*

Aim. This work is aimed at reviewing the use of biomonitoring of benzene exposure focusing on those changes in the occupational limit values and in air pollution that have prompted the research of new biomarkers.
Methods. A bibliographic review on biomedical databases has focused on papers dealing with biomonitoring of benzene in the last 40 years.
Results. In the forties of the last century occupational limit values for benzene were in the order of hundreds of ppm; one of the outcome of the epidemiological studies conducted on the Pliofilm Goodyear cohort in the eighties, was that benzene was recognised as leukemogenic to human and the limits were lowered to about 1 ppm. Until that time the biological monitoring of the exposure to benzene was performed by measuring urinary phenol, accounting for about 70% of the adsorbed dose; however a major drawback was its poor specificity.
Industrial toxicologists were forced to identify new bioindices to assess the lower occupational exposure: in the early nineties, urinary t,t-muconic acid (MA, 3 - 18 % of the absorbed dose), urinary S-phenylmercapturic acid (SPMA, < 1%), blood benzene (< 1%), and urinary benzene (<0.1%) were introduced and biological limit values were proposed. These indices were first applied to assess exposure in petrochemical workers, coke oven workers, shoe makers, and rubber workers. Almost in the same years, in Europe and US a new blended petrol, enriched with a mixture of aromatic hydrocarbons, including benzene, was introduced. Airborne benzene level increased, mostly due to auto vehicle exhaust fumes, and it became a pollutant of the living environment and a chemical of public concern. New biomarkers were then used to assess exposure to low levels of benzene in petrol station attendants, traffic policemen, bus and taxi drivers and in the general population; studies investigated the specificity and sensibility, practical and ethical implications, analytical issues, and optimal application ranges of these biomarkers. The results highlighted that MA is increased by sorbic acid in diet, so that it is not useful to assess low exposure; on the other hand blood benzene, although specific, requires an invasive sampling; urinary benzene and SPMA are non-invasive biomarkers, they are specific and useful to trace the lowest exposures.
Conclusions. Biomonitoring of benzene has been challenged by several changes toward low exposures; for this reason, it is ready to face upcoming lower occupational exposure limits as well as biomonitoring programme in the general population.

Mo-SY-B3.4

Elemental speciation in biological samples - occupational exposure experiences for better risk characterisation

Jackie Morton, *Health & Safety Laboratory, Buxton, United Kingdom*

Background
Elemental speciation analysis adds another dimension to the analysis of biological samples. Simply, it allows the different forms of an element to be individually identified and quantified. This is useful because different species of the same element vary both in toxicity and bioavailability which when separated allows better interpretation of exposure. The work reported here summarises the developments made in the area of speciation analysis for occupational monitoring and addresses the further research issues still remaining.
Methods
The Health and Safety Laboratory (HSL) has developed and published methods for mercury, arsenic and chromium speciation which have formed the backbone of these analytical developments. Using liquid chromatography coupled with inductively coupled plasma mass spectrometry, analytical techniques have been optimised so that biological samples (urine, serum, exhaled breath condensate and hair) can be analysed for different elemental species.
Results
Initial work started by developing a method for mercury species in hair samples, however, whilst showing potential in some scenarios this method proved that overall hair is not a useful medium for mercury speciation.
Initial arsenic speciation studies showed that there were significantly higher concentrations of the more toxic inorganic arsenic species in urine samples from timber treatment workers (n=49) than in either controls (n=31) or semi-conductor workers (n=46). Since then, reference ranges for five arsenic species have been established in 95 non-occupationally exposed volunteers and this is turn has allowed the interpretation of exposure in UK semiconductor workers over a five year period.
The newest speciation method developed at HSL is to allow the determination of hexavalent chromium in exhaled breath condensate. This novel analysis has shown that hexavalent chromium is present in the breath of electroplaters and though further work is necessary to better understand the kinetics of this matrix, already a better understanding of the workplace exposures has been gained from this methodology.
This area of research allows us to better interpret exposure to the more toxic species of arsenic and chromium. Whilst improvements and more developed methods are necessary in the field of elemental speciation, the work presented here helps to show that it is a vital component when assessing risk in the workplace. It also has a role in environmental biomonitoring, for example distinguishing between geological and dietary exposures thus enhancing risk characterisation.

Mo-SY-B3.5

Using PCB signatures and enantiomer fractions for source identification and to age date exposure

David Megson, *University of Toronto, Toronto, ON, Canada*
Jean-Françios Focant, *University of Liège, Liège, Belgium*
Donald Patterson, *Exponent, Atlanta, United States*
Matthew Robson, *Brock University, St. Catharines, Canada*
Maeve Lohan, *Plymouth University, Plymouth, United Kingdom*
Paul Worsfold, *Plymouth University, Plymouth, United Kingdom*
Sean Comber, *Plymouth University, Plymouth, United Kingdom*
Robert Kalin, *University of Strathclyde, Glasgow, United Kingdom*
Eric Reiner, *Ontario Ministry of the Environment and Climate Change, Toronto, Canada*
Gwen O'Sullivan, *Mount Royal University, Calgary, Canada*

Polychlorinated biphenyls (PCBs) are a group of 209 chlorinated organic compounds that were widely used throughout the 20th century. While PCBs have been largely phased out of commercial/industrial use, they remain an important legacy contaminant and can still be found in closed systems in some countries as dielectric fluids in electrical equipment and transformers. Many transformers containing PCBs are in the process of being replaced which presents a potential for human exposure. In these instances it is important to not only determine the extent of any exposure and risks to human health, but to also establish the source and age date exposure.
To achieve this, detailed polychlorinated biphenyl (PCB) signatures comprised of over 80 congeners and chiral Enantiomer Fractions (EFs) of CB-95, CB-136 and CB-149 were measured for 30 workers at a transformer dismantling plant. Approximately 1.5 g of serum was extracted and PCB signatures were created through analysis by comprehensive two-dimensional gas chromatography with time-of-flight mass spectrometry (GC×GC-TOFMS), and EFs calculated following analysis by gas chromatography with high resolution mass spectrometry (GC-HRMS). A total of 84 PCBs were identified in the serum samples with concentrations of the 7 indicator PCBs ranging from 11-350 ng g-1 of serum (1.2-39 µg g-1 lipid). PCB signatures were interpreted using principal component analysis (PCA) which distinguished workers with background or recent exposure from those with prolonged occupational exposure. Occupationally exposed individuals had a similar PCB profile to Aroclor A1260. However, individuals with prolonged exposure had depleted proportions of several PCBs that are susceptible to metabolism (CB-95, CB-101 and CB-151) and elevated proportions of PCBs that are resistant to metabolism (CB-74, CB-153, CB-138 and CB-180). The results also identified a third group of workers who appeared to have been exposed to an additional source of PCBs. The results show near complete removal of the CB-95 E2 enantiomer in some samples, indicating that bioselective metabolism or preferential excretion of one enantiomer occurs in humans. By considering PCB concentrations along with detailed congener specific signatures it was possible to identify different exposure sources, and gain an insight into both the magnitude and duration of exposure.

Mo-SY-C3: Quantitative in vitro to in vivo extrapolation (QIVIVE): Advances in tools to quantify exposure-response relationships for risk assessment - II

Mo-SY-C3.1

Passive Dosing of hydrophobic organic chemicals to in vitro assays - controlling, defining and linking exposure

Philipp Mayer, *Technical University of Denmark, Kongens Lyngby, Denmark*
Dorothea Gilbert, *Technical University of Denmark, Kongens Lyngby, Denmark*
Stine N. Schmidt, *Technical University of Denmark, Kongens Lyngby, Denmark*
Kilian EC Smith, *RWTH Aachen University, Aachen, Germany*

In vitro assays are increasingly used for toxicity testing and screening of chemicals and environmental samples. The physical format of in vitro tests (24-1536 well plates) facilitates high throughput and cost-efficient testing, but creates also challenges with regards to controlling, maintaining and defining exposure in these assays. In passive dosing, a polymer is loaded with the test substance and then applied as partitioning donor for tightly controlling the exposure of the test substance in the test1. The exposure can be expressed as freely dissolved concentration, chemical activity or equilibrium lipid concentrations1,2, which can all facilitate quantitative in vitro to in vivo extrapolations (QIVEVE).

This presentation will give an overview on passive dosing of in vitro tests and outline challenges and needs for further development. For testing in 24 well plates, silicone O-rings have been successfully applied as donor in several studies, which was practical and provided good performance and new results1,3-5. It is now important to (1) reduce the preparative steps prior to the tests, (2) develop passive dosing for 96-1536 well plates and (3) combine passive dosing with closed-well testing for in vitro testing of (semi)volatile chemicals. New application areas include in vitro assays directed at mixtures2,6,7 (level & composition) and biotransformation parameters.

References

1 Smith KEC et al. 2010. Passive dosing for producing constant and defined exposures of hydrophobic organic compounds during in-vitro toxicity tests. Chemical Research in Toxicology 23: 55-65.
2 Schmidt SN et al. 2013. Passive dosing of PAH mixtures to terrestrial springtails - linking mixture toxicity to chemical activities, equilibrium lipid concentrations, and toxic units. ES&T 47, 7020–7027.
3Smith KEC et al. 2013. The dosing determines mutagenicity of hydrophobic compounds in the Ames II assay with metabolic transformation: Passive dosing versus solvent spiking. Mutation Research, Genetic Toxicology and Environmental Mutagenesis 750: 12-18.
4 Oostingh GJ et al. 2015. Differential immunomodulatory responses to nine polycyclic aromatic hydrocarbons applied by passive dosing. Toxicology in vitro 29, 345-351.
5Vergauwen L et al. 2015. A high throughput passive dosing format for the Fish Embryo Acute Toxicity test. Chemosphere 139: 9-17.
6 Gilbert D et al. 2015. Endocrine activity of persistent organic pollutants accumulated in human silicone implants - Dosing in vitro assays by partitioning from silicone. Environment International 84, 107-114.
7Jahnke A et al. 2016. Strategies for Transferring Mixtures of Organic Contaminants from Aquatic Environments into Bioassays. In press in ES&T DOI 10.1021/acs.est.5b04687

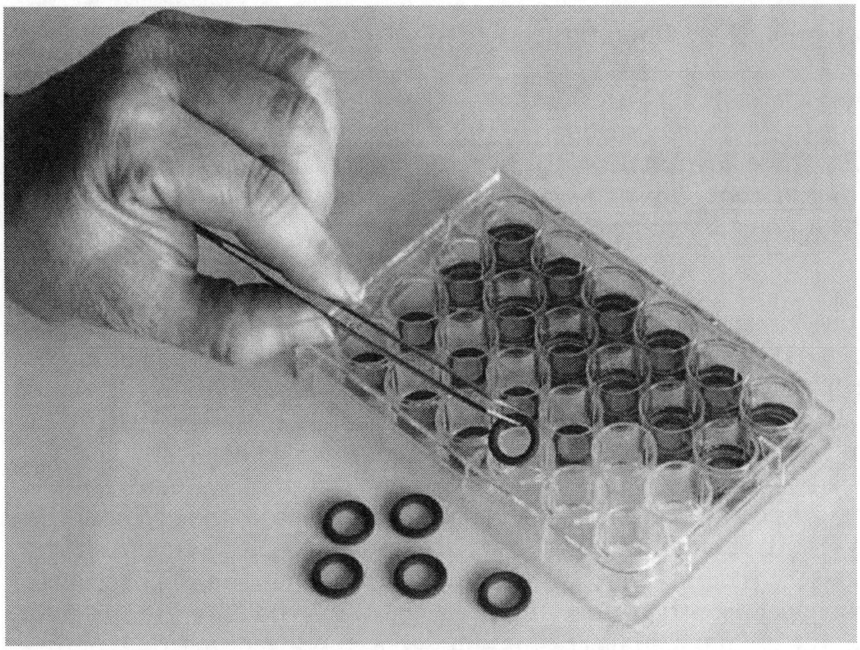

Passive dosing of in vitro tests with silicone O-rings

Mo-SY-C3.2

Examining underlying assumptions when translating in vitro bioassay results to in vivo conditions

Jon Arnot, *ARC Arnot Research and Consulting, Toronto, Canada*
James Armitage, *AES Armitage Environmental Sciences, Toronto, Canada*
Todd Gouin, *Unilever, Sharnbrook, United Kingdom*
Michelle Embry, *ILSI Health and Environmental Sciences Institute (HESI), Washington, United States*

In vitro testing is creating information that can be used to improve mechanistic understanding of toxicity pathways and for chemical assessments. The data can be used for hazard-based prioritization or they can be combined with exposure data for risk-based prioritization. Chemical potency comparisons require a consistent exposure metric ("x-axis") corresponding to the observed response ("y-axis"). Translating chemical concentrations from one system under specific conditions to a different system with different conditions requires consistent metrics and units. For example, the corroboration of in vitro data with in vivo data and comparisons of exposure and hazard concentrations for risk estimation require consideration of various sub-phase volumes and their sorptive capacities as well as chemical property information, e.g. partition coefficients. The freely dissolved concentration (Cfree; nmol/L) has been proposed as a relevant metric to translate concentrations across and within systems (e.g., blood-tissues). The objectives of this study were to examine commonly applied assumptions (e.g., Cblood, in vivo = Cnominal, in vitro) and alternative assumptions when translating in vitro bioassay results to in vivo conditions. Mass balance models and equilibrium partitioning theory were used to examine assumptions for translating in vitro test assay data to in vivo systems. A series of representative in vitro bioassay conditions are simulated with a suite of neutral hypothetical chemicals capturing a relevant range of partitioning properties (e.g. octanol-water partition coefficient, Kow). The first in vitro system (1) represents a cell-free assay. The second in vitro system (2) represents a cell-based test in which no serum is present during the test assay and the third in vitro system (3) represents a cell-based test in which 10% serum is present during the test assay. Differences in Cfree and Cblood / Cnom are relatively small for lower Kow chemicals (log Kow < 2) independent of the assay. However, as Kow increases, Cfree decreases. All else being equal, i.e., the magnitude of response in the assay for Cnom is the same for all chemicals, the higher Kow chemicals appear to be more potent than the lower Kow chemicals because they elicit the same response at lower Cfree. The difference in Cfree can be as much as 4-5 orders of magnitude lower. Extending the translation to Cblood and comparing against Cnom, the opposite interpretation would be made, i.e., the hydrophobic chemicals appear to be "less potent" because the concentration in blood required to elicit the same response in vitro (i.e., Cnom) is higher.

Mo-SY-C3.3

A range of approaches for interpreting in vitro toxicity data: PBPK, PK, mass balance and biomonitoring

Sean Hays, *Summit Toxicology, Lyons, CO, United States*
Lesa Aylward, *Summit Toxicology, Falls Church, Virginia, United States*

In vitro toxicity data can be interpreted using a wide range of approaches and in different contexts. First, the data can be interpreted in the context of existing exposures or in the context of existing risk assessments. When interpreting in an exposure context, biomonitoring data of chemicals in blood in humans provides a direct interpretation in the context of existing exposures amongst the population. To interpret in vitro toxicity data in the context of existing risk assessments, methods are required to convert the external dose based risk assessments into internal dose measures. Numerous approaches exist for making this extrapolation, ranging in complexity from simple measures of chemicals in animals at the points of departure of interest, to simple pharmacokinetic modeling to complex physiologically based pharmacokinetic (PBPK) modeling. This talk will review the various approaches and provide case studies showing a range of approaches for interpreting in vitro toxicology data in both an exposure and risk assessment context.

Mo-SY-C3.5

QIVIVE APPROACHES TO EVALUATE INTERINDIVIDUAL TOXICOKINETIC VARIABILITY

Barbara Wetmore, *ScitoVation, LLC, Research Triangle Park, NC, United States*

Toxicokinetic (TK) variability across life-stages and populations can significantly impact the amount of chemical available systemically to elicit an effect despite similar external exposures. This variability is driven by physiologic (e.g., liver weights, blood flow rates, etc.), ontogenetic (e.g., immature or developing metabolic enzyme capabilities) and genetic (e.g., polymorphisms) differences. Recent advances in experimental tools, in vitro - in vivo extrapolation (IVIVE) and in silico modeling approaches have laid the groundwork for the development of strategies that can quantitate chemical-specific TK variability that may be present across different populations. Metabolic clearance of nine ToxCast chemicals were measured in vitro using 13 cytochrome P450 (CYP) and 5 UDP-glucuronosyltransferase (UGT) isozymes that were recombinantly expressed. Together with plasma protein binding data, these isozyme-specific clearance rates were used in an IVIVE modeling approach that incorporates known differences in xenobiotic metabolizing isozyme abundances among various life-stage or ethnic-based populations to estimate the resulting systemic chemical steady-state concentrations, thus providing a strategy to quantitate TK variability. CYPs 3A4, 3A5, 2C9, and 2C19 were the most active isozymes, contributing to the clearance of all of the chemicals tested. Chemicals metabolized primarily by CYP1A2 displayed the greatest TK variability across the populations assessed. Children, in particular the newborn to 6 months of age life-stage, displayed the highest steady-state levels given a similar external exposure, identifying them as a sensitive population. Next, these steady state values were incorporated with ToxCast in vitro bioactivity concentrations to estimate the daily oral dose for each population, called the oral equivalent dose, necessary to produce steady-state in vivo blood concentrations equivalent to these in vitro bioactivity values. These external, population-specific oral equivalent doses were then compared against life-stage or population-specific external exposure estimates to provide a margin of exposure assessment that could be applied in risk-based prioritization. This study demonstrates the feasibility and value of using isozyme-specific clearance data to tailor dosimetric values for a wide range of populations. Moreover, such strategies may enable a shift away from default uncertainty factors toward chemical-specific safety factors.

Mo-SY-D3: 15 years of Human Biomonitoring in Flanders: surveillance feeding policy and research - II

Mo-SY-D3.1

Determinants of metal exposure in the biomonitoring campaigns of the Flemish Environment and Health Study (FLEHS)

Sam De Craemer, *Vrije Universiteit Brussel, Brussels, Belgium*
Stefaan De Henauw, *Ghent University, Ghent, Belgium*
Ben Nemery, *University of Leuven, Leuven, Belgium*
Ilse Loots, *University of Antwerp, Antwerp, Belgium*
Tim Nawrot, *Hasselt University, Diepenbeek, Belgium*
Greet Schoeters, *Flemish Institute for Technological Research (VITO), Mol, Belgium*
Eva Govarts, *Flemish Institute for Technological Research (VITO), Mol, Belgium*
Vera Nelen, *Provincial Institute of Hygiene, Antwerp, Belgium*
Liesbeth Bruckers, *Hasselt University, Hasselt, Belgium*
Michelle Plusquin, *Hasselt University, Diepenbeek, Belgium*
Willy Baeyens, *Vrije Universiteit Brussel, Brussels, Belgium*

Background: Identification of determinants of exposure to pollutants can be a tool for targeted policy actions to reduce exposure. In the Flemish Environment and Health studies (FLEHS), biomonitoring data are coupled to questionnaires in order to identify significant determinants of exposure for the monitored pollutants.
Objective: The aim of this study is to explore determinants of exposure to metals in adolescents, newborn-mother pairs and adults in the latest cycle of FLEHS (2012-2015). These determinants will be compared with results from the previous cycles (2002-2006, 2007-2011), to identify those that are consistent over the different campaigns of FLEHS. Part of the population exceeds existing health guidelines for internal exposure to cadmium (adults and adolescents) and arsenic (adolescents), so special focus is given to these metals. Furthermore, certain species of arsenic are considered separately, since the toxicity of arsenic heavily depends on its form.
Methods: Pollutant concentrations in biological matrices like blood and urine are the cumulative reflection of all exposure. Statistical regression techniques are performed on the biomonitoring results in each study population to identify determinants of pollutants.
Results: Some associations with significance in the latest cycle were consistent with earlier cycles. Like the positive association between blood and cord blood cadmium and smoking in adolescents and mothers respectively ($p<0.001$ and $p=0.03$), lead in cord blood and maternal age ($p<0.001$) and total arsenic in blood and seafood consumption in adults and adolescents (both $p<0.001$). Most associations were consistent with the literature. Additionally, we found that determinants differ between arsenic species.
Conclusions: Several determinants are consistent with literature and earlier FLEHS cycles. These include determinants of metals that exceed health guidelines, (passive) smoking for cadmium in adolescents and fish consumption/education for arsenic in adolescents and adults. However, determinants of arsenic differ between species, indicating that biomarkers of total arsenic exposure may have limited use for steering policy action.

Mo-SY-D3.2

Looking at human biomonitoring results through an environmental justice lens: the case of Flanders (Belgium)

Bert Morrens, *University of Antwerp, Antwerp, Belgium*
Liesbeth Bruckers, *Hasselt University, Hasselt, Belgium*
Ann Colles, *Flemish Institute for Technological Research, Mol, Belgium*
Greet Schoeters, *Flemish Institute for Technological Research, Mol, Belgium*
Vera Nelen, *Provincial Institute for Hygiene, Antwerp, Belgium*
Tim Nawrot, *Hasselt University, Hasselt, Belgium*
Nik Van Larebeke, *Ghent University, Ghent, Belgium*
Willy Baeyens, *Free University Brussels, Brussels, Belgium*
Ilse Loots, *University of Antwerp, Antwerp, Belgium*

BACKGROUND: Environmental justice research suggests that inequalities in the distribution of environmental quality systematically disadvantage the lower social strata of society. The effects of these inequalities on the human exposure to chemical pollution remain however to a large extend unknown. The Flemish Environment and Health Study (FLEHS), a large-scaled human biomonitoring program in Belgium measuring exposure to multiple pollutants in blood and urine of different age groups, can shed some new light on the environmental justice hypothesis.
OBJECTIVE: The objective of this study is to assess social gradients in human biomonitoring results of representative samples of Flemish newborns and adolescents from 2002 to 2015, and determine whether these gradients can be explained by specific underlying factors related to both exposure and social background. The hypothesis is to find negative social gradients in body concentrations: lower socioeconomic status having higher exposure.
METHODS: We investigate the associations between individual socioeconomic status (SES), measured by educational attainment, and different biomarkers of exposure, using multiple regression models.
RESULTS: Depending on the (type of) pollutant, people with lower socioeconomic status can either have higher or lower body concentrations. Exposure to some heavy metals (lead, cadmium and copper) is associated with lower SES, while exposure to persistent organic compounds (PCBs, chlorinated pesticides and flame retardants) is associated with higher SES. Social gradients in exposure remained after correcting for proximity to (suspected) pollution sources, but largely disappeared after correcting for socially constructed factors, such as dietary and lifestyle habits (smoking, breastfeeding, fish consumption). These results indicate that exposure is not only an environmental but also a social process.
CONCLUSIONS: We conclude that when assessing body concentrations of pollutants, more complex patterns of social stratification emerge than can be assumed on the basis of the environmental justice hypothesis. It therefore remains important to consider the chemical environment in relation to the social environment when monitoring environmental health risks.

Mo-SY-D3.3

Emerging contaminants in the Flemish Environment and Health biomonitoring Surveys (FLEHS)

Adrian Covaci, *University of Antwerp, Wilrijk-Antwerp, Belgium*
Nathalie Lambrechts, *VITO, Mol, Belgium*
Govindan Malarvannan, *University of Antwerp, Wilrijk-Antwerp, Belgium*
Gudrun Koppen, *VITO, Mol, Belgium*
Liesbeth Bruckers, *Hasselt University, Diepenbeek, Belgium*
Eva Govarts, *VITO, Mol, Belgium*
Vera Nelen, *Provincial Institute of Hygiene, Antwerp, Belgium*
Tim Nawrot, *Hasselt University, Diepenbeek, Belgium*
Ilse Loots, *University of Antwerp, Antwerp, Belgium*
Isabelle Sioen, *Ghent University, Ghent, Belgium*
Willy Bayens, *Free University Brussels, Brussels, Belgium*
Greet Schoeters*VITO, Mol, Belgium*

Besides classical pollutants measured in individual blood or urine samples, the Flemish and Environment Surveys (FLEHS) have also investigated human exposure to emerging contaminants. An emerging contaminant is defined as a chemical that is newly used in products, can distribute into the environment, and for which there are no or insufficient data available. Chemicals that have been used for a while, but for which we lack data regarding their occurrence, fate or toxicity are also considered emerging contaminants. These compounds have received increasing attention the last years. Exposure originates mostly from current usage of consumer products, such as plastics, food contact materials, personal care and household products, furniture, etc, and nowaday behaviour patterns. We will give an overview of past activities related to the analysis of emerging contaminants in the FLEHS campaigns and compare these with similar activities worldwide. We will then assess the (most) relevant emerging contaminants to be included in future biomonitoring studies.
For exploratory purposes, most of the work on emerging contaminants in the previous FLEHS campaigns has been done in a limited number of pooled urine or blood. While in FLEHS I, only classical pollutants were included, the following campaigns (FLEHS II and III) have included brominated flame retardants, such as polybrominated diphenyl ethers, hexabromocyclododecane and tetrabromobisphenol-A, personal care products (parabens, UV filters, triclosan) and bisphenol-A.
The newest campaign, FLEHS 4 (2016-2020) focuses on a number of new themes, such as use of open space and eco-behaviour, and some more defined investigation of emerging contaminants in adolescents. We will aim for the identification of emerging contaminants and their metabolites by using an array of analytical approaches, such as A) untargeted screening, B) suspect screening, and C) targeted measurements of emerging contaminants and/or their metabolites identified via screening. We will furthermore identify life style-specific exposure profiles: compounds that may differ in relation to specific behavioural patterns. This should give guidance towards more accurate prevention measures that protect against exposure to ubiquitous environmental toxicants and their substitutes in new materials. A number of emerging contaminants have been already identified, such as plasticizer substitutes, bisphenol-S and other bisphenols, organophosphate flame retardants, and new perfluorinated compounds. It is also expected that several new compounds will be identified through non-targeted screening. Lastly,further identification

and selection of emerging contaminants could be based on the analyses of environmental samples (indoor air, dust and products).

Mo-SY-D3.4

Early-life exposure to multiple environmental contaminants and birth outcomes: pooled analysis in four Flemish birth cohorts

Eva Govarts, *VITO, Mol, Belgium*
Lützen Portengen, *Institute for Risk Assessment Sciences (IRAS), Utrecht University, Utrecht, Netherlands*
Nathalie Lambrechts, *Flemish Institute for Technological Research (VITO), Mol, Belgium*
Liesbeth Bruckers, *Hasselt University, Diepenbeek, Belgium*
Adrian Covaci, *University of Antwerp, Antwerp, Belgium*
Vera Nelen, *Provincial Institute of Hygiene (PIH), Antwerp, Belgium*
Tim Nawrot, *Hasselt University, Diepenbeek, Belgium*
Ilse Loots, *University of Antwerp, Antwerp, Belgium*
Isabelle Sioen, *Ghent University, Ghent, Belgium*
Willy Baeyens, *Free University of Brussel (VUB), Brussels, Belgium*
Bert Morrens, *University of Antwerp, Antwerp, Belgium*
Roel Vermeulen*Institute for Risk Assessment Sciences (IRAS), Utrecht University, Utrecht, Netherlands*
Greet Schoeters, *Flemish Institute for Technological Research (VITO), Mol, Belgium*

Background:
Prenatal chemical exposure has frequently been associated with reduced fetal growth although results have been inconsistent. Most studies associate single pollutant exposure to these health outcomes, even though this does not reflect real life situations as humans are exposed to thousands pollutants during their life time. Human biomonitoring shows that complex mixtures of xenobiotic chemicals are present in the prenatal environment.
Aim:
The objective of this study is to investigate the association between prenatal exposure to a mixture of environmental chemicals and birth weight.
Methods:
We used exposure biomarker data obtained from cord blood samples of 2033 women from three Flemish birth cohorts (FLEHS I, II & III) and a regional birth cohort in the Flemish regions Dessel, Mol and Retie (3xG). The common set of available and detectable exposure measures in these cohorts are the organochlorine compounds (three PCB congeners (138, 153 and 180), HCB, p,p'-DDE) and the heavy metals Cd and Pb. Birth weight was assessed as a proxy for reduced fetal growth. In a first step, the exposure-response associations were investigated by single pollutant linear regression models adjusted for gestational age, newborn's sex, smoking of the mother during pregnancy, parity, maternal age and prepregnancy BMI. Next, elastic net regression was used to assess the effect of multipollutant exposure on birth weight.
Results:
In the pooled database, birth weight ranged from 1245 to 5575 grams with a median of 3430 grams. The median contaminant levels in cord blood were: 26 ng/g lipid for PCB 153, 15 ng/g lipid for PCB 138, 18 ng/g lipid for PCB 180, 90 ng/g lipid for p,p'-DDE, 17 ng/g lipid for HCB, 0.045 µg/L for Cd and 9 µg/L for Pb. In single pollutant models, the three PCB congeners were significantly associated with reduced birth weight. The correlations between the different pollutants are low to moderate (r = 0.11 - 0.59), except for the three PCB congeners being highly correlated with Pearson correlation coefficients ranging

from 0.74-0.84. From all exposure measures, the two PCB congeners PCB 153 and PCB 180 were most consistently associated to birth weight based on elastic net regression.

Conclusions:
Assessing health risk of combinations of exposure biomarkers reflects better real world situations. The findings allow more effective risk assessment as addressing the critical chemical in a mixture of pollutants is pivotal for risk assessment.

Mo-SY-E3: Assessing exposure to SVOCs in dust

Mo-SY-E3.1

Residential exposures to SVOCs: Identifying efficient and effective exposure measurements

Robin Dodson, *Silent Spring Institute, Newton, MA, United States*
Ruthann Rudel, *Silent Spring Institute, Newton, MA, United States*

Aim: Semivolatile organic compounds (SVOCs), occurring both in the gas and condensed phase, readily redistribute from their original source over time to indoor air, house dust, and other indoor surfaces. Their distribution in the indoor environment determines sampling approaches, how people are exposed, and strategies to reduce exposure. To inform efficient and effective exposure measurement strategies for SVOCs, we rely on extensive empirical data as well as chemical-physical properties and feasibility.
Methods: We have analyzed over 100 SVOCs in indoor air and house dust samples from 170 U.S. homes as part of our Cape Cod and California Household Exposure Studies. More recently, we have targeted 60 SVOCs in 115 indoor air samples and surface dust wipes collected in subsidized housing units as part of the U.S. Green Housing Study.
Results: SVOCs are often, although not always, correlated between indoor air and house dust. Of the 34 SVOCs in our California Household Exposure Study, 25 were significantly correlated in indoor air and house dust, although without a clear pattern across chemical class and chemical property. Ratios of measured dust and indoor air concentrations span 6 orders of magnitude, with SVOCs with lower octanol-air partitioning values (Koa) values having smaller ratios and SVOCs with higher log Koa values having higher ratios. We observed a moderately strong correlation (rho ~ 0.8) between log Koa and ratio of dust to air concentrations, and SVOCs with higher log Koa values (>10) generally had lower detection frequencies in air. We observed lower correlations between indoor air and surface wipes, likely a result of the variability of the wipe, which is not as standardized and may reflect surface material rather than settled dust more relevant for exposure.
Conclusions: SVOCs with large Koa values (>10), especially if they are not found at relatively high concentrations like phthalates, may be best measured using dust sampling or air sampling that captures both the gas and particulate phases. In contrast, chemicals with lower Koa values (<10) would be readily measured in air samples. For many SVOCs with Koa values between 5 and 10, measurements in one media (e.g. air) can be used to predict concentrations in other media (e.g. dust) reasonably well. If a researcher is interested in a large number of chemicals with a wide range of physical-chemical properties, either air or dust sampling may be able to provide information on exposure source concentrations and route-specific exposures.

Mo-SY-E3.2

Modelling the Relevance of Dust as an Exposure Pathway

Christiaan Delmaar, *RIVM, Bilthoven, Netherlands*
Vilma Sukiene, *ETHZ, zurich, Switzerland*
Andreas Gerecke, *EMPA, zurich, Switzerland*
Martine Bakker, *RIVM, Bilthoven, Netherlands*
Konrad Hungerbuhler, *ETHZ, Zurich, Switzerland*
Natalie Von Goetz, *ETHZ, Zurich, Switzerland*

Background
For semi-volatile organic compounds (SVOCs) used in consumer products, human exposure via dust is often suggested to be a potentially important exposure pathway. Various models are available to estimate the emission of substances from products into air. Other models can estimate the resulting distribution of the substances in the indoor environment. However, a model platform for estimating the concentration of substance in house dust from its concentration in the source is lacking.
Objectives
In this project, funded by CEFIC-LRI, a platform of dynamic models was developed to estimate the emission of chemicals from consumer products and the subsequent transfer of these substances into air, airborne particulate matter and dust. This platform was used to investigate under which circumstances the ingestion of a substance via house dust is important and subsequently, to define when the house dust pathway should be included in a risk assessment of a consumer product.
Methods
The model platform, called the DustEx model, includes models published in literature that jointly describe transport of SVOCs from products to indoor media. The DustEx model was used to evaluate the relative importance of the inhalation, dermal absorption and dust ingestion pathways. On the basis of the calculated indoor air and dust concentrations and further conservative parameters the exposure of consumers via the dust pathway was calculated and compared to other exposure pathways. This inter-pathway comparison was performed (1) in a generic way by exploring the chemical property space and (2) by selecting realistic case studies for different types of consumer products.
Results
The results of the inter-pathway comparison were summarized in a conservative decision tree giving guidance on when to consider the dust pathway in a risk assessment of a substance released by a single consumer product. For example, for products leading to indirect exposure (i.e. no mouthing and no contact of the product with skin) the relevance of the dust pathway is dependent on the skin permeability (kp_g) and the octanol-air partitioning coefficient (Koa) of the substance.
In the case studies it was shown that compared to conservative, lower tier calculations of direct exposure to a product, the dust pathway was always negligible (lower by more than one order of magnitude). Only in higher tier assessments with refined parameters for other pathways, the pathway of dust ingestion may be important.

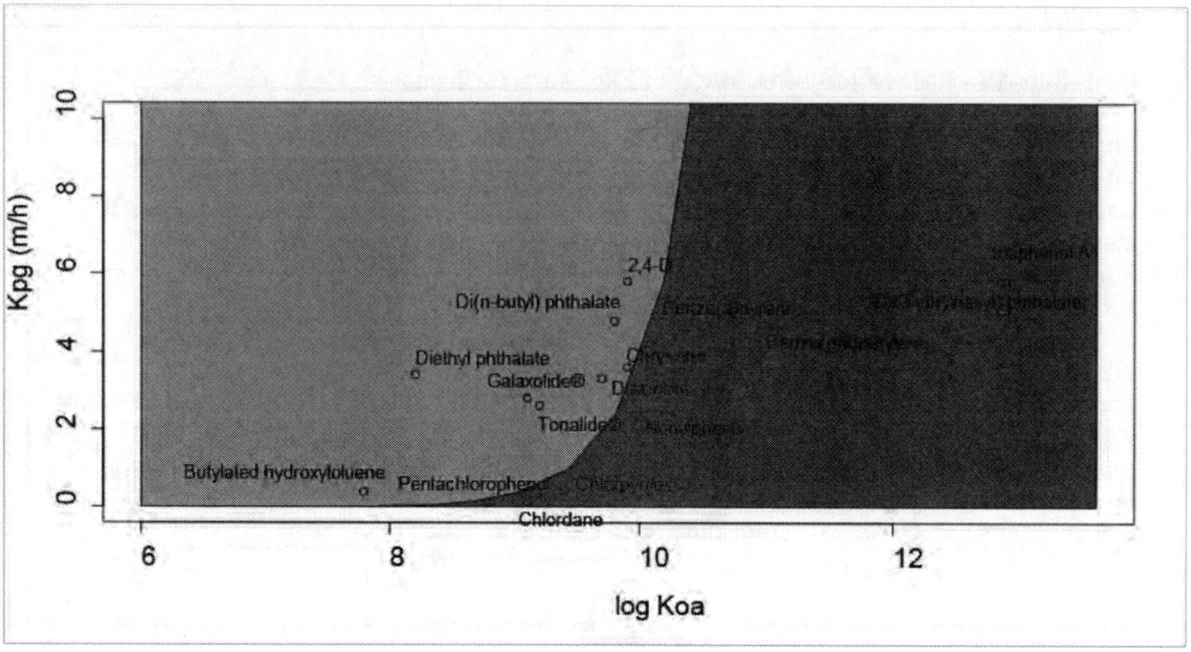

Contour plot for when to include dust as an exposure pathway, for indirect exposure. Green area: ignore dust pathway; red area: include dust pathway.

Mo-SY-E3.3

SVOCs Transfer from Products into Dust: Model vs Measurements

Vilma Sukiene, *Swiss Federal Institute of Technology, (ETH Zurich), Zurich, Switzerland*
Andreas Gerecke, *Swiss Federal Laboratories for Material Science and Technology (EMPA), Dübendorf, Switzerland*
Christiaan Delmaar, *National Institute for Public Health and the Environment (RIVM), Utrecht, Netherlands*
Martine Bakker, *National Institute for Public Health and the Environment (RIVM), Utrecht, Netherlands*
Konrad Hungerbühler, *Swiss Federal Institute of Technology, (ETH Zurich), Zurich, Switzerland*
Natalie von Goetz, *Swiss Federal Institute of Technology, (ETH Zurich), Zurich, Switzerland*

Background
Various models are available to estimate the emission of substances from products into air. Other models can estimate the resulting distribution of the substances in the indoor environment. However, a model platform for estimating the concentration of substance in house dust from its concentration in the source is lacking. Moreover, the number of experimental studies providing substance concentrations in both products and indoor dust required for model validation is limited.
Objectives
In the DustEx project a model was developed that describes the transfer of substances from consumer products into house dust. In order to test and validate the model a field study was designed that follows the transfer of labeled SVOCs from the product to the dust.
Methods
The DustEx model includes models published in literature that jointly describe transport of semi-volatile organic compounds (SVOCs) from products to indoor media. To support the model an experimental study under controlled conditions was designed. In a small-scale field study the transfer of SVOCs from consumer products to indoor air and settled dust was analyzed. Altogether two measurement campaigns were conducted. The first field study had enrolled five apartments for twelve weeks. During the second field study three apartments selected from the previous five were investigated for eight weeks. Eight deuterium-labelled target SVOCs (phthalates and adipates) were synthesized and introduced into artificial plastic consumer products and a carpet. These products were used in different scenarios with several emission processes (e.g. evaporation, mechanical stress etc.). To study consumer sprays, the participants of the study were supplied with an insecticide spray, which they used for a defined period of time. During the study indoor dust and air samples were collected and analyzed regularly.
Results
The dynamic DustEx model was used to predict concentrations of target substances in the indoor air, settled dust and the products over time. The measured concentrations in indoor media were compared with the results of the model. The most sensitive parameters influencing substance distribution indoors were identified: Octanol-air and material-air partitioning coefficients, mass transfer coefficient and organic matter fraction of dust. The values of these parameters given in literature vary largely for the single substances

and different estimation methods. Thus, the model was calibrated by optimizing these parameters within the ranges of uncertainties.

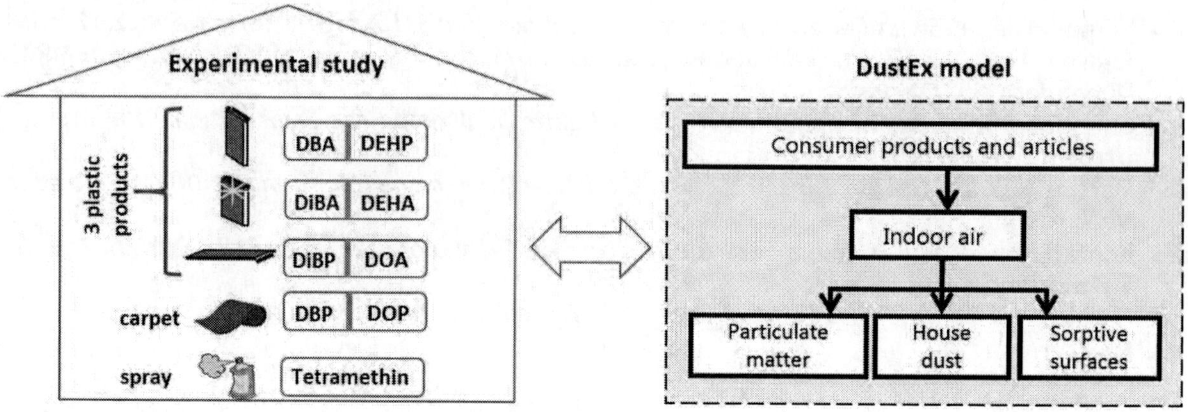

DustEx model and the experimental study setup

Mo-SY-E3.4

Exploring house dust as a path of exposure

Stefanie Klenow, *Federal Institute for Risk Assessment, Berlin, Germany*
Astrid Heiland, *Federal Institute for Risk Assessment, Berlin, Germany*
Gerhard Heinemeyer, *Federal Institute for Risk Assessment, Berlin, Germany*

Background: The intake of house dust can contribute to the exposure of humans to e.g. SVOC and needs to be considered in a comprehensive exposure assessment. However, many aspects of house dust intake are not sufficiently explored, resulting in a high degree of uncertainty in exposure assessment. In particular, assumptions on the quantity of house dust intake are very uncertain due to missing data.
Objectives: The project presented here was therefore triggered by the question whether it is possible to achieve more realistic and reliable assumptions on the amount of ingested house dust that may also appropriately reflect the variation of the amounts ingested. It is also aimed at determining what kind of studies is needed to shed additional light on this issue.
Methods: The available literature on house dust intake was compiled and evaluated. Toxicokinetic modelling based on available data was carried out. A concept was developed for a study targeting the investigation of house dust intake.
Results: The literature review revealed many open questions related to a general definition of house dust in the context of an exposure assessment, to house dust-intake scenarios which go beyond hand-mouthing in children, and to adequate sampling methods to assess exposure. Toxicokinetic modelling based on available data did not result in satisfying assumptions because no study was conducted to determine house dust intake, and consequently data essential for assessing the quantity are missing. Therefore, a concept for a study is proposed to investigate main parameters necessary for a sound toxicokinetic modelling of house dust intake rates in different age groups and to derive realistic reference values of house dust intake. To assess additional exposure via food intake, the duplicate diet method is an integral part of this concept as no substance could be identified for which this pathway could be excluded. A complete collection of urine over several days is also an essential requirement. The key prerequisite for this study is, however, the identification of suitable tracer substances. At the same time, additional aspects potentially affecting house dust intake are recorded, and, if necessary, have to be characterized in more detail in further studies.
Conclusion: Such a comprehensive study has the potential to improve the assessment of exposure via house dust, and will thus allow more valid risk assessments of substances in house dust.
Acknowledgement: This study was funded by the Federal Ministry for the Environment, Nature Conservation, Building and Nuclear Safety.

Mo-SY-E3.5

SVOCs in Dust: Motivation and challenges for a cumulative approach to exposure & risk assessment

Philippe Glorennec, *EHESP - School of Public Health, Rennes, France*
Kevin Fournier, *EHESP - School of PUblic Health, Rennes, France*
Nathalie Bonvallot, *EHESP - School of Public Health, Rennes, France*

Aim. Semi Volatile Organic Compounds (SVOCs) such as flame retardants, plasticizers, pesticides...present in dust are more generally present in indoor environment because they partition between vapor phase, airborne particulate and dust. Many indoor SVOCs have a common mode of action regarding effects on human health, so that a cumulative risk assessment is indicated rather or in complement of a contaminant by contaminant approach.
Methods. The first step of the risk assessment process, i.e. hazard identification, can in this structured in three sub-steps:(1a) Identification of contaminants people are exposed to, (1b) identification of effects and mechanisms of action of these contaminants, (1c) grouping of contaminants according to similarity of their mechanism of action and health effects. Based on this exposure-based grouping we can derive "multi-contaminant" toxicity reference values, in the "dose-response assessment" step. The third step consists of exposure assessment by combining indoor concentrations in dust and air with dust ingestion, breathing and skin permeation rates in order to calculate exposure doses. The final step consists in risk calculating from multi-contaminant exposure doses and multi-contaminant toxicity reference values.
Results. The application of this risk assessment framework on indoor SVOCs in France revealed that the main challenge is still to gather toxicological - or epidemiological - data on the largest possible number ofcontaminants that are relevant to human exposure. There is at this step a trade-off between the number of contaminants we wish to consider and the accuracy level for dose-additivity. Indeed, because of the accuracy of available information, the more contaminants we want to include, the less likely it is that the relative potency factor approach (RPF) can be used. In the case of less availability of toxicological data, a point of departure index (PODI) approach is likely to be promoted, emphasizing effects rather than mechanisms. As an illustration, considering indoor SVOCs and the reproductive and nervous systems, 7 groups of contaminants can be considered for a cumulative risk assessment: 5 groups for a RPF approach, and 2 for a PODI approach. They are composed of 5 to 12 contaminants from different chemical families in most cases. Adapted from Fournier et al. 2014, Environ Res.; 130, 20-28.
Conclusions. Although resource consuming, a cumulative approach beyond chemical families is relevant for exposure and for risk assessment.

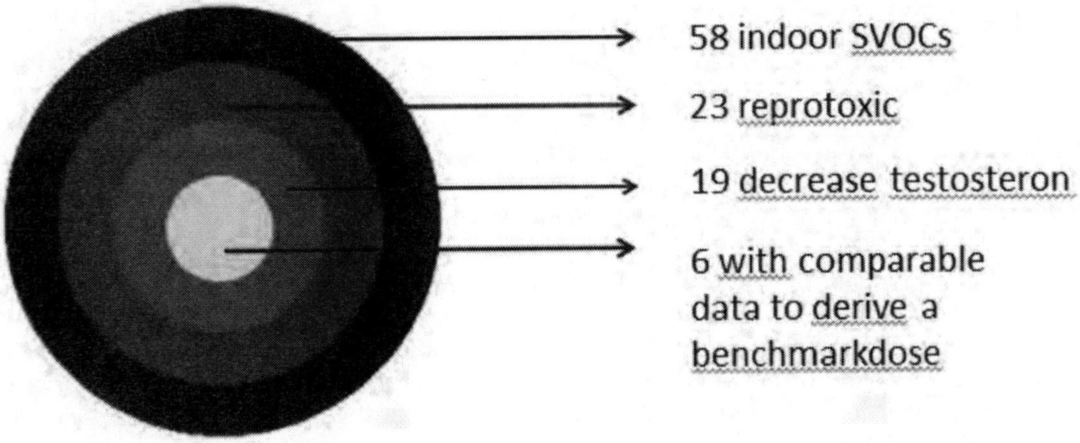

Cumulative risk assessment of indoor SVOCs. Completeness versus accuracy.

Mo-SY-F3: Thresholds of Toxicological Concern: an exposure-driven approach to risk assessment

Mo-SY-F3.1

Thresholds of Toxicological Concern (TTC) - Introduction to the tiered concept

Kirstin Kosemund, *Procter & Gamble, Schwalbach, Germany*

The TTC (Threshold of Toxicological Concern) Concept is an important pragmatic first-tier risk assessment tool for evaluating low level human exposures to chemicals lacking sufficient toxicity data to support a chemical-specific risk assessment. The origins of TTC come from work by the U.S. Food and Drug Administration to develop an approach for assessing the safety of chemicals that could migrate at low levels from food packaging materials. Since then, TTC has evolved into a tiered tool that incorporates knowledge of chemical structure to bin untested chemicals into different potency categories. Reviews by major regulatory bodies have acknowledged the scientific foundation of TTC and its wide applicability for evaluating low level exposures as well as risk prioritisation for a wide range of uses to appropriately focus societal resources and to avoid unnecessary animal testing. This talk will provide an introduction to the tiered TTC concept and an overview of its current and potential uses within current regulatory landscapes.

Mo-SY-F3.2

Ecological Threshold for Toxicological Concern (eco-TTC) - Assessing the potential of a new tool for environmental hazard assessment

Michelle Embry, *ILSI Health and Environmental Sciences Institute, Washington, DC, United States*
Amy Beasley, *The Dow Chemical Company, Midland, MI, United States*
Mace Barron, *United States Environmental Protection Agency, Gulf Breeze, FL, United States*
Scott Belanger, *The Procter & Gamble Company, Cincinnati, OH, United States*
Jessica Brill, *The Procter & Gamble Company, Cincinnati, OH, United States*
Dick De Zwart, *Mermayde, Groet, Netherlands*
Brianna Farr, *ILSI Health and Environmental Sciences Institute, Washington, DC, United States*
Aude Kienzler, *European Commission, Ispra, Italy*
Hans Sanderson, *Aarhus University, Roskilde, Denmark*
Marlies Halder, *European Commission, Ispra, Italy*
Ruth Hummel, *United States Environmental Protection Agency, Washington, DC, United States*
Teresa Norberg-King*United States Environmental Protection Agency, Duluth, MN, United States*
Robin Sternberg, *United States Environmental Protection Agency, Washington, DC, United States*
Jamie Suski, *United States Environmental Protection Agency, Washington, DC, United States*
Peter Wilson, *Sanofi, Bridgewater, NJ, United States*

The Threshold for Toxicological Concern, or TTC concept, is well-established for assessing human safety of indirect food-contact substances and has been reapplied for a variety of endpoints including carcinogenicity, teratogenicity, and reproductive toxicity. Recently, we have proposed an extension to the human safety TTC concept for application in environmental situations, termed the ecological TTC or eco-TTC. Eco-TTCs summarize the wealth of ecotoxicological information as Predicted No-Observed Effect Concentrations (PNECs) on diverse chemical substances in the form of statistical (probability) distributions. Eco-TTCs can be developed that allow prediction of untested chemicals based on structural attribute (category), mode of action, or functional use. The approach may be useful for assessing chemicals at early tiers of the risk assessment process, providing hazard perspective on chemicals that lack QSARs, guiding product development discussions, and assisting read across or category justifications. The eco-TTC approach has the potential to reduce the need for vertebrate testing (e.g., fish) in many situations.
A database consisting of approximately 110,000 unique ecotoxicological records has been developed based on recent assessments of published data and international chemical management programs. This toxicity data is associated with physical chemistry data and curated taxonomic information for the organisms tested. A process to conclude acute and chronic effects as well as identify the PNEC for exposed ecosystems based on depth and breadth of data have been devised, with the 5th percentile of PNECs for a compound group defined as the ecological TTC. Several mode of action schemes are being assessed to devise a best approach for grouping compounds. Chemicals that are categorized as neutral organics are the most abundant in the dataset, therefore are candidates for an

initial in-depth assessment of eco-TTC attributes. Approximately 500 chemicals in the database are included in this analysis at this time with approximately one third having complete acute or chronic data sets (all three taxa). The eco-TTC for non-polar and polar narcotics is explored in depth, with additional categories under development (phenols, esters, reactive compounds, surfactants, pesticides, and pharmaceuticals). Eco-TTCs look to be a promising addition to the toolkit of hazard assessment.

Mo-SY-F3.3

Progress in the Development of Internal TTC Approaches

Harvey Clewell, *ScitoVation, Durham, North Carolina, United States*
Corie Ellison, *Procter and Gamble, Cincinnati, Ohio, United States*

The Threshold of Toxicological Concern (TTC) is an important risk assessment tool that establishes acceptable low level exposure values to be applied to chemicals with limited toxicological data. The concept relies on knowledge of the range of toxicological hazard/potency for structurally relevant classes of chemicals for which good toxicity data exist. The non-cancer TTC databases consist of distributions of oral No Observed (Adverse) Effect Levels (NO(A)ELs) identified from toxicity studies. The TTC threshold limits were established by identifying the 5th percentile NO(A)EL value from the database and applying an appropriate uncertainty factor. Given the fact that the data comprising the TTC are from external oral exposures, the corresponding threshold limits are representative of external exposures.
Recently, there has been increased discussion regarding the possibility of internal TTCs (i.e. TTCs based on internal exposure such as chemical concentration in blood) and the advantage these would provide for use in risk assessment. One such example includes the most recent version (9th revision) of the SCCS Notes of Guidance for the testing of Cosmetic Ingredients and their Safety Evaluation which states that "for cosmetic ingredients, the TTC approach should be based on internal doses." However, while there has been mention of internal TTCs being developed and their application to risk assessment, the fact of the matter is that the few currently available internal TTCs are not adequate for risk assessment and more work beyond this will be necessary before there are 'actionable' internal TTCs that can be used in risk assessment. The development of internal TTCs requires a significant amount of data and computational tools (e.g. PBPK modeling) that can be used to convert the chemical specific external doses (oral NOAELs) in the TTC database to an estimated internal exposure to the relevant toxicant (i.e., parent or metabolite) that is associated with the critical effect for each chemical.
The current presentation will introduce recent work that is underway to develop internal TTCs that can be used in risk assessment. The presentation will include a summary of that state of the science, challenges associated with this work and the criteria needed for success. This work is part of the Cosmetics Europe Long Range Science Strategy Research Program 2016-2020 with the goal of promoting approaches for safety assessments without generating animal data.

Mo-SY-F3.4

Assessment of co-exposures based on combination of TTC and specific data

Heli M Hollnagel, *Dow Europe GmbH, Horgen, Switzerland*

One of the areas of risk assessment which is most impacted by a lack of toxicological data is the assessment of environmental co-exposures. Depending on how broad an analytical setup is chosen, environmental samples contain dozens to hundreds of natural and manmade chemicals. Especially for natural chemicals and metabolites of manmade chemicals, toxicological data are scarce.

The objective of this talk is to show in which cases the Concept of the Thresholds of Toxicological Concern (TTC) is a useful tool in human health risk assessment of co-exposures, how the application of TTC influences the outcome of the assessment, and to discuss which uncertainties are associated with the application of TTC in co-exposure risk assessment.

Basic principles of co-exposure risk assessment will be introduced and case studies on non-cancer repeated dose endpoints presented, comparing evaluations based on specific data, TTC only, and a combination of both approaches. From such studies, it is determined that the use of TTC in co-exposure risk assessment relies on conservative health protective assumptions. It offers a pragmatic approach to a screening level assessment allowing the identification of environmental co-exposures requiring further investigation also in the absence of specific toxicity data on some or all components.

Mo-SY-F3.5

Development and application of a new TTC for oral exposure to proteins

Carrie Fleming, *Dow AgroSciences, Indianapolis, IN, United States*
Sabitha Papineni, *Dow AgroSciences, Indianapolis, IN, United States*
Amy Beasley, *The Dow Chemical Co., Midland, MI, United States*
Manoj Aggarwal, *Dow AgroSciences, Abingdon, United Kingdom*

The threshold of toxicological concern (TTC) concept is a valuable tool for risk assessment of chemicals with low exposure potential, such as food contact materials or pesticide impurities or metabolites. Because toxicological data on proteins was not included in the original derivation of TTC values for the Cramer chemical classes, proteins are excluded from current regulatory frameworks on TTC-based risk assessment. Dietary exposures to regulated proteins, such as newly expressed proteins in genetically-modified (GM) crops, are often very low, and would lend themselves well to TTC-based risk assessment approaches. For instance, proteins in GM crops are often incorporated at very low concentrations (often less than 500 ppm) in the plant and are destroyed by extreme temperatures and pH used in food processing and preparation prior to consumption, thereby limiting human exposure potential. Application of the TTC approach in such cases could help to prioritize toxicological testing on proteins with higher exposure potential, and to significantly reduce the number of animals used in the risk assessment and registration of products containing new proteins. This presentation will present a review of the information available for derivation of a TTC for proteins, describe the development of such values, and provide examples of how this could be applied to reduce the reliance on animal studies in the risk assessment of GM crops.

Mo-SY-G3: The role of analytical chemistry within exposure science.

Mo-SY-G3.1

Current analytical tools for wide-scope screening of organic pollutants in environmental samples

Felix Hernandez, *University Jaume I, Castellon, Spain*
Lubertus Bijlsma, *University Jaume I, Castellon, Spain*
Maria Ibañez, *University Jaume I, Castellon, Spain*
Tania Portoles, *University Jaume I, Castellon, Spain*
Juan Vicente Sancho, *University Jaume I, Castellon, Spain*

The instrumental improvements and advances of high resolution mass spectrometry (HRMS) has revolutionized modern analytical chemistry and notably increased applications to environmental chemistry and exposure fields. In this work, we illustrate that recent coupling of gas and liquid chromatography to HRMS, using analyzers such as QTOF and Orbitrap. These hyphenated techniques have made the screening of large number of organic pollutants in the environment possible. Using both chromatographic techniques coupled to HRMS many different organic pollutants can be detected and identified with high reliability, from volatile/non polar compounds to non-volatile/highly polar ones. Different approaches can be applied, from non-target screening (discovery of unknowns) to target screening (with our without reference standards). After identification of the pollutants present in the samples, analytical efforts can then be directed towards the most relevant compounds by applying target quantitative methods, commonly based on tandem MS using triple quadrupole analyzers. Advantageously and thanks to the accurate-mass full-spectrum acquisition, HRMS allows retrospective analysis i.e. searching for additional compounds not considered in the initial analysis without the need of new injections. Searching for metabolites and transformation products that share common fragments with the parent molecule is another interesting possibility that may allow discovering many compounds that otherwise would remain ignored in the analyses.

Mo-SY-G3.2

Wastewater based epidemiology: recent advances

Lubertus Bijlsma, *University Jaume I, Castellon, Spain*
Pim de Voogt, *KWR Watercycle Research Institute, Nieuwegein, Netherlands*
Sara Castiglioni, *Istituto di Ricerche Farmacologiche Mario Negri, Milan, Italy*
Adrian Covaci, *University of Antwerp, Antwerp, Belgium*
Erik Emke, *KWR Watercycle Research Institute, Nieuwegein, Netherlands*
Felix Hernandez, *University Jaume I, Castellon, Spain*
Barbara Kasprzyk-Hordern, *University of Bath, Bath, United Kingdom*
Christoph Ort, *Eawag aquatic research, Dübendorf, Switzerland*
Malcolm Reid, *NIVA, Oslo, Norway*
Kevin Thomas, *NIVA, Oslo, Norway*
Alexander van Nuijs, *University of Antwerp, Antwerp, Belgium*

Relevant real-time information about lifestyle habits, public health and wellbeing can be obtained from the chemical analysis of urban wastewater. This approach, called wastewater based epidemiology (WBE), uses the analysis of specific human metabolic excretion products (biomarkers) in wastewater as an indicator of consumption or exposure of the population served by the sewer network under investigation. WBE has successfully been applied as suitable approach for the estimation of illicit drugs consumption, but it has also been exploited to other lifestyle factors such as alcohol, nicotine, caffeine and new psychoactive substances yielding satisfactory results. Its great potential also opens up the possibility of expanding the application of WBE to other human biomarkers in order to provide information about diet, health, disease or environment. For example by linking exposure to substances present in the environment or in food with disease outcomes such as higher prevalence of diabetes or cancer.
Chemical analysis of biomarkers in wastewater is the foundation of the WBE approach. Advanced analytical techniques and expertise is required to obtain accurate quantitative data. The generally low analyte concentrations in combination with the complexity and unknown composition of the wastewater matrix may hamper not only the accurate quantification but also sound identification. Hyphenation of chromatography with mass spectrometry, commonly LC-MS/MS, is the best suited approach to obtain the sensitivity, selectivity and identification requirements in chemical analysis directed towards WBE.
In this work we discuss the recent advances in WBE with emphasis on the analytical aspects and difficulties encountered.

Mo-SY-G3.3

Application of LC-full scan HRMS for non-targeted measurement of urinary biomarkers of exposure to modern pesticides

Hans Mol, *RIKILT - Wageningen UR, Wageningen, Netherlands*
Rosalie Nijssen, *RIKILT - Wageningen UR, Wageningen, Netherlands*
Paul Zomer, *RIKILT - Wageningen UR, Wageningen, Netherlands*

Background: human exposure to pesticides may occur through various routes, often dominated by food consumption. Human biomonitoring is an alternative to monitoring of residues in food and can bring added value for chemical risk assessment because it can reduce the assumptions regarding consumption rates and it integrates exposures from other sources (e.g. house-hold use) [1]. Most modern pesticides are rapidly metabolised and excreted through urine, which is a frequently used non-invasive matrix for human biomonitoring. Detection of biomarkers of exposure (typically metabolites) by targeted analysis methods such as LC-MS/MS is often not possible because analytical reference standards are not available. Non-targeted measurement by LC-full scan HRMS has substantially improved in the past years in terms of sensitivity and selectivity and offers improved possibilities to screen urine samples for pesticide biomarkers.
Objectives: development of a generic non-targeted method for the (tentative) detection of biomarkers of exposure to modern pesticides.
Methods: for non-targeted measurement a LC-Q-Orbitrap system was used. Acquisition was done using multiple scan events, to generate data with and without fragmentation (full scan/vDIA). Measurement was performed in positive and negative mode in separate injections. Urine samples were analysed with and without deconjugation, with ultrafiltration as only clean up. The raw data were searched for biomarkers using an exact mass database of ions of potential human metabolites. To demonstrate the feasibility of the approach, urine samples were analysed from a volunteer before and after consumption of strawberries known to contain multiple pesticides.
Results: through consumption of the strawberries, the volunteer was exposed to a cocktail of seven pesticides, the amounts varied from 0.17 to 0.86 mg. Metabolites, were tentatively detected in urine for five pesticides. In most cases, multiple metabolites were detected. Deconjugation resulted in (enhanced) detection of the aglycon, supporting the tentative detection of the conjugates. For three metabolites a reference standard was available which allowed full confirmation. The detectability varied widely for the different metabolites and ranged from 2 to 700 times the estimated LOD.
The results show that detection of pesticide biomarkers of exposure in urine by non-targeted measurement is feasible, even without availability of reference standards. Obviously, for full identification and quantitative analysis, a reference standard is required. Information on the detectability aids in the selection of the best human biomarker for synthesis of a standard.
[1] Choi J, Aarøe Mørck T, Polcher A, Knudsen LE, Joas A, 2014. EFSA supporting publications 2015:EN-724. [321 pp.]

Mo-SY-G3.4

Suspect screening of REACH's chemicals in environmental samples

Thomas Ter Laak, *KWR watercycle research institute / Wageningen University, Nieuwegein, Netherlands*
Rosa Sjerps, *KWR watercycle research institute, nieuwegein, Netherlands*
Ton van Leerdam, *KWR watercycle research institute, Nieuwegein, Netherlands*
Theo Traas, *RIVM, Bilthoven, Netherlands*
Emiel Rorije, *RIVM, Bilthoven, Netherlands*
Annemarie van Wezel, *KWR watercycle research institute /Utrecht University, Nieuwegein, Netherlands*

Over 310.000 chemical substances are registered and regulated via national and international authorities globally, and new substances enter the market continuously. Most of these chemicals are not monitored in the environment. Only 8% of the chemicals that are produced in volumes over 1000 tons per year are part of Dutch monitoring and screening programs and for chemicals with production volumes of 100-1000 tons, this fraction is even lower (3%). For priority substances, monitored fractions are higher, but still less than half is covered (22-47%). Chemicals' legislation in the European Union aims to control risks to humans and the environment to ensure safe use of chemicals. Currently, the assessment of chemicals is based on the ratio of predicted environmental concentrations (PEC) and predicted no effect concentrations (PNEC). The objective of this study is to test if and how broad screening analyses of environmental samples can be applied to provide additional information on occurrence of chemicals for further evaluation or risk management.
Broad screening data of environmental samples obtained with liquid chromatography coupled to high resolution accurate mass spectrometry [2] were matched with 2402 chemicals with 1846 unique elemental compositions. This matching can be considered "suspect screening" [3]. The selected chemicals had production volumes of 100-1000 tons or were part of priority lists. The selected samples consisted of nine surface waters and effluents of wastewater treatment plants all sampled in spring and summer of 2014.
The appraoch resulted in 402 matching elemental compositions in the samples. The matching elemental compositions of surface waters were qualitatively and quantitatively different that the wastewater effluents (Figure 1). We were not able to identify the observed suspects within the current study, but the applied screening approach can guide further evaluation of chemicals by triggering in-depth studies of specific REACH dossiers as well as further identification and quanntification efforts of the observed suspects.

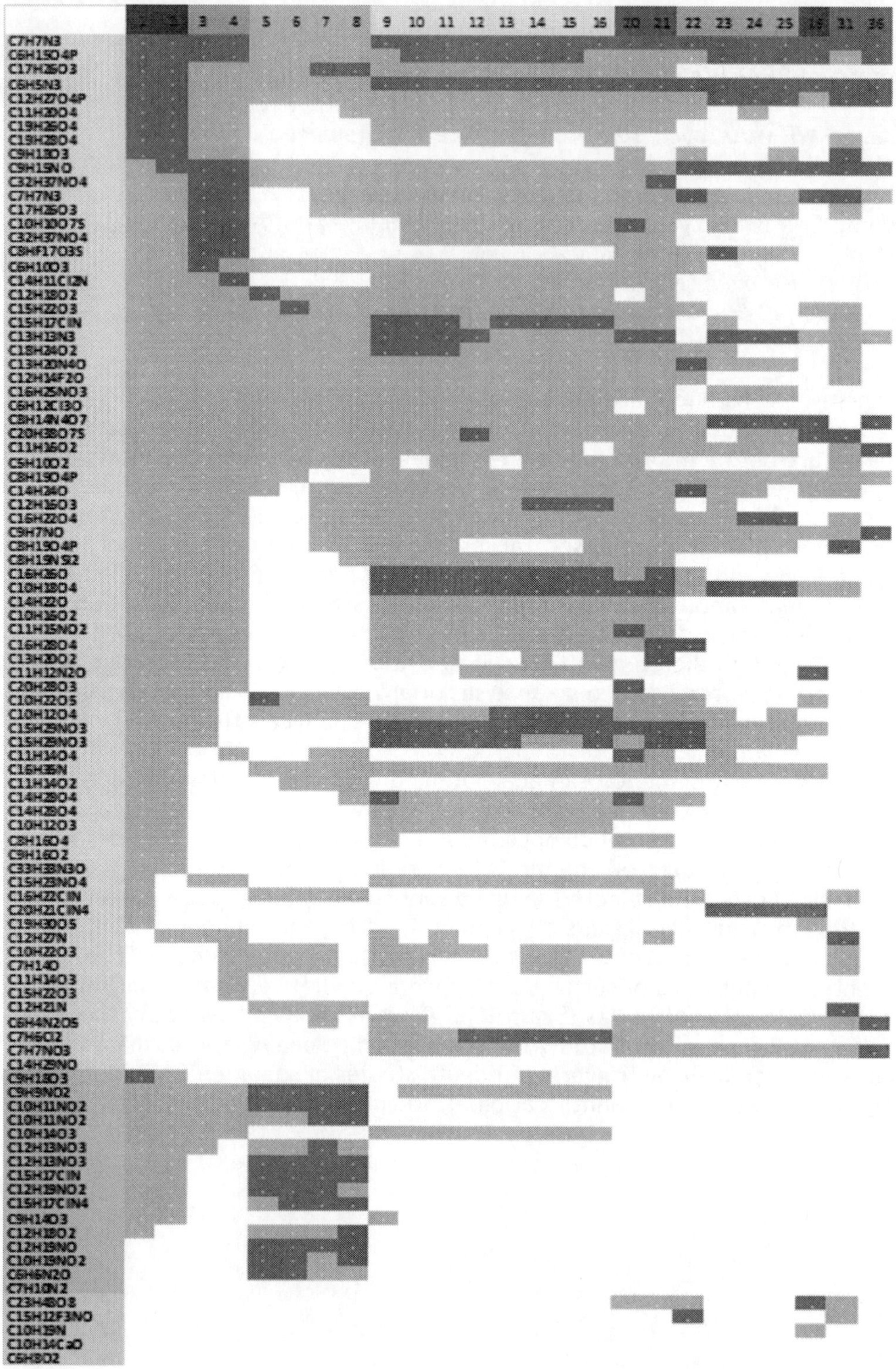

Figure 1: Heat map of the presence of accurate masses corresponding to listed elemental compositions in the different samples. Horizontally, samples from airport concentrate (red), airport effluent (dark orange), industrial effluent (orange), communal eff

Mo-SY-G3.5

Application of Effect Directed Analysis to Identify Mutagenic Nitrogenous Disinfection Byproducts of Advanced Oxidation Drinking Water Treatment

Annemieke Kolkman, *KWR Watercycle Research Institute, Nieuwegein, Netherlands*
Dennis Vughs, *KWR Watercycle Research Institute, Nieuwegein, Netherlands*
Kirsten Baken, *KWR Watercycle Research Institute, Nieuwegein, Netherlands*
Bram Martijn, *PWN Technologies, Andijk, Netherlands*
Annemarie van Wezel, *KWR Watercycle Research Institute, Nieuwegein, Netherlands*
Pim de Voogt, *KWR Watercycle Research Institute, Nieuwegein, Netherlands*

Advanced oxidation processes are important barriers for organic micropollutants (e.g., pharmaceuticals, pesticides) in (drinking) water treatment. Studies have indicated that medium pressure (MP) UV/H2O2 treatment leads to a positive response in Ames mutagenicity tests, which is then removed after granulated activated carbon (GAC) filtration. The potentially mutagenic substances formed have been scarcely identified. Earlier research showed that many substances result from the reaction of photolysis products of nitrate with (photolysis products of) natural organic material (NOM).
Using an innovative approach to trace the formation of disinfection byproducts (DBPs) of MP UV water treatment, based on stable isotope labeled nitrate combined with high resolution mass spectrometry, we showed that multiple nitrogen containing substances were formed in artificial water during MP UV treatment. Part of these DBPs were also detected in full scale water samples. In these samples, both chemical analysis and the Ames fluctuation test showed an increased response after MP UV/H2O2 treatment. For 14 DBPs the structure of the N-DPB was elucidated using in silico fragmentation tools and confirmed with analytical reference standards.
Further identification of the detected N-DBPs, using effect directed analysis to pinpoint the source of the mutagenicity or individual testing of these substances in Ames tests, would provide more insight into the relation of the N-DBPs with the observed mutagenicity. To this end, fractions of MP UV treated and untreated water extracts were prepared using preparative HPLC. These fractions were each concentrated and tested in the Ames fluctuation test. In addition, high resolution mass spectrometry was performed in all fractions to assess the presence of N-DBPs. After evaluating the results, a correlation was observed between the detection of byproducts in the fractions and the mutagenic response. Based on toxicity data and read across analysis, we could indicate five N-DBPs that are potentially genotoxic and were present in relatively high concentrations in the fractions in which mutagenicity was observed.
The results of this study offer opportunities to further evaluate the identity, potential health concern and relevance for full scale drinking water treatment plants and varying process conditions of N-DBPs formed during MP UV drinking water treatment.

Mo-SY-G3.6

Towards higher throughput in Effect-Directed Analysis: development of integrated platforms for micro-fractionation, suspect screening libraries and non-target analysis

Marja Lamoree, *Institute for Environmental Studies, Vrije Universiteit, Amsterdam, Netherlands*

Aim: In the past decades, Effect-Directed Analysis (EDA) - in which chemical analytical techniques are combined with (in vitro) bioassays to identify environmental contaminants capable of causing adverse effects - has developed into a promising tool for the identification of contaminants in various matrices such as sediment, surface and effluent waters, indoor dust and samples from mammalian origin. EDA has a long tradition but technological developments in the last decade have significantly contributed to the maturation of EDA as a powerful approach for nontarget analysis. In the past, especially the low throughput and the limited identification success rate have hampered the acceptance of EDA. In this presentation, solutions will be presented for the realization of higher throughput in EDA and compound identification.
Methods: LC and LCxLC fractionation strategies into ≥ 96 well plates were developed for combination with miniaturized bioassays to improve the bioactivity-to-identity correlation. For evaluation and interpretation of the high resolution ToF-MS data, dedicated suspect screening lists aiming at either a specific toxicological endpoint (e.g. thyroid hormone disruption) or a specific matrix (e.g. indoor dust) were designed to maximize the identification success rate.
Results: The composition of complex environmental samples covering a wide range of matrices such as waste water treatment plant (WWTP) effluents, indoor dust and dryer lint was unraveled using 1D-LC and 2D-LC fractionation in combination with bioassays for assessment of acetylcholinesterase inhibition and thyroid hormone disruption. For toxicant identification, a compound library containing thousands of chemicals known for their use in consumer products was used.
Conclusions: The implementation of comprehensive LCxLC coupled to ToF-MS adds another dimension to nontarget analysis and EDA by the inclusion of the double confirmation of compound/suspect retention times in the orthogonal separation system. The 2D-LC system was successfully used for high resolution fractionation into 4x96 or 384 well plates and bioassay testing in parallel with ToF-MS. Environmental contaminants were identified according to their accurate masses and isotopic patterns, and further confirmed by two dimensional retention alignment as well as their bioactivities in the assay. From the perspective of human exposure assessment, depending on their distribution and fate in the indoor and outdoor environment, these compounds may become of interest for future inclusion in human biomonitoring programmes.

Mo-SY-H3: Aggregate exposure assessment of contact allergens in consumer products

Mo-SY-H3.1

Aggregate Exposure to Contact Allergens: Application of a Method of Quantitative Risk Assessment

Christiaan Delmaar, *RIVM, bilthoven, Netherlands*
Bas Bokkers, *RIVM, bilthoven, Netherlands*
Martine Bakker, *RIVM, bilthoven, Netherlands*
Janine Ezendam, *RIVM, Bilthoven, Netherlands*

Background
Users of consumer products may come in contact with skin sensitizing substances contained in these products. In case the sensitizing substance is included in multiple products, the risk of dermal sensitization depends not only on the sensitization potential of a single product, but rather on the total (aggregate) exposure of all product exposures combined. Risk assessment of aggregate dermal exposure to a sensitizing substance is hampered by the fact that the processes of the induction of skin sensitization are not fully understood. A method for quantitative risk assessment (QRA) was proposed by (Api et al., 2008), but this method is currently being revised.
Aim
Application of the QRA method to aggregate consumer risk assessment is not straightforward. It will be discussed how the QRA can be combined with an aggregate consumer exposure assessment tool (the Probabilistic Aggregate Consumer Exposure Model (PACEM)) to assess risk of skin sensitization in a population.
Methods
As an exemplifying case, the method was applied to the risk assessment of geraniol, a fragrance present in cosmetics and household cleaning agents. Dermal sensitization studies were assessed to derive the point of departure needed for the estimation of the Acceptable Exposure Level (AEL). Using monte carlo simulation, information on product concentrations was combined with product use data to assess aggregate dermal exposure to geraniol in a population. Comparing the dermal aggregate exposure with the AEL, the fraction of the population at risk was determined.
Results
In this particular example it is shown that a range of 0.02-0.86% of the population may have an aggregated exposure which exceeds the acceptable exposure level. Furthermore, it is demonstrated that personal care products contribute more to the consumer's geraniol exposure compared to household cleaning agents.
References
A. Api, et al. Dermal sensitization quantitative risk assessment (QRA) for fragrance ingredients. Regul. Toxicol. Pharmacol., 52 (2008), pp. 3-23

Mo-SY-H3.2

Individual-based aggregate exposure assessment for isothiazolinones in cosmetics: Exposure factors and a calculation strategy for sensitizers

Elena Garcia Hidalgo, *ETH Zürich, Zurich, Switzerland*
Natalie von Goetz, *ETH Zürich, Zurich, Switzerland*
Konrad Hungerbuehler, *ETH Zürich, Zurich, Switzerland*

Background:
Consumers regularly use household care and personal care products (HPCP), so that it is necessary to ensure a high level of safety for the consumption of these products. Cosmetics and household ingredients are rarely aggregated, because they are subject to different regulations and information on the use of household cleaners is scarce. However, especially preservatives, like e.g. isothiazolinones, are present in both product groups, because they are used to preserve water-containing consumer products. Therefore, it is important that the aggregate exposure assessment for such preservatives comprises both cosmetics and household cleaning products.
Isothiazolinones are known skin sensitizers, i.e. they penetrate the skin and interact with the skin proteins, triggering, under specific circumstances, an allergic contact dermatitis. Dermal exposure to sensitizers is dependent on the nature, frequency, area and duration of contact between consumer products and skin.
Objectives:
The aim of this study is to estimate aggregate human exposure to four isothiazolinones (Methylisothiazolinone, Chloromethyisothiazolinone, Benzisothizolinone, and Octylisothiazolinone) that are found in many HPCPs, which are often concurrently used by the same consumer.
Methods:
Skin sensitization is induced following dermal exposure to a sensitizer in an amount exceeding the sensitization threshold. The critical determinant of exposure for evaluating skin sensitization risks is dose per unit area of exposed skin, i.e. a model should be applied that aggregates exposure per body site.
Our dermal exposure model is based upon consumer use and co-use patterns and isothiazolinone concentration data for HPCP. The former were determined by a postal questionnaire in Switzerland, which included children and adolescents, providing for the first time in Europe information regarding cleaning products combined with personal care products. For relevant products, not only use frequency was inquired, but also which body parts were treated. Isothiazolinones were analysed in HPCP that were frequently used by the survey respondents.
Results:
Around 50% of the analysed household cleaning products contained isothazolinones. In cosmetic products, and particularily in leave-on products, the percentage is much smaller. Individual-based aggregate exposure was estimated by combining the reported individual use patterns with the isothiazolinone concentrations in the products used by the individual person and the contributions of cosmetics and household cleaners were compared. For each isothiazolinone we provide realistic distributions of exposure for specific body parts of both genders and across all age groups. The exposure factors and calculation strategies developed in this study can serve as a basis for exposure assessments to other sensitizing substances.

This page intentionally left blank.

Mo-SY-H3.3

Aggregate Exposure Assessment for Skin Sensitizing Fragrances: Using an Aggregate Exposure Model to Assign Maximum Concentration Levels.

Cian O' Mahony, *Creme Global, Dublin, Ireland*
Anne Marie Api, *RIFM, New Jersey, United States*

Consumers are routinely exposed to fragrance materials in personal care products and cosmetics, which can potentially give rise to allergenic reactions and skin sensitization. Determining a safe level of a fragrance in personal care products and cosmetics requires consideration of a number of different factors, for which detailed data is often required in order to avoid worst case assumptions. In a population of consumers, different application sites are exposed to different combinations of products, which in turn are used in different amounts with varying levels of fragrances in the products themselves. Accurately accounting for this variability is necessary to accurately determine exposure, requiring probabilistic modelling and large data sets of consumer habits and practices. At the same time, a practically applicable method is needed that be used to set a maximum concentration of a fragrance material in a range of products when the No Expected Sensitisation Induction Level (NESIL) is known, along with the appropriate Safety Assessment Factors (SAFs) for different combinations of products and application sites (amongst others).
The Creme RIFM aggregate exposure model is now routinely used for the safety assessment of fragrance materials currently on the market, and is capable of assessing dermal exposure to fragrance materials in consumer populations for all major product application sites. This model has recently be applied in the Quantitative Risk Assessment (QRA) of fragrance materials to derive a set of adjustment factors that can be used to set safe concentrations of fragrance materials in cosmetics and personal care products which take aggregate exposure into account. This presents a convenient and robust methodology for setting maximum allowable concentrations of fragrances in products that can be easily applied, includes consideration of aggregate exposure, and is protective of consumer risk to skin sensitization.

Mo-SY-H3.4

Estimating Aggregate Dermal Exposure to Preservatives for Skin Sensitization Quantitative Risk Assessment.

Sarah Tozer, *Procter & Gamble, Egham, Surrey, United Kingdom*
Jane Rose, *Procter & Gamble, Cincinnati, Ohio, United States*
Cian O' Mahony, *Creme Global, Dublin, Ireland*
Cindy Ryan, *Procter & Gamble, Cincinnati, Ohio, United States*
Petra Kern, *Procter & Gamble, Brussels, Belgium*

Background: Aggregate exposure assessment should use a tiered approach that is iterative. If, in any tier, negligible or acceptable risk cannot be demonstrated, the assessment moves to a higher tier. The risk assessment is finished if (in any tier of the approach) it has been demonstrated that the risk for the population under consideration is negligible or acceptable, or if in the highest tier the risk is not acceptable and further refinements are not possible. This approach is becoming standard for systemic exposure assessments, and is here considered for aggregate dermal exposure assessment for the purpose of skin sensitization quantitative risk assessment.
Objectives: The aim of this study is to apply a tiered approach to the aggregate dermal exposure assessment of the preservatives benzyl alcohol in cosmetic and personal care products, and to assess the impact on quantitative risk assessment of skin sensitisation, by comparing the exposures to the weight of evidence (WoE) No Expected Sensitization Induction Level (NESIL) considering the appropriate safety assessment factors (SAF).
Methods: At the low tier, maximum concentrations of benzyl alcohol when used as a preservative were assumed, and simplistic deterministic additions calculated per body part providing an overly conservative risk assessment, due to some of the exposure assumptions. At the high tier, more complex exposure datasets were modelled using the subject oriented Creme Care and Cosmetics aggregate exposure tool to refine the data, including co-use data and product composition data (such as presence probability), providing more realistic population estimates of exposure.
Results: Results are shown for estimated dermal exposure (expressed as µg/cm2/day) for key body parts where different product types are applied, for the low and high tiers, and the impact on risk assessment is shown. In addition to providing a refined assessment of dermal (local/topical) total benzyl alcohol exposure, this work can be used as a case study on how to approach future dermal aggregate exposure questions for skin sensitization quantitative risk assessment.

Mo-SY-H3.5

Aggregate exposure assessment to contact allergens from essential oil consumption- A loophole in the risk assessment?

Nicolas DORNIC, *LERCCO (laboratoire d'évaluation du rsique chimique pour le consommateur - Laboratory of chemical risk assessement for the consumer), brest, France*
Anne-Sophie FICHEUX, *LERCCO (laboratoire d'évaluation du risque chimique pour le consommateur - Laboratory of chemical risk assessment for the consumer), Brest, France*
Audrey BERNARD, *LERCCO (laboratoire d'évalation du risqu chimique pour le consommateur - Laboratory of chemical risk assessement for the consumer), Brest, France*
Gregoire CHEVILLOTTE, *LERCCO (laboratoire d'évalation du risqu chimique pour le consommateur - Laboratory of chemical risk assessement for the consumer), Brest, France*
Alain-Claude ROUDOT, *LERCCO (laboratoire d'évaluation du rsqiue chimique pour le consommateur - Laboratory of chemical risk assessment for the consumer), Brest, France*

To date, risk assessment on skin sensitization is based on a quantitative risk assessment, also known as the « QRA » method. The current scientific consensus is to take into account cumulative exposure, especially resulting from the consumption of personal care products to assess the Consumer Exposure Level. However, raw aromatherapy products, i.e. essential oils, although as being applied voluntarily on the skin by consumers are not taken into account, due to the lack of knowledge on their consumption and the relative lack of regulation. Our previous consumption survey revealed that a significant part of the French population is exposed to essential oils and that consumption per dermal route might be an important source of co-exposure for contact allergens.
The aim of our study was to determine the allergenic substances composing essential oils, both qualitatively and quantitatively, in order to better understand the risks of sensitization due to the consumption of aromatherapy products.
By crossing chromatography data available in the literature and the recent list of the Scientific Committee on Consumer Safety of established contact allergens in humans, we developed a database compiling more than 500 compositions of 11 different types of essential oils intended to be applied on the skin.
Our results revealed that 6 established contact allergens are systematically found in essential oils, i.e. limonene, linalool, pinenes, terpinolenes, α-terpineol and β-caryophyllene. As a consequence, aromatherapy can be considered as a source of co-exposure as such for some allergens. Not taking into account exposure from aromatherapy could be thus a loophole in the risk assessment. Moreover, two allergens present in essential oils, i.e. limonene and linalool are found as the most occurring allergens that are required to be labelled in cosmetics. Although these are considered as weak sensitizers both by the International Fragrance Association and the Research Institute for Fragrance Materials, the prevalence of allergy could be more the result of the daily consumption of these types of products more than the inherent potency of these molecules to cause skin sensitization. The consumption of aromatherapy combined with cosmetics could be thus of concern regarding skin sensitization.

Mo-PL-I3: Risk Assessment

Mo-PL-I3.1

A conceptual framework to support exposure science research and complete the source-to-outcome continuum for risk assessment

Justin Teeguraden, *Pacific Northwest National Laboratory, Richland, Washington, United States*
Yu-Mei Tan, *U.S. Environmental Protection Agency, Durham, NC, United States*
Stephen Edwards, *U.S. Environmental Protection Agency, Durham, NC, United States*
Peter Egeghy, *U.S. Environmental Protection Agency, Durham, NC, United States*
Jeremy Leonard, *Oak Ridge Institute for Science and Education, Oak Ridge, TN, United States*
Kim Anderson, *Department of Environmental and Molecular Toxicology, Corvallis, OR, United States*
Richard Corley, *Pacific Northwest National Laboratory, Richland, WA, United States*
Anna Harding, *Oregon State University, Corvallis, OR, United States*
Molly Kile, *Oregon State University, Corvallis, OR, United States*
Staci Simonich, *Oregon State University, Corvallis, OR, United States*
Dave Stone, *Oregon State University, Corvallis, OR, United States*
Robert Tanguay*Oregon State University, Corvallis, OR, United States*
Katrina Waters, *Pacific Northwest National Laboratory, Richland, WA, United States*
Stacey Harper, *Oregon State University, Corvallis, OR, United States*
David Williams, *Oregon State University, Corvallis, OR, United States*
Jennifer Orme-Zavaleta,*U.S. Environmental Protection Agency, Durham, NC, United States*
Tina Bahadori, *U.S. Environmental Protection Agency, Durham, NC, United States*

Driven by major scientific advances in analytical methods, biomonitoring, computational tools, and a newly articulated vision for a greater impact in public health, the field of exposure science is undergoing a rapid transition from a field of observation to a field of prediction. A necessary element of this evolved paradigm is an organizational and predictive framework for exposure science that is analogous to the "systems-based approaches" used in the biological sciences. To enable such a "systems-based approach", we proposed the "Aggregate Exposure Pathway (AEP)" concept to organize data and information emerging from an invigorated and expanding field of exposure science. The AEP framework is a layered structure that describes the elements of an exposure pathway, as well as the relationship between those elements. The AEP has similar elements of the Adverse Outcome Pathway (AOP) in that the basic building blocks of an AEP retain the naming conventions used for AOPs: Key Events (KEs) to describe the measurable, obligate steps through the AEP; and Key Event Relationships (KERs) describe the linkages between KEs. The AEP offers an intuitive approach to organize exposure information from sources to internal site of action, setting the stage for forecasting exposure at an internal target site. We envision the AEP as a natural and complementary companion in exposure science to the Adverse Outcome Pathway (AOP) concept used in the toxicological sciences. The direct link between AEPs and AOPs will complete the source to outcome continuum to support more efficient integration of exposure science and toxicity testing information.

Together, the AEP and AOP frameworks form and inform a decision-making framework with thé flexibility for risk-based, hazard-based, and exposure-based decision making.

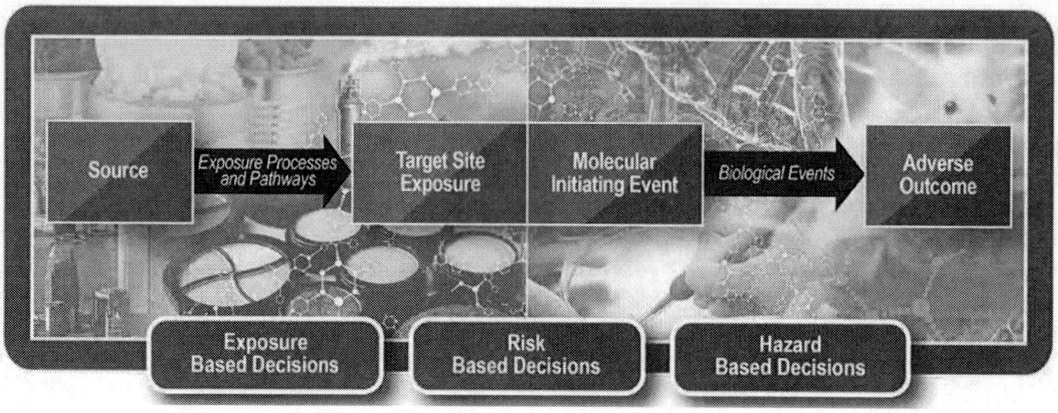

Aggregate Exposure Pathway Framework Integration with Adverse Outcome Pathway Framework

Mo-PL-I3.2

Comparative Analysis of Human Biomonitoring Values and Health-Based Guidance Values Used in Risk Assessment of Chemicals in Food: An Update

Judy Choi, *BiPRO GmbH, Munich, Germany*
Lisbeth E. Knudsen, *University of Copenhagen, Copenhagen, Djibouti*
Anke Joas, *BiPRO GmbH, Munich, Germany*

Aim: In risk assessment, authorities generally use toxicological/epidemiological data to derive health-based guidance values (HBGVs), such as acceptable/tolerable daily intake and reference doses, to control potential risks from dietary intake, but there is uncertainty on the reliability of these exposure estimates. Human biomonitoring (HBM) is an ideal tool to measure human body burden, which reflects internal exposure in an individual or a population group. HBM allows the establishment of reference values (RVs), defined as the 95th population percentile along with the 95% confidence interval of a substance's concentration, which provide valuable information about body burden at a certain point of time but do not inform directly about potential health risks or exceedance of HBGVs. In order to translate measured body burdens into HBGVs and allow interpretation of HBM data in a risk context, the German HBM Commission and experts from the USA develop health-relevant biomonitoring values such as HBM-values and biomonitoring equivalents (BEs), respectively. The aim of this work is to provide the most current analysis on the overlaps and non-overlaps among HBGVs, RVs, HBM-values and BEs of chemicals with potential dietary sources.
Method: National HBM programmes were reviewed for RVs for the general population of environmental chemicals potentially taken up by humans via the food chain. Next, HBGVs established by the World Health Organisation (WHO), European Food Safety Authority (EFSA) and US's Environmental Protection Agency (EPA), HBM-values and BEs for these chemicals were identified. The collected data were then compiled to determine the degree of overlaps and non-overlaps among the four values.
Results: We identified a total of 168 RVs, 12 HBM-values, 21 BEs and 191 HBGVs for environmental chemicals related to the food chain (see Figure). Metals such as cadmium and mercury, some phthalates and bisphenol A are some of the few chemicals that have RVs, HBM-values/BEs and HBGVs. Otherwise, there remain significant non-overlaps among these values, especially between RVs and HBGVs. This primary reason behind these non-overlaps is due to lack of HBM data on several classes of pesticides and mycotoxins or insufficient toxicological/epidemiological data to derive HBGVs.
Conclusions: This work demonstrates a current obstacle in the efficient use of HBM in risk assessment. Priorities for future method development to improve the applicability of HBM in food risk assessment are warranted.
Acknowledgement: We would like to thank EFSA for funding this project. More outcomes from this project can be found in a EFSA supporting publication (2015:EN-724).

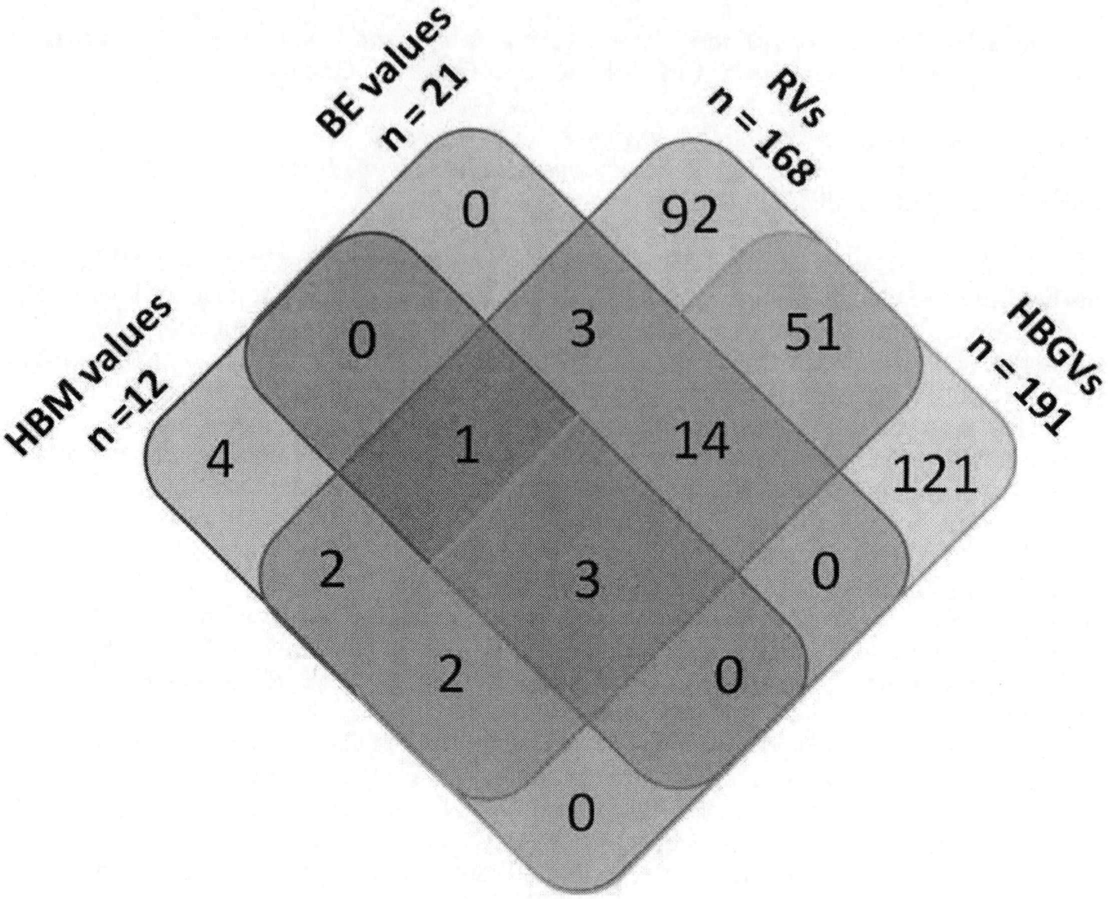

Figure: Venn diagram showing the overlaps and non-overlaps among HBM-established reference values (RVs), HBM-values, biomonitoring equivalents (BEs) and health-based guidance values (HBGVs) established by the authorities.

Mo-PL-I3.3

Rapid Environmental Assessment of Pesticide-Contaminated Sites - An Expert System For Effective Preliminary Risk Assessment With Limited Resources

John Keith, *Pure Earth, New York, United States*
Bret Ericson, *Pure Earth, New York, United States*
Barbara Jones, *Pure Earth, New York, United States*
Nick Albergo, *Pure Earth, New York, United States*
Petr Sharov, *Pure Earth, New York, United States*
Jack Caravanos, *Pure Earth, New York, United States*
Russell Dowling, *Pure Earth, New York, New York, United States*

Misuse and spillage of pesticides related to poor storage and handling practices has resulted in pesticide-contaminated sites in countries throughout the world. Many of these sites present significant public health risks, particularly where persistent pesticides were released in or close to populated areas. Many sites are in low- and moderate-income countries, where resources for investigation and remediation of sites are often limited, and access to technical expertise is inadequate. There is a need to perform effective preliminary risk assessment of sites at a low cost in such countries to allow decision-makers to prioritize resources for sites representing the most significant public health risk. Pure Earth, under contract from the UN Food and Agriculture Organization (FAO), developed an expert system that facilitates risk assessment of small- to medium-sized pesticide contaminated sites using limited resources. This Rapid Environmental Investigation (REA) system uses investigation methods that can be easily taught to investigators with limited experience in the field, but who are scientifically literate. The system incorporates available GIS data and other information about sites into a common database to enable efficient investigations. Trained investigators then perform investigations, typically in 2 days or less in the field and employ only a few analytical samples. The system is computer-based such that it can be loaded on laptop computers, and includes quality control protocols. The products of the system are 1) a database of information about sites including risk-relevant factors, and 2) semi-quantitative risk scores based on algorithms developed by experts in risk assessment, assuring consistency of risk assessments. Scores are developed for risks related to sources (amount and hazard of pesticides released); pathways (migration to populated areas and exposure routes); and receptors (number (and age) of people potentially exposed and severity of exposure.) The REA system was first developed for Vietnam then modified and expanded in 2014 and 2015 for use in former Soviet Union countries as part of a larger FAO pesticide initiative. Experience there demonstrated the effectiveness of the program and its potential for use elsewhere in the world. In 2016 the system was introduced in Cameroon. The system has been favorably received by government agencies in numerous countries, and the results of the REA investigations have enabled effective prioritization of pesticide-contaminated sites for further detailed investigation and risk management plan development.

Mo-PL-13.4

Human Biomonitoring of Endocrine-disrupting Phthalates Exposure in Children and Cumulative Risk Assessment

Maryam Zare Jeddi, *Institute for Environmental Research (IER), Tehran University of Medical Sciences, Tehran, Iran*
Noushin Rastkari, *Institute for Environmental Research, Tehran University of Medical Sciences, Tehran, Iran*
Reza Ahmadkhaniha, *School of Public Health, Tehran University of Medical Sciences, Tehran, Iran*
Masud Yunesian, *School of Public Health, Tehran University of Medical Sciences, Tehran, Iran*

Background: Phthalates are a group of environmental endocrine disruptors and have been ubiquitously applied in a multitude of common consumer products. Through contact with such products in daily life, people are frequently exposed to phthalates, which are suspected to contribute to adverse health effects, particularly in the reproductive system as "phthalate syndrome". However, limited information is available on phthalate exposure and its associated cumulative risk among school age children.
Method: In the present study, one spot urine sample was collected from 56 healthy children and adolescents (range 6-17yrs). Five major urinary phthalate metabolites (mono butyl phthalate (MBP), monobenzyl phthalate (MBzP), mono(2-ethyl-hexyl) phthalate (MEHP), mono-(2-ethyl-5-hydroxyhexyl) phthalate (MEHHP), and mono-(2-ethyl-5-oxohexyl) phthalate (MEOHP)) of three parent phthalates were analyzed by gas chromatography-mass spectrometry (GC-MS). Daily intake of phthalates was estimated using urine metabolite levels, extrapolating to ingest 'dose' using the creatinine correction approach. Based on the calculated daily intakes derived from urinary metabolite levels and acceptable levels of exposure for each individual phthalate, a cumulative risk assessment was performed for the anti-androgenic phthalates di-(2-ethylhexyl) phthalate (DEHP), dibutyl phthalate (DBP), and butylbenzyl phthalate (BBzP) related to the Reference Dose for Anti-Androgenicity (RfD AA).
Results: The analyzed metabolites were detected in almost all urine samples, and this shows that children are simultaneously exposed to multiple phthalates. The median concentrations of MBP, MBzP and the sums of di-(2-ethylhexyl) phthalate metabolites (ΣDEHPm) were 42.87, 2.2 and 35.36 µg/L, respectively. Girls were more exposed than boys. Among the phthalate metabolites determined, concentration of ΣDEHPm was the highest, with respective maximum concentration of 108.9 µg/L. The estimated daily intakes of DEHP, DBP, and BBzP were 16.42, 1.6, and 0.11 µg/kgbw-day, respectively. The execution of a cumulative risk assessment for combined phthalate exposure demonstrated that the HI was 85% RfDAAs (HI=0.6) in school age children, which might constitute to the risk of anti-androgenic effect during puberty, which is a sensitive stage regarding hormonal changes and the development of reproductive organs. Exposures to DEHP are major contributors to cumulative exposure to the phthalates assessed, thus to the overall hazard index.
Although on the basis of this study's results children exposure didn't exceed the reference values, it needs to be considered that children are exposed to other various environmental chemicals as endocrine disrupters/anti-androgens. However, taking attention to cumulative risk assessment of phthalates - and other chemicals with same endpoint and mechanism is recommended.

This page intentionally left blank.

Mo-PL-I3.5

The future of risk assessment and toxicity testing for chemical mixtures - Report from the EFSA-RIVM Symposium 18-19 May 2016

Jacob Van Klaveren, *RIVM, Bilthoven, Netherlands*
Stef Bronzwaer, *EFSA, Parma, Italy*
Jean-Lou Dorne, *EFSA, Parma, Italy*
Caroline Merten, *EFSA, Parma, Italy*
Bernadette Ossendorp, *RIVM, Bilthoven, Netherlands*
Tobin Robinson, *EFSA, Parma, Italy*

Every day, we are exposed to a mixture of substances via food, the air and our skin. Much progress has been made in developing methods for assessing the risk to public health of such exposure to multiple chemicals. EFSA and the Dutch National Institute for Public Health and the Environment (RIVM) will hold a symposium in May that aims to present and discuss the state of the art of such methods.
Currently, test strategies for assessing the safety of chemicals consider one compound at a time. Such strategies depend heavily on animal testing. We face two inter-related problems: 1) there is no efficient test strategy for mixtures; and 2) we need to develop a new test strategy that will meet public demands to reduce animal testing. Scientific experts, risk assessors and risk managers have been supporting research programmes to further develop and improve new methodologies for mixture testing.
EFSA, supported by RIVM and other European partners, has taken a major step forward in its work on assessing the risks from exposure to multiple pesticides, and implementation of these risk assessment methodologies is ongoing at EFSA and at the European Commission. However, further work is still needed to continue to refine the methodologies and to apply them to other groups of substances (e.g. metals and other contaminants).
Against this background, RIVM and EFSA are organising an international symposium to reflect on the state of play in research with regard to the risk assessment of chemical mixtures. The meeting will bring together Dutch researchers, scientists, stakeholders and policy-makers, and their European and international peers to discuss scientific progress in the field of evaluating exposure to mixtures of pesticides and to address new challenges in mixture testing for a broad range of chemicals. The aim is to report on scientific progress in the whole area of chemical mixture risk assessment and to discuss the feasibility and prioritisation for future testing in Europe and beyond.
Symposium topics

- The state of play regarding implementation of cumulative risk assessment of pesticides in Europe
- Enhancing knowledge on efficient toxicity testing of chemical mixtures while reducing animal testing
- How new testing technology will improve understanding of mixture toxicity and human risk assessment

At ISES 2016, the outcome of the symposium that will take place on 18 + 19 May will be presented.

Mo-PL-I3.6

Assessing Exposure Risks from Synthetic Biology

Andrew Gillespie, *US Environmental Protection Agency, Research Triangle park, NC, United States*
Mark Bagley, *US Environmental Protection Agency, Cincinnati, Ohio, United States*

Advancing knowledge and technology in applied genetic engineering are causing an exponential growth in development of new synthetic biology applications with potentially significant beneficial impacts on human well-being. Examples include the potential to reengineer mosquitos to eliminate infectious diseases such as malaria and Zika; to reengineer algae to produce high-efficiency carbon-neutral biofuels; to design bacteria capable of metabolizing and remediating hazardous waste and spills; and to build new organisms to produce pharmaceuticals and commercial chemicals. The trajectory of technological advancement is to move past current Genetically Modified Organisms (GMOs) into a realm of entirely de novo organisms containing DNA which has never before existed. New applications for this technology are emerging rapidly, presenting an urgent and growing need for new tools and methods to evaluate the risks of human and ecological exposure to never-before-seen organisms. Our current analytical toolkit was designed for simpler technology and a slower rate of development of new applications. Newer technologies enable simple editing of existing genomes to add, delete, or change current genetic traits in a single generation, which is greatly increasing the number of new organisms and applications available for evaluation and use. There is a concomitant growing number of applications by businesses who want to commercialize these advances, requiring government regulators to increase their throughput in evaluating such applications for potential adverse outcomes.

Understanding and managing exposure risk in such a new and complex field of biology will require a wide array of traditional exposure science disciplines such as molecular biology, ecology, epidemiology, and informatics, and will require the integration of disciplines which may not have historically been engaged as much in exposure science, such as genetic and chemical engineering and lifecycle analysis.

The US is currently updating a Coordinated Framework for the Regulation of Biotechnology to guide how US federal agencies charged with regulating synthetic biology will balance the different objectives of developing promising technologies while managing risks to protect human health and the environment. The challenge is made more complex because the regulatory framework must exist within a legal framework which has not kept up with the fast pace of change in the science and technology. Development of improved frameworks and methods for characterizing and assessing synthetic biology exposure pathways and risks would be of substantial help and interest to policymakers working in this rapidly changing field.

Mo-SY-A4: The Children's Health Exposure Analysis Resource

Mo-SY-A4.1

Accelerating Child Health Exposure Research with the CHEAR Data Center

Patricia Kovatch, *Icahn School of Medicine at Mount Sinai, New York, New York, United States*
Deborah L. McGuinness, *Rensselaer Polytechnic Institute, Troy, New York, United States*
Chris Gennings, *Icahn School of Medicine at Mount Sinai, New York, New York, United States*
Susan L. Teitelbaum, *Icahn School of Medicine at Mount Sinai, New York, New York, United States*

The United States' National Institute of Environmental Health Sciences has established an infrastructure, the Children's Health Exposure Analysis Resource (CHEAR), to provide the extramural research community access to laboratory and statistical analyses aimed at adding or expanding the inclusion of environmental exposures in their research. The CHEAR project is a $50 million multiunit infrastructure composed of a coordinating center, a network of laboratories and a data center tasked to provide researchers access to comprehensive exposure assessment for NIH funded studies of children's health. The CHEAR Data Repository, Analysis and Science Center (Data Center) is located at the Icahn School of Medicine at Mount Sinai in collaboration with Rensselaer Polytechnic Institute in New York. The goal of the data center is to catalyze new scientific insight from the co-location, integration and advanced statistical and data science analysis of multimodal data sets. The data center provides the intellectual and logistical support for the validation, interpretation, curation, and maximum reuse of data generated by the laboratory network. We aim to provide access to tools and services that incorporate and extend exposure analysis on an exposome scale (i.e., to study complex environmental influences on health) by providing a strong data, knowledge, and analytic infrastructure. We are developing semantic infrastructure for support in consistent modeling, unambiguous interpretation, and enhanced integration. For the investigators that utilize CHEAR for studies of children's environmental health using the data generated within and outside the network, the Data Center provides: 1) data repository and management; 2) statistical consultation and analysis services; 3) collaborative research support; 4) statistical and analytical methods development; and 5) data science resources, including semantic infrastructure and services powered by a family of child health exposure ontologies. In this presentation, we will discuss the opportunities for advancing the study of early life environmental exposures and later life health consequences with advanced statistical and data science approaches including the use of knowledge graphs and ontologies. We will also review our initial lessons learned from building the data repository and developing its accompanying policies for data sharing.

Mo-SY-A4.2

Interlaboratory Harmonization and External Quality Assessment for Quantifying Toxic Metals/Metalloids in Human Body Fluids

Patrick Parsons, *New York State Dept of Health, Albany, NY, United States*
Aubrey Galusha, *New York State Department of Health, Albany, NY, United States*

Background:
A key goal of the NIEHS Children's Health Exposure Analysis Resource (CHEAR) is to provide the extramural research community with access to analyses for toxic metals/metalloids in children's body fluids. The analytical data produced by the six CHEAR hub laboratories need to be accurate, precise, traceable to international standards, and possess a high degree of inter-laboratory agreement. Well-validated, sensitive analytical methods are required to achieve reliable results for toxic metals/metalloids in body fluids at trace and ultra-trace levels.
Aim
The primary objective of the current project was to establish CHEAR laboratory method validity for toxic metals/metalloids against international reference materials, and to organize ongoing assessments of CHEAR laboratory performance.
Methods
Initial validation was based on analyses of NIST Standard Reference Materials (SRM) for toxic metals/metalloids in blood (SRM955c Toxic Metals in Caprine Blood) and urine (SRM 2668/3668 Toxic Elements in Frozen Human Urine). Thereafter, CHEAR laboratories participated in a special proficiency test (PT) event for trace elements in whole blood, serum and urine, based on 5 levels of archived samples provided by the Wadsworth Center.
Results and Conclusions
Data were analyzed from 6 CHEAR laboratories for Pb (Fig 1), Cd, Hg and As in 955c, and up to 20 elements in 2668/3668. Results provided important information on assay accuracy (compared to NIST values), intra-laboratory precision, inter-laboratory precision, and internal quality assurance (IQA) practices. Results for PT samples were reported for up to 25 elements in urine, and 24 elements in whole blood and serum. At low levels, the bias for blood Pb in 955c (0.424 µg/dL) ranged from +20.5% to -6.4%, while within-lab precision varied from 3.8 - 17% RSD. For blood As, the "true" value may be lower than NIST's "reference" value (2.07 µg/L). All six laboratories reported values for urinary As (2668) but only two were within NIST's stated uncertainty. Detection limits and IQA practices varied by orders of magnitude among laboratories. The data reported for Pb, Cd, Hg and As in blood/urine were graded based on the NY State Department of Health criteria. Most results were judged satisfactory for urinary As, Pb, Cd and Hg, and for Cd and Pb in blood. One lab stated that 3/5 urinary Hg values exceeded their calibration range. Blood As proved a more challenging assay, with only 2/4 labs achieving satisfactory results. Only 3 laboratories reported analyzing serum and of those, only two achieved satisfactory scores for Cu and Se.

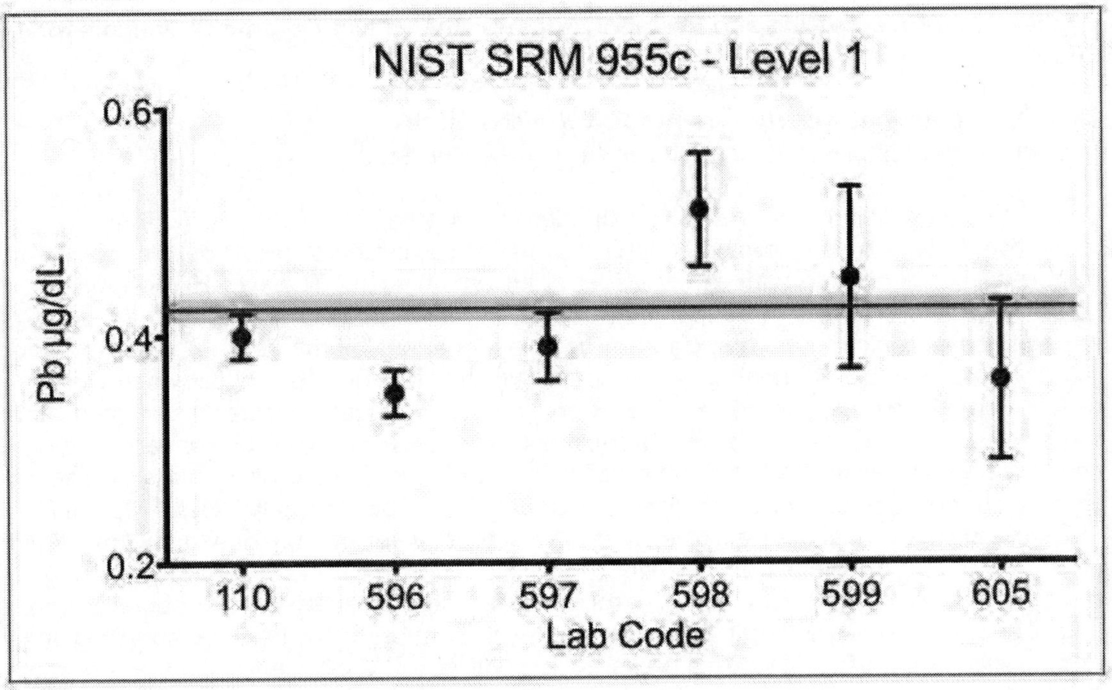

Analysis of NIST SRM 955c-level 1 certified for blood Pb at 0.424 µg/dL (±0.011 U): mean values reported by 6 CHEAR labs. error bars = within-lab uncertainty (U)

Mo-SY-A4.3

Quality Assurance Testing and Harmonization of Analysis of Organic Chemicals for CHEAR

Dana Barr, *Emory University, Atlanta, GA, United States*
Andrea Hjerpe, *Emory University, Atlanta, GA, United States*

One of the goals of the Children's Health Exposure Assessment Resource (CHEAR) laboratory hubs is to is to provide analytic tools for researchers to estimate exposure to targeted toxic chemicals which may affect children's health. The six CHEAR hub laboratory have a variety of capabilities to measure organic chemicals ranging from environmental toxicants such as phthalates to endogenous biomarkers. A series of working groups were established including members from the six laboratory hubs which were tasked with various aspects of the CHEAR program such as data reporting standards and quality assurance (QA). One of the first tasks of the QA working group for targeted chemicals was to evaluate the baseline agreement of the six laboratory hubs for the organic chemical classes of phthalates, pesticides and tobacco biomarkers. NIST SRMs 3672 and 3673 were used as accuracy bases. Further, a set of five individual, unspiked urine pools were developed and distributed to the laboratory hubs. A data reporting template was developed and all laboratories were asked to analyze each pooled sample and NIST SRM material in triplicate then to export these data to the Data Coordinating Center secure sharefile site. This presentation will detail the baseline evaluation process, results and will provide our plans for moving forward with laboratory analyses.

Mo-SY-A4.4

Integration of Targeted, Untargeted, and Biological Response Data

Robert Wright, *Mount Sinai School of Medicine, New York, New York, United States*

With rapid technological advances in genomics have come large-scale studies examining the genetic etiology of complex diseases. Although genome-wide association studies revealed genetic factors and biological networks that advanced our understanding of biology, it is increasingly recognized that the new morbidities of childhood (complex diseases such as asthma, autism, metabolic syndrome, etc) are largely not genetic, but instead reflect interactions between genetic susceptibility and environmental toxins or stressors. Unlike the genome, which is static, environmental exposures and our biological response to exposure changes at each life stage, and this variability in response must be factored in to our research. The exposome framework considers multiple exposures (e.g., chemicals, dietary intake/nutrition, social stressors), the biological and/or physiological response to exposure and how these processes change by life stage. Developmental programming results from toxicant-induced shifts in a number of molecular, cellular, and physiological processes and their interacting systems. Environmental toxicants affect many overlapping biological response processes including immune response/inflammation, response to reactive oxygen species, hormonal metabolism (sex steroids, hypothalamic-pituitary-adrenal axis), autonomic reactivity, DNA damage response, epigenetic alterations, & mitochondrial function Understanding the life stage during which exposure affects health (i.e. susceptibility windows) can also inform the underlying biology.
If we are to understand the exposome's role in biology, we must develop methods to link it to biology. Logical integrations of untargeted, targeted and biological response data could take several forms depending on the nature of the study question, the nature of the biological response data, the nature of the exposure data and the research design. One approach would be analogous to genomics in which untargeted assays are rank ordered with respect to health outcomes in a research discovery stage. The top ranked "hits" would then be addressed either singly or as a mixture in a replication stage. Biological response data, such as DNA methylomics, could then be integrated as a mediator of the top hit chemicals or their mixture. This "genomics like" approach however does not factor in exposure timing. Because effects of exposure vary by life stage, future research will need to address the likelihood of multiple "discovery stages" to explore the role of exposomic data with epigenomic or other "omic" like response data. This means there will be different sets of top hits for each life stage. Replication research will be likewise multi-dimensional. The ability of temporally resolved exposure biomarkers or models to address these complexities will be discussed.

Mo-SY-A4.5

Multi-laboratory Harmonization of Untargeted Exposomic Analyses

Susan Sumner, *RTI International, Research Triangle Park, North Carolina, United States*

The NIEHS Children's Exposure Analysis Resource (CHEAR) has established six laboratories in the United States to serve as analytical hubs. These analytical hubs use a wide variety of sample preparation, chromatography, NMR, Mass Spectrometry, and data analysis approaches for the detection and quantitation of endogenous and exogenous compounds in biospecimens. This presentation will provide a overview of the different methods being used by these analytical hubs, and highlight our approaches to achieve multi-platform and multi-laboratory harmonization of metabolomics for the analysis of both endogenous and xenobiotic compounds in biological samples.

Mo-SY-B4: New Biomarkers for Human Biological Monitoring in Occupational Health

Mo-SY-B4.1

Multi-residue analysis of pesticides in human hair

Hans Mol, *RIKILT - Wageningen UR, Wageningen, Netherlands*
Rosalie Nijssen, *RIKILT - Wageningen UR, Wageningen, Netherlands*
Paul Zomer, *RIKILT - Wageningen UR, Wageningen, Netherlands*

Background: many chemicals are incorporated in hair upon exposure. Hair analysis can reveal long-term exposure and is being used for this purpose in forensic toxicology. Advances in instrumental analysis lead to opportunities for development of new hair-based biomarkers of occupational exposure to chemicals which occur at much lower levels compared to e.g. drugs of abuse in forensic applications. The advances also improve possibilities for simultaneous detection of multiple chemicals. This is relevant in agriculture where during the season various pesticides are being applied.
Objectives: development of a generic method for the simultaneous detection of multiple pesticides in human hair, suited for research into occupational and non-occupational exposure assessment through human biomonitoring using hair as non-invasive matrix.
Methods: a multi-residue method for 20 frequently used pesticides was developed based on liquid chromatography with tandem mass spectrometry (LC-MS/MS). The method was validated at the 1 and 5 pg/mg level according to SANTE/11945/2015. The applicability was tested by analysis of hair samples from volunteers.
Results: during validation, adequate recoveries (70-120%) and repeatability's (<20%) were obtained for most pesticides down to the 1 pg/mg level and for all at the 5 pg/mg level. Hair samples from 14 individuals, not occupationally exposed, were analysed. Pesticides were detected in most hair samples, with multiple detects in several cases. Azoxystrobin, boscalid, imazalil, imidacloprid, and thiabendazole were most frequently detected. Levels were in the range 0.7-15 pg/mg.
The results show that multi-residue analysis in hair is feasible and has potential as tool for exposure assessment. Comparison of levels of exposure between different populations (e.g. occupational vs non-occupational) is in principle possible. For linking levels in hair to actual exposure, however, much more research on factors affecting incorporation, and toxicokinetics is needed.

Mo-SY-B4.2

Use of unmetabolized xenobiotics as biomarkers of exposure

Laura Campo, *University of Milan and Fondazione IRCCS Ca' Granda Ospedale Maggiore Policlinico, Milan, Italy, Italy*

Background: Using the unmetabolized (parent) chemicals as biomarker of exposure has some advantages over metabolite analyses: unmetabolized chemicals are very specific, their excretion is not influenced by interindividual differences in metabolism, and they are easily extracted from matrix.
Aim: To provide a review of recent studies on urinary benzene (BEN-U), methyl tert-butyl ether (MTBE-U), toluene (TOL-U), and PAHs (U-PAHs) as biomarkers of occupational or environmental exposure.
Methods: The studies performed by our group on exposure to benzene, MTBE, toluene, and PAHs are presented with emphasis on the intrinsic characteristics of unmetabolized compounds, their relationship with the environmental levels, and the comparison with the corresponding metabolites.
Results: In petrol station attendants, petrochemicals workers, and individuals not occupationally exposed, BEN-U was superior to the metabolites for the better relationship with personal exposure and the lower interindividual variability. A biological limit equivalent for the biomonitoring of occupational exposure to benzene in non-smokers was proposed.
MTBE-U was proposed as a biomarker of traffic exposure for its association with both airborne CO and the duration of exposure to traffic fumes in urban traffic policemen, and as a biomarker of exposure to MTBE and in general to petrol vapours in petrol station attendants. Biological exposure equivalents for MTBE-T have been calculated. As MTBE-U is not influenced by smoking habits, it was proposed as a surrogate biomarker of benzene exposure in smoking petrol station attendants.
A comparison between TOL-U and o-cresol in rotogravure printing workers has shown that TOL-U is not influenced by cigarette smoking and has higher specificity and sensitivity, lower background values, and a better correlation with airborne exposure than o-cresol. The kinetics of excretion of TOL-U has been studied to find out about the exposure period represented by TOL-U and to establish the best time for sample collection.
Researches on U-PAHs have focused firstly on high occupational exposure (i.e. coke oven workers), then on lower occupational exposure (i.e. asphalt workers), and recently, thanks to advance in analytical methods, on environmental exposure. Significant relationships have been found between U-PAHs and both environmental exposure and monohydroxylated PAH metabolites. Moreover, for the first time urinary benzo[a]pyrene and other carcinogenic PAHs were quantified in urine samples from both occupationally and environmentally exposed subjects.
Conclusions: The results of these studies show that BEN-U, MTBE-U, TOL-U, and U-PAHs are useful biomarkers to face the challenge of low occupational or even environmental exposure to toxics.

Mo-SY-B4.3

Urine collection methods for non-toilet trained children in environmental exposure assessment

Arné Oerlemans, *Radboud university medical centre, Nijmegen, Netherlands*
Gwendolyn Beckmann, *Radboudumc, Nijmegen, Netherlands*
Paul Scheepers, *Radboudumc, Nijmegen, Netherlands*

Background: Young children differ from adults in their exposure and susceptibility to environmental chemicals because of various factors such as biometry, physiology, behavior and diet. Their heightened vulnerability to environmental stressors makes it important to obtain appropriate urine samples for exposure assessment. However, collecting urine from non-toilet trained children has been shown to be methodologically and practically challenging. Collection methods should not introduce contamination or affect the integrity of the sample and must be acceptable to the participants.
Objective: The aim of this study was to evaluate various urine collection methods for non-toilet trained children which could be applied in a non-clinical setting to obtain biomonitoring data.
Methods: Selected methods for urine collection include a disposable polyacrylate diaper, a urine bag, a collection pad containing a hygroscopic polymer and the clean catch method. Advantages and limitations of these methods were evaluated with respect to minimum required sample volume, potential for contamination, timing of collection, and burden on participants. The success rate was defined as the percentage of suitable samples from the total number of sample collection attempts. An attempt was considered successful if it yielded a urine sample with a volume of at least 5 mL and free of faeces contamination. In addition, the user rating of each method was evaluated on a 0-10 scoring range.
Results: In total, 24 samples were obtained for each of the urine collection methods. The success rates were 67%, 21%, 17% and 4% for the disposable diaper, urine bag, collection pad and clean catch, respectively. The average user rates were 9.0, 4.7, 7.3 and 2.5, respectively. This indicates that a disposable polyacrylate diaper is a proper urine collection method among non-toilet trained children and therefore this method will be further evaluated for the storage stability of the analytes of interest, including clinical parameters such as creatinine and osmolality.
Preliminary results showed that urine including solutes stored in the polyacrylate granules of disposable diapers could be extracted using an aqueous solution of 15% calcium chloride. The recovery of creatinine was 92% to 95%, thereby showing that the disposable diaper is a promising method to assess environmental exposure of chemicals in non-toilet trained children. This research will be continued by analysis of the recoveries from the diaper for a range of xenobiotic metabolites.

Mo-SY-B4.4

Non-Invasive Dosimetry of Volatile Compounds: A Breath-Taking Experience

Jacqueline Biesterbos, *Radboudumc, Nijmegen, Netherlands*
Frans Russel, *Radboudumc, Nijmegen, Netherlands*
Ad Ragas, *Radboud University, Nijmegen, Netherlands*
Nel Roeleveld, *Radboudumc, Nijmegen, Netherlands*
Paul Scheepers, *Radboudumc, Nijmegen, Netherlands*

Introduction - Most volatile substances that are primarily taken up by inhalation will also be excreted via the lungs, unchanged or sometimes after biotransformation. Analysis of exhaled air can therefore be considered as an alternative to analysis of body fluids, such as blood or urine. In this contribution we focus on the use of exhaled air analysis in exposure assessment.
Objective - Provide an overview of the state-of-the-art and present opportunities for new uses of exhaled air as a medium for human biological monitoring.
Methods - Literature search in PubMed and Web of Science, combining 'exposure assessment', 'breath analysis' and 'exhaled air analysis' as search terms.
Results - Possibilities for analysis in exhaled air depend on substance properties. Substances with a vapor pressure of 1010 (for carbon monoxide) down to 0.3 Pa (mercury vapor) have been analyzed in human breath. We found evidence to demonstrate that exhaled air sampling is a good alternative to collection of blood or urine samples for many substances. However, widespread use of exhaled air analysis is currently hampered by a lack of a framework for interpretation, more specifically ranges of levels of xenobiotics observed in the general population and reference values for substances that also have an endogenous source. Nevertheless, the method has great potential to be used widely, e.g. in sample collection by self-administration, on-line sampling and real-time analysis.
Conclusion - There is concern that researchers may be caught in a fixed mindset, linking biomonitoring to (only) analysis of body fluids. We would like to inspire them to consider the option of breath analysis, particularly if less invasive methods are preferred, such as in children and elderly. Due to the increased use of breath analysis for clinical purposes it is expected that the technology will become more user friendly, and that interpretation will be supported by advanced software as a framework for interpretation.

Mo-SY-B4.5

New applications of human biological monitoring in petrochemical industry

Peter J. Boogaard, *Shell International bv, The Hague, Netherlands*

Human biological monitoring (HBM) has proven to be a highly effective approach for integrated assessment of workers' exposures and has been applied as such for decades in industrial settings. Historically, industrial biomonitoring programmes aimed to assess exposure of workers in situations where (personal) air monitoring proved unreliable or inappropriate. This would, for instance, be the case with intermittent airborne exposures of short duration and also in situations where significant dermal exposure might occur. Typical examples of the use of biomonitoring in such situations would be to assess whether there is exposure above a specified level and/or whether personal protective equipment is adequate to prevent exposure above a specified level. More recently, developments in bioanalytical chemistry have allowed extending the application of HBM to the general population.

HBM allows a more personalized health risk assessment than is possible with external exposure monitoring methods since inter-individual differences in working behaviour, in life style, and also in biological and physiological parameters, are accounted for. In addition, HBM has specific features that may allow more efficient hazard and risk assessment. Firstly, it may support a direct human link to in vitro test results. Effects observed in vitro at a certain concentration of a substance may be translated to internal human concentrations, which may improve human health risk assessment and at the same time reduce the need of animal testing. Secondly, it can be applied to improve risk assessment and management under REACH using the Biomonitoring Equivalent (BE) concept specified for the Derived No-Effect Level (DNEL). A BE translates an established reference value into a biomarker concentration using biokinetic data. If the result of an exposure assessment using human biomonitoring indicates that the levels measured are below the DNEL-based BE (BEDNEL), it would indicate that the combined exposure via all potential exposure routes is unlikely to pose a risk to human health and, as a consequence, health risk management measures might not be required.

These approaches require biokinetic data, which may be a challenge since for many substances reliable biokinetic information is currently lacking. There are two approaches to solve this issue. The first would be to adapt the current design of toxicity testing in animals. More specifically, bio-analysis of blood and/or urine should be included in repeated-dose studies to obtain biokinetic data. The second incurs the development of generic physiologically-based biokinetic models, which allow estimation of biomarker concentrations based on physicochemical properties.

Mo-SY-C4: Quantitative in vitro to in vivo extrapolation (QIVIVE): Advances in tools to quantify exposure-response relationships for risk assessment - III

Mo-SY-C4.1

High Throughput PBTK: Evaluating EPA's Open-Source Data and Tools for Dosimetry and Exposure Reconstruction

John Wambaugh, *U.S. E.P.A., Research Triangle Park, NC, United States*
Robert Pearce, *U.S. Environmental Protection Agency, Research Triangle Park, NC, United States*
Caroline Ring, *Oak Ridge Institute for Science and Education (ORISE), Research Triangle Park, NC, United States*
Woodrow Setzer, *U.S. Environmental Protection Agewncy, Research Triangle Park, NC, United States*
Barbara Wetmore, *US EPA, Research Triangle Park, NC, United States*

Thousands of chemicals have been profiled by high-throughput screening (HTS) programs such as ToxCast and Tox21; these chemicals are tested in part because most of them have limited or no data on hazard, exposure, or toxicokinetics (TK). While HTS generates in vitro bioactivity data for characterizing potential chemical hazards, TK models are needed to inform in vitro to in vivo extrapolation (IVIVE) to real world situations. The U.S. Environmental Protection Agency has created a new tool (R package "httk") for building, simulating, and evaluating TK and physiologically-based TK (PBTK) models for both IVIVE and exposure inference from biomonitoring data (i.e., reverse dosimetry). We are now able to rapidly parameterize generic PBPK models using in vitro data to allow IVIVE for 543 chemicals. Our high throughput toxicokinetics (HTTK) tools were implemented in the R statistical platform in part to allow statistical analysis of both IVIVE and our TK models. We have statistically evaluated our TK predictions using in vivo measurements of human steady-state serum concentrations, rat serum concentrations, rat tissue partition coefficients, and human volumes of distribution. We find that for many chemical classes our methods and models perform reasonably, and that we can begin to identify chemical classes for which our methods perform poorly. Our PBTK models are parameterized with not only chemical-specific parameters derived from in vitro measurements and predicted from chemical structure; but also with physiological parameters for a virtual population. We simulate population physiological parameters based on data from the most recent U.S. Centers for Disease Control National Health and Nutrition Examination Survey (NHANES), which describe distributions of demographic and anthropometric quantities in the modern U.S. population. A Monte Carlo approach, accounting for the correlation structure in physiological parameters, can be used to estimate margins between IVIVE predicted bioactive doses and estimates of exposure for the most sensitive portion of the population. While these new models are expected to have limited accuracy due to their simplicity and generalization of assumptions, the confidence in the predictions can be in part assessed using our comparison to TK in vivo data. Ultimately, we are working to identify the chemicals for which these new tools may be used with confidence, and to identify those chemicals where alternative approaches are needed. This abstract does not necessarily reflect U.S. EPA policy.

Mo-SY-C4.2

High Throughput Modeling of the Effects of Mixtures of ToxCast Chemicals on Steroid Hormone Cycles in Women

Frederic Bois, *INERIS, Verneuil en Halatte, France*
Nazanin Golbamaki-Bakhtyari, *INERIS, Verneuil en Halatte, France*
Simona Kovarich, *S-IN Soluzioni Informatiche Srl, Vincenza, Italy*
Emmanuel Lemazurier, *INERIS, Verneuil en Hallate, France*

Exposure to mixtures of chemicals is an increasing toxicological concern. The availability of exposure data for thousands of chemicals through ExpoCast project, together with the ToxCast results for the hundreds of high throughput in vitro assays, offers the opportunity to explore the toxicity of the chemical mixtures in realistic scenarios. We used computer modeling to predict the size of potential effects of random mixtures of aromatase inhibitors on women's menstrual cycle. We had previously investigated the impact of mixtures on steroidogenesis by a systems biology model for aromatase inhibition in adult female rats. In current work, to consider a larger number of events involved to hormonal balance disruption, we adapted a mathematical model of the hypothalamus-pituitary-ovarian control of estradiol and progesterone concentrations in blood. We used the model (coupled to a pharmacokinetic model of intake and disposition) to predict the effects of a million of chemical mixtures sampled by Monte Carlo simulations. To simulate a realistic exposure scenario, the exposures were also sampled from statistical distributions provided by the ExpoCast database (see illustrated work-flow). We find that a sizable fraction of the mixtures generated led to more than 20% inhibition of estradiol production. In contrast, exposures to chemicals considered individually almost never reach such effect sizes. Those results are discussed in light of the approximations and assumptions made, but demonstrate the possibility to address large scale mixture questions in a predictive toxicology framework, suitable for high throughput risk assessment of endocrine perturbation.

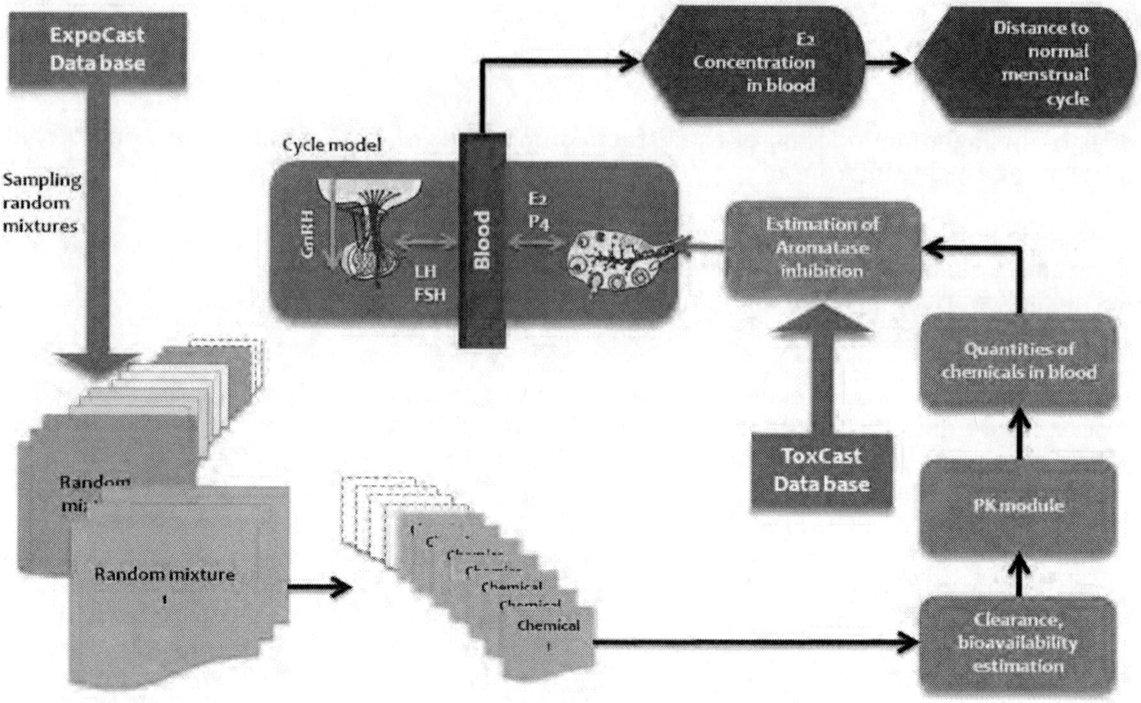

Computational model work flow.

Mo-SY-C4.3

Endocrine activity of POPs accumulated in human silicone implants - transferring in vivo exposure into in vitro bioassays

Dorothea Gilbert, *Danmarks Tekniske Universitet, Kgs. Lyngby, Denmark*
Philipp Mayer, *Danmarks Tekniske Universitet, Kgs. Lyngby, Denmark*
Mikael Pedersen, *Danmarks Tekniske Universitet, Søborg, Denmark*
Anne Marie Vinggaard, *Danmarks Tekniske Universitet, Søborg, Denmark*

Persistent organic pollutants (POPs) accumulated in human tissues may pose a risk for human health by interfering with the endocrine system. A new partitioning-controlled dosing approach from silicone was applied in the in-vitro H295 steroidogenesis assay to test whether POPs in a mixture composition and at concentrations as found in human silicone implants can interfere with steroidogenesis. Silicone served here as a reference partitioning phase to link in vivo to in vitro exposure. In the in vitro assay, silicone functioned as donor of a mixture of POPs while it also acted as sorptive sink for lipophilic hormones produced by the cells. The new dosing method from silicone was compared to conventional solvent-dosing. A mixture of PCBs, PBDEs, HCB and DDT as identified in human silicone breast implants was tested by spiking it into 24-well plate adopted silicone disks. When dosed from silicone, this chemical mixture in which individual POPs were in a freely dissolved concentration below the femtomolar range, increased the production of progestagens and androgens in the in vitro assay. However, no changes were observed when when co-solvent dosing was applied. The new silicone-based dosing approach allowed linking actual human POP levels in silicone (in vivo) to altered hormone production in a human adrenal cell line in vitro.

Mo-SY-C4.4

Combining Fish In Vitro Systems with Computational Modelling to Predict Chemical Accumulation and Altered Growth in Fish

Julita Stadnicka-Michalak, *EPF Lausanne, School of Architecture, Civil and Environmental Engineering; Eawag, Lausanne, Switzerland*
Kristin Schirmer, *Eawag; EPF Lausanne, School of Architecture, Civil and Environmental Engineering; ETH Zürich, Department of Environmental Systems Science, Dübendorf, Switzerland*

In vitro systems, i.e. cells cultured in a well-controlled environment in a petri dish, have revolutionized our mechanistic understanding of fundamental physiological processes as well as of stress adaptation and toxicity. This holds especially for cells obtained from mammals but increasingly so also for fish. Indeed, permanent fish cell cultures allow to decipher processes which are difficult or impossible to study in a fish. Yet, to fully exploit such in vitro systems, we need to devise strategies that allow us to quantitatively link in vitro and in vivo. We are working on such a strategy having three important processes in mind: bioaccumulation and alteration of survival and growth upon chemical exposure of fish. We base our approach on the assumption that the same chemical concentration in fish internal organs and in cells in culture would cause the same changes in cell survival and growth. Therefore, we develop mathematical models that aid in predicting chemical concentrations in fish tissue as well as in a culture dish. Focusing first on a fish Physiologically Based Toxicokinetic (PBTK) model, we determine in vitro biotransformation rates in order to predict the internal chemical concentrations in fish. We use fish cell cultures of tissues comprising entry routes for chemicals (gill, intestine) and liver as a major site of biotransformation. Indeed, taking the example of benzo(a)pyrene, we predicted a bioconcentration factor (BCF) of 1005 (+/-53) compared to a BCF of 920 measured in rainbow trout (Schirmer et al., in preparation). Next, we use a cell culture TK model to determine chemical concentrations in cell culture medium that would lead to intracellular chemical concentrations corresponding to the internal organ concentration. Applying the thereby determined in vitro exposure concentrations, we measure cell survival and proliferation of fish cell populations in the culture dish. Obtained live cell numbers are used to inform a mathematical model for organism growth. In this way we were able to predict reduced fish growth, resulting from several weeks of exposure to two fungicides, with very high accuracy (Stadnicka-Michalak et al., Science Advances, 2015:e1500302). Thus, our approach comprises a very promising step toward alternatives to whole organism toxicity testing, especially considering the simplicity, rapidity and low costs of this method.
We would like to acknowledge our colleagues who contributed to the two studies listed here as examples: BaP bioaccumulation - Frederic Weiss and Melanie Knöbel (Eawag); fish growth - Roman Ashauer (now University of York).

Mo-SY-C4.5

Dishing Up Nanoparticle Risks: Exposure-Based Computational Translation of In Vitro Toxicity Data to Human Risk

Justin Teeguarden, *Pacific Northwest National Laboratory, Richland, Washington, United States*
Brian Thrall, *Health Effects and Exposure Science, Richland, Washington, United States*
Dennis Thomas, *Computational Biology, Richland, Washington, United States*
Jordan Smith, *Health Effects and Exposure Science, Richland, Washington, United States*

Nanomaterials are an immense class of largely untested materials undergoing toxicity testing, principally in in vitro systems. This talk will introduce the principles for understanding how exposures in cell culture systems can be understood in the context of rodent and human exposures and demonstrate their application for iron oxide nanoparticles. The last five years has seen the rapid maturation of the fundamental concepts regarding the physical, chemical and biological features and processes that control the exposure of nanomaterials to cells in vitro and in vivo. Paralleling is evolution, increasingly sophisticated computational tools and experimental methods for acquiring dosimetry information have continued to emerge. This talk will present the importance of measuring
cellular dosimetry of nanoparticles, the implications for the fields of exposure science, toxicology and risk assessment, highlighting several mature approaches for obtaining dosimetry information in vitro and extrapolating those findings to in vivo systems with specific examples related to human health risk assessment (calculating a Permissible Exposure Limit for iron oxide nanoparticles).

Mo-SY-D4: Evidence-Based Research on Interventions to Reduce Personal Exposures to Environmental Pollutants

Mo-SY-D4.1

Bridging the Ineffective Lead Treatment Gap: Lessons from Hurricane Katrina in New Orleans, USA for Establishing an Effective Lead Intervention Strategy

Howard Mielke, *Tulane University School of Medicine, New Orleans, Louisiana, United States*
Christopher Gonzales, *Tulane University School of Medicine, New Orleans, Louisiana, United States*
Eric Powell, *Tulane University School of Medicine, New Orleans, Louisiana, United States*

Aim: Regarding reducing children's blood lead exposure, current medically prescribed treatments are ineffective. In addition the contribution from soil lead to childhood exposure and disease is deemed insufficiently characterized by the public health community. On 29 August 2005 Hurricane Katrina flooded and disrupted habitation in New Orleans. Soil and blood lead were mapped prior to Katrina. This unique study addresses soil and blood lead conditions pre- and ten years' post-Katrina and considers the effectiveness of low lead soil for lead exposure intervention. The purpose was to compare soil and blood lead levels pre- and ten years' post-Katrina to assess the impact of flooding on soil lead and blood lead levels in New Orleans.
Methods: This unique data-set includes soil lead (n=3314 and 3320, pre- vs. post-Katrina), blood lead (n=39,620 and 17,739, pre- vs. post-Katrina), distance, and changes in percent pre-1940 housing. Post-Katrina soil and blood lead surveys were stratified by the same census tracts (n=176) as pre-Katrina surveys. Statistical analysis entailed permutation procedures and Fisher's Exact Tests.
Results: Pre- vs. ten years post-Katrina soil lead medians decreased from 280 mg/kg to 132 mg/kg, blood lead medians decreased from 5 µg/dL to 1.8 µg/dL, respectively. Percent pre-1940 housing did not change significantly. Soil and blood lead shows a decrease with distance from the center of New Orleans. Except for age-of-housing results, P-values were extreme (<10-12). In the higher soil lead communities of the city, the percentage of children with median blood lead ≥ 5 µg/dL underwent a remarkable decrease from 64% before Katrina to 18.9% ten years' post-Katrina; still too many children are still being excessively exposed to lead.
Conclusions: Ten years after Katrina, profound changes in soil lead and children's blood lead occurred in New Orleans. Decreasing the lead on soil surfaces reduces children's interaction with lead dust, thus underscoring soil as a major of source of exposure. Not all communities saw the same changes and these results indicate a path forward for an effective intervention method to reduce children's lead exposure.

Mo-SY-D4.2

Home Air in Agriculture - Pediatric Intervention Trial (HAPI)

Catherine Karr, *Univ WA, Seattle, WA, United States*
Griselda Arias, *Yakima Valley Farmworkers Clinic, Toppenish, WA, United States*
Elizabeth Torres, *Northwest Community Education Center, Granger, WA, Wake Island*
Victoria Breckwich Vasquez, *University of Washington - Bothell, Bothell, WA, United States*
Michael Yost, *University of Washington, Seattle, WA, United States*

This university -community partnership identified poor air quality (particulate matter, ammonia) as a local risk factor for asthma morbidity among low income Hispanic farm worker children in a rural agricultural region of Washington State, USA. In response, a randomized trial was designed augmenting home-based asthma and environment education with placement of portable high efficiency particulate air (HEPA) cleaners containing a prefilter to reduce ammonia in the child's home. Using a rolling recruitment protocol, 30 of the total intended sample of 75 children have been recruited. All have poorly controlled asthma and live within 400 meters of crop production or dairy operations. Children participate for a one year follow up period with 2 week home indoor air contaminant measurement before randomization and approximately one year later. Asthma health is characterized twice prior and twice post randomization using both objective and subjective metrics. Validated questionnaires of asthma control, clinical utilization, exhaled nitric oxide concentration, spirometry, and urinary leukotriene concentrations are assessed. Enrolled children are all Hispanic, age 6-12 years and most (80%) are atopic based on positive skin prick testing to common aeroallergens. To date, the mean (SD) two week indoor sleeping area PM2.5 and PM10 concentrations observed are 11.8 (7.3) mcg/m3 (range 3.8-25.3) and 23.0 (13.3) mcg/m3 (range 6.7-43.9) respectively. Samples for analysis of indoor and outdoor NH3 have also been collected. Other common indoor asthma triggers of concern identified include rodent pests (69% of homes), use of irritant chemical cleaners (100% of homes), and wood burning for heat (25%). Community partners (Yakima Valley Farm Workers Clinic) and Northwest Community Education Center play a primary role in field based data collection of both health assessments and air sampling with training and oversight by University partners. Using a community engaged approach, this study is addressing a priority environmental health concern in a difficult to reach and understudied vulnerable population. Results will be incorporated into asthma education and management programs in the community.

Mo-SY-D4.3

Effect of a Randomized, Blinded Organic Diet Intervention on Pesticide Exposure among Pregnant Women

Cynthia Curl, *Boise State University, Boise, Idaho, United States*
Jessica Porter, *Boise State University, Boise, Idaho, United States*

Background: Food is certified "organic" when grown according to specific requirements, including the absence of most synthetic pesticides. The organic food market is among the fastest growing sectors of the agricultural industry, primarily due to a perception that organically-grown food is "healthier" than food that is conventionally-grown. However, there is almost no data to support (or refute) this perception. Most researchers do agree that an organic diet can reduce exposure to agricultural pesticides. It is also known that maternal exposure to some classes of pesticides during pregnancy is associated with subsequent decrements in cognitive, behavioral, and neurological outcomes in children. However, it is not known whether the amount of pesticide exposure resulting from a conventional diet is substantial enough to cause these decrements, and whether any such effects could be prevented with an organic diet.
Objectives: The purpose of this pilot project was to develop methodology for a future study of the effect of a maternal organic diet during pregnancy on subsequent cognitive outcomes in children. We aimed to: 1) design an effective recruitment strategy and a means to deliver organic and conventional produce to study participants in a blinded and scalable fashion, and 2) assess compliance with and efficacy of the dietary intervention through repeated measurement of urinary metabolites.
Methods: We partnered with Women, Infants, and Children (WIC) clinics in Idaho's Treasure Valley to recruit 20 women during their first trimester of pregnancy. Eligible women were 18-35 years of age, non-smoking, and were not expected to have a high-risk pregnancy. All participants reported eating exclusively conventionally-grown food at the time of recruitment. An equal number of participants were randomized to the organic and conventional groups; based on group assignment, each woman received weekly deliveries of either organic or conventional produce throughout their second and third trimesters. Participants also provided weekly urine samples and completed food diaries using a mobile phone app or written diaries.
Results: Weekly urine samples are pooled to represent monthly exposures throughout the second and third trimester of pregnancy. Pooled samples are analyzed for metabolites representing exposure to one herbicide (2,4-dicholorphenoxyacetic acid), seven organophosphate insecticides (diazinon, parathion, methyl parathion, ethyl parathion, chlorypyrifos, chlorpyrifos-methyl, and malathion), and eight pyrethriod insecticides (cyfluthrin, cyhalothrin, cypermethrin, deltamethrin, fenpropathrin, flumethrin, permethrin and tralomethrin). We will compare pesticide exposures across the organic and conventional groups and will investigate individual produce items associated with the highest concentrations of urinary pesticide metabolites.

Mo-SY-D4.5

The UGAAR Randomized Controlled Trial of HEPA Filter Air Cleaner Use and Residential Fine Particulate Matter in Ulaanbaatar, Mongolia

Enkhjargal Gombojav, *Mongolian National University of Medical Sciences, Ulaanbaatar, Mongolia*
Prabjit K. Barn, *Simon Fraser University, Burnaby, British Columbia, Canada*
Chimedsuren Ochir, *Mongolian National University of Medical Sciences, Ulaanbaatar, Mongolia*
Bayarkhuu Laagan, *Sukhbaatar District Health Center, Ulaanbaatar, Mongolia*
Bolor Beejin, *Public Health Institute of Mongolia, Ulaanbaatar, Mongolia*
Buyantushig Boldbaatar, *Mongolian National University of Medical Sciences, Ulaanbaatar, Mongolia*
Jargalsaikhan Galsuren, *Mongolian National University of Medical Sciences, Ulaanbaatar, Mongolia*
Bat-Undes Baasansuren, *Mongolian National University of Medical Sciences, Ulaanbaatar, Mongolia*
Altan-Ochir Khurelbaata, *Mongolian National University of Medical Sciences, Ulaanbaatar, Mongolia*
Tsogtbaatar Byambaa, *Public Health Institute of Mongolia, Ulaanbaatar, Mongolia*
Craig Janes, *University of Waterloo, Waterloo, Canada*
Bruce P. Lanphear*Simon Fraser University, Burnaby, Canada*
Young Man Roh, *Korea University, Seoul, Korea, South*
Tim Takaro, *Simon Fraser University, Burnaby, Canada*
Scott A. Venners, *Simon Fraser University, Burnaby, Canada*
Glenys M. Webster, *Simon Fraser University, Burnaby, Canada*
Christopher D. Palmer, *New York State Department of Health, Albany, United States*
Patrick J. Parsons, *New York State Department of Health, Albany, Albany, United States*
Ryan W. Allen, *Simon Fraser University, Burnaby, Canada*

Background: Portable high efficiency particulate air (HEPA) filter air cleaners can reduce indoor PM2.5 concentrations, but their efficacy has not been evaluated in highly polluted settings. In Ulaanbaatar, Mongolia, residential coal use contributes to annual average ambient PM2.5 concentrations of approximately 70 μg/m3.
Objectives: To assess the efficacy of portable HEPA filter air cleaners to reduce indoor residential PM2.5 in homes of non-smoking pregnant women in Ulaanbaatar, Mongolia.
Methods: We randomly assigned 540 participants to an intervention (1-2 HEPA air cleaners) or control (no air cleaners) group. For each participant, we conducted home visits at approximately 12 and 30 weeks gestation to make one-week continuous PM2.5 measurements using optical particle counters (Dylos DC1700). In a subset (n=82), one-week gravimetric PM2.5 concentrations were also measured and an equation for converting particle counts to PM2.5 concentrations was derived from co-located measurements. Whole blood samples collected at around 30 weeks gestation were analyzed for lead, mercury, and cadmium. Data on housing characteristics and personal behaviours were obtained via questionnaires. Additionally, 24-hour ambient PM2.5 concentrations were obtained from a centrally-located government monitoring station.
Results: One-week optical particle counts were highly correlated with indoor gravimetric PM2.5 concentrations ($r = 0.86$; $p<0.001$). Ambient 24-hour PM2.5 concentrations were

highest in winter [geometric mean (GSD): 87.8 (1.6) µg/m3] and lowest in summer [14.3 (1.6) µg/m3]. Indoor PM2.5 concentrations followed the same seasonal pattern, with geometric mean (GSD) concentrations of 62.1 (1.9) µg/m3 and 23.0 (1.4) µg/m3 in winter and summer, respectively. Living with a smoker did not contribute substantially to indoor PM2.5 concentrations. On average, PM2.5 concentrations were 26 % lower in intervention homes. Season-specific filter efficacy, comparing geometric mean PM2.5 concentrations in intervention and control homes, ranged from 24-43 % for measurements in early pregnancy, and 5-20 % in later pregnancy. Effectiveness was highest in winter for both visits. Geometric mean (GSD) blood cadmium levels were lower among intervention [0.20 (1.69) µg/L; n= 222] compared to control [0.23 (1.74) µg/L n= 198] participants (p0.05). Conclusions: HEPA filter air cleaner use was associated with lower indoor PM2.5 and blood cadmium concentrations among pregnant women. Air cleaner effectiveness varied seasonally and decreased over duration of the study.

Mo-SY-D4.6

Effects of Low Emission Zones (LEZ) on air quality in Germany

Josef Cyrys, *Helmholtz Zentrum Muenchen, Neuherberg, Germany*
Jianwei Gu, *University of Augsburg, Augsburg, Germany*
Veronika Deffner, *Ludwig-Maximilians-Universität Munich, Munich, Germany*
Helmut Kuechenhoff, *Ludwig-Maximilians-Universität, Munich, Munich, Germany*
Jens Soentgen, *University of Augsburg, Augsburg, Germany*
Annette Peters, *Helmholtz Zentrum München, Neuherberg, Germany*

Background. Low Emission Zones (LEZs) were implemented as a measure for improving air quality of ambient air, especially in cities where the European limit values for PM10 (particulate matter with an aerodynamic diameter < 10 µm) were exceeded. By end of 2015 almost 50 LEZs were introduced in Germany. In the framework of the ACCEPTED project (Assessment of changing conditions, environmental policies, time-activities, exposure and disease, www.acceptedera.eu/) we evaluated the effects of LEZs on the air quality in three German cities: Augsburg, Munich and Berlin.
Methods. To assess the effectiveness of LEZ we used general additive models adjusted for PM10 levels at reference site (located in regional background), wind direction, public holidays, day of the week, and time of the day. Because of the seasonal variation in PM10 concentrations, we modelled LEZ and time effects for summer and winter separately by introducing of an indicator function.
Results. We observed clear seasonal differences regarding the magnitude of the effect. The reduction of PM10 levels was in general more pronounced in summer season compared to winter season. A clear reduction of PM10 levels was observed in Munich and Berlin, whereas the results for Augsburg were not consistent. The PM10 reduction was especially large for Berlin, depending on the monitoring site and the active stage of the LEZ. The decrease of total carbon (TC=OC+EC) concentrations was clearly larger than the corresponding decrease of PM10 levels. Whereas PM10 concentrations at a traffic site decreased after the implementation of the LEZ by 16% in summer and 9% in winter, the corresponding reduction of TC was 24% (summer season) and 16% (winter season).
Conclusions. In sufficiently large and strictly regulated LEZs a reduction of PM10 concentration between 5 and 10 % (at traffic site partially up to 20 %) can be expected. The reduction of PM10 levels is in general more pronounced for the summer season compared to the winter season. It means that LEZs are proving successful as a measure for air pollution control. Moreover, they decrease not only PM10 but, to a much higher degree, the health-relevant components (such as diesel soot) contained in PM10. Therefore, the effect of LEZs on air quality could be much better estimated by additional monitoring of diesel soot and elemental carbon in PM10 and the benefit on human health is by far greater than it is presently visible from measurements of PM10.

Mo-SY-E4: Toward an Understanding of Indoor exposures

Mo-SY-E4.1

The added value of time-use studies in exposure science in the built environment

Jelle Laverge, *Ghent University, Gent, Belgium*
Hal Levin, *Building Ecology, Santa Cruz, United States*

Most people in our modern society spend more than 90% of their time in indoor environments. At the same time, countless studies show that the exposure in indoor environments is dominated by pollutants that are typical for that specific environment. We are shielded from some outdoor pollutants, while we or the indoor space itself are sources of pollutants that do not exist outdoors. The whereabouts of the studied subjects are therefore crucial to get a realistic estimation of their exposure.
Preferably, this information is very detailed, since the concentration of pollutants can vary with as much as an order of magnitude between spaces within the same building.
In this contribution, we look at the information that is typically gathered in time-use studies to describe the 'typical' human subject. This information is usually activity based in stead of location based, and if location is included, it is most likely limited to the functional description of the building (home, work, school). We will examine what information is available, what we can learn from it and how this information can be useful in exposure studies.

Mo-SY-E4.2

Integrated Indoor and Outdoor Exposure Assessment Framework for Fine Particulate Matter Pollution

Thomas McKone, *University of California, Berkeley, California, United States*
Natasha Hodas, *California Institute of Technology, Pasadena, California, United States*
Joshua Apte, *University of Texas, Austin, Texas, United States*
Matti Jantunen, *National Institute for Health and Welfare, Helsinki, Finland*
Olivier Jolliet, *University of Michigan, Ann Arbor, Michigan, United States*
John Evans, *Harvard School of Public Health, Boston, Massachusetts, United States*
Peter Fantke, *Technical University of Denmark, Kongens Lyngby, Denmark*

The 2010 Global Burden of Disease report demonstrates that fine particulate matter (PM2.5) pollution is the major environmental contributor to mortality. Exposures outdoors (ambient) and indoors (household) contribute almost equally to this burden. Unfortunately, the health impacts from exposure to PM2.5 are often excluded from life cycle impact assessment (LCIA) used for characterizing environmental performance of products and services. This is in large part because of the lack of well-vetted harmonized guidance about how to consistently assess the exposures and impacts of indoor and outdoor emissions of PM2.5 and its precursors. We present a modeling framework for calculating exposure factors for indoor and outdoor emissions of primary PM2.5 and secondary PM2.5 precursors, and a roadmap for further refining this modelling framework for operational use in LCIA. The framework was developed over the last three years by a task force convened under auspices of the Society of Environmental Toxicology and Chemistry (SETAC)/United Nations Environment Program (UNEP) Life-Cycle Initiative to provide guidance and methods for estimating the health impacts associated with PM2.5 exposure and to recommend PM2.5 characterization factors for application in life cycle assessment. The framework involves three stages--analyzing PM2.5 fate and exposure (including indoor and outdoor urban/rural environments), modeling exposure-response, and the integration of exposure-response and PM2.5 exposure reflecting population and location characteristics. We introduce the overall framework and present key components of the exposure assessment underlying the health impact characterization factors. The exposure metric at the center of this analysis is the population intake fraction (iF). Our exposure model is organized as a mass balance matrix that tracks the global fate of primary PM2.5 and secondary PM2.5 precursor emissions (both indoors and outdoors) as an embedded system of compartments including urban environments, rural environments, and indoor environments within urban and rural areas. The fate modeling system provides PM2.5 concentrations that are linked with human activity patterns and population geographical distribution patterns to determine intake fractions. After presenting the model structure, we will review initial results and will present geographic variability, discuss key uncertainties, and evaluate our model using results from other models and concentration measurements.

Mo-SY-E4.3

VOC exposures indoors: focus on VOCs most often found indoors

Hal Levin, *Building Ecology Research Group, Santa Cruz, California, United States*
Matti Jantunen, *National Institute for Health and Welfare, Kuopio, Finland*

AIMS: The purpose of the abstract is to put indoor exposures to VOCs in context of total exposures to assist researchers and policy-makers to address exposures and their sources appropriately.
METHODS: Selective literatue review and summary with comparisons of indoor exposures to total exposures.
RESULTS:
Human exposure to VOCs is strongly influenced, most often dominated by indoor exposures. Proper characterization of these exposures is a necessary component of any effort to assess the health relevance of exposures to any VOC and especially in application of total exposure assessment practices. (TEAM, 1989) The example below was part of the work that led to the conclusion that indoor exposures, at least for exposures to VOCs, are the dominant exposures in a total exposure assessment framework.

Figure 1. (Wallace, Annu. Rev. Energy Environ. 2001. 26:269-301)
Lance Wallace wrote: The most important sources of exposure are small and close to the person (generally right under one's nose). Rhodes has shown this in his 1991 Indoor Air journal paper. This is true for smokers' exposure to benzene. (It is also the case for smokers' exposure to styrene.) Other well documented examples include chloroform, the dominant source of exposure is chlorinated water. Chlorination kills microbes, but the excess chlorine reacts with organic materials to form chloroform. Since all the water going to our homes is chlorinated (except for persons with private wells or on groundwater systems), all the water used in washing machines, dishwashers, sinks, bathtubs, toilets, etc., is constantly emitting chloroform to the air.
These findings led to the formation of ISEA, now ISES.
CONCLUSIONS:
Indoor and outdoor exposures to numerous other health-relevant VOCs are described in terms of indoor/outdoor concentration ratios and calculations of intake fractions are useful in improving understanding health relevance of the various exposures and assisting policy-makers in addressing the sources of the exposures.

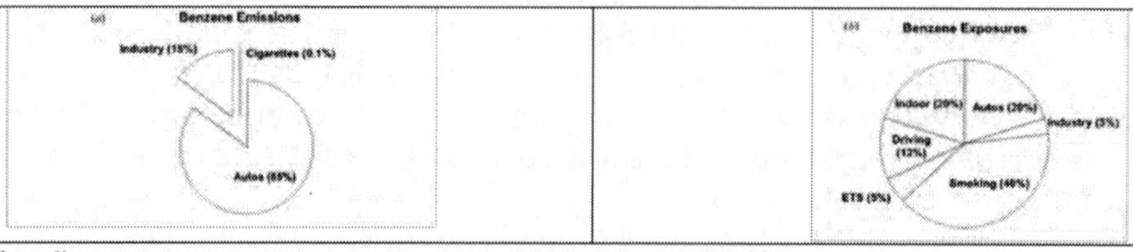

Figure 1. (Wallace, *Annu. Rev. Energy Environ.* 2001. 26:269–301)

Figure 1. (Wallace, Annu. Rev. Energy Environ. 2001. 26:269-301)

Mo-SY-E4.4

SVOC Exposure Indoors

Charles Weschler, *Rutgers University, Piscataway, New Jersey, United States*

Aim: This presentation will present an overview of the contribution of indoor pathways - inhalation, dermal absorption and dust ingestion - to total SVOC exposure, with a focus on SVOCs that are manmade. (Note that there is an entire symposium at ISES 2016 titled: "Exposure to SVOCs in the Indoor Environment - Products, Emissions, Exposure, Pharmacokinetics and Biomarkers".)
Methods: The processes and results discussed will be based on relevant peer-reviewed literature published over the past decade, emphasizing more recent studies.
Results: Under equilibrium conditions, if we know an SVOC's concentration in one compartment, we can predict it in other indoor compartments using partitioning theory. Based on actual indoor measurements, acceptable predictions have been demonstrated between
SVOC levels in the gas phase & in dust; in the gas phase & in surface films; in settled dust & in airborne particles; and in the gas phase & in skin surface lipids. Such predictions work best for central tendencies in large datasets. Several studies have shown that for many SVOCs, dietary intake is insufficient to explain the levels found in serum and/or blood. Airborne SVOCs can be inhaled or dermally absorbed. SVOCs in settled dust can be ingested. SVOCs on surfaces can be transferred to human skin via contact and subsequently dermally absorbed or ingested (hand-to-mouth activity). Intake budgets, based on concentrations measured in different indoor compartments, coupled with generally accepted exposure factors, are providing indirect evidence that indoor exposures are responsible for a large fraction of total SVOC exposures. More direct evidence is coming from a growing number of studies that have reported statistically significant associations between SVOCs in an indoor compartment and the same SVOC (or its metabolite) in blood or urine. These include significant associations between four common phthalate esters in indoor dust and corresponding metabolites in urine; between selected phthalates in skin wipes and corresponding urine metabolites; between ten lower-chlorinated PCB congeners in indoor air and their levels in serum; between polyfluorinated compounds in indoor office air and their levels in serum; and between certain PBDEs in indoor dust or handwipes and their levels in serum.
Conclusions: Numerous SVOCs, including manmade SVOCs not even produced in commercial amounts just two generations ago, are measured in bodily fluids. Indoor pathways contribute substantially to these observed body burdens.

Mo-SY-E4.5

Microbial (fungal, viral, and bacterial) exposures indoors - The Indoor Microbiome

Hal Levin, *Building Ecology Research Group, Santa Cruz, California, United States*

KEYWORDS, and THEME:
Microbes, indoor microbiome, Fungi, Bacteria, Viruses

Humans are constantly exposed to millions of bioaerosols, These include exposures to whole microorganisms, which can have both beneficial and detrimental effects. There is growing understanding of the indoor microbiome as a result, among other things, of the dramatic (approximately thousandfold) reduction in the cost of DNA sequencing during the past 10 years and significant investments in characterizing the indoor microbiome including but not limited to approximately USD 50,000,000 by the Albert P Sloan Foundation during the past 5 years. Much of the recent work has focused on characterizing the various sources of airborne microorganisms and the relative contribution of each. We have identified the following eight major categories of sources of airborne bacteria, viruses, and fungi in the built environment: humans; pets; plants; plumbing systems; heating, ventilation, and air-conditioning systems; mold; dust re-suspension; and the outdoor environment.

This work has shown that some species are associated with specific sources. The potential for source characterization and source apportionment can be extended substantially as a part of indoor environmental exposure and total exposure characterization. This potential is currently unrealized. Future studies of indoor exposures will quantify detailed emission rates of microorganisms from each source and identify the relative contributions of each source to the complete indoor microbiome. "This information could then be used to probe fundamental relationships between specific sources and human health, to design interventions to improve building health and human health, or even to provide evidence for forensic investigations." (Microbiome (2015) 3:78).

Mo-SY-F4: Exposure-Based Toxicity Testing

Mo-SY-F4.1

Exposure Based Testing- Introduction to the Symposium

Nicholas Ball, *The Dow Chemical Company, Horgen, Switzerland*
Heli M Hollnagel, *Dow Europe GmbH, Horgen, Switzerland*
Sean Hays, *Summit Toxicology, Lyons, CO, United States*

Is there a feasible way to use the potential for exposure to a chemical as the basis for the generation of hazard data?
The field of regulatory toxicology is continuing to move toward a registration approach where data requirements are determined based on manufacturing or import volume with little regard for the actual potential for exposure to a substance. In essence, the assumption exists that where the manufacturing volume is high, the potential for exposure is also high and so more should be known about the chemical in question. This can hold true in some cases, but there are many examples of where the manufacturing or import volume of a substance does not correlate with actual exposure (for example chemical intermediates). Consequently the actual exposure to a chemical is rarely factored in to the hazard assessment, leading to the generation of data that does little to improve the overall safety of workers, consumers and the environment.
With the advances being made in terms of exposure modelling and biological activity screening, it is time to consider if there is a more effective way to assess the risks posed by chemicals without resorting to extensive testing programs using large numbers of animals where the need for testing is based on arbitrary criteria rather than a more flexible and realistic approach?
So the questions for this symposium are:

- How can exposure information be factored into toxicity study selection to adequately protect against likely exposures while limiting excessive testing that neither informs risk assessment or risk management?
- How can we improve current regulatory policy and guidance on chemical safety assessments, as well as convince the public of the adequacy of using exposure information, to achieve this important objective?

Each of the speakers in this symposium will cover a range of topics around how exposure can become a cornerstone of chemical safety testing, including regulatory and industry perspective. At the end of the symposium, the audience will be encouraged to participate in a polling exercise (SciPinion) that will explore the audience's opinion about what they've heard and their thinking on this important issue.

Mo-SY-F4.2

A Data-Driven Framework for Incorporating New Tools for Toxicity, Exposure, and Risk Assessment

John Wambaugh, *U.S. E.P.A., Research Triangle Park, NC, United States*
Richard Judson, *U.S. Environmental Protection Agency, Research Triangle Park, NC, United States*
Barbara Wetmore, *Scitovation, Research Triangle Park, NC, United States*
Russell Thomas, *U.S. Environmental Protection Agency, Research Triangle Park, NC, United States*

Non-targeted screening of human biological samples indicates the presence in small quantities of thousands of chemicals that mostly have limited or no data on hazard, exposure, or toxicokinetics (TK). Thomas et al. (2013) proposed a pragmatic framework for risk-based and animal-sparing approaches to evaluating chemical safety. The tiers of this framework were guided by estimation of the margins between predicted human bioactive doses and estimates of exposure for the most sensitive portion of the population. The first tier integrates data from three key sources. The first source is high-throughput screening (HTS) in vitro assays, such as the U.S. Federal Tox21 consortium and the U.S. Environmental Protection Agency (EPA) Toxicity Forecaster (ToxCast) program, which have screened thousands of chemicals, in some cases across more than a thousand assay endpoints. These data separate chemicals based on their relative selectivity in interacting with biological targets and identify the concentration at which these interactions occur. The second source is toxicokinetic (TK) data that allow in vitro-to-in vivo extrapolation (IVIVE) from any activities observed in vitro to actual human dose rates (i.e., mg/kg bodyweight/day). These TK data are largely drawn from in vitro assays amenable to hundreds of chemicals, and have themselves been organized into tiers of confidence based on comparison to in vivo evaluation data. The third source of data is first-pass exposure predictions (mg/kg/day) that can be obtained from rapidly parameterized exposure models or from heuristic models that have been empirically calibrated to human biomonitoring data. A recent application of the first tier is the U.S. Endocrine Disruptor Screening Program (EDSP), which must consider approximately 10,000 chemicals for potential to impact function of human estrogen, androgen, steroid, and thyroid pathways. Coupled with a pathway-based model for integrating in vitro data, the EDSP is using Tier One testing to identify those chemicals most of interest for additional study. The second tier of the Thomas et al. (2013) framework involves short-term in vivo studies, expanded TK evaluations, and refined human exposure estimates. Only in the third tier are more traditional animal studies used to assess chemical safety. This tiered approach holds promise to significantly increase the efficiency in which chemical testing occurs. This abstract does not necessarily reflect U.S. EPA policy.

Mo-SY-F4.3

Exposure based testing for risk assessment

Theo Vermeire, *RIVM, Bilthoven, Netherlands*
Yuri Bruinen de Bruin, *RIVM, Bilthoven, Netherlands*

Exposure of humans or organisms in the environment is one of the decision elements in integrated testing strategies on which testing decisions are based. In other words: some tests may be waived if it can be shown that there are situations when human or environmental exposures are so low that there is a very low probability that the acquisition of additional effect information may lead to an improvement in the ability to manage risk (Exposure Based Waiving, EBW). This will reduce the number of experimental animals needed. In contrast, high exposures may trigger extra testing using experimental animals (Exposure Based Triggering, EBT). Extensive exposure knowledge through modelling or monitoring is essential for both aspects of exposure based testing. This applies to all relevant life cycle stages of a substance, from production to the waste stage. Based on such knowledge, exposure can judgedto be relevant or not. This concerns exposure of humans, directly via consumer products or at the workplace and indirectly exposed via the environment, and exposure of environmental organisms.
The presentation will report about investigations on how EBW and EBT can be used as a factor in testing strategies and how EBW can be used to reduce testing of experimental animals.

Mo-SY-F4.4

Expossure based testing, an industry perspective on current situation and where to fo next

Nicholas Ball, *The Dow Chemical Company, Horgen, Switzerland*
Heli M Hollnagel, *Dow Europe GmbH, Horgen, Switzerland*
Sean Hays, *Summit Toxicology, Lyons, CO, United States*

The concept for exposure driven testing requirements for regulatory submissions is not new. Under the EU REACH regulation, toxicity data requirements are driven primarily by the amount of a substance manufactured or imported per year (a crude proxy for exposure potential), with the theoretical possibility to avoid/reduce testing for substances where there is essentially no potential for exposure. Observations from the first years of REACH show that while legally permitted, exposure-based adaptations to the testing scheme are foreseen, regulatory practice does not use this approach for chemicals. For food contact materials (in the EU), toxicological assessment is driven by the potential for migration into food and subsequent human exposure. However while these frameworks allow for exposure based testing there is no harmonized, globally accepted approach for exposure based testing and often the possibility to reduce the need for data based on an exposure based argument is extremely difficult. Is there a way through this? What are the needs to develop a more consistent toxicological assessment of chemicals that uses the potential for exposure as the driver for data requirements? This presentation will provide a brief overview of the state of play from an industry perspective and present thoughts on how to push towards a true exposure based assessment paradigm.

Mo-SY-F4.5

Results of audience debate

Sean Hays, *Summit Toxicology, Lyons, CO, United States*

A real-time polling website (www.scipinion.com) will be used by the audience to respond to a predetermined set of charge questions written by the session organizers. The final presentation will provide the results of the poll and lead a discussion amongst the audience and speakers.

Mo-SY-G4: Advanced mass spectrometric techniques for the analysis of environmental organic contaminants

Mo-SY-G4.1

High resolution mass spectrometry provides novel insights into products of human metabolism of organophosphate and brominated flame retardants

Mohamed Abdallah, *University of Birmingham, Birmingham, United Kingdom*
Jinkang Zhang, *University of Birmingham, Birmingham, United Kingdom*
Mark Viant, *University of Bfrmingham, Birmingham, United Kingdom*
Kevin Chipman, *University of Birmingham, Birmingham, United Kingdom*
Kyle D'silva, *ThermosFisher Scientific, Bremen, Germany*
Maciej bromirski, *TherfmoFisher Scientific, Bremen, Germany*
Stuart Harrad, *University of Birmingham, Birmingham, United Kingdom*
Aristide Ganci, *University of Birmingham, Birmingham, United Kingdom*

Very little is known about the biotransformation of flame retardants in indoor dust by human liver cells. Furthermore, very little is known about the metabolic behavior of human liver cells upon concomitant exposure to multiple stressors which mimic the real life situation. In this study, Human HepG2/C3A cells were cultured in William's E medium supplemented with 5% heat-inactivated fetal bovine serum and incubated at 37°C with humidified air containing 5% CO2. HepG2/C3A cells were seeded into 6 well plates at density of 2 million cells/well and exposed to the extract of 12 mg NIST SRM2585 dust/well (relevant to the high end exposure of a 12.3 kg toddler ingesting 200 mg/day of indoor dust) for 24 hours. Samples were spiked with 13C-α, β, γ-HBCDs and TPP-d15 as internal standards prior to extraction with ethyl acetate:hexane (1:1 v/v) mixture according to a QUECHERS-based method using successive steps of vortex-mixing, ultra-sonication and centrifugation. Chromatographic analysis was achieved using a dual pump Ultimate 3000TM (ThermoScientific) UHPLC equipped with an Ultimate 3000TM XRS autosampler. Analyte separation was performed on Accucore RP-MS column (100 x 2.1, 2.6 µm, ThermoScientific) using a mobile phase of 1mM ammonium acetate (mobile phase A) and Methanol (mobile phase B). Identification of target analytes (HBCDs, TCEP, TCIPP and TDCPP) and their metabolites was performed on the ExactiveTM Plus OrbitrapTM mass spectrometer (Thermo Scientific, Bremen, Germany) using an ESI source operated in both +ve and -ve ionisation modes. The Automated gain control (AGC) was set to 3 x 106 ions at a mass resolution of 70,000 FWHM. Due to the lack of reference standards, the high resolution full scan mass spectra were used for identification of metabolites. The identity of potential metabolites were confirmed via all ion fragmentation (AIF) using the higher collisional dissociation (HCD) cell. The fragmentation patterns obtained from each metabolite provided additional information for structural elucidation. Further confirmation of metabolite structures were achieved via MS/MS analysis using the parent (nominal) mass from the Orbitrap full scan and the most predominant fragment obtained from the AIF analysis. Results revealed several hydroxylated, dehalogenated and conjugated metabolites for HBCDs, TCEP, TCIPP and TDCPP. More studies are required to investigate the toxic implications of these metabolites which may pose more risk to human health than their parent compounds.

This page intentionally left blank.

Mo-SY-G4.2

Untargeted identification of novel BFRs and their degradation/transformation products in environmental samples

Aristide Ganci, *University of Birmingham, Birmingham, United Kingdom*
Khanh Hoang Nguyen, *University of Birmingham, Birmingham, United Kingdom*
Leon Peters, *Uiversity of Birmingham, Birmingham, United Kingdom*
Thomas Moehring, *Thermo Fisher Scientific, Bremen, Germany*
Stuart Harrad, *University of Birmingham, Birmingham, United Kingdom*

Due to legislative restrictions on manufacture and use of some brominated flame retardants (BFRs), several new chemicals (NBFRs) have been developed. To explore their presence in different environmental compartments and ultimately understand their environmental fate, analytical methods for targeted analysis are required. Classically these compounds are determined by GC-based instrumental methods. In recent years, LC-based methods coupled to low resolution mass spectrometers have also been developed. Advances in high resolution mass spectrometry facilitate accurate measurements and identification of unknowns, as well as possible degradation and transformation products. Moreover, bromine isotopic pattern analysis and the use of mass defect plots and filters, helps identify relevant substances, with such techniques starting to be more commonly used in environmental science.
In this work the potential of high-resolution accurate mass (HR/AM) quadrupole Orbitrap benchtop technology will be exploited for targeted/untargeted detection and identification of NBFRs in environmental samples such as dust, soil and sediments, along with possible degradation and transformation products.
Extraction of samples was conducted using a Thermo Scientific™ Dionex™ ASE™ 350 accelerated solvent extractor and in-cell cleanup. The use of different solvent mixtures, including hexane, dichloromethane and acetone in the accelerated solvent extractor was investigated to obtain the optimal extraction results. Further in-cell clean up using silica and Florisil™ was performed to reduce matrix interferences. Final extracts were separated on a Thermo Scientific Accucore™ RP-MS 100x2.1mm, 2.6µm column on a Thermo Scientific UltiMate® 3000 HPLC system using a gradient elution program with water (mobile phase A) and methanol (mobile phase B) at a flow rate of 400 µl/min. A HPLC gradient elution program and APCI values were optimized based on the measurement of reference standard solutions. Samples were analyzed on a Q Exactive™ Plus mass spectrometer with an APCI source. Raw data files were processed using Thermo Scientific Compound Discoverer™ version 2.0 software. In addition, mass defect plots were created using Microsoft® Excel to visualize the presence of brominated compounds.
Initially, Full Scan experiments were conducted to obtain a general overview of the presence of compounds of interest in the samples. The use of high-resolution accurate mass (HRAM) instrumentation, together with powerful software tools, facilitates identification of targeted compounds and unknowns by means of selectivity, elemental compositions and isotopic pattern scoring. Later, confirmation of compounds was conducted using MS2 fragmentation spectra and measured reference standards where available. Several BFRs and NBFRs were identified in the samples.

Mo-SY-G4.3

Explore the potential of state of the art Mass Spectrometry and dedicated software in identification of NBFRs and their metabolite/degradation products

Khanh Hoang Nguyen, *University of Birmingham, Birmingham, United Kingdom*
Mohamed Abdallah, *University of Birmingham, Birmingham, United Kingdom*
Aristide Ganci, *University of Birmingham, Birmingham, United Kingdom*
Leon Peters, *University of Birmingham, Birmingham, United Kingdom*
Stuart Harrad, *University of Birmingham, Birmingham, United Kingdom*

In this research, we aim to explore the potential of a fast and reliable method to identify compounds of interest and their metabolites by a totally unknown approach using High Resolution Mass Accuracy (HRAM) data and a dedicated software. The method was applied to data from an in-vitro experiment.
Briefly, mouse liver microsomes (MLM) were pre-incubated with sterile Milli-Q water, William's E medium and EH-TBB at two concentrations (1 and 10 µM) for 10 minutes at 37oC. NADPH regenerating system was added to make a final volume of 1 mL in each well. The plate was then incubated at 37oC for 1 hour and stopped by ice-cold methanol. Samples were extracted by a proper procedure then analyzed by an UPLC-Orbitrap-HRMS system in full scan (-)APCI mode (Thermo Fisher Scientific, Bremen, Germany). Triplicate experiments were performed at both dosing levels together with non-enzymatic sample, heat-inactivated sample, solvent blank and EH-TBB standard.
Compound Discoverer 2.0 software (Thermo Fisher Scientific, Bremen, Germany) was used to interpret the data. Briefly, the software extracted spectra from input MS data then elucidated the element compositions for each peaks in every single files and grouped them based on retention time across all files. Further analysis nodes were used such as elemental composition prediction and bromine isotope pattern scoring. Finally, a "Differential Analysis" node was used to provide some simple differential statistic (PCA and ANOVA).
Overall, 1429 features were found in two groups: treated and untreated samples. ANOVA results showed 100 features with significant differences between two groups (adjusted-P < 0.1, confidence level 90%). Log2 fold changes of peak areas were then calculated. Among 100 significant features, there were 28 features with a positive log2 fold change meaning they have higher concentrations in treated samples. Finally, a bromine pattern scoring filter was applied, which resulted only 1 feature matched 4-Bromine pattern. The proposed ion composition was: [C8H4O3Br3]- which might derive from 2,3,4,5-tetrabromomethylbenzoate (C8H4O2Br4) by common (-)APCI ionization mechanism [M-Br+O]-. An increase in log2 fold change between 10 µM and 1 µM samples was also observed: 9.71 versus 2.59. This implied [C8H4O3Br3]- indeed a metabolite formed during in-vitro experiments.
In conclusion, by using HRAM and Compound Discoverer software, we were able to identify one metabolite of EH-TBB by MLM through an unknown approach. This method can be applied for any other compound and potentially real samples if LC/(-)APCI-Orbitrap library provided.

Mo-SY-G4.4

The comprehensive characterisation of diesel exhaust nanoparticles using variable ionisation mass spectrometry

Mohammed Alam, *University of Birmingham, Birmingham, West Midlands, United Kingdom*
Christopher Stark, *University of Birmingham, Birmingham, West Midlands, United Kingdom*
Roy Harrison, *University of Birmingham, Birmingham, West Midlands, United Kingdom*

Despite intensive research over the last 20 years, major questions remain concerning the composition of primary vehicle exhaust aerosol and its contribution to secondary organic aerosol (SOA) formation. These uncertainties relate especially to the semi-volatile component of the particles. Semi-Volatile Organic Compounds (SVOC) are compounds which partition directly between the gas and aerosol phases under ambient conditions. In engine exhaust the SVOC are typically hydrocarbons in the C15-C35 range and are largely uncharacterised, other than the homologous series of the n-alkanes. This is due to the drawbacks of monitoring techniques available, as the SVOC are unresolved by traditional gas chromatography and form a large hump in the chromatogram referred to as Unresolved Complex Mixture (UCM).
In this study, we exploit 2D Gas-Chromatography coupled to variable ionisation Time-of-Flight Mass-Spectrometry (GC×GC-ToF-MS) to characterise the composition of SVOC from diesel exhaust emission. Samples were collected from the exhaust of a diesel engine using both filter and impaction. The GC×GC-ToF-MS technique has been demonstrated capable of resolving specific components of the UCM, which typically makes up 95% of the area of chromatogram using conventional 1D separation. Samples were collected from the exhaust of a diesel engine with and without abatement devices fitted. Preliminary analyses indicate the separation of ~ 13,000 peaks, of which many homologous series' are identified. These include the homologous series' such as n-alkanes, alkyl-cyclohexanes, alkyl-cyclopentanes and aromatics; similar to both fresh lubricating oil and fuel. Although we can identify that many of the chromatographic peaks belong to a particular homologous series, it is often difficult to determine the positioning and degree of branching of many isomeric hydrocarbon compounds. Using variable electron energy ionisation, mass spectrometry, the positioning and degree of branching of isomeric hydrocarbon compounds are also identified and presented. We identify the mono-, di-, tri- and tetra- substituted methylalkanes and alkylcycloalkanes, which are in many cases more abundant in terms of mass than n-alkanes and n-alkylcycloalkanes.

Mo-SY-G4.5

The Thermo Scientific Q Exactive GC Orbitrap for Multi Dimensional Position Specific Isotope Analysis of Trace Organic Environmental Contaminants

Leon Peters, *Thermofisher Scientific and The University of Birmingham, Birmingham, United Kingdom*
Khanh Hoang Nguyen, *Thermofisher Scientific and The University of Birmingham, Birmingham, United Kingdom*
Stuart Harrad, *University of Birmingham, Birnimgham, United Kingdom*

The position specific stable isotopic composition (PSIA) of organic environmental contaminants holds the potential to categorically identify pollutant sources, understand global and local transportation processes and determine degradation pathways and kinetics.
Traditional isotopic measurements conducted using magnetic sector type isotope ratio mass spectrometers (IRMS) and analyte gasses derived by chemical conversion from the original sample of interest; do not retain the intra-molecular isotopic distributions critical for these applications. Further, even the most sensitive of conversion processes requires sample concentrations on the order of ng on-column, which for the most part, prohibits the isotopic measurement of trace-level organic environmental contaminants.
In an attempt to retain the isotopic integrity of analytes as well as to reduce limits of detection, methods using single quadrupole mass spectrometers (qMS) operating in selected ion monitoring mode (SIM) have been developed. These have been successful in describing the degradation kinetics and pathways of several organic contaminants, under laboratory conditions and in field studies of highly contaminated sites where sufficient concentrations for analysis could be extracted and purified. However, these approaches suffer in that they often require a separate compound-specific δ13C analysis to calculate offsets present as isobaric interferences at low mass resolution. Also these measurements require multiple injections and long integration times- increasing the mass of compound required for analysis.
This study aims to provide a comprehensive solution for PSIA of organic contaminants by high resolution-accurate mass (HR/AM) full scan analysis on the Thermo Scientific Q Exactive GC Orbitrap system. Methodology has been successfully developed which is capable of mass resolving isobaric interferences and providing position-specific isotopic data of organic moieties at trace concentrations (as low as 10 ppb).
This full scan technique is presented and discussed as a comparison against SIM-qMS and IRMS methodology for δ13C, δ37Cl and δD of Hexachlorocyclohexane gamma isomer (lindane) as well as δ81Br, δ13C and δD of several polybrominated diphenol ether (PBDEs) and brominated dioxin and furan (PBDD/Fs) congeners. Accuracy and precision of the measurements was found to be highly dependent upon mass spectrometric and chromatographic conditions, however showed reproducible accuracy of between 0.1 - 1.5‰ (δ37Cl of lindane) at resolutions ranging from 15 - 120k.

These results show for the first time the capability of the GC-Orbitrap platform to perform PSIA for a range of organic molecules at concentrations relevant to environmental conditions, enabling a new generation of applications in the field of environmental contaminants research.

Mo-SY-H4: Analysis of Patterns of Co-Exposure: Methodologies and Applications

Mo-SY-H4.1

Analyzing patterns of co-exposure in exposure space

Thomas Webster, *Boston University School of Public Health, Boston, MA, United States*

Aim: Exposure science plays a crucial role in the understanding of the health effects of mixtures: we need to know what mixtures populations are exposed to. Developments in analytical chemistry mean that large amounts of exposure data will be routinely generated. Analysis of the distribution of compounds in exposure space provides useful information. Large areas of this space will be empty and do not need to be investigated by toxicologists. Epidemiologists need to understand the patterns of co-exposure in order to consider confounding as well as colinearity and other statistical issues. Epidemiologists will also need to consider generalizability, including the similarity of patterns of correlations between populations.
Methods: Consider a p-dimensional space where each perpendicular axis denotes a specific exposure. We are interested in the distribution of data in this space and correlations. For example, exposures that are very highly correlated will form rays. 1) To examine correlations, Spearman correlation coefficients provide a robust statistic that does not require assumptions about data distributions. Correlation matrices provide one method for portraying such data, but this can become unwieldy when p is large. An alternative method constructs dendrograms ("family trees") using hierarchical clustering where relatedness depends on the correlation. We constructed such trees by a) defining distance (dissimilarity) as one minus the absolute value of the Spearman correlation coefficient; b) constructing the tree using average linkage. 2) A second problem involves comparison of patterns of correlations between populations. One can compare dendrograms (e.g., via cophenetic correlation) or the underlying correlation matrix. The latter can be readily done via the Mantel statistic: compute the Spearman correlation of the elements in the lower triangular forms of the dissimilarity matrices. P-values are computed via a permutation test (as elements are not independent). We illustrate these procedures using serum concentrations of a set of persistent organic pollutants from a Boston cohort, comparing the pattern of correlations with NHANES.
Results and Conclusions: In the Boston cohort, serum concentrations showed two major clusters: PentaBDEs and other compounds: organochlorine pesticides (OCs) and PCBs. The latter contained two main subclusters, lower and higher molecular weight PCBs. The 2003-4 NHANES measured the PCBs in a separate sample from the other compounds. The comparison of the Boston data and NHANES was therefore restricted to the PBDEs and OCs. The Mantel test yielded an overall correlation of 0.7 ($p=0.002$), indicating a strong similarity between the two.

Mo-SY-H4.2

Co-variation in circulating levels of 45 environmental contaminants from different chemical classes in a human population.

Lars Lind, *Uppsala University, Uppsala, Sweden*
Erik Lampa, *Uppsala University, Uppsala, Sweden*
Samira Salihovic, *Uppsala University, Uppsala, Sweden*
Bert van Bavel, *Örebro University, Örebro, Sweden*
Monica Lind, *Uppsala University, Uppsala, Sweden*

Aim: When evaluating effects of mixtures in humans it is essential to know to what extent high levels of environmental contaminants co-exist in he individual. If so, some selected contaminants could be used as markers of exposure. The aim of the present study is to investigate co-variation of multiple contaminants in the circulation in a large population-based sample.
Method: In almost 1,000 subjects in the Prospective Study of the Vasculature in Uppsala Seniors (PIVUS) study (50% women, all aged 70 years), 45 different environmental contaminants from different classes (PCBs, dioxin, brominated flam retardant, pesticides, BPA, phthalates, PFAS, metals) were measured. Hierarchical cluster analysis and principal component analysis were used.
Results: Except for one cluster of low-chlorinated PCBs (congeners 74, 99, 105, 118, 138, and 153) and one cluster of highly chlorinated PCBs (congeners 156, 157, 170, 180, 189, 194, 206, and 209), and one cluster consisting of some of the PFAS, no major co-variation was seen in the investigated contaminants.
Conclusion: Except for some PCBs, and some PFAS, very little co-variation was seen amongst a large number of environmental contaminants from different classes. Thus, a large number of contaminants have to be measured on order to evaluate mixture effects in humans.

Mo-SY-H4.3

Development of correlation globes to map out environment-wide associations and to determine the multiplicity burden of association tests

Chirag Patel, *Harvard Medical School, Boston, MA, United States*

The phenomenon of environmental exposure is complex and humans are not exposed to one or a handful factors but potentially hundreds factors throughout their lives. The exposome, the totality of exposures encountered from birth, is hypothesized to consist of multiple inter-dependencies, or correlations, between individual exposures. These correlations may reflect how individuals are exposed. Currently, we lack methods to comprehensively identify robust and replicated correlations between environmental exposures of the exposome, the comprehensive battery of exposures encountered from birth to death. Further, we have not mapped how exposures associated with disease identified by environment-wide association studies (EWAS) are correlated with other exposures. To this end, we implement methods to describe a first "exposome globe", a comprehensive display of replicated correlations between individual exposures of the exposome. First, we describe overall characteristics of the dense correlations between exposures, showing that we are able to replicate 2,656 correlations between individual exposures of 81,937 total considered (3%). We document the correlation within and between broad a priori defined categories of exposures (e.g., pollutants and nutrient exposures). We also demonstrate utility of the exposome globe to contextualize exposures found through two EWASs in type 2 diabetes and all-cause mortality, such as exposure clusters putatively related to smoking behaviors and persistent pollutant exposure. The exposome globe construct is a useful tool for the display and communication of the complex relationships between exposure factors and between exposure factors related to disease status. We will demonstrate open-source tools to estimate and visualize exposome globes and their role in deciphering associations between exposures and phenotype.

Mo-SY-H4.4

Identifying Robust Co-Occurrence Patterns in Personal Care Product Purchases

Rogelio Tornero-Velez, *U.S. EPA, Research Triangle Park, NC, United States*
Kristin Isaacs, *U.S. EPA, Research Triangle Park, NC, United States*
Michael Nye, *U.S. EPA, Washington DC, DC, United States*
Paul Price, *U.S. EPA, Research Triangle Park, NC, United States*
Timothy Buckley, *U.S. EPA, RTP, NC, United States*

Personal care products (PCPs) are used for beautification and personal hygiene, and because they are applied to the skin, hair, and mouth, they provide an efficient delivery vehicle for chemicals into our bodies. Although efforts have been made to enumerate the chemicals in individual PCPs and understand their health risks, little is known on the combined chemical exposures and risks that occur from the co-use of PCPs. To address this need, we employed association rules mining, a method with origins in market basket analysis, to assess patterns of co-occurrence in PCP purchases. We applied this method to an anonymized database of consumer product transactions for sixty thousand households collected over one year, provided by a major market research firm. PCP categories included hair care, oral hygiene, cosmetics, soap and bath products, fragrances, and toiletries. To identify co-occurring sets of products, we applied a mathematical technique, the Apriori algorithm, which finds nested combinations of increasing order as long as all satisfy a minimum 'support' (occurrence) threshold. We further examined robustness of co-occurrence patterns by demographic variables. Identifying assemblages of products is a prerequisite to assessing risks from the chemicals in multiple products. We demonstrate that this approach is an efficient framework to consider consumer product co-occurrence to inform chemical exposure and risk assessment. This abstract does not necessarily reflect U.S. EPA policy.

Mo-SY-H4.5

Quantifying Associations between Environmental Stressors and Demographic Factors

Hongtai Huang, *Environmental Protection Agency (ORISE), Durham, North Carolina, United States*
Rogelio Tornero-Velez, *U.S. Environmental Protection Agency, Research Triangle Park, North Carolina, United States*
Timothy Barzyk, *U.S. Environmental Protection Agency, Research Triangle Park, North Carolina, United States*

Association rule mining (ARM) [1, 2], also known as frequent item set mining [3] or market basket analysis [2], has been widely applied in many different areas, such as business product portfolio planning, intrusion detection infrastructure design, gene expression analysis, medical diagnosis, and drug prescription pattern. In recent years, ARM has also been used to analyze relationships between environmental stressors and adverse human health effects [4, 5].

We employed ARM to identify and quantify associations within and between ambient pollutants (environmental stressors) and demographic factors such as age, poverty, race/ethnicity, and education attainment. Specifically, we linked the 2011 NATA (National-Scale Air Toxics Assessment) U.S. Census tract-level air pollutant exposure concentration data with the 2010-2014, 5-Year Summary Files in the American Community Survey (ACS), and created relevant chemical and demographic variables. Association rules were generated based on the merged data (NATA Data and ACS 5-Year Summary Files) and filtered with specific criteria or measurements to enhance understanding of the relationships between multiple chemical stressors and socio-demographic factors. We also utilized a graph-based visualization tool [6] to depict the interacting relations among all the stressors or factors that play active roles in the resultant rules. Our main aim is to demonstrate the ability of using unsupervised data mining methods to identify associations among multiple stressors (e.g., to find the underlying structure of and the relationship[s] between the stressors), which can be useful for assessment of co-exposure to chemical and non-chemical stressors, and informative for public health decision-making, especially when it comes to addressing environmental justice issues and social disparities.

References:

1. Agrawal, R., T. Imieliński, and A. Swami. Mining association rules between sets of items in large databases. in ACM SIGMOD 1993.
2. Hastie, T., R. Tibshirani, and J. Friedman, The elements of statistical learning: data mining, inference and prediction. 2005, Springer.
3. Borgelt, C., Frequent item set mining. Wiley Interdisciplinary Reviews: Data Mining and Knowledge Discovery, 2012. 2(6): p. 437-456.
4. Bell, S.M. and S.W. Edwards. Building associations between markers of environmental stressors and adverse human health impacts using frequent itemset mining. in Society for Industrial and Applied Mathematics (SIAM) International Conference on Data Mining. 2014.
5. Bell, S.M. and S.W. Edwards, Identification and Prioritization of Relationships between Environmental Stressors and Adverse Human Health Impacts. Environ Health Perspect, 2015. 123(11): p. 1193-9.
6. Hahsler, M. and S. Chelluboina, Visualizing association rules: Introduction to the R-extension package arulesViz, in R project module. 2011. p. 223-238.

Mo-PL-I4: VOCs and SVOCs

Mo-PL-I4.1

Use of Indoor Dust Levels to Reconstruct Exposure to Semivolatile Organic Compounds: Evaluation with NHANES biomarkers

Hyeong-Moo Shin, *University of California-Davis, Davis, California, United States*
Thomas McKone, *Lawrence Berkeley National Laboratory/University of California-Berkeley, Berkeley, California, United States*
Deborah Bennett, *University of California-Davis, Davis, California, United States*

Background/Aims: House dust can serve as a marker of exposure. Many studies have found that polybrominated diphenyl ether (PBDE) concentrations in indoor dust are correlated with concentrations in biological samples. However, the role of indoor dust levels on predicting human body burden for other chemical classes remains mostly unexplored. Our objective is (1) to quantify exposures from indoor dust levels for a suite of semivolatile organic compounds (SVOC) with indoor sources and (2) to better understand how much indoor dust levels can inform human body burden depending on chemical properties and indoor source strength.
Methods: Compounds of interest include PBDEs, phthalates, polycyclic aromatic hydrocarbons (PAHs), and bisphenol A (BPA). We couple measured levels in indoor dust with the partitioning relationships among the gas phase, airborne particles and settled dust to estimate concentrations in the gas phase and airborne particles. We model intake rates (iR) using the exposure concentrations (i.e., dust levels, gas phase and airborne particles) and recommended exposure factors (e.g., inhalation rate, dust ingestion rate). We then compare the modeled total iRs with those inferred from the measured urine or blood levels in the NHANES survey.
Results: The contribution of individual exposure pathways (e.g., inhalation, dermal uptake, dust ingestion) to the total iRs is similar to the results in previous studies. For compounds with large octanol-air partition coefficients (log Koa >9), the primary exposure pathway is dust ingestion (>70%), but our methods underestimate the total iRs (only 2 to 16% of the total iRs from biomonitoring data). It is likely that exposure for these compounds is driven by non-indoor exposure pathways or that the gas-phase concentrations are not equilibrated with the settled dust levels. We also found that our results are very sensitive to the Koa, which involves in partitioning relationships between the gas phase and the particle phase as well as between the gas phase and settled dust.
Conclusion: The results from this study indicate that reconstructed exposure from measured indoor dust levels and estimated gas- and particle-phase concentrations can better inform human body burden than that from the dust levels only, especially for SVOCs with small Koa. Our methods can provide reliable exposure estimates in a high-throughput manner when biomarker measurements are lacking but indoor dust levels are known.

Mo-PL-I4.3

Reactive and Water-Soluble Organic Gases inside Several Residences in New Jersey and North Carolina

Sara Duncan, *Rutgers University, New Brunswick, New Jersey, United States*
Barbara Turpin, *University of North Carolina, Chapel Hill, North Carolina, United States*
Kenneth Sexton, *University of North Carolina, Chapel Hill, North Carolina, United States*
Ron Lauck, *Rutgers University, New Brunswick, NJ, United States*

Volatile organic compounds (VOCs; alkanes, alkenes, aromatics) are frequently found at much higher concentrations inside than outside of homes. However, while some water-soluble oxidized VOCs have been measured indoors, the concentration and composition of water-soluble organic gases (WSOG) in indoor spaces are poorly understood. We hypothesize that WSOGs also exist in higher concentrations indoors than outdoors and that many water-soluble gases are present indoors that have not yet been identified. These compounds can be directly emitted from sources (e.g., cooking) and formed when VOCs are oxidized in indoor air. In damp homes, liquid water (on surfaces, in wet particles, and in the respiratory tract of occupants) is a sink for WSOG and could possibly be an important medium for chemistry, altering dermal and inhalation exposures indoors. WSOGs were collected inside and outside of 13 homes in central New Jersey and the Triangle region of North Carolina in summer and fall, 2015 using four mist-chamber sampling devices in parallel (two sampling indoors and two sampling outdoors). These devices scrubbed water-soluble gases out of the air into water with an air flow rate of 25 liters per minute and a refluxing collection volume of 25 milli-liters of water. In each home, two sequential 2-hour samples were collected and composited. Total organic carbon (TOC) concentrations in the indoor samples were 6 - 27 times higher than concurrent outdoor concentrations (542 to 1,387 micro-molar carbon indoors; 28 to 107 micro-molar carbon outdoors). On average, at least 86 percent of water-soluble organic carbon collected indoors originated from indoor sources and indoor formation, rather than from outdoor-to-indoor transport (This lower-bound estimate assumes outdoor-to-indoor transport was 100 percent efficient).
Samples were reacted with OH radicals in water to identify collected WSOGs that are particularly reactive in the aqueous phase. Elemental composition and structural information for reactive WSOGs were provided by electrospray ionization mass spectrometry (ESI-MS) (soft ionization, 40 V fragmentor voltage) and MS-MS. In each home, several compounds were reactive with OH radicals, including reduced nitrogen compounds and polyols. This work will help to enable future investigation into how chemistry in liquid water indoors may alter indoor gas-phase chemical makeup in homes and consequent exposure risks.

Mo-PL-I4.4

Comparison of three biomarkers for benzene exposure during turnaround works and derivation of an assessment value for urinary benzene

Michael Bader, *BASF SE, Ludwigshafen, Germany*
Thomas Jaeger, *BASF SE, Ludwigshafen, Germany*
Sandra Baecker, *BASF SE, Ludwigshafen, Germany*
Tom Van Weyenbergh, *BASF Antwerpen N.V., Antwerpen, Belgium*
Elke Verwerft, *BASF Antwerpen N.V., Antwerpen, Belgium*
Thomas Schettgen, *RWTH Aachen, Aachen, Germany*
Thomas Kraus, *RWTH Aachen, Aachen, Germany*

Benzene is one key compound for exposure monitoring during turnaround works in chemical production plants such as steamcrackers and aromatic hydrocarbon synthesis. At two large industrial sites in Germany and Belgium, benzene biomonitoring is carried out routinely during turnarounds. While exposure is generally low and well-controlled, the results of the biomonitoring programs can be used to compare the validity and applicability of different biomarkers. In particular, urinary benzene is currently under discussion, and one aim of the investigations was to evaluate the correlation between urinary benzene and the established biomarkers trans,trans-muconic acid (ttMA) and S-phenylmercapturic acid (SPMA).
Several turnaround campaigns since 2006 in chemical plants with potential exposure of the maintenance workers to benzene were monitored. ttMA and SPMA as well as urinary benzene were analyzed in parallel, according to procedures recommended by the German Research Foundation (DFG). The parameters were continuously certified by successful participation in the German External Quality Assessment Scheme (G-EQUAS). The limits of detection were 50 µg/L for ttMA, 1 µg/L for SPMA, and 0.02 µg/L µg/L for urinary benzene. The Biological Exposure Indices (BEI) of the American Conference of Governmental Industrial Hygieinists (ACGIH) were used as internal action values: 500 µg ttMA/g creatinine, 25 µg SPMA/g creatinine (no assessment value available for urinary benzene).
In four campaigns, altogether 1003 samples were collected from 215 employees, and analyzed for at least two biomarkers. In two campaigns, all three biomarkers were analyzed. Additionally, 79 urine samples from employees without occupational exposure to benzene were analyzed as controls. Action value excursions were observed in less than 5 % of the samples. The median values were up to 210 µg/g creatinine for ttMA, 1.7 µg/g creatinine for SPMA and 0.6 µg/L for urinary benzene. All three biomarkers showed close correlations between each other, and the relation between the BEI values (500 µg ttMA/g creatinine vs. 25 µg SPMA/g creatinine) was confirmed. At that level, the urinary benzene concentration is 4.5 µg/L.
The biomonitoring programs have shown the very low overall exposure to benzene during turnaround works in chemical plants, but they provided nevertheless a solid database for a comparison of the three biomarkers. Urinary benzene shows a good correlation to the established biomarkers ttMA and SPMA, it is relatively easy to analyze, and an action value of 5 µg/L, corresponding to 0.6 ppm benzene in air, could be derived.

Mo-PL-I4.5

Bioaccessibility of semi-volatile organic compounds (SVOCs) in settled dust

Gaelle Raffy, *EHESP - School of public health / Irset Inserm UMR 1085- Research Institute for Environmental and Occupational Health, Rennes, France*
Fabien Mercier, *EHESP - School of public health / Irset Inserm UMR 1085- Research Institute for Environmental and Occupational Health, Rennes, France*
Philippe Glorennec, *EHESP - School of public health / Irset-Inserm UMR 1085- Research Institute for Environmental and Occupational Health, Rennes, France*
Corinne Mandin, *CSTB - Scientific and technical center for building / OQAI - French indoor air quality observatory, Rennes, France*
Barbara Lebot, *EHESP - School of public health / Irset-Inserm UMR 1085- Research Institute for Environmental and Occupational Health, Rennes, France*

Humans are exposed to a wide range of indoor chemical pollutants including semi-volatile organic compounds (SVOCs), which are suspected of adverse health effects such as reprotoxic and neurotoxic effects. Dust ingestion is a non-negligible pathway of human exposure to several of these compounds. To improve this human exposure assessment, it is necessary to consider the bioaccessibility of SVOCs, i.e. the fraction of pollutants released in the digestive tract following the ingestion of dust. The present work reviews the literature for the methods, measured values, and influencing factors related to the bioaccessibility of SVOCs in indoor dust.
Reported bioaccessibility measurement methods simulate the gastrointestinal tract, with dust samples being successively submitted to synthetic gastric and intestinal fluids. Models were sometimes extended to include the role of saliva or colon. Milk proteins, Tenax® beads or Caco-2 cells could also be added for a better physiological relevance.
So far, SVOC bioaccessibility in dust has not been well documented in the scientific literature. However, the available articles show that measured bioaccessibilities ranged from < 20% for bromodiphenylether (BDE) 209, polycyclic aromatic hydrocarbons (PAHs) or high molecular weight phthalates, up to > 60% for organophosphorous flame retardants (OPFRs), low molecular weight polybromodiphenylether (PBDEs), hexabromocyclododecanes or tetrabromobisphenol A.
SVOC bioaccessibility is influenced by several factors related to the matrix. An increase of organic carbon content was often associated with a decrease of bioaccessibility: this was observed for PBDEs or the more hydrophobic OPFRs and phthalate esters, but no effect was noticed on the less hydrophobic ones. A decrease of dust particle size or an increase of dust pore volumes lead to a larger specific surface area which was linked to an increase of bioaccessibility.
SVOC bioaccessibility was also influenced by factors related to the compound itself. For example, substances with higher octanol-water partition coefficients (Kow) were less bioaccessible. For PBDEs, congeners that had integrated dust by adsorption were more bioaccessible than BDE 209 whose presence in dust is suspected to originate from material abrasion. Bioaccessibility was not influenced by the pollutant's concentration.
Bioaccessibility data are useful for a better quantification of SVOC exposure doses, which could otherwise be overestimated when the pollutant's total concentration is considered. However, more studies are needed, which should be compared to in-vivo studies for validation.

Poster sessions Monday October 10, 2016

Biomonitoring

Mo-Po-01

Measurement of Urinary Phthalate Metabolites in a Pilot Study of Nail Salon Workers and Comparison to a Sample of the U.S. Population

Julia Varshavsky, *UC Berkeley, Berkeley, United States*
Suhash Harwani, *Cancer Prevention Institute of California, Berkeley, United States*
Martin Snider, *Department of Toxic Substances Control, Berkeley, United States*
Syrago-Styliani Petropoulou, *Department of Toxic Substances Control, Berkeley, United States*
June-Soo Park, *Department of Toxic Substances Control, Berkeley, United States*
Myrto Petreas, *Department of Toxic Substances Control, Berkeley, United States*
Peggy Reynolds, *Cancer Prevention Institute of California, Berkeley, United States*
Tuan Nguyen, *State Compensation Insurance Fund, Berkeley, United States*
Thu Quach, *Cancer Prevention Institute of California, Berkeley, United States*

California's nail care industry is rapidly growing, with over 100,000 registered nail technicians in the state. Many are low-income Vietnamese immigrants who have limited access to health care, labor protections, and chemical health and safety information. Because they use phthalate-containing products on a regular basis, we predicted these nail salon workers to be a uniquely exposed and vulnerable occupational group. To characterize phthalate exposure in this subpopulation, we collected post-shift urine samples from 17 Vietnamese American workers at six Bay Area nail salons. We analyzed the urine samples for four primary phthalate metabolites: monoethyl-, monoisobutyl-, monobutyl- , and mono-(2-ethylhexyl)-phthalates (MEP, MiBP, MBP, and MEHP, respectively) using a modified method we validated from existing literature. We then compared our findings to the 2011-2012 Asian American National Health and Nutritional Examination Survey (NHANES) population. Nail salon worker geometric means (GMs) for each phthalate metabolite were higher than Asian American NHANES population averages, and some of the differences were statistically significant. We found MBP, MiBP, MEHP, and MEP to be 2.5 ($p=0.0003$), 1.6 ($p=0.015$), 2.6 ($p<0.0001$), and 0.25 ($p=0.4445$) times higher in nail salon workers compared to NHANES. Additionally, our results show that some workers are exposed to very high phthalate levels, well above the NHANES 95th percentile. This pilot study provides suggestive evidence that nail salon workers may be disproportionately exposed to certain phthalates, warranting further investigation into these findings.

Mo-Po-02

Measurement of urinary environmental chemicals in a convenience sample of 3 to 5 year old American children: a pilot study for NHANES

Mary Mortensen, *U.S. Centers for Disease Control, Atlanta, GA, United States*
Kathleen Caldwell, *Centers for Disease Control and Prevention, Atlanta, GA, United States*
Maria Morel-Espinosa, *Centers for Disease Control and Prevention, Atlanta, GA, United States*
Benjamin Blount, *Centers for Disease Control and Prevention, Atlanta, GA, United States*
Robert Jones, *Centers for Disease Control and Prevention, Atlanta, GA, United States*
Lanqing Want, *Centers for Disease Control and Prevention, Atlanta, GA, United States*
Christine Pfeiffer, *Centers for Disease Control and Prevention, Atlanta, GA, United States*

Background: The National Health and Nutrition Examination Survey (NHANES) collects data and biospecimens from Americans aged one year and older to evaluate their health and nutritional status. A spot urine obtained from participants ages 6 years and older is used to assess exposure to environmental chemicals; information is scarce for children younger than 6.
Methods: NHANES developed a pilot study to collect urine from children 3-5 years old during 4 months of a 2-year cycle. We analyzed the urine for metals, iodine, perchlorate, nitrate, thiocyanate, phytoestrogens, and several novel tobacco exposure biomarkers using mass spectrometry. The standard methodology for analyzing urine from children 6-11 was applied to the urine samples from children 3-5 years.
Results: Approximately 120 children provided urine. We measured 21 metals and iodine; perchlorate, nitrate, and thiocyanate (anions); 6 phytoestrogens; 5 tobacco-specific N-nitrosamines; and 6 volatile N-nitrosamines. We detected 16 metals, iodine, 3 anions, 6 phytoestrogens, and the tobacco biomarker NNAL in more than 50% of the children. Concentrations spanned several orders of magnitude, depending on the biomarker and were largely within the ranges reported previously for NHANES participants 6-11 years old.
Conclusion: based on a sample of children 3-5 years old, we detected 21 metals and iodine, 3 anions, and 6 phytoestrogens in most of these children. Most tobacco biomarkers were not detected, with the exception of NNAL, which was detected in more than 60% of children and suggestive of second-hand smoke exposure.

Mo-Po-03

Urinary 1-nitropyrene metabolites as markers of exposure to diesel exhaust in an underground mine.

Christopher Simpson, *University of Washington, Seattle, WA, United States*
Joemy Ramsay, *University of Washington, Seattle, WA, United States*
Erin Riley, *University of Washington, Seattle, WA, United States*
Dale Stephenson, *Boise State University, Boise, ID, United States*
Terry Spear, *Montana Tech, Butte, MT, United States*
Michael Paulsen, *University of Washington, Seattle, WA, United States*
Emily Zamzow, *University of Washington, Seattle, WA, United States*
Emily Carpenter, *University of Washington, Seattle, WA, United States*
Lianne Sheppard, *University of Washington, Seattle, WA, United States*
Noah Seixas, *University of Washington, Seattle, WA, United States*

Elevated exposure to diesel exhaust (DE) is widespread and has been linked to adverse health outcomes including respiratory irritation, cardiovascular disease, immune dysfunction and lung cancer. Underground miners experience amongst the highest exposures to DE of any occupation. Thus, miners are at high risk for suffering adverse health effects associated with DE exposure. MSHA currently mandates measurement of elemental carbon (EC) and total carbon (TC) to assess workers exposures to DE in underground metal/nonmetal mines. However, limitations in the specificity and reliability of these metrics of DE exposure hamper quantitative evaluation of links between DE exposure and adverse health outcomes. The DE-specific chemical 1-nitropyrene (1-NP) has been suggested as a potential alternative marker of exposure to DE, and 1-nitropyrene metabolites in urine may serve as useful biomarkers of exposure to DE.
In the current study we measured DE exposures in a cohort of 20 workers at a large underground metal mine. Diesel powered equipment at this mine uses a B70 biodiesel blend fuel. Full shift personal air samples were collected on up to eight occasions from each worker using an MSHA compliant SKC DPM impactor downstream of a GS-1 cyclone pre-filter. 103 of these samples were analyzed for EC and 1-NP. A total of 535 urine samples were collected pre- and post shift from the workers - of which 170 were analyzed for two specific metabolites of 1-NP - 6-hydroxy-1-nitropyrene (6-OHNP) and 8-hydroxy-1-nitropyrene (8-OHNP). The geometric mean (GM) metabolite levels were 0.014 pg/mg creatinine for 6-OHNP and 0.007 pg/mg creatinine for 8-OHNP. Metabolite levels were lowest in the pre-shift sample on the first day of the work week, and increased progressively throughout the work week. Consistent with this observation, metabolite levels did not show a significant association with personal exposure to 1-NP in the single work-shift preceeding collection of the urine sample. However, urinary metabolite concentrations were associated with cumulative 1-NP exposure across the four days prior to collection of the urine sample.
These data indicate that in this workplace where DE is anticipated to be the only source of 1-NP, urinary metabolites of 1-NP show promise as a biomarker of occupational exposure to DE.

Cross-week variation in metabolite levels

Mo-Po-04

Metabolomic Indicators of Primary Traffic Exposures in the Dorm Room Inhalation to Vehicle Emissions (DRIVE) Study

Donghai Liang, *Emory University, Atlanta, Georgia, United States*
Rachel Golan, *Ben-Gurion University of the Negev, Givat Brenner, Israel*
Jennifer Moutinho, *Georgia Institute of Technology, Atlanta, Georgia, United States*
Tianwei Yu, *Emory University, Atlanta, Georgia, United States*
Chandresh Ladva, *Emory University, Atlanta, Georgia, United States*
Roby Greenwald, *Georgia State University, Atlanta, Georgia, United States*
Rodney Weber, *Georgia Institute of Technology, Atlanta, Georgia, United States*
Stefanie Sarnat, *Emory University, Atlanta, Georgia, United States*
Armistead Russell, *Georgia Institute of Technology, Atlanta, Georgia, United States*
Vishal Verma, *University of Illinois at Urbana-Champaign, Urbana, Illinois, United States*
Dean Jones, *Emory University, Atlanta, Georgia, United States*
Jeremy Sarnat*Emory University, Atlanta, Georgia, United States*

Introduction. Traffic pollution health studies increasingly focus on identification of sensitive, biologically-relevant indicators of exposure and response. Environmental metabolomics, where metabolites associated with endogenous and exogenous processes can be quantitated, holds promise as a powerful tool to improve internal exposure estimation to complex air pollution mixtures, including primary traffic emissions. Methods. We conducted a panel-based study to measure an extensive suite of pollutants at ambient and indoor sites ranging from 0.01 to 2.3 km away from a major highway artery. In addition, 54 students living in dormitories either near (20 m) or far (1.4 km) from the highway conducted personal sampling and contributed weekly biomonitoring (plasma and saliva). Plasma and saliva metabolites were analyzed using high-resolution mass spectrometry (Q Exactive Highfield Orbitrap). We used targeted and untargeted metabolomics-wide association analyses to examine associations between primary traffic and corresponding metabolomics profiles in the panel. Results. Ambient traffic pollutants level were substantially higher outside and inside the near dorm as compared to the far dorm. Exposures to traffic pollution were different between students living in the near and far dorms. GIS analyses indicated that, the mean time-weighted distance to the traffic hotpot for the near dorm participants was 0.5 km, which contrast with 2.2 km for participants living in the far dorm. Despite this, personal PM2.5 exposures was comparable in participants from both dorms (8.3 and 7.5 ug/m3, respectively). A total of 20,766 metabolites were reliably extracted from plasma samples and 29,013 from saliva samples. Linear random effects models were conducted to examine associations between metabolite intensity (relative concentration) and student dorm location, controlling for multiple covariates. In all, 221 metabolites were robustly identified and significantly different in the near dorm metabolic profiles compared to those in far dorm ($p < 0.05$, Benjamini-Hochberg FDR correction). Untargeted functional analyses indicate that 30% of these metabolites significantly associated with 19 reliable modules and 21 known pathways, of which include redox biology, the carnitine shuttle, and oxidative stress. Conclusions. This study is among the first to examine the metabolic response to complex traffic exposures. Comprehensive pathway analysis and chemical validation is currently being conducted to identify specific metabolite patterns and further develop biologically-relevant indicators to primary traffic exposures for panel-based epidemiologic studies.

Mo-Po-05

Exposure assessment of multiple chemicals starting from biomonitoring data

Evangelos Handakas, *Aristotle University of Thessaloniki, Thessaloniki, Greece*
Dimosthenis Sarigiannis, *Aristotle University of Thessaloniki, Thessaloniki, Greece*
Spyros Karakitsios, *Aristotle University of Thessaloniki, Thessaloniki, Greece*
Alberto Gotti, *Aristotle University of Thessaloniki, Thessaloniki, Greece*

Aim: The current study aims at the estimation of external and target tissue exposure to 15 different chemicals, including both rapidly (BPA, DEHP, triclosan) and non-rapidly (PCBs, BDEs, HCB, DDT) metabolized compounds, starting from human biomonitoring (HBM) data.
Methods: The simulations were carried out in the INTEGRA platform, a software that provides realistic exposure scenarios coupled with a generic physiologic based bio-kinetic (PBBK) model and numerical "reversal" techniques for exposure reconstruction. The exposure reconstruction algorithm is based on the Markov chain Monte Carlo and dynamic evolution Monte Carlo techniques. The process starts from ancillary exposure-related data that are fed into the exposure model taking into account multiple exposure routes. The results are evaluated against the biomonitoring data distributions, aiming at the reduction of uncertainty in back-calculating doses, by minimizing the error between the predicted and the actual biomonitored data. Parameterization of the model for a large chemical space is facilitated by quantitative structure-activity relationship (QSAR) models. HBM data were obtained from cohort and biomonitoring studies from Mediterranean Countries. Ancillary exposure parameters were obtained from the INTEGRA database. The study focused on perinatal and childhood exposure.
Results: The results showed that the predicted intake dose is commensurate with intake estimates found in literature for both short and long term exposure scenarios of the European population. In all cases external intake estimates (e.g. for BPA 0.4 μg/kg_bw/d) were significantly lower than the respective tolerable daily intake (TDI). The estimated internal dose and the respective concentration in breast milk of BPA, DEHP and triclosan was very low because of their rapid metabolism. Similar were the results on exposure of neonates and infants. On the contrary, fetuses and newborns are highly exposed to POPs through trans-placental transfer during pregnancy and through maternal milk during lactation. Despite the fact that use of these chemicals is regulated, people are still exposed to low doses due to their environmental persistence and continuous transfer through the food web.
Conclusions: Exposure reconstruction offers unique capabilities for the utilization of the continuously growing amount of available biomonitoring data in Europe and the world. In this way, biomonitoring data can be mechanistically linked to both external and internal exposure, effectively supporting the screening and prioritization process for assessing chemical risk.

Mo-Po-06

CDC's National Biomonitoring Program: State Biomonitoring 2016 update

Lovisa Romanoff, *Centers for Disease Control and Prevention, Atlanta, GA, United States*
Kristin Dortch, *Centers for Disease Control and Prevention, Atlanta, GA, United States*
Antonia Calafat, *Centers for Disease Control and Prevention, Atlanta, GA, United States*

Biomonitoring can provide unique and valuable information on human exposure to environmental compounds by measuring chemicals or their breakdown products in people's blood or urine. The Centers for Disease Control and Prevention (CDC) uses biomonitoring to conduct an ongoing assessment of the U.S. population's exposure to environmental chemicals; however, CDC's biomonitoring data do not provide exposure information by specific state or locality.
Since 2003, CDC has granted competitive funding awards to selected states to increase the capability and capacity of state public health laboratories to a) conduct high-quality biomonitoring science and b) use biomonitoring to assess chemical exposures of concern in their communities. States use funding to purchase laboratory equipment and supplies; hire and train specialized staff; and conduct fieldwork and data analysis, while CDC provides training and technology transfer, quality assessment services, and technical assistance to awardees. Currently, nine states receive funding for projects not limited to assessing exposures to: multiple environmental chemicals in a state-wide survey; arsenic from geographic sources in people with private wells; combustion products in firefighters; and perfluoroalkyl substances in communities located near industrial sites.
Since their inception, the CDC-supported state programs have increased analytical capabilities for state laboratories conducting exposure assessment to environmental chemicals, expanded state population-based biomonitoring surveillance, and completed studies of exposure assessment in target populations. Other successes include increased collaboration between biomonitoring programs; advances in communication resources for data reporting to study participants; increased capability to identify emerging chemicals of concern; and increased harmonization of laboratory analyses used for biomonitoring efforts.

Environmental/Human Health

Mo-Po-07

Risk assessment of dietary exposures to aflatoxin for corn tortilla consumption in Veracruz city

Hiram Alejandro Wall Martínez, *Université de Bretagne Occidentale, Brest, France*
Nicolas Dornic, *Université de Bretagne Occidentale, Brest, France*
Anne Sophie Ficheux, *Université de Bretagne Occidentale, Brest, France*
Alejandra Ramirez-Martínez, *Universidad Veracruzana, Veracruz, Mexico*
Nathalie Wesolek, *Université de Bretagne Occidentale, Brest, France*
Catherine Brabet, *Centre de Coopération Internationale en Recherche Agronomique pour le Développement, Montpellier, France*
Guadalupe del Carmen Rodríguez Jimenes, *Instituto Tecnológico de Veracruz, Veracruz, Mexico*
Miguel Ángel García Alvarado, *Instituto Tecnológico de Veracruz, Veracruz, Mexico*
Marco Antonio Salgado Cervantes, *Instituto Tecnológico de Veracruz, Veracruz, Mexico*
Víctor José Robles Olvera, *Instituto Tecnológico de Veracruz, Veracruz, Mexico*
Alain-Claude Roudot, *Université de Bretagne Occidentale, Brest, France*

Human exposure to mycotoxins is a public health issue worldwide (Brera, Miraglia, & Colatosti, 1998). Although international variations in diet and cancer indicate that diet is an important risk factor for many cancers, it has been difficult to ascribe a clear role in cancer causation to exposure to specific individual chemicals or mixtures of chemicals, only alcohol intake and food contaminated with aflatoxins have been documented as risk factors in humans (Key et al., 2002). Since aflatoxins are classified as carcinogenic and genotoxic contaminants, the ALARA (As Low As Reasonable Achievable) approach is recommended (JECFA, 2001). Based on JECFA and SCF reports (JECFA, 1999 and SCF, 1994), even a very low exposure level to aflatoxins (1 ng kg−1 bw day−1) may induce liver cancer cases. The aim of this study was to assess the probabilistic risk of mycotoxin ingesting for corn tortilla consumer in Veracruz. One hundred twenty samples of tortilla corn were randomly collected in 3 season: October 2013, October 2014, and February 2015 to estimate the intake of aflatoxin in Veracruz City, México. The quantification of aflatoxin was performed by high-performance liquid chromatography with fluorescence detection and electrochemical derivatization. Tortilla corn consumption was evaluated in the population of Veracruz City through dietary intake questionnaires. A daily consumption questionnaire was used to determine the consumption of corn tortillas. Descriptive statistics and Probability Density Functions (PDF) of the daily consumption were determined and analyzed using @Risk6 (Palisade, Inc.). Descriptive statistics of daily consumption include the mean, median, standard deviation and the 95th percentile. Furthermore, PDF for aflatoxin concentration, body weight and consumption of the inhabitants of Veracruz City were generated. Calculation of PDF is based in the Monte Carlo simulation method with 10,000 iterations. The estimated daily intake PDF led to a mean of 3.24 ng kg−1 bw day−1 with a standard deviation 4.40 and a 95th percentile of 11.47 ng kg−1 bw day−1 for Octubre 2013 (season with the higher contamination), this represent a risk of 69.7 % based on JECFA data (1 ng kg−1 bw day−1).

Mo-Po-09

Exposure to Indoor Wood Smoke as Measured by Low-cost Air Quality Monitor

NADĚŽDA ZÍKOVÁ, *Clarkson University, Potsdam, New York, United States*
Thomas Twomey, *Clarkson Univesity, Potsdam, New York, United States*
Philip Hopke, *Clarkson University, Potsdam, NY, United States*
David Chalupa, *University of Rochester Medical Center, Rochester, New York, United States*
David Rich, *University of Rochester Medical Center, Rochester, New York, United States*
Andrea Ferro, *Clarkson University, Potsdam, New York, United States*

Particulate matter (PM) has been linked to adverse effects on respiratory and cardiovascular health (e.g. Pope et al. 2009). In urban settings, human exposure to PM is often connected to sources emitting black carbon (Naeher et al 2007), both outdoors and indoors (Smith et al. 2010), such as residential wood combustion for heat. Evidence suggests that wood combustion particles have similar toxicity to urban PM (Naeher et al. 2007). However, the emissions and thus chemical composition of wood smoke particles depend strongly on the combustion device, fuel, and operating conditions (Bolling et al. 2009). The current regulatory measurement network cannot cover large spatial and temporal variability in wood smoke (WS) concentrations, nor predict indoor exposures.
We conducted a measurement campaign in Rochester, NY (ca 210,000 inhabitants) using 52 low cost PM monitors (Speck; Airviz Inc., Pittsburgh, PA, USA), located at 26 sampling sites with wood burning appliances. Indoor/outdoor and spatial-temporal relationships of PM in the area, and the contribution of WS to personal PM exposures will be examined.
At each location, between November 2015 and March 2016, one indoor and one outdoor monitor concurrently measured 1-minute particle number concentrations, estimated particle mass concentrations of PM between 0.5 and 3 µm, and temperature. Additionally, a CO monitor was placed inside each house to distinguish between combustion and non-combustion sources of indoor PM. The study participants completed a survey on house type and age, heating fuel, and other activities influencing indoor air quality data.
Preliminary results show strong dependence of indoor activities on low-cost monitor concentrations, and differences in indoor and outdoor PM concentrations at the homes during periods when wood smoke is expected to be present (e.g., during the holidays).
Several additional results will be available after the heating season. The indoor/outdoor ratio will be calculated and compared to house type, ventilation etc. The spatial variability of outdoor PM concentrations in the county will be described, and the proportion of wood smoke in the outdoor and indoor PM concentrations will be estimated.
This work was supported by the New York State Research and Development Authority (NYSERDA) under grant 63040.

References:

Bolling, A.K. et al. (2009). Part. Fibre Toxicol. 6.1:1;
Naeher, L.P., et al. (2007) InhalToxicol 19.1: 67-106;
Pope III, C.A. et al. (2009) New Engl J Med 360.4: 376-386;
Smith, K.R. et al. (2010) J Expo Sci Env Epid 20.5: 406-416;
Wallace, L.A. et al.(2011) J Expo Sci Env Epid 21:49-64.

Mo-Po-10

Estimating Exposure to DDTs and Potential Carcinogenic and Non-carcinogenic Risks among Breast-fed Infants

Maryam Zare Jeddi, *Institute for Environmental Research (IER), Tehran University of Medical Sciences, Tehran, Iran*
Noushin Rastkari, *Institute for Environmental Research (IER), Tehran University of Medical Sciences., Tehran, Iran*
Reza Ahmadkhaniha, *School of Public Health, Tehran University of Medical Sciences, Tehran, Iran*
Masud Yunesian, *School of Public Health, Tehran University of Medical Sciences, Tehran, Iran*

Background: Breast milk consumption is the primary route of infant exposure to certain lipophilic toxicants such as organochlorine pesticides (OCPs). Dichlodiphenyl-trichloroethane (DDT) as the main type of OCPs has been associated with endocrine disruption and several cancers in many experimental and epidemiological studies.
Objectives: Because of the potentially serious health risk posed to newborn infants by the exposure to DDTs through breast milk feeding at a time of rapid development and growth, the present study has attempted to estimate the extent of carcinogenic and non-carcinogenic risk.
Methods: In the present study, breast milk samples were collected from 50 lactating mothers in the first week of post-partum from Tehran on May, 2015 and evaluated for DDT metabolites including p,p'-DDT, p,p'-DDD, p,p'-DDE, o, p'-DDE using GC-MS method. Subjects' mean age was 29±6 years, ranging from 18 to 42 years. All infants were exclusively breast-fed. For exposure assessment, daily intake was estimated based on the concentration of analytes and the assumption of a daily breast milk consumption rate of 700 g/day for a neonate weighing 5 kg. For the non-carcinogenic risk assessment purposes, hazard quotients were calculated. For carcinogenic risk assessment estimation, the cancer benchmark concentrations (CBC) were derived using oral slope factors and their hazard ratios (HR).
Results: p,p'-DDT, p,p'-DDD, p,p'-DDE, o, p'-DDE were detected in all breast milk samples, with a ΣDDTs mean concentration of 0.026±0.004 μg/g lipid wt, and ranged between 0.004-0.03 μg/g lipid wt, suggesting past usage and long-term accumulation of DDTs in humans. In addition, the DDT/DDE ratio was lower than one (mean: 0.021), which also indicates a historical exposure to this pollutant. A daily intake estimate of ΣDDTs, through breast milk consumption was 0.032 μg/kg per day. The highest intakes were recorded for p,p'-DDE (0.02 μg/kg-day) whilst the lowest was for p,p'-DDD, (0.003 μg/kg-day). The HQ for average DDTs concentrations was found to be low (0.002). HQ lower than 1 was noted for ΣDDTs, indicating these compounds were not a threat to the health of any breast-fed infant. Also, hazard ratio obtained for p,p'-DDE, p,p'-DDT, p,p'-DDD in breast milk could not pose potential carcinogenic risk to breast-fed infants since all HRs were lower than 1. Our study results reveal infants' exposure to low levels of DDTs through breast milk consumption. However, infants as the vulnerable group might be subject to the potential additive and/or synergistic health effects from the other xenobiotic chemicals present in breast milk.

Mo-Po-12

Current knowledge on the health benefits and risks of indoor air ionization

Paul Scheepers, *Radboudumc, Nijmegen, Netherlands*
Ilse Ottenbros, *Radboudumc, Nijmegen, Netherlands*
Arné Oerlemans, *Radboudumc, Nijmegen, Netherlands*

Background - Outdoor air is to some extent ionized (200-400 ions/cm3), depending on atmospheric conditions and weather systems. In addition, air treatment technologies designed for indoor settings can also produce ionized airborne particles. These ambient particles of various origin may accumulate surface charge, including microorganisms such as bacteria, bacterial spores and mites. Extremely high ionization (10^5-10^6 ions/cm3) was reported to reduce viability of bacteria and mites in laboratory studies. Some studies also describe the influence of ionization on diffusion and increased soiling on surfaces.
Aim - To provide an update on the potential health implications of indoor air ionization.
Methods - A search strategy was prepared for Pubmed and Medline using surface charge, charged particle, charged molecule, anion, cation, air ionization, ionized air and air ions as search terms. This set was combined with the search terms toxicokinetics, health effects, reproduction or neoplasms and resulted in 1.051 reviews. Further selections were made based on the abstracts by two experts independently resulting in 56 papers that were included in the review.
Results - There is much information on the toxicity of charged particles at the molecular and cellular level. Cell membranes and organelles were found to be affected by charged particles, as well as organ system, e.g. circulation, central nervous system and lungs. Differences in toxicity were observed between particles carrying positive or negative charge and neutral particles. The higher cytotoxicity observed in phagocytic cells for anionic nanoparticles could be related to their sensitivity in response to negative surface charge normally carried by pathogens. In contrast, for non-phagocytic cells it was suggested that interaction of cations with negatively charged membranes explains adverse effects to the membrane, embedded proteins and ion channels. Only few studies have addressed ionization on an organism level. Some theoretical and modelling studies predicted increased deposition in the higher airways. So far, human volunteer and population studies have failed to provide strong evidence to support benefits for treatment of (mainly asthma) patients. Nor have they demonstrated a convincing positive or negative effect in healthy populations.
Conclusion - In vitro studies indicated higher cellular toxicity of positively charged particles relative to neutral and negatively charged nanoparticles. So far, no studies were identified that show how airborne charged particles are taken up by inhalation, how their charge may change on the air-liquid interface in the lungs and how charged particles are distributed and reach target tissues in vivo.

Mo-Po-13

CITI-SENSE Edinburgh -The empowerment potential of participatory tools for environmental monitoring of air quality

Susanne Steinle, *Institute Of Occupational Medicine, Edinburgh, United Kingdom*
Karen S. Galea, *Institute Of Occupational Medicine, Edinburgh, United Kingdom*
Alena Bartonova, *Norwegian Institute for Air Research, Kjeller, Norway*

CITI-SENSE is a Collaborative Project partly funded by the EU FP7-ENV-2012 under grant agreement no 308524. The project involves nine cities across Europe which are defining and implementing the concept of a Citizens' Observatory. The aim of the project is to collect, collate and convey a multitude of environmental health information with participatory practices. This poster reports on the feedback from Edinburgh where people evaluated the empowerment potential of a variety of tools regarding urban air quality. Empowerment meaning enhancing an individual's or group's capacity to make effective choices, effective in the sense of enabling them to transform those choices into desired actions and outcomes.

A web portal that hosts information on all tools developed and applied within the frame of the project has been created and contains a Citizens' Observatory toolbox (http://co.citi-sense.eu/). In Edinburgh three tools have been applied and evaluated with members of the general public in combination, these being the:

1. The Little Environment Observatory (LEO) - a personal sensor that measures NO, NO2 and O3 as well as temperature and relative humidity. The LEOs work in combination with an app and transmits data to a visualization webpage (item 3). Participants carried the LEOs with them for a week collecting air quality data as well as gaining user experience.
2. The CityAir app - this gives citizens the opportunity to log their perception of air quality at any given point in time and space using a colorimetric indicator. Additional comments can be left every time a new perception is logged. Participants were asked to log their perception regularly during the 1 week trial with the LEOs.
3. The visualization web page - designed to allow citizens to view data from the LEOs as well as from the CityAir app. In addition it allows users to access data from other CITI-SENSE tools should they wish to do so.

Upon participation citizens asked to provide feedback on the usability and empowerment potential of the tools via questionnaire and a short interview. User evaluation of the outlined tools will take place during April - May 2016, with the analysis of the collected information following thereafter. Analysis will be based on empowerment potential guidelines developed by the project team aiming at usefulness of the tools, opportunities and barriers for improving air quality based on the tools. Our poster will present the findings of the evaluation process that took place with the participating Edinburgh citizens.

Mo-Po-14

Estimation of health risks associated with trace elements emitted from cooking with electric stove

Aiymgul Kerimray, *School of Engineering, Nazarbayev University, Astana, Kazakhstan*
Soudabeh Gorjinezhad, *Middle East Technical University Northern Cyprus Campus, Guzelyurt, Turkey*
Mehdi Amouei Torkmahalleh, *School of Engineering, Nazarbayev University, Astana, Kazakhstan*
Melek Keles, *Abant Izzet Baysal University, Bolu, Turkey*
Fatma Ozturk, *Abant Izzet Baysal University, Bolu, Turkey*
Philip Hopke, *Clarkson University, New York, United States*

Background and objective
Cooking is one of the main sources of indoor particulate matter (PM). Cooking even with cleaner sources of energy (gas, electricity) may also pose risks to human health. PM emissions from cooking contain trace elements, some of them being toxic and/or carcinogenic (As, Cd, Ni, Cr). The main objective of this study is to perform health risk assessments as a result of exposure to trace elements generated from cooking.
Method
A controlled study was conducted to understand the contribution of each cooking component including corn oil and beef meat on trace elements in PM emissions. No mechanical ventilation and natural ventilation existed on the sampling site. In each experiment, corn oil, corn oil with table salt and beef meet were heated using an electric stove. Each set of experiments lasted 20 minutes (14 minutes heating and 6 minutes cooling for corn oil experiments, 18 minutes grilling and 2 minutes cooling for beef experiments).
An Eight-Stage, Non-Viable Andersen Impactor was employed to collect the generated particles ranging from 0<0.43 μm to 3.3 μm (six cut sizes) on 81 mm quartz fiber filters including a backup filter for collecting PM0.43. Metal analyses were performed using an Inductive Coupled Plasma-Mass Spectrometry (ICPMS).
A human health risk assessment was performed as a result of chronic exposure to As, Cd, Co, Cr, Ni, Pb, Mn, Ba. The methodology used for assessing health risk is described in a Risk Assessment Guidance for Superfund (Part F, Supplemental Guidance for Inhalation Risk Assessment) (US EPA, 2011). Exposure time was assumed to be 2 hours per day, exposure frequency (EF) was assumed 365 day, and exposure duration 40 years.
Results
The results of health risk assessment indicate that carcinogenic risk from As, Co, Cr exceed the acceptable level (1 × 10-4) (Table 1). Cr poses the highest carcinogenic risk, with the risk exceeding the acceptable level by two orders of magnitude. Carcinogenic risk from Cd, Ni and Pb is within the acceptable level. Non-carcinogenic risk (HQ) values from all elements (except for Cd in oil with salt experiment) is higher than the safe level (= 1), with total risk exceeding the safe level 53-115 times. Ni poses the highest non-carcinogenic risk, with the HQ value 7.74-46.22 depending on the experiment. Thus, professional cooks or housewives, which are exposed to toxic elements at chronic levels from cooking activities under poor ventilation may have significant health outcomes.

Table 1 – Carcinogenic and non-carcinogenic risk from exposure to trace elements

	Non-carcinogenic			Carcinogenic		
	Oil	Oil with Salt	Meat	Oil	Salt	Meat
As	9.75	3.96	6.98	3.60E-04	1.46E-04	2.57E-04
Cd	2.12	0.08	1.06	2.18E-05	8.63E-07	1.09E-05
Co	11.39	18.06	12.81	3.52E-04	5.57E-04	3.95E-04
Cr	6.67	7.22	6.79	3.20E-02	3.47E-02	3.26E-02
Ni	46.22	7.74	36.52	8.88E-05	1.49E-05	7.01E-05
Pb	6.12	6.47	5.06	2.24E-05	2.36E-05	1.85E-05
Mn	28.22	7.02	19.12			
Ba	4.43	2.70	4.07			
Sum	114.94	53.26	92.40	3.28E-02	3.54E-02	3.34E-02

Mo-Po-15

Frequency, duration and severity of air pollution events: implications from repeated exposure to moderate

Yuli Huang, *National Kaohsiung First University of Science and Technology, Kaohsiung, China, Republic of (Taiwan)*
Yi-Chun Chen, *Ishou University, Kaohsiung, China, Republic of (Taiwan)*

Despite decades of efforts, air pollution remains a serious environmental health issue. Although a long-term decreasing trend was observed, bouts with very high pollutant concentrations have also been found. The occurrence and characteristics of these short-term events with high air pollutant levels have not been well examined. In this study, the frequency, duration and levels of ambient PM10 and PM2.5 concentrations were examined. Ambient air quality monitoring data between 2005 and 2015 were obtained from Taiwan EPA for a municipal monitoring station in southern Taiwan. Hourly concentrations for ozone, particulate matter (PM10) and fine particulate matters (PM2.5) were extracted. For ozone, the hourly and 8-hour average concentrations were analyzed, and ozone events were identified as having 1-hour level above 120 ppb or the 8-hour average level above 60 ppb. Any consecutive hours above the screening level would be treated as the same pollution event. For PM10, 24-hour average concentrations were calculated sequentially, and a 24-hour average concentration above 125 µg/m3 would be considered an event. Similarly, a high-PM2.5 event was identified as having 24-hour average or weighted 12-hour average concentration above 35µg/m3 (criteria limit for air quality standard and for pollution indicator value, respectively). Over the 11-year period analyzed, the annual average concentrations for the pollutant concentrations have decreased steadily: PM10 from 86.7 down to 72.6 µg/m3, and PM2.5 from 47.6 down to 25.4 µg/m3. By using the criteria limit as screening value, the 24-hour PM10¬ concentration exceeded the criterial limit 11.6% of the time. For PM2.5, the 24-hour average concentration exceeded the 35 µg/m3 limit 55.7% of the time, and the 12-hour weighted concentration also occurred 52.3% of the time. During an event, the pollutant concentrations were substantially above the screening level: 131 ppb for 1-hour ozone level, 72.8 ppb for 8-hour ozone level, 148.6 µg/m3 ¬for PM10, 56.4 µg/m3 for the 12-hour weighted PM2.5 concentration, and 55 µg/m3 for 24-hour average PM2.5 concentrations. The event length also varied substantially by pollutant type and averaging time, and a distinct seasonal pattern could be observed for these air pollution events. Implications of these repeated air pollution events on health outcomes warrant further analyses.

Mo-Po-16

NICU-Based Phthalate Exposure Impacts Early Neurodevelopmental Performance

Annemarie Stroustrup, *Icahn School of Medicine at Mount Sinai, New York, NY, United States*
Jennifer Bragg, *Stamford Hospital, Stamford, CT, United States*
Syam Andra, *Icahn School of Medicine at Mount Sinai, New York, NY, United States*
Paul Curtin, *Icahn School of Medicine at Mount Sinai, New York, NY, United States*
Chris Gennings, *Icahn School of Medicine at Mount Sinai, New York, NY, United States*
Christine Austin, *Icahn School of Medicine at Mount Sinai, New York, NY, United States*

Background: Each year, over 300,000 newborns in the US are cared for in a neonatal intensive care unit (NICU) where they are exposed to a chemical-intensive hospital environment. In utero phthalate exposure at this point in development can alter neurodevelopmental outcomes in healthy fetuses. Neurodevelopmental disorders among NICU grads are common and incompletely predicted by degree of prematurity or neonatal illness. The NICU-Hospital Exposures and Long-Term Health (NICU-HEALTH) study evaluates the impact of early life environmental exposures on preterm infants. The earliest neurodevelopmental assessment of NICU-HEALTH participants is the NICU Network Neurobehavioral Scale (NNNS), a structured physical exam of infant neurobehavioral organization, neurologic reflexes, motor development, active and passive tone, and signs of stress performed by a trained examiner prior to NICU discharge.
Objective: To evaluate associations between NICU-based phthalate exposure and short-term neurodevelopment.
Design/Methods: Urine specimens were non-invasively collected from NICU-HEALTH study infants multiple times during the NICU stay and frozen for batch analysis. We analyzed each specimen for a panel of 15 phthalate metabolites using high-performance liquid chromatography coupled to tandem mass spectrometry. Next, we used weighted quantile sum (WQS) regression, adjusting for specimen concentration, clinical, and demographic parameters to evaluate the association between patient averaged biomarkers of phthalate exposure and performance on each of the NNNS summary scores.
Results: 104 urine specimens from 46 hospitalized preterm infants were available for analysis. 10 phthalate monoesters were detected in >95% of specimens and were included in statistical analyses. The WQS Index of 10 phthalate monoesters indicated a negative association of phthalate exposure with NNNS quality of movement (beta=-0.12, p=0.01), a positive association between phthalate exposure and arousal (beta=0.07, p=0.017) and a negative association between phthalate exposure and lethargy (beta=-1.97, p=0.05). We also identified the indication of an association between phthalate exposure and attention; the inclusion of a quadratic term in this model provided a better fit than a simple linear model (two degrees of freedom contrast test, p=0.08).
Conclusions: NICU-based exposure to phthalates may be associated with worse performance on the NNNS at NICU discharge. These findings are significant as abnormal performance on NNNS quality of movement, arousal, lethargy and attention summary scores is associated with worse motor, cognitive, and behavioral function in later childhood.

Mo-Po-18

Housing, indoor air quality, and pediatric asthma in a low income multifamily housing site in Boston - a systems science approach

M. Patricia Fabian, *Boston University School of Public Health, Boston, Massachusetts, United States*
Lindsay Underhill, *Boston University School of Public Health, Boston, Massachusetts, United States*
Kimberly Vermeer, *Urban Habitats Initiative Inc., Boston, Massachusetts, United States*
Gary Adamkiewicz, *Harvard T.H. Chan School of Public Health, Boston, Massachusetts, United States*
Megan Sandel, *Boston University School of Medicine, Boston, Massachusetts, United States*
Jonathan Levy, *Boston University School of Public Health, Boston, Massachusetts, United States*

Aim. Asthma is a complex disease affecting over 20 million children in the United States, with a disproportionate impact on low income urban populations. The causes of asthma exacerbations are multifactorial, and include exposure to residential indoor environmental contaminants such as allergens and combustion pollutants. Housing characteristics and building interventions (e.g. energy-saving retrofits) impact asthma outcomes by modifying exposures to these indoor environmental contaminants, but quantifying the impact of these changes is challenging given complex airflows, variable pollution source characteristics, and unknown resident behavior.
Methods. Using a systems science approach, we applied a pediatric asthma discrete event simulation model (DEM) to evaluate the health impacts of energy-saving building retrofits in a low-income multi-family housing complex in Boston, MA. Indoor environmental conditions and pollutant concentrations ($NO2$ and $PM2.5$) were modeled using CONTAM, a multi-zone airflow and contaminant transport analysis program. The resulting air exchange rates and pollutant concentrations were used in conjunction with allergen data to parameterize the DEM for a large simulated population of asthmatic children. Simulations were run for 100,000 children in pre and post retrofit scenarios, with outputs such as pollutant concentrations and pediatric asthma outcomes evaluated over 10 years.
Results. Across all simulated households, retrofits led to a 2% increase in PM2.5 concentrations from environmental tobacco smoke, and a 46% and 22% decrease from cooking activities and outdoor infiltration, respectively. NO2 concentrations post retrofit also decreased by 45% and 8% for cooking and outdoor infiltration respectively. Cockroach allergen was reduced 82% post retrofit. These differences are due to a combination of factors, including post retrofit changes in air exchange rates, exhaust fan installation and use, and filter efficiency improvements in the mechanical ventilation system. Post retrofit, we estimated that there were on average 6% fewer days with asthma symptoms, and a 16-19% reduction in serious asthma events including clinic visits, emergency room visits, and hospitalizations. In addition, there were significant reductions in the percentage of children progressing to a more severe asthma classification over the simulation period (from 21% to 8%).
Conclusion. Our simulation models indicated that building retrofits targeting energy savings resulted in a decrease in asthma outcomes, although with variable impacts as a function of resident behaviors such as smoking and exhaust fan use. Our study illustrates the utility of a systems science approach to evaluate the complex tradeoffs between building retrofits, indoor air quality and pediatric asthma outcomes.

Mo-Po-19

Cumulative Environmental Effects: Expanding Research with the Hopi Tribe

Mary Kay O'Rourke, *The University of Arizona, Tucson, Arizona, United States*
Robin Harris, *The University of Arizona, Tucson, Arizona, United States*
Gayl Honanie, *The Hopi Tribe, Kykotsmovi, Arizona, United States*
Lorencita Joshweseoma, *The Hopi Tribe, Kykotsmovi, Arizona, United States*
Jefferey Burgess, *The University of Arizona, Tucson, Arizona, United States*

The "Hopi Environmental Health Project" is a component of the newly funded Center for Indigenous Environmental Health Research (CIEHR) aimed at eliminating environmental health disparities. The Hopi Project objectives include: (1) Characterization of the magnitude of environmental exposures to particulate matter (PM), As species, U and other contaminants from air, water, and food in households; (2) Evaluation of exposure moderation social determinants of health, social capital and community resilience; and (3) Expansion of the Hopi Tribe's capacity to address areas of environmental concern that can inform programs and policy. Based on earlier survey work with the Tribe, the research team identified tribal health concerns addressing asthma and diabetes prevalence, while the Hopi Environmental Office indicated the need for ambient PM sampling, a concern regarding solid waste/ash disposal, arsenic in the water, and impact of contaminated water on people and crops. Self-reported asthma on the Hopi reservation affects 24% of the people as compared to a national self-reported asthma rate of 10.5%. This represents a clear health disparity that may be associated with housing type (traditional/stone masonry; "modern"/primarily block; and manufactured) combined with use of wood (38%) and coal (35%) to heat homes from late October through early April. Over the 3 years of the project 90 homes will be sampled for PM10 & 2.5 by operating a PDR 1500 Personal Data Ram for a 24 hour period during the heating and non-heating season. Water samples, dietary information and surveys describing behaviors will be evaluated. Homes will be selected with replacement by randomizing members listed on the tribal roles, characterizing homes by construction and heating mode and filling a grid defined by housing type and heating fuel utilized. The Hopi Environmental Protection office has identified other concerns related to homes. These include ash disposal, arsenic in drinking water, moisture/mold problems in homes, as well as radon and formaldehyde within buildings. These concerns will be ranked by the Community Advisory Board for analysis in conjunction with PM. Household sampling will begin in late October of 2016. This describes the background, rationale and study design proposed to evaluate environmental health disparities experienced by the Hopi Tribe of Arizona, USA.

Measuring/monitoring/strategy

Mo-Po-20

Americans' Exposure to the Insect Repellent N,N-diethyl-m-toluamide (DEET)

Maria Ospina, *Centers for Disease Control and Prevention, Atlanta, Georgia, United States*
Lee-Yang Wong, *Centers for Disease Control and Prevention, Atlanta, Georgia, United States*
Samuel Baker, *Centers for Disease Control and Prevention, Atlanta, Georgia, United States*
Amanda Bishop, *Centers for Disease Control and Prevention, Atlanta, Georgia, United States*
Pilar Morales-Agudelo, *Centers for Disease Control and Prevention, Atlanta, Georgia, United States*
Liza Valentin-Blasini, *Centers for Disease Control and Prevention, Atlanta, Georgia, United States*
Antonia Calafat, *Centers for Disease Control and Prevention, Atlanta, Georgia, United States*

Background: N,N-diethyl-m-toluamide (DEET) has become the most effective and ubiquitous insect repellent in the USA since its development in 1946. DEET repels disease-carrying vectors such as deer ticks associated with Lyme disease, and mosquitoes that can transmit malaria, encephalitis, Dengue fever, and West Nile and Zika viruses. Understanding exposure to DEET is of public health interest.
Methods: We studied DEET exposure, using urinary concentrations of DEET and two of its metabolites, 3-(diethylcarbamoyl)benzoic acid (DCBA) and N,N-diethyl-3-hydroxymethylbenzamide (DHMB) in 5,348 Americans 6 years and older from the 2007-2010 National Health and Nutrition Examination Survey. We used multiple regression to examine associations between several variables (e.g., age, sex, race/ethnicity, household income, season of year) and concentrations of DCBA, the most detected DEET biomarker.
Results: We detected DEET at concentration ranges (>0.1-45.1 µg/L) much lower than those of DCBA (>0.5-30,400 µg/L) and DHMB (>0.1-332 µg/L). DCBA was the most frequently detected biomarker (84%) while DHMB and DEET were detected in 15% and 3% of samples, respectively. DCBA adjusted geometric mean concentrations were higher in May-Sep than in Oct-Apr. Non-Hispanic whites in warm months were more likely than in colder months [adjusted odds ratio = 10.49; 95% confidence interval, 3.23-34.10] to have DCBA concentrations above the 95th percentile (an arbitrary value selected as an example of higher than average concentrations).
Conclusions: Almost 85% of Americans are exposed to DEET. However, reliance on DEET as the sole biomarker would underestimate the extent of exposure; instead, DEET oxidative metabolites are adequate exposure biomarkers. Higher concentrations of DEET biomarkers in the warm season agree with increasing use of DEET when people usually spend time outdoors for recreational activities; also pests are more abundant with higher seasonal temperatures, thus likely additional protection against vector borne diseases in warmer weather is needed.

Mo-Po-21

Exploring metallome risk of gestational diabetes mellitus on the context of meconium internal chemical environmental changes: A systems approach

Heqing Shen, *Institute of Urban Environment, Chinese Academy of Sciences, Xiamen, China, People's Republic of*

Exposomics, the systems environmental exposure and health risk assessment approach, has been suggested by Wild and the other scientists [1, 2] to investigate the complex environmental health issue. Although some strategies had been suggested to practice this perfect concept [2, 3], there are still many challenges to include the complete set of environmental factors in one study design. Here we aim to report the potential sub-group exogenous factors from meconium metallome as the gestational diabetes mellitus (GDM) and their linkages to the rest part of the internal chemical environmental changes, i.e., the endogenous response of metabolomics changes, in uterus during the pregnancy.
A cross-sectional population of 1359 pregnant women has been recruited and the meconium samples from their newborn babies were collected; a subset subjects with the 142 GDM (12.21% occurrence) and the 197 health controls have been selected to set as a nested case-control study. Perversely we had found that some toxic heavy metals, esp. arsenic, could be the risk factors of GDM [4], and ten most potential metabolites from meconium metabolome also showed their potential prediction of GDM. In the same time, these metabolites also implied the interrupted metabolic path ways in the GDM cases when comparing the health controls [5]. In the present work, the meconium metallome including 23 metal and metalloid elements were measured, their correlations with the metabolites were investigated in the systems way. Exogenous factor network and metabolic response network were constructed by calculating the covariance of each pairs of metals from the potential GDM risk factors and each pairs of metabolites from the potential GDM biomarkers, in respectively; in further we assessed relationships between the exposure matrix of exogenous factors and response matrix of endogenous metabolites.
The core part of metabolic network suggested some mode of actions, esp., the adenosine/L-arginine/nitric oxide (ALANO) pathway may be deeply involved the GDM risk; in further, some metal pollutants induced reactive oxygen species (ROS) may couple to the nitrosative stress, the later produce more reactive nitrogen species (RNS) such as peroxynitrite (ONOO-) and induced the adverse effect of vascular endothelial cells (VECs) in placenta and cord. This work also contributes the establishment of the systems approach by linking the exposure matrix to the response matrix through coinciding covariance analysis and topological network plotting.
Mining their network instead of the risk factor and response biomarker along may address the extra biological information of GDM and its environmental risk.

UTRECHT
THE NETHERLANDS
OCTOBER 9-13
ISES 2016

Mo-Po-22

Potential Dermal and Inhalation Exposure of Workers During Pest Control of Oak Processionary Moth by Spray Application

Thomas Göen, *University of Erlangen-Nuremberg, Erlangen, Germany*
Ralph Hebisch, *Federal Institute for Occupational Safety and Health (BAuA), Dortmund, Germany*
Anja Schäferhenrich, *University of Erlangen-Nuremberg, Erlangen, Germany*
Anja Baumgärtel, *Federal Institute for Occupational Safety and Health (BAuA), Dortmund, Germany*
Frank Burgmann, *Federal Institute for Occupational Safety and Health (BAuA), Dortmund, Germany*
Karin Ludwig-Fischer, *Federal Institute for Occupational Safety and Health (BAuA), Dortmund, Germany*
Michael Roitzsch, *Federal Institute for Occupational Safety and Health (BAuA), Dortmund, Germany*

Background: The larvae of the oak processionary moth (OPM) form irritating hairs which can cause severe skin, eye and respiratory irritation as well as allergic reactions to humans. For protection of human health, the OPM is commonly controlled by biocide spray application. Biocidal products require authorisation according Regulation (EU) No 528/2012 of the European Parliament. For authorisation, a risk assessment based on exposure estimation is performed for all intended uses of the respective product. Detailed information on the exposure of operators and of casual bystanders was lacking until now. Considering the height of oak trees and the locations where OPM is controlled, it can be assumed that the exposure patterns of biocide applications might be significantly different from those in agricultural spray applications, for which several models are available.
Objectives: This study investigated the dermal and inhalation exposure of pest control operators and bystanders during spray application against the OPM (biocide: Diflubenzuron, DimilinTM). Task-specific exposure data for knapsack as well as vehicle mounted sprayers were collected, covering activities such as weighing and portioning of the biocidal product, on-site preparation and application of the spray liquid and cleaning of the equipment.
Methods: The suitable dosimeters used to quantify the potential dermal exposure were whole body polyethylene coveralls and cotton gloves. Quantified diflubenzuron levels were related to the amount of active substance applied or to the duration of the task. Inhalation exposure was measured using an individual sampler provided with a glass fiber filter with organic binder (capture of the particulates) and a portable pump (SG10/2; 10 L/min).
Results: 51 overalls and 83 pairs of gloves were obtained during OPM control in 2014 and 2015. The total dermal exposure to diflubenzuron during weighing and portioning was 6.05±3.31 µg/g (n=2), for application using knapsack sprayers 3420±4840 µg/g (n=18), for application using vehicle mounted sprayers 72.5±81.8 µg/g (n=17) and for cleaning of the equipment 1520±1290 µg/min (n=3).
For the determination of inhalation exposure 51 personal air samples were taken. During weighing and portioning, pest control operators were exposed to 484±297 µg/m^3 (n=2) diflubenzuron by inhalation, during knapsack spraying to 5.22±5.35 µg/m^3 (n=18), during

application using vehicle mounted sprayers 1.58±2.70 μg/m^3 (n=17) and during cleaning of the equipment 2.36±1.17 μg/m^3 (n=3).
Cluster analysis of dermal and inhalation exposure data allow a clear distinction of the different tasks. The data presented here provide a reliable database for the authorisation of biocidal products according to the EU-Regulation.

Mo-Po-25

From consumer use surveys of personal care products to chemical emission estimates at wastewater treatment plant level

Mélanie Douziech, *Radboud University, Nijmegen, Netherlands*
Antonio Franco, *Unilever, Colworth, United Kingdom*
Jan Hendriks, *Radboud University, Nijmegen, Netherlands*
Mark Huijbregts, *Radboud University, Nijmegen, Netherlands*
Henry King, *Unilever, Colworth, United Kingdom*
Rik Oldenkamp, *Radboud University, Nijmegen, Netherlands*
Rosalie Van Zelm, *Radboud University, Nijmegen, Netherlands*

Personal care products (PCPs) are part of our daily lives. Their use results in the release of chemicals into the environment. Estimations of the amount of PCPs used are required to identify areas with potentially higher environmental risks. Such emission estimates are scarce, uncertain, and variable. Nowadays, regulatory assessments rely primarily on total tonnage estimates collated from industry confidential tonnage data (ECHA in Europe) or generated by market research organisations (e.g. Euromonitor). Unfortunately, these methods are not entirely transparent and the associated uncertainty is difficult to quantify.
The objective of this research was to develop a method to predict the yearly amounts of different personal care products used within countries, by means of consumer surveys. This bottom-up method is more systematic and transparent than total market estimates currently in use.
Our analysis considered published articles about consumer use habits of PCPs [1-6]. We focused on "down-the-drain" PCPs, namely hair styling products, shampoo, conditioner, shower gel, and toothpaste. Published data were reviewed and selected based on the quality of the information available to derive per capita daily use. The mean daily consumption data published in [4, 5] for the American female population, and in [2] for the South Korean population were used directly. In [3] the probability distributions of the frequency of use and amount per application were multiplied to derive per capita daily use. The mean frequency of use from [1] was multiplied with the mean amount per application from [6]. Mean values of per capita daily product use were multiplied with the prevalence of use and the population number corresponding to the user categories differentiated upon the study (e.g. gender). Finally, these yearly country-level estimates were compared to available total market estimates.
Figure 1 shows a comparison of the computed values for the use of some PCPs and those provided by Euromonitor. Both methods show similar estimates for the Netherlands, but vary significantly for other countries, especially the USA where only the female population is considered [4].

References

[1] Ficheux, A.S., et al., Food Chem Toxicol, 2015. 78
[2] Park, J.Y., et al., Food Chem Toxicol, 2015. 77
[3] Biesterbos, J.W., et al., Food Chem Toxicol, 2013. 55
[4] Loretz, L., et al., Food Chem Toxicol, 2006. 44(12)
[5] Loretz, L.J., et al., Food Chem Toxicol, 2008. 46(5)
[6] Ficheux, A.S., et al., Food Chem Toxicol, 2016. 90

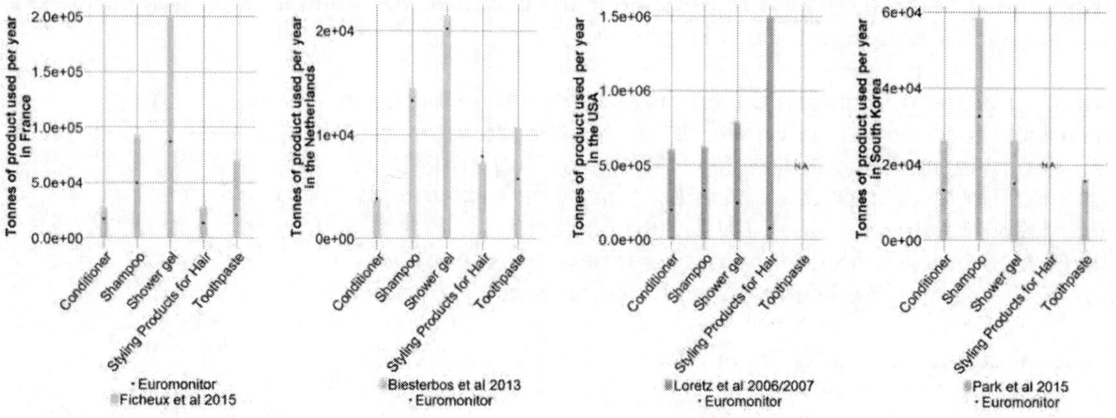

Figure 1 - Total amount of product used per country derived from consumer use studies and compared to Euromonitor value. The totals are computed considering the entire population for France, the Netherlands, and South Korea and considering only the female

Mo-Po-26

Harmonization of Analysis of Real-Time Monitoring Data from RTI MicroPEM™ Through Open-Source Software

Maëlle SALMON, *ISGlobal, Center for Research in Environmental Epidemiology (CREAL), Barcelona, Spain*
Zheng Zhou, *Mailman School of Public Health, Columbia University, New York, United States*
Vakacherla Sreekanth, *CMR Institute of Technology, Bangalore, India*
Julian D. Marshall, *University of Washington, Seattle, United States*
Cathryn Tonne, *ISGlobal, Center for Research in Environmental Epidemiology (CREAL), Barcelona, Spain*

Aims: RTI MicroPEM is a small PM exposure monitor, increasingly used in research in developed and developing countries. The output of one measurement session is a csv file, including a header section with information on instrument settings and a table of tens of thousands observations of time-varying variables such as PM2.5 concentration, relative humidity, temperature and accelerometer measures over three axis. Such files need processing which currently is not done using an harmonized approach. Our objective was to develop software to improve quality control and ease of processing MicroPEM output files.
Methods: We developed an open-source R package for reading data from MicroPEM output files and generating clean datasets with time-varying variables and instrument setting parameters for further analysis. We chose R for implementation because it contains many packages for fast and easy data wrangling, and so that the prepared data can directly be analyzed using R inference packages, before being communicated using R visualization and report producing commands. We also wanted our package to be open-source for better reproducibility, easier involvement of new contributors, and free use, particularly in developing countries.
Results: We applied the package in two research projects, including a large number of measurements. The functionalities of our package are three-fold: allowing conversion of files including batch conversion of all files in a directory, empowering easy data quality checks, and supporting data cleaning through e.g. a function for correcting PM2.5 measures when there is a shift in the baseline.
The package enables easy inspection of MicroPEM output files. The package allows outputting a table comparing parameters from a set of files for checking whether all measures of a study are comparable. The plotting function also permits a quick overview of an individual PM2.5 time series. Moreover, we developed an 'R Shiny' interface which automatically produces graphs and tunable alarms such as "Nephelometer slope was not 3". The interface can be used in the field for easy, inspection of data quality in individual MicroPEM files.
Conclusions: The package can be easily installed from the free web-based repository hosting service Github. We use unit tests and continuous integration to ensure good code quality and cross-platform compatibility. Our R package will contribute towards optimization and harmonization of MicroPEM data processing and analysis across studies.

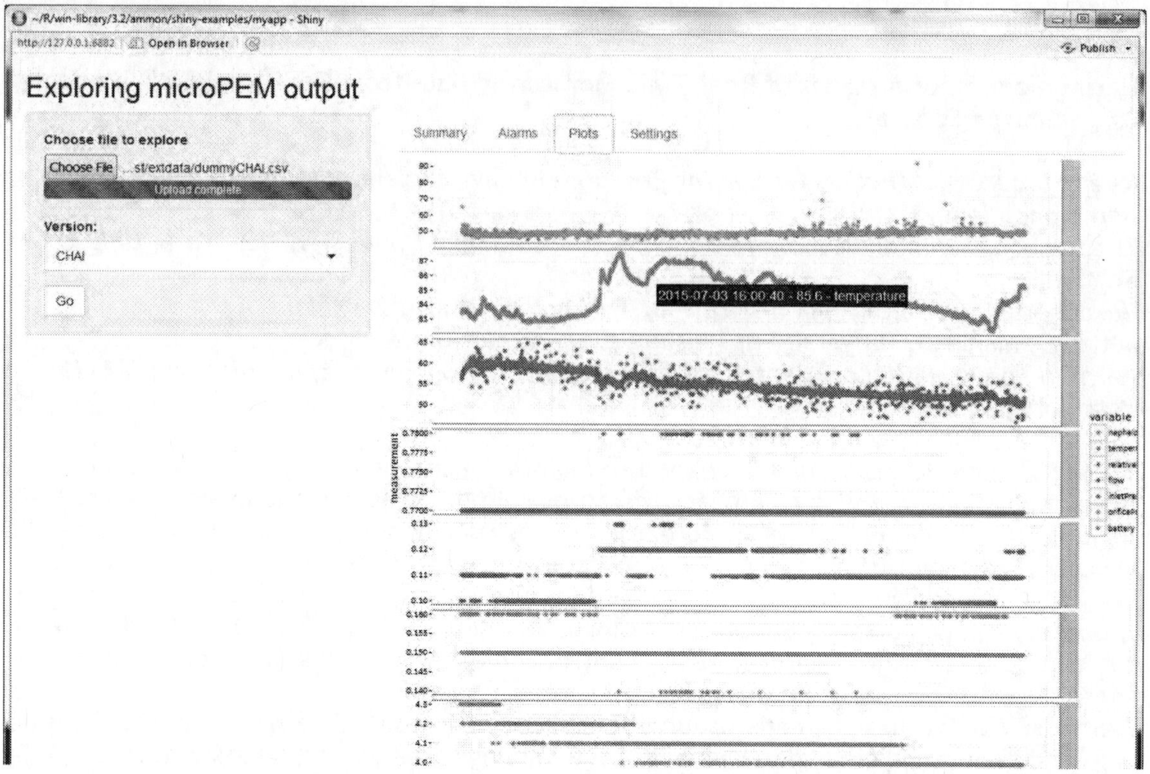

Screenshot of the Shiny app interface implemented in the R package ammon. The plot tab offers an interactive plot of the measurements time series. The interface can be used with no R experience.

Mo-Po-27

Policy Implications of the Health Co-benefit Assessment of Alternative GHG mitigation Strategies in Suzhou, China

Miaomiao Liu, *School of the Environment, Nanjing University, Nanjing, China, People's Republic of*
Yining Huang, *School of the Environment, Nanjing University, Nanjing, China, People's Republic of*
Zhou Jin, *School of the Environment, Nanjing University, Nanjing, China, People's Republic of*
Xingyu Liu, *School of the Environment, Nanjing University, Nanjing, China, People's Republic of*
Jun Bi, *School of the Environment, Nanjing University, Nanjing, China, People's Republic of*
Matti Jantunen, *National Institute for Health and Welfare, Kuopio, Finland*

Suzhou, 70 km West of Shanghai, is a historic and one of the economically most advanced cities in China, with 10.6 million people. As one of the "low carbon development pilot cities", Suzhou is committed to not increase its carbon emissions after 2020 and to reduce the carbon intensity of its economy by 54% from 2005 (SDRC, 2013). The rapid growth of its export-oriented high-tech industry is a challenge to these targets.
Our objective is to evaluate the CO2 and health benefits of three policy scenarios against the business as usual scenario (BAU) in 2020, (i) industrial structure dominated (ISD), (ii) technology dominated (TD), and (iii) integrated carbon reduction (ICR, which combines ISD and TD).
We apply the IEA Energy Balance Tables based GHG-PAM model to develop these four scenarios into balanced energy supply and use tables and to assess the GHG and PM emission, and burden of disease (BoD) impacts of each scenario.
The key results are presented in the attached Table. Compared to the 2010 Background both BAU and ISD 2020 scenarios would double the emissions and BoD and even the TD and ICR scenarios would increase them by 20...25 %. The BoD reduction from the BAU to ICR scenario is 51.5 kDALY/a or 44%, 43% due to industrial and only 1% to all other source reductions. The BoD reduced per CO2 emission reduction [DALY/MtCO2] varies greatly between the different industries. This could have local policy implications in Suzhou, because per unit CO2 reductions the health co-benefits in Paper, pulp & printing, Food & tobacco, and Wood products industries would be 2..3 times higher than in, e.g., Machinery, Non-ferrous metals, and Transport equipment industries.
Acknowledgements: URGENCHE study funded by EU FP7 contract n:o 265114

Scenario	GHG emission	Burden of disease	BoD benefit vs BAU 2020
	$MtCO_2/a$	kDALY/a	$DALY/MtCO_2$
2010 Background	148	55,6	
2020 BAU	297	116,8	-
2020 ISD	284	110,7	469
2020 TD	180	68,5	413
2020 ICR	173	65,3	415
Economic sector	**Emission reduction: 2020 ICR vs 2020 BAU**	**BoD reduction: 2020 ICR vs 2020 BAU**	
Iron & Steel	**36,7**	**17,90**	488
Chemical & petroleum	**19,4**	**8,20**	423
Paper, pulp & printing	14,4	**7,70**	**535**
Machinery	**28,5**	5,90	207
Non-metallic minerals	4,1	1,76	430
Food & tobacco	1,3	0,81	**620**
Non-ferrous metals	0,9	0,23	260
Wood products	0,3	0,18	**590**
Transport equipment	0,2	0,05	230
Transportation	5,5	**1,50**	**273**
Resid & commercial	1,5	0,20	133

Suzhou GHG policies comparison

Mo-Po-28

Development and application of traffic density-based parameters for studying near-road air pollutant exposure

Shi Liu, *US EPA National Exposure Research Laboratory, Durham, North Carolina, United States*
Jianping Xue, *US EPA National Exposure Research Laboratory, Durham, North Carolina, United States*
Fu-lin Chen, *US EPA National Exposure Research Laboratory, Durham, North Carolina, United States*

Increasingly human populations are living and/or working in close proximity to heavily travelled roadways. There is a growing body of research indicating a variety of health conditions are adversely affected by near-road air pollutants. To reliably estimate the health risk associated with near-road air pollutants, one of the greatest challenges is accurately understanding exposures in these microenvironments because actual monitoring data on the air pollutants are often not available. To overcome this limitation, we performed a metadata analysis on data compiled from literature search and obtained from EPA-sponsored measurement studies to evaluate air pollutants' concentration decay rate (R) which equals Ln (Ca/Ci) / D where Ci and Ca represents concentration initially and after a distance (D), respectively. We also compared correlation of near-road air pollutants' concentrations with conventional traffic indicators such as nearest distance (ND) to major road and total length (TL) of major road as well as three traffic density-based indicators we developed: Major-Road Density (MRD), All-Traffic Density (ATD) and Heavy-Traffic Density (HTD). Our metadata analysis yielded mean decay rates (fraction of relative concentration decrease per meter) for black carbon (BC), carbon monoxide (CO), nitrogen dioxide (NO2) and nitrogen oxides (NOX) as 0.0026, 0.0019, 0.0004, and 0.0027, respectively, while no such decay was noticeable for particulate matter (PM). Traffic density-based indicators MRD, ATD) and HTD yielded respective average correlation coefficients of 0.26, 0.18 and 0.48 for BC measured in the Near-road Exposures and Effects of Urban Air Pollutants Study (NEXUS). In contrast, average correlation coefficients of -0.31 and 0.25 were generated for BC when considering its relationship to two commonly used traffic indicators: ND and TL. Further analysis of MRD and ATD with ambient concentration data for ozone (O3), CO, NOx, sulfur dioxide (SO2), lead (Pb), and PM retrieved from US Environmental Protection Agency (EPA)'s Aerometric Information Retrieval System (AIRS) showed correlation coefficients in range of 0.22 to 0.47 for the mobile source related pollutants CO and NOX but less than 0.19 for O3, SO2, Pb, and PM. Thus, the traffic density-based parameters may be more specific indicators than conventional traffic indicators for near-road exposure to air pollutants from mobile source emissions and may be useful in assessing traffic contribution to health risk on human populations living and/or working in near-road environments.

Mo-Po-29

Impact of inter-coder differences in occupation and industry classification coding on exposure estimates obtained via job-exposure matrix: example of gasoline engine emissions in CANJEM

Thomas RÉMEN, *CRCHUM, MONTREAL, Canada*
Jack SIEMIATYCKI, *CRCHUM, Montréal, Canada*
Jérôme LAVOUÉ, *CRCHUM, Montréal, Canada*
Marc-André Verner, *Université de Montréal, Montreal, Canada*

The purpose of this study is to determine how inter-coder differences in coding jobs impact the exposures subsequently assigned by a JEM.
1,000 jobs were selected from among the job histories reported by subjects in a case-control study of lung cancer conducted in Montreal. Two coders independently each job to four different occupation classifications (OC) and three different industry classifications (IC) which can be linked to the CANJEM job exposure matrix. For jobs coded differently by the two experts, CANJEM was used to obtain various metrics of exposure to gasoline engine emissions (GEE): exposed or unexposed (based on three cut-points of probability of exposure: 5%, 25% or 50%), exposure intensity (categorical) and frequency-weighted intensity (FWI - continuous). Interrater agreement between the exposure metrics to GEE was measured using Kappa statistics for categorical metrics and Intra-class Correlation Coefficients (ICCs) for FWI.
Depending on the classification used, at the highest level of resolution, the proportion of jobs coded differently by the two experts varied from 46.8% to 64.3% for OC and from 21.5% to 36.8% for IC. Based on jobs coded differently, Cohen's kappa statistic for exposure status ranged from 0.32 to 0.49 when using OC and from -0.02 to 0.50 when using IC depending on the cut-point used. The corresponding numbers for intensity of exposure ranged from 0.34 to 0.48 for OC and from 0 to 0.44 for IC. When restricting analysis to jobs considered as exposed using both codes, ICC for FWI varied from -0.24 to 0.35 (OC) and from -0.13 to 0.68 (IC).
Similar to other studies, a quite high proportion of jobs were coded differently by the two experts, especially for OC. Jobs with different codes had null to moderate agreement in exposure estimates. This exercise highlights the importance of improving and standardizing coding of occupations and industries.

Mo-Po-32

New approach to study the real exposure to fungi in cork industry: nasal swabs mycobiota investigation coupled with screening on fungal resistance to azoles

Carla Viegas, *ESTeSL-IPL; ENSP-UNL, Lisbon, Portugal*
Anália Clérigo, *ESTeSL - IPL, Lisbon, Portugal*
Cátia Pacífico, *ESTeSL-IPL, LIsbon, Portugal*
Tiago Faria, *ESTeSL-IPL, Lisbon, Portugal*
Raquel Sabino, *INSA; ESTeSL-IPL, Lisbon, Portugal*
Mariana Francisco, *INSA, Lisbon, Portugal*
Cristina Verissimo, *INSA, Lisbon, Portugal*
Susana Viegas, *ESTeSL-IPL; ENSP-UNL, Lisbon, Portugal*

The permanent contact with cork may lead to constant exposure to fungi, raising awareness as a potential occupational hazard in the cork industry. In fact, the presence of fungi belonging to the Penicillium glabrum complex has been associated with the development of respiratory diseases such as suberosis, one of the most prevalent diseases among workers from cork industries, besides occupational asthma.
The aim of this work was to characterize fungal distribution and to evaluate the incidence of isolates resistant to azoles present in nose swab samples from the cork industry workers. Nose swab samples were collected from 305 workers from two cork industries. Samples were plated onto malt extract agar (MEA) media (for morphological identification of the mycobiota present) and also onto screening media of Itraconazol and Voriconazol (to detect azole-resistant Aspergillus isolates). Plates were incubated at 27°C during 5 to 7 days.
Using macro- and microscopic analysis of the colonies, fungal contamination was evident in 267 from the 305 samples collected (87.5%). From 267 contaminated samples, 109 presented countless units of Chrysonilia sytophila (40.8%). The most prevalent genus found was Penicillium sp. (73.6%), followed by Cladosporium sp. (9%), Chrysonilia sp. (2.9%) and Acremonium sp. (2.6%). From the 305 samples collected, 161 belong to workers from the cork yard. Penicillium sp., Cladosporium sp., Acremonium sp. and A. niger complex were the most commonly found species. Itraconazole resistant isolates of the Aspergillus genus were found in nose samples from two workers of the cork yard, belonging to the Nigri and Fumigati sections.
This approach allowed knowing the real contact with fungi of the workers of cork industry. Additionally, was possible to obtained data regarding the fungal mycobiota present that can result in negative health effects. Ongoing studies on this population are being performed in order to assess possible health implications related to fungal exposure in this occupational setting.

Mo-Po-33

An algorithm for quantitatively estimating non-occupational pesticide exposure intensity for spouses in the Agricultural Health Study

Nicole Deziel, *Yale School of Public Health, New Haven, CT, United States*
Laura Beane Freeman, *National Cancer Institute, Rockville, MD, United States*
Jane Hoppin, *North Carolina State University, Raleigh, NC, United States*
Kent Thomas, *National Exposure Research Laboratory, Research Triangle Park, NC, United States*
Catherine Lerro, *National Cancer Institute, Rockville, MD, United States*
Rena Jones, *National Cancer Institute, Rockville, MD, United States*
Aaron Blair, *National Cancer Institute, Rockville, MD, United States*
Barry Graubard, *National Cancer Institute, Rockville, MD, United States*
Jay Lubin, *National Cancer Institute, Rockville, MD, United States*
Dale Sandler, *National Institute of Environmental Health Sciences, Research Triangle Park, NC, United States*
Honglei Chen, *National Institute of Environmental Health Sciences, Research Triangle Park, NC, United States*
Gabriella Andreotti*National Cancer Institute, Rockville, MD, United States*
Michael Alavanja, *National Cancer Institute, Rockville, MD, United States*
Melissa Friesen, *National Cancer Institute, Rockville, MD, United States*

Purpose: Women living or working on farms may be exposed to pesticides from direct occupational use of agricultural pesticides and from non-occupational pathways, such as take-home exposure from skin, clothes and shoes of farmworkers, drift from nearby fields, and pest treatments in the home/yard. Failure to account for non-occupational pathways may underestimate total exposure, increase exposure misclassification and reduce power to detect associations in epidemiologic analyses, particularly for women who have less occupational pesticide contact than men. We developed an active-ingredient-specific algorithm for cumulative, non-occupational pesticide exposure for female spouses of pesticide applicators in the Agricultural Health Study (AHS) that quantified exposure intensity from four pathways: bystander, take-home, agricultural drift, and residential pesticide use. Methods: We used exposure data from previous meta-analyses to develop pathway weights. We used spouse and applicator responses to questions on pesticide use, farm characteristics, and other activities to identify subject-specific contrasts in pesticide exposure intensity. Results: In our algorithm, bystander exposure was a function of time a spouse spent working in fields, take-home exposure was a function of time a spouse spent at home, and both were proportional to days and years the applicator applied an active ingredient. Exposure from agricultural drift was a function of distance between homes and treated fields and days and years the applicator applied the active ingredient. Residential pesticide exposure was a function of the combined contribution of years of multiple home pest treatments, accounting for the probability the active ingredient was used in specific treatments. Conclusion: This transparent, data-driven algorithm of cumulative, aggregate pesticide exposure intensity will facilitate etiologic analyses of health effects in the AHS and could be applied to studies with similar information.

Mo-Po-34

Exposure to the Non-Phthalate Plasticizer 1,2-Cyclohexane Dicarboxylic Acid, Diisononyl Ester (DINCH) in Portuguese Children

Luísa Correia-Sá, *REQUIMTE/LAQV - Instituto Superior de Engenharia do Porto do Instituto Politécnico do Porto, Porto, Portugal*
André Schütze, *IPA- Institute for Prevention and Occupational Medicine of the German Social Accident Insurance, Institute of the Ruhr-University Bochum, Bochum, Germany*
Sónia Norberto, *CINTESIS - Centro de Investigação em Tecnologias e Sistemas de Informação em Saúde, Centro de Investigação Médica, Porto, Portugal*
Conceição Calhau, *CINTESIS - Centro de Investigação em Tecnologias e Sistemas de Informação em Saúde, Centro de Investigação Médica, Porto, Portugal*
Valentina F. Domingues, *REQUIMTE/LAQV - Instituto Superior de Engenharia do Porto do Instituto Politécnico do Porto, Porto, Portugal*
Holger M. Koch, *IPA- Institute for Prevention and Occupational Medicine of the German Social Accident Insurance, Institute of the Ruhr-University Bochum, Bochum, Germany*

Di-iso-nonyl-cyclohexane-1,2-dicarboxylate (DINCH) is used as substitute for High Molecular Weight (HMW) phthalate plasticizers like di (2-ethylhexyl) phthalate (DEHP) and the diisononyl phthalate (DiNP). The available data suggests that DINCH, contrary to phthalates such as DEHP and DINP, is neither an endocrine disruptor nor a reproductive toxicant. Similar to the phthalates, DINCH is only dissolved in the polymer and not chemically bound to it. Thus, due to the ongoing substitution process DINCH exposures of the general population seems likely to occur. The aim of this study was to evaluate the exposure of a group of 112 children from Portugal, divide in two groups: i) diagnosed for obesity/overweight (cases) and ii) healthy weight (controls). Samples were collected during the years 2014 and 2015. Oxidized DINCH metabolites (OH-MINCH, cx-MINCH and oxo-MINCH) were analyzed after enzymatic hydrolysis via on-line HPLC-MS/MS with isotope quantification. In this study, a detection rate of 100% for OH-MINCH and cx-MINCH, and of 99% for oxo-MINCH was achieved. The median creatinine adjusted (95th percentile) values were of 1.16 µg/g (8.79) for cx-MINCH, 1.09 µg/g (7.22) for oxo-MINCH, and 2.14 µg/g (17.25) for OH-MINCH. For the unadjusted concentrations, the median (95th percentile) values were of 1.08 µg/L (7.33) for cx-MINCH, 1.10 µg/L (7.54) for oxo-MINCH, and 2.14 µg/L (15.91) for OH-MINCH. No significant differences were observed between the exposure levels of the case and the control. To the best of our knowledge, the levels of DINCH urinary metabolites have not been yet reported for the Iberian population. The results show a widespread exposure in the participating children, indicating a possible similar scenario for the overall Portuguese population. Acknowledgments: L. Correia-Sá is grateful to FCT by the grant (SFRH/BD/87019/2012), financed by POCH, subsidized by Fundo Social Europeu and Ministério da Ciência, Tecnologia e Ensino Superior.

Mo-Po-35

Assessment of indoor temperature, relative humidity, carbon dioxide, noise and illuminance level in two general hospitals

Boram Lee, *Seoul National University, Seoul, Korea, North*
Kiyoung Lee, *seoul national university, seoul, Korea, South*
kyoosang Kim, *Seoul Medical Center, seoul, Korea, South*

Aim: Healthy and comfortable indoor environment in a hospital has a major effect on patient well-being as well as on the work efficiency of the hospital staff. Although thermal comfort such as temperature and humidity is one major factor in indoor comfort, noise, lights and air quality are also important for indoor comfort. The purpose of this study was to assess the indoor temperature, relative humidity(RH), carbon dioxide, noise and illuminance by places. And also propose the control plan of indoor environment management.
Methods: Various indoor environmental conditions were measured in two general hospitals during a year (April 2014- April 2015). Each hospital was measured for 1 month, after which the monitoring instruments were moved to the other hospital also for 1 month. The indoor air temperature, relative humidity, carbon dioxide, noise level and illuminance level were measured at the same time.
Results: In hospital A and B, There was no difference between the temperatures of the each place during spring, summer and fall. In winter, the lowest temperature was the health screening centers ($19 \pm 2.3^{\circ}C$) in hospital A, lobby ($18.9 \pm 2.4^{\circ}C$) in hospital B and the highest was the nurse station ($27.3 \pm 0.6^{\circ}C$) in hospital A, nurse station ($27.2 \pm 0.4^{\circ}C$) in hospital B. Indoor relative humidity was the highest during summer (hospital A: $78 \pm 6\%$, hospital B: $73 \pm 10\%$) and the lowest during winter (hospital A: $18 \pm 6\%$, hospital B: $20 \pm 4\%$) in both two hospitals. There were no significant difference between RH of the five places in each hospitals. CO_2 levels of two hospitals were maintained less than 1000ppm during a year except nurse station and there are no seasonal differences of the CO_2 levels. The noise level of the two hospitals had exceeded WHO standards in hospitals of 30 LAeq dB during a year. Even the background noise during non-operating hours in the two hospitals exceeded 30dB. The results from other hospital noise level researches had similar results. Therefore, the WHO noise level guideline for hospitals seems hard to be achieved. The ISO 8995-1:2002 recommend indoor work place illuminance should not be less than 200lx. The seasonal average of Lobby, injection room and nurse station in hospital A were the only places that had illuminance of >200 lx.
Conclusion: Two hospitals were not satisfy the recommendations of indoor relative humidity, noise and illuminance levels. These factors that can affect the health and work efficiency of occupants such as patients and hopital staff shoud be managed.

Summer (Jun-Sep)	Hospital	Lobby	Injection room	Health screen center	Nurse station	Emergency room	Waiting room
Temperature	A	25.4±1.0	24.9±0.7	26.7±0.8	26.7±0.7	NA	27.0±0.9
(℃)	B	26.1±1.1	26.5±1.1	27.3±1.2	26.7±0.8	25.9±1.1	NA
RH	A	80.3±4.2	83.6±6.2	78.2±7.5	75.8±6.0	NA	73.0±6.8
(%)	B	74.0±9.5	75.1±10.0	74.0±9.5	67.4±5.6	77.4±11.4	NA
CO_2	A	451±102	527±113	562±130	598±37	NA	555±155
(ppm)	B	493±78	541±116	516±81	715±164	428±80	NA
LAeq	A	58.2	40.9	40.0	49.6	NA	49.1
(dB)	B	54.0	45.9	51.4	49.1	53.2	NA
Illuminance	A	248±146	264±271	59±67	359±38	NA	130±115
(Lux)	B	129±73	55±84	30±50	127±80	179±80	NA

Winter (Nov-Jan)	Hospital	Lobby	Injection room	Health screen center	Nurse station	Emergency room	Waiting room
Temperature	A	21.0±16	24.6±1.4	18.9±2.1	27.3±0.6	NA	19.3±2.3
(℃)	B	18.9±2.4	23.9±1.8	19.9±1.8	27.2±0.4	21.8±1.0	NA
RH	A	15.4±5.7	14.7±5.4	20.6±7.2	17.4±4.1	NA	20.4±6.3
(%)	B	25.2±13.2	17.8±7.2	22.3±9.1	13.6±2.1	19.8±8.0	NA
CO_2	A	539±72	565±121	507±53	843±66	NA	557±142
(ppm)	B	550±81	552±111	548±77	780±98	597±46	NA
LAeq	A	56.9	52.2	45.8	47.4	NA	43.5
(dB)	B	49.1	42.9	48.6	50.1	53.8	NA
Illuminance	A	185±133	93±174	60±69	217±40	NA	85±101
(Lux)	B	26±30	58±87	10±11	148±9	171±58	NA

Indoor temperature, RH, CO2, noise level and illuminance in two general hospitals.

Mo-Po-36

Bioavailability of plasticizers in dust and food after oral administration in pigs

Veronika Plichta, *Bavarian Health and Food Safety Authority, Munich, Germany*

Bioavailability of plasticizers in dust and food after oral administration in pigs
Veronika Plichta, Hermann Fromme
Department of Chemical Safety and Toxicology, Bavarian Health and Food Safety Authority, Munich, Germany
Background: According to the US Environmental protection agency (US-EPA), infants have an unintended daily uptake of 60-100 mg dust. Dust contains among other things a various amount of pollutants. Especially phthalates are present in high amounts. This could pose a potential health risk. At the current state of knowledge, there is no information about the bioavailability of pollutants in the matrix dust. Therefore, in risk assessments, the bioavailability is assumed to be 100%. This could lead to an overestimation because in vitro digestions tests indicate that the bioaccessability of phthalates in house dust range only between 2.4%-32%. Phthalates are widely used as plasticizers in many consumer products e.g. food packaging, toys, clothing and personal care products. Phthalates are not chemically bound to the polymer matrix, so they can easily be released into the surrounding by leaching or migration. Humans are exposed to phthalates on a daily basis via ingestions, inhalations or dermal uptake. Phthalates act as endocrine disruptors and target mainly the reproductive system. They do not accumulate in the body and are rapidly metabolized and excreted via urine within 48 hours.
Objectives: The oral bioavailability of certain Phthalates and Diisononylcyclohexa-1,2-dicarboxylat (DINCH) in house dust will be determined absolutely and relative to the bioavailability after intravenous administration or by ingestion.
Methods: Eight 5-week old piglets were fed once with five different dust samples from daycare centers and one food sample. Additionally, the target compounds were given by intravenous injection. The urine was collected over a period of 38 hours. The metabolites were quantified in the urine by using an HPLC-MS/MS method.
Results: The preliminary experiment (three piglets and three dust samples) showed a median uptake rate of 30% DEHP (range 24-43%) and 27% DINCH (range 23-31%) of the given dose. Those results indicated that the bioavailability is higher than in the in vitro digestion tests (DEHP: 2.24-12.6%) but also not 100%. The results of the main experiment are in progress.

Mo-Po-37

Quantitative material releases from articles containing manufactured nanomaterials

Joonas Koivisto, *National Research Centre for the Working Environment, Copenhagen, Denmark*
Alexander C. Ø. Jensen, *National Research Centre for the Working Environment, Copenhagen, Denmark*
Kirsten I Kling, *National Research Centre for the Working Environment, Copenhagen, Denmark*
Asger Nørgaard, *National Research Centre for the Working Environment, Copenhagen, Denmark*
Anna Brinch, *COWI, Kongens Lyngby, Denmark*
Frans Christensen, *COWI, Kongens Lyngby, Denmark*
Keld A Jensen, *National Research Centre for the Working Environment, Copenhagen, Denmark*

Introduction
Environmental and human risk assessment models are critical for estimating the impact of products containing engineered nanomaterials (NMs). Risk assessment is based on quantitative exposure levels and hazard estimation. In this review study, we extracted quantitative releases (e.g. mg/mL, mg/kg/wash), or emission rates (e.g. mg/m^2/year, mg/s) as well as properties of released particles from 320 different scenarios, including mechanical treatment, UV irradiation, washing, leaching, and spraying.
Methods
A total of 89 peer-reviewed scientific publications were identified as relevant studies considering release of NMs from consumer products. Quantitative releases or emission rates were calculated from the measured average concentration levels and volumes of immersion fluid or dilution air by assuming full mixing and insignificant sampling losses. This provides first order approximation for the quantitative releases or emission rates.
Conclusions
This review provides a basis for an NM emission library, which ideally should take into account the quantitative release, properties of the released particles, release scenario, and potential fate of the released NMs. We were able extract quantitative releases from 33 studies while the majority of the studies did not provide sufficient conceptual information. Emissions varied many orders of magnitude depending on the stress applied to the article, environmental conditions, and article properties (Table 1). NMs were mainly released as agglomerates coated by matrix. Release of NM free of matrix was minor or not reported. In future, the studies should take into account losses in experimental setup (e.g. deposition, sampling efficiency) and influence on the release characteristics (e.g. coagulation, flocculation). In airborne emission rate assessments, characterization should include size-resolved particle concentrations from 10 nm to 10 µm, respirable mass concentration and total material removal rate.
Acknowledgements
This work was supported by the European Union Seventh Framework Programme [FP7/2007-2013] under EC-GA No. 604305 'SUN', a grant from the Danish Environmental Protection Agency, and by a grant from the Danish Centre for Nanosafety (grant agreement no. 20110092173/3). The abstract template is adopted from the European Aerosol Conference.

Table 1. Quantitative emission ranges from articles containing NMs.

Release scenario	Range of release
Artificial weathering	10^1 to 10^5 mg m^{-2} at UV dose of ca. 150 MJ m^{-2}
Mechanical treatment	9×10^4 to 3.1×10^{10} s^{-1} / 52 to 258 μg s^{-1}
Mechanical treatment of artificially weathered articles	<1 to 2.7×10^6 s^{-1}
Pump sprays	1.8×10^7 to 2.9×10^9 g^{-1}
Propellant sprays	2.0×10^9 to 9.0×10^{10} g^{-1}
Fabrics after 1st wash and rinse	Containing Ag: 0.5 to 35 % Containing TiO_2: 0.01 to 3.4 %

Mo-Po-38

Exposure to Carbon Monoxide during the Operation of Recreational Watercraft - a Public Health Hazard with Potentially Lethal Outcomes

Dale Stephenson, *Boise State University, Boise, Idaho, United States*
Megan Smith, *Boise State University, Boise, Idaho, United States*
Robert Vercellino, *Boise State, Boise, Idaho, United States*
Lea Duffin, *Boise State University, Boise, Idaho, United States*
Tim Burke, *Boise State University, Boise, Idaho, United States*

Carbon monoxide (CO) is a colorless and odorless gas generated from incomplete combustion of hydrocarbon-based fuels. Human exposure to elevated CO concentrations causes serious injury or even death. Of increasing concern are CO-related poisonings and fatalities associated with the use of recreational watercraft. According the US Coast Guard, from 1981 to 2009 there were 879 poisonings and 160 fatalities as a result of exposure to CO during the operation of recreational watercraft. The US EPA has established CO exposure thresholds of 420 ppm and 1700 ppm, which, if the former is exceeded, impairs the ability to self-escape and, if the latter is exceeded, has the potential for life-threatening effects. Further, it is well-recognized by medical experts that instantaneous exposures exceeding 2000 ppm can be rapidly fatal. This study is intended to promote awareness of this issue in anticipation of policies and practices that will be developed to protect the public from adverse exposure to CO during the operation of recreational watercraft.

We monitored real-time CO concentrations on and adjacent to six ski boats, three with inboard engines and three with outboard engines. While the engines were at idle, real time CO grab samples were collected from thirty monitoring locations on and adjacent to each boat. We also collected real time CO grab samples from the breathing zone of an individual situated on a floatation device being towed behind each boat while the engine was under load. Finally, we fixed a data-logging CO monitor to estimate exposures received by a boat operator while the engine was both at idle and under load.

Results from grab samples collected at engine idle (n=180) showed that CO concentrations at the swim ladder and swim platform locations were highest, ranging from 1800 ppm to 5000 ppm. Results of breathing zone samples collected while an individual was being towed five feet behind each boat (n=60) ranged from 80 ppm to 500 ppm. Estimation of exposures received at the passenger seat location while a boat was both at idle and traveling 5-10 miles an hour (n=240) ranged from 5 ppm to 1000 ppm. These data suggest that the maximum CO concentration values found in this study have the potential to exceed US EPA exposure thresholds and, in some instances, exceed concentrations thought to be rapidly fatal.

Mo-Po-39

Improving Personal Exposure Assessment for Trace Metals

Ashley Newton, *Johns Hopkins Bloomberg School of Public Health, Baltimore, Maryland, United States*
Berrin Serdar, *University of Colorado at Denver, Denver, Colorado, United States*
Kirsten Koehler, *Johns Hopkins Bloomberg School of Public Health, Baltimore, Maryland, United States*

AIM: Airborne particulate metals exposure leads to many adverse health outcomes, including multiple cancers and depression of both the cardiovascular and respiratory systems. However, the high cost involved in the sampling and analysis of trace metals has led to few personal exposure assessments being performed for the exposed populations. Additionally, conventional samplers measure the inhalable fraction of particulate matter, a dose that corresponds most strongly with external exposure. However, it is the fraction of particles deposited in the lung that contributes most to overall body burden. This lung-deposited fraction can differ from the inhalable fraction by a factor between two and six. We aim to validate a low-cost analysis method for trace metal particulate matter using microfluidic paper-based analytical devices (μPADs) paired with an inexpensive lung deposition sampler (LDS). The LDS uses polyurethane foam as the collection substrate and is engineered to collect only the lung-deposited fraction. This system not only reduces the cost of personal exposure assessment for our metals of interest, but it also provides a measure of internal dose as a more physiologically relevant exposure metric.
METHODS: Metal fumes were generated inside an air chamber and side-by-side filter and LDS samples were collected in order to capture both the inhalable and lung-deposited fractions. Collection substrates were extracted using microwave-assisted acid digestion, and a portion of the extract was analyzed via ICP-MS to serve as the reference. μPADs were then used to quantify metals concentrations with both the conventional filter and the LDS for iron, nickel, copper, and total chromium.
As it is solely capturing the deposited dose, the LDS/μPAD system shows systematically lower results than the filter/μPAD system previously validated by Cate et al. Thus, in order to meaningfully compare the two systems, the total particulate metal concentration was calculated as the sum of the metals concentration measured by the LDS and a secondary filter behind the LDS to capture the mass not deposited on the foam plug (i.e. the exhaled fraction).
RESULTS and CONCLUSIONS: We have validated a low-cost system to analyze the lung-deposited fraction of metals in particulate matter. Good agreement was observed between ICP-MS and the μPAD metals concentrations using the LDS substrate, meeting NIOSH guidelines for method equivalence.
This combination of sampler and analytical method has the potential to reduce costs for personal exposure assessments by several orders of magnitude, leading to an increase in hazard surveillance in the occupational and environmental settings.

Mo-Po-40

Colorimetric Paper-Based Biosensing Device for the Assessment of Bisphenol A in Indoor Dust.

Alan Rossner, *Clarkson University, Potsdam, NY, United States*
Ramiz Alkasir, *Clarkson University, Potsdam, NY, United States*
Silvana Andreescu, *Clarkson University, Potsdam, NY, United States*

Objective: The aim of our work was to evaluate a new design for a paper-based sensing device for the colorimetric detection of Bisphenol A (BPA) in household dust. Bisphenol A (BPA) is found in poly-carbonate plastic and epoxy resin and is used in variety a of commercial and consumer products. The leaching of BPA consumer products can result in human exposure via inhalation, ingestion, and dermal routes. As a result, humans have been exposed in their home and work environment to BPA. Several studies have reported detectable levels of total urinary BPA in the majority of individuals in a number of populations, in the United States and other locations around the world. To reduce human exposure we need improved sensing devices to allow for quick, effective and inexpensive screening of our living and working environment. Methods: A rapid procedure for dust collection is used with a sensitive method for BPA detection, based on the formation of a greenish color, on the test zone of the sensing device. The system employs interchangeable low-cost paper-based enzyme sensors as a test zone (0.6 cm in diameter) for BPA detection interfaced with an air-sampling cassette as a sample collection area. The color results from the formation of a Schiff base compound, quinine-imine, formed by reaction of chitosan with the enzymatic product of tyrosinase o-quinone on paper coated in a layer-by-layer (LbL) assembly approach. Colorimetric response was concentration dependent with a detection limit of 0.28 µg/g. The color started to appear within the first 60 s and stabilized after 30 min. Replicate samples were run on a Gas Chromatography (GC) as means of validating the colorimetric data. Field sampling was conducted in a series of homes where dust specimens were collected from different homes and a day care center. Results: Results between the GC and colorimetric sensor showed a linear regression (R2 = 0.9743) for samples measured by both of the colorimetric and GC methods. In this work, BPA ranged in concentration from 0.05 to 3.87 µg/g in 57 samples of household dust when both methods were used. Conclusions: While the sample set was relatively small (n=57), the correlation between the colorimetric sensor and GC method was excellent, thus we feel the sensors is promising as a quick, inexpensive means of measuring BPA in settled dust in the home and work environment.

Mo-Po-41

Radionuclides in Contaminated Soils as Possible Source of Inhalation Exposure

Olorunfemi Adetona, *The Ohio State University, Columbus, Ohio, United States*
Chieh-ming Wu, *The Ohio State University, Columbus, Ohio, United States*
Brian Viner, *US Department of Energy, Aiken, South Carolina, United States*
Tim Jannik, *US Department of Energy, Aiken, South Carolina, United States*
Daniel Stone, *US Department of Energy, Aiken, South Carolina, United States*
Allan Hepworth, *US Department of Agriculture, Aiken, South Carolina, United States*
Teresa Eddy, *US Department of Energy, Aiken, South Carolina, United States*
Luke Naeher, *University of Georgia, Athens, Georgia, United States*
Anna Adetona, *University of Georgia, Athens, GA, United States*

Release of radionuclides into surface waters due to past production of nuclear materials resulted in the incorporation of radionuclides in the forest soils at the Savannah River Site (SRS), South Carolina, United States. As part of an effort to model possible exposure of emergency response workers including forest firefighters during hypothetical fires in the contaminated wetlands of the site, concentrations of natural and artificial radionuclides were determined by gamma spectrometry in mineral and surface (fresh and decomposed vegetative litter from trees) soil samples collected along four streams with historical inflow from nuclear facilities at the site. The soil samples were collected from twelve paired locations downstream of historical sources of radionuclide inflow for each stream. Each paired location comprised of one sample site in a predetermined contaminated area and another in a predetermined uncontaminated area. The contamination status of the areas was determined by gamma overflight with ground validation of gamma emission levels. Linear mixed effect models were used to compare radionuclide activities across the different streams and sample locations, while Spearman rank correlation was run to determine the association between the radionuclides. Generally, concentrations of artificial radionuclides including 137Cs, 60Co 238Pu and 239Pu, and gross beta were higher at contaminated locations in both the mineral and surface soil ($p<0.05$). Such differences were not observed with naturally occurring radionuclides of uranium and 40K. 137Cs, which was the predominant radionuclide in discharged cooling water, had the highest radioactivity in both mineral and surface soils among the radionuclides, and had higher radioactivity upstream in mineral soil. The highest model estimated stream average of 137Cs observed was 593 Bq/kg, and is within the range previously reported for the site. There was moderate correlation between mineral and surface soil concentrations of 137Cs ($r=0.74$) and gross beta ($r=0.50$). Additionally, 137Cs was moderately correlated with gross beta in mineral ($r=0.70$) and surface ($r=0.66$) soil samples. These results suggest that 137Cs in discharged water from nuclear production contributes substantially to radioactivity and gross beta emissions in the soil at SRS. Additionally, the results are an evidence that 137Cs is readily taken up by vegetation from the wetland soil as indicated by the extent of correlation of 137Cs radioactivity between mineral and surface soil. The measured radionuclide concentrations in this study will be input into an already developed model that incorporates emissions and plume dispersion to estimate potential exposure to radioactivity in contaminated areas of SRS.

Mo-Po-42

An Assessment of Legionella pneumophila and Mycobacterium avium in Residential and Commercial Structures

Maura Donohue, *US EPA, Cincinnati, Ohio, United States*
Lisa Melnyk, *US EPA, Cincinnati, OH, United States*

Legionella pneumophila and Mycobacterium avium are water-borne pathogens. I investigated the occurrence of L. pneumophila and M. avium in potable water at residential and commercial structures. This was performed to understand which locality may pose the highest risk for infection based on the occurrence and observed concentrations of these microorganisms in potable water.
Between the years of 2009-2014, potable water from 136 taps (62 residential and 74 commercial) across the United States were monitored for the presence of these microorganisms. Water was collected at three independent time points for a total of 408 samples. In general, three liters of water were filtered through a polycarbonate membrane. The DNA was extracted from the captured material. Quantitative Polymerase Chain Reaction (qPCR) was then used to detect and measure the concentrations of these microorganisms in the final extract.
Both microorganisms were detected in the potable water from both residential and commercial buildings. Of the 62 taps that were located in a residence, 27% were positive for L. pneumophila, and 23% were positive for M. avium. The 74 taps that were located in a building, 41% and 39% were positive for M. avium and L. pneumophila, respectively. Longitudinal data revealed the L. pneumophila persisted for longer lengths of time in water from residences and at higher concentrations (avg: 10,830 genomic targets/L, Max: 82,250 genomic targets/L). For M. avium, higher concentrations (avg: 715 genomic targets/L, Max: 148,270 genomic targets/L) were detected in buildings. These findings will help craft a better understanding as to possible exposure routes to these microorganisms that are most relevant for infection to occur.
Disclaimer: The views expressed in this abstract are those of the authors and do not necessarily reflect the views or policies of the U.S. EPA. Mention of trade names or commercial products does not constitute endorsement or recommendation for use.

Mo-Po-43

Particle inhalation rate as a metric for ambient air pollution exposure

Laura Corlin, *Tufts University, Medford, MA, United States*
Mark Woodin, *Tufts University, Medford, MA, United States*
Jaime Hart, *Channing Division of Network Medicine, Boston, MA, United States*
Matthew Simon, *Tufts University, Medford, MA, United States*
David Gute, *Tufts University, Medford, MA, United States*
Harsha Amaravadi, *Tufts University, Medford, MA, United States*
Doug Brugge, *Tufts University, Boston, MA, United States*
John Durant, *Tufts University, Medford, MA, United States*

Aim: Many air pollution epidemiology studies use ambient residential air pollutant concentrations to represent exposure. Nevertheless, ambient concentrations may not accurately reflect even relative exposure levels since factors such as age, sex, weight, and physical activity differentially affect intake of air pollutants. We developed and tested a new exposure metric to try to better approximate the biologically relevant dose.
Methods: Using data from a longitudinal study of 812 adults (69% female) in Boston (MA, USA), we first modeled participants' annual residential average exposure to ultrafine particulate matter (UFP, <0.1 µm diameter, measured as particle number concentration or PNC). We then multiplied PNC estimates (particles/L) by hourly respiratory volume (L of air inhaled/hr) for each participant to obtain the average particle inhalation rate (PIR, particles inhaled/hr). Respiratory volume was estimated using published estimates of sex-, age-, and physical activity-adjusted ventilation rates together with data on how many hours per day participants engaged in defined levels of physical activity. We compared the distributions of PNC and PIR, considered whether associations between UFP exposure and cardiovascular disease (CVD) risk factors differed for PNC and PIR, and examined how sensitive the PIR effect estimates were to factors such as physical activity and respiratory medication use.
Results: While the PNC distribution was slightly left skewed (mean=23,000, median=24,000, min=10,000, max=32,000 particles/cc), the PIR distribution was slightly right skewed and had greater variability (mean=13x10^9, median=12x10^9, min=3.7x10^9, max=54x10^9 particles inhaled/hr). Distributions were stable over the five year study period. By design, distributions for the PIR strongly reflected physical activity patterns (r=0.7 p<0.001) even in this population with generally low physical activity levels. Notably, among those with highest PIR (greater than 90th percentile), 6% had low PNC exposure (<10th percentile).
As may be expected based on the different distributions, PNC and PIR showed different associations with CVD risk factors. We found that PNC was more strongly associated with increases in systolic blood pressure, pulse pressure, and high sensitivity C-reactive protein (a biomarker of systemic inflammation) while the PIR was more strongly associated with increases in diastolic blood pressure. Our PIR results did not change substantially when we conducted sensitivity analyses excluding participants taking respiratory medications (25% of participants) or excluding participants reporting the highest physical activity levels.
Conclusions: Our findings suggest that adjusting ambient UFP exposure estimates for inhalation rate affects the shape and variability of the exposure distribution and alters effect estimates for the association with CVD risk factors.

This page intentionally left blank.

Mo-Po-44

Concentration-Based High-Throughput Exposure Screening of Chemicals in Flooring Materials

Lei Huang, *University of Michigan, Ann Arbor, Michigan, United States*
Olivier Jolliet, *University of Michigan, Ann Arbor, Michigan, United States*
Deborah Bennett, *University of California, Davis, Davis, California, United States*
Hyeong-Moo Shin, *University of California, Davis, Davis, California, United States*
Jon Arnot, *ARC Arnot Research and Consulting, Toronto, Ontario, Canada*
Vy Nguyen, *University of Michigan Medical School, Ann Arbor, MI, United States*

Recently, increasing attention has been paid to the near-field human exposure to chemicals formulated in consumer products. Our previous Tier 1 project provided screening-level exposure estimates for chemicals in consumer products using total production volume and simple models/assumptions. The current Tier 2 effort aims at more refined exposure estimates using more advanced models and actual chemical concentrations in products. This study provides Tier 2 high-throughput exposure estimates for chemicals encapsulated in flooring materials, a major passive indoor emission source. A total of 186 chemicals were identified as ingredients in flooring materials based on the Pharos building material database, of which 136 have chemical content data in different types of flooring (e.g., wood or vinyl), constituting a total of 264 chemical-flooring combinations (cases). For each chemical-flooring type combination, we calculated typical concentration ranges using median and maximum possible content. A parsimonious model was developed to predict the diffusive chemical emissions from materials, subsequent losses by ventilation as well as flooring surface concentrations. This model is solely based on explicit equations, suitable for high-throughput calculations. Quantitative structure-activity relationships were developed to predict the key model parameters, diffusion coefficient Dm and material-air partition coefficient Kma, and were applied to the 264 cases. We then calculated the inhalation and dermal product intake fractions (PiFs), determining the fraction of the chemical in the flooring that is taken in by adults and children during its use phase.
The median chemical contents in the 264 cases vary by several orders of magnitude, from 0.001% to 31.5%. Likewise, the model-predicted fractional mass emitted from flooring to air over 3 years ranges from 2E-8 to 1. VOCs generally have a low content in flooring and a high fractional mass emitted, while the opposite is typically observed for SVOCs. For children under five, the inhalation PiF ranges from 4E-12 to 2E-4 and the dermal PiF from 1E-13 to 0.1 (Figure 1). For adults, the inhalation PiF ranges from 1E-10 to 1E-3, no dermal contact with flooring being assumed for adults. The daily intake dose is typically higher for children than for adults. The children's inhalation doses are in the range of 1E-4 to 1E4 µg/kg-day, with dermal dose of similar order of magnitude as for inhalation. Several chemicals that may lead to high exposures are identified for further investigations. These results suggest that the near-field exposures to chemicals in flooring can be substantial, especially for children crawling on the floor.

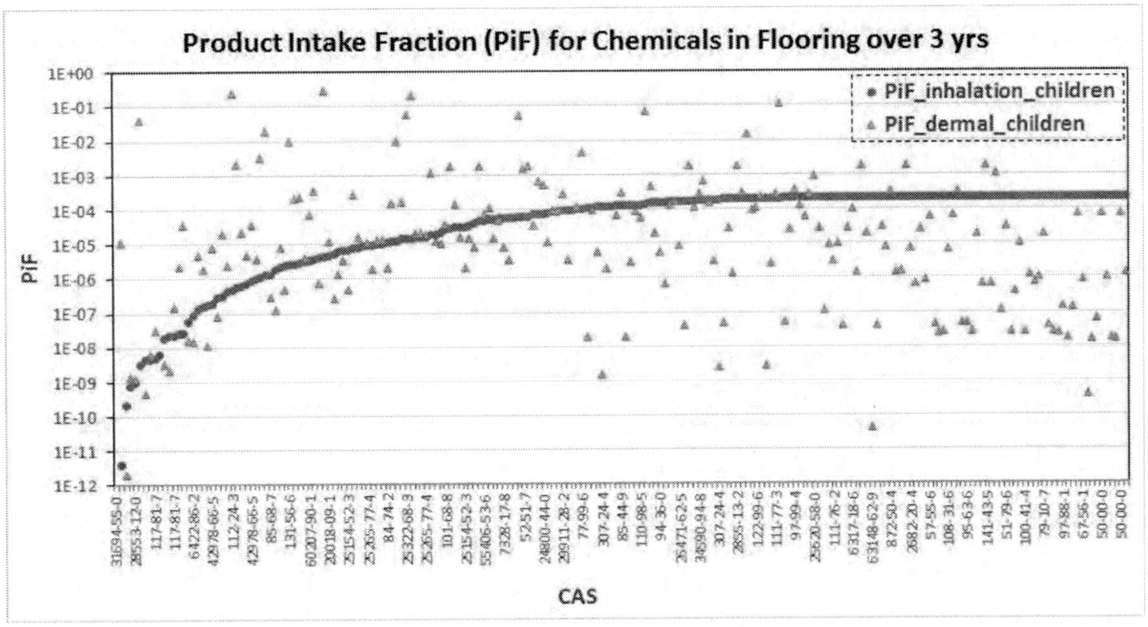

Figure 1. Product Intake Fraction (PiF) for Chemicals in Flooring over 3 yrs

Mo-Po-46

Exposure to PM2.5 and Blood Lead Level in Two Populations in Ulaanbaatar, Mongolia

Undarmaa Enkhbat, *Mongolian National University of Medical Sciences, School of Public Health, Ulaanbaatar, Mongolia*
Ana Rule, *Johns Hopkins Bloomberg School of Public Health, Baltimore, Maryland, United States*
Carol Resnick, *Johns Hopkins Bloomberg School of Public Health, Baltimore, Maryland, United States*
Chimedsuren Ochir, *Mongolian National University of Medical Sciences, School of Public Health, Ulaanbaatar, Mongolia*
Purevdorj Olkhanud, *Mongolian National University of Medical Sciences, School of Public Health, Ulaanbaatar, Mongolia*
D'Ann Williams, *Johns Hopkins Bloomberg School of Public Health, Baltimore, Maryland, United States*

Introduction: The World Health Organization (WHO) has ranked Ulaanbaatar as the second most polluted city in the world and much attention has been given to the issue of outdoor air pollution.6 The World Bank reports that particulate matter is the main air pollutant in Ulaanbaatar, mainly due to coal burning from individual gers. Through this research study we aim to test our hypothesis that women living in gers (traditional dwillings) burning coal in traditional stoves for cooking and heating during the winter are exposed to higher concentrations of airborne PM2.5 than women living in apartments in Ulaanbaatar, Mongolia, and this exposure may include exposures to lead in coal with effects on blood lead levels.

Methods: This cross-sectional study recruited a total of 50 women, 40-60 years of age living in apartments and gers. Air sampling was carried out during peak cooking and heating times, 5:00 p.m.-11:00 p.m., collecting direct-reading measurement (TSI SidePak™) gravimetric methods. Blood lead level (BLL) was measured using a LeadCare II rapid field test method.
Results: Measured PM2.5 geometric mean (GM) concentrations using the SidePak™ in the apartment group was 31.5 (95% CI:17-99) µg/m3, and 100 (95% CI: 67-187) µg/m3 in ger households ($p < 0.001$). The GM integrated gravimetric PM2.5 concentrations in the apartment group were 52.8 (95% CI: 39-297) µg/m3 and 127.8 (95% CI: 86-190) µg/m3 in ger households ($p = 0.004$). The correlation coefficient for the SidePak™ PM2.5 concentrations and filter based PM2.5 concentrations was $r = 0.72$ ($p < 0.001$). Blood Lead Levels were not statistically significant different between apartment residents and ger residents ($p = 0.15$). The BLL is statistically significant different ($p = 0.01$) when stratified by length of exposures outside of the home. This statistically significant difference in increased BLL could be due to occupational or frequent exposure to outdoor air pollution that were not measured. This study was first to measure BLLs among an adult population in Mongolia.
Conclusion: The results suggest that elevated BLLs may not be a major public health issue for older adult women who use coal however, the elevated PM2.5 levels observed inside ger homes are concerning and continue to be a significant risk to public health.

Mo-Po-47

Seasonal variation of time activity pattern characteristics in Korean population

Sewon Lee, *Seoul National University Graduate School of Public Health, Seoul, Korea, South*
Kiyoung Lee, *Seoul National University, Seoul, Korea, South*

The time-activity patterns distribution is necessary to calculate population exposure. This objective of this study was to characterize time-activity pattern of Korean population in three different seasons. The time-activity data of a total of 26,988 subjects older than 10 years was provided by the Time Use Survey of the Statistics Korea (KOSTAT). The survey was conducted in summer, fall and winter. The subjects were selected using probability proportional extraction method from 800 administrative areas of Korean in 2014. Each subjects completed a diary with 10-minutes interval for two consecutive days regardless of day of week. The location codes of the diary were classified into 5 microenvironments: indoor at home, workplace/school, restaurant/bar, transportation and other indoor/outdoor locations. They also completed a questionnaire including demographic, socioeconomic and family information. This study analyzed only weekday data. The time-activity pattern was analyzed by gender, age-related, urban-rural difference in summer, fall, and winter seasons. The times spent in different microenvironments were similar in the three seasons. The most time was spent at home (62-64%) followed by workplace/school (18-20%), transportation (6-7%), bar/restaurant (2%) and other indoor/outdoor locations (9-10%) in all seasons. The times spent at home and workplace/school were different by gender. Male spent 10-11% less time at home than female. However, male spent 7-8% more time at workplace/school than female. The times spent at home, workplace/school and other locations were different by age group. The teenage group spent 52-61% at home and the senior group spent 74-79% at home. The other age groups spent 57-62% at home. The time spent at workplace/school was 5-6% in the senior group, while times at workplace/school were 22-34% for the teenage group and 19-24% for the other age group. The times spent at different microenvironments were not different by urban-rural status. The difference of time spent at each microenvironment was less than 2%. Overall time location data of Korean population were different by age and gender but not urban/rural status in three seasons. However, further analysis is necessary to compare daily time activity pattern by the variables and determine interaction between the demographic variables on time activity patterns. Since time activity pattern is associated with exposure, population exposure assessment shall include difference by age and gender.

Mo-Po-48

Assessment of Personal Exposure to Black Carbon and Nitrogen Dioxide in Contrasting Urban (Road Traffic) and Industrial (Fracking) Environments

Eliani Ezani, *University of Strathclyde, Glasgow, United Kingdom*
Iain Beverland, *University of Strathclyde, Glasgow, United Kingdom*

Aim: This study used real-time monitors to assess and compare environmental and occupational exposure to black carbon (BC), nitrogen dioxide (NO_2), and ozone (O_3) in contrasting urban and industrial environments.
Methods: Pedestrian exposures to traffic-related air pollution in central Glasgow were compared to occupational exposures to diesel engine exhaust from industrial fracking equipment at an experimental hydraulic fracturing (HF) test site in Poland. Mobile measurements at varying distances from sources in both of the above locations were made using portable real-time micro-aethalometers (AE51) for BC and Aeroqual series-500 monitors for NO_2 and O_3 carried by researchers. Duplicate BC measurements were compared with NO_2 observations, after correction of the sensor response for the latter for O_3 interference effects.
Results: The highest 1-minute average BC concentrations measured at the HF test site were approximately five times higher than equivalent 1-minute average concentrations observed in central Glasgow (51.2 µg/m^3 and 10.0 µg/m^3 respectively). Similarly, maximum 1-minute average NO_2 concentrations measured at the HF test site exceeded equivalent 1-min average observations in Glasgow (292.3 µg/m^3 and 108.1 µg/m^3 respectively). The overall BC mean exposure on roadside walking routes in Glasgow was higher than the overall mean exposure observed across multiple locations, upwind and downwind of diesel engine sources, at the HF test site (2.0 µg/m^3 and 1.6 µg/m^{3}respectively). Mean NO_2 concentrations across Glasgow central roadside locations (64.7 µg/m^3) was also higher than overall mean occupational exposure to NO_2 at the HF test site (43.9 µg/m^3). Duplicate BC instruments provided very similar real-time measurements, which in turn were relatively highly correlated with NO_2 observations at 1-minute temporal resolution at the HF experimental site (R^2 = 0.65) and the central Glasgow walking route (R^2 = 0.88).
Conclusion: Marked elevations of BC and NO_2 concentrations were observed in downwind proximity to traffic and industrial fracking equipment sources. Exposure to diesel engine exhaust emissions from fracking equipment may present a significant risk to people working on HF sites over extended time periods. The short time resolution of the portable instruments used, enabled identification of sources of occupational and environmental exposure to combustion-related air pollutants.

Mo-Po-49

Residential PM2.5 concentrations in Ger, traditional residence in Mongolia

Hyunkyung Ban, *Seoul National University, Seoul, Korea, South*
Chimedsuren Ochir, *Mongolian National University of Medical Sciences, Ulaanbaatar, Mongolia*
Undarmaa Enkhbat, *Mongolian National University of Medical Sciences, Ulaanbaatar, Mongolia*
Yanjinlkham Bold, *Mongolian National University of Medical Sciences, Ulaanbaatar, Mongolia*
Kiyoung Lee, *Seoul National University, Seoul, Korea, South*

Aim: Significant proportion of population in Ulaanbaatar (UB), the capital city of Mongolia, is living in traditional ger home. Residents in ger can be exposed to high concentration of PM2.5 from burning coal and biomass fuels. The purposes of this study were to measure indoor PM2.5 levels in gers and determine effects of fuel combustion for cooking and heating on indoor PM2.5 level.
Method: This study was conducted in 28 gers in UB in January 2016. PM2.5 concentration, temperature and relative humidity were simultaneously measured by real time monitors during daytime for about 7 hours. During the monitoring, two researchers recorded stove type, fuel type, fuel usage and the number of putting fuels into stove.
Results: The average PM2.5 concentration, temperature and relative humidity in 28 gers were 208±176 µg/m3 (n=28), 23±4 °C (n=27, one home is missing due to malfunction of the monitor) and 21±8 % (n=28), respectively. Overall, the PM2.5 concentration was higher in the morning and then decreased constantly until late afternoon. Five temporal profiles of indoor PM2.5 concentration in 28 gers were determined (Fig 1). Average PM2.5 concentrations in the five groups were 384±171 µg/m3, 171±71 µg/m3, 123±56 µg/m3, 74±27 µg/m3 and 27±20 µg/m3. In the first group, indoor PM2.5 concentrations were very high and extreme peak of about 8500 µg/m3 was observed. In the second and third groups, indoor PM2.5 concentrations in the morning were 600 and 400 µg/m3 respectively and decreased gradually. In the second group, there was peak of about 1100 µg/m3 in the afternoon. In the third group, less peak was observed. In the fourth group, indoor PM2.5 concentrations in the morning were 200 µg/m3. There was peak of about 180 µg/m3 in the afternoon. In the fifth group, PM2.5 concentration was maintained below 50 µg/m3 in the afternoon. The peaks in daily profile of PM2.5 concentration were coincided with smoking, burning herbs or candle, opening stove for refueling, cooking fried food and using vacuum cleaner.
Conclusions: The indoor PM2.5 concentrations in gers were very high with peak concentration up to 8500 µg/m3. However, our measurement might underestimate 24 hour exposure due to lack of monitoring during night and early morning. Further study is necessary to measure indoor concentration for 24 hours and identify factors associated with daily indoor exposure.

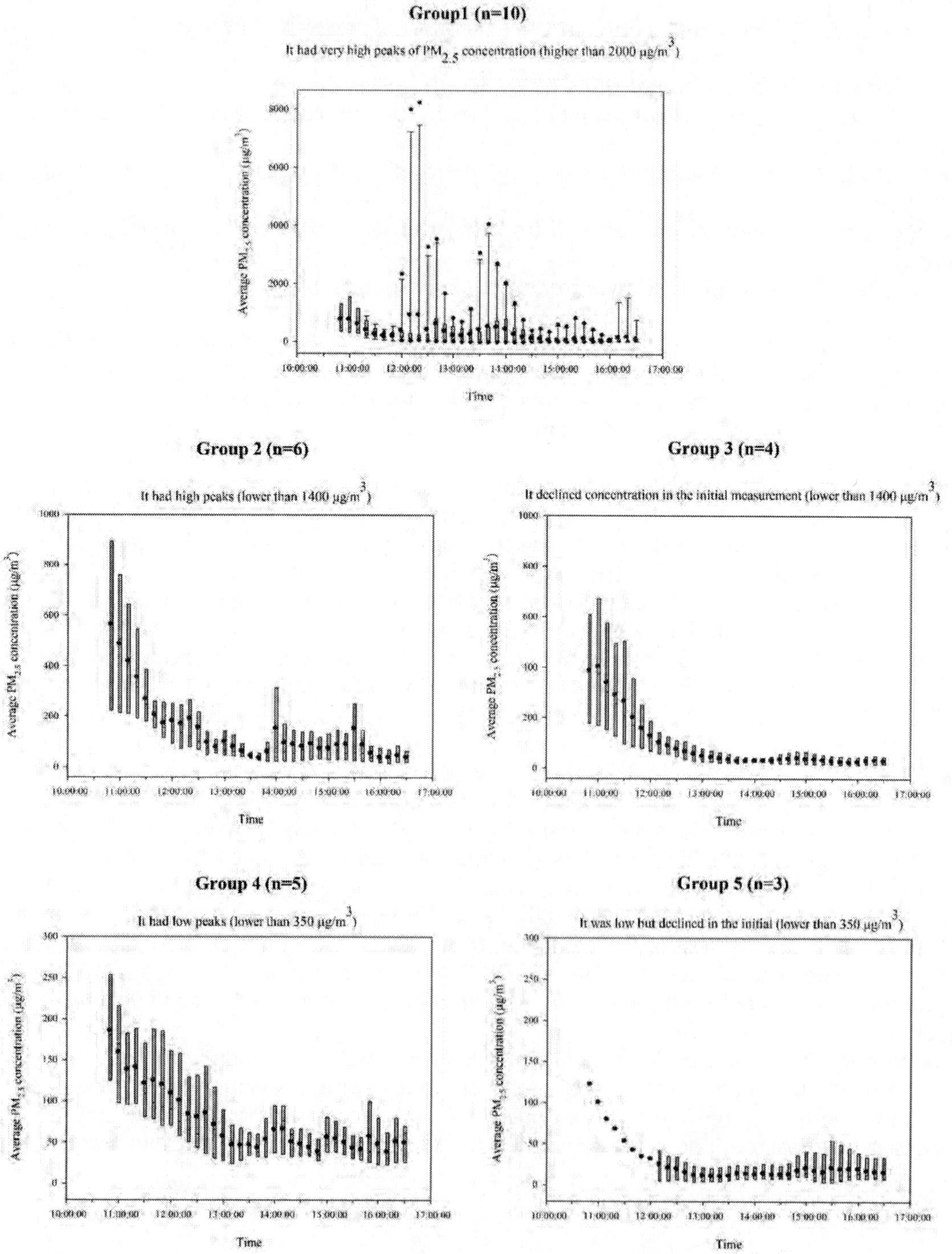

Fig. 1. Five types of temporal profiles of $PM_{2.5}$ concentration in 28 gers

Mo-Po-50

Microenvironmental Exposure to Ultrafine Particles Among Adolescent Children Characterized by A Personal Sensor with High Spatial and Temporal Resolution

Christopher Wolfe, *Cincinnati Children's Hospital Medical Center, Cincinnati, OH, United States*
Grace LeMasters, *University of Cincinnati, Cincinnati, OH, United States*
James Lockey, *University of Cincinnati, Cincinnati, OH, United States*
Sang Young Son, *University of Cincinnati, Cincinnati, OH, United States*
Patrick Ryan, *Cincinnati Children's Hospital Medical Center, Cincinnati, OH, United States*

Background: Ultrafine particles (UFPs) exhibit high spatiotemporal variation, which makes assigning individual level exposures challenging. Additionally, the dynamic time-activity patterns of individuals in daily life further complicate exposure assessment and necessitate exposure methods with high spatial and temporal resolution. To address this need, a personal UFP sensor (PUFP C100) capable of measuring GPS location and UFP concentration at 1 second intervals has been developed (Enmont, LLC). Based on user feedback from previous field testing of the PUFP C100 model, significant improvements have been made while maintaining sensor performance. The purpose of this report is to detail the results of additional field testing to assess the performance and user satisfaction with the modified sensor (PUFP C200).
Methods: Adolescent children (n = 10; ages 13 - 17) will wear the second generation UFP sensor for 3 hours after school for 7 consecutive days in Spring 2016. During sensor operation, participants will document their activity patterns including location and exposure sources (tobacco smoke, cooking, etc.). We will use the documented activity patterns to validate GPS location and to develop an algorithm to characterize UFP concentrations within different exposure microenvironments (home, transit, school). We will also examine the inter/intra-personal variation of UFP concentration by microenvironment.
Results: The PUFP C200 has significant reductions in size (40%; 910 cc vs. 1500 cc) and weight (25%; 0.75kg vs. 1 kg). A new software interface (EView™) complete with data visualization tools for real-time UFP monitoring via Bluetooth has also been developed. Furthermore, the software translates GPS location data into Google Earth compatible files with formatting options to indicate the degree of UFP exposure. Participant recruitment has been completed, and field testing is currently underway. Once field testing has concluded in Spring 2016, over 200 hours of sampling data will be collected, analyzed prepared and for presentation.
Conclusion: We hypothesize that sensor improvements will increase user satisfaction and that personal exposures will differ significantly by study participant and microenvironment. In the future, the PUFP C200 will be deployed in a larger epidemiological study to characterize UFP exposure. The goal of this prospective study will be to examine the association between UFP exposure and pulmonary function in adolescents with and without asthma.

Mo-Po-55

Measuring concentrations and sources of flame retardants and phthalates indoors

Miriam Diamond, *University of Toronto, Toronto, Ontario, Canada*
Shelley Harris, *Cancer Care Ontario, Dalla Lana School of Public Health, Toronto, Ontario, Canada*
Congqiao Yang, *University of Toronto, Toronto, Ontario, Canada*
Liisa Jantunen, *Environment Canada, Egbert, Ontario, Canada*
Dina Tsirlin, *Cancer Care Ontario, Toronto, Ontario, Canada*
Bruce Fraser, *Health Canada, Ottawa, Ontario, Canada*
Joseph Okeme, *University of Toronto, Toronto, Canada*

Indoor concentrations of and exposures to flame retardants and plasticizers are influenced by poorly characterized emission sources. Furthermore, the physical-chemical characteristics of individual compounds vary greatly, influencing their indoor partitioning and residence time, and hence indoor levels and patterns.
Here we report on a case control study in Ontario in which a subsample of 51 women from the Greater Toronto Area and Ottawa were enrolled. Our goal was to characterize indoor concentrations, indoor fate and exposure to participants of brominated and phosphorus-based flame retardants and phthalate plasticizers and to estimate exposure. In the bedroom and most used room we measured air concentrations in participants' homes using two passive air samplers (Harner-type polyurethane foam and polydimethylsiloxane), collected floor dust and sampled window films. We also wiped the surfaces of electronic products to assess their potential contribution to indoor levels, as well as hand wipes of participants. A questionnaire was developed and administered to gather personal demographic, and lifestyle data, and environmental/household characteristics including the presence and patterns of use of electronic devices, types of furniture and mattresses and presence of foam padding. Analysis of passive air samplers (PDMS n=100), showed that the flame retardant TDCPP was the most abundant novel-BFRs followed by ATE, PBBz, and PBT, with PBEB and TBB randomly detected; BDE-47 was the major PBDE congener followed by BDE-28, -99, BDE-49, and -17, with BDE-49, and -153 randomly detected. Phthalate esters were found at concentrations orders of magnitude higher than flame retardants.

Mo-Po-59

Determination of dietary patterns prior to exposure assessment of populations of Benin, Cameroon, Mali and Nigeria to harmful residues and contaminants

Luc Ingenbleek, *Centre Pasteur of Cameroon, Yaounde, Cameroon*
Sara Eyangoh, *Centre Pasteur of Cameroon, Yaounde, Cameroon*
Bruno Le Bizec, *LABERCA/Oniris, Nantes, France*
Philippe Verger, *INRA/Met@risk, Paris, France*
Jean-Charles Leblanc, *ANSES, Maisons Alfort, France*

Aim: In the framework of a WTO/FAO funded 3-year (2014-2017) project aimed at assessing the dietary exposure of populations of Sub-Saharan Africa to toxic contaminants, we are developing a cost effective way of determining the food consumption patterns in developing countries, prior to sampling food in a representative way.
Method: The latest official household budget surveys validated by national institutes for statistics of Benin, Cameroon, Mali and Nigeria representing food expenditure data of more than 70.000 households were processed with SPSS software, with the inclusion of edible fraction conversion factor and processing yield factor to express the average food daily intake and percentile 95 as consumed.
Key steps of the food intake calculation were: i) setting up of a dedicated food classification system including 3-level hierarchy and a correspondence list of each national food item recorded by national institutes for statistics with energy content (kcal/100g), edible conversion factor and yield factor ii) conversion of each household member into adult male equivalent (AME) iii) selection of normally reporting households based on energy daily intake (1200-5100 kcal/AME/day)iv) data processing to obtain food intakes as consumed.
Food list covered 95% of the total diet.
Sampling was designed with 12 subsamples by individual composite.
Food samples were prepared as consumed.
Results: The food consumption patterns of Benin, Cameroon, Mali and Nigeria were determined in cost effective manner, which is replicable to other developing countries. Although the utilization of 4 different datasets showed different average food intake profiles both within each country of the scope of the study and between the 4 countries, reflecting different customs and cultures, the total intake was consistent with an average intake of 1804g (SD=5%).
North versus South food consumption patterns appeared to be significantly different within the same country, with staple food used as diet markers, shifting progressively from cereal-based diet to tuber-based diet, which is probably resulting in differential exposure patterns, in particular as far as mycotoxins are concerned.
Conclusion: exposure assessment is not yet performed but will be available in 2017. As for now, the total diet study for Sub-Saharan Africa has focused on designing a replicable method suitable for developing countries and based on statistically meaningful already existing household budget surveys initially designed for following up poverty and living standards indicators.

Mo-Po-58

Exposure Measurement Error Reduced by Personal Air Pollutant Exposures Monitoring in an Active Young Adult Cohort

Donghai Liang, *Emory University, Atlanta, GA, United States*

Aim: We conducted a multipollutant exposure and health risk evaluation to assess a link between traffic pollution and health effects using a panel cohort: the Dorm Room Inhalation to Vehicle Emission (DRIVE) study. A sub-aim of this study is to assess the error between different exposure measurement approaches. Method: We measured PM2.5 concentrations at the personal level for 54 college students who live in two dormitories of different proximities from a highway (20 m (NR) (n=25) vs. 1.4 km (FR) (n=29)) with similar demongraphic characteristics including firsthand and secondhand tobacco smoke exposure, one indoor location of each dormitory, and six outdoor locations of different proximities from the highway connector (from 10 m to 2.3 km). Two consecutive 48-hours personal PM2.5 concentration measurements using RTI MicroPEM provided real-time data and filter samples with personal real-time location information by GPS. Stationary monitors of GRIMM and TEOM collected real-time data. Black carbon (BC) as marker of traffic emission was optically analyzed in real-time by MicroAeth and at an integrated level from the personal filter samples by reflectance. Results: Outdoor sample data showed a sharp decrease of BC between 10 m and 20 m from the highway (1.7 vs. 0.9 µg/m3). However, other locations further from the highway, including the background site (2.3 km) showed similar level of BC and BC/PM2.5 ratio, and PM2.5 between NR and FR were similar, suggesting a contribution by local traffic. Indoor samples showed a lower PM2.5, and higher BC and BC/PM2.5 ratio for NR compared to FR. Personal samples showed negligible differences in PM2.5 between NR and FR (8.3 vs. 7.5 µg/m3). Participants spent at least 40% of their time not in their dormitory. GPS data showed the averaged time-weighted distances of each participant from the highway are 509 m for NR participants and 2174 m for FR participants. This suggests a potential exposure misclassification by classifying exposure by residential location, especially for the NR. Conclusions: Accurate exposure assessment using personal exposure monitors, such as the MicroPEM, are feasible for use with healthy adults with diverse activity patterns. Stationary monitors inaccurately estimate personal exposure to PM, as they do not capture the exposure during the time when participants are away from the stationary monitoring sites.

Mo-Po-60

Pesticide residue monitoring programs: valuable tools for refined dietary exposure assessment and support of minor crop agriculture

Carrie Fleming, *Dow AgroSciences, Indianapolis, IN, United States*
Manoj Aggarwal, *Dow AgroSciences, Abingdon, United Kingdom*

Pesticide residue monitoring programs exist within various regulatory bodies around the world, and vary in the breadth of sampling, the transparency of their process, and the publication of the data. Information from these programs can be highly valuable in gaining more realistic estimates of human dietary exposure to pesticides. Such information is used in regulatory risk assessments in some geographies, while in others, conservative assumptions such as residues at the legal maximum residue limit (MRL) and 100% of food treated are employed. The use of such disparate assumptions to estimate dietary exposures can result in large discrepancies in exposure and risk estimates between countries unrelated to actual differences in exposures, and can create major challenges for global food trade. Furthermore, the use of conservative overestimates of exposure in place of existing data can limit the availability of key agricultural tools, particularly in specialty crops. This presentation will provide a survey of available data from residue monitoring programs and discuss their value in supporting pesticide dietary exposure and risk assessments. Real world examples will be provided demonstrating that the use of these data results in more realistic estimates of human exposure and the impacts that these programs have on specialty crop agriculture.

Late Breaking Abstracts

Mo-LBA-02

Young adult street vendors and reported health outcomes affected by measured exposures to near-roadway traffic-related air pollution in Bangkok, Thailand

Saisattha Noomnual, *Rutgers School of Public Health, Piscataway, New Jersey, United States*
Derek Shendell, *Rutgers School of Public Health, and Environmental and Occupational Health Sciences Institute, Piscataway, New Jersey, United States*

Aims and Introduction/Background: This study estimated adverse respiratory health outcomes among street vendors, in particular young adults, in Bangkok, Thailand, using a self-report questionnaire developed from previous validated questions from the peer-reviewed literature and air quality data. Air pollutants of concern were traffic-related air pollution (TRAP), including particulate matter in respirable coarse and fine size fractions (PM10 and PM2.5). There are no available data about knowledge, awareness and attitudes among Thai street vendors about using proper respiratory masks.
Methods: A cross-sectional, repeated measures field study conducted spring 2016 (March 14-23). Street vendors from selected study locations, i.e., roadsides at Chong Nonsi, Yannawa, Bangkok, Thailand, who were either literate in English or can communicate in both English and Thai language for clarification as needed, plus not currently smoking and not having known diagnosed respiratory diseases, were recruited. Additionally, real-time PM10 and PM2.5 were collected from the study locations using TSI AM510 particle counters (flow rates calibrated at Rutgers then daily zero calibration conducted in Bangkok). These field data were then compared to the air quality data from the closest station of the Pollution Control Department (PCD), Thailand.
Results: Overall, of 45, 30 participating street vendors (10 males, 20 females) were asked to complete the questionnaire, including demographic and respiratory masks use; eight males and seven females (n=15) declined participation. The percentage of participated street vendors who reported they have lower respiratory, upper respiratory, and other symptoms were 50, 37, and 70, respectively. Also, 53%of participants have never used respiratory protective mask. Among those who have used mask, 71% used hygiene mask and 29% used anti-dust/cotton mask.
Conclusions: The results suggested outdoor TRAP is associated with adverse health effects. The knowledge, awareness and attitudes of health in using proper respiratory protective masks among street vendors in Bangkok Thailand need to be increased.

Mo-LBA-03

Target and nontarget screening of chemicals in the indoor environment for human exposure assessment - SHINE

Marja Lamoree, *Institute for Environmental Studies, Vrije Universiteit, Amsterdam, Netherlands*
Katleen De Brouwere, *Flemish Institue for Technological Research (VITO), Mol, Belgium*
Stuart Harrad, *University of Birmingham, Birmingham, United Kingdom*
Cynthia De Wit, *Stockholm University, Stockholm, Sweden*
Adrian Covaci, *University of Antwerp, Antwerp, Belgium*
Pim Leonards, *Institute for Environmental Studies, Vrije universiteit, Amsterdam, Netherlands*
Jacob De Boer, *Institute for Environmental Studies, Vrije Universiteit, Amsterdam, Netherlands*

Worldwide, people are spending more time indoors in well-insulated buildings and are more heavily engaged with multiple electronic devices. Various types of emerging chemicals such as dirt and water repellents, flame retardants (FRs), and plasticizers can be emitted from construction materials, electronic equipment, carpets, textiles, flooring and furniture through evaporation (off gassing) or abrasion (small particles breaking off from foam, textile fibers, etc.). The indoor environment is rather complex as there are several sources of substances and some of the substances can even have multiple functions. For example, some compounds used as FRs can also be applied as plasticizers, e.g. in sealants or as additives in waxes. Other compounds reported in the literature vary from personal care products to siloxanes and from dirt and moisture repellents, like perfluoroalkyl substances (PFASs) in textile and clothing to nanoparticles. These multiple sources and dual functionality of some chemicals contribute to the total indoor exposure for humans. Chemical concentrations in indoor dust and air can correlate with body burden, but until now it is unknown if inadvertent dust ingestion (especially for young children who have regular hand-mouth contact), inhalation or dermal uptake is the main route of exposure.

The last couple of years have seen substantial efforts expended on the development of sophisticated, high tier modelling of integrated exposure, however, these models have not yet been thoroughly tested for a wide range of chemicals, including these new chemicals. As many of them will be more polar or possess other properties than previously investigated ('classical') compounds, the existing models also need to be checked for their applicability to these emerging chemicals.

Within the SHINE project (2016-2019) we will carry out sampling and targeted analysis of emerging contaminants, e.g. brominated/organophosphate flame retardants, novel plasticizers and polymer additives, in dust and air of schools/daycare centers, homes and offices in various European countries and conduct non-target screening to identify additional contaminants and combinations of chemicals. To facilitate the nontarget high resolution mass spectrometry screening, existing information on chemicals found indoors will be compiled from various sources, e.g. the ECHA chemicals database, scientific literature. For the newly identified chemicals, the applicability and use of existing exposure models will be verified and modifications to the models will be proposed - if needed. Finally, the measured and modelled data will be compared to biomonitoring data from the literature and ongoing projects.

Mo-LBA-04

RESIDENTIAL ATTACHED GARAGE INTERVENTION STUDY: REDUCING POLLUTANT TRANSFER FROM GARAGES THROUGH IMPROVED BUILDING ENVELOPE AIR TIGHTNESS

Daniel Aubin, *National Research Council of Canada, Ottawa, Ontario, Canada*

Aim: An intervention study was conducted in Ottawa, Canada to evaluate interventions aimed at improving indoor air quality (IAQ) by reducing the transfer of harmful pollutants from attached garages into adjoining living spaces. This study was a joint-research project between Health Canada (HC) and the National Research Council of Canada (NRC). Previous HC and NRC field studies identified that attached garages, when present, are a major source of benzene in homes due to the emissions from the presence of vehicles, internal combustion engine equipment, and solvents.
Methods: The two interventions were tested in the winter and included; 1) running an exhaust fan in the garage (n=33); and 2) improving the airtightness of the house-garage interface wall to reduce the movement of air from the garage into the home (n=29). This presentation will report the results from the sealing intervention. IAQ relevant parameters (VOC's, aldehydes, CO, CO2, relative humidity, air exchange rate, and building envelope air tightness) were measured before and after each intervention. Questionnaires capturing information relating to the indoor environment and occupant behavior were administered to the participants. The sealing intervention was undertaken by two teams of weatherization contractors. All construction products to be used were tested for their VOC profile by dynamic head-space analysis with a TD-GC-MS to ensure that there would be no cross-contamination.
Results: The sealing interventions generally took less than 3 hours to complete. The most common leakage points identified and addressed included: the weather stripping and door frame, the gap between the drywall and concrete foundation, and the electrical outlets. The average amount of material used per house included 1 tube of caulking, 0.5 can of expansion foam, and 1 set of weather stripping and bumper threshold. After the sealing intervention the median airtightness (ACH50) increased by 5.2% and 14.7% for the home and garage respectively.
Conclusions: More than half of Canadian single-family homes have an attached garage which represents a significant exposure route to BTEX, CO, and other harmful pollutants. This study provided practical, effective, and field validated techniques to reduce the transfer of pollutants from attached garages into adjoining living spaces. This study will enable Health Canada to provide evidence-based advice to Canadians to reduce their exposure to harmful pollutants and allow NRC to support the construction industry with field validated intervention techniques.

Mo-LBA-05

Re-analysis of the ETEAM Database for the ECETOC TRAv3 Model

Chris Money, *Cynara Consulting, Brockenhurst, United Kingdom*
Neil Barone, *ExxonMobil, Clinton, New Jersey, United States*
Dook Noij, *Dow Chemical, Terneuzen, Netherlands*
Detleff Keller, *Henkel, Dusseldorf, Germany*
Rosalie Tibaldi, *ExxonMobil, Clinton, New Jersey, United States*
Gerald Bachler, *Shell, The Hague, Netherlands*

Aim
The ETEAM study (BAuA, 2015) sought to evaluate the performance of a range of REACH worker exposure models. ECETOC was given the opportunity to examine the ETEAM database with an aim to characterise the extent to which the findings could help inform what revisions might be implemented for improving the reliability of TRA worker exposure estimates as, the ETEAM analysis was unable to address all the scientific comments originally submitted by ECETOC.
Methods
The contents of the ETEAM database was first examined to confirm it contained sufficient contextual information to make accurate comparisons of the measured data to TRA estimates. Preliminary review indicated a more comprehensive analysis could be undertaken.
1. Datasets within the TRA's domain were identified. Applicability and accuracy (of coding) was confirmed using a consensus based approach to reduce user variation and account for the recent EU REACH Guidance on Use Descriptors (ECHA, 2015).
2. The data were then re-assembled to re-construct the ETEAM analyses but with the TRA comparator being the mean of the P75 values for relevant datasets (as the TRA predicts exposures for the population of interest).
Results
The ETEAM database does not contain sufficient data to address all the situations expected to be covered under REACH. Specifically, for volatile liquids, sufficient data appears to be available for only 6 settings (PROCs) compared with the 30 covered by REACH. For solid materials only 4 scenarios are supported by representative data. Further examination indicates that a limited number of datasets may 'bias' the ETEAM findings for some PROCs. For example, for some PROCs up to 96% of the samples derive from only 4 datasets (see Figure below). Frequent coding errors have also been identified concerning how information on vapour pressures, risk management measures and operational conditions has been used to derive TRA estimates.

A series of updated comparisons against the original ETEAM findings have been developed. These indicate that while some exposure predictions of the TRA do underestimate those obtained from measurements, they appear infrequent..
Conclusions
The ETEAM database contains substantial data and supporting information. A number of errors have been identified which raise questions concerning the integrity of the original ETEAM findings as they relate to the TRA.
Accounting for these deficiencies, updated analyses still suggest that in some cases the TRA may under-predict exposure. However these occurrences appear to be less severe than implied by the ETEAM findings.

Contribution of Different Types of Dataset in ETEAM Database (volatiles)

Mo-LBA-06

Solid-Phase Microextraction Procedure to Measure Endocrine Disruptors in Personal Care Products

Shih-Wei Tsai, *National Taiwan University, Taipei, China, Republic of (Taiwan)*
Wei-Ju Tseng, *National Taiwan University, Taipei, China, Republic of (Taiwan)*

Aim: Emerging environmental pollutants have caused concerned in recent years. For example, a variety of chemical components such as parabens, triclosan, bisphenol A, and synthetic musks (e.g., galaxolide (HHCB) and tonalide (AHTN)), which have been detected from personal care products, may be released into the environment. Among them, synthetic musk comprises a broad variety of chemically heterogeneous compounds. In addition, the USEPA classified polycyclic musk as one of the High Production Volume (HPV) chemicals. From in vivo and in vitro studies, there are evidences that some synthetic musks are considered as the potential endocrine disruptors. To assess the possible health effects, a method for the analysis of synthetic musks in personal care products was developed in this research. The concentrations in various samples were also determined.
Methods: Synthetic musks were prepared in mixtures as standard solutions. The samples were first diluted by water and equilibrated for 5 minutes before the extraction. Hence, the extraction was performed at 25°C for 30 minutes with 500 rpm. The SPME fiber was exposed to the headspace over the samples. After adsorption equilibrium has been reached, the SPME fiber was inserted into the injector of the gas chromatography with tandem mass spectrometry for thermal desorption and further analysis. Effects from various factors, including the types of the SPME fibers, extraction time, extraction temperature, stirring velocity, desorption time and desorption temperature, were all evaluated.
Results: The SPME procedure coupled with GC/MS/MS analysis for the determinations of synthetic musks in the samples of personal care products was established in this study. No carry-over effect was observed from the thermal desorption of the sample. The linear range of all compounds ranged from 0.005 to 0.05 µg ml-1, and the method detection limits were 0.00015 to 0.00075 µg ml-1. Among the personal care products tested in this study, including perfume, body lotion, hair care product, facial cleanser, body wash, and shapoo, HHCB were all detected with concentration as high as 2871.02 µg/g sample. Besides, DPMI, ADBI, AHMI, and AHTN were also detected with concentrations ranged from N.D. to 780.85 µg/g sample. In addition, nitromusks, such as MA, MX, MM and MK, were not found in the samples.
Conclusions: The SPME procedure was applied in this study, while advantages over conventional methods, such as solve-free and time-saving, were reached. Besides, the sensitivities of the method for different compounds were low enough to determine the concentrations from personal care products.

Mo-LBA-07

Metabolomics reveals metabolic disorders in mice exposed to thirdhand tobacco smoke

Noelia Ramirez, *Instituto d'investigacio Sanitaria Pere Virgili, Tarragona, Spain*
Sara Samino, *University Rovira i Virgili, Tarragona, Spain*
Sonia Torres, *Institute of Sanitary Investigation Pere Virgili-URV, Tarragona, Spain*
Beatrix Paton, *University Rovira i Virgili, Tarragona, Spain*
Neema Adhami, *University of California Riverside, Riverside, California, United States*
Manuela Martins-Green, *University of California Riverside, Riverside, California, United States*
Xavier Correig, *University Rovira i Virgili, Tarragona, Spain*

Thirdhand tobacco smoke (THS) is a novel and poorly understood pathway of tobacco exposure that is produced by the deposition and ageing of tobacco smoke particles and toxicants in surfaces and dust. This aged tobacco smoke becomes increasingly toxic with age, re-emitted into the air or react with other chemicals in the environment to yield new toxicants, including carcinogens. Furthermore, THS remains in indoor environments long after smokers move out, which makes THS a serious health problem, especially for children with smoking parents, which are the most vulnerable population to this pathway of tobacco exposure. Although the increasing evidences of THS hazards, the specific cellular and molecular consequences of exposure to THS remain to be fully elucidated. To address this, here we present the first non-targeted metabolomics approach applied to THS-exposed animal model: C57BL/6 mice, exposed to THS under conditions that mimic exposure of humans in homes of smokers. THS-exposed mice showed alterations in multiple organ systems including non-alcoholic fatty liver disease, inflammation in the lung, poor healing of cutaneous wounds, hyperactivity, hyperglycemia and insulin resistance in the form of non-obese type II diabetes (NODII) through oxidative stress. Urine samples from THS-exposed mice and control ones were analyzed using GC-QTOF and reverse phase HPLC-QTOF in both ionization modes. Accurate multivariate analysis revealed hundreds of differently expressed metabolites. Interestingly, some of the altered metabolites coincide with those reported in metabolomics studies of current smokers. Nevertheless, this study also reveals for the first time other metabolic pathways altered by this passive way of tobacco exposure, not related to tobacco until now. Furthermore, biomarkers of tobacco exposure have been also detected in the urine of the THS-exposed animals using a target HPLC-QQQ approach in concentrations similar to those reported in children exposed to second hand smoke.
This study demonstrates the power of metabolomics for identifying the health hazards of THS exposure and, if confirmed in humans, would have a major impact on current tobacco health and environmental policies.

Mo-LBA-09

A spatial analysis approach combining multi-media and human models to map the lead exposure of children in a French region

Ségolène Siméon, *INERIS, Verneuil en Halatte, France*
Julien Caudeville, *INERIS, Verneuil en Halatte, France*
Cleo Tebby, *INERIS, Verneuil en Halatte, France*
Céline Brochot, *INERIS, Verneuil en Halatte, France*

Context and objectives: The last decade has witnessed an increasing interest in assessing health risks caused by contaminants present in several environmental media, i.e. soil, air, and water. To that end, mathematical models describing the fate of compounds in the environment (multi-media models) and in the human body (toxicokinetic models) can be combined to simulate realistic exposure scenarios of the human populations. These models can also be integrated in a Geographic Information System (GIS) to produce maps of exposure and reveal spatial patterns. The aim of this study was to develop a spatial stochastic multimedia and human exposure model for detecting vulnerable populations and analyzing exposure determinants at a fine resolution and regional scale. This approach was applied to the exposure to lead of children in the Region Nord-Pas-de-Calais in France.
Methods: A GIS-based modeling platform for quantifying human exposure to chemical substances (PLAINE: environmental inequalities analysis platform) was used to build and discretize environmental and population variables collected from different sources on a 1 km2 regular grid. Inhalation and ingestion exposure via contaminated food, drinking water, and soil were taken into account. Either direct observations or multi-media models were used to compute daily intakes for different reference groups (age, dietary habits, and the fraction of food produced locally). In each cell of the grid, Monte-Carlo simulations were performed to generate a sample of 100 daily intakes. These latter were then used as input in a physiologically based pharmacokinetic (PBPK) model for children to simulate the associated blood lead levels since birth.
Results and Conclusion: Maps of the simulated blood lead levels (BLL) for children aged from 1 to 11 years old were generated for year 2015. The 95th percentile of the distribution of BLL for each age was used to compute the risk of exceeding the reference value of 5µg/dL. Our results showed that the majority of the predicted BLL fell under the reference value, and several hotspots were detected as a former industrial site and Lille, the capital of the region. Drinking water and surface soil were identified as the main determinants of the children BLL. To conclude, exposure maps are a valuable tool in risk assessment to explore changes in disease patterns potentially associated with changes in environment quality and to better characterize the links between the sources of pollution and health effects. Future developments will consider the integration of health data in our approach.

Tuesday, October 11, 2016

Plenary Address 3: The exposome: moving from concept to reality

Chris Wild, International Agency for Research on Cancer (IARC), Lyon, France

Tu-SY-A1: The Exposome: From concept to practice - I

Tu-SY-A1.2

Early observations from HELIX - building the Human Early-Life Exposome

Martine Vrijheid, *CREAL, Barcelona, Spain*
for the HELIX Consortium, *CREAL, Barcelona, Spain*

Aim
Environmental hazards form a heterogeneous, interacting, multi-level part of disease causation, but epidemiology has almost uniquely focused on single exposure-outcome relationships. The "exposome" concept was proposed to encompass the totality of exposures from conception onwards, complementing the genome. The general aim of HELIX is to measure a range of chemical and physical environmental hazards in food, consumer products, water, air, noise, and the built environment, in pre and postnatal periods, and link these with molecular omics profiles and child health outcomes. As a first step, we aim to describe correlations and variabilities within estimates of the chemical and physical environment, as well as the determinants of exposure patterns.
Methods
In six existing birth cohort studies in Europe, HELIX estimated prenatal and postnatal exposures. Exposure models for the outdoor exposome (air pollutants, noise, meteorological factors, and natural and built environment characteristics) were developed for a total of 28,000 mother-child pairs. Exposure biomarkers (for persistent organic pollutants, metals, phthalate metabolites, phenolic compounds and organophosphate pesticides) were measured in a subset of 1,200. Nested repeat-sampling panel studies (N=150) collected data on biomarker variability for non-persistent chemicals (phthalates and phenolic compounds). We assessed correlations between exposures, derived principal components, and associated exposures with determinants (country, education, ethnicity, and other). Further we calculated intraclass-correlation coefficient (ICC) to assess variability components.
Results
We find strong levels of correlations within families of exposure (grouped by structure or source) and weak to moderate correlations between exposure families. Education and ethnicity are important determinants of the outdoor exposome with more disadvantaged social groups more exposed to unhealthy urban environments within some European cities. Exposure biomarkers are less determined by sociodemographic factors and more by the country of residence. Between-day ICCs ranged from 0.52 for bisphenol-A to 0.87 for benzophenone-3, and between-season ICCs ranged from 0.18 for bisphenol A to 0.73 for mono-2-ethylhexyl phthalate (MEHP).
Conclusion
The exposome concept provides an important new framework to improve knowledge how multiple environmental exposures co-exist and impact on health, with early life as a crucial developmental period. Information on the structure of the exposome in terms of correlations, determinants and variability, will aid interpretation of reported findings from epidemiological studies in general and inform future analyses in HELIX.

Tu-SY-A1.3

We do exposome as much as we can: Japan Environment and Children's Study

Shoji Nakayama, *National Institute for Environmental Studies, Tsukuba, Japan*

Japan Environment and Children's Study, JECS shortly, is a national birth cohort study launched in 2011 by the Ministry of the Environment. JECS is designed to evaluate the effect of the environment on children's health and development. A total of 103,000 mother-child pairs were registered. Biological samples, such as blood, urine, cord blood, breast milk and hair were collected during pregnancy, at birth and a month after birth from mothers, children and fathers. Questionnaires have been administered to obtain the exposure and health outcome information. Numerical models are used to estimate air pollutants and physical environment (e.g. noise, radiation).
JECS considers the concept of 'exposome' seriously. Every exposure during pregnancy and childhood could affect children's health and development. JECS takes 'exposure' broadly to capture not only exposures that can be measured by so called -omics techniques but also those need to be estimated by questionnaires, interviews, direct observations, personal sampling, sensing and simulation models. The exposure information should be gathered periodically throughout children's life course.
It is a huge challenge for JECS. Techniques such as personal sampling and sensing are yet to come. No standardised questionnaires are available for exposome measurements. JECS finished its recruitment in March 2014. The first recruited children are becoming 5 years old. Children never stop growing. We are desperate to find out the best ways to incorporate 'exposome' into JECS study as soon and much as possible.

Tu-SY-A1.4

Early Observations from CHEAR

Robert Wright, *Mount Sinai School of Medicine, New York, New York, United States*

The mission of CHEAR is to accelerate cutting-edge transdisciplinary research utilizing life course-informed models of health. A critical need is the development of methods to understand the time-sensitive and dynamic nature of perinatal and childhood environmental influences on the developmental health trajectories that are interrelated antecedents of chronic disease . Our CHEAR program has developed methods that integrate developmental biology, temporally resolved exposure metrics and advanced statistics to define the critical windows by environmental exposures predict long term health. The CHEAR Network can incorporate these principles into existing studies with a wide variety of designs, child health phenotypes, and archived biological samples. The pursuit of life course-informed research within the CHEAR structure represents a shift in focus from single cause-single disease inquiries to research that addresses multiple causes linked through shared vulnerabilities. The life-course approach also emphasizes the need to understand the physical, biological, genetic, and social/behavioral factors that independently, cumulatively, and interactively influence health and development, starting prenatally. CHEAR will address these challenges in the following ways: first, we will measure environmental exposures and their timing more accurately and comprehensively. We will emphasize mixtures of chemicals, as well as non-chemical (social, nutritional) stressors, to more fully understand their health risks. We will help delineate the nature of sensitive-period effects by incorporating innovative exposure measures that can be integrated with a host of biological response indicators to assess underlying mechanisms and to provide a set of intermediate molecular profiles linked with the chronic diseases of childhood (e.g., obesity, asthma, attention deficit disorder, etc.). I will present preliminary data illustrating these methods from the Mexico City PROGRESS (Programming Research in Obesity, GRowth, and Social Stressors) cohort, using both a novel tooth based biomarker of exposure and a satellite based model of ambient particulate matter. The value of exposure methods that can objectively predict the dose and timing of past exposures and biological responses will be as critical to the larger Exposome effort as methods to generate untargeted assays, EWAS statistical approaches, and data harmonization tools. By developing such methods, CHEAR can begin to assess the complex, time varying nature of the exposome with its complex interactions that may even occur across time, rather than from joint exposure. The long-term goal of CHEAR is to pool/meta-regress the findings from multiple studies to better characterize the exposome's impact on health. This presentation will present a roadmap towards this goal.

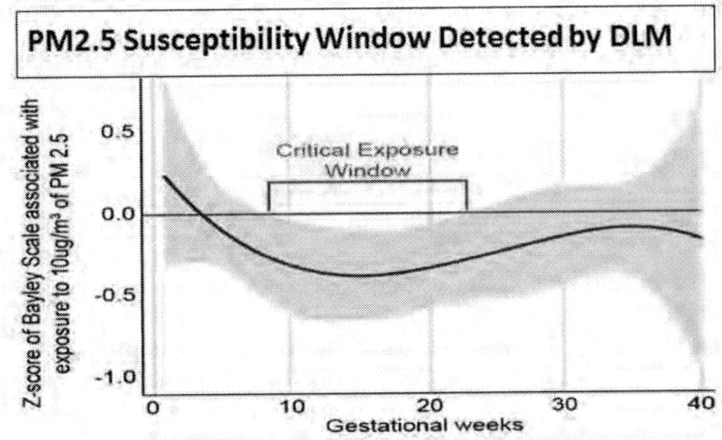

Y axis is Beta coefficient, X axis is weeks of gestation in pregnant women. Graph shows change over time in beta between PM and Bayley Scale infant development. Inverse Betas are significant between 8-22 weeks, suggesting that the sensitivity window is here.

Figure of Distributed Lag Model showing critical window for PM in pregnancy and infant development: i.e time point when PM exposure is most toxic.

Tu-SY-B1: Uncertainty in scientific assessments: Recent efforts by governmental bodies to develop guidance for assessors

Tu-SY-B1.1

Why do we need to improve the treatment of uncertainty in exposure and risk assessment?

Andy Hart, *Fera Science Ltd., York, United Kingdom*

The need to address uncertainty in risk assessment has long been recognised at international level. The Codex Working Principles for Risk Analysis, first established in 2003, say uncertainties should be explicitly considered at each step in risk assessment, documented transparently and quantified to the extent that is scientifically achievable. They also say that responsibility for resolving the impact of uncertainty on the risk management decision lies with the risk manager, not the risk assessors. The first part of this presentation will discuss the rationale and meaning of these requirements, and why they are important, illustrated by practical examples.
Substantial efforts have been invested in developing practical guidance on how to implement these principles. Guidance on methods for addressing uncertainty in human exposure assessment was published by EFSA in 2006 and IPCS/WHO in 2008, while ECHA's 2008 guidance also included hazard, risk and environmental assessments. In 2014-15, IPCS/WHO published guidance on uncertainty in risk characterisation, accompanied by a spreadsheet calculator, and EFSA published draft guidance on addressing uncertainty in all areas of its work. Further work has been undertaken by national authorities. Some of these initiatives will be presented later in this session.
The focus is now shifting to applying these principles and guidance in the day-to-day practice of exposure and risk assessment. This raises a number of challenges for assessors, for the interaction between assessors and decision-makers, and for the wider process of risk communication. The second part of this presentation will discuss these challenges, outline some potential solutions, and set the scene for the following detailed presentations.

Tu-SY-B1.2

EFSA's approach to uncertainty analysis in scientific assessment

Andrea Germini, *European Food Safety Authority (EFSA), Parma, Italy*
Anthony Richard Hardy, *EFSA Scientific Committee, York, United Kingdom*
Andy Hart, *Food and Environment Research Agency (FERA), York, United Kingdom*
Caroline Merten, *European Food Safety Authority (EFSA), Parma, Italy*

The commitment to provide EU risk managers with objective, reliable and transparent scientific advice is one of the key values of the European Food Safety Authority. As part of this commitment, identifying and characterising uncertainties, and explaining their implications for assessment conclusions, is an important element of EFSA's risk assessment process.
Following its earlier guidance focussing on the characterization of uncertainties in exposure assessment, EFSA decided to develop a harmonised framework on how to characterise, document and explain uncertainties in the various steps of its risk assessment that would be applicable to all relevant working areas of the Authority.
In order to produce a Guidance Document that would be applicable to all EFSA's activity, a working group was created with representatives from all the EFSA Panels and the support of relevant external experts. During the development of the guidance consultations took place with all EFSA's relevant stakeholders including EC risk managers, national authorities, EU sister agencies, and international organisations. The draft guidance was published for public consultation and later revised in light of the comments received. The revised version of the draft guidance is currently being tested by all EFSA Panels on selected scientific opinions. Following feedback on the Panels' testing phase and an impact assessment of its implementation the guidance document will be further revised and finalised.
The draft guidance document provides a harmonised, but flexible methodological framework that should be applicable to all areas of EFSA, and all types of uncertainty affecting scientific assessment. The document provides guidance on general principles and a menu of different qualitative and quantitative methods which can be used to help assessors to systematically identify, characterise, explain and account for sources of uncertainty at different stages of the assessment process. The approach aims to be sufficiently flexible to adapt to the needs of the different EFSA Panels and to the circumstances of each assessment, e.g., from urgent advices to longer-term comprehensive reviews of all available scientific knowledge.
The Guidance Document is expected to contribute to further improve the scientific assessment process carried out by EFSA, increase its transparency and strengthen the basis for an informed decision-making process in the area of food safety.
EFSA acknowledges the following experts that contributed to the development of the Guidance Document:
J.Alexander, D.Benford, P.Craig, A.DiDomenico, A.Havelaar, M.Jeger, R.Luttik, A.Martin, L.Martino, A.Mortensen, O.Mosbach-Schulz, K.Nielsen, B.Nørrung, B.Ossendorp, D.Renshaw, J.Schlatter, A.Smith, H.H.Thulke, N.VonGoetz and D.Wölfle.

Tu-SY-B1.3

The BfR-guidance on uncertainty assessment for exposure modelling

Gerhard Heinemeyer, *Federal Institute for Risk Assessment, Berlin, Germany*
Michael Schümann, *Chairman of BfR Expert Commitee "EXPO", Berlin, Germany*
Matthias Greiner, *BfR, Berlin, Germany*
Olaf Mosbach-Schulz, *EFSA, Parma, Italy*

Uncertainty analysis (UA) has gained increased importance in risk assessment. Several guidance documents have been developed by national and international bodies in connection with regulatory work. On the basis of these documents, a working group at the Federal Institute for Risk Assessment (BfR) has prepared a guidance document focusing on a structured qualitative uncertainty analysis. This document has the objective to improve the comprehensibility and coherence of scientific opinions, to support the use of harmonized terminology in the field of risk assessment, and to ensure the provision of the best-possible advice in the various areas of activity of the BfR. The guidance is embedded in a general concept of BfR guidance document for risk assessment. The procedures are tailored to meet the requirements at different levels of refinement using a tiered approach.

As a first iteration step of UA, a qualitative procedure has been proposed. This first level of the analysis focusses on qualitative uncertainty analysis, a step which is generally accepted in international guidance documents. This supports the quality of uncertainty evaluation in exposure assessments.

The BfR approach combines the tiered approach of exposure analysis with a tiered step by step procedure of the uncertainty analysis. It uses a structured questionnaire list which covers all the possible issues of uncertainties. The questionnaire is structured according to the four basic pillars of an uncertainty analysis:

1. goal and question formulation of the analysis
2. scenario uncertainty
3. model uncertainty and
4. parameter uncertainty

Answer formulation enables the assessor to gain an overview about the state of scientifific knowledge and its limitations. Other sources of uncertainties including interpretation of available data and expert opinion should be taken into account. Furthermore it supports semi-quantitative sensitivity analyses. The result of this qualitative check list gives hints for necessary further quantitative evaluations and supports risk communication.

The final implementation is foreseen to take place after a pilot phase in 2016. First results will be presented at the meeting.

Developed by "Sub-Committee for Statistics and Uncertainty Analysis of the Committee for Exposure Assessment and Exposure Standardisation" (Olaf Mosbach-Schulz, Lothar Kreienbrock, Michael Schümann, Gerhard Heinemeyer, Matthias Filter, Matthias Greiner, Matthias Herzler, Oliver Lindtner, Stephanie Kurzenhäuser, Bettina Roeder)

Tu-SY-B1.4

The Anses-guidance on evaluation of weight of evidence and uncertainty analysis

Moez SANAA, *The French Agency for Food, Environmental and Occupational Health & Safety (ANSES), Maisons Alfort, France*
Claire Bladier, *Anses, Maisons Alfort, France*
Sandrine FRAIZE-FRONTIER, *Anses, Maisons Alfort, France*
Eve Feinblatt, *ANSES, Maisons Alfort, France*

With the aim of improving the transparency of its risk assessment process, ANSES requested its permanent working group of Risk Assessment Methodology to conduct a critical analysis of approaches used in the evaluation of weight of evidence and uncertainty analysis. The working group organized the work in three steps: i) inventory of ANSES current practices, ii) review of the scientific literature and approaches used by national and international agencies iii) recommendations on how to conduct evaluation of weight of evidence and uncertainty analysis iv) assessment of the recommendations through case studies.
The report provides definitions of key terms and recommends structuring the process of weight of evidence evaluation and uncertainty analysis in common and specific steps to the evaluation of weight of evidence or analysis of uncertainty. The planning step (first and common step) aims to define the scope of the risk assessment, to reveal the public health and decision issues and conclude with the appropriate approaches to be undertaken for the risk assessment. At this step, the needs for the evaluation of weight evidence and uncertainty analysis are discussed. A clear problem formulation should help to define explicitly the hypotheses that need to be evaluated and risk metrics that need to be assessed. Furthermore, the complexity and resources allocated to the risk assessment and especially to the weight of evidence evaluation and uncertainty analysis are decided at this step. The extensiveness of weight of evidence and uncertainty analysis should be proportionate to the importance of the public health issue and fit the purpose of the risk assessment. The specific steps of the evaluation of weight of evidence are i) establishment of line of evidence (including search and selection of studies, data extraction, evaluation of the quality of the studies and analysis of similar type of studies to provide the lines of evidence; ii) integration of the lines of evidence to establish the weight of evidence; and iv) expression and communication of weight of evidence. Uncertainty analysis is organized in five steps: i) Identify and describe all the uncertainties; ii) evaluate the main uncertainties; iii) assess the combined impact of uncertainty on the risk assessment output; and iv) prioritize uncertainty sources according to their contribution to the overall uncertainty (optional step). The expression and communication of weight of evidence and uncertainty analysis constitute the last step of the process.

Tu-SY-B1.5

Considering Uncertainty based on International Experience

Bette Meek, *McLaughlin Centre for Risk Science, Ottawa, Ontario, Canada*

Increasing experience internationally in tiered assessment strategies has important implications for the consideration of uncertainty, variability and sensitivity in exposure and hazard assessment. These strategies, which are designed to increase efficiency through inclusion of various levels of priority setting and assessment have underscored the importance of early consideration and communication of envisaged methodological approaches to characterize uncertainty in an assessment strategy tailored to meet specific objectives. This is critical within tiered approaches in order to characterize and communicate relative degrees of conservatism (and associated uncertainty) of, for example, screening versus full assessments.

Specific consideration of sensitivity is also important as a basis to inform risk managers, stakeholders and the research community concerning the most important aspects influencing outcome as a basis in part, to identify data critical for generation for higher tiered assessments.

The nature of this evolving experience internationally and its implications for guidance on uncertainty assessment will be considered. This includes delineating considerations for selection of approaches from a continuum of increasingly complex qualitative to quantitative methods in planning of the assessment following formal problem formulation. Formal planning takes into consideration the appropriate focus and methodology based on consideration of, for example, the objectives and importance of the question at hand (and potential consequences), available resources (including data) and previous work.

Tu-PL-C1: Aggregate and Cumulative Exposure Evaluations

Tu-PL-C1.1

Methods to evaluate housing and neighborhood-related environmental health disparities using public databases

M. Patricia Fabian, *Boston University School of Public Health, Boston, Massachusetts, United States*
Anna Rosofsky, *Boston University School of Public Health, Boston, Massachusetts, United States*
Jonathan Levy, *Boston University School of Public Health, Boston, Massachusetts, United States*

Aim: It is generally acknowledged that environmental health disparities disproportionately affect populations of certain sociodemographic groups, patterned around individual racial and economic characteristics. Beyond individual characteristics, many stressors can add to the cumulative burden of environmental health disparities in these populations, including the housing, neighborhoods, and the social networks in which they live and function. Quantifying the burden of these multiple sources of disparities and their effects on vulnerable communities has been limited by the ability to obtain and analyze a diverse set of datasets, and their availability across large geographical regions. However, recent advances in the availability of public data, methods for handling big data, and geographical information systems allow us the opportunity to quantify these disparities at relevant geographic resolutions.
Methods: In this study we developed a methodology to construct environmental health disparity indexes related to housing, neighborhood and social characteristics by linking geospatial and demographic datasets from a variety of public sources. For example, an outdoor pollution infiltration index was constructed from meteorology, individual housing, and aggregated neighborhood characteristics, and used to describe infiltration disparities across Massachusetts. Disparities were quantified using the Atkinson index and the Cumulative Environmental Hazard Inequality Index, which were evaluated at multiple geographic resolutions to determine the most relevant to the stressor under study.
Results: Results revealed patterns of pollution infiltration-related disparities across Massachusetts that differed from traditional race/income disparities analysis. Results will be validated in our case communities of Chelsea and Dorchester, Massachusetts by comparing field study measurements to the environmental health disparities indexes.
Conclusions: This method can be implemented across multiple environmental health disparity dimensions and geographical locations where public data exists.

Tu-PL-C1.2

An automated protocol for assigning address-level air pollution exposure for longitudinal birth cohort studies

Daniela Fecht, *Imperial College London, London, United Kingdom*
Kevin Garwood, *Imperial College London, London, United Kingdom*
John Gulliver, *Imperial College London, London, United Kingdom*

Aim: To assess environmental exposures in longitudinal cohort studies, a better understanding is needed about the effects of incomplete or inaccurate records of residential mobility on exposure measurement error.
Methods: We developed an ALGorithm for Generating Address Exposures (ALGAE). It is a generic, automated protocol that uses residential address histories to assign air pollution exposures. The protocol corrects gaps and overlaps in the address histories and creates sensitivity variables to assess exposure measurement error. We tested the protocol using address records and historic daily air pollution estimates (particulate matter ≤10 µm [PM10]) for the Avon Longitudinal Study of Parents and Children (ALSPAC). ALSPAC recruited 14,541 pregnant women between 1990 and 1992 in the South West of England. For each pregnancy, we modelled trimester-specific exposure estimates based on PM10 concentrations a) at residential address at birth and b) using reconstructed address histories to account for mobility during pregnancy.
Results: We reconstructed residential address histories for 14,027 pregnant women, 4,059 of whom moved out of the study area and were excluded from the analysis. We were able to assign trimester-specific PM10 exposures for the remaining 9,968 pregnancies. Average PM10 exposure were 28.8 µg/m3 (StDev. 4.56 µg/m3), 28.5 µg/m3 (StDev. 4.68 µg/m3) and 26.9 µg/m3 (StDev. 4.58 µg/m3) for first, second and third trimester, respectively. 1,447 women (14.5%) changed address during their pregnancy. For those women, we identified a difference of up to a 5.4 µg/m3 (StDev. 0.75 µg/m3) between the exposure assessed using the address at birth compared to the exposure assessed using the complete cleaned address history.
Conclusions: Accounting for residential mobility in the exposure assessment of birth cohort studies has the potential to substantially reduce the exposure measurement error. ALGAE provides an open-source solution to clean addresses stored in the cohort contact database and assigns life-stage specific exposures. Its generic code base means that it is adaptable for other cohort studies.

Tu-PL-C1.3

Time Varying Associations between Air Pollution and Birth Weight: Results from the MIREC Cohort Study in Canada

Markey Johnson, *Health Canada, Ottawa, Ontario, Canada*
Hwashin Shin, *Health Canada, Ottawa, Ontario, Canada*
Eric Roberts, *Public Health Institute, Richmond, California, United States*
Sunny Sun, *Health Canada, Ottawa, Ontario, Canada*
Eric Lavigne, *Health Canada, Ottawa, Ontario, Canada*
Nina Clark, *Health Canada, Ottawa, Ontario, Canada*
Mandy Fisher, *Health Canada, Ottawa, Ontario, Canada*
Randall Martin, *Dalhousie University, Halifax, Nova Scotia, Canada*
Aaron van Donkelaar, *Dalhousie University, Halifax, Nova Scotia, Canada*
Perry Hystad, *Oregon State University, Corvallis, Oregon, United States*
William Fraser, *Sainte-Justine University Hospital, Montreal, Quebec, Canada*
Tye Arbuckle*Health Canada, Ottawa, Ontario, Canada*

Adverse birth outcomes have been linked with increased risk for infant mortality, as well as developmental, cardiovascular, and respiratory health problems in childhood and adulthood; and have consistently ranked in the top ten global causes for disability-adjusted life years. Previous studies have reported associations between air pollution and adverse birth outcomes. However, results have been inconsistent, and largely focused on preselected periods such as trimesters. This study used a Bayesian random date selection approach to examine time-varying vulnerability to air pollution during fetal development. Analyses were conducted using data from Maternal-Infant Research on Environmental Chemicals (MIREC) study, a prospective pregnancy cohort spanning 10 Canadian cities. Models were limited to term pregnancies with data available for birth weight, covariates and > 75% daily pollution values. Of the 1600 mother-baby pairs, approximately 1300 were included in the models.
Daily pollution estimates were assigned to cohort members based on their forward sortation area (FSA) where FSA size < 20 km2. Daily air pollution estimates were calculated using measurements at ground level monitoring sites within 30 km of the FSA centroid combined with long term concentrations from a national land use regression model for nitrogen dioxide (NO_2) and satellite derived fine particulate matter ($PM_{2.5}$). Critical periods were identified using random selection of date pairs (start and end dates) from 86 days before conception through delivery. Date pairs with selection density > 75th percentile of the empirical distribution were identified as critical windows. Bayesian estimates relating air pollution to birth weight were calculated for each critical window. Models were adjusted for gestation day, infant sex and birth season; as well as maternal age, ethnicity, smoking, education, marital status, household income, parity, and alcohol use.
Critical windows were identified for both NO_2 and $PM_{2.5}$ during late pregnancy (NO_2: day 184-273, $PM_{2.5}$: day 251-274), and for $PM_{2.5}$ during early pregnancy (day 74-139). Exposure to ambient pollution during those periods was associated with decreased birth weight (approximately 3-4 g per 1 μg/m3 change in $PM_{2.5}$, 10 g per 1 ppb change in NO_2). We identified periods during pregnancy when exposure to air pollution was more highly associated with decreased birth weight. Two critical periods were identified for $PM_{2.5}$ compared with one for NO_2. This may be due to greater variation in daily $PM_{2.5}$ versus

daily NO2 estimates. These results suggest the need for further research and targeted interventions.

Tu-PL-C1.4

Screening and identification of high priority compounds in textile products

Iris Meuwissen, *Avans University of Applied Sciences, Breda, Netherlands*
Ben de Rooij, *Avans University of Applied Sciences, Breda, Netherlands*
André Muller, *RIVM, Bilthoven, Netherlands*
Theo Noij, *Avans University of Applied Sciences, Breda, Netherlands*
Yuri Bruinen de Bruin, *RIVM, Bilthoven, Netherlands*

During the production of textile products like clothing, different chemicals are used, including colorants and anti-wrinkle agents. By wearing these clothes humans potentially could be exposed to these chemicals. Because not all chemicals in clothing are fully known or sufficiently regulated, there is a concern that this exposure may bear unwanted health effects. For this reason a priority list of compounds was proposed.
The goal of this project was to develop an analytical screening method to determine whether compounds can be released from clothing using artificial sweat to mimic real life situations.
In this project a non-targeted approach and a targeted approach were used. The non-targeted approach was an extraction of several different pieces of children's-clothing, executed with different types of artificial sweat: acid sweat (pH 4.3), neutral sweat (pH 5.5) and basic sweat (pH 8.0). After the extraction the samples were analyzed by screening LC-MS and GC-MS methods
In the targeted approach, clothing was spiked with colorants Malachite Green, Reactive Orange 16 and Acid Red 66, extracted by artificial sweat and analyzed by HPLC_DAD.
As expected, the non-targeted method showed a variety of substances. The results of different types of artificial sweat were reflected by the intensity of the extracted substances. The intensity and composition of the extracted patterns reflected the different types of artificial sweat. After comparison of the theoretical molecular masses of the priority list with the extracted masses in LC-MS, two potential matches were found for Reactive Orange 16 and Acid Red 66. Further research is required to confirm the potential matches.
In the GC-MS method several compounds were tentatively identified using the NIST library. None of these compounds were part of the priority list.
Conclusion: the extraction of commercially obtained clothing by different types of artificial perspiration resulted in different chromatographic patters as visualized by LC-MS and GC-MS. Comparisons were made to a list of priority compounds and the NIST library. The comparisons resulted in suggested compounds. These suggestions need further research and confirmation/
References:
1. KEMI Report 6/14, Swedish Chemicals Agency (2014): Chemicals in textiles, risks to human health and the environment.
2. European standard: NEN ISO 105-B07: 2009: Textiles - Tests for color fastness - Part B07: Color fastness to light of textiles wetted with artificial sweat.
Acknowledgement:
Financial support by National Institute for Public Health and the Environment (RIVM), the Netherlands and by the Dutch Food and Consumer Product Safety Authority (NVWA) is gratefully acknowledged.

Tu-PL-C1.5

Assessment of Children's Residential Exposure to Agricultural Pesticides: the PIAMA Birth Cohort.

Joseph Bukalasa, *Utrecht University, Utrecht, Netherlands*
Bert Brunekreef, *Utrecht University, Utrecht, Netherlands*
Maartje Brouwer, *Utrecht University, Utrecht, Netherlands*
Roel Vermeulen, *Utrecht University, Utrecht, Netherlands*
Johan C. de Jongste, *Erasmus University Medical Center/Sophia Children's Hospital, Rotterdam, Netherlands*
Lenie van Rossem, *University Medical Center Utrecht, Utrecht, Netherlands*
Judith M. Vonk, *University of Groningen, Groningen, Netherlands*
Alet H. Wijga, *National Institute for Public Health and the Environment, Bilthoven, Netherlands*
Anke Huss, *Utrecht University, Utrecht, Netherlands*
Ulrike Gehring, *Utrecht University, Utrecht, Netherlands*

Aim: Agricultural pesticides are frequently used for crop growth and food production. However, residents living in close proximity to treated fields may be exposed to pesticides through primary spray drift or after application (e.g. trough volatilization of pesticide residues from crops or soil or wind erosion of soil particles). There is some indication that children living near agricultural fields have an increased risk of developing asthma and decreased lung function. The aim of this study was to evaluate the use of geographical information on the presence of crops and data on pesticide use on specific crops for the assessment of residential pesticide exposure.
Methods: For a cross-sectional study, we investigated the potential for residential pesticide exposure at the 14-year home addresses of 2291 participants of a birth cohort study living in the North, West, and Center of the Netherlands. Using ArcGIS and geographic data on presence of crops for the year 2012, we created circular buffers with radii of 50, 100, 500 and 1000 meters around these addresses and calculated the surface area for specific crops (cereals, corn, potatoes, open field vegetables, open field floriculture and bulbs, orchard crops, and commercial crops like beets, chicory, rapeseed, flax and hemp fibre) within those buffers. Information on agricultural pesticides use in the specific crops from a Dutch survey in 2012 was used to estimate amounts of pesticides applied (in gram) within the buffers. Pesticides with potential relevance for respiratory outcomes were identified from the Pesticides Properties Database (PPDB) and The Pesticides Manual (2009).
Results: Any of the specific crops were found in 3%, 7%, 40% and 65% of the 50, 100, 500 and 1000 metres buffers around the addresses, respectively. The most prevalent crops were corn, cereals and potatoes. About 3%, 7%, 39% and 63% of the addresses likely had pesticides with potential relevance for respiratory outcomes applied within 50, 100, 500 and 1000 meter buffers, respectively.
Conclusions: A small proportion of children participating in this study lives at close proximity to agricultural fields. However, as pesticides may be transported over longer distances and as children may play in treated fields, exposure to children living at larger distances cannot be excluded.

Tu-PL-D1: Land Use Regression Modeling - I

Tu-PL-D1.1

Land Use Models for Elemental Components of Particulate Matter in an Urban Environment: A Comparison of Regression and Random Forest Models

Cole Brokamp, *Cincinnati Children's Hospital Medical Center, Cincinnati, United States*
Roman Jandarov, *University of Cincinnati, Cincinnati, United States*
MB Rao, *University of Cincinnati, Cincinnati, United States*
Grace LeMasters, *University of Cincinnati, Cincinnati, United States*
Patrick Ryan, *Cincinnati Children's Hospital Medical Center, Cincinnati, United States*

Background: Exposure assessment for elemental components of particulate matter (PM) using land use modeling is a complex problem due to the high spatial and temporal variations in pollutant concentrations at the local scale. Land use regression (LUR) models often fail to capture complex interactions and non-linear relationships. The increasing availability of big spatial data and machine learning methods present an opportunity for improvement in PM exposure assessment models.
Objectives: Our objective was to develop a novel land use random forest (LURF) model and compare its accuracy and precision to a LUR model for elemental components of PM in the urban city of Cincinnati, Ohio.
Methods: Total PM and eleven elemental components were measured at 24 sampling stations from the Cincinnati Children Allergy and Air Pollution Study (CCAAPS). Over 50 different predictors associated with transportation, physical features, community socioeconomic characteristics, greenspace, land cover, and emission point sources were used to construct LUR and LURF models. Cross validation was used to quantify and compare model performance.
Results: Successful LURF and LUR models were created for Al, Cu, Fe, K, Mn, Ni, Pb, S, Si, V, Zn, and total PM2.5 in the CCAAPS study area. LURF utilized a more diverse and greater number of predictors than LUR in its final models. Cross validation revealed that LURF models had a lower predictive error than LUR models for all elements except Fe, Mn, and Ni. Furthermore, LUR models showed a differential exposure assessment bias and had a higher prediction error variance. LURF models predictions were approximately ten fold more precise compared to LUR model predictions. LURF will be a useful exposure assessment tool for epidemiologic studies associating elemental components of PM with health effects. Random forest and other machine learning methods should be incorporated into future land use models for more accurate and precise exposure assessment.

Tu-PL-D1.2

Comparison and Evaluation of Spatiotemporal Air Quality Exposure Fields Developed using Ten Methods

Haofei Yu, *Georgia Institute of Technology, Atlanta, Georgia, United States*
Armistead Russell, *Georgia Institute of Technology, Atlanta, Georgia, United States*
James Mulholland, *Georgia Institute of Technology, Atlanta, Georgia, United States*
M Talat Odman, *Georgia Institute of Technology, Atlanta, Georgia, United States*
Yongtao Hu, *Georgia Institute of Technology, Atlanta, Georgia, United States*
Howard Chang, *Emory University, Atlanta, Georgia, United States*

A variety of measures are currently being used to assign air pollutant concentrations to individuals in air quality exposure-related studies. The outcomes of those assessments reflect the methods used and may differ between methods. Comprehensive comparison of multiple methods to a common data set is needed to inform users of potential issues that may impact their use in health studies. Here, we compare pollutant concentration fields generated using ten methods and evaluate them regarding their use in exposure related studies. The application is the Atlanta (GA, USA) metropolitan region and the pollutants are CO, NO2, SO2, O3, PM2.5 and PM2.5 constituents elemental carbon, organic carbon and sulfate. The selected methods involve the use of central monitor data, multi-site averaging, three spatial interpolation methods (inverse distance weighting, tessellation and kriging), land use regression with satellite AOD information, chemical transport modeling (CMAQ-derived fields) combined with data assimilation/fusion methods, and fine scale dispersion modeling. The central monitor and site averaging methods are spatially, temporally and chemically incomplete, and performance varies across pollutants, depending on the number and locations of monitors. Interpolation methods are able to generate spatially resolved fields, although the spatial patterns of primary pollutants are poorly captured due to sparse monitoring networks. The results are similar between IDW and kriging, but could also vary depending on other interpolation approach applied or interpolation parameters used. The CMAQ model provides complete fields and reasonable spatial concentration patterns, but CMAQ model performance shows substantial biases and other errors that vary between pollutants and over time. Incorporating data fusion approaches improves model performance. Satellite AOD-derived fields perform similarly to using CMAQ fields with data fusion. Results of this study contribute to our understanding of the strengths and weakness of different methods regarding their application in exposure related studies.

Tu-PL-D1.3

Development of Land Use Regression models for assessment of annual average PM10 and endotoxin exposure levels in ambient air in a livestock dense area

Myrna De Rooij, *IRAS - Utrecht University, Utrecht, Netherlands*
Dick Heederik, *IRAS - Utrecht University, Utrecht, Netherlands*
Erik van Nunen, *IRAS - Utrecht University, Utrecht, Netherlands*
Isabella Oosting, *IRAS - Utrecht University, Utrecht, Netherlands*
Gerard Hoek, *IRAS - Utrecht University, Utrecht, Netherlands*
Inge Wouters, *IRAS - Utrecht University, Utrecht, Netherlands*

Aim
There is concern about the influence of livestock farms on the health of Dutch inhabitants. Results from studies on health effects of livestock farming based on exposure proxies are inconsistent. Dust exposure measurements may enable more refined exposure-response analyses. We aimed to develop Land Use Regression (LUR) models for particulate matter 10 (PM10) and endotoxin, known to be emitted from livestock farms and associated with adverse health effects.
Methods
Ambient PM10 was collected with Harvard Impactors at 61 sites (residential gardens) representing a variety of nearby livestock related characteristics. Three to four 2-week averaged PM10 samples were collected at each site over the course of 1.5 year. In addition a local reference site was set up and measured continuously to take into account temporal variation. Samples were analyzed for PM10 mass by weighing and endotoxin using the Limulus-Amebocyte-Lysate assay. LUR models were developed using temporally adjusted annual PM10 and endotoxin exposure averages and livestock-related GIS variables (distances to and number of farms /animals by animal species).
Results
More spatial variation was observed for endotoxin compared to PM10. The model explained variance was higher for endotoxin than for PM10 (R2 0.68 and 0.18 respectively; number of predictor variables 6 and 2 respectively). Predictor variables included number of farms and type of animal species kept in the surroundings, and distance to the farms.
Conclusions
In conclusion, the effect of livestock-related sources on annual average exposure levels seems considerable for endotoxin and more limited for PM10. The LUR approach used, similar to air measurement studies focusing on other air pollutants, was found to be suitable to describe spatial variation in a livestock dense area. Thus far only livestock related predictor variables were explored. The developed LUR models should be further enriched and validated, before they can be applied to predict air pollution concentrations. Different validation methods will be applied in the near future to assess robustness of the models for PM10 and endotoxin. Leave-one-out cross validation will be performed, each site will be sequentially be left out from the model while the included predictors will be left unchanged. In addition, hold-out validation methods will be performed based on stratified random exclusion of sites resulting in multiple models. Lastly, assessment of external validation will be explored with pilot measurements previously performed in a neighboring area.

Tu-PL-D1.4

Spatial and Temporal Assessment of Air Pollution in the Calgary, Alberta Air Zone

Markey Johnson, *Health Canada, Ottawa, Ontario, Canada*
Isabelle Couloigner, *University of Calgary, Calgary, Alberta, Canada*
Stefania Bertazzon, *University of Calgary, Calgary, Alberta, Canada*
Fox Underwood, *University of Calgary, Calgary, Alberta, Canada*
Keith Van Ryswyk, *Health Canada, Ottawa, Ontario, Canada*
Ryan Kulka, *Health Canada, Ottawa, Ontario, Canada*
Hongyu You, *Health Canada, Ottawa, Ontario, Canada*

Canadian air zones represent a complex mixture of urban and rural land-use impacted by diverse emissions sources. The Calgary Spatial and Temporal Exposure Modeling (CSTEM) Study was designed to provide spatial and temporal air quality information for Calgary and surrounding areas to support local air zone management strategies and air pollution health studies.

CSTEM collected two-week integrated measurements of nitrogen dioxide (NO2), volatile organic compounds, particulate matter (PM10, PM2.5), black carbon (BC), and PM-components at 125 sites in summer (August 5-19, 2015) and winter (January 20-February 3, 2016). Seasonal trends were assessed by collecting two-week integrated data every two weeks at four temporal sites across the city from March 25, 2015-April 27, 2016.

NO2 and VOCs were measured using Ogawa Passive Samplers and Organic Vapor Passive Samplers. Gravimetric PM2.5 and PM1.0 measurements were collected using Harvard Cascade Impactors with 37 mm Teflon filters. PM2.5 samples were analyzed for elemental composition using HF-nitric acid digested inductively coupled plasma mass spectrometry. BC was measured via optical scanning of gravimetric PM2.5 samples using a SootScan Model OT21 Transmissometer. Continuous BC was collected at 40 sites using microAethalometers. Analysis of collocated BC samples showed good agreement (R2>0.70) between the methods.

Air pollution data were combined with land-use information to develop land-use regression (LUR) models. Stepwise selection and regression tree methods were used to identify best predictors. The Getis-Ord Gi statistic and global Moran's I were applied to assess local variation of pollutants. Land use regressionLUR ordinary least squares (OLS) regression and geographically weighted regression (GWR) techniques. Summer results follow.

NO2 displayed greater local variation compared with PM2.5. Therefore, GWR and regional OLS models were developed for NO2. Global OLS models performed poorly, predicting only 56% of the variability in NO2. Regional OLS models performed slightly better, with R2 ranging from 0.56-0.60. GWR models performed best, explaining > 80% of the global variability in NO2, with local R2 ranging from 0.56-0.87 (Q25=0.72, Q75=0.84). ANOVA tests confirmed that GWR provided a statistically significant improvement over OLS. Industrial zoning, infrastructure and major roads were significant predictors of NO2. Industrial zoning, PM emitting facilities, and local roads were major predictors of PM2.5.

CSTEM results provide insight into best approaches for characterizing air pollution in a large, diverse air zone. Future analyses will focus on seasonal and temporal modeling; modeling BC, VOCs, and metals; and integrating data from chemical transport models, satellite remote sensing, and continuous regulatory monitoring.

Tu-PL-D1.5

Land-use regression modelling of ultrafine particles in the Augsburg Region, Germany

Josef Cyrys, *Helmholtz Zentrum Muenchen, Neuherberg, Germany*
Kathrin Wolf, *Helmholtz Zentrum Muenchen, Neuherberg, Germany*
Alexandra Schneider, *Helmholtz Zentrum Muenchen, Neuherberg, Germany*
Jianwei Gu, *University of Augsburg, Augsburg, Germany*
Thomas Kusch, *University of Augsburg, Augsburg, Germany*
Annette Peters, *Helmholtz Zentrum Muenchen, Neuherberg, Germany*

Background. Though important health relevance has been suggested for ultrafine particles (UFP), studies on long-term effects are still scarce, mainly due to the lack of appropriate spatial exposure models. In the past few years land-use regression (LUR) models have been increasingly used for assessing small-scale air pollution variability. However, only a few LUR models have been developed for UFP. Due to the cost of monitoring devices most of them have used mobile measurements or short-term measurement at several sites. We developed LUR models based upon methodology developed within the framework of the European Study of Cohorts for Air Pollution Effects (ESCAPE) to predict the spatial variability of several air pollutants and particle number concentration (PNC) as indicator for UFP in the Augsburg region, Southern Germany.
Methods. Three bi-weekly measurements (reflecting the warm, cold and one intermediate season) of PNC, particulate matter (PM10, PM2.5), soot (absorbance of PM filters) and nitrogen oxides (NOx, NO2) were performed at 20 sites in 2014/15. Annual mean and median concentration were calculated and temporally adjusted by continuous measurements from a reference site located in the urban background. As geographic predictors we offered several traffic and land-use variables, elevation, population and building density. Models were validated using leave-one-out cross-validation.
Results. Annual mean and median PNC concentrations ranged from 5,489-13,232 and 4,421-9,458 particles/cm^3. Model explained variance (R^2) was high for both mean and median PNC (0.92 and 0.94, respectively). Cross-validation R^2 was slightly lower (0.82 and 0.87, respectively) but still indicated a very good fit. PNC was moderately correlated with PM (Spearman r: 0.63-0.65), but high with nitrogen oxides and soot (r=0.81-0.89). Also, main predictors for the latter were similar including traffic and building density in the close vicinity (25-50m) and green and industrial area in the medium surrounding (100-500m).
Conclusions. LUR models for PNC performed well in our study region. The performance of our model was better than the performance of a previously reported UFP LUR model developed for Amsterdam (adjusted model R2 was 0.65). To our knowledge no further UFP LUR models based upon fixed monitoring sites have been conducted so far. The high correlation with nitrogen oxides and soot and similar predictors for the LUR models indicate PNC being part of the traffic-related air pollutant mixture in the Augsburg region. We are currently applying the LUR models to the residential addresses of our KORA (Cooperative Health Research in the Augsburg Region) participants.

Tu-SY-E1: Real-time measurements and integrated models to estimate traffic exposures in complex urban environments.

Tu-SY-E1.1

Addressing confounding by noise in air pollution studies

John Gulliver, *Imperial College London, London, United Kingdom*
David Morley, *Imperial College London, London, United Kingdom*
Daniela Fecht, *Imperial College London, London, United Kingdom*

Aim. There is a need in studies of traffic-related air pollution and health to evaluate the robustness of associations in order to distinguish the effects from other factors such as noise. We present what is known about the correlation between noise and air pollution in urban settings and implications for exposure and health assessment.
Methods. We compared modelled levels of noise (LAeq16, Lnight, Lden) and air pollutants (total PM2.5 and PM10, PM2.5 and PM10 from road traffic only, NO2 and O3) for ~190,000 residential postcode centroids within different geographical units (neighbourhoods, n=5,359; 1km grid squares, n=2,171; districts, n=32; and across all postcodes) of London, UK. All comparisons were made using Spearman's correlation.
Results. Across all London postcodes we observed overall moderate correlations between noise and air pollution estimates (Spearman's rho range: |0.34 - 0.55|). Correlations, however, varied considerably depending on the spatial unit: largest ranges were seen in neighbourhoods and 1km grid squares (Spearman's rho range: |0.01 - 0.87|) and the range was less for districts (Spearman's rho range: |0.21 - 0.78|). Differences in correlations could not be explained by exposure or deprivation tertiles, or by distance from road.
Conclusion. Exploring the differences in spatial correlation between noise and air pollution is important for studies of traffic-related air pollution and health. Correlations may vary from weak to strong depending on the pollutant and geographical unit of analysis. This study suggests that although sharing the same source co-linearity between air pollution and noise cannot be assumed in health assessment.

Tu-SY-E1.2

Exposure to greenness in the urban environment: assessment and interactions with co-varying exposures

Charlotte Clark, *Queen Mary University of London, London, United Kingdom*
Emily Rugel, *University of British Columbia, Vancouver, Canada*
Hind Sbihi, *University of British Columbia, Vancouver, British Columbia, Canada*
Hugh Davies, *University of British Columbia, Vancouver, BC, Canada*
Jochem Klompmaker, *RIVM, Bilthoven, Netherlands*
Michael Brauer, *University of British Columbia, Vancouver, British Columbia, Canada*
Perry Hystad, *Oregon State University, Portland, Oregon, United States*

Background: Motor vehicle transportation not only produces air pollutants, but the associated infra-structure affects urban form via access to the natural environment, commonly quantified as a measure of "greenness" through the satellite-derived normalized difference vegetation index (NDVI). Using various buffer sizes surrounding study participants' homes, these indices may capture different aspects of natural spaces, having relevance to distinct pathways. Other exposure metrics such as greenspace access, quality, visibility, microbial diversity may capture other potentially relevant factors in disease etiology.
Methods: Greenspace exposure and two outcomes, namely asthma and type 2 diabetes, in children and adults cohorts respectively, were investigated.
Linked administrative health databases allowed for longitudinal follow-up of participants' home postal codes such to reconstruct their environmental exposures. We applied greenness measures with other spatially varying exposures (air pollution, noise, "walkability") to assess three hypothesized pathways by which natural spaces may impact health: (1) stress-reducing effect of greenspace proximity; (2) reduction of harmful environmental exposures; and 3) providing spaces for increased physical activity.
Finally, measures of greenspace access and quality were integrated into a comprehensive natural space measure and compared with NDVI.
Results: NDVI was negatively correlated with walkability (r=-0.7) while other built environment exposures were modestly correlated with this metric whether assessed in 100m and 250m buffers; thereby allowing for incremental model building of multiple exposures.
In the children's cohort, air pollution was associated with increased odds of incident asthma between birth and age 5 [adjusted OR per interquartile exposure increase (95% CI): NO: 1.06 (1.01 - 1.11), NO2: 1.09 (1.04 - 1.13)), CO: 1.05 (1.01 - 1.1)]. These associations with air pollution were partially offset by the independent protective effect of the level of residential greenness [0.96 (0.93-0.99)].
In the adult cohort, greenness was associated with a significant reduction in the odds of type 2 diabetes (0.4 (0.31 - 0.48)). Although independent of the increased risk related to noise exposure, which increased the risk of incident type 2 diabetes [1.05 (1.02 - 1.07)], this effect was attenuated after adjusting for risk factors and co-varying exposures to air pollution, walkability, and residential greenspace.
For the quality appraisal, no significant differences by neighborhood-level income were shown, yet the percentage of publicly accessible greenspace was highest in the lowest SES areas.

Conclusion: NDVI can be used to estimate greenness presence in studies of large populations. Greenness measures may modify the effects of air pollution or be independently associated with health outcomes.

Tu-SY-E1.3

The link between exposure to traffic-related air pollution and SES

Jane Clougherty, *University of Pittsburgh, Pittsburgh, PA, United States*

Socioeconomic status (SES) and associated chronic stressors (e.g., poverty, violence) are often spatially correlated with air pollution; for example, traffic-related air pollution is shown to be tightly correlated, in space and time, with traffic-related noise. Epidemiologic studies consistently link SES (variously indicated by education, income, etc.) with a wide range of health outcomes, though the "causal components" which drive SES-related susceptibility remain unclear. Separately, both air pollution and psychosocial stress have been shown to act through many of the same biological pathways (e.g., inflammation) and to influence common endpoints (e.g., hypertension). For these reasons, it has been hypothesized - and growing evidence suggests - that lower SES, and higher chronic social stressor exposures, may exacerbate air pollution health effects. In this presentation, Dr. Clougherty will discuss the current state of knowledge on the combined effects of air pollutants and chronic stressors on health, and describe methods for identifying, quantifying, and accounting for, potential confounding and effect modification by SES and chronic stressors in air pollution epidemiology.

Tu-SY-E1.4

Understanding the contribution of other sources than traffic in urban environments

Gerard Hoek, *University Utrecht, Utrecht, Netherlands*

Aim:
This presentation will provide an overview of the contribution of sources other than the regulated taipipe traffic emissions to near-road air pollution exposure such as non-tailpipe traffic emissions, local restaurants, (air)ports, and wood burning. It is important to disentangle their contribution from tailpipe emissions of ultrafine and fine particles.
Methods:
A review of published papers will be conducted, focusing on the contribution of different sources to ambient concnetrations of fine, ultrafine and black carbon particles.
Results
Although the major source of ultrafine particles (UFP) in urban areas is tailpipe emissions from motroized road traffic, recent studies have shown that other sources in urban areas may be associated with substantial increases in UFP as well. Studies in Utrecht, the Netherlands documented increases in UFP concentrations related to proximity to restaurants and increases in PM2.5 related to wood burning of a similar magnitude as proximity to major roadways. Studies near Los Angeles airport and Schiphol airport documented a signifcant contribution of airports, several kms away from the airport. Several studies in Europe, North America and Asia have documented that near-roadway contrasts of metals such as Cu and Fe related to non-regualted non-tailpipe emissions may be as large as those of ultrafine and black carbon particles. Studies in Nort-America and Europe have documented large increases of fine and balc carbon particles related to wood burning.
Conclusions
Multiple sources may affect fine and ultrafine particle concentrations in urban areas. The contribution of these sources should be accounted for in helath impact assessment of (near roadway) urban air pollution.

Tu-SY-E1.5

Aggregate Exposure Pathway (AEP) and Exposure Science in the 21st Century (ES21)

David Carslaw, *University of York, York, United Kingdom*
Vasu Kilaru, *US Environmental Protection Agency, Research Triangle Park, NC, United States*

Recent concerns over emissions from road vehicles has highlighted the importance of understanding and measuring 'real-world' emissions. The robust quantification of emissions from motor vehicles is a uniquely challenging problem. The reasons why this area is challenging are many fold but include: the sources are mobile and emissions vary in both time and space, there are millions of individual vehicles that have differing age profiles, vehicle technologies, fuel types and maintenance procedures and so on. Additionally, vehicles can be driven very differently depending on driver behaviour patterns and traffic conditions. While there are many programmes in place around the world that set emission limits for Type Approval of new vehicles, it has become clear that the emissions measured for reasons of compliance may bear little relation to the actual emissions in use. This presentation will consider these issues in detail and present the recent evidence relating to real-world emissions. The focus will be on the emissions of both nitrogen oxides (NOx) and nitrogen dioxide (NO2) for which there is currently considerable interest in Europe. Consideration will be given to the principal techniques used to measure real-world emissions including vehicle emission remote sensing and Portable Emission Monitoring Systems. The talk will summarise the most recent evidence relating to vehicle emissions of NOx and NO2 and consider the implications for air pollution in general and urban air pollution exposure specifically. Finally, the talk will outline the future challenges for real-world vehicle emission measurement and how changes to vehicle emissions legislation in Europe aims to address these challenges.

Tu-SY-F1: OECD Task Force on Exposure Assessment - Better exposure science for better lives - I

Tu-SY-F1.1

Overview on OECD activity on Exposure Assessment

Takahiro Hasegawa, *OECD, Paris, France*

This presentation will provide a brief overview on the OECD Task Force on Exposure Assessment (TFEA) activities.
The TFEA was established in 1995 to facilitate sharing, developing, disseminating, comparing and where possible, harmonising exposure assessment related information (such as emission factors), methodologies and tools (such as databases, guidance documents, harmonised templates and exposure models) for assessing the impact of the release of chemicals during their life cycle on the environment and various populations including the general population, consumers, workers and children at various age groups. This will include:

- developing and updating OECD Emission Scenario Documents (ESDs) for estimating the release of chemicals during their lifecycle including production, processing, use, service-life, recycling, treatment and disposal of chemicals;
- developing, comparing and harmonising information, methodologies and tools for assessing the exposure of chemicals to specific populations, and the exposure of specific types of chemicals to humans and the environment.
- facilitate sharing and compiling information, and develop methodologies and tools (e.g. databases, guidance documents, templates) relating to chemical exposure via products; and
- exchange and share experiences and knowledge on information gathering from stakeholders relating to the exposure assessment and exposure mitigation.

Tu-SY-F1.2

Development of Emission Scenarios

Nhan Nguyen, *US EPA, Washington DC, Virginia, United States*
Wataru Naito, *National Institute of Advanced Industrial Science and Technology (AIST), Tsukuba, Japan*
Hirai Yusuke, *National Institute of Technology and Evaluation (NITE), Tokyo, Japan*

One objective of the OECD Task Force on Exposure Assessment (TFEA) is to facilitate and support the work of OECD on the exposure assessment of chemicals and chemical products with special emphasis on sharing and developing exposure related information, methodologies and tools. To fulfill this objective, the TFEA members have on-going activities to develop, share, and exchange information on exposure assessment tools and models. An important activity of the TFEA on tools and models is development of emission scenario documents or ESDs.

This presentation will provide an overview of the development of emission scenario documents (ESDs) including the purpose of an ESD, the information included in an ESD, the steps involved in the development of an ESD, the published ESDs and the ESDs under development.

Note: The views in this abstract are those of the author and does not represent Agency policy or endorsement.

Tu-SY-F1.3

Children's health

Gerlienke Schuur, *National Institute for Public Health and the Environment (RIVM), Bilthoven, Netherlands*

Children's health
Children can be more vulnerable to chemicals than adults. Considering global concern for children's health, the OECD has been working to bring together knowledge and experiences to reduce risks to children's health from chemicals.
Following an OECD-wide survey of methodologies and tools used to assess the risk of chemicals to children's health in 2011-2012 (OECD, 2013) and a workshop on children's exposure to chemicals held in the Netherlands in 2013 (OECD, 2014), the following projects are currently ongoing within the OECD-Task Force on Exposure Assessment:
1. Children's exposure decision tree
2. Mouthing tool

Children's exposure decision tree
Children's exposure to chemicals from consumer products may deviate from the exposure of adults due to age related behaviour and characteristics such as hand-to-mouth behaviour, object-to-mouth behaviour, crawling/ playing, incidental oral ingestion (swallowing), and sleep pattern. For certain products/product categories or certain substances the exposure and risk assessment for adults may cover the risk for children, but for some products/product categories a separate exposure assessment for children may be indicated.
Currently, there is no structured approach to determine the need to carry out a separate children's exposure assessment within risk assessment. Therefore the OECD-TFEA is developing a decision tree on when to assess children's exposure. This decision tree is intended to guide in determining when children's exposure should be addressed separately in a risk assessment.
A preliminary Children's exposure decision tree was presented at the OECD workshop (OECD, 2014). Work is ongoing to refine this decision tree, and to include case studies to test and improve the decision tree.
The guidance contains a checklist with issues relevant for the direct/indirect exposure to children and is based on a list containing the following questions:

- Is the product/article (category) intended for use by consumers?
- Does this concern direct or indirect use?
- Is the product specifically meant for children?
- Could consumers come into contact with the product? Is this exposure directly or indirectly?

Based on a careful check of the decision tree, a specific exposure assessment for children should be incorporated in the exposure assessment for a certain consumer product.

References
OECD (2013). ASSESSING THE RISK OF CHEMICALS TO CHILDREN'S HEALTH: AN OECD-WIDE SURVEY. Series on Testing & Assessment No. 192. ENV/JM/MONO(2013)20. http://www.oecd.org/officialdocuments/publicdisplaydocumentpdf/?cote=env/jm/mono(2013)20&doclanguage=en
OECD (2014). REPORT ON OECD WORKSHOP ON CHILDREN'S EXPOSURE TO CHEMICALS Series on Testing and Assessment No. 209. ENV/JM/MONO(2014)29.

http://www.oecd.org/officialdocuments/publicdisplaydocumentpdf/?cote=env/jm/mono(2014)29&doclanguage=en

Tu-SY-F1.4

Occupational Exposure

Gudrun Walendzik, *BAuA, Federal Institute for Occupational Safety and Health, Dortmund, Germany*

This presentation will summarize the ongoing activities of the Organization for Economic Cooperation and Development (OECD) Task Force on Exposure Assessment (TFEA) subgroup on occupational exposure. The objective of this subgroup is to suggest possible project ideas / categories for discussion on a scientific basis to reach a common understanding and to identify needs for further developments in order to tackle these issues. The subgroup currently counts 13 members who are governmental officials and researchers from the OECD countries as well as representatives from non-governmental organizations and industrial sectors.

Work products of the TFEA occupational subgroup members such as overview on occupational exposure assessment related tools and models, validation activities of (Tier-1) exposure assessment tools or a proposal for a tier-1dermal exposure assessment approach were made publicly available. Presentations on ongoing research projects e.g. a German research project on occupational dermal exposure (SysDEA) are discussed during TFEA meetings to have a scientific discourse amongst the exposure experts.

Currently a starting compilation for OECD-wide project ideas for further input by the Task Force members to encourage possible collaboration on specific projects exists with the aim and objective to develop and share information and guidance on inhalation and dermal exposure between the different OECD countries.

These are for example:

- Survey on measured data to compile an inventory of databases with inhalation/dermal data in various industry sectors
- Survey of Tier 1 assessment tools used to assess the dermal exposure of chemicals to workers
- Data on efficiency of engineering controls
- Exchange on information and knowledge on occupational exposure which were identified to be useful to the work on issues on manufactured nanomaterials

The presentation will provide a high level overview of the OECD TFEA activities on occupational exposure.

Tu-SY-G1: Environmental Justice: Developing the Scientific Foundation Supporting Cumulative Exposures/Risks/Impacts and Disparate Impacts Research - I

Tu-SY-G1.1

Hand- and Object-Mouthing by Rural Bangladeshi Children 6-20 Months Old; Importance of Including Food-Related Contacts

Laura Kwong, *Stanford University, Stanford, CA, United States*
Ayse Ercumen, *University of Californa, Berkeley, Berkeley, CA, United States*
Amy Pickering, *Stanford University, Stanford, CA, United States*
Leanne Unicomb, *International Centre of Diarrhoeal Disease Research, Bangladesh, Dhaka, Bangladesh*
Jennifer Davis, *Stanford University, Stanford, CA, United States*
Stephen P Luby, *Stanford University, Stanford, CA, United States*

Background: Children are exposed to environmental contamination through mouthing of hands and objects. However, few mouthing studies quantify hand- and object-mouthing associated with eating, which may result in underestimation of children's total exposure. Eating-related exposures may be particularly important in settings where children eat or are fed by hand. We collected activity data from young children in rural Bangladesh in order to quantify their frequency and duration of mouthing, including feeding-related events. The study objective was to assess the contribution of feeding-related hand-to-mouth and object-to-mouth contacts to children's total exposure through mouthing.
Methods: We video recorded the activities of 28 rural Bangladeshi children aged 6-20 months over six consecutive hours and used computer software to record all observed mouth contact with hands and objects, including utensils used for eating. We also reviewed the videos to better understand sources of contamination associated with feeding events.
Results: All children used their hands for eating and/or were fed by hand during observation, and all but two children also used eating utensils. The median frequency of hand-to-mouth contact for children aged 6-12 months (N=19) was 43.8 contacts/hr and for children aged 12-20 months (N=9) it was 33.4 contacts/hr. Food-related contact with the child's own hand or a caregiver's hand accounted for more than one-third of total hand mouthing. During feeding events, children aged 6-12 months mouthed hands a median of 16.9 times/hr and children aged 12-20 months old mouthed hands 11.7 times/hr. Of the fourteen children who fed themselves and were also fed by a caregiver by hand, median contact frequency with the child's own hands (7.3 contacts/hr) was similar to the contact frequency with caregiver's hands (8.4 contacts/hr). Mouthing eating utensils accounted for ~15% of the frequency of mouthing all objects (children aged 6-12 months: median 9.8 utensil contacts/hr of 55.8 total object contacts/hr; children aged 12-20 months: 4.3 utensil contacts/hr of 31.2 total object contacts/hr). Video recordings showed that children did not wash hands before eating. Children were also observed eating off of earthen floors and consuming food that had fallen into the dirt.
Conclusion: In contexts where children eat or are fed by hand, excluding eating periods from mouthing exposure studies may substantially underestimate exposure due to hand-mouthing and underestimate exposure due to mouthing of potentially contaminated

utensils. Washing children's hands before eating and preventing them from consuming contaminated food could reduce foodborne exposure.

Tu-SY-G1.2

Community-Level Stressors and Their Impacts on Food Contamination

Lisa Melnyk, *US EPA, Cincinnati, Ohio, United States*
Zahra Al-Hamdani, *US EPA, Research Triangle Park, NC, United States*
Kent Thomas, *US EPA, Research Triangle Park, NC, United States*
Jacqueline MacDonald Gibson, *University of North Carolina at Chapel Hill, Chapel Hill, NC, United States*

Community programs could benefit from advances in an understanding of food resources, utilization of those resources, and how the built and natural environment impact food access and potential exposures to chemical contaminants in food. To study these factors, a research project was designed to identify stressors, such as, socioeconomic status, transportation, ethnicity, type of industry within the community, drinking water quality, food access, chemical residues on foods, and environmental conditions; to evaluate the relationships between stressors; and to explore potential interactions between food resources and chemical and non-chemical stressor exposures. By evaluating various chemical and non-chemical stressors, an understanding of the relationships between a community's food resources and potential exposures to contaminants that could lead to public health issues can be obtained. The objectives of this research are 1) to obtain information on a community's environmental exposures (chemical and non-chemical stressors) from available data sources and 2) to evaluate impacts of the various chemical and non-chemical stressors on dietary exposures that may lead to adverse public health outcomes. To capture the dynamics of the effects of chemical and non-chemical stressors on a community diet which could lead to impacts on public health, a cumulative exposure model approach was developed. Geospatial Information System (GIS) mapping of Durham County, North Carolina was used to visually describe the area. Data from the US Census Bureau, Dun & Bradstreet, the National Health and Nutrition Examination Survey, state-level food residue measurements, EPA's EnviroAtlas, and EPA's EJScreen provide the input for the analyses. Correlations between the various inputs are calculated using the statistical software, R. This research will enhance public tools, in particular, the Community-Focused Exposure and Risk Screening Tool (CFERST), which can be used by community leaders in decision making by bridging all pertinent information to inform policy. Improving our understanding about a community's potential dietary exposures and contributing factors will help in the identification and mitigation of issues that could impact public health. Community level health analyses can support protective actions, be used by communities to identify and prioritize their risks based on scientific data and ensure that resources are directed where they will provide the greatest benefit.

Tu-SY-G1.3

The cumulative MeHg and PCBs exposure and risk of tribal and US general population with SHEDS-multimedia

Jianping Xue, *EPA, RTP, NC, United States*
Valerie Zartarian, *EPA, Boston, United States*
Ken Bailey, *EPA, CIN, United States*

Studies have shown that the U.S. population continues to be exposed to methyl mercury (MeHg) and polychlorinated biphenyls (PCBs) due to the long half-life of those environmental contaminants. Fish intake of Tribal populations is much higher than the U.S. general population due to dietary habits and unique cultural practices. Large fish tissue concentration data sets from the Environmental Protections Agency's (EPA's) Office of Water, USGS's EMMMA program, and other data sources, were integrated, analyzed, and combined with recent tribal fish intake data for exposure analyses using the dietary module within EPA's SHEDS-Multimedia model. SHEDS-Multimedia is a physically-based, probabilistic model, which can simulate cumulative (multiple chemicals) or aggregate (single chemical) exposures over time for a population via various pathways of exposure for a variety of multimedia, multipathway environmental chemicals. Our results show that MeHg and total PCBs exposure of tribal populations from fish are about 3 to 10 and 5 to 15 times higher than the US general population, respectively, and that the estimated exposures pose potential health risks. The cumulative exposures of MeHg and total PCBs will be assessed to generate the joint exposure profiles for Tribal and US general populations. Model sensitivity analyses will identify the important contributions of the cumulative exposures of MeHg and total PCBs such as fish types, locations, and size, and key exposure factors. Biomarker data from NHANES will be used to evaluate SHEDS-Multimedia outputs. These exposure assessments can be used to help inform decisions regarding meal sizes and frequency, types of fish and water bodies to avoid, and other factors to minimize exposures and potential health risks from contaminated fish in Tribal populations and high exposure groups from the U.S. general population.

Tu-SY-G1.4

Community-engaged modeling of exposures to chemical and non-chemical stressors in a low-income community near a Superfund site

M. Patricia Fabian, *Boston University School of Public Health, Boston, Massachusetts, United States*
Komal Basra, *Boston University School of Public Health, Boston, Massachusetts, United States*
Jonathan Levy, *Boston University School of Public Health, Boston, Massachusetts, United States*

Cumulative risk assessment requires models of exposures to chemical and non-chemical stressors at sufficient geographic and demographic resolution to accurately identify high-risk subpopulations. Many non-chemical stressors are of considerable interest to community stakeholders, above and beyond their connections to chemical-oriented risk assessments, but relevant exposure models have not been leveraged for these purposes. In this study, we developed a methodology to provide high resolution exposure estimates, constructing detailed synthetic demographic microdata, and using these data as predictors in regression models for multiple stressors in New Bedford, Massachusetts, a low income community located near a Superfund site. Chemical exposure regression models utilized biomarker measurements from a cohort study conducted in the New Bedford area, and non-chemical stressor models leveraged data from the Behavioral Risk Factor Surveillance System (BRFSS). Through a series of meetings with community partners, behavior and health questions of interest from the BRFSS were identified, including fruit and vegetable consumption, obesity, and diabetes. Paralleling the structure of our chemical stressor models, we constructed multivariable regression models of the probability of eating fruits and vegetables, body mass index (BMI), and diabetes prevalence. Regression models were applied to the synthetic microdata and results mapped across the community to identify census tracts at high risk for these behaviors and outcomes in adults. Comparisons of geographic patterns of these stressors of interest to community partners with geographic patterns of chemical stressors identified areas of common emphasis. The maps and modeled demographic patterns will be used by community partners for city planning and policy activities such as parent support programs for people living with chronic diseases, locating new farmers markets, expansion of the fresh food voucher program, and prioritizing selection of existing brownfields to be converted to community gardens. Our study emphasizes the value of multi-stressor exposure modeling in the context of cumulative risk assessment, the insights provided by community engagement, and the opportunity for innovative exposure modeling approaches to connect with broader community concerns.

Tu-SY-G1.5

GIS-Mapping and Statistical Analyses to Identify Climate-Vulnerable Communities and Populations in New England Exposed to Contaminated Sites and Combined Sewer Overflows

Valerie Zartarian, *US EPA, Boston, MA, United States*
Jianping Xue, *EPA, RTP, NC, United States*
Tony Poulakis, *Vistronix, Boston, MA, United States*
Alex Dichter, *US EPA, Boston, MA, United States*
Michael Jahne, *US EPA, Cincinnati, OH, United States*
Jeff Yang, *US EPA, Cincinnati, OH, United States*

Climate change-related cumulative health risks are expected to be disproportionately greater for overburdened communities that are often in proximity and thus greater exposures to chemical sources and flood zones. Communities and populations vulnerable to climate change-associated impacts from contaminated sites and CSOs (combined sewer overflows) were identified using GIS mapping and statistical analyses. Databases from U.S. EPA and other Federal Agencies were combined to overlay Superfund sites by elevation, flood zones, and environmental justice related variables. Additional analyses examined disparities associated with the most vulnerable communities and particular groups based on socioeconomic, racial and ethnic factors, as well as proximity to contamination. The regional-scale screening results could be used to identify areas of potential concern, for informing climate resiliency efforts and conducting human exposure and risk analyses. Case studies in New England region, as presented here, have focused on mapping EPA Superfund sites, hazard-facility sites and CSOs vulnerable to climate-induced flooding, and better understanding the agents and key exposure factors related to increased risk of toxic chemical and microbial exposure. The study findings enhance understanding of climate vulnerability and resiliency indices used in community adaptation and risk mitigation strategies. Such information can enable vulnerable communities to take proactive and effective actions to reduce future risks from impacts of flooding, extreme weather, and sea level rise on contaminated sites, waste-storage releases, and CSOs. The methods developed can be used for other communities in the U.S..

Tu-PL-H1: Analytical Methods - I

Tu-PL-H1.1

Considerations for Stability of Environmental Samples in Storage in Children's Environmental Epidemiology Studies

Susan Viet, *Westat, Rockville, MD, United States*
Maire S.A. Heikkinen, *Westat, Rockville, United States*
Michael Dellarco, *NIH/NICHD, Bethesda, United States*

For long-term, large scale children's environmental epidemiological studies, it is advantageous to store collected environmental samples for future analysis. This permits spreading out costs for analyses over time and flexibility for analysis of samples. Samples can be analyzed for specific analytes after sufficient numbers have been collected for efficiency, subsets of samples can be analyzed in nested case-control studies for specific analytes after health outcomes of interest have occurred, or to include new target analytes not previously considered for study. Information about sample stability is critical to determine if there will be analyte loss or gain or degradation during storage. Failure to evaluate and consider analyte stability could result in inaccurate results and bias subsequent exposure assessments due to partial or complete analyte decomposition, or chemical transformation.
We reviewed the literature and organizational guidelines for guidance in developing a sample stability program for the US National Children's Study Vanguard Study. We surveyed literature in PubMed and other commercial sources to identify published sample preparation and storage information. We found 53 peer-reviewed articles and 9 reports and book chapters dealing with stability of chemical compounds, mostly pesticides and metals, in various matrices: food, animal tissue, soil/sediment, dust, water, and air. There was little documentation to support preservation of analytes during long-term storage. Our review showed variable guidance in storage temperatures and holding times for water samples and dust wipes. Similarly we found different guidance protocols regarding sample preparation procedures for water samples prior to storage.
Our results show there is a need for more data and guidelines to ensure stability of environmental samples stored over long time periods. In the interim, based on our research, a sample stability program should be integrated with sample collection and storage as a part of the study quality assurance procedures. This would enable data to be collected about analyte stability during storage and document the integrity of stored samples and the analytes of interest.

Tu-PL-H1.2

Combination of multiple analytical platforms and nontargeted approaches for comprehensive risk assessment characterization: thirdhand tobacco smoke as case study

Noelia Ramirez, *Institut d'investigacio Sanitaria Pere Virgili-URV-CIBERDEM, Tarragona, Spain*
Sonia Torres, *Instituto d'Investigacio Sanitaria Pere Virgili, Tarragona, Spain*
Xavi Domingo, *Universitat Rovira i Virgili, Tarragona, Spain*
Lara Gundel, *Lawrence Berkeley National Laboratory, Berkeley, United States*
Hugo deataillats, *Lawrence Berkeley National Laboratory, Berkeley, United States*
Rosa Maria Marce, *Universitat Rovira i Virgili, Tarragona, Spain*
Francesc Borrull, *Universitat Rovira i Virgili, Tarragona, Spain*
Xavier Correig, *Universitat Rovira i Virgili-IISPV-CIBERDEM, Tarragona, Spain*

Aim: To the date, most of the risk assessment studies are focused on the determination of specific families of toxicants in different environmental matrices and the evaluation of the human health risk associated to this exposure. Nevertheless, the risk associated to complex matrices exposure, such as indoor dust, which contains hundreds of contaminants, may lead to possible mixture-related effects – such as, antagonistic, synergistic, potentiating or additive effects – that are underestimated in targeted studies. Here, we evaluate the benefits of using different analytical platforms to achieve a more complete identification of the hazardous composition of complex environmental samples for a more holistic characterization of the human risk through the exposure to these contaminants. The developed methodology is applied to the risk evaluation of dust samples from smokers' and non-smokers' homes polluted with residual tobacco smoke, also so-called, thirdhand tobacco smoke (THS). Tobacco smoke is considered one of the major sources of inorganic and organic carcinogens in indoor environments, but more evidence of the chemical toxicity of THS and its impact on human health is necessary to improve understanding of the risks of THS-polluted environments.
Methods: We have extracted indoor dust samples from smokers' and non-smokers' homes using pressurized liquid extraction and ethyl acetate as solvent. We have used three different analytical platforms to analyze these extracts, including single and comprehensive gas chromatography (GC-QTOF and GC×GC-TOF) and liquid chromatography (UPLC-QTOF). The raw data have been processed using in-house and commercial software programs. Univariate and multivariate statistical analysis have been used for feature selection to better profile THS toxicants.
Results: The use of multiple analytical platforms allowed a more complete identification of the toxicological profile of THS. GC×GC-TOF, with enhanced separation power, provided a visual profiling of the samples and allowed the classification of the toxicants by chemical group. GC-QTOF, which provided the exact mass of the ions, favored the identification of the contaminants. UHPLC-QTOF is complementary to GC and permitted the identification of polar and semivolatile contaminants. The putative identification of the main contaminants was confirmed with the MS/MS spectrum and the use of compound standards.
Conclusions: The results presented here demonstrate the benefits of using multiple analytical platforms and non-targeted approaches to achieve a more comprehensive risk assessment of complex environmental mixtures. The application of this holistic

methodology in future studies will help to better understand the chemical risk of polluted environments.

Tu-PL-H1.3

Sample extraction strategies for target and non-target analysis of xenobiotics in biological fluids

Christine Baduel, *University of Queensland, Brisbane, Queensland, Australia*
Jochen F. Mueller, *University of Queensland, Brisbane, Queensland, Australia*
Henghang Tsai, *Queensland Government, Brisbane, Australia*
Maria Jose Gomez Ramos, *University of Queensland, Brisbane, Australia*

Our developed, highly technological society has more than 100,000 chemical substances registered with several hundred new chemicals being introduced and registered every year. Due to the potentially adverse environmental and health effect associated with exposure to such chemicals and their degradation products, data concerning the exposure of these chemicals in biological matrices is needed. For several decades, target analysis have been mainly use to analyse contaminants in biological matrices. The drawbacks of this technique is that only predeterminated chemicals can be detected and quantified meaning that all the other chemicals present, potentially toxic and/or in high concentration are overlooked.
Aims: Stepping towards a more comprehensive assessment of human and biota exposure to contaminants, we present in this work the development and optimization of an analytical strategy that combines efficient and reliable extraction, purification and analysis of a broad range of polar and non-polar target analytes in fatty biological matrices such as breast milk.
Methods: To extract a wide range of chemicals, the partition/extraction procedure used for the QuEChERS (Quick, Easy, Cheap, Effective, Rugged and Safe) was modified for the extraction. The method development was done by using analytes with different physico-chemical properties (log Kow ranges from -0.3 to 10) from highly polar pesticides and personal care products (PPCPs) to highly lipophilic chemicals such as polybrominated diphenyl ethers (PBDEs), polychlorinated biphenyls (PCBs), polycyclic aromatic hydrocarbons (PAHs). The proposed method combines targeted multiresidue analysis using gas chromatography triple quadrupole mass spectrometry (GC-QqQ-MS/MS) and a multi-targeted analysis complemented with non-target screening using liquid chromatography coupled to a quadrupole time of flight mass spectrometry (LC-QTOF-MS/MS).
Results: The method was fully validated for samples of breast milk through the evaluation of recoveries, matrix effects, limit of quantification, linearity and precision (interday and intraday). Mean recoveries (n=5) were between 70% and 120% with relative standard deviations (RSD) less than 20% in most of the cases. To demonstrate the applicability and suitability of the validated method, 5 breast milk samples were analysed. The results showed that of the 77 target compounds monitored, a total of 29 were quantified in the samples. The present work also demonstrated the feasibility of discovering untargeted chemicals such as the transformation products of widely used insecticides, a flame retardant, fluorosurfactant and preservatives and its metabolites.
Conclusion: This strategy has shown to be suitable for target and non-target screening given a more comprehensive view of the true overall contaminants present in the samples.

Tu-PL-H1.4

Screening persistent polar contamination in drinking water with UHPLC-QTOF: focus on reverse osmosis applied to riverbank filtrate

Pim De Voogt, *Univeristeit van Amsterdam - Institute for Biodiversity and Ecosystem Dynamics, Amsterdam, Netherlands*
Vittorio Albergamo, *University of Amsterdam - Institute for Biodiversity and Ecosystem Dynamics, Amsterdam, Netherlands*
Rick Helmmus, *University of Amsterdam - Institue for Biodiversity and Ecosystem Dynamics, Amsterdam, Netherlands*

Incomplete removal of polar organic contaminants by riverbank filtration and reverse osmosis (RO) treatment has been discussed in scientific literature. These compounds tend to be highly mobile within the water cycle and can accumulate in the environment due to their high water solubility. Therefore it is necessary to assess their presence in the source water and final product. In the case of RO it is also necessary to screen for moderately/less polar compounds, as they might persist RO treatment due to adsorption onto RO membranes and diffusion to the permeate side.The characterization of organic contaminants in the watercycle is essential to assess the potential ecotoxic and toxic effects of individual chemicals and mixtures. In the present work, we developed an analytical method for the screening of small polar MPs with UHPLC-ESI-QTOF-MS. We have developed a LC method which employs a novel core-shell biphenyl stationary phase that provides satisfactory retention of polar contaminants having low molecular weights (<420 Da) and covering a broad range of physicochemical properties, such as acesulfame (Log Kow= -1.33) and PFOA (Log Kow= 6.30). We apply the UHPLC-QTOF method for the suspect screening of finished drinking water and its sources from a drinking water treatment plant (DWTP) in The Netherlands. This plant will employ a standalone reverse osmosis (RO) treatment to riverbank filtrate from the Rhine basin before delivering a remineralized RO permeate to its customers.
Suspect screening is performed by acquiring accurate mass fragmentation data from riverbank filtrate and RO permeate samples in broadband collision-induced dissociation (bbCID) MS/MS full scan mode. Following examination of mass accuracy, isotopic patterns, adducts and chromatographic behavior, a preliminary candidate list is obtained. The m/z ratios are added to a database incorporated in an automated screening software tool for the fast processing of large batches of samples. A subset of the samples are then re-analyzed in auto MS/MS full scan mode to elucidate the fragmentation pattern of candidates if not yet known. One or more fragments are subsequently used as qualifier ions for the tentative suspect identification. When possible, identity is confirmed with a reference standard.
This approach proved to be successful in identifying polar organic micropollutants in riverbank filtrate as well as in qualitatively assessing chemical removal by RO treatment. We believe that our strategy adds valuable insight to the global efforts towards profiling emerging and unknown water contamination with HRMS and towards a better comprehension of contaminant removal by RO during drinking water production.

Tu-PL-H1.5

Contamination, Exposure and Risk Assessment: Pyrazol Case in the Dutch surface water

Erik Emke, *KWR Watercycle Research Institute (KWR), ieuwegein, Netherlands*
Annemieke Kolkman, *KWR Watercycle Research Institute (KWR), Nieuwegein, Netherlands*
Dennis Vughs, *KWR Watercycle Research Institute (KWR), Nieuwegein, Netherlands*
Kirsten Baken, *KWR Watercycle Research Institute (KWR), Nieuwegein, Netherlands*
Patrick Bäuerlein, *KWR Watercycle Research Institute (KWR), Nieuwegein, Netherlands*
Annemarie van Wezel, *KWR Watercycle Research Institute (KWR), Nieuwegein, Netherlands*
Pim de Voogt, *KWR Watercycle Research Institute (KWR), Nieuwegein, Netherlands*

In the Netherlands around 40% of the drinking water originates from surface water mainly from the rivers Rhine and Meuse. Therefore several early warning monitoring stations upstream from the intake points safeguard the quality of the surface water by an array of sensors, daphnia, algae, mussels and instrumental techniques like HPLC Diode Array Detection (DAD). The different monitoring stations agreed on a common best practice protocol based on HPLC-DAD screening, the so-called UV fingerprint screening which is performed daily. Known and unknown compounds are followed by using their retention time index, their UV spectrum, and internal standard equivalents. In the summer of 2015 a daphnia sensor and the mussel monitor were triggered. A sample was measured by the UV fingerprint screening showing a large broad peak emerging with a relative short retention index indicating a contaminant with a highly polar nature not present in the UV database. This resulted in a closedown of the water intake for the production of drinking water in large parts of the Netherlands.
The aim of the present study was to identify this new emerging compound by hyphenating the HPLC-DAD to the LTQ-FT-Orbitrap and employing different ionisation techniques. The effluent of the HPLC-DAD screening system was transferred directly to the LTQ-FT-Orbitrap without splitting. Initial experiments were done by using a heated electrospray interface (HESI) and acquiring both in positive and negative ionisation mode and no corresponding peak was detected in the MS. Further experiments were done by exchanging the HESI for the Atmospheric Pressure Chemical Ionisation (APCI) interface which showed a significant improvement in ionization efficiency. Two possible suspects were selected based on structure and log Kow namely imidazole and pyrazole. Since no fragmentation was observed it was obligatory to confirm the suspect with a standard. Pyrazole resulted in a perfect match and through establishing a calibration curve the concentration in the alarm sample was found to be 100µg/L. Pyrazole is widely used as a starting product for the synthesis of pharmaceuticals and pesticides and a known industrial by-product. The origin of the discharge into the environment was located and the company informed. Daily monitoring revealed that the discharge was happening frequently and measures were taken to prevent pollution of the surface water. Since almost no information was available on the toxic properties of pyrazole a provisional guideline value was established at 15 µg/L for drinking water.

Tu-PL-I1: Quantitative Methods

Tu-PL-I1.1

Integrated approach for external and internal exposure assessment: 2 case studies with benzene and chlorpyrifos

Joost Westerhout, *TNO, Zeist, Netherlands*
Henk Goede, *TNO, Zeist, Netherlands*
Wouter Fransman, *TNO, Zeist, Netherlands*
Rob Stierum, *TNO, Zeist, Netherlands*
Anjoeka Pronk, *TNO, Zeist, Netherlands*
Hans Marquart, *Triskelion B.V., Zeist, Netherlands*
Rianda Gerritsen-Ebben, *TNO, Zeist, Netherlands*

Introduction: Current human health risk assessment of chemicals is fragmented, using conservative factors to account for uncertainty in all separate steps, without proper balance. Consequently, the risk assessment is often over conservative, resulting in unnecessary stringent management and disapproval of chemicals.
The aim of this project was to develop an aggregated and cumulative model of three exposure routes (inhalation, dermal, ingestion) that integrates both external exposure and the resulting internal exposure, providing more realistic exposure estimates for risk assessment purposes.
Methods: An interactive tool was created using R software, allowing the prediction of blood and (target) organ concentrations on the basis of a variety of exposure scenarios (e.g., single or multiple exposure routes, short or long duration, single or daily exposure). The underlying physiologically-based toxicokinetic (PBTK) model includes children and adults of all ages and sizes and uses generic physicochemical properties and default values for human physiological parameters for its predictions. Benzene and chlorpyrifos were selected as model compounds to test the functionality of the tool. Several experimental studies, in which an exposure scenario was clearly defined (e.g. tasks performed, task duration, use of respirators or gloves) and the amounts absorbed and/or blood or expired air concentrations were reported, were derived from literature for both benzene and chlorpyrifos. The exposure scenarios were then applied to the tool and predicted amounts absorbed and/or blood or expired air concentrations were compared to the reported values.
Results: For benzene, results indicate that the PBTK model, using generic physicochemical properties for predictions, overestimates blood concentrations for different exposure scenarios up to 3-fold, while expired air concentrations are overestimated up to 2-fold, thus resulting in a conservative estimation. For chlorpyrifos, results indicate that dermal absorption is overestimated up to 3-fold with the PBTK model using generic physicochemical properties, whereas the estimated exposure via inhalation is only minimal.
Discussion: Overall, the PBTK model allows for the bridging of the knowledge gap between single-route 'potential' external exposure models and kinetic internal models with multiple organ endpoints. Furthermore, the interactive tool allows for end users to make predictions of blood and organ concentrations for any chemical without requiring programming skills. As our understanding of human toxicology is improving by investigating adverse outcome pathways (AOPs) and the molecular initiating events (MIEs) leading to the

adverse outcome, the human health risk assessment process could further be improved by linking these combined internal exposures to AOPs and MIEs.

Tu-PL-I1.2

Assessing Tobacco Smoke Exposure Categories from Continuous Biomarker Measurements Using Cumulative ROC Curve Analysis

Rey DeCastro, *U.S. Centers for Disease Control, Atlanta, GA, United States*

AIM: Discriminating three or more exposure categories (e.g., low-medium-high) from continuous measurements, such as biomarker concentrations, has been hindered because standard receiver operator characteristic (ROC) curve analysis is limited to binary outcomes. A new method free of this binary restraint -- cumulative ROC curve analysis -- is used to determine cutpoints discriminating three categories of exposure to tobacco smoke. METHODS: Cotinine in blood serum is a biomarker specific to tobacco smoke exposure. Cumulative ROC curve analysis was used to identify cutpoint concentrations of serum cotinine discriminating three ordinal categories of self-reported exposure to tobacco smoke: non-exposed, exposed to second-hand smoke, and active smoker. Serum cotinine measurements were obtained from participants in the United States National Health and Nutrition Examination Survey (NHANES). Cumulative ROC curve analysis comprises a two-stage, semiparametric approach combining conventional cumulative logit regression with a cumulative version of ROC curve analysis. RESULTS AND CONCLUSIONS: Cumulative ROC curve analysis estimated serum cotinine cutpoints at 2.4 ng/ml (discriminating non-exposed vs. second-hand smoke exposed) and 10.4 ng/ml (discriminating second-hand smoke exposed vs. active smoker). Cumulative ROC curve analysis may prove useful when it may be of interest to estimate intermediate exposure levels.

Tu-PL-I1.3

Models to estimate asbestos exposure of brake mechanics without sampling

María Fernanda Cely-García, *Universidad de los Andes, Bogotá, Colombia*
Frank C. Curriero, *Johns Hopkins University, Baltimore, MD, United States*
Mauricio Sánchez-Silva, *Universidad de los Andes, Bogotá, Colombia*
Patrick N. Breysse, . *Work conducted while affiliated with the Department of Environmental Health Sciences, Johns Hopkins University, Bloomberg School of Public Health, Baltimore, MD, USA; currently employed by the Centers for Disease Control and Prevention. Patrick Breysse, Baltimore, MD, United States*
Margarita Giraldo, *Universidad de los Andes, Bogotá, Colombia*
Lorena Méndez, *Universidad de los Andes, Bogotá, Colombia*
Carlos Torres-Duque, *Fundación Neumológica Colombiana, Bogotá, Colombia*
Mauricio Durán, *Fundación Neumológica Colombiana, Bogotá, Colombia*
Mauricio González-García, *Fundación Neumológica Colombiana, Bogotá, Colombia*
Patricia Parada, *Fundación Neumológica Colombiana, Bogotá, Colombia*
Juan Ramos-Bonilla, *Universidad de los Andes, Bogotá, DC, Colombia*

Introduction: Decades ago it was common among brake mechanics in high-income countries to manipulate asbestos containing brake products, work-tasks that are no longer performed. Air asbestos samples were rarely collected during these manipulations, limiting our ability to understand asbestos exposure of brake mechanics, something crucial for epidemiological studies. Brake mechanics that work in brake repair shops (BRS) in Bogotá, Colombia still manipulate asbestos containing brake products.
Objective: Propose models to estimate personal asbestos exposures of brake mechanics that manipulate asbestos containing brake products, based on the results of asbestos sampling campaigns conducted in BRS located in Bogotá, Colombia.
Methods: Since 2010, personal air asbestos concentrations were determined on 28 riveters that work in 18 brake repair shops (BRS). Work-shift personal asbestos exposures were determined based on both 30-min personal samples collected during manipulation activities of brake products, and personal samples collected during non-manipulation activities. Longitudinal based linear regression models and Monte Carlo simulations were used to construct models to estimate work-shift personal asbestos concentrations, based on the tasks performed by workers during the work-shift. Spearman correlations were used to evaluate the relationship between the actual and the model-estimated work-shift personal asbestos concentrations.
Results: Three hundred and twelve (312) short-term asbestos personal samples collected during manipulation activities, and 289 asbestos personal samples collected during non-manipulation activities, were used to calculate 103 work-shifts personal asbestos concentrations. Longitudinal based linear regression models showed that asbestos personal exposures in a work-shift increase 5.25-folds with at least one manipulation activity of asbestos containing brake products ($p=0.001$), increase 5.52-folds if the worker remains inside of the manipulation area during non-manipulation activities ($p=0.014$), and increase 4% per each additional asbestos containing product manipulated daily ($p=0.150$). Monte Carlo simulations conducted with different numbers of manipulations (i.e., 0 to 16 sets of brake products manipulated), showed that the estimated work-shift personal asbestos exposure mean ranges between 0.02 to 0.85 f/cc. The Spearman correlation between the estimated and the actual personal asbestos work-shift exposure was ~0.7 ($p=0.000$).

Conclusion: The models proposed were constructed based on real sampling data, and can be applied to estimate asbestos exposures without sampling campaigns. The models can be useful for both current and retrospective asbestos exposure studies of brake mechanics. Caution is required because of potential underestimation of the real work-shift exposure. The next step in the process will be to have an external validation of the models.

Tu-PL-I1.4

Bayesian Approach for Summarizing Uncertainties Related to a Prevalence Estimate

Matthias Greiner, *Federal Institute for Rissk Assessment (BfR) and Veterinary University Hannover (TiHo), Berlin, Germany*
Matthias Flor, *Federal Institute for Risk Assessment (BfR), Berlin, Germany*
Chrsitine Müller-Graf, *Federal Institute for Risk Assessment (BfR), Berlin, Germany*

The generic concept of prevalence is applicable to quantify the proportion of any binary trait in a population such as proportion of contaminated food items or proportion of consumers with a certain exposure behavior. Major sources of uncertainty of prevalence estimates are related to the study design and sampling issues and are usually described using the concepts of precision (statistical parameter uncertainty of the estimate) and bias. A typical source of (information) bias is diagnostic misclassification.
Aim: We aimed at developing a statistical approach for prevalence estimation and characterization of all uncertainties that are associated with the study data and relevant meta-data such as diagnostic method performance characteristics.
Methods: We developed a user-friendly, universally applicable Bayesian version of a prevalence estimator that accounts for diagnostic misclassification (based on Rogan and Gladen, 1978; Am J Epidemiol 107.1: 71-76). The model requires input data for the (apparent) prevalence estimate (AP) along with prior information about the diagnostic sensitivity (Se) and specificity (Sp) that are available from validation studies of the test or instrument and a prior estimate for the true prevalence. Priors may also be elicited from experts. We illustrate using a simple hypothetical example (AP = 12/100, Se = 80/120, Sp = 79/80, and a uniform, non-informative prior for prevalence).
Results: Using the example input we obtain a prevalence estimate of 0.158 (95 % credible interval 0.056-0.281), which is adjusted for misclassification and deviates from the unadjusted estimate (AP = 0.12). Further uncertainties can be handled using repeated analyses with alternative priors.

Conclusions: Prevalence estimations can easily be adjusted for diagnostic misclassification if the diagnostic performance (sensitivity and specificity) of the test or instrument has been characterized. The use of a Bayesian model is a transparent and flexible approach for quantifying the combined uncertainties of all model parameters which can be informed by empirical data or expert opinion.

Tu-PL-I1.5

Lifetime exposome modeling

Olivier Jolliet, *University of Michigan, Ann Arbor, Michigan, United States*

"Tell me when and where you lived and worked, what you consumed and ate... and I will tell you who you are (your exposome!). In the era of high-throughput biology and biomedicine, genome-wide associations need to be complemented by exposome wide associations, thus the importance to develop high-throughput exposure strategies. Complementary to biomarker oriented approaches, this presentation explores how recent databases enables us to trace and predict the lifetime exposure of individuals to thousands of chemicals on the basis of the time evolution of their home location, sector workplace, consumption patterns and eating habits.

We use measured concentrations from the OSHA Chemical Exposure Health Database to characterize exposure to 250 chemicals per blue collars hour worked in 880 NAICS industrial sectors. These occupational exposures are confirmed by significantly higher biomarker levels measured in NHANES for blue collar workers compared to the rest of the population. Coupling of household product databases with near field exposure models with near-field modeling of product intake fractions (PiFs) provide first estimate of ranges of exposure to cosmetics, building material or other consumer products. A spatialized multiscale multimedia model (Pangea) provides improved resolution to better assess environmental and food contaminants.

This presentation will show how these recent developments in exposure data and modeling enable us to predict lifetime exposure across multiple chemicals. The approach will be illustrated through a case study in Michigan determining the lifetime dioxin exposome of hundreds of habitants in a contaminated area, based on their dioxin blood measurements and phamacokinetic modeling. The study identifies as major factors of influence the number of years lived in the contaminated area, the distance to the local incinerator and the consumption of contaminated food. At a wider scale we will finally shortly discuss how the exposome relates to the age response of multiple NHANES biomarkers.

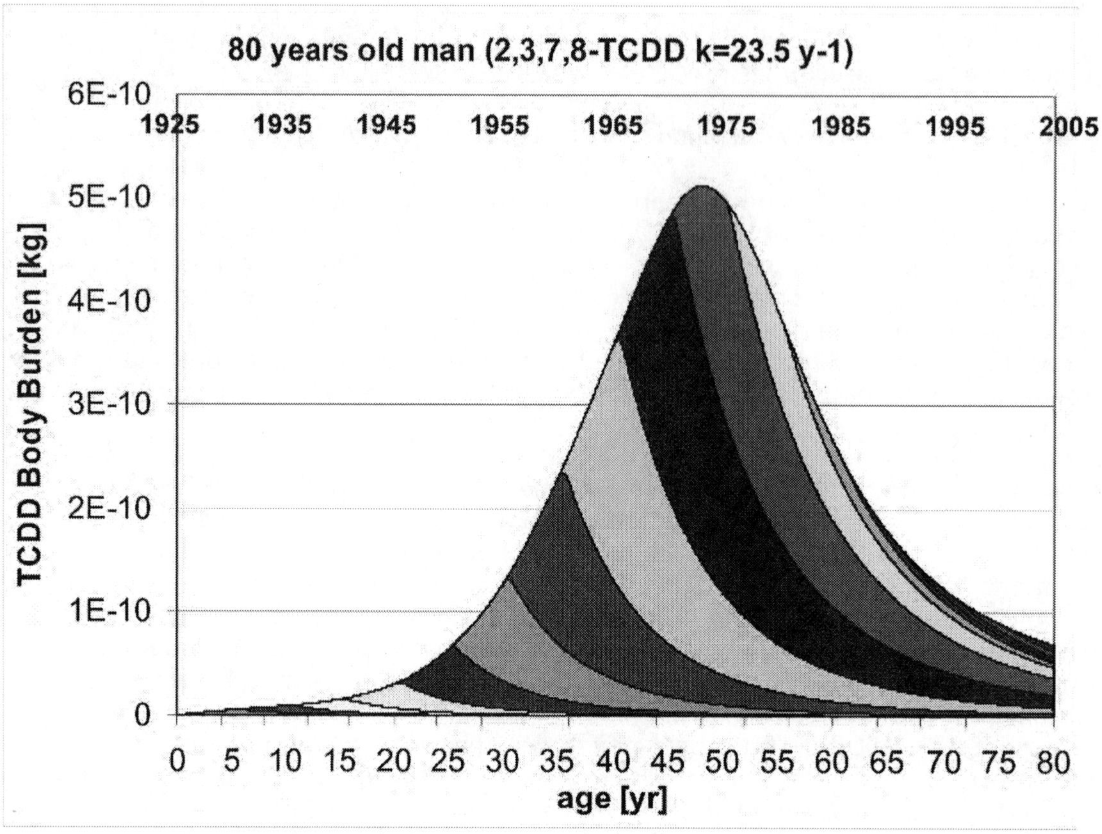

TCDD lifetime exposome of a 80 years old inhabitant in a contaminated are. in 2005 a large part of its measured blood concnetration is due to exposure prior to 1983

Tu-SY-A2: The Exposome: From concept to practice - II

Tu-SY-A2.1

Early Observations from EXPOSoMICS

Marc Chadeau-Hyam, *Imperial College London, London, United Kingdom*
Paolo Vineis, *Imperial College London, London, United Kingdom*
Roel Vermeulen, *Utrecht University, Utrecht, Netherlands*

EXPOSOMICS is one of the first European large-scale project aiming at the characterisation of the exposome compiling data and molecular profiles from bio-banked blood samples as well as using existing cohort resources. Adopting a 'top down' approach, the project aims to identify molecular signatures of several prioritised (air and water) exposures.
In order to capture the complex nature of the exposome the project design focused on the exploration of acute, short-term and long-term effects of these exposures by using, respectively, experimental study designs, performing personal exposure monitoring using sensors, and modelled long-term exposures. The study also accounts for the dynamics of the exposome by covering several critical life stages by including mother-child pairs, children, young adults, and adult cohorts.
In this presentation, we will describe the rich exposome dataset representing both the external and internal environments that the project has generated. We will describe its amplitude and complexity, and will present the overall approach we have adopted to fully exploit this unique resource.
In order to improve statistical power, Exposomics has made use of specific study designs designs (e.g. cross-over designs, longitudinal data) which are not routinely used in an OMIC-screening context. This raises specific methodological challenges which adds to the typical high dimensional situation observed in computational biology. We developed efficient models to exploit these data and to provide interpretable results. We will describe these approaches based on a few examples.
In addition to these proof-of-principle examples, we will describe on-going work, and short-term perspectives envisaged to exploit such information-rich exposome datasets. From theses examples we will draw conclusions on the state of the art of 'real-word' exposome research, and will highlight both the advances it has already yielded as well as the potential for further improvements. The latter will call upon an accelerated and trans-disciplinary effort to devise scalable and interpretable methods to explore and integrate these high resolution/high quality OMIC data sets.

Tu-SY-A2.2

Early observations from HEALS

Dimosthenis Sarigiannis, *Aristotle University of Thessaloniki, Thessaloniki, Greece*

Aim: The exposome represents the totality of exposures from conception onwards. Unraveling it requires simultaneously identifying, characterizing and quantifying the exogenous and endogenous exposures and modifiable risk factors that predispose to and predict diseases throughout a person's life span. Using the exposome to identify causes of human disease implies that environmental exposures and genetic variation are reliably measured in tandem and linked through mechanistic analysis of toxicity pathways rather than only phenotypically associated.
Methods: To better understand the interaction between environmental exposure and disease, we need to; (a) capture the biological perturbations initiated by exposure to environmental stressors; and (b) identify which of these perturbations overcome the homeostasis barrier, resulting in observed alterations of the cell/tissue environment and eventually to pathologic phenotypes. Exposure biology provides the methodological elements for the surveillance of changes at different levels of biological organization through the use of multiple -omics and post-omics technologies including epigenetics. Starting from untargeted transcriptomics and metabolomics we proceed with joint analysis of biological and metabolic processes induced by exposure to xenobiotics. In HEALS already existing cohort samples and data are re-analyzed to derive new insight on disease causality. Prenatal exposure of Polish mothers to phthalates was determined by (a) untargeted metabolomics analysis (using a 600 MHz NMR) and (b) targeted metabolomics measuring 11 phthalate metabolites in urinary samples from (n=165) mothers during the third trimester of pregnancy (prenatal exposure) using HPLC-MS/MS. Genetic variability and psychomotor development was assessed in their children at the age of 2 years by the Bayley Scales of Infant and Toddler Development.
Results: Results showed that the mothers with higher exposure to phthalates have completely different metabolic profiles compared to the ones with lower exposure levels. Child motor development was inversely associated with concentrations of several phthalates ($\beta = -2.5$; 95% CI -4.1 to -0.9) in the urine collected from mothers during pregnancy.
Conclusions: Metabolic pathway analysis using Agilent GeneSpring revealed that alterations in urine metabolites are related to the TCA cycle, suggesting impaired mitochondrial respiration; the latter is central to energy metabolism and cellular signaling and plays fundamental roles in synthesis of nucleotides and active transport processes. Inhibition of mitochondrial oxidative phosphorylation could also cause defective mitochondrial energy production during fetal formation and development. Impaired mitochondrial respiration and energy generation seem to affect early life motor development. These observations suggest a plausible set of biological process pathways causing neurodevelopmental disorders.

Tu-SY-B2: Uncertainty in scientific assessments: Recent efforts by governmental bodies to develop guidance for assessors

Tu-SY-B2.1

Application of Quantitative Methods for Uncertainty Assessment in Chemical Risk Assessment: 2-Dimensional-Monte Carlo Method for PBDEs in Food

David Trudel, *Arcadis Schweiz AG, Zurich, Switzerland*
Natalie von Goetz, *ETH Zurich, Zurich, Switzerland*

Background
Quantitative methods for uncertainty analysis in chemical risk assessment are available, but not yet routinely used. Among these, the probabilistic method of 2-dimensional Monte Carlo (2D-MC) is one of the most advanced methods for refined exposure assessment: It allows attributing the variation of exposure estimates to either variability or uncertainty in the underlying parameters, which helps with management decisions and the allocation of analytical resources.
Objectives
Polybrominated diphenyl ethers (PBDE) were selected for exemplarily showing the benefits of using a refined quantitative method for uncertainty assessment. They are flame retardants, which are widespread environmental contaminants and, thus, regularly monitored in a number of food items. Due to endocrine disrupting properties of some PBDEs refined risk assessments are of interest. We used 2D-MC for exposure assessment of PBDEs in food in order to explore the uncertainties in the underlying dataset and identify the most effective strategy for future sampling of food items.
Methods
The probabilistic aggregate exposure estimation was based on the extensive Irish database on PBDE concentrations in food and on food consumption data for the Irish adult population. The calculations were performed for nine important PBDE congeners: BDE-28, -47, -49, -99, -100, -153, -154, -183, and -209. Only foods of animal origin and vegetable oil were modeled, because all data points from other sample matrices (such as fruits) were below the analytical limit of detection.
Results
The uncertainty analysis showed that dairy fat and lean fish were the most important contributors to PBDE exposure. The dominating congeners were BDE-47, -99, and -209. Table 1 shows the effect of using 2D-MC on the exposure estimates as compared to 1D-Monte Carlo. The HQD is the high-quantile exposure estimate taking into account variability and uncertainty at Q97.5, respectively.
Table 1: Exposure estimates for 1D-MC and 2D-MC
see attachment
Most of the HQD from 2D-MC are about twice as high as the high-quantile exposure Q97.5 from 1D-MC. Thus, the safety margins in risk assessments will be reduced correspondingly, but at the same time the confidence in the risk assessments will considerably be increased, since a high quantile for uncertainty (Q97.5) is covered.

Congener	1D-MC		2D-MC	Congener	1D-MC		2D-MC
	Mean	Q97.5	HQD		Mean	Q97.5	HQD
BDE-28	0.022	0.059	0.12	BDE-153	0.045	0.096	0.17
BDE-47	0.24	1.0	1.8	BDE-154	0.028	0.096	0.16
BDE-49	0.053	0.28	0.52	BDE-183	0.035	0.073	0.07
BDE-99	0.14	0.36	0.80	BDE-209	0.27	0.53	0.87
BDE-100	0.056	0.23	0.42	∑BDE	0.70	2.1	3.6

Table 1: Exposure estimates for 1D-MC and 2D-MC

Tu-SY-B2.2

APROBA-Plus: An Excel tool for an approximate probabilistic risk assessment taking uncertainties into account

Wout Slob, *RIVM, Bilthoven, Netherlands*

In 2014 WHO published guidance on evaluating and expressing uncertainties in human health hazard characterization (HC). In this approach, the outcome of a HC is expressed as an interval or distribution rather than the usual deterministic point estimate (such as RfD, ADI), thereby communicating potential uncertainties more clearly. Risk management protection goals, in particular the acceptable magnitude of effect (M) and associated incidence (I) in the human population are first made explicit. The goal is to estimate the „HDMI", i.e. the „true" human dose associated with M and I (e.g. body weight decreased by ≥ 10% (M) in 5% (I) of the population). The WHO approach provides an uncertainty distribution for the HDMI, which is calculated by probabilistically combining the individual uncertainties in the various aspects of the HC (such as POD, inter-, and intraspecies extrapolation). WHO also published a user-friendly Excel tool that can do those probabilistic calculations in an approximate but quick and easy way. This tool is called APROBA, and can be downloaded from
http://www.who.int/ipcs/methods/harmonization/areas/hazard_assessment/en/
Recently, the tool was extended by an option to enter exposure estimates, with an uncertainty range. This extended tool (APROBA-plus) plots the uncertainty in the HDMI against the exposure uncertainty, resulting in an ellipse that transparently indicates the uncertainty about the distance between the HDMI and the exposure, given the information available.
The use of APROBA-Plus was recently evaluated by applying it to 19 different substances, showing that APROBA-Plus can indeed be used as a quick tool for risk assessment while making the (approximate) uncertainties in both the hazard and the exposure visible in a single plot. By making the uncertainties visible, the outcome from a risk assessment becomes more transparent and informative than the more usual deterministic approaches, so that risk managers can make better-informed decisions, e.g. directly taking measures or asking for refinement of the risk assessment. If the latter, APROBA-Plus can help in showing which aspects in the risk assessment contributed most to the overall uncertainty, as an indication what type of refinement would be most effective. This tool could easily serve as a standard extension of routine risk assessments.

Tu-SY-B2.3

EFSA target audience research project on communicating scientific uncertainties

Anthony Smith, *European Food Safety Authority, Parma, Italy*

Background
EFSA is developing guidance on uncertainty in scientific assessments to apply across the work of its scientific panels, staff and partners in EU member states. EFSA's Scientific Committee considers effective communication of uncertainties vital in this process but the literature is equivocal about the best methods to do so. Further, there is a lack of empirical data on the best approaches for communicating scientific uncertainties to non-technical audiences. EFSA therefore commissioned a project to generate such data through target audience research.

Objectives
Although EFSA regularly communicates the scientific uncertainties related to its assessments, it has not developed a model that is applied consistently across the organisation. The project, therefore, aims to test effective methods, approaches and tools for communicating scientific uncertainties to European and national decision-makers, stakeholders and the general public. The results will be used to update and finalise the guidance document under development at EFSA and also to establish best practice at EFSA on communicating scientific uncertainties to the Authority's broad range of stakeholders, including decision-makers and the general public.

Methods
EFSA intends to differentiate more systematically the level of scientific technicality in the communications messages on uncertainties intended for different target audience. EFSA identified five main target audience groups for the project: technical policy-makers, political decision-makers, civil society stakeholders, food chain operators and informed members of the public. Individual representatives of these groups were selected from both European and national levels. In focus groups of 6 to 8 people, participants complete two tasks:

An individual task to read and evaluate nine short statements containing different presentations of uncertainty - including quantitative and qualitative expressions of uncertainty - relating to a case study, then answer a brief questionnaire.

A collective task involves a discussion to explore the issue of uncertainty and the case study. The conversation is guided to cover the different ways of expressing uncertainty used in the statements. It explores matters of clarity, usefulness, and gives participants the opportunity to express any relevant considerations. The group discussion is recorded, transcribed and analysed.

Results
The responses contribute to EFSA's understanding of how different ways of expressing uncertainty might be received by individuals and in groups. This provides a rich and nuanced understanding of how participants respond to uncertainty information, as individuals and as a collective. Salient themes are identified, as well as commonalities and differences within the group.

Tu-SY-C2: Use of Agent Based Models in Exposure Assessment

Tu-SY-C2.1

Towards an in silico Experimental Platform for Air Quality: Houston, TX as a Case Study

Bianica Pires, *Virginia Tech, Arlington, VA, United States*
Gizem Korkmaz, *Virginia Tech, Arlington, VA, United States*
Katherine Ensor, *Rice University, Houston, TX, United States*
David Higdon, *Virginia Tech, Arlington, VA, United States*
Sallie Keller, *Virginia Tech, Arlington, VA, United States*
Bryan Lewis, *Virginia Tech, Blacksburg, VA, United States*
Aaron Schroeder, *Virginia Tech, Arlington, VA, United Kingdom*
Paul Price, *US EPA, RTP, NC, United States*

It has been demonstrated that localized specific exposures to ozone can dramatically increase health risks for cardiac events and asthma. Many studies, however, use 12- or 24-hour activity summaries. Aggregating time to daily periods misses important details such as variations in ozone levels across physical space and time, which can significantly impact individual and population exposure levels. For example, it was found that an increase of 20 parts per billion (ppb) in ozone over a period of one to three hours is associated with a 4.4% increased risk of having an out-of-hospital cardiac arrest, for which 90% of cases result in death (Ensor et al., 2013). We seek to estimate exposure levels at a higher granularity by coupling a spatiotemporal air quality model of ozone concentration levels with a synthetic information model of the Houston Metropolitan Area. The synthetic population includes socio-demographically relevant activity sequences and geo-spatially mapped locations for these activities, thus we estimate the movements of each individual in the population and their location second-by-second. This population contains 4.9 million individuals, grouped into 1.8 million households, who perform activities that occur in 1.2 million physical locations. We then match the resolution of time intervals obtained from the 47 monitors that measure ozone across Houston. While traditional approaches often aggregate the population, activities, or concentration levels of the pollutant across space and/or time, this research utilizes high performance computing and statistical learning tools to maintain the granularity of the data, allowing specific exposure levels to be attached to the synthetic individuals. Furthermore, the heterogeneous exposure levels of the population across time are more accurately reflected, allowing for increased sensitivity to detecting the variation of exposure across the population. Several scenarios of the model were run at different levels of resolution, one in which individuals were assumed to stay home all day. While average hourly exposures to ozone across the population were similar across the scenarios, when we maintain the granularity of the data, the variation of exposure could reach an increase of 20 ppb over a short period of time, which could be particularly important if experienced by sensitive populations. This results in varying levels of exposure among individuals in the same zip code, neighborhood, block, and even household depending on their activity patterns throughout the day.
Ensor, K.B., Raun, L.H., and Persse, D. (2013). A case-crossover analysis of out-of-hospital cardiac arrest and air pollution. Circulation, AHA-113.

Tu-SY-C2.2

Using agent-based modelling for interpreting the individual exposome

Dimosthenis Sarigiannis, *Aristotle University of Thessaloniki, Thessaloniki, Greece*
Dimitris Chapizanis, *Aristotle University of Thessaloniki, Thessaloniki, Greece*
Spyros Karakitsios, *Aristotle University of Thessaloniki, Thessaloniki, Greece*
Alberto Gotti, *Aristotle University of Thessaloniki, Thessaloniki, Greece*

Aim: This paper outlines a prototype methodology for exploring and interpreting the exposome concept at the individual level using multi-sensor data fusion and agent-based modelling to capture behavioural determinants of exposure profiles. In this way, cumulative exposure to major potential health stressors over critical windows of life is estimated. The integral of cumulative exposure over one's life course would then define the external exposome.
Methods: Time series of data from wearable sensors used for tracking personal location, activity intensity, food consumption and consumer product use patterns are fused using artificial neural networks (ANNs) to derive time-activity models. Data from food consumption and consumer product use are statistically associated to gender, age and socio-economic status (SES) of the individuals participating in the study. Coupling time-activity models with location data allows us to derive individual space-time trajectories. These individual trajectories together with gender, age and SES are used as input to a spatially explicit agent-based model (ABM). The studied population is clustered by age, gender and SES. Through stochastic interactions among model agent clusters space-time trajectories of population subgroups emerge from the ABM simulations. These trajectories are then superposed onto high resolution indoor and ambient air quality maps. Food consumption and consumer products use data are linked to residue levels (of pesticides and other industrial chemicals) of food items and consumer products respectively. Age, gender and intensity of activity data are taken into account for estimating population subgroup exposure.
Results: By coupling data from multiple location, activity tracking and environmental sensors with an urban ABM, complex exposure was estimated for a city of a population of 1.000.000. Exposure to various pollutants was found to vary significantly based on SES conditions. In principle, people with lower income tend to live in areas characterized by higher levels of air pollution (especially with regard to combustion-related pollutants) and to consume food with higher content of pesticides. On the other hand, people with higher income were exposed to higher levels of PBDEs and heavy metals.
Conclusions: Agent-based modelling proved to be a powerful tool for exposome research. It allows the derivation of statistically robust exposure estimates at the population level from a limited number of individual exposure profiles, while avoiding the inherent bias of probabilistic exposure modelling based on Bayesian statistics. ABMs incorporate explicitly socio-economic determinants of exposure and support the enhanced use of multi-sensor systems for exposome characterization at high degree of granularity.

Tu-SY-C2.3

A framework for the use of agent based modeling to simulate inter- and intra-individual variation in human behaviors

Paul Price, *US EPA, RTP, North Carolina, United States*
Namdi Brandon, *US EPA, TRP, North Carolina, United States*
Kathie Dionisio, *US EPA, RTP, North Carolina, United States*
Rogelio Tornero-Velez, *US EPA, RTP, North Carolina, United States*
Kristin Isaacs, *US EPA, RTP, North Carolina, United States*

Simulation of human behavior in exposure modeling is a complex task. Traditionally, inter-individual variation in human activity has been modeled by drawing from a pool of single day time-activity diaries such as the US EPA Consolidated Human Activity Database (CHAD). Here, an agent-based model (ABM) is used to simulate population distributions of longitudinal patterns of four macro activities (sleeping, eating, working, and commuting) in populations of adults over a period of one year. In this ABM, an individual is modeled as an agent whose movement through time and space is determined by a set of decision rules. The rules are based on the agent having time-varying "needs" that are satisfied by performing actions. Needs are modeled as increasing over time, and taking an action reduces the need. Need-satisfying actions include sleeping (meeting the need for rest), eating (meeting the need for food), and commuting/working (meeting the need for income). Every time an action is completed, the model determines the next action the agent will take based on the magnitude of each of the agent's needs at that point in time. Different activities advertise their ability to satisfy various needs of the agent (such as food to eat or sleeping in a bed or on a couch). The model then chooses the activity that satisfies the greatest of the agent's needs. When multiple actions could address a need, the model will choose the most effective of the actions (bed over the couch). In addition, multiple activities can be linked to a single decision (e.g., commuting must precede and follow working). An agent's needs and the rate at which the needs increase over time are varied across agents and are correlated with the agents' fixed personal attributes (e.g., age, gender, etc.) and household physical characteristics (distance between residence and work). Model parameters such as individuals' rates of need increases are informed using data from CHAD. The advantage of ABM is that, unlike CHAD, it can provide information on human activity over periods of time longer than one day. We will present predictions for a population of adults for the four activities and compare the model outputs to the CHAD data. In future work we propose to extend this "need-based" framework to model usage of consumer products. For example, each agent is assigned personal hygiene and home cleanliness needs which drive their use of personal care products and household cleaning supplies.

ISES 2016 UTRECHT THE NETHERLANDS OCTOBER 9-13

Tu-PL-D2: Land Use Regression Modeling - II

Tu-PL-D2.1

Effect of Monitoring Network Design on Land Use Regression Model for Estimating Residential NO2 Concentration

Hao Wu, *University of Edinburgh, Edinburgh, United Kingdom*
Stefan Reis, *NERC Centre for Ecology and Hydrology, Penicuik, United Kingdom*
Mathew Heal, *University of Edinburgh, Edinburgh, United Kingdom*

Land-use regression (LUR) models are increasingly used to estimate exposure to air pollution in urban areas. An appropriate monitoring network is an important component in the development of a robust LUR model. In this study concentrations of NO₂ were simulated by a dispersion model at 'virtual' monitoring sites in 54 networks of varying numbers and types of site, using a 25 km² area in Edinburgh, UK, as an example location. Separate LUR models were developed for each network. These were then used to estimate NO₂ concentration at all residential addresses, which were evaluated against the dispersion-modelled concentration at these addresses. The improvement in predictive capability of the LUR models was insignificant above ~30 monitoring sites, although more sites tended to yield more precise LUR models. Monitoring networks containing sites located within highly populated areas better estimated across all residential NO₂ concentrations. LUR models constructed from networks containing more roadside sites better characterised the high end of residential NO₂ concentrations but increased errors when considering the whole range of concentrations. No particular composition of monitoring network resulted in good estimation simultaneously across all residential NO₂ concentration and of the highest NO₂ levels. This evaluation with dispersion modelling has shown that previous LUR model validation methods may have been optimistic in their assessment of the model's predictive performance at residential locations. Dispersion modelling has proven to be a useful tool for designing an effective network for LUR model development and evaluation.

Tu-PL-D2.2

European models incorporating satellite and chemical transport modelling with local variables in LUR

Kees De Hoogh, *Swiss Tropical and Public Health Institute, Basel, Switzerland*
John Gulliver, *Imperial College London, London, United Kingdom*
Aaron van Donkelaar, *Dalhousie University, Halifax, Canada*
Randall V Martin, *Dalhousie University, Halifax, Canada*
Julian D Marshall, *University of Washington, Minneapolis, United States*
Danielle Vienneau, *Swiss Tropical and Public Health Institute, Basel, Switzerland*
Gerard Hoek, *University of Utrecht, Utrecht, Netherlands*

Introduction. To quantify effects on health outcomes over background air pollution levels, it is necessary to undertake large epidemiological studies and/or pool data from multiple cohorts. Air pollution exposure estimates over large geographic areas at sufficient spatial resolution are thus needed. Satellite-derived (SAT) and chemical transport model (CTM) estimates of PM2.5 and NO2 are increasingly used in combination with Land Use Regression (LUR) models to accomplish this. We aimed to compare the contribution of SAT and CTM data to the performance of LUR PM2.5 and NO2 models for Europe.
Methods. Four sets of models, all including local traffic and land use variables, were compared (LUR without SAT or CTM, with SAT only, with CTM only, and with both SAT and CTM). LUR models were developed using two monitoring data sets: PM2.5 and NO2 ground level measurements from the European Study of Cohorts for Air Pollution Effects (ESCAPE) and from the European AIRBASE network.
Results. LUR PM2.5 models including SAT and SAT+CTM explained ~60% of spatial variation in measured PM2.5 concentrations, substantially more than the LUR model without SAT and CTM (adjR2: 0.33-0.38). For NO2 CTM improved prediction modestly (adjR2: 0.58) compared to models without SAT and CTM (adjR2: 0.47-0.51). Both monitoring networks are capable of producing models explaining the spatial variance over a large study area.
Conclusions. SAT and CTM estimates of PM2.5 and NO2 significantly improved the performance of high spatial resolution LUR models at the European scale for use in large epidemiological studies.

Tu-PL-D2.3

Modeling the Intraurban Variation in Traffic Exposure in Urban Areas in Kathmandu Valley, Nepal

Anobha Gurung, *Yale University, New Haven, Connecticut, United States*
Johnathan Levy, *Boston University, Boston, United States*
Michelle Bell, *Yale University, New Haven, United States*

Aim: With growing urbanization, traffic pollution has become one of the main sources of air pollution in Nepal. Understanding the impact of air pollution on health requires estimation of exposure. While many exposure assessment studies are available in the developed world, such assessments are relatively scarce in developing countries. Landuse regression (LUR) modeling is widely used to investigate intraurban variation in air pollution, specifically traffic pollution, for Western cities. We developed LUR modeling towards understanding intraurban variation of traffic pollution in urban areas of Kathmandu Valley, Nepal, one of the fastest urbanizing areas in South Asia.
Methods: Over the study area, 135 monitoring sites were selected using stratified random sampling based on building density and road density and purposeful sampling. In 2014, four sampling campaigns were performed, one per season, for two weeks each where nitrogen dioxide (NO2) was measured using duplicate Palmes tubes at 135 sites and Ogawa badges at 28 sites. Ogawa badges were used to measure NO2 and nitrogen oxides (NOX). Geographical variables (e.g., road network, landuse, built area) were used as predictor variables LUR modeling. Predictor data were unavailable outside of the study area except for Landsat data. Predictor variables were estimated for buffers 25-400m around each monitoring site. When portions of buffers were outside the study area, values of predictor variables with missing data were interpolated.
Results: Annual average NO2 by site ranged from 1.62 to 349ppb for the study area. High completion rate by campaign was observed for Palmes tubes (90.4-95.6%) and Ogawa badges (78.6-100%). Sensitivity analysis showed that interpolation of predictor data outside of the study area performed well. The final model selected accounted for 51% of the variance in NO2 levels. In the final model, length of major road, built area, and industrial area were positively associated with NO2 concentration while normalized difference vegetation index (NDVI) was negatively associated with NO2 concentration. Cross validation of the results confirmed the reliability of the model.
Conclusions: Findings demonstrate that NO2 annual average concentration was higher in the Village Development Committees (VDCs) of Kathmandu and Lalitpur than in Kirtipur, Thimi, and Bhaktapur with variability present within each VDC. LUR modeling allows understanding of intraurban variation of traffic pollution for better exposure estimation for future epidemiological studies.

Tu-SY-E2: Exposure science meets social science: Tools for the effective communication of the health risks associated with air pollution exposure and implications for policy development

Tu-SY-E2.1

Personalised Air Quality Data Gathered Through Community-Based Projects as a Tool For Communicating Air Pollution as a Public Health Risk.

Diana Silva, *King's College London, London, London, United Kingdom*

Background: It is widely recognised that air pollution is a public health concern accountable for numerous health problems and tens of thousands of premature deaths per year in the UK. Despite this evidence, awareness of the issue is low in comparison to other public health risks. Improved methods for engaging with the public and communicating this risk are required.
Aim: To investigate the impact of an air pollution community-based project on participants' perceptions and attitudes towards air quality issues.
Methods: The methodology used is rooted in Community-Based Participatory Research (CBPR) and uses observation, surveys and interviews. A range of community groups took part in this study, including Primary school children, senior citizens, patients from COPD recovery groups and parents from a mother and baby group. All members of these groups participated in an information session on air pollution causes and effects; then using portable exposure monitors and GPS watches, a subset of individuals from each group measured the air pollution they are exposed to as they go about their normal day.
Results: Most of the participants expressed the view that being able to collect air pollution data themselves was a key motivator for deciding to take part in the project.
The majority of the participants stated that having access to personalised environmental information they themselves gathered increased their air pollution awareness and their desire to identify ways in which they could reduce their air pollution exposure. Additionally, some participants expressed their desire to use the air pollution data collected to raise awareness and persuade local government to address the issue. Finally, the personal nature of some of the results and the issues around dealing with unexpected exposures revealed some of the complexities surrounding the benefits and limitations of communicating personal exposure data to individuals.
Conclusion: This project appears to have raised awareness of the risks of poor air quality by supplementing information provision with active collection of personalised exposure data. Citizens' responses included personal behaviour change in order to reduce air pollution exposure as well as a desire to use data to lobby local government for change. This project also highlights the importance of suitable personal exposure data reporting, tailored to specific settings and social surroundings.

Tu-SY-E2.2

Environmental exposure and citizen sensing: New modes of monitoring, new modes of politics

Helen Pritchard, *Goldsmiths, University of London, London, United Kingdom*
Jennifer Gabrys, *Goldsmiths, University of London, London, United Kingdom*

A number of environmental sensing technologies and practices are emerging that seek to enable citizens to use DIY and low-tech monitoring tools to understand and act upon environmental problems such as air pollution. These "citizen sensing" projects intend to gather data sets, which can indicate environmental change and give rise to political action. This presentation will discuss citizen-sensing efforts related to monitoring PM2.5 in the gas fields of northeastern Pennsylvania, and the ways in which citizen-gathered data has generated new insights for understanding environmental exposure.

Tu-SY-E2.3

Evidences of a social sciences pilot research from the FP7 SEFIRA Socio-economic implications for individual responses to Air Pollution Policies in EU +27

Benjamin Barratt, *King's College London, London, United Kingdom*

The SEFIRA project investigated with a pilot research the preferences, behaviour and responses of individuals and stakeholder groups that can influence the uptake of or resistance to EU air quality policies. Qualitative and quantitative methods have been combined for a pilot study based on a survey involving 16,100 European citizens from 7 countries (Austria, Belgium, Germany, Italy, Poland, Sweden, United Kingdom) testing their environmental behaviour and preferences regarding a selection of air quality policies. In addition, qualitative research in the four metropolitan areas of Antwerp, Malmö, Milan and Warsaw has been carried out, consulting 12 focus groups each with about 10 citizens and interviewing 38 top experts and policy makers. From 2014 to 2016 SEFIRA has organized public meetings for the consultation of public stakeholders in Antwerp, Warsaw, Malmö, Milan, London and Vienna that saw the involvement of more than 200 participants from a wide range of civil society organizations, NGOs and public institutions. Evidence from the project shows a high degree of environmental consciousness of air quality from a wide section of the European population, at least in urban areas, while a mismatch between scientific knowledge and perceptions among the general population still exists. Citizens show different attitudes when they are asked to express their preferences towards possible air quality policies, according to where they live and their socio-economic characteristics. Overall they appear to be more willing to reduce car use and polluting consumption patterns rather than accepting additional visible costs and environmental taxation and point out the need for a transition to a re-organization of mobility and work-life balance in order to reduce car use in urban areas. The impact of European policies is often reduced in the policy translation when regional and municipal governments face strong resistance from important economic sectors or because of trade-offs between environment and economic growth accentuated by the 2008 economic crisis. The study confirms the necessity of a more effective integration between social sciences and atmospheric and exposure sciences in order to improve the design and effectivity of air quality policies on the different european, national and local scales.

Tu-SY-F2: OECD Task Force on Exposure Assessment - Better exposure science for better lives - II

Tu-SY-F2.1

Combined Exposure Assessment

Takahiro Hasegawa, *OECD, Paris, France*

Human and ecological receptors are continuously co-exposed to multiple chemicals; however, chemicals have traditionally been assessed and managed on a chemical-by-chemical basis. As a result, there is a possibility that there are instances where chemicals that independently do not pose a risk to human health and the environment may do so when considered in combination.
OECD is discussing technical aspects of performing a hazard and exposure assessment for a cumulative risk assessment to provide further guidance in the following areas:

- Development of problem formulation guidance on prioritization/triggers/scope for assessment of combined exposures.
- Considerations regarding hazard characterization to inform assessment of combined exposures.
- Considerations regarding co-exposure characterization to inform assessment of combined exposures.
- Considerations regarding risk assessment of combined exposures using various approaches and capturing and communicating uncertainties in findings.

This presentation will provide an overview of the OECD's development of guidance to support the assessment of risks from the combined exposures to multiple chemicals.

Tu-SY-F2.2

Development of Internationally Harmonized Use Codes

Cathy Fehrenbacher, *U.S. Environmental Protection Agency, Washington, DC, United States*
Charles Bevington, *U.S. Environmental Protection Agency, Washington, DC, United States*
Franklyn Hall, *U.S. Environmental Protection Agency, Washington, DC, United States*
Andreas Ahrens, *European Chemicals Agency, Helsinki, Finland*

This presentation will provide an update on the OECD Task Force on Exposure Assessment's (TFEA) development of internationally harmonized codes for reporting information on uses of chemicals to inform exposure assessment. Many existing approaches to categorize uses are broad in nature, when applied to large numbers of chemicals. Broad categorization results in a greater number of chemicals present in any given category, and may present challenges when used in assessing exposure. However, a very fine level of granularity might limit the international harmonization and ease of use if chemical use categories are too specific. Thus, it is important to balance the need for granularity with the burden of reporting the information.
This OECD TFEA project developed proposed internationally harmonized use codes based on a review of existing functional use and product categories, and bilateral discussions between the United States and the European Chemicals Agency (ECHA). The proposed use codes include about 120 functional use codes, about 140 product use codes, and 8 material types and 7 article categories. A comprehensive approach was taken for development of the functional use codes, with a goal towards defining codes which are most helpful in evaluating exposure to chemicals throughout the whole lifecycle. This includes functions in industrial processes, in products and in articles. The product and article use categories are intended to focus on the end-use application of chemicals within products and articles, rather than upstream manufacturing and processing.. The proposed internationally harmonized use codes have undergone extensive review within the TFEA, and will be made available by the OECD for use in developing models, databases, exposure assessments, etc.
Disclaimer: The views expressed in this abstract are those of the authors and do not represent United States Environmental Protection Agency policy or endorsement.

Tu-SY-F2.3

An industry perspective on future exposure science needs to support chemical risk assessment

Oliver Price, *Unilever, Sharnbrook, United Kingdom*
Richard Becker, *American Chemistry Council, Washington, DC, United States*
Bruno Hubesch, *European Chemical Industry Council - Cefic, Brussels, Belgium*

The International Council of Chemical Associations' (ICCA) Long-Range Research Initiative (LRI), is tasked with supporting the development of high quality scientific research to better understand the potential impacts of chemicals on human health and the environment. The LRI develops new science to address challenging issues such as the appropriate use of new chemical hazard data (e.g. high-throughput screening data), a need for more accurate exposure data on chemicals used in commerce, increased public demands for safe products and concerns about animal welfare. The LRI global research strategy, 'Advancing Chemical Safety Assessment for the 21st Century', directly addresses these challenges and consists of three priority areas (1) Innovating Chemical Testing, (2) Understanding Everyday Exposures to Chemicals and (3) Translating research outcomes for product safety.

Understanding exposure to chemicals remains a challenge across multiple chemical sectors and for academics and regulators alike. Detailed understanding of how chemicals are used and distributed is critical for assessing the potential human and environmental health risks and for informing decisions about new innovations. LRI research in exposure science fosters initiatives to develop predictive models for estimating environmentally-relevant exposures to chemicals, supports the development of novel biomarkers, and advances approaches for interpreting new and available exposure data. This presentation will provide an overview of current industry research challenges that are tasked with improving our understanding of consumer exposure and the incorporation of novel hazard data to assess risks from chemicals. We present exposure science challenges to realising quantitative in vitro to in vivo (QIVIVE). Case studies that, combine human exposure and dose information generated using PBTK models with hazard data from high-throughput assays to support decision making are discussed. We outline remaining uncertainties in the approaches, such as the model applicability and performance, characterisation of chemical-response relationships, robust use/relevance of ADME data, that once addressed will further increase our confidence in risk-based decision making using QIVIVE.

Tu-SY-G2: Environmental Justice: Developing the Scientific Foundation Supporting Cumulative Exposures/Risks/Impacts and Disparate Impacts Research - II

Tu-SY-G2.1

Quantifying Exposure and Risk Disproportionality in Environmental Justice Populations

Timothy Barzyk, *U.S. EPA, Research Triangle Park, United States*
Hongtai Huang, *Oak Ridge Institute for Science and Engineering, Research Triangle Park, United States*
Lawrence Martin, *U.S. Environmental Protection Agency, Washington, D.C, United States*
Liem Tran, *The University of Tennessee, Knoxville, United States*
Don Catanzaro, *Oneida Total Integrated Enterprises, Milwaukee, United States*

Disproportionate risk estimates can indicate a predisposition within an individual or population to be either differentially exposed or differentially affected by a given stressor or combination of stressors, which can be especially prevalent in Environmental Justice (EJ) communities. Research gaps remain in accurately quantifying disproportionate risk, including: 1) the extent to which varying stressors can be combined into a single health risk score, and 2) determining rigorous, standardized scientific methods to compare the relative risk of each stressor in relation to the others, and to develop cross-cutting solutions to address them. This work focuses on the latter. Stressors can be chemical or non-chemical in origin and serve as a link between to human and ecological health. A primary research topic is how chemicals and chemical mixtures (i.e., pollution/contamination), interact with those most susceptible and vulnerable to adverse health effects associated with these exposures and how that might affect health risk estimates (e.g., how dose-response functions might be impacted). While associations have been well-documented (i.e., that low-income, minority populations are highly exposed or disproportionately impacted), less has been done to determine the relative risk associated with particular stressors in order to establish priorities for policy development and implementation. Until quantitative techniques become available to combine varying stressors, comparative or relative risk assessment provides an opportunity to include data, expert advice, and local knowledge to make informed decisions about addressing risk. These methods have been applied to multiple case studies in EPA Regions throughout 2014-2015. The process we followed was informed by stakeholder input and real-world testing, and has been developed into a web-based software application called the Community Cumulative Assessment Tool (CCAT), a broadly designed tool appropriate to use for evaluating wide-ranging risk scenarios, yet retaining some user flexibility to meet the unique needs of each assessment. We present an overview of CCAT, along with case study results and lessons learned, to highlight the utility of comparative risk assessment (and resulting risk management actions), especially in the absence of methods that combine risk from varying stressors.

Tu-SY-G2.2

Connecting the dots: Linking quantifiable environmental justice indicators to exposure assessment methodologies

Hongtai Huang, *Environmental Protection Agency (ORISE), Durham, North Carolina, United States*
Timothy Barzyk, *National Exposure Research Laboratory, U.S. Environmental Protection Agency, Durham, North Carolina, United States*
Rogelio Tornero-Velez, *National Exposure Research Laboratory, U.S. Environmental Protection Agency, Durham, North Carolina, United States*

Cumulative risk assessment (CRA) offers a unique context for addressing Environmental Justice (EJ) issues from scientific perspectives, especially when it comes to examining combined effects of multiple environmental stressors1. Not only chemical stressors (e.g. radon, toluene and particulate matter) but also non-chemical stressors (e.g. smoking, noise and violence) can be evaluated2-4. EJ indicators, used as a tool to assess and quantify some of these non-chemical factors, include health, economic, and social indicators such as vulnerability and susceptibility5. Many studies have identified the associations between EJ indicators and chemical stressors6-10, but fewer focused on the interrelation between stressors and indicators and their effects on health. In this study, we utilized both established and novel quantitative methods in order to better understand how the interaction of multiple chemical and non-chemical stressors affect population level exposure assessment. Established approaches include the Average Daily Dose (ADD) model11 that has been commonly used in exposure assessment within the conventional risk assessment framework12, while novel techniques encompass unsupervised data mining methods such as association rules mining13. A major intention of this work is to quantify what are often considered qualitative EJ indicators to provide a more accurate representation of environmental exposures and impacts.

Tu-SY-G2.3

For Better or For Worse: Environmental Health Promotion in Support of Community Action

Eric Hall, *US Environmental Protection Agency, Durham, North Carolina, United States*
Cheryl Johnson, *People for Community Recovery (PCR), Chicago, Illinois, United States*
Brandi White, *Medical University of South Carolina, Charleston, South Carolina, United States*

OBJECTIVE/AIM: Environmental Health Education (EHE) is effective when incorporated with community member knowledge, environmental science, risk education, and health education. EHE programs must inform residents about exposures affecting their health and the attendant risks. This research assesses EHE's impact on civic engagement, knowledge of environmental exposures, associated health risks, and community health outcomes.
METHODS: The focus of the research is a public housing community surrounded by landfills, hazardous waste sites, and manufacturing facilities located in Chicago, Illinois. An environmental justice organization, People for Community Recovery (PCR), was the community partner. Data was collected from community residents during one week in March 2009 using both qualitative and quantitative research methods, including a focus group and a survey provided to two different resident groups, to understand attitudes/beliefs about environmental exposures, hazardous wastes, landfills, lead, and EHE preferences.
RESULTS: There were 42 community who residents participated [97.7% Black/African American; 61.9% female; mean age: 45.1 (SD±13.5) years]. Most (79%) were concerned about physical and social hazards. They felt that no one provided clear and pertinent information about physical hazards. Approximately 60% believe there are too few laws regulating environmental risks, and 64.2% do not believe government will help them with serious health community problems. Trusted sources of EHE information include their community-based environmental justice organization and community health workers.
DISCUSSION: This study has been useful in organizing community efforts and fostering collaborations to improve community health. EHE initiatives can improve community residents' knowledge of environmental exposures and inform researchers on the broader environmental factors that community residents consider important including socio-economic and psychosocial stressors.

Tu-PL-H2: Analytical Methods - II

Tu-PL-H2.1

Towards a metrological validation of gas sensors for exposure assessment

Gertjan Kok, *VSL, Delft, Zuid-Holland, Netherlands*
Stefan Persijn, *VSL, Delft, Zuid-Holland, Netherlands*
Annarita Baldan, *VSL, Delft, Zuid-Holland, Netherlands*
Hugo Ent, *VSL, Delft, Zuid-Holland, Netherlands*

Recent years showed a strong increase in the world-wide use of new low-cost sensors in medium and large scale environmental monitoring studies, particularly carried out in cities. Such sensor systems are considered by many to be the future direction for ambient air quality monitoring, and even personal monitoring. Next to their low costs other advantages are that they can deliver real-time, spatially dense, pollution data.
For the uptake of this technology the quality of the measurement data needs to be carefully considered. Therefore independent assessment of the sensor performance is required. As Dutch National Metrology Institute VSL is currently developing a gas sensor test facility based on VSL's extensive expertise on gas mixture preparation and gas analysis, in particular using highly selective infrared spectroscopy. This facility will allow gas sensor manufacturers and end users to assess sensor performance and cross sensitivities at different temperatures, pressures, relative humidities and wind speeds. This facility will be unique in its kind, because gas reference concentrations are measured in-situ and at a high time resolution. A large variety of molecular species can be generated and analyzed including many reactive species such as formaldehyde and hydrogen chloride down to the ppb level. The built up expertise is based on the outcome of MACPoll project and on the partnership in testing of gas sensors in the KEY-VOCs project (www.key-vocs.eu).
We will show the design of the test facility, together with our capabilities in gas generation and analysis. Results of pilot experiments performed during the realization of this facility will be presented. In conclusion it is anticipated that the development of this new VSL gas sensor test facility will contribute to the harmonization of exposure assessment.

Tu-PL-H2.3

A Novel Method for the Multi-Element Analysis of Dried Blood Spots

Jessica Pawly, *McGill University, Montreal, Canada*
Niladri Basu, *McGill University, Montreal, Canada*

Objective: Dried blood spots (DBS), capillary blood collected on specialized filter paper by pricking an individual's finger or heel, are a minimally invasive and cost effective alternative to venipuncture for elemental analysis. Exposure assessments using DBS can help overcome certain logistical challenges faced by researchers, particularly in terms of data collection efforts in low- and middle-income countries as well as newborn screening programs. Although previous studies have used DBS for elemental measurements, technological and practical hurdles exist (e.g. detection limits, sample volume) and there is currently no widely accepted standard method of analysis. The objective of this research is to develop and validate a novel method of quantifying select essential elements (copper, zinc, and selenium) and lead, a prevalent environmental contaminant, in DBS using Total Reflection X-Ray Fluorescence (TXRF).
Methods: A TXRF-based method for elemental analysis was established by studying DBS from different human blood standard reference materials (Institut National de Santé Publique du Québec, INSPQ; n=7) with varying and known concentrations of elements. Percent recoveries, calculated by comparing DBS to known values, and coefficients of variation were analyzed for accuracy and precision. Stability of analyses was assessed by comparing results over 16 batch runs. Additional work is underway to address other factors, such as DBS storage time and temperature, as well as to increase the number of elements that can be quantified.
Results: Percent recoveries for copper, zinc, and selenium using an entire 25uL DBS were 106.7+/-10%, 97.9+/-11.9%, 105.5+/-9.2%, respectively. Percent recoveries for copper, zinc, and selenium using a 3mm diameter DBS sub-sample were 99.5+/-3.4%, 102.6+/-4.4%, 100.2+/-6.9%, respectively. Preliminary results show that percent recovery of lead is accurate at higher concentrations (>10 µg/dL).
Conclusions: These results indicate that elemental analysis of DBS using TXRF can accurately quantify select elements using both entire and sub-sampled DBS, though additional research is needed to improve accuracy of lead analysis. These results can help establish a new method for elemental analyses of DBS, overcoming some of the challenges associated with other methods.

Tu-PL-H2.4

Proficiency tests for external quality assurance of human biomonitoring data

Thomas Göen, *University of Erlangen-Nuremberg, Erlangen, Germany*
Barbara Schaller, *University of Erlangen-Nuremberg, Erlangen, Germany*
Hans Drexler, *University of Erlangen-Nuremberg, Erlangen, Germany*

Background: The data derived in human biomonitoring (HBM) of chemical exposure have extensive consequences in occupational as well as in environmental medicine. Therefore measurements of quality assurance are a must to guaranty the reliability of these data. Because adequate certified reference materials are not available for the most HBM parameters, the participation in an inter-laboratory comparison tests is usually the exclusive feasibility to receive information on comparability and accuracy. To comply with this request a proficiency test programme for HBM parameters (GEQUAS) was started in 1982, which was adjusted over the time to toxicological as well as international requirements.
Objectives: The presentation displays the concept of the proficiency test programme, the requirements and limitation for the adjustment of the programme as well as an extract of the results.
Methods: Parameter scheme, participation quota, target values, tolerance ranges and rates of success (certificate number/participant number) were extracted from the data and results of the last 10 runs of the proficiency test programme. Each parameter was provided in two different concentrations. Generally, the target values and the tolerance ranges were estimated by the results of several so-called reference laboratories. A successful participation was certified for a parameter if the results of the participant were found to be inside the tolerance ranges of both concentrations.
Results: 156 HBM parameters were provided in the last run (7 metals in blood in occupational exposure range (OER), 3 metals in blood in environmental exposure ranges (EER), 29 inorganic parameters in urine in OER, 11 metals in urine in EER, 16 organic parameters in urine in OER, 20 organic parameters in urine in EER, 10 mercapturic acids in urine, 10 phenolic compounds in urine, 5 amines in urine, 11 solvents in blood for headspace technique, 12 solvents in urine for headspace technique, 15 halogenated hydrocarbons in serum, 11 metals in plasma, 5 N-terminal valine adducts in globin). In last runs about 200 laboratories participated, of which three-fourth were located outside Germany. The rate of success ranged for most parameters between 60 and 100 %. Parameters with poor rates of success in the last runs were phenol in urine, 4,4'-methylenedianiline in urine, inorganic arsenic in urine in EER, alpha-hexachlorocyclohexane in plasma, PCB52 in plasma, and some headspace parameters. For the phenolic compounds comparability and rate of success decreased when glucuronide conjugates were used in the preparation of the testing material.

Tu-PL-I2: Close Contact: Contaminants in Clothing

Tu-PL-I2.1

Accumulation of SVOCs in clothing from air

Tunga Salthammer, *Fraunhofer WKI, Braunschweig, Germany*
Erik Uhde, *Fraunhofer WKI, Braunschweig, Germany*
Tobias Schripp, *Fraunhofer WKI, Braunschweig, Germany*
Deniz Varol, *Fraunhofer WKI, Braunschweig, Germany*
Glenn C. Morrison, *Missouri University of Science and Technology, Rolla, United States*

BACKGROUND: Semi-volatile organic compounds (SVOCs) are generally found in quite low concentrations in indoor air. Nevertheless, they may be of importance to human health and wellbeing, as there are exposure pathways beside inhalation, e.g. food or skin contact to surfaces. Clothing could be of specific relevance here, since it is a porous material with good sorption properties, and it can have close skin contact over an extended period of time.
OBJECTIVES: Current research indicates that the human exposure towards SVOCs may be significantly influenced by adsorption/desorption-process in clothes. Fabrics can act as effective sinks for SVOC during production, treatment, but also during plain storage in a closet. Especially when they are stored for a long time the uptake of SVOC like phthalates, flame retardants, pesticides could be considerable. Due to the close contact with the skin, a quick and effective transfer followed by dermal uptake may happen. Studying such processes in detail, however, is difficult: The relevant target substances do typically appear in very low concentrations in the indoor environment, and the equilibration may take weeks to months. Also, the concentration levels to be expected in different fabric types are still unclear.
METHODS: To study the adsorption processes a test set-up was designed. Artificial sources provide humidified air with a defined, low contamination of several SVOC substances (4-nonylphenol, pentachlorophenol, parathione, HHCB, PCB-28, anthracene, di-n-butyl-phthalate, di-n-pentylphthalate, di-n-hexylphthalate). From each source a controlled air flow is led into two small environmental test chambers (see Figure 1). In total, 9 different fabrics were studied at different temperatures (2x cotton, 2x polyester, linen, polyester/46% cotton, cotton/2% elasthane, viscose/5% elasthane, and polyester/viscose). The complete system is located in a climate room to ensure a good temperature stability of all tubing and connections. Air samples were taken from each chamber and from the air supply (source) lines to monitor SVOC concentrations. Every 5 weeks, fabric samples were removed and extracted to determine the SVOC content.
RESULTS: For most of the studied substances, the equilibration time exceeded 2 months. Afterwards, fairly stable concentrations were found for a period of five months. In the source lines, steady-state concentrations were between 0.1 $\mu g/m^3$ (di-n-hexylphthalate) and 35 $\mu g/m^3$ (4-nonylphenol). In the exposure chambers, the concentrations were found to be lower. Analysis of exposed fabric samples indicated relevant accumulation of the target compounds in the material. Therefore dermal exposure via clothing is considered as a relevant uptake route for SVOCs.

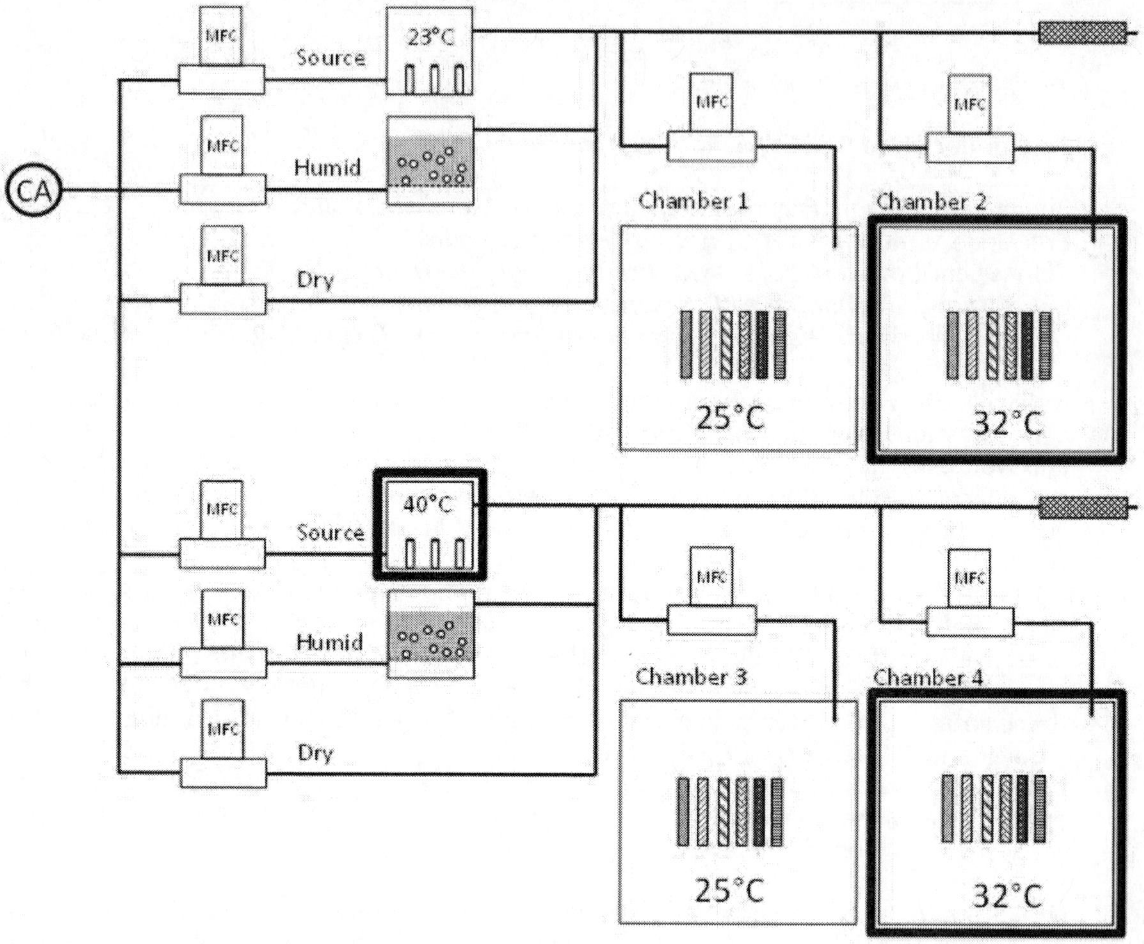

Figure 1: Schematic of the experimental setup

Tu-PL-I2.2

Effects Of Weathering On PFASs Used In Durable Water Repellence Of Textiles

Ike Van der Veen, *VU University Amsterdam, Amsterdam, Netherlands*
Anne-Charlotte Hanning, *Swerea, Mölndal, Sweden*
Jana Weiss, *Stockholm University, Stockholm, Sweden*
Pim Leonards, *VU University Amsterdam, Amsterdam, Netherlands*

Introduction and Aim
Per- and polyfluoroalkyl substances (PFASs) are used in textiles for their durable water and soil repellent (DWR) properties. Because PFASs with long perfluorinated chains have been shown to be persistent, bioaccumulative and (eco)toxic, the textile industry is phasing-out the long-chain PFASs and is replacing those with alternative chemistries to deliver the desired DWR effect.
The aim of our research in the SUPFES (Substitution in practice of prioritised fluorinated compounds for textile applications) project is to assess the alternative DWRs to (i) their structural properties, (ii) the loss and degradation processes, and (iii) the hazard profile for the emitted substances.
As part of SUPFES the influence of weather conditions on DWR treatments are assessed. Here the effect of weathering on the PFASs concentrations in outdoor clothing is presented.
Methods
Nine samples of outdoor clothing were exposed to UV radiation, humidity, and temperature in an aging device for 300 h, which is equivalent to the life time of the clothing. The textile samples were, before and after aging, extracted with methanol and analysed by LC-MS/MS.
Results
The original nine textile samples contained different levels of PFAAs with different patterns. After aging the results are different for all samples. In two samples odd-chain length PFAA appeared. In 5 samples the mean concentration of PFAAs increased 5 times or more, and in one of the samples it amplified more than 100 times.
An explanation of the forming of those PFAAs might be the transformation of the precursors fluorotelomer alcohols (FTOHs) and fluorotelomer acrylates (FTACs), which are used for the formation of polymers used for DWR, or by the degradation of the polymers themselves.
The FTOH and FTAC concentrations in the textile samples before and after aging will be discussed to get a comprehensive overview of the transformation and degradation pathways by exposure to weather conditions.
Conclusions
Weather conditions, like sunlight, high temperature, or humidity have an effect on PFASs used in DWR of outdoor clothing. Concentrations of PFAA increased and other PFAA were formed during exposure to weather conditions.

Tu-PL-I2.3

Measurements of Dermal Uptake of Nicotine Directly from Air and Clothing

Gabriel Bekö, *Technical University of Denmark, Lyngby, Denmark*
Glenn Morrison, *Missouri University of Science and Technology, Rolla, MO, United States*
Charles Weschler, *Rutgers University & Technical University of Denmark, Piscataway, United States*
Holger Koch, *Institute for Prevention and Occupational Medicine of the German Social Accident Insurance, Bochum, Germany*
Tunga Salthammer, *Fraunhofer WKI, Braunschweig, Germany*
Tobias Schripp, *Fraunhofer WKI, Braunschweig, Germany*
Jørn Toftum, *Technical University of Denmark, Lyngby, Denmark*
Geo Clausen, *Technical University of Denmark, Lyngby, Denmark*

Recent studies suggest that dermal uptake of certain semivolatile organic compounds directly from air can be a significant exposure pathway. This has been experimentally confirmed for two phthalates. An experiment has been conducted to investigate if dermal uptake of nicotine directly from air or from clothing may be similarly important. Two subjects wearing only shorts and a third subject wearing clean cotton clothes were exposed to environmental tobacco smoke (ETS) for three hours in a 55m3 chamber while breathing clean air from hoods they wore. The ETS was generated by mechanically "smoking" cigarettes, with three lit at any given time. The resulting average nicotine concentration (475 µg/m3) is comparable to the highest levels reported for smoking sections of pubs. Urine samples were collected immediately before exposure. For the subjects wearing only shorts, all urine was collected for the 60 hours post-exposure. These samples were pooled for the first 12 h, 12-36 h and 36-60 h post-exposure. For the clothed subject, urine samples were collected until the next morning. After collecting a new pre-exposure urine sample, this subject entered the chamber for another three-hour exposure wearing a hood and clothes, including a shirt that has been exposed for 5 days to elevated nicotine levels (>200 µg/m3). The urine samples were analyzed for nicotine and two metabolites -- cotinine and 3OH-cotinine. Following exposure, the subjects who wore only shorts excreted a significant amount of nicotine and nicotine metabolites. Assuming that 90% of nicotine and its metabolites are excreted via urine and that nicotine, cotinine and 3OH-cotinine constitute 85% of what is excreted via urine, the back-calculated minimum amount of dermally absorbed nicotine was 570 µg for the bare-skinned subjects. For the subject wearing clean clothes, it was 20 µg, and while wearing a shirt previously exposed to nicotine, it was 80 µg. Peak cotinine and 3OH-cotinine concentrations in the urine of the bare-skinned subjects were an order of magnitude higher than for non-smokers who avoid ETS exposure and comparable to levels measured among non-smokers in hospitality environments before smoking bans. This study indicates that meaningful dermal uptake of nicotine can occur from exposure to environmental tobacco smoke. This is especially important for children in homes where smoking or vaping occurs. Fresh clothing can significantly limit dermal uptake, but clothing can also be a source if it was pre-exposed to cigarette smoke.

Tu-PL-I2.4

SPME-based C-history method, accurate measurement of important parameters for assessing SVOC dermal exposure: diffusion and partition coefficients of SVOCs adsorbed by clothing

Jianping Cao, *Tsinghua University, Beijing, China, People's Republic of*
Yinping Zhang, *Tsinghua University, Beijing, China, People's Republic of*

Aim: It has been recently recognized that the impact of clothing requires to be taken into consideration when estimating dermal exposure to SVOCs. The knowledge of the SVOC's partition coefficient between the clothing and air, K, and its diffusion coefficient in the clothing, D, is a prerequisite for estimating this impact. In this study, an approach, named SPME-based C-history method is developed to accurately measure these two parameters.
Methods: We designed a sealed experimental chamber as illustrated in the attached figure. Solid phase Micro-extraction (SPME) devices are employed to monitor the time profile of SVOC concentrations in the chamber air. Analysing SVOC mass transfer in the chamber, the SVOC concentration in the chamber air (Ca) can be expressed as: Ca=a·(1-b·exp(-c·t)); where a is a constant, t is the time, b and c are functions of K and D. b and c can be obtained by fitting this expression to Ca measured at a series of times by SPME. Then K and D can be calculated directly because they are functions of b and c, and we have two equations with two unknown parameters.
Results: K and D of three kinds of phthalates (DiBP, DnBP and DEHP) adsorbed by two kinds of clothes (pure cotton T-shirt and jeans) are measured at 25 °C and 32 °C. Results show that, for DiBP and DnBP, D and K are on the order of 10e-11 and 10e5, respectively; while for DEHP, D and K are on the order of 10-13 and 10e7, respectively. Comparing the present K with that in the literature of the same SVOC-clothing pairs, small deviations between them support the accuracy of the present method.
Conclusions: The proposed method together with the measured data should be useful in more accurately estimating dermal exposure to gas-phase SVOCs.

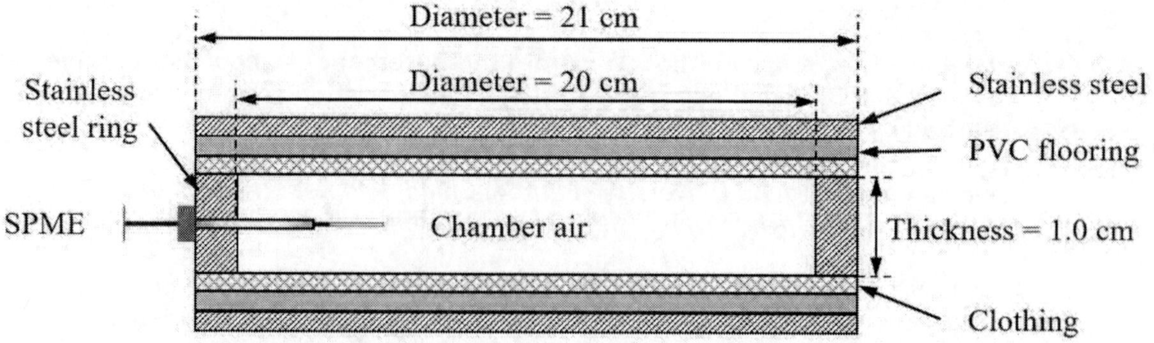

Schematic of the experimental chamber

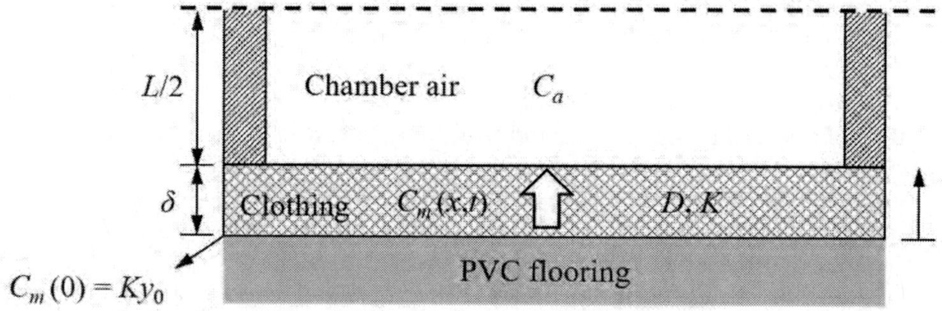

Schematic of SVOC mass transfer in the chamber

Expression: $$C_a = y_0 - 2y_0 \frac{e^{-Dq^2\delta^{-2}t}}{q/\sin q + \cos q} \quad \text{with } q \tan q = \frac{K\delta}{L}$$

Schematic of the experimental chamber and SVOC mass transfer in the chamber

Tu-SY-A3: The Exposome: From concept to practice - III

Tu-SY-A3.1

Sensor Technologies and the Exposome

Miranda Loh, *Institute of Occupational Medicine, Edinburgh, United Kingdom*
Denis Sarigiannis, *Aristotle University of Thessaloniki, Thessaloniki, Greece*
Alberto Gotti, *Aristotle University of Thessaloniki, Thessaloniki, Greece*
Spyros Karakitsios, *Aristotle University of Thessaloniki, Thessaloniki, Greece*
Anjoeka Pronk, *TNO, Utrecht, Netherlands*
Eelco Kuijpers, *TNO, Utrecht, Netherlands*
Isabella Annesi-Maesano, *Universite Pierre et Marie Curie, Paris, France*
Nour Baiz, *Universite Pierre et Marie Curie, Paris, France*
Joana Madureira, *University of Porto, Porto, Portugal*
Eduardo Oliveira Fernandes, *University of Porto, Porto, Portugal*
Michael Jerrett, *University of California, Los Angeles, Los Angeles, CA, United States*
John Cherrie*Institute of Occupational Medicine, Edinburgh, United Kingdom*

The exposome lays out a challenge to characterize exposures and exposure-related attributes over a person's lifetime. Historically, exposure assessment methods relied on measurements over short periods of time or models based on short-term data collection. Wireless technologies, downsizing and low-cost sensors provide the potential for longer-term and more detailed data collection, which will help us better define the holism of the exposome. Such monitoring technologies put exposure science at a new frontier with new challenges. Several projects are exploring the use of wireless sensors in assessing the exposome. As an example, the EU-funded HEALS project uses a variety of apps, in-home and personal sensor devices to measure external exposure and related factors. Researcher and participant experiences will be discussed, along with challenges in data quality, data quantity, privacy, data analysis and interpretation and cost. Currently, sensor-based technologies provide good opportunities to better characterize exposure information, but technology for personalized quantification of exposure to environmental chemicals and other hazards is limited, particularly in relation to ease of use and measurement sensitivity / specificity requirements. Combination of sensor-based data collection with exposure modelling to estimate personal level exposures will provide us an ability to improve our understanding of the exposome. In the longer term, personalized monitors such as for air quality, and even biomarkers may become a reality.

Tu-SY-A3.2

A Time Geography of the Exposome

Michael Jerrett, *University of California, Los Angeles, Los Angeles, California, United States*

Background and Objectives: Geographers and others interested in spatial analyses have long recognized the potential for locational tracking in exposure science. von Haggerstrand first introduced the idea of the "Time Geography" in which an individual would follow different hazard paths of exposure through a time-space prism. This idea can be seen as a prelude and compliment to the exposome, which is broadly defined as the totality of the environmental exposures an individual faces through the course of their lifetime. The objective of this presentation will be to better link the conceptual and modeling aspects of time geographies with the exposome. I will draw on several ongoing and recently completed studies in Europe and North America to illustrate specific points.
Methods: With the evolution of cellular phone technologies to mobile sensing devices, huge volumes of data can be collected on geographic position, physical activity, emotional state, and ambient environmental conditions on smart devices carried by literally billions of individuals globally. Smart phones can also serve as base stations for receiving and transmitting data from other external devices either worn on the person or embedded into urban infrastructure (e.g., pollution monitoring, heat sensing).
Results and Discussion: Although promising, tracking the time geography of exposure with ubiquitous devices poses numerous technical, analytical, and computational challenges. First, the data obtained from these devices is likely to be very large. For example, in a recent study in Barcelona with 180 participants carrying a smart phone that logged geographic position and physical activity every 10 seconds, more than 10,000,000 observations were generated in one week. Second, the data are likely to be autocorrelated in time and space, which violates the fundamental premise for statistical inference that the data be independent. Dealing with autocorrelation creates computational problems that exceed the capacity of common computing platforms. Third, the data are likely to be spatially biased because signal loss from the global positioning system will follow distinct gradients in the urban domain based on building obstructions and indoor environments. Temporal bias will also occur depending on wear time. For instance, a recent study showed that only 3 days of valid data on physical activity were obtained from phones that study participants wore for 5 days. Understanding these sources of bias will be important for scaling up the exposure estimates. I will conclude with some possible near and long-term resolutions to better characterize the time geography of the exposome.

Tu-SY-A3.3

Metabolomics for Environmental Biomonitoring

Dean Jones, *Emory University, Atlanta, GA, United States*

Background and Objectives: A broad range of environmental exposures occurs globally due to >30,000 chemicals used in commerce. Surveillance is limited to a relatively small number of known hazards and at-risk populations because analytic costs preclude comprehensive monitoring. We developed high-resolution metabolomics (HRM) with ultra-high resolution mass spectrometry coupled to liquid chromatography to provide advanced blood chemistry for healthcare. In optimization and application to disease cohorts, the results showed routine detection of over 10,000 metabolites and environmental chemicals. This provided a basis to explore use of HRM for high-throughput biomonitoring of environmental exposures.
Methods: Ultra-performance liquid chromatography with electrospray ionization (ESI) and ultra-high resolution mass spectrometry is performed using commercial instruments with Fourier-transform ion-cyclotron resonance or orbitrap detectors. Wide scan ranges are used depending upon the instrument, typically including the range of 85-850 m/z, with options from 50 m/z to 2000 m/z. The instruments are configured with switching valves to enable dual chromatography (one column being washed and re-equilibrated while samples are analyzed on a parallel column). Alternate ESI polarity with orthogonal chromatography on the columns provides optimal performance with +ESI/HILIC and -ESI/C18 as an optimal pair; protein is removed by addition of acetonitrile:sample (2:1),including a mixture of stable isotopic internal standards, and samples are analyzed with acetonitrile and formic acid gradients. Run times are typically 10 min/analysis, and each sample is analyzed with three technical replicates. Data are extracted using xMSanalyzer with apLCMS, and quality control and assurance are based upon mass accuracy and reproducibility of internal standards and inclusion of pooled reference samples before and after each batch of 20 samples. Reference standardization was used to quantify individual environmental chemicals.
Results and Discussion:
Results establish that metabolomics analyses directed toward measurement of nutrients and intermediary metabolism in healthy adults also provides quantitative information on environmental chemicals with identities confirmed by ion dissociation mass spectrometry (MS/MS) and co-elution with standards. This included metabolites of DDT and other persistent halogenated chemicals (chlorobenzoic acid, chlorophenylacetic acid), plasticizers (dibutylphthalate, di(2-ethylhexyl)adipate, di-isononylphthalate, dipropylphthalate), insecticides (methomyl, pirimicarb), commercial chemicals (styrene, tetraethylene glycol), and flame retardants (octylphenol, triethylphosphate, triphenylphosphate). Accurate mass matches to a broad range of other chemicals were correlated with chlorophenylacetic acid, including dichloroacetate, chloroallylaldehyde, chlorobenzoate, nitrosonaphthalene, naphthalenesulfonic acid, cyanofenphos, pyraclofos, 1-chloro-2,2-bis(4'-chlorophenyl)ethane, and trichloroethanolglucuronide.
Targeted metabolome wide association studies (MWAS) of specific environmental chemicals shows that perturbations of metabolism can be directly linked to the environmental chemicals through a dose-response relationship. The results establish that widely available ultra-high resolution mass spectrometry instrumentation can be used with triplicate analyses and rigorous standard operating procedures to detect a wide range of environmental chemicals within an analytical structure that is also useful for measurement

of many nutrients and intermediary metabolites. The results indicate that costs for environmental biomonitoring may be controlled by incorporation of biomonitoring into routine health analyses involving blood or urine. Although numerous obstacles wound need to be addressed to operationalize such an approach, the capabilities are available today and could be especially useful for undiagnosed health conditions of possible environmental origin.

Tu-SY-A3.4

Transcriptomics: at the interface of exposure and biological response

Jelle Vlaanderen, *IRAS, Utrecht University, Utrecht, Netherlands*
TW Wang, *Boston University School of Medicine, Boston, MA, United States*
W Hu, *National Cancer Institute, Rockville, MD, United States*
N Rothman, *National Cancer Institute, Rockville, MD, United States*
Q Lan, *National Cancer Institute, Rockville, MD, United States*
A Spira, *Boston University School of Medicine, Boston, MA, United States*
RCH Vermeulen, *IRAS, Utrecht University, Utrecht, Netherlands*

In recent years characterization of the human transcriptome has contributed to the refinement of clinical diagnosis of disease and the development of diagnostic and prognostic markers. As the human transcriptome is highly sensitive to external stressors and to inter-individual variability in biological processes, it is a promising tool for individualized assessment of biological perturbations due to environmental stressors. Studies of tobacco smoke have established associations between smoking patterns and specific gene expression profiles in bronchial, buccal, and nasal epithelial cells. Showing both reversible and irreversible transcriptomic changes. In a recent indoor air pollution study among Chinese non-smoking women, we observed perturbations in gene expression profiles in buccal epithelial cells due to exposure to smoky coal that were highly similar to the gene expression profiles associated with smoking in Caucasians. We observed a similar overlap with gene expression profiles assessed in peripheral blood that were associated with long-term exposure to ambient air pollution. In a study among European workers we assessed whether occupational exposure to carbon nanotubes induced perturbations of the transcriptome in nasal epithelial cells. While we observed limited evidence for significant perturbations of the abundance of individual transcripts due to exposure to carbon nanotubes, changes across the transcriptome overlapped significantly with those that were associated with smoking. The degree of overlap in transcriptome profiles between these studies indicates shared biological pathways between these sources of particulate matter exposure.
Successful application of transcriptomics in human studies can provide physiologically relevant, molecular insights reflecting the interface of exposure and biological response.

Tu-PL-B3: Urinary Biomarkers

Tu-PL-B3.1

Three-Year Temporal Variability in Urinary Concentrations of Environmental Chemicals among a Multi-Ethnic Cohort Of Girls In The United States.

Susan L. Teitelbaum, *Icahn School of Medicine at Mount Sinai, New York, New York, United States*
Ashley Pajak, *Icahn School of Medicine at Mount Sinai, New York, New York, United States*
Gayle C. Windham, *California Department of Health, Richmond, California, United States*
Susan M. Pinney, *University of Cincinnati, Cincinnati, Ohio, United States*
Antonia M. Calafat, *Centers for Disease Control and Prevention, Atlanta, Georgia, United States*
Xiaoyun Ye, *Centers for Disease Control and Prevention, New York, New York, United States*
Manori Silva, *Centers for Disease Control and Prevention, Atlanta, Georgia, United States*
Michael Rybak, *Centers for Disease Control and Prevention, Atlanta, Georgia, United States*
Lawrence H. Kushi, *Kaiser Permanente, Oakland, California, United States*
Frank M. Biro, *Cincinnati Children's Hospital Medical Center, Cincinnati, Ohio, United States*
Mary S. Wolff, *Icahn School of Medicine at Mount Sinai, New Yokr, New York, United States*

Environmental chemicals, including phthalates, phytoestrogens and phenols, have been widely studied in relation to human health. Exposure can be quantified using urinary biomarkers, but the time frame that these biomarkers represent is known to be brief. The underlying sources of exposure can be relatively constant and these biomarkers have been shown to be predictive of exposure levels up to one year in children. Data on the temporal variability of environmental chemical biomarkers for longer time periods are sparse, especially among children. The objective of this investigation was to evaluate intra-individual temporal variability in 16 urinary biomarkers (8 phthalates, 3 phytoestrogens and 5 phenols) among a multi-ethnic cohort of girls.
Healthy girls (N=309; 6-8 years at baseline) provided 3 urine samples over a 3 year period. To assess temporal variability we used two statistical methods, intraclass correlation coefficient (ICC) and surrogate category analysis, used to determine how well tertile categories based on a single measurement represented ranking on a 3-year mean concentration.
Surrogate category analysis suggested that a single sample provides reliable ranking for all biomarker classes; all 3 of the surrogate samples predicted the 3-year mean concentration ranking. Of the 16 analytes, the ICC was >0.4 (range: 0.4-0.7) for 4 analytes: 2,5-dichlorophenol; enterolactone; benzophenone-3; and mono-ethylphthalate. The ICC for 10 analytes was between 0.2-0.4, while the ICC was <0.2 for 2.
Important considerations for the use of phthalate, phytoestrogen and phenol biomarkers include prevalence of exposure and accurate reflection of the exposure of interest. These results indicate that a single urine sample measurement demonstrates reasonable ranking of exposure over a 3 year period, possibly longer, for several of these environmental

chemical biomarkers; and provide additional support for the use of these biomarkers in children's environmental health research.

Tu-PL-B3.2

Antibiotic internal exposure levels of School Children in East China based on biomonitoring Study

Ying Zhou, *Fudan University, Shanghai, China, People's Republic of*
Hexing Wang, *Fudan University, Shanghai, China, People's Republic of*
Qingwu Jiang, *Fudan University, Shanghai, China, People's Republic of*

To explore the antibiotic internal exposure levels of Chinese children, a biomonitoring method was developed to detect total urinary concentrations (free and conjugated) of 18 representative antibiotics (5 macrolides, 2 ß-lactams, 3 tetracyclines, 4 quinolones, and 4 sulfonamides) based on ultraperformance liquid chromatography coupled to quadrupole time-of-flight mass spectrometry. By usage of this proposed method, the urinary samples collected from 1064 school students, which were recruited from 3 areas in East China area in 2013, were measured. The detection frequencies of all 18 antibiotics ranged from 0.4% to 19.6%, and 58.3% of the whole urine samples were found with target antibiotics. Of them, 47.8% of the urine samples had a sum of mass concentration of all antibiotics between 0.1 (minimum) and 20.0 ng/mL, and 8 antibiotics had their concentrations of above 1000 ng/mL in some urine samples. At least one antibiotic was found in more than 50% of the whole urine samples and the proportion reached at 74.4% in one study area. Three veterinary antibiotics, 4 human antibiotics, and 11 human/veterinary antibiotics were detected overall in 6.3, 19.9, and 49.4% of urine samples, respectively. The results suggested that Chinese school children were extensively exposured to antibiotics and contaminated food or environment might be potential exposure sources.

Table 1. Descriptive analysis of target antibiotics in all subjects (n=1064)

Antibiotics	Usage	n (%)[a]	Percentile					Maximum
			50th	75th	90th	95th	99th	
Macrolides[b]		205 (19.3)	-	-	4.9 (12.0)[c]	39.9 (99.9)	3092.8 (5759.1)	19608.6 (113458.9)
Azithromycin	Human	174 (16.4)	-	-	3.8 (8.5)	31.5 (81.5)	2401.7 (4684.9)	7056.8 (18025)
Clarithromycin	Human	18 (1.7)	-	-	-	-	1.2(3.2)	1240.7 (25063)
Erythromycin	Human/veterinary	11 (1.0)	-	-	-	-	0.6 (1.5)	19605.2 (111721.1)
Roxithromycin	Human	4 (0.4)	-	-	-	-	-	6.8(19)
Tylosin	Veterinary	11 (1.0)	-	-	-	-	0.7 (1.2)	4.1 (8.8)
β-Lactams[b]		72 (6.8)	-	-	-	1.2 (1.8)	61.4 (118.9)	42689.5 (78005.5)
Ampicillin	Human/veterinary	47 (4.4)	-	-	-	-	4.9 (5.3)	42689.5 (78005.5)
Cefaclor	Human	29 (2.7)	-	-	-	-	35.3 (69.9)	25758.2 (69254.6)
Tetracyclines[b]		40 (3.8)	-	-	-	-	4.6 (10.5)	2682.6 (9478.8)
Oxytetracycline	Human/veterinary	24 (2.3)	-	-	-	-	2.0 (6.0)	2626.7 (9281.5)
Chlortetracycline	Veterinary	13 (1.2)	-	-	-	-	1.1 (1.4)	361.2 (184.3)
Tetracycline	Human/veterinary	7 (0.7)	-	-	-	-	-	55.8 (197.3)
Quinolones[b]		281 (26.4)	-	0.3 (0.6)	1.4 (3.5)	3.2 (6.4)	25.2 (51.6)	499.8 (323.3)
Ofloxacin	Human/veterinary	142 (13.3)	-	-	0.6 (1.1)	1.1 (2.3)	4.9 (9.1)	499.8 (323.3)
Ciprofloxacin	Human/veterinary	113 (10.6)	-	-	0.3 (0.7)	0.7 (2.2)	5.5 (7.5)	42.3 (60.0)
Enrofloxacin	Veterinary	45 (4.2)	-	-	-	-	2.1(5.8)	10.6 (16.6)
Norfloxacin	Human/veterinary	37 (3.5)	-	-	-	-	7.1(10.1)	45.2 (79.1)
Sulfonamides[b]		321 (30.2)	-	1.1 (1.9)	3.8 (8.5)	6.7 (12.9)	21.2 (44.9)	3021.1 (6510.2)
Sulfamethazine	Human/veterinary	209 (19.6)	-	-	3.2 (6.3)	4.9 (10.0)	13.3 (24.4)	23.6 (66.7)
Trimethoprim[d]	Human/veterinary	107 (10.1)	-	-	0.2 (0.1)	0.5 (1.3)	3.9 (8.3)	1268.8 (2734.0)
Sulfamethoxazole	Human/veterinary	49 (4.6)	-	-	-	-	3.9 (11.5)	1750.0 (3771.0)
Sulfadiazine	Human/veterinary	5 (0.5)	-	-	-	-	-	66.2 (201.2)
All antibiotics[e]		620 (58.3)	0.9 (1.5)	4.4 (9.1)	22.8 (51.5)	87.5 (208.7)	3585.7 (8927.7)	42689.8 (113460.3)

[a]Number of positive detection (detection frequency, %). [b]Mass sum of antibiotics in corresponding antibiotic categories. [c]Volume-based concentration value, ng/mL (creatinine-corrected concentration value, μg/g creatinine). [d]Due to usually mixed use with sulfonamides in practical application, trimethoprim was together analyzed with sulfonamides. [e]Mass sum of all antibiotics. -: <Limit of detection (LOD).

Descriptive analysis of target antibiotics in all subjects (n=1064)

Tu-PL-B3.3

Plasticizer monitoring in the urine of 2 to 6 year old children from North Rhine-Westphalia, Germany - exposure trends within a period of 4 years

Martin Kraft, *North Rhine Westphalian State Agency for Nature, Environment and Consumer Protection, Recklinghausen, North Rhine Westphalia, Germany*
Susanne Rudzok, *North Rhine Westphalian State Agency for Nature, Environment and Consumer Protection, Recklinghausen, North Rhine Westphalia, Germany*
Silvia Sievering, *North Rhine Westphalian State Agency for Nature, Environment and Consumer Protection, Recklinghausen, North-Rhine Westphalia, Germany*
Yvonni Chovolou, *North Rhine Westphalian State Agency for Nature, Environment and Consumer Protection, Recklinghausen, North Rhine Westphalia, Germany*
Wolfgang Völkel, *Bavarian Health and Food Safety Authority, München, Bavaria, Germany*
Hermann Fromme, *Bavarian Health and Food Safety Authority, München, Bavaria, Germany*

Background: Plasticizer like phthalates or diisononyl-cyclohexane-1,2-dicarboxylate (DINCH) are found in many consumer products used in daily life. When present in consumer products plasticizers can be released and incorporated by consumers. To estimate the body burden metabolites of Phthalate and DINCH can be used for human biomonitoring.
Aim: There is little information on the exposure to phthalates and DINCH in children under the age of six years. Therefore, the urine of 2 to 6 year old children was examined on selected plasticizers in winter/spring 2011/2012 and in a follow up in winter/spring 2014/2015. These surveys are part of the LUPE III project (Länderuntersuchungsprogramm).
Methods: About 250 children participated in each human biomonitoring study. The urine samples were taken throughout the day while attending kindergarten. Overall, 11 phthalate metabolites and 3 DINCH metabolites were measured in the urine. In addition, information on anthropometric data, on lifestyle and environmental factors were collected via questionnaires.
Results: The phthalate metabolites mono-n-butyl phthalate (MnBP) and mono-iso-butyl phthalate (MiBP) were found most frequently and displayed the highest median urinary concentrations in the examined children, followed by mono-ethyl phthalate (MEP), Mono(2-ethyl-5-hydroxyhexyl)phthalate (OH-MEHP) and Mono(2-ethyl-5-oxohexyl) phthalate (oxo-MEHP). From 2011/2012 to 2014/2015, a statistically significant decrease was found for the median of OH-MEHP and oxo-MEHP by 37 % and 42 %, respectively. There was a statistically significant increase in median measured urinary MEP concentrations by 24 %. Considering health based assessment criteria, the urinary phthalate concentrations were within tolerable limits, except for MnBP and MiBP. Here, 2.3 % and 5.1 % of the examined children exceeded the tolerable daily intake (TDI).
The mediane urinary concentrations of the DINCH-metabolites cyclohexane-1,2-dicarboxylic acid-mono(hydroxy-isononyl) (OH-MINCH) and cyclohexane-1,2-dicarboxylic acid-mono(carboxyisooctyl) (cx-MINCH) were significantly increased by 100 % and 41 %, respectively. However, the obtained values are still low compared to the most measured phthalate metabolite concentrations and fall markedly below the health based assessment criterion.
Conclusion: The carried out human biomonitoring studies serve as a valuable tool by monitoring the present exposure and the possible trends of plasticizers in urine of

children. The results of this study show statistically significant declining exposure levels of DEHP-metabolites and an increase of DEP- and DINCH-metabolites within a period of four years.

Tu-PL-B3.4

Assessing the impact of a single biomarker measurement to reconstruct the exposure of pregnant women to Di(2-ethylhexyl) phthalate

Doris Tan, *INERIS, Verneuil-en-Halatte, France*
Florence Zeman, *INERIS, Verneuil-en-Halatte, France*
Maribel Casas, *CREAL, Barcelona, Spain*
Amrit K. Sakhi, *Norwegian Institute of Public Health, Oslo, Norway*
Berit Granum, *Norwegian Institute of Public Health Oslo, Oslo, Norway*
Cyntia Manzano, *CREAL, Barcelona, Spain*
Martine Vrijheid, *CREAL, Barcelona, Spain*
Céline Brochot, *INERIS, Verneuil-en-Halatte, France*

Background and Objectives: Large-scale human biomonitoring surveys typically gather biomarker measurements from single time points. For non-persistent compounds (i.e., with a biological half-life of hours), the biomarker levels are known to greatly vary within the same individual over a relatively short time period. In this work, we aimed at assessing the impact of a single biomarker measurement to reconstruct the exposure of individuals to a non-persistent compound (a phthalate) by testing several sampling scenarios. In particular, we studied the exposure of pregnant women to Di(2-ethylhexyl) phthalate (DEHP) using the urinary concentrations of four DEHP metabolites. Phthalates are a family of chemicals that can be found in a wide array of products, and are suspected to induce reproductive and developmental toxicity.
Methods: Thirty women from two countries (Spain and Norway) were followed over a week during the second trimester of pregnancy. A protocol was designed specifically to assess the intra-individual variability over the week. Two spot samples per day (first and last voids) and the pool of the urines collected over the week were analyzed. A toxicokinetic model was applied to back-calculate the exposure levels to DEHP from the urinary metabolites concentrations. Additional individual data were collected from questionnaires and integrated in the model such as the bodyweight and the times of urination.
Results and conclusion: The measured concentrations of DEHP metabolites in urine exhibit high variability between women and within each woman. The daily intakes (DI) were estimated according to several scenarios based on different types of biomarkers (spot or pool) and exposure patterns. For a continuous exposure over the week, the individual DI were estimated between 0.49 to 7.17 µg/kg bw/d using the pool samples. On average, the DI estimated using the 14 spot samples were lower by 14%, but discrepancies were observed between the women. More realistic exposure scenarios were tested to improve the model adjustment to the data when all the samples are analysed. A constant exposure during the day and no exposure during the night were assumed. Using this scenario, the extreme (i.e., the lowest and highest) concentrations of the metabolites were better estimated than using a constant exposure and the estimated DI were slightly higher (10% on average). Our results show the need to define realistic exposure scenario to describe adequately the time evolution of the urinary concentrations over the week. Future steps will consist in refining the scenarios using individual information.

UTRECHT
THE NETHERLANDS
OCTOBER 9-13
ISES 2016

Tu-PL-B3.5

Urinary concentrations of parabens in young children - A human biomonitoring study from north Rhine Westphalia, Germany

Yvonni Chovolou, *North Rhine-Westphalian State Agency for Nature, Environment and Consumer Protection, Essen, Germany*
Susanne Rudzok, *North Rhine-Westphalian State Agency for Nature, Environment and Consumer Protection, Essen, Germany*
Silvia Sievering, *North Rhine-Westphalian State Agency for Nature, Environment and Consumer Protection, Essen, Germany*
Holger Koch, *Institute for Prevention and Occupational Medicine of the German Social Accident Insurance, Institute of the Ruhr-Universität Bochum (IPA), Bochum, Germany*
Martin Kraft, *North Rhine-Westphalian State Agency for Nature, Environment and Consumer Protection, Essen, Germany*

Background: Parabens are alkyl esters of para-hydroxybenzoic acid and are widely used as antimicrobial preservatives in personal care products but also in food, beverages, food, and pharmaceutical preparations. Although they are considered to be slightly toxic to humans, the use of parabens has raised concern due to their possible endocrine disrupting activities as demonstrated in several in vivo and in vitro studies.
Aim: Only a few data are available on the magnitude of children's exposure to chemicals present in many consumer products. The aim of this study was to determine concentrations and profiles of parabens in young children from North Rhine-Westphalia, Germany. These new biomonitoring data were compared to previous data from urine samples collected in 2011/2012 as part of the LUPE III project (Länder-untersuchungsprogramm).
Methods: Nine Parabens were measured in spot urine specimens from 255 young children aged from 27 to 98 months. Samples were collected between December 2014 and Mai 2015.
Results: Among the nine parabens analyzed, methylparaben (MeP) and ethylparaben (EtP) were the parabens found in most urine samples. They were detected in concentrations above the limit of quantification (LOQ) in 100 % (MeP) and 86 % (EtP) of the urine samples. n-Propylparaben (n-PrP) was above the LOQ in 38 % of the samples. iso-butylparaben, benzylparaben and n-butylparaben were only detectable in < 6 % of the samples. Iso-propylparaben, pentylparaben and heptylparaben could not be quantified in any of the urine samples. The mean concentration of MeP was 159 µg/L, followed by n-PrP (10.4 µg/L) and EtP (3.18 µg/L). The maximal paraben levels found were 7700 µg/L for MeP, 699 µg/L for
n-PrP and 189 µg/L for EtP. A significant decrease in paraben concentrations for MeP and n-PrP was found between our study population examined 2011/2012 and 2014/2015. The median concentration for MeP and n-PrP decreased over a 3-year period by 85 % and 70 %, respectively. The measured median urinary concentrations for MeP (6.91 µg/L),
n-PrP (< 0.50 µg/L) and EtP (1.02 µg/L) in our study population were in range with results reported by several European and North American studies surveying children in various age groups.
Conclusion:
Parabens (in particular MeP, EtP; n-PrP) were frequently found in urine from young children indicating ubiquitous exposure to these compounds. We found decreasing mean

urinary concentrations for MeP and n-PrP over a period of 3-4 years, which probably reflects a reduced use of these compounds in personal care products.

Tu-SY-C3: Health effects of air pollutant

Tu-SY-C3.1

Global burden of diseases, injuries and risk factors, a bridge between disease epidemiology, risk assessment and public health policy

Mohammad Forouzanfar, *IHME, Seattle, Washington, United States*

The Global Burden of Diseases, Injuries and Risk Factors study (GBD) is a systematic, scientific effort to quantify the comparative magnitude of health loss. In GBD 2013 the health loss due to 240 causes of death and 301 causes of morbidity, and 79 risk factors was estimated in terms of deaths, years of life lost due to premature death and years lived with disability. Estimates were generated for 188 countries and for subnational locations in China, Mexico, and the United Kingdom by age and by sex and 5 years intervals from 1990 to 2015. More than 1400 collaborators provided feedback and oversaw the project. Moreover, GBD is equipped with a rich scientific team as well as a strong computational infrastructure, visualizations, and communication experts for the global dissemination of results and engagement.
In GBD 2013, we estimated that 802 million deaths were related to exposure to environmental and occupational risks in 2013. More than 5.5 million deaths were caused by air pollution and about one million by lead and residential exposure to radon. Satellite measurements, a chemical transport model and ground monitoring measurements of air particulate matter concentration were analyzed to estimate the annual average of fine particle (PM2.5) at $0.1^\circ \times 0.1^\circ$ spatial resolution. Concentration response curve analysis provided the effect size of air pollution on the health outcomes. In 2013, 3.0% (95% CI: 2.5 - 3.5%) of the total burden (in terms of disability adjusted life years, DALYs) was lost due to the exposure to PM2.5 pollution. Globally, 14.6% of cardiovascular deaths, 4.2% of lower respiratory infection, 4.0% of chronic respiratory diseases deaths, and 3.6% of cancer deaths can be attributed to exposure to PM2.5 of more than 8.7 μg/m3. Exposure to ambient air pollution is the top environmental risk in Western Europe. The age-standardized mortality rate in southern Europe including Italy, Spain and France is about three-folds higher than Nordic countries except the Netherlands and Denmark with 24.8 and 15.5 deaths per 100,000 in 2013, respectively.
In this presentation, I will present an overview of the methodology for estimating disease and risk factor burden in GBD and discuss results and challenges of quantifying ambient air pollution adverse health effects.

Tu-SY-C3.2

Disease Burden Estimates for Ambient Air Pollution in Finland and Related Parametric and Model Uncertainties

Heli Lehtomäki, *National Institute for Health and Welfare (THL), Kuopio, Finland*
Arja Asikainen, *National Institute for Health and Welfare, Kuopio, Finland*
Isabell Rumrich, *National Institute for Health and Welfare, Kuopio, Finland*
Otto Hänninen, *National Institute for Health and Welfare, Kuopio, Finland*

Background: Ambient air pollution is known to cause adverse health effects on humans. National disease burden estimates for ambient air pollution have earlier mainly focused on fine particles (PM2.5) and ozone (O3). WHO working group published new recommendations on concentration response functions for health impact evaluations in 2015. Those recommendations included besides updates for fine particles and ozone also concentration response functions for PM10 and NO2.
Objective: (i) To calculate updated disease burden estimates for ambient air pollution in Finland for PM2.5, PM10, NO2 and O3 and (ii) to analyse the main model and parametric uncertainties.
Methods: Based on established environmental burden of disease methods (Prüss Üstün et al., 2003) and using the World Health Organization's (WHO) Global Health Estimates 2012 as a background disease burden data we calculated estimates for deaths, years of life lost and years lived with disability attributable to air pollution. Air pollution exposure estimates covered the whole Finland for 2013 (Korhonen et al., 2015) and concentration-response (C-R) functions were adopted from WHO working group's recommendations (Heroux et al., 2015).
Results: The disease burden caused by the four selected air pollutants was 5,500 DALYs per million people in Finland (population 5.4 M) in 2013. PM2.5 had the largest share (70 %; 3,800 DALY) of the total disease burden caused by the selected air pollutants. For PM2.5 uncertainty given in confidence intervals related to C-R functions was 2,300 DALY (lower CI -1,100 DALY; upper CI +1,200 DALY). There are also other large uncertainties in the disease burden estimates, for instance, in relation to the shape of C-R curve, averaging of exposures and choice of endpoints.
Conclusions: The projects results show that PM2.5 has the biggest share of the disease burden caused by the selected air pollutants. Even though uncertainties are large, we can still conclude that healthwise PM2.5 is the most important air pollutant.
References:
Héroux, M. E., Anderson, H. R., Atkinson, R., Brunekreef, B., Cohen, A., Forastiere, F., ... & Künzli, N. (2015). Quantifying the health impacts of ambient air pollutants: recommendations of a WHO/Europe project. International journal of public health, 60(5), 619-627.
Korhonen A, Asikainen A, Rumrich I & Hänninen O, 2015. Ilmansaasteiden altistustasot Suomessa. ISTE-report, National Institute for Health and Welfare.
Prüss-Üstün, A., Mathers C., Corvalán, C. & Woodwards, A. Introduction and methods: assessing the environmental burden of disease at national and local levels. Geneva, World Health Organization, 2003. (WHO Environmental Burden of Disease Series, No. 1).

Tu-SY-C3.3

External cost of air pollution in Nordic countries evaluation using latest evidence in EVA-model and development needs

Mikael Skou Andersen, *Aarhus University, Aarhus, Denmark*

The objective is to demonstrate how interdisciplinary collaboration between health sciences, atmospheric science and socio-economic expertise can provide policy-support in the context of air pollution control and climate mitigation policies.
Health impacts from air pollution and the related external cost have been estimated with the integrated EVA model system. The EVA system is based on the impact-pathway methodology, where site-specific emissions, via atmospheric transport and chemistry, provide the basis for a concentration distribution, which together with detailed population data, is used to estimate the population-level exposure. Using exposure-response functions from health sciences and economic valuations, the exposure impacts are analyzed with regard to human health impacts (mortality and morbidity) and the related external costs. As a basis for the EVA system lifetable methodology is applied to estimate the years of life lost by air pollution victims due to long-term chronic exposures. It is shown that using established mortality risk ratios the average loss of life expectancy is in the range of 10-12 years, whereby the 'harvesting' effect hypothesis regarding air pollution can be dismissed for chronic exposures. Sensitivity analysis is conducted to explore implications of this finding for the final valuation of external costs, as different methodologies are preferred for valuation of statistical lives in USA and EU respectively. The statistical life (VSL) approach as preferred in USA produces higher external costs than the lifeyear (VOLY) approach preferred in EU, but for our loss of life expectancy results differences are less dramatic than conventionally assumed. Nevertheless it is cautioned that for policy support purposes the VSL approach seems less credible than the VOLY approach, considering the age profile of air pollution victims.
Although Nordic countries with their relatively low population densities and high environmental standards generally experience modest external costs from air pollution, their populations in larger urban centers are relatively more exposed. Factoring in the external health costs in appraisals of energy technologies tend to even out the cost advantages of fossil fuels relative to renewables and deserve more attention in the context of climate mitigation policies.
The presentation is part of the NordicWelfAir project funded by NordForsk, the Research Program of the Nordic Council of Ministers.

Tu-SY-C3.4

Air pollution and fetal growth- a study on ultrasound measures of Swedish children

Ebba Malmqvist, *Lund University, Lund, Sweden*
Zeyan Liew, *UCLA, Los Angeles, United States*
karin Källén, *Lund University, Lund, Sweden*
Ralf Rittner, *Lund University, Lund, Sweden*
Lars Rylander, *Lund University, Lund, Sweden*
Beate Ritz, *UCLA, Los Angeles, United States*

Background: Air pollution has been suggested to affect fetal growth. Most studies have relied on birth weight from birth records. Recent studies have been using ultrasound measures of fetal growth to assess air pollution effects. All of these studies have, however, been conducted in smaller study populations. Here, we have the opportunity to further advance our knowledge regarding the influence of air pollution exposures on fetal growth in a population based cohort of more than 48 000 pregnancies in women from Southern Sweden. The large study population allows us to perform sensitivity analyses.

Methods: In this study we used outcome data from Swedish medical birth registers and an ultrasound database for around 55 000 pregnancies with two ultrasound measurements (one early in pregnancy and one late in pregnancy). We estimated exposures during different parts of pregnancies for all the pregnancies in the ultrasound database using geocoded residential adresses. The measures of air pollution exposure were obtained through dispersion modelling with input data from an emissions database (NOx) with high resolution (100m grids). We had knowledge from registers on potential confounders/interfering factors and effect modifiers (e.g.parity, sex, smoking, age and education of the mother).

Results: In this large cohort of Swedish born children, we consistently estimated negative effects for NOx on most measures of fetal growth late in pregnancy. As an example, we show that our modelled NOx-exposures exhibit an effect on birth weight reducing birth weight by approximately 10g per 10 µg/m3 increment of NOx.

One of the challenges of the regulators is to assess whether or not there is a threshold for pollution under which we do NOT expect to see any further effects and what that might be. This study provides some insight on air polluton effect in the lower end of exposure assessment.

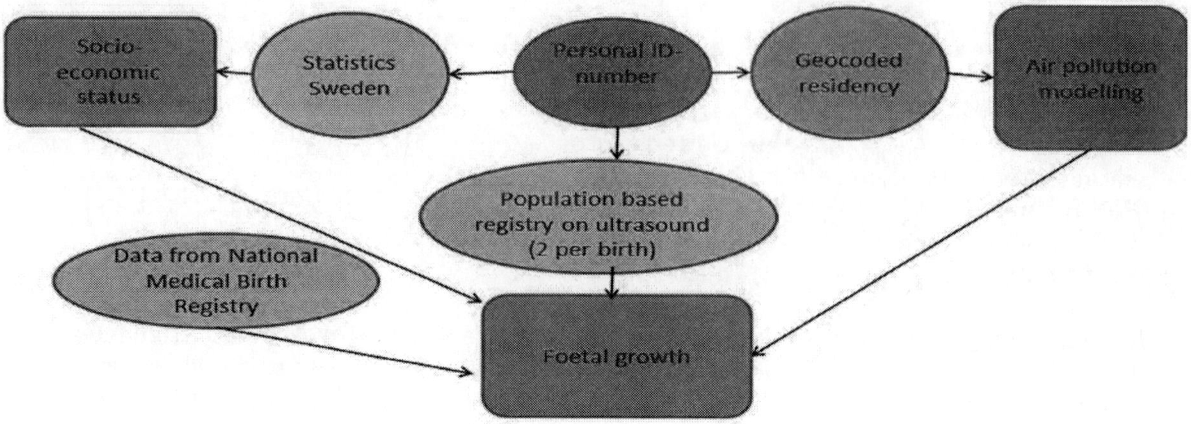
Socio-economic status
Statistics Sweden
Personal ID-number
Geocoded residency
Air pollution modelling
Population based registry on ultrasound (2 per birth)
Data from National Medical Birth Registry
Foetal growth

Tu-SY-C3.5

Implications of Nordic building stock on concentration-exposure and -respiratory uptake relationships

Otto Hänninen, *National Institute for Health and Welfare, Kuopio, Finland*

Background. Ambient particulate matter has been proposed to be the twelfth most harmful environmental exposure globally. Health impact assessment is based on outdoor concentrations, even though almost all populations in developed countries spend majority of their time indoors. Especially in some situations like in Nordic countries due to the cold climate, the buildings separate people indoors from outdoor pollution. Several studies have shown that ventilation system characteristics are associated with differences in C-R relationship. The aim of this paper is to discuss the possible interpretations of the existing evidence on the impact of building stock and building characteristics on the health effects of outdoor air pollution.
Methods. We conducted a literature search and discuss various possible hypotheses to explain heterogeneity in C-R relationships observed in epidemiological studies such as compositional differences, population structure, and behaviour and building stock. We developed a simple single compartment complete mixing mass-balance -based modelling environment to characterize the infiltration process in buildings of various tightness and ventilation. The building characteristics were partly based on existing building stocks and partly on energy efficiency driven policies for future buildings. In Europe all new buildings are expected to be nearly zero energy buildings (NZEB) by 2020 and the whole building stock in 2050.
Results. The Nordic building stock represents globally high fraction of mechanically ventilated buildings applying intake air filtration in practically all new buildings and a large fraction of the whole building stock. Moreover, the building tightness and energy performance is planned to be substantially increased further limiting the penetration of ambient particulate matter indoors. This seems to be the most significant regional difference on which we have hard data on. It will be interesting and relevant to compare these indoor exposure factors with compositional differences in PM as well as differences in population characteristics such as age structures and population ageing.
Changes of anthropogenic emissions for the current century projected in climate changes models suggest also substantial reductions. Together with the tightening buildings, mechanical ventilation with filtration, and reduction of combustion engines in urban transportation systems may lead to removal of the PM problem by the end of 21st century. Nevertheless, over the coming decades correct targeting of policies provides substantial opportunities for improved public health and avoided health care costs.

Tu-SY-D3: How can knowledge of toxicokinetics, mode of action and biomonitoring help you in human exposure risk assessment of chemicals?

Tu-SY-D3.1

How can knowledge of toxicokinetics, mode of action and biomonitoring help you in human exposure risk assessment of chemicals?

Manoj Aggarwal, *Dow AgroSciences, Abingdon, United Kingdom*

Risk assessment involves the integration of both exposure assessment and hazard assessment. In recent years, hazard assessment of chemicals has been greatly improved with the use of advanced technology and tools available in toxicology. The vision for 21st century toxicity testing is to move away from traditional animal toxicity testing and rely on in silico, in vitro and 'omics' tools for hazard characterization. Mode of action (MoA) and adverse outcome pathways (AOPs) have also been incorporated into hazard assessment, especially to better understand human relevance of hazards identified in toxicity studies. At the same time, exposure assessment has been migrating towards the use of internal (and absorbed dose) measures of exposure via biomonitoring. These advances in both hazard and exposure assessment is necessitating a greater need for understanding toxicokinetics of chemicals in animals and humans. Unique opportunities exist to incorporate toxicokinetic measurements in toxicity studies to give better understanding of 'internal dosimetry' and identification of biomarkers e.g. for biomonitoring studies. Despite all these developments, one of the key issues remains is to how to integrate all this knowledge into routine human exposure and risk assessments. The objective of this symposium is to discuss current knowledge on these advanced tools and technology, and present some case studies to demonstrate their use in integrated human risk assessments of chemicals that achieve the goals of 21st century toxicity assessment, while reducing uncertainties in human health risk assessments.

Tu-SY-D3.2

Use of animal toxicokinetic and human biomonitoring data in human risk assessments

Sean Hays, *Summit Toxicology, Lyons, CO, United States*
Lesa Aylward, *Summit Toxicology, Falls Church, Virginia, United States*

Biomonitoring is increasingly being used to assess human's exposures to chemicals. Uncertainties in exposure assessments can be significantly reduced by the use of biomonitoring data. However, translation of existing risk assessments into a dose metric consistent with those measured via biomonitoring data is required to interpret the biomonitoring data in the context of existing risk assessments. An understanding of the toxicokinetics (TK) of chemicals in humans and/or animals provides the ability to translate the existing risk assessments into Biomonitoring Equivalents (BEs), which can then be used to interpret human biomonitoring data in a risk assessment context. By having this knowledge of TK, the uncertainties in a risk assessment can be substantially reduced. This talk will highlight how the use of TK data can reduce the uncertainty in a risk assessment and the importance of collecting additional TK data in animal studies. Case studies are provided to highlight these issues.

Tu-SY-D3.3

Innovative Strategies for Agrochemical Safety Assessments: Use of Toxicokinetic Data for ArylexTM and RinskorTM

Sabitha Papineni, *Dow AgroSciences, Zionsville, Indiana, United States*
Manoj Aggarwal, *Dow AgroSciences, Abingdon, United Kingdom*

Integrating toxicokinetics (TK) into toxicity studies, without use of additional animals, provides valuable data on metabolism, systemic exposure and dose response for observed toxicity. Guidance documents, such as the OECD GD 116, highlight the importance of TK for dose level selection. Increasingly, TK data are being used to provide insights on MoA, study design, and in vitro to in vivo extrapolation in human health risk assessments. Described here are case studies, with the herbicides, ArylexTM (Halauxifen-methyl) and RinskorTM (XDE-848 Benzyl Ester), highlighting the use of integrated TK approaches in novel testing strategies to explore a MoA, or implement kinetically-derived maximum dose (KMD) strategies, respectively. In the MoA case study, Halauxifen-methyl is rapidly hydrolyzed in rodent liver to a single primary metabolite, Halauxifen-acid, and induces rodent liver effects via nuclear receptor (NR) activation. Halauxifen-acid does not activate NR and in vitro assays in rat and human blood, liver S9, and gastric fluid evaluated species differences in hydrolysis rates. TK and hydrolysis data were incorporated into a PBPK model for rat and human systemic exposure to Halauxifen-methyl, and supported non-human relevance of the Halauxifen-methyl liver MoA. In the KMD case study, XDE-848 Benzyl Ester is hydrolyzed to a single primary metabolite, XDE-848 acid, which displays non-linear kinetics based on integrated TK from 28- to 90-day toxicity studies. Interestingly, these studies indicated no toxicity up to the limit dose (1000 mg/kg/day) and the TK data provided strong justification for use of a KMD approach on the OECD 453 and 416 studies in the rat, and 1-year chronic toxicity study in the dog. A high dose of 300 mg/kg/day was chosen in contrast to traditional toxicity testing paradigms using a maximum tolerated dose approach where the limit dose is chosen as the high dose in the absence of toxicity. Taken together, integration of TK into guideline toxicity studies can increase understanding of the intrinsic properties of a molecule to develop human health risk assessments that utilize the best available science.

Tu-SY-D3.4

Kinetics in vitro versus in vivo in the context of quantitative in vitro in vivo extrapolation (QIVIVE)

Nynke Kramer, *Utrecht University, Utrecht, The Netherlands*

A thorough understanding of toxicokinetic processes goes a long way to explaining the variation in toxicity between chemicals, species and population groups. By using physiologically based biokinetic models (PBBK), quantitative in vitro-in vivo (dose) extrapolations (QIVIVE) take these processes into account, extrapolating an in vitro effect concentration to a human relevant toxic dose. Kinetic processes in in vitro cell-based assays has received less attention, but may also be essential for QIVIVE as the amount of a chemical reaching the cell in an in vitro assay may be different between cell assays and between in vitro and in vivo systems, even if the nominal or total concentration in exposure medium is the same.

Tu-SY-D3.5

Use of mode of action (MoA)/adverse outcome pathways (AOP) un human health risk assessments

Bette Meek, *McLaughlin Centre for Risk Science, Ottawa, ON, Canada*

A MOAs/AOP comprises a series of key events, each of which has its own dose-response relationship. Integration of AOP/MOA into risk assessment requires quantitative assessment of the likely progression of key events, enabling a more realistic estimate of risk and of population variability.

Tu-SY-E3: The Effects of Climate Change on Human Exposures to Air Pollution

Tu-SY-E3.1

Climate change impacts on human exposures to air pollution

Lisa Baxter, *United States Environmental Protection Agency, Research Triangle Park, North Carolina, United States*
Kathie Dionisio, *United States Environmental Protection Agency, Research Triangle Pare, North Carolina, United States*

Exposures to air pollutants both indoors and outdoors are influenced by a wide range of air quality, meteorological, behavioral, land-use, and housing-related factors. Many of these factors will be affected by climate change. Any changes in air quality and human exposures due to changes in climate will in turn impact human health. The objective of this presentation is to present an overview of approaches to characterize human's exposures to air pollutants and resulting health impacts in the context of climate change. This presentation will also describe the overall organization of the symposium as well as the importance of this topic.
Simulations of future changes in air quality from climate models are essential in projecting human exposures and future health impacts due to climate change. It is therefore necessary to calibrate these model outputs and to better quantify their uncertainties. The meteorological data associated with these projections such as temperature, rainfall, and humidity are critical determinants to human exposures to air pollutants as they impact human activities (e.g. exercise), locations (e.g. outdoors versus indoors), and behaviors (e.g. opening of windows and use of air conditioning). In addition to altering ambient air quality, climate change may also alter indoor air quality. A changing climate could affect the indoor environment in a number of ways such as changing infiltration and ventilation patterns, leading to changes in indoor exposure to outdoor air pollutants. Finally, air pollution exists as a complex mixture with various pollutants having their own spatial and temporal patterns. Changes in meteorology can have different effects depending on the pollutant altering the air pollution mixture that the population is exposed to.
The relationships between climate change and air pollution exposures is becoming increasingly important to understand. These relationships are complex, highly variable and depend on local conditions. An improved understanding of the impacts of climate change on air pollution exposures will allow for the more accurate estimation of future health risks.

Tu-SY-E3.2

A changing climate: impacts on human exposures to O3 using an integrated modeling methodology

Kathie Dionisio, *U.S. EPA, Research Triangle Park, NC, United States*
Chris Nolte, *U.S. EPA, Research Triangle Park, NC, United States*
Tanya Spero, *U.S. EPA, Research Triangle Park, NC, United States*
Kristin Isaacs, *U.S. EPA, Research Triangle Park, NC, United States*

Predicting the impacts of changing climate on human exposure to air pollution requires future scenarios that account for changes in ambient pollutant concentrations, population sizes and distributions, and housing stocks. An integrated methodology to model changes in human exposures due to these impacts was developed by linking climate, air quality, land-use, and human exposure models. This methodology was then applied to characterize changes in predicted human exposures to O3 under multiple future scenarios. Regional climate projections for the U.S. were developed by downscaling global circulation model (GCM) scenarios for three of the Intergovernmental Panel on Climate Change's (IPCC's) Representative Concentration Pathways (RCPs) using the Weather Research and Forecasting (WRF) model. The regional climate results were in turn used to generate air quality (concentration) projections using the Community Multiscale Air Quality (CMAQ) model. For each of the climate change scenarios, future U.S. census-tract level population distributions from the Integrated Climate and Land Use Scenarios (ICLUS) model for four future scenarios based on the IPCC's Special Report on Emissions Scenarios (SRES) storylines were used. These climate, air quality, and population projections were used as inputs to EPA's Air Pollutants Exposure (APEX) model for 12 U.S. cities. Probability density functions show changes in the population distribution of 8 h maximum daily O3 exposure by age, gender, and city for each of the three future climate scenarios. Of the 12 cities analyzed, some cities see an increase in the number of exceedances (e.g., Los Angeles), while others see a decrease (e.g., Chicago). In contrast, results show that there is minimal change in exposure distributions across future population scenarios. Thus we expect the change in ambient air quality concentrations in future climate scenarios to have a greater impact on future exposure distributions than potential scenarios of population change.

Tu-SY-E3.3

Quantifying recent associations between meteorology and multipollutant day types to inform future air quality projections

John Pearce, *Medical University of South Carolina, Charleston, South Carolina, United States*
Howard Chang, *Emory University, Atlanta, Georgia, United States*
Stefanie Sarnat, *Emory University, Atlanta, Georgia, United States*
Jennifer Runkle, *NOAA's National Centers for Environmental Information, Asheville, North Carolina, United States*
James Mulholland, *Georgia Institute of Technology, Atlanta, Georgia, United States*
Ian Rumsey, *College of Charleston, Charleston, South Carolina, United States*
Lance Waller, *Emory University, Atlanta, Georgia, United States*

Background: Changes to the climate system will impact air quality and related health effects through changes in exposure patterns. Objective: Establish influence of meteorology on air pollution using recently observed pollutant levels and estimate potential changes in pollutant mixtures experienced due to expected changes in climate. Methods: We obtained four years (2011-2014) of daily average CO, NOy, SO2, O3, PM2.5 EC, PM2.5 OC, PM2.5 NO3, PM2.5 NH4, and SO4 for seven cities across the American Southeast from USEPA's NCore network and corresponding meteorological conditions for each NCore site from National Center for Environmental Prediction (NCEP) reanalysis data. Generalized additive models (GAMs) were constructed for each pollutant in order to establish present day associations between daily pollution and daily maximum temperature and daily precipitation rate. Future climate conditions were obtained for each NCcore site for the years 2030-2040 from downscaled Coupled Model Intercomparison Project Phase 5 (CMIP5) projections using the Community Climate System Model (CCSM4.1) and the forcing pathway of the Representative Concentration Pathway (RCP4.5) emissions trajectory. Fitted GAMs were then used to predict corresponding responses of daily pollution levels. Finally, self-organizing maps (SOMs) identify categories of days based on multipollutant conditions (i.e., multipollutant day types (MDTs)) and establish differences in present day and future day type frequencies. Results: We found that twelve MDT profiles well explain the nature of pollutant combinations presently experienced on days across our cities. MDTs conditions ranged from relatively clean days, high single pollutant days (e.g., SO2), to high combination days (e.g., CO, NOy, EC). Using our present day SOM to classify days under our future climate scenario revealed that the largest increases in MDTs frequencies occurred on days characterized by moderate-to-high O3 pollution (21% increase), days dominated by relatively moderate-to-high CO, NOy, and EC (10.5% increase), and days with elevated SO2 and OC. The largest decreases occurred for relatively clean days (10.7% decrease); we show that these days transition to a similar profile with higher O3 in the future. Conclusion: In reality, future air quality will depend on both emissions in the future as well as changes in meteorology; however, isolating the influence of meteorology enables us to determine which combinations of pollutants should be expected to experience penalties driven solely by future weather conditions. We find combining multipollutant day typing (SOM), GAMs, and future climate predictions provides a complementary suite of tools for investigating potential air quality changes driven solely by future meteorological conditions.

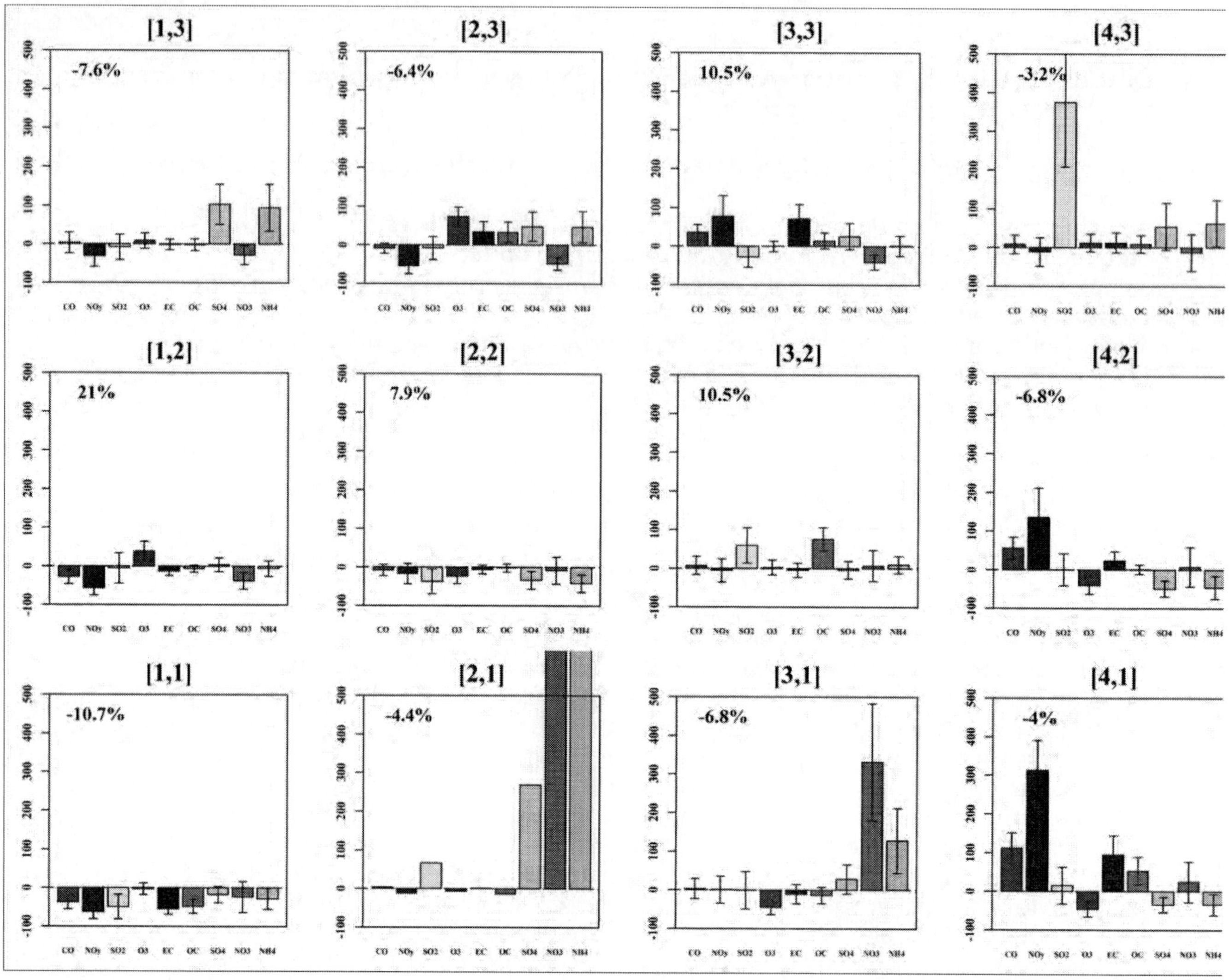

A 4x4 SOM illustrating twelve multipollutant day types (MDTs) identified from EPA NCore data obtained from 7 cities in the American South during the years 2011 to 2014. Bars represent overall mean-centered individual pollutant concentrations under each MD

Tu-SY-E3.4

Statistical Projections of Future Ozone Levels and Their Health Impacts in 5 US Cities

Howard Chang, *Emory University, Atlanta, GA, United States*
Xinyi Zhao, *Emory University, Atlanta, GA, United States*
Stefanie Sarnat, *Emory University, Atlanta, GA, United States*
Yang Liu, *Emory University, Atlanta, GA, United States*

Various meteorological conditions are drivers of ambient air quality. Hence, there is increasing interest in quantifying the impacts of climate change on future air pollution levels and their associated health effects. We describe a statistical modeling framework for projecting future ambient ozone levels. Previous studies have typically utilized outputs from numerical models for projecting future ozone levels; however, these models are computationally expensive and provide only deterministic projections. In contrast, a statistical approach, driven by meteorology and precursor levels, can flexibly incorporate various sources of uncertainties in the future projections, which may be useful to inform public health risk assessment. We first develop statistical models for predicting daily maximum 8-hour average ozone levels in Atlanta, Baltimore, Chicago, Houston, and Los Angeles based on observed daily levels of volatile organic compounds and nitrogen oxides. The models account for non-linear associations between precursor levels and meteorology, and achieve an average out-of-sample prediction R2 of 0.60. We then perform future ozone projections using bias-corrected climate model simulations of meteorology and changes in precursor levels. We describe a multivariate bias-correction method to account for the complex dependent structure in meteorological variables that are often not present in climate model outputs. Projections of health impacts, as measured by annual excess mortality and hospital admissions, are also conducted. Finally, we quantify the relative uncertainties in the health impact projections that are associated with heterogeneity in ozone health effects, ozone projection uncertainty, error in climate model bias-correction, and impacts of emission scenario on ozone precursor levels.

Tu-SY-E3.5

Impact of Ambient Temperature on Pollutant Infiltration and Exposure Processes: How Current Field Studies Inform Future Climate Change Effects

Donghai Liang, *Emory University, Atlanta, Georgia, United States*
Wan-Chen Lee, *Harvard University, Boston, Massachusetts, United States*
Joy Lawrence, *Harvard University, Boston, Massachusetts, United States*
Jeremy Sarnat, *Emory University, Atlanta, Georgia, United States*
Stefanie Sarnat, *Emory University, Atlanta, Georgia, United States*
Petros Koutrakis, *Harvard University, Boston, Massachusetts, United States*

Aim. Rising temperatures associated with climate change are expected to influence future air pollution exposures through changes in home air exchange rates, altering contributions of indoor and outdoor particle sources to indoor air quality. Using data from current field studies of homes in two US cities with different climatic conditions, we examine associations between indoor concentrations of particles of outdoor and indoor origin and ambient temperature to inform future air pollution exposure and health.
Methods. We assembled a large database of two retrospective cohorts (321 homes) in the Boston Area and a prospective cohort (840 homes) in Atlanta. Given that generally there is no indoor sulfur sources, indoor-outdoor sulfur ratios were used as a surrogate of total particle infiltration for PM2.5. We used linear mixed-effects models to examine the sulfur ratio-temperature relationship on both the whole population and a subset of naturally ventilated homes, using archived samples in Boston. Projected meteorological values, obtained from an ensemble of 15 Coupled Model Inter-comparison Project Phase 5 (CMIP5) models, were incorporated to predict sulfur ratio for 20 years in the future (2046-2065) and the past (1981-2000).
Results. The average sulfur ratio in the cohorts in Boston was 0.55 ± 0.19, with a 0.04 lower sulfur ratio in homes (N=43) without air conditioning (AC) compared to those (N=278) with AC. Temperature was the only meteorological factor found to significantly predict sulfur ratio ($p < 0.05$) in both population scenarios (whole population and naturally ventilated house only). A positive linear relationship was found between temperature and sulfur ratio for the whole population, with every Celsius degree increase in temperature associated with an increase of 0.006 in sulfur ratio. The predicted future summer-winter difference in sulfur ratio was as high as 54% for naturally ventilated homes and 30% for the whole population, using winter as the baseline. In contrast, the long-term difference was small with a maximum of 7% and 2 % increase in sulfur ratio in summer for the populations, respectively.
Conclusion. Substantial increment in sulfur ratio was found particularly in summer or the 20 years in the future. Ongoing analyses on the prospective cohort in Atlanta will be compared to the sulfur ratio-temperature relationship obtained from the cohorts in Boston. Together these analyses can help minimize exposure misclassification in future epidemiologic studies of PM2.5, as well as provide a better understanding of the potential influence of climate change on PM2.5 associated health effects.

Tu-SY-F3: Current opportunities and challenges in exposure surveillance to implement prevention strategies at the national and European scale

Tu-SY-F3.1

Occupational exposures to some chemical carcinogens by gender in France (from Matgéné program).

Corinne Pilorget, *French Institute for Public Health Surveillance / University of Lyon, Lyon, France*
Marie Houot, *French Institute for Public health surveillance, Saint Maurice, France*
Laurène Delabre, *French Institute for public health surveillance, Saint Maurice, France*
Stéphane Ducamp, *French Institute for public health surveillance / University of Bordeaux, Bordeaux, France*
Loïc Garras, *French Institut for public health surveillance, Saint Maurice, France*
Dephine Jezewski-Serra, *French Institute of Public Health Surveillance, Saint Maurice, France*
Jean-Baptiste Pelletan, *French Institute for public health surveillance / University of Lyon, Lyon, France*
Danièle Luce, *French Institute for Public health surveillance, Saint Maurice, France*
Mounia El Yamani, *French Insitute for public surveillance, Saint Maurice, France*

Matgéné program aims to create job-exposure matrices (JEMs) that give exposure indices for all the French workers by period. Linking JEMs, developed for carcinogenic, agents with occupational data allows a description of this occupational exposure to carcinogens by gender and activity at different dates for the current job or for all the occupational life in the French working population.
JEMs were developed for carcinogens classified in group1 or 2A by the International Agency for Research on Cancer. These JEMs were linked with 2 dataset: a representative sample of the French population in 2007 including complete occupational calendars and the 1999 French Census data. For each carcinogen, exposure prevalence in 1999 and in 2007 and lifetime exposure prevalence were estimated among men and women. These results are described by activity in order to identify the most exposed sector by gender. Carcinogens assessed by JEMs are solvents (trichloroethylene, perchloroethylene, methylene chloride, benzene), dusts (leather, silica) and asbestos fibers. The 1999 and 2007 exposure prevalence's are generally low. They are higher for men (from 0.07% to 7.4%) than for women (from 0.02% to 0.38%), except for leather dust and perchloroethylene more frequent among women. The lifetime exposure prevalence's are very high (i.e 5.6% of exposed men to silica in 2007 vs 15.3% of exposed men in their career) but remain lower among women (0.75% women exposed during their career to silica).
The highest exposed activity sectors are different by gender; the exposed men to trichloroethylene are working mostly in shop and automobile repair (31%) while exposed women are working mostly in rubber and plastic industry (29%).
These results highlight heterogeneous occupational exposure by gender due to different jobs occupied by men and women. They indicate that gender is an important element to define and prioritize prevention actions. Finally, they point out the interest to record all occupational jobs occupied during lifetime for men and women in order to facilitate past

exposure assessment that is essential for occupational cancers recognition or for estimating attributable risk fractions.

Tu-SY-F3.2

Component of the Esteban study 2014-2016 to describe exposures to environmental contaminants in the French general population

Clémence FILLOL, *French Insitute for Public Health Surveillance, Saint Maurice, France*
Alexis BALICCO, *French Institute for public health surveillance, Saint Maurice, France*
Laura BOSCHAT, *French institute for public health surveillance, Saint Maurice, France*
Amivi OLEKO, *French Institute for public health surveillance, Saint-Maurice, France*
Abdessatar SAOUDI, *Franch Insitute for public health surveillance, saint maurice, France*
Emmanuelle SZEGO, *French Institute for public health surveillance, saint maurice, France*
abdelkrim ZEGHNOUN, *French Insitute for Public Health surveillance, Saint Maurice, France*

BACKGROUND/OBJECTIVES
As part of the French biomonitoring program, a cross-sectional survey called Esteban, combining health exams and a nutrition study with biomarker dosages is ongoing. Its aims in the field of exposure biomonitoring are to describe and establish reference values for the levels of biomarkers of exposure to chemical agents present in the environment (incl. food); to analyze the determinants of exposure biomarkers levels and to compare with results from studies conducted abroad.
METHODS
Esteban is a cross-sectional study conducted in the continental French population (adults aged 18-74 and children aged 6-17), which took place during 2 years between April 2014 and March 2016, to take into account seasonality of exposure to environmental and food substances. The sampling design of the Esteban study has been defined as a stratified plan with three degrees (primary units, households and one member per household). Each participant answers a « questionnaire survey » (sociodemographic data, use of medical care, dietary habits, exposures to environmental pollutants), partially conducted face-to-face during 2 visits at home and partially self-administered and via telephone, and undergo a «biological and clinical exam» including collection of biological samples (blood, urine and hair, collected either at home during a visit by a nurse, or in an health center). Biological samples are prepared and sent to a biobank, for long term conservation at -80°C.
RESULTS
Inclusions have begun in April 2014, and full size implementation of data collection continues until the end of March 2016. As of March 4th, 2016, 2, 350 adults and 1, 058 children have been included in the study. An iterative consensus process (adapted from Delphi) among experts was used to obtain a prioritized list of biomarkers' families (Metals, Benzene, Perfluorochemicals, Pesticids, Polybrominated compounds, bisphenols, parabens...). More than 100 biomarkers were prioritized by this method.
CONCLUSION
This survey will offer a unique opportunity to assess the levels of impregnation of the French population by many chemicals. It will then allow comparison across time, and will

be complementary to other national initiatives (e.g. biomonitoring in children as part of the Elfe cohorte). The results of Esteban will be compared with surveys conducted abroad (in other European biomonitoring programs). Perspectives are to monitor time trends in biomarkers levels, when previous results are available and to monitor the impact of public health policies and regulations aiming to reduce environmental exposures to chemicals.

Tu-SY-F3.3

Using human biomonitoring to assess chemical exposures in French mothers of newborns, 2011: results obtained in the framework of the French biomonitoring program

Clémentine Dereumeaux, *French Institute for Public Health Surveillance, Saint-Maurice, France*
Abdesattar Saoudi, *French Institute for Public Health Surveillance, Saint-Maurice, France*
Marie Pecheux, *French Institute for Public Health Surveillance, Saint-Maurice, France*
Bénédicte Bérat, *French Institute for Public Health Surveillance, Saint-Maurice, France*
Vérène Wagner, *French Institute for Public Health Surveillance, Saint-Maurice, French Guiana*
Sarah Goria, *French Institute for Public Health Surveillance, Saint-Maurice, France*
Serge Brunel, *French Institute for Public Health Surveillance, Saint-Maurice, France*
Perrine De Crouy-Chanel, *French Institute for Public Health Surveillance, Saint-Maurice, France*
Clémence Fillol, *French Institute for Public Health Surveillance, Saint-Maurice, France*
Amivi Oleko, *French Institute for Public Health Surveillance, Saint-Maurice, France*
Corinne Delamaire, *French Institute for Public Health Surveillance, Saint-Maurice, France*
Alain Le Tertre*French Institute for Public Health Surveillance, Saint-Maurice, France*
Sébastien Denys, *French Institute for Public Health Surveillance, Saint-Maurice, France*
Agnès Lefranc, *French Institute for Public Health Surveillance, Saint-Maurice, France*
Laurence Guldner, *French Institute for Public Health Surveillance, Saint-Maurice, France*

Aim: As part of the French human biomonitoring program, the perinatal component provides a description of internal exposure to priority environmental contaminants among pregnant women having given birth in continental France in 2011. The study also aims: (i) to compare the biomarker levels with those observed in previous surveys conducted in France and abroad (e.g. in other European HBM programs); (ii) to identify and quantify some determinants of exposure.
Methods: Exposure biomarkers were measured in biological samples (urine, blood, cord blood and hair) collected from pregnant women randomly selected among the participants in the clinical and biological component of the Elfe cohort (n=4,145). Biomarkers analyzed in the perinatal component of the French HBM program were both well-known pollutants (e.g. toxic metals) and emerging substances (e.g. phthalates, bisphenol A, pesticides). For each biomarker, the geometric mean and percentiles of the levels distribution were estimated, taking into account the sampling design, in order to obtain estimates representative of the pregnant women having given birth in continental France in 2011. Multivariate analyses were conducted to search for determinants of impregnation levels.

Results: Comparisons with results previously obtained in the French population highlight a downward temporal trend in exposure to some chemicals (e.g. lead, mercury) that may be related to regulations taken to limit exposure to these chemicals. On the opposite, comparisons with results obtained abroad provide insight of some higher exposures in France than in other European countries or northern America (e.g. mercury, pyrethroids) that may partly be related to lifestyle characteristics and regulatory specificities. Methodological aspects that might impact biomonitoring results such as progressive improvement in measurements of emerging substances (bisphenol A, phthalates) have also to be considered in these comparisons.
Contributors to biomarker levels in pregnant women identified in this study are mainly consistent with those highlighted in previous studies. However interpretation of these findings warrants caution in relation with, on the one hand the short half-life of some biomarkers (e.g. bisphenol A, phthalates, pesticides), and on the other hand the fact that questionnaire-collected consumption behaviors (e.g. food, tobacco) may not reflect behaviors immediately preceding the spot biological collection used in the present study.
Conclusions: The results of the perinatal component of the French HBM program underline the necessity to study exposure to substitutes of some chemicals (e.g. bisphenol S, bisphenol F) and to develop multi-disciplinary approaches to assess exposure to chemicals.

Tu-SY-F3.4

Improved Risk Assessment through the integration of toxicokinetic modelling to connect external exposure to internal dose: the case of persistent chemicals

Camille Béchaux, *ANSES, Maisons-Alfort, France*
Laurent Bodin, *ANSES, Maisons-Alfort, France*
Stéphan Clémençon, *TelecomParisTech, Paris, France*
Amélie Crépet, *ANSES, Maisons-Alfort, France*

Biomonitoring data has many advantages for exposure assessment since it represents an integration of exposure from all sources and routes, takes into account the accumulation of the chemical in the body and represents direct measurements of the dose of the chemical substance that is really taken up from the environment (internal dose). Although the availability of human biomonitoring data can greatly improve the risk assessment especially for persistent chemicals, their interpretation often remains their main weakness because of the inability to identify the different sources of exposure, their relative contribution to the total exposure or to interpret it in terms of individual behaviors. Consequently, biomonitoring data needs to be linked with external exposure for interpretation through the integration of toxicokinetic modelling.
The objective of this work is thus to propose a method which makes it possible to combine all the information available to assess the exposure and to link external and internal exposure using relevant toxicokinetic models and integrating inter and intra individual variability when possible.
Two case studies were conducted: dioxins with dioxin-like PCBs and cadmium. Focusing on cadmium, concentration in the urine of the French population from French Nutrition and Health Survey (ENNS) were used. Dietary and smoking habits recorded in the ENNS study were combined with contamination levels in food and cigarettes to assess individual exposures. A physiologically based toxicokinetic (PBTK) model was used in a Bayesian population model to link this external exposure with the measured urinary concentrations. In this model, the level of the past exposure was corrected thanks to a scaling function which account for a trend in the French dietary exposure. It resulted in a modelling which was able to explain the current urinary concentrations measured in the French population through current and past exposure levels. Risk related to cadmium exposure in the general French population was then assessed from external and internal critical values corresponding to kidney effects.
This work illustrates how the combination of different data in a population modelling can improve risk assessment for persistent chemical. It does not aim to convince that all this data are needed to evaluate the risk but rather to investigate what can be learned when combining all the data.

Tu-SY-G3: Advancing human exposure metrics in Life Cycle Assessment (LCA) and Chemical Alternatives Assessment (CAA) - I

Tu-SY-G3.1

Framework for Multi-Pathway Cumulative Exposure for Comparative Assessments

Tom McKone, *University of California, Berkeley, California, United States*
Peter Fantke, *Technical University of Denmark, Lyngby, Denmark*

Efforts to assess human and ecosystem exposure to contaminants released to multiple environmental media have been evolving over the last decades. In this talk, we summarize the development and evolution of the multimedia mass-balance approach combined with multi-pathway exposure assessment as a framework for comparative assessment of chemicals, products, and services. We first review the development and evolution of the multimedia mass-balance approach to pollutant fate and exposure evaluation and illustrate some of the calculations used in multimedia, multi-pathway exposure assessments. The multimedia approach requires comprehensive assessments that locate all points of chemical release to the environment, characterize mass-balance relationships, and track contaminants through the entire environmental system to exposure of individuals or populations or specific ecosystems. For use in comparative risk assessment, life-cycle assessment (LCA), and chemical alternatives assessment (CAA), multimedia fate and exposure models synthesize information about partitioning, reaction, and intermedia-transport properties of chemicals in a representative (local to regional) or generic (continental to global) environment with information about larger scale populations rather than specific individuals or vulnerable subgroups. Although there can be large uncertainties in this approach, it provides insight on how chemical properties and use patterns map onto population-scale metrics of exposure, such as intake fraction for characterizing human intake per unit emission and aquatic or terrestrial ecosystem exposure concentrations per unit emission. We next discuss the reliability with which fate models at different levels of geographic scale--from near field indoor scales to urban, regional, continental and even global scale--can be used to determine cumulative human exposure and/or ecosystem exposure from multiple pollutants and emissions sources. The key question here is whether the results of cumulative assessments can provide sufficient insight for decision makers who are concerned with life-cycle impacts and chemical alternatives. We present a regional case study for pesticide alternatives in an agricultural valley of California to assess the opportunities and future prospects for the multi-pathway cumulative framework in LCA and CAA. This case reveals that the relative contributions to cumulative pollutant intake via different exposure pathways depend on (a) persistence of chemicals at different levels of integration (regional, urban-scale, food-web, indoors), (b) basic chemical properties, (c) the retention of chemicals in food webs, and (d) the retention of chemicals by indoor surfaces.

Tu-SY-G3.2

What are the elements for considering exposure in alternative assessments?

Ann Mason, *American Chemistry Council, Washington, DC, United States*
Brett Howard, *American Chemistry Council, Washington, DC, United States*

The alternative assessment community faces a variety of challenges when considering alternative chemistries or products. Specifically, assessors work to balance sometimes competing goals to 1) reduce or eliminate the risk of chemical ingredients; 2) reward chemicals and products that are designed using the principles of green engineering and chemistry; and 3) assure functional and economic performance. Overall, any alternative chemical or product must be measurably safer for the user and the environment, perform similarly or better than its predecessor, and have a cost comparable to the chemical or product it replaces. For some, the assessment includes the belief that one must move from 'older' chemicals and products toward newer ones and therefore the original should not be included during the assessment of alternatives. For others, inclusion of the original is a necessary 'proof point' to justify the resources when making the case to redesign a chemical or product or to move to a new alternative. Given the broad spectrum of functional roles for chemicals and products, there is currently no "best" solution that will work in all circumstances. Full alternative assessments as currently practiced involve a substantial amount of expert resources, data, and time so it is important that they result in clear and lasting benefits. The goal of this work is to suggest a set of elements for a screening-level evaluation that directly responds to the question posed by the assessor and that considers hazard and exposure within the use context of the chemical or product. This evaluation process sets out six elements to aid the alternative assessment practitioner during the initial comparative hazard/exposure evaluation step for a chemical or product in their use applications. These elements allow the practitioner to consider the results of the screening-level evaluation and decide whether a more complete alternative assessment is needed. Specifically, this proposed screening evaluation defines the criteria, considerations and approaches needed when using any hazard approach and any exposure model to provide a screening-level hazard/exposure (risk) evaluation.

Tu-SY-G3.3

What are the elements required to improve exposure estimates in life cycle assessments?

Alexi Ernstoff, *DTU, Lyngby, Denmark*
Ralph Rosenbaum, *Irstea, Montpelier, France*
Manuele Margni, *École Polytechnique de Montréal, Montreal, Canada*
Peter Fantke, *Technical University of Denmark, Lyngby, Denmark*

In this study we aim to identify and discuss priority elements required to improve exposure estimates in Life cycle assessment (LCA). LCA aims at guiding decision-support to minimize damages on resources, humans, and ecosystems which incur via providing society with products and services. Potential human toxicity and ecosystem toxicity of chemicals posed by different product life cycle stages are characterized in the life cycle impact assessment (LCIA) phase. Exposure and effect quantification as part of LCIA toxicity characterization faces numerous challenges related to inventory analysis (e.g. number and quantity of chemicals emitted), substance-specific modelling (e.g. organics, inorganics, nano-materials) in various environments and time horizons, human and ecosystem exposure quantification (e.g. exposed organisms and exposure pathways), and toxicity end-points (e.g. carcinogenicity). There are many relevant areas for improving exposure quantification in LCIA.
We explore prioritising future work based on investigating existing mitigation efforts, observed damages, and potential for (irreversible) harm to ensure LCIA covers at least the most relevant concerns faced by societies today regarding chemical exposure and harmful effects. Thereby, we structure this study of key elements identified as areas of elevated public, industrial, regulatory, and scientific concerns.
We found the majority of missing elements are directly related to the definition of exposed populations (both ecosystems and humans). For example, current LCIA human toxicity methods focus on exposure of the general population via chemical emissions. Occupational and consumer exposure to chemicals is of elevated concern for various stakeholders and leads to millions, if not billions, of dollars of damages yearly (e.g. through mesothelioma). Although consumer and occupational exposures often occur at magnitudes far greater than exposure mediated via environmental emissions, they are notably missing from current LCIA methods. As another example, recommended LCIA ecotoxicity methods focus on freshwater ecosystems. A significant amount of resources has been spent to mitigate damages on marine and terrestrial organisms such as fishes, bees, and birds. However, recommended methods are currently unavailable in LCIA to consider these organisms to evaluate the sensitivity of terrestrial and marine ecosystems. Microbes are another elevated concern due to the rise of antibiotic resistant organisms due to microbial exposure to disinfectants, antimicrobials and antibiotics etc. Yet, both the microbial exposure to chemicals and human exposure to microbes (and other disease vectors) are entirely missing from current LCIA exposure frameworks.
In all, defining exposure sub-populations and developing suitable methods can improve exposure methods in LCIA and capture major societal concerns.

Tu-SY-G3.4

The Current State of Alternatives Assessment in REACH - An Evaluation of 100+ Authorization Applications

Steffen Foss Hansen, *Technical University of Denmark, Kgs. Lyngby, Copenhagen, Denmark*
Rune Hjorth, *Technical University of Denmark, Kgs. Lyngby, Copenhagen, Denmark*
Molly Jacobs, *University of Massachusetts Lowell, Lowell, Massachussetts, United States*
Joel Tickner, *University of Massachusetts Lowell, Lowell, Massachussetts, United States*

Completion of an "Analysis of Alternatives" (AoA or AA) is a key component of manufacturers/importers request to obtain REACH authorization of their substances.
As of March 2016, 68 opinions on applications for REACH authorization had been adopted and 40 applications are undergoing consultations. In this analysis, we investigate how these AAs were completed in the regulatory applications and whether they addressed key elements of an AA based on the recent literature of current practices as well as guidance developed by the European Chemicals Agency (ECHA).
Based on the Guidance on Authorisation Applications developed by ECHA (2011), the work by Geiser et al. (2015; DOI: 10.1111/risa.12507) and Jacobs et al. (2016; http://dx.doi.org/10.1289/ehp.1409581), we identified a list of 25 basic AA elements that it seems relevant to investigate whether these have been included in the 108 applications for authorization. Elements reviewed included: what types of alternatives (e.g. chemical, material, product, or system alternatives) were considered; what analytical methods (e.g., hazard assessment, life-cycle assessment, exposure assessment) are used; and how different criteria e.g. hazard, technical feasibility and economic considerations were weighted in the AA.
Our analysis reveals that, while the applications compare alternatives based on technical feasibility, economic feasibility and risk, many of the applications do not include even the most basic elements of what one would normally considered a robust AA based on the emerging regulatory science policy literature on AA. A minority of applications provides an extensive identification of alternatives and although ECHA (2011) does provide guidance on what factors to consider in the assessment, a key problem seems to be that it is not always clear how to include environmental, safety and health considerations and technical feasibility when considering numerous alternatives. In general, most of the AA conclude that there is no alternative available and it could seem that there is a problem in the fact that the AA is performed by companies required to do them in order to justify continued use of the substance that they manufacturer. The ECHA SEAC committee needs to be more rigorous when it comes to assessing this aspect of the applications.

Tu-SY-G3.5

Identifying critical hazard and exposure information for Chemical Alternatives Assessment (CAA) decision-making

Cal Baier-Anderson, *US Environmental Protection Agency, Washington, District of Columbia, United States*

Background: Exposure considerations can add significant value to chemical alternatives assessment (CAA) and assessors can benefit from a targeted approach to collecting data to streamline the assessment and focus on relevant factors with available data. Objectives: This presentation will focus on the use of problem formulation to identify critical exposure information, defined as factors that may substantively change routes, patterns and/or levels of exposure, and discuss how exposure information can be used to inform or modify decision-making (e.g., compare exposure levels, identify differential routes of exposure, identify critical data gaps, etc.). Methods: The conceptual model for the use of chlorinated phosphate ester flame retardants (CPE FR) in consumer products illustrates relevant exposure pathways and receptors in the home. Comparing conceptual model exposure pathways and receptors with available hazard and exposure data or models shows which exposure pathways can be quantified and which pathways contain critical data gaps. Results: EPA/OPPT is evaluating three CPE FR chemicals [i.e., ethanol, 2-chloro-, phosphate (TCEP), 2-Propanol, 1-chloro-, 2,2',2''-phosphate (TCPP) and 2-Propanol, 1,3-dichloro-, phosphate (TDCPP)] for potential risks to human health and the environment. These chemicals exhibit low to moderately volatility and there may be multiple exposure pathways, including inhalation, ingestion of particles and dust, and dermal contact. None of the chemicals have sufficient toxicological data to evaluate each of the pathways identified, leaving data gaps. Knowing which exposure pathways are existing (or likely) can help to prioritize needed toxicological data. Substitute chemicals with different physical chemical characteristics (such as polymeric flame retardants) demonstrate how the relevant exposure pathways may change, leading to different priorities for toxicological data. Options for incorporating exposure data into the alternatives assessment are not always straight forward. Comparing relevant pathways, key hazards, receptors and data gaps can provide clues to how the exposure information can be used to inform decision-making and what trade-offs are likely. The views expressed in this presentation are those of the author and do not represent Agency policy or endorsement.

Tu-SY-G3.6

Panel discussion "Ideas and debate around current challenges and gaps in addressing exposure in LCA and CAA"

Peter Fantke, *Technical University of Denmark (DTU), Kgs. Lyngby, Denmark*
Kathie Dionisio, *U.S. Environmental Protection Agency, Research Triangle Park, United States*

This is a special panel discussion slot to efficiently receiving input from a wider audience in a lively, useful discussion of ideas and debate around current challenges and gaps in addressing exposure in LCA and CAA focused on the following aligned talks:
1) Framework for multi-pathway cumulative exposure for comparative assessments
2) What are the elements for considering exposure in alternative assessments?
3) What are the elements required to improve exposure in life cycle assessments?
4) The (questionable) use of CAA when it comes to granting authorization of Substances of Very High Concern (SVHCs) under the REACH regulation
5) Identifying critical exposure information for CAA decision-making

Tu-PL-H3: Pesticides

Tu-PL-H3.1

Influence of pesticide toxicokinetic parameters on the association between plasma and hair concentration

Caroline Chata, *LIH, Luxembourg, Luxembourg*
Nathalie Grova, *LIH, luxembourg, Luxembourg*
Emilie Hardy, *LIH, luxembourg, Luxembourg*
Brice Appenzeller, *LIH, Luxembourg, Luxembourg*

Aims: Mainly used for the detection of medical and illicit drugs, hair analysis is increasingly used for the assessment of human exposure to pollutants thanks to recent progresses in analytical techniques which allowed the detection of low levels of concentration. Although the relationship between chemicals intake and resulting concentration in hair remains incompletely elucidated, the transfer from blood to hair bulb is generally considered the main route of incorporation in hair.
Mechanisms of absorption, distribution and metabolization into blood should modulate the transfer of chemicals depending on their toxicokinetic parameters, resulting in different "blood to hair concentration" ratios.
The present work investigated the correlation between "blood and hair concentration" and "toxicokinetic parameters" of more than twenty pesticides from different chemical classes in animal models submitted to controlled exposure.
Methods: Two animal experiments were performed on the same strain (Lister Hooded). The first one was conducted to observe the relationship between hair and blood concentration. Hence, rats were administered pesticides by gavage over a 90 days-period, 3 times per week, at 7 different levels plus one control group. Each level of exposure consisted of n=8 animals. Animals' hair was collected at the end of the experiment by shaving. The second one provided toxicokinetic parameters of pesticides into blood. Rats were administered pesticides by gavage of a single dose and blood was sampled at different times using a catheter in the caudal vein. Toxicokinetic was established with 20 different time points with a 12-repetition for each in order to reduce individual variability.
After hair sample decontamination, pulverization and extraction, both parents and metabolites were analyzed by GC-MS/MS. Blood was immediately turned into plasma, and after extraction, the same compounds were analyzed also by GC-MS/MS.
Results: The data obtained for all the investigated compounds demonstrated significant association between plasma and hair concentrations (P value of 2.97E-45 and Rpearson of 0.875), with the exception of 3 outliers. For all the target compounds, toxicokinetic parameters (such as Cmax, tmax, Cmin, elimination half-life, area under the curve) were investigated in order to understand the influence of these parameters on outlier's specific behavior.
Conclusions: Our results support that the concentration of chemicals in hair depends on the respective concentration in plasma and suggest that for most pesticides, the transfer from blood to hair would not represent a limiting step in the incorporation. Results will however be analyzed more in detail in regard to compound toxicokinetic parameters.

This page intentionally left blank.

Tu-PL-H3.2

Matphyto: a French program for retrospective pesticide exposure assessment

Laura Chaperon, *French Institute for Public Health Surveillance, Lyon, France*
Céline Gentil, *French Institute for Public Health Surveillance, Fort de France, France*
Johan Spinosi, *French Institute for Public Health Surveillance, Lyon, France*
Mounia El Yamani, *French Institute for Public Health Surveillance, Paris, France*

OBJECTIVES:
Acute effects of pesticides are well known, but information on delayed effects (e.g. cancers) is lacking. Detailed knowledge of past occupational exposures to pesticides is required for epidemiologic studies or monitoring survey. Retrospective assessment of occupational pesticides exposure is complex. Because no reporting system for pesticides use existed in France, Matphyto program has been initiated to remedy this knowledge gap. This program consists in developing crop exposure matrices (CEMs) to pesticides, to reconstruct historical pesticides use patterns in France. Concurrently, Matphyto lists all active substances which are registered in France since 1961, for each crop and each use in agriculture, in a database named CIPA.
METHODS:
The Matphyto program aims at developing CEMs for each main crop since the 60's in both metropolitan and overseas France. For each crop, data of the pesticides used are collected and compiled. Different periods and geographical areas are characterized. Three exposure indicators for each active substance and/or chemical family are defined for each period and each geographical area: the probability, the frequency and the intensity. CEMs are validated by agronomists.
The CIPA database was developed from Acta plant protection products indexes. This yearly books presents data about all the active substances registered in France since 1961 and is published by Acta, the head of the agricultural technical institutes network. In order to ensure continuous monitoring of the registered uses of pesticides over time, an important work of interpretation, homogenization and data input has been made. The choices made were approved by experts from technical institutes.
RESULTS:
First, Matphyto focuses on the main French crops: straw cereals, potatoes, corn and wine-growing. Wine-growing CEM lists the use of pesticides on 9 vineyards with 50 herbicidal active substances, 108 insecticides and 94 fungicides from respectively 41, 25 and 35 chemical families. CEMs can be useful to the occupational medicine and to epidemiological studies to assess more precisely pesticides exposure.
CIPA provides data on pesticides as:

- 1053 individual forms (one form for each substance) with the characteristics and the agricultural uses of the substance,
- An Access® database which allows advanced queries about the active substances,
- Graphs from the main retrievals of the CIPA database.

CONCLUSION:
Matphyto program covers most of the crops in France. The data will be widely and freely usable to support assessment of occupational exposure to pesticides. The first CEMs are already available. The CIPA database is downloadable for free on the website http://index-matphyto.univ-lyon1.fr.

This page intentionally left blank.

Tu-PL-H3.4

Relevance of hair analysis for the biomonitoring of pesticide exposure - comparison with blood and urine in an animal model

Brice M.R. Appenzeller, *Luxembourg Institute of Health, Esch-sur-Alzette, Luxembourg*
Emilie Hardy, *Luxembourg Institute of Health, Esch-sur-Alzette, Luxembourg*
Nathalie Grova, *Luxembourg Institute of Health, Esch-sur-Alzette, Luxembourg*
Caroline Chata, *Luxembourg Institute of Health, Esch-sur-Alzette, Luxembourg*
Olivier Briand, *French Ministry of Agriculture, Agrifood, and Forestry, Paris, France*
Henri Schoeder, *Université de Lorraine, Vandoeuvre-lés-Nancy, France*
Radu C. Duca, *Luxembourg Institute of Health, Esch-sur-Alzette, Luxembourg*

Aim: Increasing interest is currently observed in hair analysis for the assessment of human exposure to organic pollutants. Nevertheless, only few studies have been published in this specific field so far, and several aspects still have to be investigated for optimum use of the information hair analysis can provide. In particular, relationships between individuals' exposure intensity and pollutants concentration in hair were only little investigated.
The present work, based on animal experimentation, was to investigate to what extent hair concentration was associated with level of exposure to a series of pesticides, and to compare results obtained from hair analysis to urine and plasma, which are more classically used in biomonitoring.
Methods: Rats (Lister Hooded, bicolor animals) were administered pesticides by gavage over a 90 days-period, 3 times per week, at 7 different levels plus one control group (no exposure). Each level of exposure consisted of n=8 animals. Pesticide mix consisted of twenty different compounds (organochlorines, organophosphates, pyrethroids, carbamates and other pesticides). Animals' hair, urine and plasma were collected at the end of the experiment. Black hair and white hair were analyzed separately in order to assess any possible effect of melanin on the incorporation of chemicals in hair. After samples decontamination, pulverization, and extraction, both parent compounds and metabolites were analyzed by GC-MS/MS.
Results: For most of the 27 target chemicals, the level of exposure was significantly associated with the concentration in the matrix, with coefficients of determination (R2) >0.9 for several compounds. In hair and in plasma, the number of chemicals detected ranged from 19 in controls to 25 in the group of highest exposure. In urine, the number of detected chemicals ranged from 8 to 18. The comparison between white and black hair demonstrated limited influence of pigmentation in the concentration of pesticides in hair. The possibility to reattribute animals to their correct group of exposure based on pesticide concentration in hair, plasma and urine demonstrated the superiority of hair, which provided the best score for 13 out of the 27 chemicals.
Conclusions: The significant association observed here between level of exposure and chemical concentration in hair supports the relevance of hair analysis for the assessment of human exposure to pollutants.

Tu-PL-H3.5

Detection of Glyphosate and its metbolite AMPA in the urine of 2- to 6-year children from the German state North Rhine-Westphalia

Susanne Rudzok, *North Rhine Westphalian State Agency for Nature, Environment and Consumer Protection, Recklinghausen, Germany*
Silvia Sievering, *North Rhine Westphalian State Agency for Nature, Environment and Consumer Protection, Recklinghausen, Germany*
Yvonni Chovolou, *North Rhine Westphalian State Agency for Nature, Environment and Consumer Protection, Recklinghausen, Germany*
Martin Kraft, *North Rhine Westphalian State Agency for Nature, Environment and Consumer Protection, Recklinghausen, Germany*

Aim: Human biomonitoring studies are effective tools to detect the load on the body with harmful substances. Glyphosate is a non-selective active herbicide ingredient contained in a number of pesticides used worldwide. Currently, various health evaluations of Glyphosate have been published. The question came up, if and to what extent 2- to 6-year old children are exposed to Glyphosate-residues.
Methods: Glyphosate and its metabolite AMPA were determined in spot urinary samples of 250 probands attending kindergarten in the German State North Rhine-Westphalia. The urine was collected between December 2014 and Mai 2015 as part of the epidemiological human biomonitoring study program in North Rhine-Westphalia. Additionally, urine collected over a time period of 24 hours was analyzed for 50 probands. Anthropometric data were collected by questionnaires.
Results: In 63 % of the samples Glyphosate concentrations were measured above the detection limit of 0.1 µg/L. The median Glyphosate concentration amounted to 0.14 µg/L and the 95th percentile was 0.97 µg/L. Based on the urinary concentration a daily intake was back calculated and compared to the Acceptable Daily Intake (ADI) of 0.5 mg per kg body weight per day derived newly by the European Food Safety Agency (EFSA). The estimated daily Glyphosate intake amounted to 0.004 % and the maximum to 0.11 % of this ADI. Thus, health related effects are presumably not expected for the measured urinary Glyphosate concentrations.
For AMPA, 58 % of the samples were above the limit of determination, with median concentration of 0.13 µg/L and a 95th percentile of 0.44 µg/L. The urinary concentrations of Glyphosate and AMPA did not display a strong correlation suggesting different exposure sources.
Furthermore, spot urinary samples were compared to corresponding 24 h urinary samples (n = 50). The results display that the first morning urine is an appropriate survey parameter to evaluate the daily intake of Glyphosate residues.
Conclusion: This study is the first epidemiological study determining the Glyphosate exposure to 2- to 6-year children worldwide. Although, the results showing no presumably expected health effects, it is undeniable, that there is a quantifiable Glyphosate body burden in the examined children.

Tu-PL-I3: Understanding Exposure Measurement Error

Tu-PL-I3.1

Assessment of Multi-Pollutant Indicators of Primary Traffic Pollution in a Near-Road Setting

Donghai Liang, *Emory University, Atlanta, Georgia, United States*
Jennifer Moutinho, *Georgia Institute of Technology, Atlanta, Georgia, United States*
Rachel Golan, *Ben-Gurion University of the Negev, Givat Brenner, Israel*
Howard Chang, *Emory University, Atlanta, Georgia, United States*
Roby Greenwald, *Georgia State University, Atlanta, Georgia, United States*
Rodney Weber, *Georgia Institute of Technology, Atlanta, Georgia, United States*
Stefanie Sarnat, *Emory University, Atlanta, Georgia, United States*
Armistead Russell, *Georgia Institute of Technology, Atlanta, Georgia, United States*
Vishal Verma, *University of Illinois at Urbana-Champaign, Urbana, Illinois, United States*
Jeremy Sarnat, *Emory University, Atlanta, Georgia, United States*

Background. Numerous approaches have been proposed to model multi-pollutant air pollution exposures and corresponding health response. For near road settings, these approaches are critical given the heterogeneity of the pollution mix. A better understanding of exposure measurement error surrounding these approaches is needed for the design and interpretation of the many observational studies linking traffic pollution and adverse health. We conducted the Dorm Room Inhalation to Vehicle Emissions (DRIVE) study to examine differential errors associated with using single- and multi-pollutant primary traffic indicators in health effect studies. Methods. We measured traditional single- and multi-pollutant traffic indicators at 8 monitoring sites (2 indoor and 6 outdoor) ranging from 0.01 to 2.3 km away from a major highway artery. An extensive suite of traffic pollutants was measured at each site, selected to characterize the heterogeneous composition of primary traffic emissions, including nitrogen oxides, carbon monoxide, particle mass and number concentration, elemental and organic carbon and particulate oxidative potential, expressed as water soluble dithiothreitol (ws-DTT) activity. We also conducted personal exposure sampling and metabolomics analysis for 54 students living in dormitories near (20 m) or far (1.4 km) from the highway. We quantified sources of spatiotemporal variability in the various traffic indicators using correlation and mixed effect modeling. Results. Spatial gradients varied substantially and differed by pollutant; median carbon monoxide, nitric oxide, and particle number concentrations were 109%, 100%, and 67% higher outside of the near-road dorm compared to the far-road dorm. Roadside versus near-dorm temporal correlations for all traffic indicators were strong (Spearman's r: 0.7-0.9); correlations were lower between the roadside and far dorm (r: 0.4-0.6), with substantial diurnal variability. In contrast to single-pollutant indicators of traffic, ws-DTT levels were more homogeneously distributed across the study domain, exhibiting a moderate, inverse gradient with respect to the highway source (median outdoor ws-DTT/volume levels at the highway roadside, near dorm, and far dorm sites = 87, 107, and 124 pmol/min/m3, respectively), trends similar to overall organic aerosol and PM2.5 mass concentrations. Conclusions. This study is among the first to comprehensively evaluate the use of single- and multi-pollutant indicators of primary traffic exposure. Early results suggest differences between single- and multi-pollutant indicators that have implications for their use and interpretation in traffic-related health effects studies.

UTRECHT
THE NETHERLANDS
OCTOBER 9-13
ISES 2016

This page intentionally left blank.

Tu-PL-I3.2

Evaluation of a Novel Approach to Refine Exposure Assessment in Epidemiological Studies of Chemicals with Short Half-Lives

Marc-André Verner, *Université de Montréal, Montreal, Quebec, Canada*
Fraser Gaspar, *SomaLogic, Denver, Colorado, United States*
Karl Forest-Bérard, *Université de Montréal, Montreal, Quebec, Canada*
Pierre Ayotte, *Université Laval, Quebec city, Quebec, Canada*
Jonathan Chevrier, *McGill University Faculty of Medicine, Montreal, Quebec, Canada*
Maryse Bouchard, *Université de Montréal, Montreal, Quebec, Canada*

Background: Epidemiological studies of prenatal exposure to environmental chemicals with short biological half-lives usually rely on urinary metabolite levels in spot samples from pregnant women to estimate exposure. However, studies have shown that within-person variability in levels measured in spot samples is often greater than between-person variability. Failure to adequately characterize between-person variability in exposure is a major obstacle to the evaluation of exposure-outcome associations. Because collecting repeated urine samples in large epidemiological studies is impractical, new tools are needed to circumvent this methodological challenge.
Aim: To evaluate whether predictive models based on spot sample measurements and sociodemographic variables can help refine the estimation of gestational exposure to organophosphate pesticides.
Methods: We recruited 50 pregnant women during their first prenatal care visit (gestational week 11), 43 of which provided 10 spot urine samples over 10 consecutive days. Six dialkylphosphate (DAP) metabolites of organophosphate pesticides were measured in the first spot urine sample and in a pool of the 10 collected samples. Linear regression and the deletion/substitution/addition (DSA) machine-learning algorithm were used to develop predictive models of the sum of DAPs (nmol/g creatinine) in the 10-sample pool (dependent variable) using DAP levels in the first spot sample and data gathered from a questionnaire (independent variables).
Results: All samples had detectable concentrations of DAP metabolites. In a univariate regression model, the sum of DAPs in the first urine sample alone explained 23% of the variability in the 10-sample pool sum of DAP levels. When we included reported determinants of DAP levels in the linear regression model (i.e., maternal age, pets, fruit and juice consumption), the model explained 38% of the variability in the 10-sample pool sum of DAP levels. On the other hand, the DSA algorithm, which performs 10-fold cross-validation to develop the predictive model, only selected the sum of dimethyl phosphate metabolites in the first urine sample to predict DAP levels in the 10-sample pool.
Conclusions: In the context of our study, we did not find questionnaire data to explain additional variance of pooled DAP concentrations in our cross-validated models. Studies evaluating other approaches to refine exposure assessment for chemicals with short biological half-lives are warranted.

Tu-PL-I3.3

3D variability of different particle metrics in urban areas: findings from the "supersito" project in bologna (Italy)

Stefano Zauli Sajani, *Arpae Emilia Romagna, Modena, Italy*
Dimitri Bacco, *Arpae Emilia Romagna, Bologna, Italy*
Stefano Marchesi, *Arpae Emilia Romagna, Modena, Italy*
Arianna Trentini, *Arpae Emilia Romagna, Bologna, Italy*
Isabella Ricciardelli, *Arpae Emilia Romagna, Bologna, Italy*
Vanes Poluzzi, *Arpae Emilia Romagna, Bologna, Italy*
Sabrina Rovelli, *University of Insubria, Como, Italy*
Gianluigi De Gennaro, *University of Bari, Bari, Italy*

Aim
The study aims at assessing the 3D spatial variability of various particle metrics within urban areas and the potential for misclassification in epidemiological studies.
Methods
Several monitoring campaigns were conducted in different sites and seasons during the period 2012-2015. Measurements included size distribution, mass and chemical composition of PM2.5. Chemical speciation were carried out in terms of a number of metals and ions, and organic and elemental carbon. The measurement activities were conducted in the city of Bologna (Italy), located in one of the most urbanized, industrialized and polluted areas of Europe. Gravimetric and optical monitors were used for PM2.5 mass. Size distributions were provided by spectrometers (FMPS-TSI). Monitoring sites included high and low traffic areas, front and back of buildings placed near busy streets, and high rise building with measurements at different floors. Some measurements were also performed in terms of NO2 and BTEX.
Results
A comparison between high and low traffic sites showed a very little variability of PM2.5 concentrations (< 15%). On the contrary very strong variability was found in the finer fractions of the size distribution and Ultrafine Particle Concentrations (3.5 times higher concentrations at the traffic site). Measurements at the front and rear of a building which fronted onto a major urban road showed differences very similar to those found comparing high and low traffic areas. Marked differences in relations to traffic proximity was found for some chemical species such as Elemental Carbon, Iron, Manganese and Tin. Little vertical variation was found for PM2.5 while no linear trend was found for UFP. Large vertical gradients resulted for NO2 and BTEX.
Conclusions
The results of the study provided the basis for a ranking of potential for exposure misclassification in epidemiological studies in relation to some different particle metrics.

Tu-PL-I3.4

Data Assimilation for Improved Exposure Modeling of Source Impacts on PM2.5 for Continental United States

Cesunica Ivey, *Georgia Institute of Technology, Atlanta, United States*
Yongtao Hu, *Georgia Institute of Technology, Atlanta, United States*
James Mulholland, *Georgia Institute of Technology, Atlanta, United States*
Armistead Russell, *Georgia Institute of Technology, Atlanta, United States*

Air quality models can provide temporally and spatially resolved fields for exposure analyses as they incorporate emissions and meteorological variations and can provide information on fields that are less routinely monitored. Further, such models can directly link air quality impacts to sources, including impacts on ozone and secondary PM2.5 species such as sulfate, nitrate, ammonium and organic carbon (SANOC). However, such models are potentially subject to significant biases and errors. Results from chemical transport models (e.g., CMAQ) are influenced by uncertainties in modeled secondary formation processes, such as chemical mechanisms, volatilization, and condensation rates. SANOC constitutes the majority of PM2.5 mass, and reducing bias in estimated concentrations has benefits for exposure assessments that use air quality data for health assessments. In this work, a method for assimilating speciated observation and modeled data is used to improve source impact estimates on PM2.5 and SANOC, which provides quantitative estimates of source contributions and reduces bias in modeled concentrations compared to observations. The approach uses sensitivities of both primary and secondary source impacts from CMAQ and observations of metals, elemental carbon, and SANOC; accounts for uncertainties in the optimization; and is applied both spatially and temporally. We apply the method over the continental United States to provide daily source impact fields for health analyses. The normalized mean bias for initial CMAQ-modeled concentrations compared to observations is -0.28 (OC), 0.11 (NO3), 0.05 (NH4), -0.08 (SO4), before applying the secondary bias correction. The bias is effectively zero after applying the method. Further, the correlation of PM2.5 observations and modeled concentrations improved after applying the secondary bias correction (observation and CMAQ: r = 0.49; observation and secondary-adjusted: r = 0.89). Multiple years of fields are being developed, and results for 2006 find that modeled source impacts and concentrations better reflect observations.

Tu-PL-I3.5

Toddlers' inhalation exposure to pyrethroids in homes

Jiaqi Zhou, *Rutgers, The State University of New Jersey, Piscataway, United States*
Clifford Weisel, *Environmental and Occupational Health Sciences Institute, Rutgers, The State University of New Jersey, Piscataway, United States*

Application of pyrethroid insecticides in residential settings results in children's exposures to these chemicals and possible adverse health effects. Household dust is a recognized reservoir for pyrethroids and a potential medium for multi-route pyrethroid exposure. Dust resuspension resulting from human's activities can increase human's inhalation exposure to pyrethroids. Young children (one to three years old) move and play in a manner that contribute to the resuspension of dust and since their breathing zone is closer to the floor, will have higher inhalation exposure to pesticide laden dust than other age groups. Their physical development stage may enhance their vulnerability to exposure to toxicants, such as pyrethroids. Directly measuring a toddler's exposure to household dust presents many logistic challenges, so few studies have reported exposure levels to resuspended dust for this age group. We simulated the dust resuspension by a toddler using a commercially available robot, which also serves as a platform to collect air samples at the breathing zone height. We tested 5 sets of experiments on carpet flooring. Our results showed that pyrethroids concentrations in floor dust, stationary air samples and mobile air samples were <1.0mg/g, 1.0-2.0mg/g, 1.0-4.0mg/g, respectively. Mean pyrethroids air concentrations in stationary and mobile samples were 0.037μg/m3 and 0.061μg/m3, respectively. We further characterized particle size distribution in stationary and mobile sampling zone. Mean particle concentrations in stationary and mobile samples were 27μg/m3 and 40μg/m3, respectively. Ratio of particle counts between mobile samples and stationary samples was 1.10—1.40 in different size ranges. Thus, the use of floor dust samples and stationary air samples may underestimate by approximately a factor of 5 and 2, respectively, a toddler's personal inhalation exposure to pyrethroids in residential houses.

Tu-SY-A4: The Exposome: From concept to practice - IV

Tu-SY-A4.1

Statistical Inference from Multipollutant Models in Exposome Studies

Lützen Portengen, *Institute for Risk Assessment Sciences (IRAS), Utrecht University, Utrecht, Netherlands*

Background: the exposome constitutes a promising framework for better understanding the effect of the totality of an individual's exposure history, in terms of both the compounds involved and the timing of the exposures. Although statistical methods are available that account for the multiplicity of hypotheses being tested, those based on simple univariate testing are unable to cope with the strong correlations that are likely to be present between different exposures or between different exposure windows for a single exposure, while multi-variable methods may suffer from low power due to the resulting high collinearity. More sophisticated dimension reduction and variable selection methods developed for machine learning tend to focus on the quality of the model predictions, rather than the implied model structure.
Methods: we will illustrate some of the deficiencies of existing methods, and highlight some recent developments that attempt to address these, using examples from simulation studies.
Discussion: statistical inference in the context of a full-fledged exposome study poses several challenges that have not been fully resolved. The dense correlation structure of the exposome results in low power for fully agnostic analyses of single exposure effects and interactions, which suggests that these may require large sample sizes. For smaller samples a more realistic goal may be to identify the joint effect of groups of (correlated) exposures, or to use stronger a-priori information on exposure effects.

Tu-SY-A4.2

Bioinformatics methods to enable exposome-based discovery

Chirag Patel, *Harvard Medical School, Boston, MA, United States*

The exposome concept, defined as the comprehensive battery of environmental exposures encountered from birth to death, promises to be a “big data” endeavor if realized. For example, the exposome is partitioned into the "external" and "internal" components, connoting ascertainments that occur inside and outside of the human. We claim new analytical tools, methods, and databases will need to be built to support and enable exposome-based discovery. Specifically, methods and databases will need to enable the deconvolution of the internal and external exposome and associate the longitudinal exposome with human and health and disease. In this talk, we will give a short review of the current day computer infrastructure and new tools to accelerate the association of high-throughput measures of environmental exposures with disease. Second, in the spirit of reproducible and collaborative science, we will exemplify new methods and tools with hands-on demonstrations of software packages and data for participants to try on their own.

Tu-SY-A4.3

Causal Assessment of the Exposome

Roel Vermeulen, *Utrecht University, Utrecht, Netherlands*

The Exposome has been identified as a unifying framework. The Exposome is a bold conceptual leap, seeking to bridge the role of the environment in human disease over multiple continua including: from populations to individuals, from external to internal environments, from discreet exposures to life course, and from single stressors to multiple determinants. Due to this inherent multi-dimensional complexity, exploitation of the "Exposome" is the ultimate Big-Data opportunity to improve public and patient health. In recent years, several studies have embraced the Exposome concept and more-and-more studies are generating multi-dimensional data to characterize the Exposome and associated health effects. However, the interpretation of these data is not straightforward due to the large number of variables and general lack of biological hypothesis. Unfortunately, a statistical and causal inference framework that fully accounts for the inherent complexities of a comprehensive exposome analysis is only just emerging. Until now researchers have chosen to simplify the problem by choosing easier parameters (i.e. bivariate associations), applying dimension reduction techniques, and multiple testing corrections to get a more restricted (tractable) number of bivariate associations. The fundamental problem with this approach is that it assumes that important causal relationships can be estimated via bivariate assumptions which due to its asymptotic assumptions may not (always) hold and that proposed solution may lack direct interpretability. As such there is a need for integrating casual assessment in the interpretation of exposome data which can be achieved through i) causal models (e.g. counterfactual and structural equation models) and through causal graphs; ii) the use of prior biological and empirical knowledge; and iii) replication/validation. In this presentation we will focus on a heuristic framework of statistical inference and Causal assessment of Exposome data.

Tu-PL-B4: Occupational Exposures

Tu-PL-B4.1

Differences in Fine Particulates and Estimated Pulmonary Ventilation Rate with Respect to Work Tasks of Wildland Firefighters: A Repeated Measures Study

Anna Adetona, *University of Georgia, Athens, Georgia, United States*
Olorunfemi Adetona, *The Ohio State University, Columbus, Ohio, United States*
Ryan Chartier, *RTI International, Research Triangle Park, North Carolina, United States*
Michael Paulsen, *University of Washington, Seattle, Washington, United States*
Christopher Simpson, *University of Washington, Seattle, Washington, United States*
Stephen Rathbun, *University of Georgia, Athens, Georgia, United States*
Luke Naeher, *University of Georgia, Athens, Georgia, United States*

Background: Wildland firefighters are exposed to a mixture of chemicals found in woodsmoke and emissions from non-woodsmoke sources such as gasoline or diesel. Improved exposure assessment approaches capable of distinguishing among the sources and estimating internal dose are needed for determining accurate exposure response relationships in epidemiological studies.
Objectives: We investigated compositional differences in fine particulate matter associated with various work tasks employed during prescribed burning. We used dual functioning personal aerosol monitors to collect personal PM2.5 (particulate matter with aerodynamic diameter of 2.5 microns and less) and accelerometry data on wildland firefighters. The PM2.5 samples are being analyzed for black carbon and levoglucosan.
Methods: Repeated measures on ten wildland firefighters employed by the United States Forest Service-Savannah River, SC and two individuals certified to work on prescribed burns were collected on burn and non-burn days during January-July 2015. Personal monitoring consisted of real-time and gravimetric PM2.5 (MicroPEM, RTI International, Research Triangle Park, NC) and carbon monoxide measurements (Dräger Pac III, DrägerSafety Inc., Pittsburgh, PA). The MicroPEM's accelerometer data were correlated with participant activity level to estimate average person-day ventilation rates. Estimated ventilation rates were then applied to calculate inhaled total PM2.5 dose. Primary analyses consisted of linear mixed-effects models.
Results: Least square means and corresponding 95% upper and lower confidence limits [95% CLs] showed gravimetric PM2.5 did not statistically differ between firefighters who managed fire boundaries (holders) and firefighters who conducted lighting (lighters) ($p=0.486$; $n=16$, 290 [162, 520] µg/m3; $n=30$, 250 [148, 424] µg/m3, respectively). Lighters had significantly higher ventilation rates compared to holders ($p<0.0001$; $n=30$, 30.7 [28.7, 32.8] L/min; $n=16$, 13.8 [12.7, 15.0] L/min, respectively). Additionally, lighters had marginally significantly higher inhaled total dose compared to holders ($p=0.0751$; $n=30$, 1310 [561, 3054] µg; $n=16$, 841 [344, 2054] µg, respectively).
Conclusion: Our study is the first to apply accelerometry data to estimate ventilation rates and inhaled particulate matter dose of wildland firefighters. We observed no difference in PM2.5 exposure concentrations by work task; however, after adjusting for ventilation rates, estimated inhaled total PM2.5 dose resulted in significant (marginally) differences. Even with some uncertainty, the ability to account for ventilation rate to estimate inhaled dose instead of the traditionally used air exposure concentration would result in more accurate assessment of the associations between occupational exposures and biological responses observed among wildland firefighters.

Tu-PL-B4.2

Occupational Heat Exposure in Washington State Roofers: A Comparison of Area, Personal, and Core Body Temperatures.

Miriam Calkins, *University of Washington, Seattle, Washington, United States*
June Spector, *University of Washington, Seattle, Washington, United States*

Aim: In Washington State (WA), the construction industry, and roofers in particular, experience high injury rates from falls and other mechanisms and also have the highest rate of heat-related illness (HRI) in the state. These outdoor workers are often working in conditions where they are not only exposed to solar radiation, but also additional sources of heat inherent to the roofing process. The optimal use of area, personal ambient, and internal temperature metrics for assessing the risk of heat health effects in this population has not to date been extensively studied. The objective of this study is to evaluate the association between occupational heat exposure, heat stress, and heat strain, in a sample of roofing workers.
Methods: Using a repeated measures design, this study measures personal heat exposure in a sample of roofers in the greater Seattle area during peak summertime activities. Days are identified as either exposed or unexposed to extreme heat using thresholds selected a priori from regionally appropriate historic trends in temperature. For this study, ambient temperature is measured both in terms of dry temperature was well as the apparent temperature, which accounts for humidity, solar radiation, and other meteorological conditions. Area temperature, personal temperature, core body temperature, heart rate, and activity level is collected continuously for each worker for an entire shift on both an exposed and unexposed day using QUESTempTM portable wet bulb globe temperature (WBGT) monitors, Thermochron iButtons®, CorTempTM ingestible temperature sensors, Polar® chest band monitors, and personal ActiGraph monitors, respectively. Additional risk factors, including job task and clothing, are collected through observations and survey questions.
Results: This study reports the relationship between area- and personal-level ambient temperature exposure, personal temperature and core body temperature, and personal temperature and physiological strain using the physiological strain index (PSI), taking into account other factors that influence heat strain, in a sample of roofers in the greater Seattle area.
Conclusions: This study provides insight into the most appropriate approach for quantifying occupational heat exposure, as related to the risk of heat strain, in a high-risk occupation where area level measures may not adequately characterize potential interpersonal variability resulting from different activities and multi-source exposures. These results are anticipated to provide valuable information for future heat exposure and health research and also for targeting HRI prevention efforts. Future analyses of these data will additionally evaluate the relationship between occupational heat exposure and heat-related injury risk.

Tu-PL-B4.3

Airborne dioxins, furans and polycyclic aromatic hydrocarbons exposure to military personnel in Iraq

Mauro Masiol, *Clarkson University, Potsdam, NY, United States*
Philip Hopke, *Clarkson University, Potsdam, NY, United States*
Timothy Mallon, *Uniformed Services University of the Health Sciences, Bethesda, MD, United States*
Mark Utell, *University of Rochester Medical Center, Rochester, NY, United States*

Objectives: The objective was to utilize the ambient concentrations of polycyclic aromatic hydrocarbon (PAH), polychlorinated dibenzo-p-dioxins (PCDD) and polychlorinated dibenzofurans (PCDF) measured at Joint Base Balard in Iraq in 2007 to identify the spatial patterns of exposure to these species and to identify and apportion the likely sources.
Methods: The ratios of the measured species were compared to literature data for open burning of simulated military wastes and other likely sources. Using the multiple site measurements on specific days, contour maps have been drawn using inverse distance weighting (IDW). Positive Matrix Factorization (PMF) was applied to the data to obtain the chemical profiles and contribution time series of the PAHs, PCDDs, and PCDFs. Conditional probability function (CPF) analyses were performed to assess the source directionality relative to the monitoring sites
Results: These analyses point to multiple sources of the PAH and PCDD/PCDF compounds including the burn pit (primarily a source of PCDD/PCDFs), the transportation field (primarily as source of PAHs) and other sources of PAHs that might include aircraft (fixed wing and helicopter), space heating, and diesel power generation. These three source types were identified and apportioned. The CPF plots were consistent with the assigned source types.
Conclusions: The ambient PAH and PCDD/PCDF concentration data provided insights into the identities and locations of their multiple sources. The PCDDs and PCDFs originated primarily from the burn pit and were highly focused in that area. Higher molecular weight PAHs were associated with vehicle emissions while the aircraft emissions were enriched in low molecular weight PAHs.

Tu-PL-B4.4

Fungi spores dimension matters in health effects: A methodology for more detail fungi exposure assessment

Carla Viegas, *ESTeSL-IPL; ENSP - UNL, Lisbon, Portugal*
Tiago Faria, *ESTeSL- IPL, Lisbon, Portugal*
Raquel Sabino, *ESTeSL -IPL; INSA, Lisbon, Portugal*
Susana Viegas, *ESTeSL-IPL; ENSP-UNL, Lisbon, Portugal*

Health effects resulting from dust inhalation in occupational environments may be more strongly associated with specific microbial components, such as fungi, than to the particles. The aim of the present study is to characterize the occupational exposure to the fungal burden in four different occupational settings (two feed industries, one poultry and one waste sorting industry), presenting results from two air sampling methods - the impinger collector and the use of filters. In addition, the equipment used for the filter sampling method allowed a more accurate characterization regarding the dimension of the collected fungal particles (less than 2.5 µm size). Air samples of 300L were collected using the impinger Coriolis µ air sampler. Simultaneously, the aerosol monitor (DustTrak II model 8532, TSI®) allowed assessing viable microbiological material below the 2.5 µm size. After sampling, filters were immersed in 300 mL of sterilized distilled water and agitated for 30 min at 100 rpm. 150 µl from the sterilized distilled water were subsequently spread onto malt extract agar (2%) with chloramphenicol (0.05 g/L). All plates were incubated at 27.5 °C during 5-7 days. With the impinger method, the fungal load ranged from 0 to 413 CFU.m-3 and with the filter method, ranged from 0 to 64 CFU.m-3. In one feed industry, Penicillium genus was the most frequently found genus (66.7%) using the impinger method and three more fungi species/genera/complex were found. The filter assay allowed the detection of only two species/genera/complex in the same industry. In the other feed industry, Cladosporium sp. was the most found (33.3%) with impinger method and three more species/genera/complex were also found. Through the filter assay four fungi species/genera/complex were found. In the assessed poultry, Rhyzopus sp. was the most frequently detected (61.2%) and more three species/genera/complex were isolated. Through the filter assay, only two fungal species/genera/complex were found. In the waste sorting industry Penicillium sp. was the most prevalent (73.6%) with the impinger method, being isolated two more different fungi species/genera/complex. Through the filter assay only Penicillium sp. was found. A more precise determination of occupational fungal exposure was ensured, since it was possible to obtain information regarding not only the characterization of fungal contamination (impinger method), but also the size of dust particles, and viable fungal particles, that can reach the worker´s respiratory tract (filters method). Both methods should be used in parallel to enrich discussion regarding potential health effects of occupational exposure to fungi.

Tu-PL-B4.6

Influence of Genetic Variance on Occupational Exposure Assessment for 1,6-Hexamethylene Diisocyanate

Kathie Y Sun, *University of North Carolina at Chapel Hill, Chapel Hill, North Carolina, United States*
John E French, *University of North Carolina at Chapel Hill, Chapel Hill, North Carolina, United States*
Jayne C Boyer, *University of North Carolina at Chapel Hill,, Chapel Hill, North Carolina, United States*
Leena A Nylander-French, *University of North Carolina at Chapel Hill, Chapel Hill, North Carolina, United States*

Significant differences in systemic response to xenobiotic exposure result from inter-individual genetic variation, but this variation is not included as a predictor of outcome in current exposure assessment models. We developed an approach to investigate and identify individual differences in genetic variation that influence biomarkers of exposure levels. Urine biomarker 1,6-hexamethylene diamine (HDA) was measured as a quantitative biological phenotype in a well-characterized population of automotive spray painters exposed to 1,6-hexamethylene diisocyanate (HDI). Exposure measurements were conducted over the course of an entire workday for up to three separate workdays that were spaced approximately one month apart. Inhalation exposure was measured using personal breathing-zone sampling while skin exposure was measured using tape-strip sampling. Urine samples were collected throughout the workday. Our innovative statistical modeling approach contains whole-genome markers along with other exposure predictors to determine the contribution of individual genetic variants to the observed urine biomarker levels among the exposed workers. The workers (n=33) were genotyped using genome-wide Affymetrix 6.0 microarrays, which feature ~1.8 million genetic markers. PLINK was used for the candidate-gene and genome-wide association analysis, and the regression model that captured the most significant SNPs included population substructure, current smoking status, HDI exposure in the worker's breathing zone, and HDI skin exposure level as covariates. We identified 26 significant genome-wide variants that were associated with the urine HDA levels. Associations were adjusted for multiple comparisons at a false discovery rate (FDR) $p<0.05$. We also performed a candidate gene analysis using 19 genes encompassing 296 genetic markers associated with HDI exposure and occupational asthma selected from published literature. No candidate genes were significant at a FDR<0.05. The genetic contribution to variation observed in urine biomarker levels was determined using linear mixed-effects models (LMM) that accounted for personal HDI exposure, individual exposure determinants, and whole-genome polymorphic markers. The biological relevance of significant variants was determined through predictive network analysis using GeneMania, Ingenuity Pathway Analysis, and Metacore. The most significant genetic marker, rs2697962 (p=9.0E-09), was also highly significant in the LMM ($p<0.0001$). This marker is located in the 3'-UTR region of PRDM2, which encodes a zinc finger protein that is a member of a nuclear histone/protein methyltransferase family. We found significant associations between urine HDA levels and genetic markers in this worker population. The results indicate that we can incorporate genetic markers with other exposure covariates in predictive exposure assessment models to better identify individual differences in biomarker levels.

Tu-SY-C4: Advanced methods for characterizing air pollution exposures at community scale

Tu-SY-C4.1

The evaluation of advanced human exposure models using personal exposure measurements

Sean Beevers, *King's College London, London, United Kingdom*
James Smith, *King's College London, London, United Kingdom*

Aim: Results from the London Human Exposure Model LHEM (Smith et al., 2016), which estimates human exposure in indoor and outdoor micro environments, shows exposure indoors, travel on the London Underground (LU) and within vehicles to be important contributors to daily PM2.5 and NO2 exposure. There is a need to evaluate the LHEM's predictions with personal exposure measurements, which has previously been difficult due to a lack of data. However, personal monitoring undertaken by KCL on the London Underground, and as part of the research council funded "Characterisation of COPD Exacerbations using Environmental Exposure Modelling COPE" study in London, provides opportunities to do so.
Methods:
London Underground: Exposure on the LU is currently modelled by the LHEM using a mean of 94 µg m-3, calculated from a small number of measurements on one line using a TSI Sidepak AM510, which requires calibration factors for this environment of iron rich particles. Consequently, KCL have undertaken high time resolution measurements to calculate calibration factors, to apply them to 34 hours of personal PM2.5 measurements and to model exposure across the entire LU network.
Indoor Environment: Exposure indoors in the LHEM is currently modelled using indoor/outdoor ratios applied to outdoor concentrations, without accounting for indoor sources. As part of the COPE study, up to 160 patients will provide PM and NO2 exposure data, which when paired with monitoring outside of their homes, will provide observed indoor/outdoor ratios.
Results: We have found large variations in PM2.5 concentrations on different LU lines including; 29 µg m-3 on the Hammersmith & City, 103 µg m-3 on the Metropolitan, 189 µg m-3 on the Bakerloo, and 326 µg m-3 on the Piccadilly lines. Within line variation was also important, with concentrations rapidly decreasing to ambient levels on outdoors sections of the network. The PM2.5 indoor/outdoor ratios assumed in the LHEM ranges between 0.45 and 0.63. Comparisons with the COPE data will establish how robust these assumptions are in modelling exposure indoors, as well as the magnitude of indoor sources of NOX and PM.
Conclusions: The LHEM model shows that personal exposure is poorly correlated with exposure outdoors at the residential address (Smith et al., 2016). Combining personal monitoring with indoor and transport micro environmental exposure models represents an important step forward for these methods.
References
Smith JD et al., 2016. Developing the London Hybrid Exposure Model (LHEM) for estimating human exposure to NO2 and PM2.5. Submitted ES&T

This page intentionally left blank.

Tu-SY-C4.2

Modeling Air Pollution Exposure Metrics for the Diabetes and Environment Panel Study (DEPS)

Michael Breen, *US Environmental Protection Agency, Research Triangle Park, North Carolina, United States*
Yadong Xu, *US Environmental Protection Agency, Research Triangle Park, United States*
Alexandra Schneider, *Helmholtz Zentrum München - German Research Center for Environmental Health (GmbH), Neuherberg, Germany*
Robert Devlin, *US Environmental Protection Agency, Chapel Hill, United States*

Air pollution health studies of fine particulate matter (PM) often use outdoor concentrations as exposure surrogates. To improve exposure assessments, we developed and evaluated an exposure model for individuals (EMI), which predicts five tiers of individual-level exposure metrics for ambient PM using outdoor concentrations, questionnaires, weather, and time-location information. We linked a mechanistic air exchange rate (AER) model to a mass-balance PM infiltration model to predict residential AER (Tier 1), infiltration factors (Finf, Tier 2), indoor concentrations (Cin, Tier 3), personal exposure factors (Fpex, Tier 4), and personal exposures (E, Tier 5) for ambient PM. In this study, we applied EMI to predict daily PM exposure metrics (Tiers 1-5) for the 21 participants in a cohort health study in central North Carolina called Diabetes and Environment Panel Study (DEPS). Using literature-reported parameters for the PM infiltration model, individual predictions were compared to 76 daily measurements of Fpex based on ratio of personal to home-outdoor sulfate concentrations from the 21 participants. Median difference between measured and modeled Fpex was 14% (25th and 75th percentiles of 7% and 34%, respectively). Using EMI, we predicted house-to-house and temporal variability of AER, Finf, and Cin (Tiers 1-3); and person-to-person variability of Fpex and E (Tiers 4-5). The capability of EMI could help reduce uncertainty of ambient PM exposure metrics used in health studies, such as DEPS, in support of improving health risk estimates.

Tu-SY-C4.3

Detailed near-port dispersion modeling for exposure assessments in Norfork, VA

Elizabeth Smith, *US EPA, Durham, NC, United States*
Timothy Barzyk, *US EPA, Durham, NC, United States*
Vlad Isakov, *US EPA, Durham, NC, United States*
Saravanan Arunachalam, *University of North Carolina, Chapel Hill, NC, United States*

This presentation will describe the use of reduced-form modeling for community-scale applications that explore exposures under a suite of future scenarios to inform actions to improve community resiliency. US Port communities are faced with a number of opportunities and challenges related to future port operations that complicate the assessment of trade-offs associated with decisions about 1) whether to expand to accommodate larger ships (e.g. post-Panamax), 2) how to anticipate impacts from potential changes in both use and distribution of alternative energy sources, and 3) how global trade might affect numbers and sizes of ships and other transportation infrastructure (rail, roadways) necessary to carry goods to and from other parts of the country. Decision-makers need flexible, simple-to-use tools to assess potential changes in exposures from a suite of air toxics associated with these multi-modal transportation sources, and to design mitigation strategies where exposures reach unacceptable levels. The C-PORT (Community air-dispersion model for Ports) model is designed to meet this need. A web-based tool that can be used for mapping and analyzing exposures to traffic-related air pollution in selected port communities, C-PORT requires no experience with GIS or model parameterization to view and modify dispersion of emissions related to port activities. The US EPA is working closely with several port communities across the country to provide such broadly-accessible tools that respond to community concerns so that management activities that improve resiliency and sustainability can be identified. Newport News, Virginia is an early case study that has several major seaport terminals nearby including the Newport News Marine Terminal (NNMT) and Norfolk International Terminals (NIT). Together, these Port of Virginia terminals represent the east coast's deepest harbors thus will likely see some of the biggest changes in shipping resulting from the expansion of the Panama Canal. These Virginia seaports also have the largest percentage of rail-based cargo transport in the US, and include transportation of the largest volume of coal in the country resulting in dispersion of significant amounts of coal dust. An illustrative example of how the C-PORT model can be applied in the Norfolk area to evaluate how exposures might change in the future will be presented.

Tu-SY-C4.4

Web-based models for exposure assessment on a community scale

Saravanan Arunachalam, *University of North Carolina at Chapel Hill, Chapel Hill, North Carolina, United States*
Brian Naess, *University of North Carolina at Chapel Hill, Chapel Hill, North Carolina, United States*
Catherine Seppanen, *University of North Carolina at Chapel Hill, Chapel Hill, North Carolina, United States*
Vlad Isakov, *U.S. Environmental Protection Agency, RTP, North Carolina, United States*
Timothy Barzyk, *U.S. Environmental Protection Agency, RTP, North Carolina, United States*

This presentation will describe the web-based, easy-to-use community tools (e.g. C-LINE and C-PORT) to study air pollution exposures due to various sources at a community scale. C-LINE is designed to study air pollution due to traffic-related sources while C-PORT is designed to study air pollution due to port-related sources such as ocean-going vessels, on-terminal stationary sources, rail and trucks. The power of such tools is to be able to make these assessments in a fairly rapid time, and to assess what-if scenarios on-demand. These scenarios are created by changing input parameters related to activity, emissions or even meteorological parameters and understanding changes in associated air quality and health risk at community scales. The targeted user-community is expected to be non-technical and less sophisticated with modeling expertise, and hence the web-based approach to keep things more intuitive and easy to use for planning purposes. The algorithms in these tools are reduced-form versions of other established models, and after extensive validation, are optimized for quick execution through the web-based interface. These tools have been developed and applied to areas within the U.S. to date. However, such web-based, easy-to-use modeling systems are of potential interest to global mega-cities that are seeing extensive growth in mobile source emissions and associated increases in exposures to traffic-related air pollutants in the near-source environment. Community groups around the world are becoming increasingly active in local initiatives that seek to mitigate potentially harmful environmental conditions. However, there is a lack of tools that can be applied to study near-source pollution in an easy manner, and explore the benefits of improvements to air quality and exposures - either due to voluntary or mandatory programs. There is interest in developing such community-scale modeling capabilities in places where emission inventories are not readily available (e.g. India, South America). This presentation will discuss the data / infrastructure needs for easy adaptation and implementation of the community tools for new cities, and will show an illustration for extending C-LINE to other cities of the world.

Tu-SY-D4: Human Biological Monitoring Following Chemical Incidents

Tu-SY-D4.1

The role of human biological monitoring in civil protection in Germany

Michael Mueller, *University Medical Center Goettingen, Goettingen, Niedersachsen, Germany*

The release of chemical, biological and radio-nuclear (CBRN-) agents creates special scenarios which can involve the exposure of disaster relief forces and the general public. Human biological monitoring (HBM) may be used in these incidents to determine the absorbed and biologically active internal dose of C-agents in the human body.
In a research project the compendium "Human biomonitoring in civil protection" was designed as a guideline for medical personnel to cover exposure of disaster relief forces and the general public after a CBRN incident. The compendium builds on HBM procedures, to be applied after exposure to chemical agents. HBM analysis methods were evaluated and basic toxicity data (including biological reference and threshold values) are given for 50 agents, previously identified as relevant in German civil protection. It also describes the sampling of human specimen to be analyzed for biological agents and radio-nuclear target isotopes, in a single sampling approach, thus limiting the burden for potentially exposed persons and facilitating comparison of their individual exposures to different CBRN agents (Müller et al., 2014).
In a second step, the compendium was implemented by a workshop of end users, training courses for medical personnel were designed and optimized by evaluation of the participant's feedback. In addition, an internet application was set up, including a periodically updated list of national and international HBM laboratories.
This development comprises a German approach for HBM in civil protection based on the obligate collection of human specimen and their subsequent analysis. Parallel developments in the Netherlands and Belgium use the concept of a transparent decision process for the application of HBM based on ambient monitoring data, simple dispersion modeling and toxicokinetic modeling (Scheepers et al., 2014). Both concepts have advantages and limitations. Nevertheless, these national approaches may serve as examples of a European concept to be developed for the application of HBM with ample and with limited resources of the respective countries.
References
Müller M, Schmiechen K, Heselmann D, Schmidt L, Göen T (2014) Human biological monitoring - A versatile tool in the aftermath of a CBRN incident. Toxicol Lett 231: 306 - 314.
Scheepers PT, van Brederode NE, Bos PM, Nijhuis NJ, van de Weerdt RH, van der Woude I, Eggens ML (2014) Human biological monitoring for exposure assessment in response to an incident involving hazardous materials. Toxicol Lett. 231: 295 - 305

Tu-SY-D4.2

Human biological monitoring following chemical incidents. Use of a guideline to support decisions to use or not use biomonitoring in The Netherlands.

Paul Scheepers, *Radboudumc, Nijmegen, Netherlands*
Gwendolyn Beckmann, *Radboudumc, Nijmegen, Netherlands*
Rob Anzion, *Radboudumc, Nijmegen, Netherlands*

Introduction - In the Netherlands public health advisors hazmat (GAGS) expressed their interest in a potentional role of human biological monitoring (HBM) in the response to chemical incidents. In 2012 a guideline was established to address this issue. End-users with a GAGS background as well as physicians, toxicologists and researchers with a background in exposure sciences and epidemiology contributed to the preparation of this guideline. Since its establishment HBM was considered in a number of incidents. This experience is used to evaluate the guideline.
Objective - To review the experience with a new guideline for decision-making regarding the use of HBM in the context of chemical incidents.
Methods - Requests received from professional users were collated into an overview and analyzed.
Results - Over the past three years HBM was considered in 12 chemical incidents. In five cases experts recommended not to use HBM to resolve the questions related to exposure and health effects. Of the cases that were followed up, three cases involved industrial exposures (fumigants, toxic metals and organic solvent) and in one incident first responders were involved (illicit drugs laboratory). In three remaining incidents residents were involved (mercury, polyurethane and metals) and two of these cases involved young children. The strength of this guideline for HBM related to chemical incidents is that it stimulated awareness for potential pitfalls and could use a framework to take decision in an early stage of the incident. In all of these cases this decision was made after consultation of a toxicologist in the region or from the national poison center. An inherent weakness of applying HBM in this setting is that HBM is often not perceived as a priority in an early stage of an incident and some of the requests for support were too late to be able to respond adequately.
Conclusion - HBM was employed in about half of the incidents and contributed to resolving questions concerning exposure. In most cases HBM was used in an assessment that could be used for reassurance of the involved individuals regarding exposure and the potential health consequences.

Tu-SY-D4.3

Assessing exposures to PCBs after an accidental fire in an electric transformer in France: (ir)relevance of human biomonitoring

Philippe Pirard, *InVS, Saint-Maurice, France*
Christine Ricoux, *InVS, Montpellier, France*
Marie-Laure Bidondo, *InVS, Saint-Maurice, France*
Dominique Mestre-Pujol, *ARS Languedic-Roussillon-Midi-Pyrénées, Carcassonne, France*
Franck Golliot, *InVS, Montpellier, France*

Introduction
In June 2013, a fire occurred in an electrical transformer in process of dismantling. The local health authorities considered to launch a program of prevention of contamination, environmental measurements, and health monitoring for the following reasons: there were doubts on pyralene remnants in the transformer and the fire could not be extinguished exposing to black smoke residents who refused evacuation.

Objective
The objective is to review the different steps of the procedure dealing with a power transformer fire: from exposure identification, to potential contamination calculation, selection of candidates for measurements and ways of differentiating acute exposure to the fire from chronic exposure to a contaminated soil.

Methods
A field investigation was initiated with geographical grid mapping and questionnaires for residents in order to assess the environmental and health impact. A step by step environmental contamination (PCB-L, PCB-NL, dioxins and furans) measuring plan was launched and a biological monitoring study (blood and urine samples, with specific blood sampling and conservation in case of necessity of dioxine measurements) was proposed to the symptomatic residents.

Results
Only the supposed most exposed quarter could be checked since the population ("Roma community") concerned was difficult to approach. Recommendations to prevention dietary contamination and tracking were scrupulously respected and no traces from the fire were observed upon visual inspection. Most of the interviewed residents declared headaches, and symptoms of irritation. Four residents were admitted to hospital due to a suspected of carbon-monoxide intoxication which revealed to be compatible with smoking habits. Measurements in soil highly exposed to fire-fighting operations and in the near field of the transformer did not reveal elevated concentrations (PCDD-TEQ = 2,7 ng/kg - PCB-TEQ 8.0 ng/kg). It turned out there was no demand for any biological monitoring.

Conclusion
Due to the high expense and technical difficulties to perform dioxine monitoring the local authorities stopped preparations for the monitoring process study.Targeting on PCB-congeners (including "lower chlorinated") for biomonitoring in combination with soil measurements could have highlighted more easily the exposure to the transformer and the acute exposure from the past ones. Nevertheless the acceptability of biomonitoring measurements by the exposed population remains an issue as well as the balance between advantages and disadvantages of such an action on a public health point of view.

Tu-SY-D4.4

Human biomonitoring for emergency responders - A valuable tool for exposure analysis and feedback on protection measures

Michael Bader, *BASF SE, Ludwigshafen, Germany*
Thomas Jaeger, *BASF SE, Ludwigshafen, Germany*
Sandra Baecker, *BASF SE, Ludwigshafen, Ghana*
Stefan Webendoerfer, *BASF SE, Ludwigshafen, Germany*
Gert Van Bortel, *BASF SE, Ludwigshafen, Germany*
Elke Verwerft, *BASF Antwerpen N.V., Antwerpen, Belgium*
Tom Van Weyenbergh, *BASF Antwerpen N.V., Antwerpen, Belgium*

Human Biomonitoring (HBM) is a well-established tool for the analysis and assessment of exposure to hazardous substances under regular working conditions. However, its potential for investigations after short-term exposure and chemical incidents is a relatively new and fast developing field of application. In particular, emergency responders like firefighters are potentially prone to contact with hazardous substances after chemical spills and accidents or work in high contaminated areas. However, data on individual exposure of firefighters to hazardous substances are still scarce. To address this issue, HBM has been carried since more than 15 years among professional firefighters at two major chemical production sites in Europe.
Post-exposure sampling is offered and carried out for altogether 38 different substances, including (poly-cyclic) aromatic hydrocarbons, isocyanates, amines, phenols, alkylating chemicals and metals. The analyses are carried out according to scientifically recommended procedures (such as the German DFG standard operating procedures) under quality-controlled conditions (German External Quality Assessment Scheme). Sampling material (urine beakers, blood tubes) as well as storage capacities (refrigerators, freezers) are available for larger sampling campaigns, if required. The professional firefighters are regularly informed about results and new developments, the purpose, the scope and practical aspects of the HBM programs. Detailed reports and individual feedback on the operations are either collected by a standardized questionnaire (Ludwigshafen) or by interview (Antwerpen).
In the course of the past 10 years, more than 1,000 samples from up to ten incident campaigns per year were analysed, the severity of the incidents ranging widely from small spills to larger product releases and fires. Most samples were analyzed for aromatic hydrocarbons such as benzene and toluene. In general, excursions of occupational limit values are rare (< 5 %) and often associated with post-campaign contaminations (skin, clothes). In example, excursions of the internal action value for the polycyclic aromatic hydrocarbons (PAH) marker 1-hydroxypyrene (1 µg/g crea.) were observed in 4 samples (3 - 45 µg/g crea.) after an operation with dermal exposure to contaminated surfaces being the most likely route of uptake.
The biomonitoring programs for emergency responders in Ludwigshafen and Antwerp address the most important and critical aspects of HBM after chemical incidents: preparedness and fast response, standard operating procedures, toxicological and medical assessment, and communication. A particular added value of these programs is the feedback on the efficacy of the protection equipment which is a practical support for the post-campaign evaluation of safety and health protection measures.

This page intentionally left blank.

Tu-SY-D4.5

Human biomonitoring as a tool of objective exposure assessment: A case-study of a major train accident with acrylonitrile in Belgium

An Van Nieuwenhuyse, *Scientific Institute of Public Health Belgium, Brussels, Belgium*
Tom De Smedt, *Scientific Institute of Public Health Belgium, Brussels, Belgium*
Koen Simons, *Scientific Institute of Public Health Belgium, Brussels, Belgium*
Christophe Stove, *Ghent University, Ghent, Belgium*
Peter De Paepe, *Ghent University Hospital, Ghent, Belgium*
Ben Nemery, *Katholieke Universiteit Leuven, Leuven, Belgium*
Michael Bader, *BASF SE, Ludwigshafen, Germany*
Christiane Vleminckx, *Scientific Institute of Public Health Belgium, Brussels, Belgium*
Ilse Van Overmeire, *Scientific Institute of Public Health Belgium, Brussels, Belgium*
Sébastien Fierens, *Scientific Institute of Public Health Belgium, Brussels, Belgium*
Birgit Mertens, *Scientific Institute of Public Health Belgium, Brussels, Belgium*
Koen De Cremer*Scientific Institute of Public Health Belgium, Brussels, Belgium*
Thomas Göen, *University of Erlangen-Nuremberg, Erlangen, Germany*
Thomas Schettgen, *RWTH Aachen University, Aachen, Germany*
Herman Van Oyen, *Scientific Institute of Public Health Belgium, Brussels, Belgium*
Joris Van Loco,*Scientific Institute of Public Health Belgium, Brussels, Belgium*

Background: Following a train derailment, several tons of acrylonitrile (ACN) exploded, inflamed and part of the ACN ended up in the sewage system of the village of Wetteren (Belgium).
Objectives: The objectives of the present study were: 1) To assess the human exposure to ACN in the populations with the highest suspected exposure, i.e. the local population and the emergency responders; 2) To investigate potential determinants of exposure to ACN; and 3) To explore the association between a biomarker of exposure and self-reported short-term health effects in the local population.
Methods: 242 residents and 841 emergency responders participated in the study. N-2-cyanoethylvaline (CEV), a highly specific biomarker for ACN exposure, was measured in blood. To account for potential influence of smoking, cotinine was determined in the urine. Participants also filled in a questionnaire including reporting of short-term health effects.
Results: In the evacuated zone, 37.3% of the non-smokers and 40.0% of the smokers had CEV concentrations above the reference values of 10 and 200 pmol/g globin, respectively, at the time of the train accident. Spatial mapping of the CEV concentrations depending on the residential address showed a distribution pattern following the sewage system. The most frequently reported symptoms were local symptoms of irritation. In the non-smokers, a dose-response relation was observed between the CEV concentrations and the reporting of short-term health effects. Overall, the value of self-reported symptoms to assess exposure was limited, with the exception of some local symptoms known to be prominent for the specific chemical exposure studied. Even then, consistent symptom reporting was observed only in case of exposures that resulted in CEV values exceeding 10 times the reference value. For the lower exposure ranges, there was no clear relationship between symptom reporting and exposure. In the emergency responders, 26% of the non-smokers exceeded the CEV reference value. ACN exposure among the non-smokers was predicted by (1) the distance to the accident, (2) the duration of exposure, and (3) the occupational function. In contrast with the local population, CEV concentrations in the emergency

responders remained relatively moderate and were comparable with background levels for a smoking population.

Conclusion: The present study is one of the first to relate accidental exposure to short-term health effects. The results of this study confirm that a critical view should be taken when considering self-reported health complaints and that ideally biomarkers are monitored to allow an objective assessment of exposure.

Tu-SY-E4: Air pollution exposure assessment getting personal: a European perspective

Tu-SY-E4.1

The use of low cost sensors to assess personal exposure to air pollution: results from the HEALS pilot study

Miranda Loh, *Institute of Occupational Medicine, Edinburgh, United Kingdom*
Susanne Steinle, *Institute of Occupational Medicine, Edinburgh, United Kingdom*
Anjoeka Pronk, *TNO, Utrecht, Netherlands*
Anne Sleeuwenhoek, *Institute of Occupational Medicine, Edinburgh, United Kingdom*
Eelco Kuijpers, *TNO, Utrecht, Netherlands*
John Cherrie, *Institute of Occupational Medicine, Edinburgh, United Kingdom*
Dimitris Chapizanis, *Aristotle University of Thessaloniki, Thessaloniki, Greece*
Thomas Maggos, *NCSR Demokritos, Athens, Greece*
Mina Stamatelopoulou, *NCSR Demokritos, Athens, Greece*
Denis Sarigiannis, *Aristotle University of Thessaloniki, Thessaloniki, Greece*

A pilot study for the HEALS (EU FP7) project is investigating the use of low-cost sensors for the assessment of personal exposure to particulate matter. For over 100 parents (with a child of <3 years) in 3 European countries (Greece, Netherlands, UK) personal time activity patterns have been captured for 5 days with the MOVES app (May 2015- May 2016). In parallel, indoor exposure to PM was assessed in the main living area of the home using the Dylos, a low-cost particle counter. Particle counts per minute were transformed into particle mass concentrations using a model developed by Semple et al. based on the relationship between Dylos small particle counts and the TSI Sidepak. Indoor mean (SD) PM2.5 levels measured by the Dylos were 8.9 (5.2) µg.m-3 with minute peaks reaching 791 µg.m-3. Outdoor levels around the same time period ranged from 3.4 - 13 µg.m-3 with a mean (SD) level of 6.4 (2.9) µg.m-3. Preliminary estimates of personal exposures for mothers in the Edinburgh study site were calculated by combining time-stamped location with air pollution data from monitoring sites and Dylos data. Further analyses will explore using higher spatial resolution models of particulate matter for assigning outdoor exposures, and estimates of in-transport PM levels, to improve the estimates of personal exposure.

Tu-SY-E4.2

Mobile-phone based air pollution exposure assessment

Audrey De Nazelle, *Imperial College London, London, United Kingdom*
Juan Pablo Orjuela Mendoza, *Imperial College, London, United Kingdom*
Markus Schläpfer, *Santa Fe Institute, Santa Fe, United States*
David Donaire, *Centre for Environmental Epidemiology, Barcelona, Spain*
Edmund Seto, *University of Washington, Seattle, United States*
Mark Nieuwenhuijsen, *Centre for Research in Environmental Epidemiology, Barcelona, Spain*
Michael Jerrett,, *UCLA, Los Angeles, United States*

Background
Methods integrating ubiquitous sensing technology have been proposed to address limitations of exposures assessments in epidemiologic and health impact assessment studies. While theoretically promising in terms of their ability to generate population-wide exposure assessments that account for activity patterns, in actual applications, smart-phone based exposure studies have so far incorporated only a handful of participants. Recent studies have scaled up these approaches with 100 or more subjects. This presentation will show results from one such study conducted as part of the Transportation Air Pollution and Physical Activities (TAPAS) project. It will also discuss progression towards game-changing use of big data from telecom companies for exposure assessment.

Methods
Over a period of one year in Barcelona, Spain, 172 commuters were fitted with a smart phone equipped with CalFit, a geo-tracking and physical activity assessment app, during one week. Activity patterns derived from CalFit were then overlaid with spatially-resolved land use regression (LUR) air pollution map, and exposures and inhalation of NO2 and PM were estimated also accounting for specific microenvironment concentration ratios. In another study, mobile phone activity data from 8 million users provided by Telecom Italia was used to estimate exposures in 7 Italian cities. In this case, activity patterns, based on presence of users every 15 minutes on a spatial grid varying from 0.1 to 30km2, were overlaid with a European-wide LUR model.

Results
Exposures in the CalFit-based study in Barcelona were shown to be 20% higher when activity patterns were accounted for (ie personal exposures) than when home based exposures were estimated. Travel activities accounted for 11% of exposures (concentrations), 20% of inhalation, but only 9% of time. The time weighted average concentrations contributions from home, work and other activities were 47%, 28% and 14%. The Italian big data showed NO2 home-based exposures were 5 to 50% higher than activity-based exposures.

Discussion
Big data provided by telecom companies on mobile phone usage have the advantage of enabling a representation of movements for a very broad population bases. However, the companies are not necessarily willing or able to provide individual-based data. Alternatively, data collected through smart phone apps such as CalFit enable detailed

information to assess personal exposure assessment (and potentially other health and socio-demographic factors). However this is still a burdensome and costly process, limiting its applicability to relatively small group sizes. Methods to bridge these two approaches need to be investigated.

Tu-SY-E4.3

Measuring personal exposure to ultra-fine particles in the EXPOsOMICS project

John Gulliver, *Imperial College, London, United Kingdom*
Gerard Hoek, *University of Utrecht, Utrecht, Netherlands*
Roel Vermeulen, *University of Utrecht, Utrecht, Netherlands*
Erik van Nunen, *University of Utrecht, Utrecht, Netherlands*
Kees Meliefste, *University of Utrecht, Utrecht, Netherlands*
Jelle Vlaanderen, *University of Utrecht, Utrecht, Netherlands*
David Donaire-Gonzalez, *CREAL, Barcelona, Spain*
Ariadna Curto, *CREAL, Barcelona, Spain*
Laia Font-Ribera, *CREAL, Barcelona, Spain*
Alex Ineichen, *University of Basel, Basel, Switzerland*
Mark Davey, *University of Basel, Basel, Switzerland*
Nicole Probst-Hensch*University of Basel, Basel, Switzerland*
Paolo Vineis, *Im, London, United Kingdom*
David Morley, *I, London, United Kingdom*
Debbie Jarvis, *Im, London, United Kingdom*
Andre Amaral, *Im, London, United Kingdom*
Alessio Naccarati, *HuGeF Foundation, Torino, Italy*
Laura Zini, *HuGeF Foundation, Torino, Italy*
Sabrina Bertinetti, *HuGeF Foundation, Torino, Italy*

Aim: To assess 24-hour personal exposures to ultra-fine particles (UFP) of individuals from five areas in Europe taking part in the EU-funded EXPOsOMICS (http://www.exposomicsproject.eu) project, and to provide these data to studies relating UFP exposures to a series of omics (e.g. metabolomics, adductomics) being undertaken on blood samples that were taken immediately after each UFP exposure measurement.
Methods. Adult and child personal exposure monitoring of UFP using the DiSCmini sensor (Matter Aerosol AG, Switzerland) was undertaken in five areas in Europe (Adults: Basel (Switzerland), Amsterdam and Utrecht (The Netherlands), Turin (Italy); and Norwich (UK); Children, Sabadell (Spain)) during 2014 and 2015. 24-hour measurements were repeated up to three times on each participant in different seasons. Raw 1-second data from the DiSCmini was cleaned using R scripts developed for the project. Using a completeness threshold of > 75% in the 1-second data, a total of 495 UFP personal exposure measurements from 162 adults and 42 children were retained to calculate 24-hour mean and median exposures. Average and median 24-hour particle number counts (PNC) (particles cm-3) were produced for each individual.
Results. For adults, mean UFP PNC were highest in Turin (20434; n = 108) followed by Amsterdam/Utrecht (15429; n = 124), Norwich (14303; n = 54) and Basel (11632; n = 127). For children (Sabadell), mean UFP were 18208 (n = 82). Values of median UFP PNC were lower than means due to exposures overall being positively skewed: Turin (16627), Amsterdam/Utrecht (11308), Norwich (11112), Basel (8981) and children in Sabadell (15823). Eleven of the 24-hour mean PNC exceeded 50000 with nine of these from Amsterdam/Utrecht and Turin.
Conclusion. Mean UFP exposures varied up to about two fold between study areas. The EXPOsOMICS project has a rich dataset on UFP exposures to determine the contribution of different microenvironments (e.g. home, journeys, work) to average/total exposures and undertake omic analysis.

Tu-SY-E4.4

Using Wearables To Quantify Personal Levels Of Physical Activity And Exposure To Traffic Related Air Pollution In Three European Cities

Evi Dons, *VITO, Mol, Belgium*
Michelle Laeremans, *VITO, Mol, Belgium*
Juan Pablo Orjuela, *Imperial College London, London, United Kingdom*
Ione Avila-Palencia, *ISGLOBAL, Barcelona, Spain*
Tom Cole-Hunter, *ISGlobal, Barcelona, Spain*
Mark Nieuwenhuijsen, *ISGlobal, Barcelona, Spain*
Audrey de Nazelle, *Imperial College London, London, United Kingdom*
Luc Int Panis, *VITO, Mol, Belgium*

Background / aims: Being active in traffic implies exposure to elevated air pollution concentrations which may come with a health cost. We designed a study to collect objective and quantitative data on the health effects of being physically active in a polluted environment as part of daily routines.
Methods: This study is part of the European project (EU FP7) Physical Activity through Sustainable Transport Approaches 'PASTA'. In it, 120 healthy adults (45% male, median age 33.75y (range 18-61), 40 participants/city) wore devices to track movement, air pollution and physiological health markers in real time (a.o. GPS, Sensewear, Zephyr BioHarness). Participants were monitored for 7 days continuously on three occasions (in three different seasons: cold, warm, intermediate). Measurements took place in parallel in Antwerp BE, Barcelona ES, and London GB, from February 2015 to March 2016.
Results: Weeklong average personal exposure to black carbon was highest in participants in Barcelona (1.7 ± 0.6 $\mu g/m^3$), followed by London (1.4 ± 0.5 $\mu g/m^3$) and Antwerp (1.3 ± 0.6 $\mu g/m^3$) respectively. Seasonal variation was observed with highest exposures in wintertime and lowest in summer. The intra-class correlation within participants for weeklong personal black carbon exposure is rather low (0.17) due to variation in background concentrations and in space-time activity patterns. Overall, participants were considered sufficiently physically active according to WHO recommendations, with only about 10% not meeting these recommendations. On average, we see that higher METs (metabolic equivalents of task) correspond to higher exposures to black carbon, driven mainly by elevated exposure during active compared to non-active travel. As our participants are a rather active subgroup of the population, we estimate ventilation and inhaled concentration to more accurately reflect internal exposure. Depending on the methods and sensors used, there can be a difference of up to a factor of 2 in the estimated inhaled concentration.
Later on in the project, personal monitoring data will be related to a number of noninvasive subclinical health markers measured in the same individuals: blood pressure, heart rate variability, retinal photography, lung function and lung inflammation.
Conclusions: By using mobile devices, the PASTA project is trying to further reduce exposure misclassification by moving away from an aggregated assessment of exposure and focus on individual risks. In the PASTA project new technologies and techniques to estimate exposure are being upscaled to move from a handful of participants to larger samples measuring for longer time periods.

Tu-SY-E4.5

Moving from short-term to long-term personal exposure monitoring - the COPE study

Benjamin Barratt, *King's College London, London, United Kingdom*
Lia Chatzidiakou, *University of Cambridge, Cambridge, United Kingdom*
Elizabeth Moore, *Imperial College London, London, United Kingdom*
Sean Beevers, *King's College London, London, United Kingdom*
Jennifer Quint, *Imperial College London, London, United Kingdom*
Rod Jones, *University of Cambridge, Cambridge, United Kingdom*
Frank Kelly, *King's College London, London, United Kingdom*

The long term deployment of personal environmental sensors to patients with respiratory disease provides the opportunity to investigate associations between symptoms and environmental conditions individually, without recourse to modelling or questionnaires.
Aim: The COPE study ("Characterisation of COPD Exacerbations using Environmental Exposure Modelling") aims to develop a method of predicting COPD exacerbations utilising personal air quality sensors, environmental exposure modelling and electronic health records. Here, we present initial results from early recruits to the study.
Methods: A portable sensor unit continuously recording PM and gaseous pollution, temperature, humidity, noise, activity level and GPS position was carried by COPD patients for up to six months. During this period patients kept records of symptoms relating to their condition (such as breathlessness, cough and wheeze) on diary cards and took daily exhaled breath flow tests. An activity algorithm was developed using multi-parameter ratios and decay rates, validated against a subset of healthy patients, who kept electronic activity diaries. When complete, the full cohort will comprise 160 COPD patients.
Results: Over 100 million personal exposure data points have been gathered. Associations are being explored between COPD symptoms and exacerbations, environmental stressors and activities such as travelling, cooking and exposure to tobacco smoke. Over a six month period central monitors in London were able to describe approximately 12% and 27% of the variation in personal exposure to daily mean PM2.5 and NOX concentrations respectively. Measurements are being used to validate a hybrid time-activity model, which will then be used to scale up results to population level.
Conclusions: Long-term deployment of personal samplers has the potential to identify individual susceptibility to environmental stress. The use of multi-sensor monitoring units can tag activities to time series data without recourse to an activity diary. Associations between risk of exacerbation and activity, rather than pollutant concentration, could produce more engaging behavioural advice to respiratory patients.

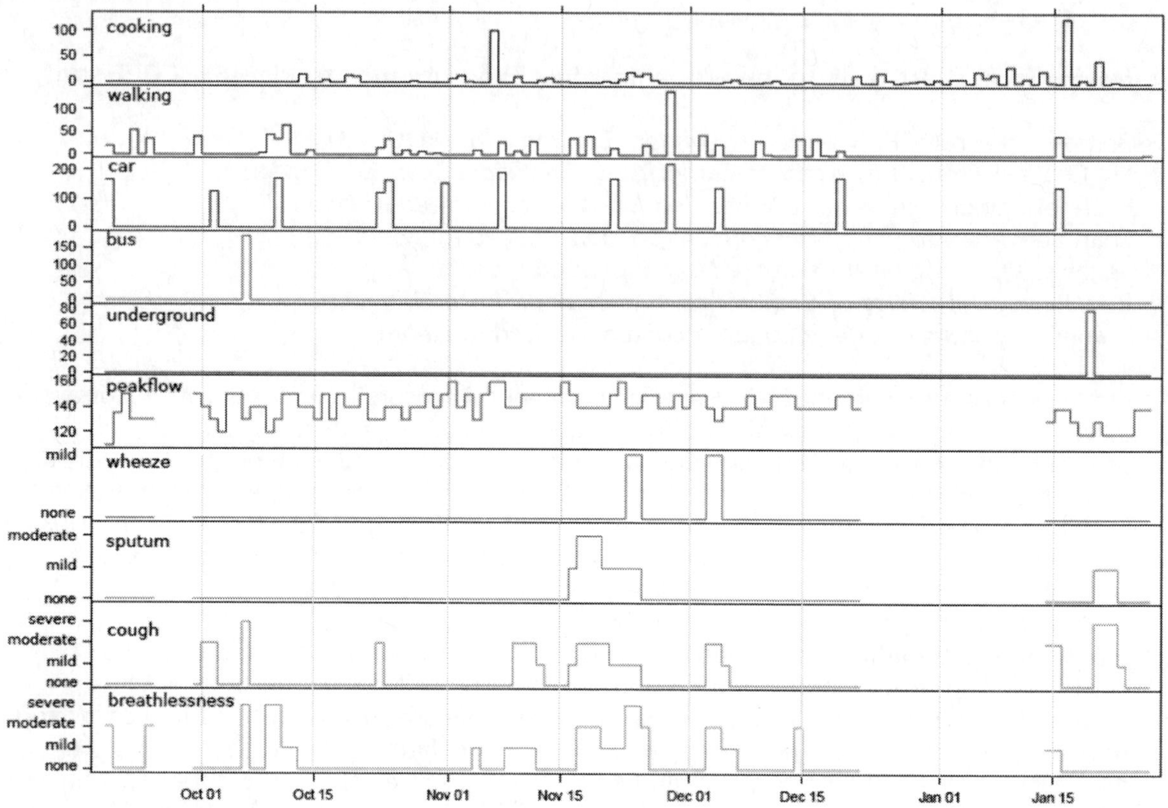

Daily time-activity-location patterns (minutes) matched with health symptoms for Subject 003

Tu-SY-E4.6

The CITI-SENSE Citizens' Observatory Toolbox: visualising and evaluating citizen-contributed environmental exposure information

Tom Cole-Hunter, *CREAL - Centre for Research in Environmental Epidemiology, Barcelona, Catalonia, Spain*
Leonardo Santiago, *Ateknea Systems, Barcelona, Spain*
Alexander Arpaci, *UBIMET GmbH, Vienna, Austria*
David Broday, *Technion, Haifa, Israel*
Núria Castell, *NILU - Norwegian Institute for Air Research, Kjeller, Norway*
Karen Galea, *IOM - Institute of Occupational Medicine, Edinburgh, United Kingdom*
Milena Jovasevic-Stojanovic, *University of Belgrade, Belgrade, Serbia*
Tania Martínez, *CREAL - Centre for Research in Environmental Epidemiology, Barcelona, Spain*
Johanna Robinson, *JSI - Jožef Stefan Institute, Ljubljana, Slovenia*
David Kocman, *JSI - Jožef Stefan Institute, Ljubljana, Slovenia*
Vlasta Svecova, *Institute of Experimental Medicine, Prague, Czech Republic*
Jonatan Moreno*Iritziak Batuz, Bilbao, Spain*
Hans Keune, *INBO - Research Institute for Nature and Forest, Brussels, Belgium*
Fintan Hurley, *IOM - Institute of Occupational Medicine, Edinburgh, United Kingdom*
Hai-Ying Liu, *NILU - Norwegian Institute for Air Research, Kjeller, Norway*
Alena Bartonova, *NILU - Norwegian Institute for Air Research, Kjeller, Norway*

Aim: CITI-SENSE, an EU-wide project involving nine cities (partly funded by the EU FP7-ENV-2012, grant agreement # 308524), is using participatory practices to develop the Citizens' Observatory Toolbox (COT) for collecting, collating and conveying a multitude of environmental exposure information. The CITI-SENSE COT should facilitate environmental health governance and contribute to future work with time-space-activity exposure classification and dynamic air quality maps. All tools within it are to be made available online upon completion of the project < http://co.citi-sense.eu/CitizensObservatoriesToolbox.aspx >.
Methods: Environmental micro-sensor nodes for measuring air quality (AQ) are being hosted by the public to monitor NO, NO_2, O_3, CO, PM2.5, PM10, and noise. Static nodes (AQMesh pods) and portable nodes were deployed with different stakeholders, including local authorities and medical outpatients, respectively. The portable nodes are coupled to the ExpoApp smartphone application (Android), which also collects geolocation and accelerometry (physical activity) data. Besides objective data, users of the CityAir smartphone application (Android, iOS) can identify perceived AQ hotspots and emission sources. Data from these tools is encrypted and then transmitted wirelessly and in near-real-time to a dedicated Spatial and Environmental Data Service (SEDS) and Web Feature Service (WFS) for processing, visualisation and evaluation by users.
Results/Conclusions: AQMesh pods with an updated algorithm are showing high correlations for NO_2 and PM2.5 when comparing 15-minute and daily averages obtained by reference instruments at municipal monitoring stations, respectively ($R^2 > 0.7$). The portable nodes, due to the challenges of micro-sensors and egomotion, are communicating real-time Air Pollution Indication (APIN) levels to users. The APIN is based upon the Common AQ Index (CAQI) but represents minute averages, showing potential to improve micro-environmental exposure classifications as well as personal AQ advice. Dynamic city-wide AQ maps have been produced using data assimilation techniques, fusing data from

the AQMesh pods with base maps obtained by land-use regression modeling. The CityAir application, so-far with around a thousand users and growing in popularity, is collecting and visualising (and sharing via social media) observations in the eight project cities, identifying emission hotspots particularly in Barcelona and Ostrava, but also with users in the United States, India and Iran. The CityAir responses are being compared (spatiotemporally) to AQ measured by the same individuals using portable AQ micro-sensor nodes, and also to stationary node network-modeled values for emission source characterisation. The products are being evaluated by focus groups of stakeholders, to assess the usefulness of the COT.

Tu-SY-F4: Advancing Exposure Science to Address Complex Environmental Issues

Tu-SY-F4.1

Advancing Exposure Science to Solve Complex Environmental Issues

Jennifer Orme-Zavaleta, *US Environmental Protection Agency, Research Triangle Park, North Carolina, United States*

Exposure science investigates the contact of humans or other organisms with chemical, physical, and biologic stressors, and their fate in living systems. Understanding exposure provides the real-world context for describing risk, along with information on the most effective ways to reduce exposure and improve health. Exposure science has become more important with the emergence of today's complex problems including climate change, security threats, population pressure, urbanization, depletion of natural resources, and increased understanding of environmentally related illness. This complexity, combined with advances in measurement and computational technologies, provides new opportunities for advancing and using exposure science to address today's wide range of health challenges.

The US established an Exposure Sciences in the 21st Century (ES21) Federal Partners Working Group that builds on a framework recommended by the National Academy of Sciences in its 2012 report on Exposure Science in the 21st Century: a Vision and a Strategy, and promotes Federal collaboration in the development of exposure science. The ES21 working group is focusing on advancing exposure science through catalyzing research including non-targeted chemical analysis, advanced GIS mapping, new biomonitoring technologies, advanced modeling, exposure analytics and informatics, integrating this information into the concepts of the exposome and advancing our understanding of cumulative risks in the real-world to rapidly and prospectively predict exposure.

The focus areas in this symposium have identified three (3) critical areas including: Advancing exposure science through technology: Focus on sensors; Advancing research on exposures to chemical stressors; Exposure Science Research Preparedness.

Tu-SY-F4.2

Advancing Exposure Science Through Technology: Focus on Sensors

Michele Penza, *ENEA; Italian National Agency for New Technologies, Energy and Sustainable Economic Development, Brindisi, Jamaica*
Tim Watkins, *United States Environmental Protection Agency, Research Triangle Park, Northe Carolina, United States*
Annette Guiseppi-Elie, *US Environmental Protection Agency, Research Triangle Park, North Carolina, United States*

There is broad international commitment to the development, evaluation, and application of novel sensor/dosimeter technologies to measure contaminants in multiple media (e.g., air, water, biological samples), as well as, related measures such as human activity or non-chemical stressors. The field of sensor technology is large and diverse. This presentation will highlight how international collaboration across agencies and sectors is advancing the field. Two examples are provided below.
In the US, Federal Agencies are combining resources for innovative technology development, including supporting sensor research and development, purchasing and using sensors, making sensor data and data products available to the public, and investigating how application of sensor technologies can help accomplish the agencies' goals. As an example, several agencies have worked on a coordinated Small Business Innovation Research (SBIR) solicitation for the development of low cost, easy to use reliable sensors. A coordinated website (Sensor Technology for the 21st Century) has been activated that will allow researchers to explore funding opportunities beyond the announcements that they might normally investigate. As part of a separate effort, EPA worked with NOAA on a series of airborne field studies (known as DISCOVER AQ) to evaluate the performance of several newly developed personal ozone monitors.
In Europe, a Concerted Action on New Sensing Technologies for Air-Pollution Control and Environmental Sustainability is a Network funded in the framework European Cooperation in the field of Scientific and Technical Research (COST). Specifically, COST Action TD 1105 EuNetAir is funded from 2012-2016. This international Network, coordinated by ENEA (Italy), includes over 120 big institutions from 31 COST Countries and 7 International Partners Countries (extra-Europe: Australia, Canada, China, Morocco, Russia, Ukraine, USA) to create a S&T critical mass in the environmental issues. The main objective of the Concerted Action is to develop new sensing technologies for Air Quality Control at integrated and multidisciplinary scale by coordinated research on nanomaterials, sensor-systems, air-quality modelling and standardised methods for supporting environmental sustainability with a special focus on Small and Medium Enterprises.

Tu-SY-F4.3

Advancing Research on Exposures to Chemical Stressors

Annette Guiseppi-Elie, *US Environmental Protection Agency, Research Triangle Park, North Carolina, United States*
Kathleen Plotzke, *Dow Corning Corporation, Midland, Michigan, United States*
James Franklin, *Environmental Chemistry Consultant, Grez-Doiceau, Belgium*
Jon Arnot, *ARC Arnot Research and Consulting, Toronto, ON, Canada*

Significant research is being under taken by regulatory agencies and industry in the area of chemical safety for sustainability. This presentation will provide research perspectives on exposure evaluation for chemical stressors of broad concern with an emphasis on advancing 21st century exposure science.
Understanding the role of exposure in assessing the actual risk for substances that are of global interest (e.g., widespread, persist in the environment and have potential for bioaccumulation) is critical in safety assessments of chemicals. However, while human and ecological health may be of broad concern, how these complex issues are addressed can be markedly different based on jurisdiction (both by agencies and countries).
As an example, persistent and bioaccummulative (PB) materials often fall into the category of chemicals of global concern but PB materials can be defined and treated differently under programs including, for example, the Toxic Substances Control Act under the United States (US) Environmental Protection Agency's New Chemical Program, the Canadian Environmental Protection Act, and the European Registration, Evaluation, Authorisation, and Restriction of Chemicals (REACH). In some circumstances, exposure may be taken into account and in others only the toxicity of the material. However, the presence or accumulation of a chemical in the environment or in an organism does not in itself represent an adverse biological effect. Meeting or exceeding the individual indicators of PB or toxicity (T) does not imply that there will be exposures leading to adverse outcomes in the environment. Such a conclusion must be based on additional, more refined assessments of the individual lines of evidence that defines what the actual exposure in the environment will be and if that exposure can lead to the potential for adverse effects. Further, there is a recognized concern related to substances that biomagnify, as these substances can accumulate in the food web in progressively higher and unpredictable concentrations and therefore threaten top predators and humans. Recent advances incorporating exposure properties allow for a more robust evaluation of these types of substances to assess if these substances will exhibit these properties in the environment and actually present a risk to organisms in the environment or to humans. The ES21 Federal Working Group has initially identified perflorinated materials (as a case example) for collaboration across agencies and other partners in advancing exposure science. The goal is to build mechanisms to address more generally constituents of global public concern with complex attendant issues.

Tu-SY-F4.4

Exposure Science Research Preparedness- The Disaster Research Response Program

Richard Kwok, *National Institute of Environmental Health Sciences, Research Triangle Park, North Carolina, United States*
Aubrey Miller, *National Institute of Environmental Health Sciences, Bethesda, Maryland, United States*
Joseph Hughes, *National Institute of Environmental Health Sciences, Research Triangle Park, North Carolina, United States*
Steve Ramsey, *Social and Scientific Systems, Inc., Durham, North Carolina, United States*

Background
After each new environmental emergency, responders, researchers and stakeholders race to determine known and unknown information on exposures, health impacts and to develop questionnaires to evaluate a situation. The tools created for studying exposures after each disaster are often difficult to find and delayed in deploying to the field by IRB and internal processes. Multiple barriers can prevent the collection of important exposure data that may inform policies and safety during response and recovery.
Objectives
To address the many barriers to rapid disaster research, presented by internal processes and external policies, the U.S. National Institutes of Health (NIH) has developed a new Disaster Research Response Program (DR2). The DR2 Program aims to develop a system of needed products, processes, and relationships to encourage rapid 'bench to trench' transdisciplinary research to better understand the human health impacts of environmental disasters for informing scientists and policymakers. Addressing issues of ethics, exposure science, tool development and integration with emergency response, DR2's goals will advance the field of exposure science and disaster research, while improving human health.
Methods
DR2 conducted a literature search to identify questionnaires used in past disasters, and collected meta-data on tool types, languages and exposures assessed. DR2 developed the Rapid Acquisition of Pre- and Post-Incident Disaster Data Study (RAPIDD) protocol to asses occupational and disaster exposures among response workers. DR2 has also engaged the NIH IRB to begin to address logistical and ethical barriers to rapidly approving a disaster protocol. DR2 has also begun work with disaster and health response agencies to explain the importance of exposure science and to begin integrating research into responses.
Results
DR2 has successfully removed barriers to rapid exposure research. Tools and resources are publically available for researchers and easy to find. The RAPIDD protocol can be customized for length of administration, and includes the ability to collect biomarkers of exposure. RAPIDD has received conditional approval from the NIH IRB, allowing a shorter turn around following an event. The NIH 'Best Practices Working Group for the Development of Special Considerations for IRB Review of Disaster and Emergency Related Public Health Research" has begun addressing institutional barriers to rapid IRB approval. Following multiple events in the US, DR2 has been called by other agencies for support. The program has bridged disciplines, both internal to research organizations and with external agencies, to improve exposure science responses to disasters and better inform response and recovery policy.

This page intentionally left blank.

Tu-SY-F4.5

Panel Discussion: Making Collaborations on Complex Environmental Issues Successful

Annette Guiseppi-Elie, *US Environmental Protection Agency, Research Triangle Park, South Carolina, United States*
Jennifer Orme-Zavaleta, *United States Environmental Protection Agency, Research Triangle Park, North Carolina, United States*
Michele Penza, *ENEA, Italian National Agency for New Technologies, Energy and Sustainable Economic Development, Brindisi, Israel*
Aubrey Miller, *National Institute of Health/National Institute of Environmental Health Sciences, Bethseda, Maryland, United States*
Kathleen Plotzke, *Dow Corning Corporation, Midland, Michigan, United States*
Tim Watkins, *United States Enviromental Protection Agency, Research Triangle Park, North Carolina, United States*
James Franklin, *Environmental Chemistry Consultant, Grez-Doiceau, Belgium*

The case example areas represented are some of a few of the complex environmental issues facing government, industry and ultimately the public. The presentations provide success stories on how to leverage resources and knowledge capital to advance solutions in innovative ways that may not be achieved without such collaboration.
The panel discussion with provide an opportunity for recommendations on such successful collaborations.

Tu-SY-G4: Advancing human exposure metrics in Life Cycle Assessment (LCA) and Chemical Alternatives Assessment (CAA) - II

Tu-SY-G4.1

Integrating Exposure into Chemical Alternatives Assessment Using a Qualitative Approach

Bill Greggs, *Soleil Consulting, LLC, Sanibel, FL, United States*
Scott Arnold, *The Dow Chemical Company, Midland, MI, United States*
Thomas Burns, *Novozymes North America, Inc., Raleigh, NC, United States*
Peter Egeghy, *U.S. Environmental Protection Agency, Durham, NC, United States*
Peter Fantke, *Technical University of Denmark, Kgs. Lyngby, Denmark*
Bonnie Gaborek, *DuPont, Newark, DE, United States*
Lauren Heine, *Northwest Green Chemistry, Juneau, AK, United States*
Olivier Jolliet, *University of Michigan, Ann Arbor, MI, United States*
Derek Muir, *Environment Canada, Burlington, ON, Canada*
Joseph Rinkevich, *SciVera LLC, Charlottesville, VA, United States*
Neha Sunger, *West Chester University, West Chester, PA, United States*
Jennifer Tanir*ILSI Health and Environmental Sciences Institute, Washington, DC, United States*
Margaret Whittaker, *ToxServices LLC, Washington, DC, United States*

Most alternatives assessments (AA) published to date are largely hazard-based rankings, and as such may not represent a fully informed consideration of the advantages and disadvantages of possible alternatives. With an assessment goal of identifying an alternative chemical that is more sustainable, other attributes beyond hazard are also important, including exposure, risk, life-cycle impacts, performance, cost, and social responsibility. Building on the 2014 recommendations by the U.S. National Academy of Sciences to improve AA decisions by including comparative exposure assessment, the HESI Sustainable Chemical Alternatives Technical Committee, which consists of scientists from academia, industry, government, and NGOs, has developed a qualitative comparative exposure approach. Conducting such a comparison can screen for alternatives that are expected to have a higher exposure potential, which could trigger a higher-tiered, more quantitative exposure assessment on the alternatives being considered.
This talk will demonstrate an approach for including chemical- and product-related exposure information in a qualitative AA comparison. Starting from existing hazard AAs, a series of four chemical-product application scenarios were examined to test the concept, to understand the effort required, and to determine the value of exposure data in AA decision-making. The group has developed a classification approach for ingredient and product parameters to support comparisons between alternatives as well as methodology to address data quality. The ingredient parameters include a range of physicochemical properties that can impact routes and magnitude of exposure, while the product parameters include aspects such as exposure pathways, use pattern, frequency/duration of use, chemical concentration in product, and use volume, accessibility, and disposal. Key learnings, challenges, and opportunities for further work will also be presented.
The views expressed in this presentation do not necessarily reflect the views or policies of the U.S. Environmental Protection Agency.

Tu-SY-G4.2

A flexible matrix-based human exposure assessment framework suitable for LCA and CAA

Olivier Jolliet, *University of Michigan, Ann Arbor, Michigan, United States*
Alexi Ernstoff, *Technical University of Denmark, Kgs. Lyngby, Denmark*
Lei Huang, *University of Michigan, Ann Arbor, Michigan, United States*
Susan Csiszar, *Oak Ridge Institute for Science and Education Research Participation Program at U.S. EPA, Oak Ridge, Tennessee, United States*
Peter Fantke, *Technical University of Denmark, Kgs. Lyngby, Denmark*

Humans can be exposed to chemicals via near-field exposure pathways (e.g. through consumer product use) and far-field exposure pathways (e.g. through environmental emissions along product life cycles). Pathways are often complex where chemicals can transfer directly from products to humans during use or exchange between near- and far-field compartments until sub-fractions reach humans via inhalation, ingestion or dermal uptake. Currently, however, multimedia exposure models mainly focus on far-field exposure pathways. Metrics and modeling approaches used in far-field, emission-based models are not applicable to all types of near-field chemical releases from consumer products, e.g. direct dermal application. A consistent near- and far-field framework is needed for life cycle assessment (LCA) and chemical alternative assessment (CAA) to inform mitigation of human exposure to harmful chemicals. To close the current research gaps, we (i) define a near- and far-field matrix-based exposure pathways framework that builds on a quantitative metric based on chemical mass in products, (ii) provide input data for the framework, e.g. chemical concentrations in products linked to functional use categories, and (iii) propose a consistent set of underlying models to populate the matrix-based framework for all relevant multimedia transfers and exposure pathways. Output is a flexible mass balance-based model structuring multimedia transfers in a matrix of first-order inter-compartmental transfer fractions. Inverting this matrix yields cumulative multimedia transfer fractions and exposure pathway-specific Product Intake Fractions defined as chemical mass taken in by humans per unit mass of chemical in a product. When the chemical mass in products is unavailable from individual studies and databases, it can be estimated from chemical-product function relationships or regulatory frame formulations. Combining Product Intake Fractions with chemical masses in products yields exposure estimates per unit mass compatible with LCA and CAA. We demonstrate how this matrix-based modeling system offers a consistent and efficient way to compare exposure pathways for different user groups (e.g. children and adults) and the general population exposed via the environment associated with product use. Our framework constitutes a user-friendly approach to test and interpret multiple human exposure scenarios in a coupled system of near- and far-field pathways and helps to understand the contribution of individual pathways to overall human exposure in various product application contexts. When combined with toxicity information this approach is a resourceful way to inform LCA and CAA and minimize human exposure to toxic chemicals in consumer products through both product use and environmental emissions.

Tu-SY-G4.3

Integrated approach for characterizing and comparing exposure-based impacts with life cycle impacts

Peter Fantke, *Technical University of Denmark (DTU), Kgs. Lyngby, Denmark*
Olivier Jolliet, *University of Michigan, Ann Arbor, Michigan, United States*

To address hazardous chemicals in consumer products, chemical alternatives assessment (CAA) is an emerging approach combining hazard and exposure assessment with technical and economic feasibility. Life cycle aspects are typically not consistently considered in CAA, but are relevant to avoid decisions that involve burden shifting or that result in only incremental improvement. Focusing in the life cycle impacts on widely accepted and applied impact categories like global warming potential or cumulative energy demand aggregating several impact categories will lead to underestimations of life cycle emissions of potentially harmful chemicals and their proposed replacements. Hence, an assessment framework is required that is able to account for near-field consumer exposure to chemicals in products during and after product use as well as population far-field exposure to chemical emissions to the environment from product-related processes along the product life cycle. We build on a flexible mass balance-based modeling system yielding cumulative multimedia transfer fractions and exposure pathway-specific Product Intake Fractions defined as chemical mass taken in by humans per unit mass of chemical in a product. When combined chemical masses in products and further with toxicity information, this approach is a resourceful way to inform CAA and minimize human exposure to toxic chemicals in consumer products through both product use and environmental emissions. We use an example of chemicals in consumer products to demonstrate how this matrix-based system offers a consistent and efficient way to compare exposure pathways for different user groups (e.g. children and adults) and the general population exposed via the environment. We further compare toxicity-related outcomes with outcomes from other life cycle impacts to compare the relevance of different impact categories for different consumer product classes. Through our examples, we will show (a) how to align assumptions used in different assessment methods in a manner that can avoid contradictory results, (b) to consistently consider and compare all relevant impacts, thereby avoiding burden shifting that could result from disregarding chemical and product life cycles, and (c) to prioritize the most relevant impacts across all life cycle stages, thereby setting the scene for a “life cycle alternatives assessment” (LCAA).

Tu-SY-G4.4

A modular Human Exposure Model (HEM) framework to characterize near-field chemical exposure in LCIA and CAA

Kathie Dionisio, *U.S. EPA, Research Triangle Park, NC, United States*
Daniel Vallero, *U.S. EPA, Research Triangle Park, NC, United States*
Peter Egeghy, *U.S. EPA, Research Triangle Park, NC, United States*
Kristin Isaacs, *U.S. EPA, Research Triangle Park, NC, United States*
Kent Thomas, *U.S. EPA, Research Triangle Park, NC, United States*
Paul Price, *U.S. EPA, Research Triangle Park, NC, United States*

Life Cycle Impact Analysis (LCIA) has proven to be a valuable tool for systematically comparing processes and products, and has been proposed for use in Chemical Alternatives Analysis (CAA). The exposure assessment portion of the human health impact scores of LCIA has historically focused on far-field sources (environmentally mediated exposures) while research has shown that use related exposures, (near-field exposures) typically dominate population exposure. Characterizing the human health impacts of chemicals in consumer products over the life cycle of these products requires an evaluation of both near-field as well far-field sources. Assessing the impacts of the near-field exposures requires bridging the scientific and technical gaps that currently prevent the harmonious use of the best available methods and tools from the fields of LCIA and human health exposure and risk assessment. The U.S. EPA's Chemical Safety and Sustainability LC-HEM project is developing the Human Exposure Model (HEM) to assess near-field exposures to chemicals that occur to various populations over the life cycle of a commercial product. The HEM will be a publically available, web-based, modular system which will allow for the evaluation of chemical/product impacts in a LCIA framework to support CAA. We present here an overview of the framework for the modular HEM system. The framework includes a data flow diagram of in-progress and future planned modules, the definition of each module including required inputs and outputs, and interactions between modules (as inputs/outputs, and via feedback loops). Planned modules for the HEM include: 1) Population generator module, simulating a population with associated family structures, age, race, and physiologic characteristics representative of the U.S. population; 2) Housing/residential characteristics module, assigning characteristics of the home such as housing type, air exchange rates, heating fuel type, etc.; 3) Product composition module, containing information on the chemical composition and associated functional use and weight fractions for products; 4) Occupational exposure module, characterizing exposures during occupational use of a product; 5) Agent-based model for simulating human behavior; 6) Dose intake module, to bring together the personal characteristics, human behaviors, housing characteristics, and product composition data to calculate population intake dose; and 7) Dosimetry module, to translate intake doses into blood concentrations.

Tu-SY-G4.5

Automated human and environmental exposure estimation to support prioritization of chemicals management actions

Joseph Rinkevich, *SciVera LLC, Charlottesville, Virginia, United States*
Patricia Beattie, *SciVera LLC, Charlottesville, Virginia, United States*
James Orchard-Hays, *SciVera LLC, Charlottesville, Virginia, United States*

New regulations and market pressures are driving companies to implement enhanced business processes that improve understanding of the human and environmental health characteristics of chemicals in products and processes. Chemicals considered for use in products and processes present varying hazard characteristics for human and/or environmental endpoints. Some chemicals may show high hazard for one endpoint while others may indicate high hazard for other endpoints. These varying characteristics offer opportunity for exploration of hazard-related trade offs between preferred alternative chemicals in specific applications. In addition to a review of chemical hazards, other attributes such as technical and functional performance, market availability, and cost factor into the viability of alternatives.
The practice of Alternatives Assessment (AA) can incorporate contextual review of chemical use and predicted exposure to facilitate prioritization of alternative chemicals for a specific application. Doing this work at scale, for dozens or hundreds of chemicals, requires automation for reasons of efficiency and economy. Screening-level exposure assessment and corresponding risk characterization, when automated via computer software, can assist in the rapid, consistent, and cost-effective processing of AAs where hazard characteristics vary across potential alternatives in a common application.
This talk will present specific examples using the cloud-based software known as SciVera Lens® to illustrate automated exposure estimation. The examples presented will show how automated screening-level exposure assessment and risk characterization can support the growing need throughout the consumer product value chain to evaluate chemicals in context of use, at scale, to enhance product and process sustainability attributes. Chemicals vary in their available hazard data. Products, materials, and processes vary in applicable exposure assumptions. SciVera Lens® offers significant efficiencies in applying hazard assessments in context, as well as varying exposure scenarios across products, materials, and processes to enable rapid screening of chemicals for hazard, exposure, and risk.

Tu-SY-G4.6

Panel discussion "Challenging and discussing the presented approaches and tools to address exposure in LCA and CAA"

Peter Fantke, *Technical University of Denmark (DTU), Kgs. Lyngby, Denmark*
Kathie Dionisio, *U.S. Environmental Protection Agency, Research Triangle Park, United States*

This is a special panel discussion slot to efficiently receiving input from a wider audience in a lively, useful discussion ofchallenging and discussing the presented approaches and tools to address exposure in LCA and CAA focused on the following aligned talks:
1) Integrating exposure into chemical alternatives assessment using a qualitative approach
2) A flexible matrix-based human exposure assessment framework suitable for LCA and CAA
3) Overall integrated matrix approach for characterizing and comparing exposure and health outcomes
4) A modular human exposure model (HEM) to incorporate near-field chemical exposure in LCA and CAA
5) Automated human and environmental exposure estimation to support prioritization of chemicals management actions

Tu-PL-H4: Kinetics

Tu-PL-H4.1

Using Exposure Bands for Rapid Decision-Making in the RISK21 Tiered Exposure Assessment

Bonnie Gaborek, *DuPont, Newark, DE, United States*
Rosemary Zaleski, *ExxonMobil Biomedical Sciences, Inc., Annandale, NJ, United States*
Hua Qian, *ExxonMobil Biomedical Sciences, Inc., Annandale, NJ, United States*
Olivier Jolliet, *University of Michigan, Ann Arbor, MI, United States*
Michael Dellarco, *National Institute of Biomedical Imaging and Bioengineering, Bethesda, MD, United States*
Neha Sunger, *West Chester University, West Chester, PA, United States*
Jennifer Tanir, *ILSI Health and Environmental Sciences Institute, Washington, DC, United States*

The Risk Assessment in the 21st Century (RISK21) project was initiated by the ILSI Health and Environmental Sciences Institute to address and catalyze improvements to human health risk assessment, with an emphasis on using problem formulation and exposure to focus the assessment. RISK21 developed a conceptual roadmap and risk matrix visualization tool to facilitate the transparent evaluation of both hazard and exposure components of risk assessment. Exposure tiers were defined to adapt readily available information to more quickly inform exposure decision-making, with increasing resource utilization and refinement at increasing tiers. In the lowest tier, Tier 0, screening level exposure predictions can be determined with minimal information regarding the substance or its applications. For this level, exposure banding, or grouping of substances based on ranges of predicted exposures was developed using physicochemical properties, exposure routes, and basic exposure models for estimating exposures to workers, consumers, and the general population indirectly exposed via the environment. In particular, look-up tables of banded exposure values were developed from publically-available exposure tools (European Centre for Ecotoxicology and Toxicology of Chemicals (ECETOC) Targeted Risk Assessment (TRA) for worker exposure, ECETOC TRA and European Solvents Industry Group (ESIG) Generic Exposure Scenario (GES) Risk and Exposure Tool (EGRET) for consumer exposure, and USEtox for indirect exposure to humans via the environment). The look-up tables provide value because they deliver rapid, screening-level exposure estimates for a wide range of substances and their applications with limited data knowledge or input. When these exposure estimates are then applied to the RISK21 risk matrix visualization tool, one can ascertain if adequate margins of safety are achieved and if a higher tier exposure assessment is necessary for further refinement of the estimates. A hypothetical case study demonstrated that the newly-developed exposure-banding methodologies provide suitable conservative exposure estimates for risk assessment purposes. Furthermore, the results of this effort showed that the RISK21 approach is useful for problem formulation, exposure estimation, and risk assessment visualization and decision-making.

Tu-PL-H4.2

Estimating the early-life exposure to two perfluorinated compounds (PFOS and PFOA) using PBPK modeling and biomarker measurements

Céline Brochot, *INERIS, Verneuil en Halatte, France*
Maribel Casas, *CREAL, Barcelona, Spain*
Line S. Haug, *Norwegian Institute of Public Health, Oslo, Norway*
Cyntia Manzano-Salgado, *CREAL, Barcelona, Spain*
Florence Zeman, *INERIS, Verneuil en Halatte, France*
Martine Vrijheid, *CREAL, Barcelona, Spain*

Context and objectives: Large-scale biomonitoring studies usually rely on biomarkers measurements from single time points to assess the human exposure to chemicals. Reverse dosimetry approaches were developed to help the interpretation of such biomarkers and aim at reconstructing the external exposure using the measured biomarkers, a physiologically based pharmacokinetic (PBPK) model accounting for the processes that the chemical undergoes in the human body and individual characteristics of the population. Such approaches are also valuable to simulate exposure between biomarker measurement time points, especially during critical windows of exposure. In this work, we propose to estimate the early-life exposure of children to two perfluorinated compounds, perfluorooctanesulfonic acid (PFOS) and perfluorooctanoic acid (PFOA). These compounds are ubiquitous contaminants that have been detected worldwide in environmental media and human tissues, and are suspected to induce adverse reproductive and developmental outcomes.
Methods: Our study involved 97 mother-child pairs, from the HELIX sub-cohort, living in the area of Barcelona (Spain). PFOS and PFOA were measured in maternal plasma or serum at the time of pregnancy (first trimester), and in child plasma at the age of 6-9 years old. Realistic exposure scenarios were defined to take into account the previous pregnancy and lactation periods of the mother. A lifetime PBPK model including childhood, pregnancy, and lactation periods was parameterized for PFOS and PFOA, and for each woman and child based on their individual characteristics (e.g., age, weight, birth weight of the child).
Results and Conclusion: First the PBPK model was run for each mother to provide the exposure estimates for the child during the pregnancy and breastfeeding period. For PFOA, the individual daily intakes of the mothers were estimated between 3.8 and 165.7 ng/day (mean 24.8 ng/day) that are similar to previous assessment. Part of the variability between the mothers was explained by their individual exposure scenario. The estimates of in utero and breastfeeding exposure were used as inputs to the PBPK model for the child together with the biomarker measurements to reconstruct his/her early-life exposure (i.e., daily intakes). Finally the internal exposure of children was simulated in target organs (brain, liver and kidneys) during critical time periods in the form of a maximal concentration or a cumulated amount. These new biomarkers of internal exposure could be used in epidemiological studies to better characterize associations between exposure to these chemicals and health outcomes.

Tu-PL-H4.3

Simple Pharmacokinetic Modeling of Infant Impacts From Exposure to PCB 153 in Mother's Milk

Matthew Lorber, *United States Environmental Protection Agency, Washington, DC, United States*
Leisa-Maree Toms, *Queensland University of Technology, Kelvin Grove, Australia*
Jochen Mueller, *University of Queensland, Coopers Plain, Australia*

Researchers in Australia have been assessing the general population body burdens of numerous persistent contaminants such as PBDEs, DDE, and PCBs. They have been pooling de-identified serum samples from routine pathology testing, including from infants. Pooled samples corresponding to half year increments of infant life starting from 0 to 0.5 year up until 4 years were measured for a suite of PCB congeners, including PCB 153. These samples were collected in 2006 and 2007 (data not published). Simple pharmacokinetic modeling was used to assess the impact of breast-feeding on infant body burden of PCB 153. The modeling followed procedures developed to model dioxin in infants by Lorber and Philips (2002; EHP 110: A325). A simple one-compartment first-order model was employed to maintain a mass balance of PCB 153 from birth until age 4. Key concepts for dioxin brought forward include: 1) elimination in infants was very much more rapid as compared to adults, and 2) mother's milk concentration declined over the course of lactation. Key parameters include: initial infant body burden, initial mother's milk concentration, a background exposure to PCB 153 from bottle-feeding and after breast/bottle-feeding, and the percent of infants in Australia who breast-feed and for how long. The infant serum data showed that infant body burdens were near 5 ng/g lipid for every 6-month increment during the first 4 years of life. Modeling showed that this body burden could not be duplicated by bottle feeding and background exposures; infant body burdens dropped to below 1 ng/g lipid independent of initial infant body burden, and only rose above 1 ng/g at age 2, to end near 2 ng/g at age 4. Breast-feeding rose infant body burdens to above 10 ng/g within 2 months, after which it declined to just below 5 ng/g at 1.5 years of age, to end near 5 ng/g at age 4. This initial rise in modeled body burden with breast feeding was not reflected in the data, but breast feeding appears to be the most plausible explanation to explain measured body burdens of PCB 153 consistently in the range of 5 ng/g lipid up until age 4. Refinements in the modeling parameters, and modeling of other PCBs and persistent contaminants in infants from the Australian data base are being pursued.
Disclaimer: The findings and conclusions in this abstract are those of the authors and do not necessarily represent the official position of the US EPA.

Tu-PL-H4.4

Using biologically motivated models for the lactating mother and nursing infant to link iodine deficiency with thyroid hormone production and hypothyroxinemia

Jeffrey Fisher, *US Food and Drug Administration/ National Center for Toxicological Research, Jefferson, Arkansas, United States*
Eva McLanahan, *Agency for Toxic Substances and Disease Registry, Atlanta, Georgia, United States*

Iodine is an essential micronutrient for thyroid hormone formation and is obtained primarily from food. Iodine deficiency remains a public health issue world-wide despite ongoing interventions. Iodide deficiency during development may lead to irreversible neurodevelopmental toxicity. The nursing infant is recognized as a sensitive subpopulation; however, no quantitative analyses of the relationship between intake of iodide and thyroid hormone homeostasis has been published for the nursing infant and lactating mother. A biologically motivated model was developed to examine the effect of iodine sufficiency and moderate insufficiency on thyroid hormone serum levels in the nursing infant and lactating mother pair. The mechanism of action for iodide insufficiency induced hypothyroxinemia was assumed to be a reduction in thyroid hormone production caused by reduced organified iodine content of the thyroid gland. Hypothyroxinemia is defined as low serum free thyroxine (fT4), while serum thyroid stimulating hormone (TSH) remains normal. Maternal hypothyroxinemia during early pregnancy is associated with neurological deficits in children. The adverse consequences of hypothyroxinemia in infants and neonates are less clear. Using published statistically-derived population reference ranges for serum TSH, free fT4, and total thyroxine (tT4) in the nursing infant, simulations were conducted for dietary iodine intake ranging from moderately low (50-100 µg/day) to sufficient (290 to 400 µg/day). Development and calibration of the models clearly established that the infant hypothalamic pituitary thyroid (HPT) axis is 'revved up' and through-put is much greater than in the adult. Interestingly, non-TSH compensatory mechanisms, may be involved in maintaining serum thyroid hormones for conditions of chronic intake of low dietary iodide, and are not well understood. Simulation of data sets from intervention studies, where iodine is added to the diet of iodine deficient populations of lactating mothers, suggests that the utilization of the iodine by the HPT is less than would be predicted for an iodine sufficient population. The reasons for these findings are unknown; however, the model was used to help interpret these data. This biologically motivated model can be linked to PBPK models for thyroid active chemicals or drugs that act on the HPT axis within the thyroid gland or elsewhere in the body such as the brain or organs involved in deiodinase metabolism or Phase II conjugation of thyroid hormones. Understanding the iodine- HPT axis status is fundamental to interpreting the effects of chemicals or drugs on the HPT axis.

Tu-PL-H4.5

Integration of environmental and human PBPK exposure models: application of MERLIN-Expo modelling tool to POPs exposure in Venice lagoon.

Artur Radomyski, *University Ca' Foscari Venice, Venice, Italy*
Elisa Giubilato, *University Ca' Foscari of Venice, Venice, Italy*
Andrea Critto, *University Ca' Foscari of Venice, Venice, Italy*
Philippe Ciffroy, *EDF R&D, Paris, France*
Céline Brochot, *INERIS, Paris, France*
Antonio Marcomini, *University Ca' Foscari of Venice, Venice, Italy*

MERLIN-Expo is a new tool for integrated exposure assessment recently developed under the FP7 project "4FUN". MERLIN-Expo incorporates advanced models simulating the fate of chemicals in the environment and in human body (PBPK model) into an easy-to-use tool. Models available in the MERLIN-Expo library are implemented on a common platform to facilitate integrated full-chain assessments for combined exposures. Models can be used to simulate fate of organic (PAHs, PCBs, dioxins) and inorganic contaminants. Software enables end-user to apply set of functionalities such as uncertainty and sensitivity analysis, dynamic deterministic and probabilistic simulations in order to address different exposure and chemical fate problems. MERLIN-Expo was applied to assess long term ecological and human exposure to PCBs and PCDDs in the Venice lagoon (Italy). Data from literature describing pollution historical trends in the lagoon, estimated in dated sediment cores, were used as time-dependent model input. In order to simulate bioaccumulation in specific aquatic food web three models were implemented in MERLIN-Expo library: Phytoplankton, Aquatic Invertebrate and Fish, allowing to represent specific organisms. Aquatic food web model was then coupled to PBPK model to simulate chemical concentration in aquatic organisms and in human serum after dietary exposure to contaminated seafood. Modelling results are then tested against available monitoring data on chemical concentrations in edible aquatic species and concentrations in serum of adult men in Venice area to assess the accuracy and applicability of the proposed tool to real complex scenarios. Full chain exposure assessment is then complemented by uncertainty and sensitivity analysis including local sensitivity methods, screening methods (e.g. Morris method), global regression methods (e.g. Standardised Regression Coefficients), and global variance based methods (e.g. FAST, EFAST, Sobol). These methods allow to follow for instance WHO (2008) recommendations to perform three stage uncertainty/sensitivity analysis, adopting qualitative, semi-qualitative, and quantitative methods. Integration of environmental and human exposure models in MERLIN-Expo allows comprehensive assessment of exposure and thus better characterisation of overall risk to human and environment especially in the case of higher tier assessment. Finally, MERLIN-Expo follows a Quality Assurance and Standardisation process for documentation in collaboration with CEN (European Committee for Standardisation). This makes the tool interesting and promising for potential applications in different regulatory domains.

Tu-PL-I4: Neurotoxicants

Tu-PL-I4.1

Outdoor Air Pollution and Brain Morphology in the Adult Health and Behavior II and Pittsburgh Imaging Project Cohorts

Sheila Tripathy, *University of Pittsburgh Graduate School of Public Health, Pittsburgh, Pennsylvania, United States*
Brett Tunno, *University of Pittsburgh Graduate School of Public Health, Pittsburgh, Pennsylvania, United States*
Drew Michanowicz, *Harvard T.H. Chan School of Public Health, Boston, Massachusetts, United States*
Ellen Kinnee, *University of Pittsburgh Graduate School of Public Health, Pittsburgh, Pennsylvania, United States*
Jessie Shmool, *University of Pittsburgh Graduate School of Public Health, Pittsburgh, Pennsylvania, United States*
Peter Gianaros, *University of Pittsburgh, Pittsburgh, Pennsylvania, United States*
Jane Clougherty, *University of Pittsburgh Graduate School of Public Health, Pittsburgh, Pennsylvania, United States*

Background: Exposure to ambient fine particulate matter (PM2.5) has been associated with indicators of brain morphology. While mechanisms remain unclear, PM2.5 may impact brain morphology through multiple pathways including systemic inflammation, disruption of the blood-brain barrier, and/or translocation via olfactory mucosa. These effects likely vary by PM2.5 composition, with metals [e.g., lead (Pb), manganese (Mn), iron (Fe), zinc (Zn)] most directly linked to adverse cognitive and neurological function. Because PM2.5 composition varies spatially, characterizing fine-scale intra-urban variation in metal constituents is critical to understanding causal mechanisms and refining exposure-response estimates.
Objectives: We aimed to examine spatial variation in metal constituents of PM2.5 by developing hybrid land use regression (LUR) models for PM2.5, Pb, Mn, Zn, and Fe, and to study relationships between PM2.5 composition and indicators of brain morphology (e.g., gray and white matter volume). We will also examine relationships between estimated pollutant exposures and markers of systemic inflammation (e.g., C-reactive protein, interleukin-6,) in two Pittsburgh based cohorts.
Methods: PM2.5 filters were collected during a monitoring campaign with 37 sites across the Pittsburgh region during summer 2012 and winter 2013. Filters were analyzed using inductively-coupled plasma mass spectrometry to determine metal concentrations. We built hybrid LUR models using GIS-based source indicators coupled with AERMOD-predicted industrial PM2.5 emissions to predict fine-scale metal concentration variation. These models were used to estimate pollutant exposures within a 300 m buffer around cohort participant addresses. Brain morphology indicators were obtained from magnetic resonance images and inflammatory markers from blood samples of participants in the Adult Health Behavior II and Pittsburgh Imaging Project Cohorts (n=750, mean age=42 years). Linear regression models were developed for pollutant exposures and outcomes, adjusting for intracranial volume (for brain outcomes), age, sex and smoking status.
Results: The majority of spatial variation in metal concentrations was explained by industrial emissions and land use. LUR R2 values of 0.52, 0.76, 0.53, and 0.75 were found

for Pb, Mn, Zn, and Fe respectively. Preliminary results show significant associations between total PM2.5, Pb, and Zn with markers of systemic inflammation (e.g., C-reactive protein). No associations were found with brain morphology measures.

Tu-PL-I4.2

Dvelopmental neurotoxicity assessment of chemical mixtures in children

Pim Leonards, *VU University, Amsterdam, Netherlands*
Milou Dingemans, *Utrecht University, Utrecht, Netherlands*

Worldwide, serious concern has arisen about the increased incidence of learning and developmental disorders in children and the potential role of exposure to neurotoxic chemicals during early brain development. DENAMIC "Developmental Neurotoxicity Assessment of Mixtures in Children" investigated developmentally neurotoxic effects of low-concentration mixtures of biocides and a number of common environmental pollutants in children. Research in DENAMIC focused on hazard characterization and epidemiology. The hazard characterisation studied the effects of neurotoxic chemicals and mixtures thereof using novel tools, testing methods and procedures for screening (mixtures of) chemicals for (developmental)neurotoxicity. Rodent studies focused on the effects of low-level exposure to biocides and mixtures on neurobehavior, cognitive and motor function, including evaluations of underlying mechanisms of observed effects with consideration of exposure timing, critical windows during neuronal development and consequences on susceptibility. In the epidemiology part, prenatal and early-childhood exposure was studied in maternal urine, breast milk, cord blood and urine of children in European cohorts. Associations with learning and developmental disorders, including ADHD, ASD and anxiety were explored. Prenatal and neonatal exposure profiles in the cohorts showed that the European population is exposed to low concentrations of neurotoxic chemicals (e.g. PCBs, organophosphates, carbamates, pyrethroids, PFAS, methylmercury). A number of associations were found between the exposure of neurotoxic chemical and neuropsychological and behavioural development in children. Factors affecting the relative differences in socio-demographic and economic impact for (developmental) neurotoxicity resulting from exposure to environmental neurotoxicants were also studied. The experimental data on mixtures concluded that additivity cannot be assumed as a default approach for risk assessment of mixtures. The risk assessment showed that the margins of safety (MOS) in the EU are sufficient for neurotoxic and neurobehavioral endpoints for carbaryl and cypermethrin, while for chlorpyrifos, endosulfan and methylmercury such MOS may not be adequate in specific populations. It is therefore recommended to investigate whether further exposure reduction measures are needed. Moreover, current risk assessment assumes that no effect levels are safe based on individual chemical toxicity studies. Some experiments done in DENAMIC indicate that combined exposure to biocides or other contaminants, that individually not resulting in any effect, may in fact cause an effect when present in mixtures. This observation brings doubt in the present approach for risk assessment, which is based on exposure to individual chemicals and may in fact result in an underestimation of the (human) risk.

Tu-PL-I4.3

Developing PBPK/PD model to characterize the mixture effect of TCDD and DEHP altering estradiol kinetic in ovary via crosstalk mechanism

Raju Prasad Sharma, *Universitat Rovira i Virgili, Tarragona, Tarragona, Spain*
Marta Schuhmacher, *Universitat Rovira i Virgili, Tarragona, Tarragona, Spain*
Vikas Kumar, *Universitat Rovira i Virgili, Tarragona, Tarragona, Spain*

The ubiquitous presence of dioxin and phthalates in the environment and raised concern over their common ovarian toxicity catch attention towards their mixture effect. Among dioxin and phthalates, 2, 3, 7, 8-Tetrachlorodibenzo-p-dioxin (TCDD) and Di(2-ethylhexyl) phthalate (DEHP), are the most toxic chemicals described. DEHP and its metabolite mono(2-ethylhexyl)phthalate (MEHP), mainly affects estrogen production and action in granulosa cell, resulting in hypo-estrogenic, polycystic ovary and anovulatory cycles, which could leads to infertility. In parallel to this, exposure of TCDD is linked with prevalence of endometriosis; reduced fecundity, reduced follicle number and anovulation. To estimate the ovarian toxicity, a physiologically-based pharmacokinetic / pharmacodynamics (PBPK/PD) model for mixture (TCDD and DEHP) was developed. This model integrates target organ (ovary) dosimetry and dynamic response (i.e. aromatase inhibition, CYP1B1 induction) describing time course of external exposure of mixture to estrogen kinetics. Model simulation was performed for single as well as mixture of TCDD and DEHP. The PBPK model was validated with previous biomonitoring study. In its turn, pharmacodynamics simulations were compared against previously published experimental in-vitro and in-vivo data. The simulation results shows that the mixture has synergistic effect in inhibition of estradiol level via crosstalk of receptors PPARγ and Ahr for the inhibition of aromatase and induction of CYP, responsible for synthesis and metabolism of estrogen respectively. Developed PBPK/PD model shows quantitatively estimates target tissue dosimetry and aromatase inhibition, CYP1B1 induction linked with estrogen kinetic. Developed model can provide basis for risk assessment of chemical mixtures causing reproductive dysfunction under different scenarios of body physiology as well as level of chemical exposure.

Tu-PL-I4.4

Use of Biomonitoring for Arsenic, Mercury and Lead to Assess Exposure and Health Risks in Children in a Northern Canadian Smelter Community

Adam Safruk, *Intrinsik Environmental Sciences Inc., Mississauga, Ontario, Canada*
Elliot Sigal, *Intrinsik Environmental Sciences Inc., Mississauga, Ontario, Canada*

Emissions from a mine and base metal smelting complex in a northern Canadian community have resulted in elevated concentrations of a number of metals and metalloids in the surrounding environment. Results from a human health risk assessment indicated that arsenic, lead, and inorganic mercury were present at concentrations in the local environment that warranted further assessment of exposure and risk. An initial biomonitoring study was undertaken for local children to examine urinary arsenic, blood lead, and urinary inorganic mercury levels. Overall, 447 children participated in the study providing 202 blood samples and 379 urine samples. The biomonitoring study was conducted to assess exposure and potential health risks, as well as to identify personal factors that may be associated with measured internal exposures of children in the community. Results demonstrated that despite elevated concentrations of arsenic and mercury in soils and other environmental media, exposure among children was low and consistent with levels measured in national surveys and reference communities. The geometric mean (GM) blood lead level (BLL) (2.73 µg/dL) was higher than national averages, with 13% of children with BLLs above 5 µg/dL, indicating atypical sources of lead exposure.

A second biomonitoring study was conducted approximately 26 months after the closure of the smelter and the implementation of several exposure reduction measures. This follow-up study included the collection of 119 blood samples and was designed to examine the impact of various environmental media, including outdoor soil, household dust, tap water and lead paint, on the BLLs of local children. Environmental samples were collected from the households of study participants, and the relationship between lead content in co-located environmental samples and children's BLLs was examined. The GM BLL was 1.41 µg/dL, representing a statistically significant reduction in BLLs between studies, with 2% of participants with BLLs greater than 5 µg/dL. Despite statistically significant relationships between BLLs and lead content in soil, household dust, and paint, the variability in BLLs was poorly explained by these factors alone ($r2= 0.07$, 0.12 and 0.06, respectively). BLLs were considered to be within the normative range and study results indicated additional intervention strategies were not likely to further influence BLLs in the community. These results have important implications on future assessment of risks associated with exposure to arsenic, lead, and inorganic mercury in soils and indicate that soil remediation or removal may not be warranted to reduce exposure.

Tu-PL-I4.5

Perinatal lead exposure and white matter microstructure in children

Megan Horton, *Icahn School of Medicine at Mount Sinai, New York, New York, United States*
Paul Curtin, *Icahn School of Medicine at Mount Sinai, New York, New York, United States*
Chris Gennings, *Icahn School of Medicine at Mount Sinai, New York, New York, United States*
Victoria Wang, *Icahn School of Medicine at Mount Sinai, New York, New York, United States*
Erika Proal, *Instituto Nacional de Salud Pública, Cuernavaca, Mexico*
Lourdes Schnaas, *Instituto Nacional de Perinatología, Mexico City, Mexico*
Martha Ma. Téllez Rojo, *Instituto Nacional de Salud Pública, Cuernavaca, United States*
Ernesto Roldan-Valadez, *Centro Nacional de Imagenología e Instrumentación Médica, Mexico City, Mexico*
Francisco Xavier Castellanos, *New York University, New York, New York, United States*
Cheuk Tang, *Icahn School of Medicine at Mount Sinai, New York, New York, United States*
Roberta White, *Boston University, New York, New York, United States*
Robert Wright*Icahn school of Medicine at Mount Sinai, New York, New York, United States*

Background
Perinatal lead (Pb) exposure is associated with adverse cognitive and behavioral outcomes that may be mediated by altered brain structure and function. Childhood Pb exposure has been associated with persistent impacts on adult white matter microstructure. The objective of this study was to assess the impact of perinatal Pb exposure on white matter microstructure in children using diffusion tensor imaging (DTI).
Methods
This study took place in the ELEMENT cohort in Mexico City. We randomly selected 20 subjects at age 6 years for a magnetic resonance imaging (MRI) pilot study. DTI images were acquired with a 3T Philips Achieva scanner using gradient echo planar imaging. A voxel-wise statistical analysis for diffusivity measures, including fractional anisotropy (FA), was performed along major white matter tracts. All 20 subjects had blood biomarkers of Pb collected during 2nd and 3rd trimesters and at delivery (umbilical cord blood). We examined correlations between Pb biomarkers and FA values. To capture associations between perinatal Pb exposure and within-brain variability at different time points, we focused on mean FA and standard deviation (SD) of FA across 48 template regions of interest (ROIs); the latter metric captures bidirectional effects that may discretely increase or decrease FA values. Linear regression models examined the association of 2nd, 3rd-trimester and cord blood Pb levels, FA and SD of FA from the ROIs.
Results
Pb levels in 2nd trimester blood were positively correlated with increased global FA after controlling for multiple comparisons ($p < 0.05$). Higher cord blood Pb was associated with increased FA ($\beta = 0.010$, $p = 0.05$) and increased variability of FA ($\beta = 0.0014$, $p = 0.07$).
Discussion
These pilot data suggest changes in white matter microstructure associated with perinatal Pb exposure. Pathological alterations can decrease or increase FA, thus our pilot findings may be consistent with neurotoxic effects of perinatal Pb exposure.

Poster sessions Tuesday October 11, 2016

Biomonitoring

Tu-Po-01

Urinary formic acid as a dose biomarker for Occupational Exposure to Formaldehyde

Giovanna Tranfo, *INAIL Research, Monte Porzio Catone (RM), Italy*
Daniela Pigini, *INAIL Research, Monte Porzio Catone (RM), Italy*
Monica Gherardi, *INAIL Research, Monte Porzio Catone (RM), Italy*
Maria Pia Gatto, *INAIL Research, Monte Porzio Catone (RM), Italy*

Aim: This paper explores the possibility to assess occupational or environmental exposure to Formaldehyde (FA) by biomonitoring, on an accessible matrix like urine. FA is a common contaminant recently recognized as a carcinogen by IARC. Occupational exposure occurs in anatomical pathology, engineering industries, furniture and plastic utensils production, beauty salons and gasoline station. The SCOEL recommends a Limit Value of 0.3 ppm (8 h TWA) with a STEL limit of 0.6 ppm. Inhaled FA is oxidized and excreted as formic acid in the urine; formic acid can derive from other metabolic sources.
Methods: A bibliographic search was performed on Scopus and PubMed, retrieving papers published from 2006 to 2015, using key words or strings pertinent to biomonitoring for formaldehyde exposure: 17 papers were selected and read and the use of urinary formic acid as biomarker is discussed. Sampling and analysis of airborne FA content were also conducted in indoor and outdoor air in non-occupational settings, according to the NIOSH method n.2016.
Results: Formic acid stays stable in urine up to 13 days at room temperature: its concentration is used as biomarker of exposure to FA in 5 papers, analyzed by headspace GC-FID or HPLC/UV. In all papers measured concentrations of airborne FA were well below the SCOEL Limit Value of 0.3 ppm, ranging from 0.009 to 0.09 ppm, while formic acid was between 17 and 42 mg/L. In exposed workers levels were statistically significant higher than in controls, and one paper reports higher values in gasoline station female workers than in males. In beauty salons a background median level of formic acid is measured of 11.05 mg/L, and a post exposure of 15.22. When if data are put together a linear relationship between airborne FA and urinary formic acid is found (figure). The formic acid concentration corresponding to the SCOEL Limit Value of 0.3 ppm can be extrapolated, being 84.42 mg/L, and a background level of 9.00 mg/L for no exposure. Airborne FA levels measured in non-occupational settings would correspond to formic acid levels very close to the calculated background.
Conclusions: Few papers were published in the last 9 years reporting the use of urinary formic acid as a biomarker for FA exposure. Urinary formic acid seems a promising biomarker for the measure of occupational exposure, but non for environmental. Further

investigations is needed on this subject, a growing need, as exposure assessment is compulsory for carcinogenic substances like FA.

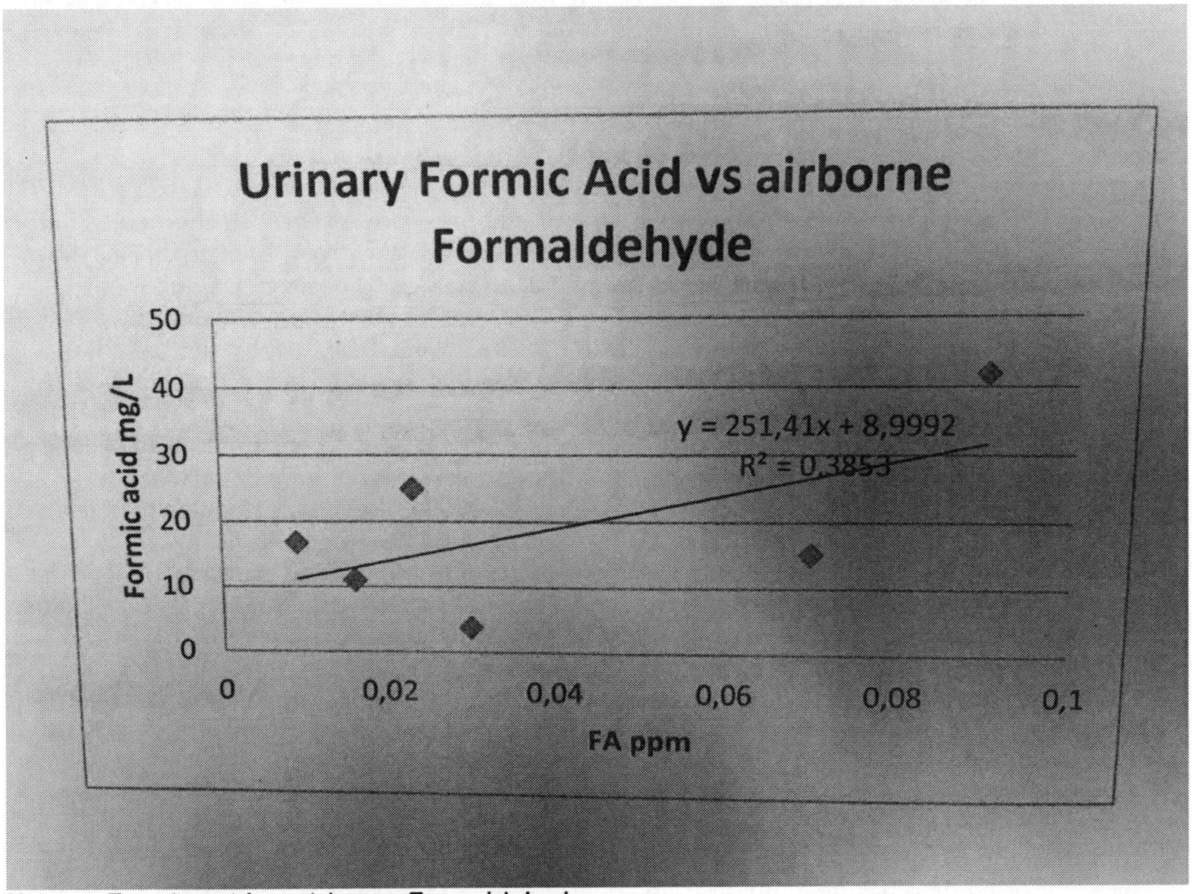

Urinary Formic acid vs airborne Formaldehyde

Tu-Po-02

A Perspective on Guidelines for Interpreting Risk at the Individual Level Derived from Biomonitoring Data for Northern First Nations

Mylene Ratelle, *University of Waterloo, Waterloo, Quebec, Canada*
Brian Laird, *University of Waterloo, Waterloo, Ontario, Canada*

First Nations country foods (e.g. fish, moose, fowl), which have been associated with lower risk factors for cardiovascular disease and diabetes, are integral to the health and food security of communities in the Northwest Territories. However, concerns regarding contaminants have led to a series of contaminant advisories to reduce fish and meat consumption. Therefore, the development of public health strategies related to contaminant exposures from country foods in the Northwest Territories need to strike a balance between risks and benefits. Biomonitoring is important to environmental health, and widely used in national surveys.
The current project uses a risk-benefit approach to country food consumption in order to improve nutrients levels while lessening contaminant levels among First Nations communities and to estimate if current advisory levels are relevant for health policies. This biomonitoring project, supported by the Northern Contaminant Program (NCP), is divided into three components: 1) the implementation of biomonitoring for heavy metals, persistent organic pollutants, essential elements, and fatty acids in blood, urine and hair and completion of dietary surveys, 2) the return of results to individual participants, comparing exposures to selected risk assessment guidelines, and 3) drafting public health messaging in collaboration with community members, local governments and stakeholders. A total of 9 communities in the Northwest Territories, Canada, were invited to participate in the project.
The interpretation of individual biomonitoring data is challenged by the lack of relevant biomonitoring guidelines in northern First Nations, due to different diet and genetic toxicokinetic parameters. Instead, the primary means of interpretation of individual biomonitoring data has been qualitative comparisons to population references: i) upper-percentiles of exposure (95th) from population biomonitoring projects, ii) the clinical guidance values for health in a medical perspective and iii) institutional/occupational guidance values to insure the safety of every individual even the most susceptible. In the specific case of a biomonitoring study designed with a risk-benefit approach, the common use of guidelines cannot characterize contaminant risk. Overall, this project will increase knowledge of the contaminant exposure levels in the north and will align further country food consumption advisories.

Tu-Po-03

Human Biomonitoring of Phthalates in Portuguese Children

Luísa Correia-Sá, *REQUIMTE/LAQV - Instituto Superior de Engenharia do Porto do Instituto Politécnico do Porto, Porto, Portugal*
Claudia Pälmke, *IPA- Institute for Prevention and Occupational Medicine of the German Social Accident Insurance, Institute of the Ruhr-University Bochum, Bochum, Germany*
André Schütze, *IPA- Institute for Prevention and Occupational Medicine of the German Social Accident Insurance, Institute of the Ruhr-University Bochum, Bochum, Germany*
Sónia Norberto, *CINTESIS - Centro de Investigação em Tecnologias e Sistemas de Informação em Saúde, Centro de Investigação Médica, Porto, Portugal*
Conceição Calhau, *CINTESIS - Centro de Investigação em Tecnologias e Sistemas de Informação em Saúde, Centro de Investigação Médica, Porto, Portugal*
Valentina F. Domingues, *REQUIMTE/LAQV - Instituto Superior de Engenharia do Porto do Instituto Politécnico do Porto, Rua Dr. António Bernardino de Almeida, 431, 4200-072 Porto, Portugal, Porto, Portugal*
Holger M. Koch, *IPA- Institute for Prevention and Occupational Medicine of the German Social Accident Insurance, Institute of the Ruhr-University Bochum, Bochum, Germany*

Phthalates are dialkyl or alkylary esters of the ortho-benzene dicarboxylic acid (phthalic acid). Because phthalates are not chemically bound to the polymer they are constantly released into the environment. The dietary source is considered the main source of population exposure to high molecular weight (HMW) phthalates. For low-molecular weight (LMW) phthalates other lifestyle-dependent exposure pathways seem to be more relevant. As a result, the general population is widely and continuously exposed to phthalates. Some phthalates, such as DnBP, DiBP, BBzP, DEHP and DiNP are developmental and reproductive toxicants. The aim of this study was to evaluate the phthalates exposure of 112 Portuguese children (obese/overweight (cases) and healthy weight (controls)). Urine samples collected in 2014/2015 were analyzed for phthalate metabolites using on-line HPLC-MS/MS with isotope dilution after enzymatic deconjugation. Most of the measured metabolites were above the limits of quantification (79 to 100 % positive detects, with the exception of MCHP, MnPEP and MnOP which were detected to a much lower degree). For the LMW phthalates the median creatinine adjusted values of MEP were the highest (65.87 µg/g for the cases and 54.84 for the controls), followed by DiBP metabolites (2ΣDiBP 22.16 and 31.68 µg/g, for cases and controls respectively), DnBP (2ΣDnBP 15.19 and 18.04 µg/g, for cases and controls respectively), MMP (around 3.00µg/g for all the population), and MBzP (around 2.00µg/g for all the population), respectively. For the HMW phthalates the median creatinine adjusted concentrations of the DEHP metabolites were the highest (4ΣDEHP for cases presented a value of 34.90 and for controls a value of 49.19 µg/g) followed by DiNP metabolites (Σ3DiNP 12.74 and 16.96 µg/g for cases and controls respectively), and DiDP (3ΣDiDP 3.19 and 3.84 µg/g for cases and controls respectively). The results showed an omnipresent phthalate exposure in the participating children and are in line with previous DEMOCOPHES data (except for MiBP, MnBP and MBzP) from Portugal and other parts of Europe. Obese/overweight children had significantly lower HMW phthalate levels than controls, which might be due to dietary consultation program set up for these children.
Acknowledgments: L. Correia-Sá is grateful to FCT by the grant (SFRH/BD/87019/2012), financed by POCH, subsidized by Fundo Social Europeu and Ministério da Ciência, Tecnologia e Ensino Superior.

Tu-Po-04

The development of a 'point of care' fluorescent immunosensor for the benzene biomarker S-PMA in human urine

Cees Koopal, *TNO, Utrecht, Netherlands*
Anjoeka Pronk, *TNO, Zeist, Utrecht, Netherlands*
Sjaak van Veen, *TNO, Utrecht, Netherlands*
Tim Meijster, *Shell, Den Haag, Netherlands*
Jan Urbaus, *Shell, Den Haag, Netherlands*
Paul Aston, *AB Biomonitoring, London, United Kingdom*

Background
Biomonitoring of occupational exposure to benzene can be done using an ELISA based method that measures S-PMA in urine. ELISA assays are usually performed in a laboratory by specialized personnel, resulting in a lag time of several weeks between urine collection on work locations and assay results. We aimed to develop a proof of concept for a fluorescent immunosensor based on an existing S-PMA ELISA assay for in the field biological monitoring of benzene with which results can be obtained within hours and analyses can be done at logistically challenging locations. In this project a point of care (POC) test, with sufficient sensitivity to detect levels below the current occupational exposure limit was developed (0.5 ppm benzene).
Development of the method
A unique sheep anti-PMA antiserum also used in the lab assay (AB Biomontoring, UK) was purified and labeled with a fluorescent probe. The resulting labeled antibodies were utilized in a fluorescent immunosensor format. The counter part of the format was a PMA-hapten, conjugated to a carrier protein and immobilized on a glass surface with low background binding properties. Based on these immunochemical building blocks, an inhibition assay was developed with S-PMA as the sample in solution. The performance of the assay was tested using both calibration and human urine samples spiked with S-PMA.
Results and conclusions
Preliminary validation results of the developed inhibition assay for S-PMA samples in human urine are promising. These will be presented together with the outline of the envisaged POC instrument format for benzene occupational biomonitoring deployable in a field situation. A POC assay can be performed on the spot by less specialized personnel such as a nurse or occupational physician, thereby providing results that are relevant for taking direct management measures.

Tu-Po-05

Human Biomonitoring of Di(2-ethylhexyl) terephthalate in Portuguese Children

Frederik Lessmann, *Institute for Prevention and Occupational Medicine of the German Social Accident Insurance, Institute of the Ruhr-Universität Bochum (IPA), Bochum, Germany*
Luísa Correia-Sá, *REQUIMTE/LAQV - Instituto Superior de Engenharia do Porto do Instituto Politécnico do Porto, Porto, Portugal*
Conceição Calhau, *CINTESIS - Centro de Investigação em Tecnologias e Sistemas de Informação em Saúde, Centro de Investigação Médica, 2° piso, edif. Nascente, Faculdade de Medicina da Universidade do Porto, Porto, Portugal*
Valentina F. Domingues, *REQUIMTE/LAQV - Instituto Superior de Engenharia do Porto do Instituto Politécnico do Porto, Porto, Portugal*
Tobias Weiß, *Institute for Prevention and Occupational Medicine of the German Social Accident Insurance, Institute of the Ruhr-Universität Bochum (IPA), Bochum, Germany*
Thomas Brüning, *Institute for Prevention and Occupational Medicine of the German Social Accident Insurance, Institute of the Ruhr-Universität Bochum (IPA), Bochum, Germany*
Holger M. Koch, *Institute for Prevention and Occupational Medicine of the German Social Accident Insurance, Institute of the Ruhr-Universität Bochum (IPA), Bochum, Germany*

Di(2-ethylhexyl) terephthalate (DEHTP) is used as a substitute for di(2-ethylhexyl) phthalate (DEHP), an ortho-phthalate based plasticizer that is classified and labeled due to its toxicity to reproduction. Due to the ongoing substitution process, increasing DEHTP exposures of the general population seem likely. Previous HBM studies have shown that children are exposed to plasticizers to a higher degree than adults.
For this study, we obtained 107 spot urine samples of Portuguese children (55 girls and 52 boys, aged 4-17) collected in 2014/2015. 68 of these children were classified as overweight/obese according to the body mass index and received specific nutritional guidance. The other 39 children were normal weight and did not receive any specific nutritional guidance. The samples were analyzed for the specific sidechain-oxidized monoester metabolites of DEHTP (5OH-MEHTP, 5oxo-MEHTP, 5cx-MEPTP and 2cx-MMHTP) by a previously published online-SPE HPLC-MS/MS method with isotope dilution.
We detected the main specific metabolite 5cx-MEPTP in almost all samples (98 % >LOQ) with a median concentration of 4.4 µg/L (maximum concentration 2220 µg/L). Compared to the, so far, only existing data of a pilot biomonitoring study with 34 German adults (median 0.9 µg/L, maximum: 38.7 µg/L), the levels determined in this study were higher. The other DEHTP metabolites correlated well with 5cx-MEPTP but were detected at lower levels and rates.
The concentrations of 5cx-MEPTP in overweight/obese children with specific nutritional guidance (median 3.1 µg/L) were lower compared to normal weight children (median 5.2 µg/L) but not significantly different (Mann-Whitney-U; p=0.11). Gender specific differences in metabolite levels could not be observed.
With this study we provide a first data set documenting the omnipresent DEHTP exposure of the participating Portuguese children.

Tu-Po-06

Concentrations of urinary biomarkers of non-persistent environmental pollutants among 316 Polish men - patients of infertility clinic.

Bartosz Wielgomas, *Medical University of Gdańsk, Gdańsk, Pomorskie, Poland*
Anna Klimowska, *Medical University of Gdańsk, Gdańsk, Pomorskie, Poland*
Joanna Jurewicz, *Nofer Institute of Occupational Medicine, Łódź, Łódzkie, Poland*
Michał Radwan, *"Gameta" Hospital, Łódź, Łódzkie, Poland*
Paweł Radwan, *"Gameta" Hospital, Łódź, Łódzkie, Poland*
Wojciech Hanke, *Nofer Institute of Occupational Medicine, Łódź, Łódzkie, Poland*

Aim
The major aim of this study was to assess exposure to non-persistent environmental pollutants among fertile men, patients of infertility clinic.
Methods
The study population consisted of 316 men who were attending an infertility clinic in Łodź, Poland for diagnostic purposes and who had normal semen concentration of 15-300 mln/mL. Methyl-, ethyl-, propyl-, butyl- and isobutyl- parabens, 3,5,6-trichloro-2-pyridinol, 4-nonylphenol, 1- and 2-naphthols, benzophenone-3, triclosan and bisphenol-A were determined in urine samples by gas chromatography tandem mass spectrometry (GC-MS/MS).
Results
Naphthols, 3,5,6-trichloro-2-pyridinol, benzophenone-3, methyl paraben and bisphenol A were detected in over 95% of the samples. 4-nonylphenol was not detected in any sample (LOD=0.5 ng/mL). Most of the studied biomarkers were present at levels observed in other populations over the world. Slightly higher levels of urinary naphthols in Polish men in comparison to other populations in Europe might be a result of air pollution and exposure to polycyclic hydrocarbons.
Conclusion
This is the first study to document widespread exposure to a number of environmental chemicals among Polish men. Further studies are needed especially to confirm the high exposure to naphthalene and identify possible sources of exposure.

Tu-Po-07a

Development of capability and capacity to conduct biomonitoring in NJ

Zhi-Hua (Tina) Fan, *New Jersey Department of Health, Trenton, New Jersey, United States*
Chang Ho Yu, *New Jersey Department Of Health, Trenton, New Jersey, United States*
Eric Bind, *New Jersey Department of Health, Trenton, United States*
C. David Riker, *New Jersey Department of Health, Trenton, United States*
Douglas Haltmeier, *New Jersey Department of Health, Trenton, United States*
Robert Servis, *New Jersey Department of Health, Trenton, United States*
Marilou Palencia, *New Jersey Department of Health, Trenton, United States*
Chris Hargrave, *New Jersey Department of Health, Trenton, United States*
Jhindan Mukherjee, *New Jersey Department of Health, Trenton, United States*
Susan Muscato, *New Jersey Department of Health, Trenton, United States*
Bahman Parsa, *New Jersey Department of Health, Trenton, United States*

New Jersey Department of Health, Public Health & Environmental Laboratories
Environmental and Chemical Laboratory Services
New Jersey (NJ) is home to the highest number and density of Superfund/National Priorities List sites compared to the rest of US, and NJ residents are disproportionally exposed to a variety of environmental pollutants. There was no biomonitoring program in NJ until the NJ Department of Health (NJDOH) received a 5-year grant from CDC to establish a state-wide biomonitoring program in 2014. Under this grant, NJDOH will increase its capability and capacity to conduct chemical analyses related to biomonitoring while working to build a sustainable program to assess State residents' exposures to harmful chemicals both within and beyond the scope of the grant. The target analytes include heavy metals, perfluorinated compounds (PFCs), and polychlorinated biphenyls (PCBs).

To achieve the primary goal of the grant of developing laboratory capability and capacity, additional staff were recruited and trained and new instrumentation, i.e. SPE-LC-MS/MS for PFCs analyses and GC-HRMS for PCBs analyses, has been acquired. The sample receiving and login method has been developed to handle logging a large number of specimens into LIMS. In addition, the CDC SPE-LC-MS/MS Method (6304.04, 2013) for PFCs testing was optimized by changing the analytical column, mobile phase composition, gradient program, and cleaning procedures. The modified method is more than ten times sensitive than the CDC method, with better resolution and a shorter run time (10' vs. 15'). To demonstrate its capabilities, NJDOH will conduct three biomonitoring projects. One of the projects is to measure toxic heavy metals (e.g., mercury, lead, arsenic, cadmium), PFCs and PCBs in remnant blood/serum or urine specimen from clinical laboratories and blood banks throughout the state. To date, 800 of the 3000 targeted whole blood samples and 200 of the 1000 targeted urine samples have been collected for metals analyses using ICP-MS, 80 serum samples have been collected of the 500 targeted for PFCs analyses using SPE-LC-MS/MS, and 1000 targeted for PCB analyses using GC-HRMS. Analyses are in progress and the body burdens of environmental contaminants for the study population will be characterized by socio-demographic strata (i.e., age, gender, race, and location by county). The results will also be compared to the general US population using the latest NHANES data. The foundation and future of the program continue to be solidified through the establishment of the NJ State Biomonitoring Commission and through the forging of partnerships with key collaborators.

Environmental/Human Health

Tu-Po-07b

From the farm to the fork: fungal occupational exposure in the swine meat supply chain

Carla Viegas, *ESTeSL-IPL, Lisbon, Portugal*
Catia Pacífico, *ESTeSL-IPL, Lisbon, Poland*
Tiago Faria, *ESTeSL-IPL, Lisbon, Portugal*
Anita Gomes, *ESTeSL-IPL; Faculty of Medicine of Lisbon, Lisbon, Portugal*
Susana Viegas, *ESTeSL-IPL; ENSP-UNL, Lisbon, Portugal*

Feed production, swine and slaughterhouses were already reported as occupational environments with high fungal contamination. This condition can ultimately lead to the development of several health effects. This study aimed at characterizing the occupational exposure to fungal burden in the three different settings.
Air samples were collected from the three different settings through an impaction method onto malt extract agar (MEA) supplemented with chloramphenicol (0.05%), alongside with surface swabs. Outdoor samples were also performed to be used as reference. All the collected samples were incubated at 27°C for 5 to 7 days. In addition, we collected air samples using the impinger method in order to perform real-time quantitative PCR (qPCR) amplification of genes from Aspergillus sections Circumdati, Flavi and Fumigati.
In the swine feed unit 1906 isolates were counted in the air, being 54.6% from Cladosporium sp. and 35.8% from Alternaria sp.. In addition Mucor sp., Rhyzopus sp., Alternaria sp. and Chrysonilia sitophyla were detected in the surface samples. Warehouse and the silage were the most contaminated. In one swine 80.6% of the 3080 isolates found in air belonged to Cladosporium sp., followed by Aspergillus ochraceus complex and Fusarium graminearum complex (3.7%). In the surfaces, countless colonies of Mucor sp. and Rhyzopus sp. were detected. The air from the other swine presented a total of 5080 isolates and Cladosporium sp. (52.7%), A. ochraceus complex (23.7%) and Penicillium sp. (11.9%) were present. Scopulariopsis candida, Penicillium sp. and Rhyzopus sp. were detected in surfaces. In the slaughterhouse, the most prevalent species in air were the ones belonging to Cladosporium sp. (48.2%), followed by Penicillium sp. (31.8%) and Aureobasidium sp. (10.6%). 51.8% from 850 isolates were present in the gutting section. No contamination was found in the surfaces. Molecular tools were only able to detect the presence of A. fumigatus complex. However, qPCR analysis successfully amplified DNA from the A. fumigatus complex in 10 out of 20 sampling sites where the presence of this fungal species was not identified by conventional methods. Although both swine units showed the highest fungal load, in all the 3 settings fungal species with toxigenic potential were present. Therefore, is important to consider interactions between fungi and mycotoxins and this should be taken into account in the risk assessment process. Importantly, the molecular tools applied allowed to target selected fungal indicators, allowing a more precise characterization of the fungal burden.

This page intentionally left blank.

Tu-Po-09

Harmonizing exposure metrics and methods for sustainability assessments of food contact materials

Alexi Ernstoff, *DTU, Lyngby, Denmark*
Olivier Jolliet, *University of Michigan, School of Public Health, Ann Arbor, MI, United States*
Monia Niero, *Technical University of Denmark, Lyngby, Denmark*
Peter Fantke, *Technical University of Denmark, Lyngby, Denmark*

We aim to develop harmonized and operational methods for quantifying exposure to chemicals in food packaging specifically for sustainability assessments. Thousands of chemicals are approved for food packaging and numerous contaminates occur, e.g. through recycling. Chemical migration into food, as a function of the chemical, food, and package properties and storage conditions, is responsible for human exposure to many chemicals of concern. In addition to complying with regulatory standards, stakeholders concerned with environmental sustainability draw on strategies such as Life Cycle Assessment (LCA) and Cradle to Cradle to support packaging design. Each assessment has distinct context and goals, but can help manage exposure to toxic chemicals and other environmental impacts. Metrics and methods to quantify and characterize exposure to potentially toxic chemicals specifically in food packaging are, however, notably lacking from such assessments. Furthermore, previous case studies demonstrated that sustainable packaging design focuses, such as decreasing greenhouse gas emissions or resource consumption, can increase exposure to toxic chemicals through packaging. Thereby, developing harmonized methods for quantifying exposure to chemicals in food packaging is critical to ensure 'sustainable packages' do not increase exposure to toxic chemicals. Therefore we developed modelling methods suitable for first-tier risk screening and environmental assessments. The modelling framework was based on the new product intake fraction (PiF) exposure metric, with units of chemical mass taken in by exposed persons versus chemical mass within a product. To model this metric, we used analytical approximations for regulatory models. We investigated model results for various chemical-package-food combinations to facilitate operation in assessments and identify combinations of priority.
Modelling results predicted with accuracy previous findings, that exposure is dependent on diffusive and partitioning behaviors according to each chemical-package-food combination. Harmonizing exposure modeling with environmental assessments, like LCA, finally facilitates including exposure to chemicals as a sustainable packaging design issue. Results were demonstrated in context of the pilot-scale Product Environmental Footprint regulatory method in the European Union. Increasing recycled content, decreasing greenhouse gas emissions by selecting plastics over glass, and adding chemicals with a design function were identified as risk management issues.
We conclude developing an exposure framework, suitable for sustainability assessments commonly used for food packaging, is feasible to help guide packaging design to consider both the environment and human exposure. Future work is required for refinement and operationality. This is the first study addressing the need for quantitative, harmonized exposure metrics and methods for food packaging within sustainability assessment frameworks.

This page intentionally left blank.

Tu-Po-12

Exposure to diesel emissions among truck drivers and consequent health risks prevention through PAH biomonitoring assessment

Ikhtiar Uddin, *PMAS Arid Agriculture University, Rawalpindi, Pakistan*
Afshan Naseem, *National University of Sciences and Technology, Islamabad, Pakistan*
Audil Rashid, *PMAS Arid Agriculture University, Rawalpindi, Pakistan*

Unavoidable occupational exposure to petrochemicals is an increasingly health risk which is marked as key environmental stressor for truck drivers. Exposure to diesel exhaust was quantified using PAH biomonitoring and associated health effects were assessed. We monitored pyrene levels in serum and its metabolite 1-hydroxy pyrene in urine. Self-reported health status of truck drivers (n=81) was noted using self-structured questionnaire item with focus on physical symptoms (e.g. skin lesions, eye redness, dryness of tongue/lips, appetite loss, acidity after meals at workplace) and neurasthenic symptoms (e.g. energy loss, fatigue, fainting, twitching, sleeplessness, irritability, body aches). These PAH exposure estimates were correlated with symptoms of health disorders to examine the occupational exposure effects. Median serum pyrene was 2 to 8-fold higher in smokers compared to non-smokers. Our logistic regression model has predicted up to 63% serum pyrene attributed by active smoking and urinary 1-hydroxy pyrene was most strongly affected work hours per day (OR=3.13, 95% CI=1.27-9.25). Neurasthenic symptoms were found in 44% of the subjects and were associated with years of involvement in job. Occupational association of continuous ten years or more as truck driver has attributed substantial development of neurasthenic effects (OR=2.79, 95% CI=1.38-5.59). These individuals rated their overall health and functional capacity significantly poorer than that of urban area general population. Our study may prove helpful in the implementation of human biomonitoring as an instrument for health risk assessment among occupational exposure to petrochemicals.

Tu-Po-13

Development of emission standards for metallurgical industry based on results of human health risk assessment

Arina Petrosian, *State Institution "O.M. Marzeyev Institute for Public Health of the National Academy of Medical Sciences of Ukraine, Kyiv, Ukraine*
Olena Turos, *State Institution "O.M. Marzeyev Institute for Public Health of the National Academy of Medical Sciences of Ukraine", Kyiv, Ukraine*
Oksana Ananyeva, *State Institution "O.M. Marzeyev Institute for Public Health of the National Academy of Medical Sciences of Ukraine", Kyiv, Kyiv, Ukraine*
Varvara Morguleva, *State Institution "O.M. Marzeyev Institute for Public Health of the National Academy of Medical Sciences of Ukraine", Kyiv, Ukraine*

Background/Aims: As part of the integration processes on Ukraine's implementation of international standards: Convention on Long-range Transboundary Air Pollution, Directive 2010/75/EU, Directive 2008/50/EU etc, the question arises as to reduce and control emissions of pollutants into the air from industrial and preventing their impact on public health.
The aim of the research is development of emission standards for PM, NO2, SO2 and CO from the metallurgical industry based on results human health risk assessment. The study included 61 stationary source emissions of iron production equipment by 9 powerful metallurgical enterprises of Ukraine.
Methods. Averaged (1, 24-hours, annual) concentrations were calculated using air pollution dispersion model ISC-AERMOD v.8.8.9 (80 receptor points with 500 m grid spacing, 10 km buffer zone). Application of this modeling algorithm allowed counting in terrain, land-use peculiarities, annual meteorological observations, source parameters and emission characteristics in calculation procedure. Computed concentrations were compared with field data in terms of consistency. Demographic data (for adult and child population) was processed by ArcGIS 10.0 tools and decoded according to the places of residence. Zones of the highest density of exposed population were identified. Risk criteria assessment was completed according to approved U.S. EPA, WHO procedure of risk assessment.
Results. According to the data obtained in applied dispersion model and field studies it was found that the highest levels of inhalation effects on peoples are characteristic of emission of NO2, SO2 and CO. Averaged concentrations were: PM (min/max, mg/m3: C1-h=0,03÷0,52, C24-h=0,004÷0,06, Can=0,001÷0,009); NO2 (min/max, mg/m3: C1-h=0,012÷0,63, C24-h=0,004÷0,095, Can=0,001÷0,005); SO2 (min/max, mg/m3: C1-h=0,011÷0,64, C24-h=0,013÷0,12, Can=0,001÷0,007) and CO (min/max, mg/m3: C1-h=0,48÷5,8, C24-h=0,26÷1,21, Can=0,031÷0,14). Calculated risks are subject to acute and chronic health effects of inhaled exposed population and implemented conservation measures to reduce emissions. Found that in case of emission standards (emission at source) on equipment for iron metallurgical enterprises at level: PM - 50 mg/m3; CO - 50-2890 mg/m3 (for various technological production units); NO2 - 120 mg/m3; SO2 - 200 mg/m3, risk to the population is at a minimum.
Conclusion. This research is allowed to unify the requirements related to air quality control and to the adoption of adequate environmental protection measures, in accordance with European standards.

UTRECHT
THE NETHERLANDS
OCTOBER 9-13
ISES 2016

Tu-Po-16

Assessing and managing infectious risk: a conceptual model for exposure scientists

Yuli Huang, *National Kaohsiung First University of Science and Technology, Kaohsiung, China, Republic of (Taiwan)*
Bing-Mu Hsu, *National Chung Cheng University, Chiayi, China, Republic of (Taiwan)*

Recent emergence in infectious disease outbreaks such as Zika virus, enterovirus, rotavirus and foodborne diseases have brought forth the importance of infectious disease prevention. Various models have been proposed for estimating the infectious risk. Quantitative Microbial Risk Assessment (QMRA), for example, uses risk assessment framework similar to that for chemical risk assessment, and predicts risk of infection under given exposure scenarios. A problem with the approach is that actual dose-response model for a pathogen may be available. In addition, it is often difficult to estimate retrospectively for potential exposure or dose level. There are two major differences between chemical and microbial risk assessment. Unlike chemical hazards which typically require a long time to develop, infection may occur as a result of single or multiple exposures, and the exposure level are often difficult to quantify. In many cases, exposure to pathogen may only be assessed qualitatively. In order to better assess risk of infection, a conceptual model is proposed that looks at likelihood of pathogen exposure as a proxy for infection risk. The conceptual model consists of three main component: pathogen, environment and host. The three components intersect with each other. In most situations, the pathogen may be present in the ambient environment, but does not present a health risk until upon exposure. Interactions between host and environment may be considered an exposure, but risk of infection would only be considered when pathogen may come in contact. Even if a host is to come in contact with the pathogen, infection may not occur unless the individual is susceptible to the pathogen. Only susceptible hosts are subject to infection upon pathogen exposure, thus the risk of infection rely on two essential conditions: exposure to pathogen and susceptibility to infection. The emergence of a disease outbreak may be considered a result of increased exposure, but in some cases an outbreak may be a result of increased susceptibility. Environmental monitoring for potential pathogen exposure may help reduce the risk of infection, and reducing susceptibility (e.g., through immunization) may help prevent infection.

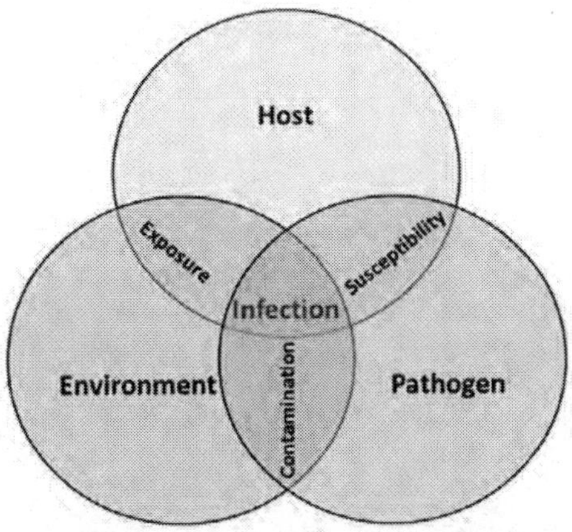

Conceptual framework for infectious risk assessment

Tu-Po-17

Heritability of Synergistic Interactions Following Co-Exposure to Anticancer Drugs in Genetically-Diverse Lymphoblastoid Cell Lines

Kyle Roell, *North Carolina State University, Raleigh, NC, United States*
John Jack, *North Carolina State University, Raleigh, NC, United States*
David Reif, *North Carolina State University, Raleigh, NC, United States*
Alison Motsinger-Reif, *North Carolina State University, Raleigh, NC, United States*

Real-world exposure scenarios for environmental chemicals or pharmacological drugs often involve significant co-exposure, wherein mixtures of several compounds are involved. However, interindividual responses are most often studied within a single-compound scenario, despite clinical evidence that compounds do not act independently when co-administered. Such synergistic (or antagonistic) effects occur between chemicals when the observed effect of the combination is more (or less) than what would be predicted from the effects of each agent working alone. Modeling synergistic interactions presents challenges in both quantitative modeling and underlying biology. There are a number of statistical methods for modeling synergy from a qualitative perspective, but rigorous quantification and detection of synergy, specifically within drug/chemical mixtures, is limited. Additionally, more investigation is needed into underlying mechanisms of synergistic effects, particularly in regards to potential genetic etiology. We have taken a novel approach to study synergy between chemotherapeutic drugs by applying known methods to show that synergy is heritable in a proven model of cytotoxic response using lymphoblastoid cell lines (LCLs). First, we used the Chou Talalay combination index approach to quantify synergy and results suggest interactions among common chemotherapy drugs. Next, we then used these results to test for a genetic component of variation in synergistic response. Importantly, we identified significant genetic components (> 50% heritability) among several drug combinations. Building on this evidence of a heritable component in synergistic response to co-exposure, we are performing genome wide association studies to identify candidate genes that explain a significant amount of synergistic variation.

Tu-Po-18a

Improving Risk assessment of Metal mixture for Neurotoxicity: in-vitro Toxicological interactions studies of metal mixture

Venkatanaidu Karri, *Universitat Rovira i Virgili, Tarragona, Tarragona, Spain*
Marta Schuhmacher, *Universitat Rovira i Virgili, Tarragona, Tarragona, Spain*
David Ramos-López, *Parque Científico de Barcelona (Barcelona Science Park), Barcelona, Barcelona, Spain*
Vikas Kumar, *Universitat Rovira i Virgili, Tarragona, Tarragona, Spain*
Raju Sharma, *Universitat Rovira i Virgili, Tarragona, Spain*

Human exposure to toxic heavy metals is a global challenge. Frequently occurring metals in environment as Pb, Cd, As and MeHg often in concurrent exposure, have a common disease outcome of cognitive dysfunction. Potential risk of metal mixture exposure could be more than single metal due to their common mode of actions which may have synergistic or additive interaction profile rather than independent as often assumed. Until now, there are no reported published studies that extend the analysis to these metals simultaneously to understand the common mechanism. Experimental data on mixture has been real constraint in development of predictive risk assessment models linking external exposure to adverse outcomes. In a recent review report, we have found that the combination of metals may produce additive/synergetic effects due to their common binding affinity with NMDA receptor (Pb, As, MeHg), Na+ - K+ ATP-ase pump (Cd, MeHg), biological Ca+2 (Pb, Cd, MeHg), and Glutamate neurotransmitter (Pb, MeHg). To further validate these assumptions and determine interaction profile of metal mixtures, in-vitro approach (cytotoxicity and proteomics) has been proposed. The specialised cell culture system can predict the toxicity of complex mixtures and to gain further insight into the mechanistic processes of these metal mixtures. The objective of this study is to develop the efficient approach for assessment of metal mixture (Pb, Cd, As, MeHg) interaction profile by using the cytotoxicity studies. We present the preliminary results of our cytotoxicity experiments.
In this study, we employed mice HT22 cell line as an in-vitro model to evaluate the interaction profile of these metal mixtures. Based on the review report, we investigated preliminary MTT/NR cytotoxicity screening by using reference IC50 values of Pb, Cd, As and MeHg for neuronal cells (in vivo) as a reference dose. With the literature reference doses of these metals, results shows more than 95% mortality of cells for both individual and combination of mixture. This indicates higher sensitivity of HT22 cell line to the in vivo metal concentrations values. Further, screening with lower doses shows the percentage of cell death decreases and with new dose-response curves we were able to establish new IC50 values for Hippocampal cells. For certain mixture combination, we have found that 50% inhibition of cells for mixture shifts to lower concentration compare to individual dose curve, which shows more than additive effects of these mixture combinations. This screening test will help us for selecting the defined concentration and mixture combination for further proteomics analysis.

Measuring/monitoring/strategy

Tu-Po-19

Creating a risk index for allergic diseases with indoor and outdoor risk factors in Seoul:

SungChul Seo, *Korea University, Seoul, Korea, South*
Dohyeong Kim, *University of Texas at Dallas (, Richardson, Texas, United States*
Ki-Youn Kim, *Catholic University of Pusan, Busan, Korea, South*
Hanjong Ko, *Korea National Open University, Seoul, Korea, South*

Background: The existing literatures on allergic diseases have identified indoor and outdoor environmental risk factors but not considered the relative weight for each factor, or considered relative weights but not linked analysis to geographic location at disaggregated levels. The lack of understanding for spatial variation of various risk factors results in uninformed decisions on allocating resources to reduce allergic disease burden. We aimed to develop a risk index and explore the spatial association of a risk index with allergic diseases.
Method: We used GIS and statistical modeling to analyze the household survey data collected from 2,147 children in kindergarten located in Seoul, including a series of indoor risk factors for allergic diseases. As for outdoor risk factors, since there are only 25 monitoring stations measuring the level of ambient air pollutants in Seoul, a spatial interpolation technique was used to match with the survey data. By integrating all relevant data, we performed statistical analysis on all data layers together.
Results: Statistical analysis reveals that each of three allergic diseases is associated with a different group of risk factors (e.g., asthma with SO2, indoor smoking, mold, atopic dermatitis with NO2, gender, and allergic rhinitis with ozone, indoor leaking, etc.). Using the coefficients from the statistical models, we created risk index for each allergic disease for every household in the survey, along with district-level priority maps coded by the index covering the entire area of Seoul. The resulting maps use weighted risk factors to spatially locate modeled risk zones and highlight critical areas for targeted intervention. This GIS-based statistical approach enables environmental and public health policymakers to design and implement programs that protect people before they suffer from allergic diseases, and advance the scientific community's understanding of the spatial distribution and magnitude of allergic diseases.

Tu-Po-20

Importance of size-selective particle sampling for assessing occupational exposures - Results from three different occupational settings

Susana Viegas, *ESTeSL-IPL, Lisbon, Portugal*
Tiago Faria, *Environment and Health Research Group - ESTeSL/IPL, Lisbon, Portugal*
Carla Viegas, *Environment and Health Research Group - ESTeSL/IPL, Lisbon, Portugal*
Elisabete Carolino, *Environment and Health Research Group - ESTeSL/IPL, Lisbon, Portugal*

Sampling the total air concentration of particulate matter (PM) only provides a basic estimate of exposure that normally not allows correlating with the observed health effects. Therefore is of extreme importance to know the particles size distribution and, in more detail, the exposure to fine particles (≤ 2.5 µm). This particles dimension corresponds to the respirable fraction, the one that can result, besides local effects, in systemic effects due to particle deposition and clearance from the lungs and transport within the organism.
This study intended to describe occupational exposure to PM2.5 in three units located near Lisbon: swine and poultry feed production and waste management. It was performed a size-selective particle sampling in three to five workplaces of each unit with an aerosol monitor (DustTrak II model 8532, TSI®).
Data showed poultry feed unit with higher values, with statistical significant differences from the others units (p's <0.05). In swine feed values range was 0.007 to 0.143 mg/m3 (0.054 + 0.042), being the reception room the workplace with higher values, in poultry feed the values were between 0.028 and 0.198 mg/m3 (0.098 + 0.061) with the bagging line as the workplace with higher values and, finally, in waste management values ranged from 0.036 to 0.059 mg/m3 (0.046 + 0.006) being the sorting cabinet the workplace with higher values (Figure 1)
This data allow a better estimation of particle penetration into the thoracic and respiratory regions of the respiratory tract and a better prediction of PM exposure health effects. Additionally, allows also to identify the workplaces where investment to prevent and control exposure should be prioritize.

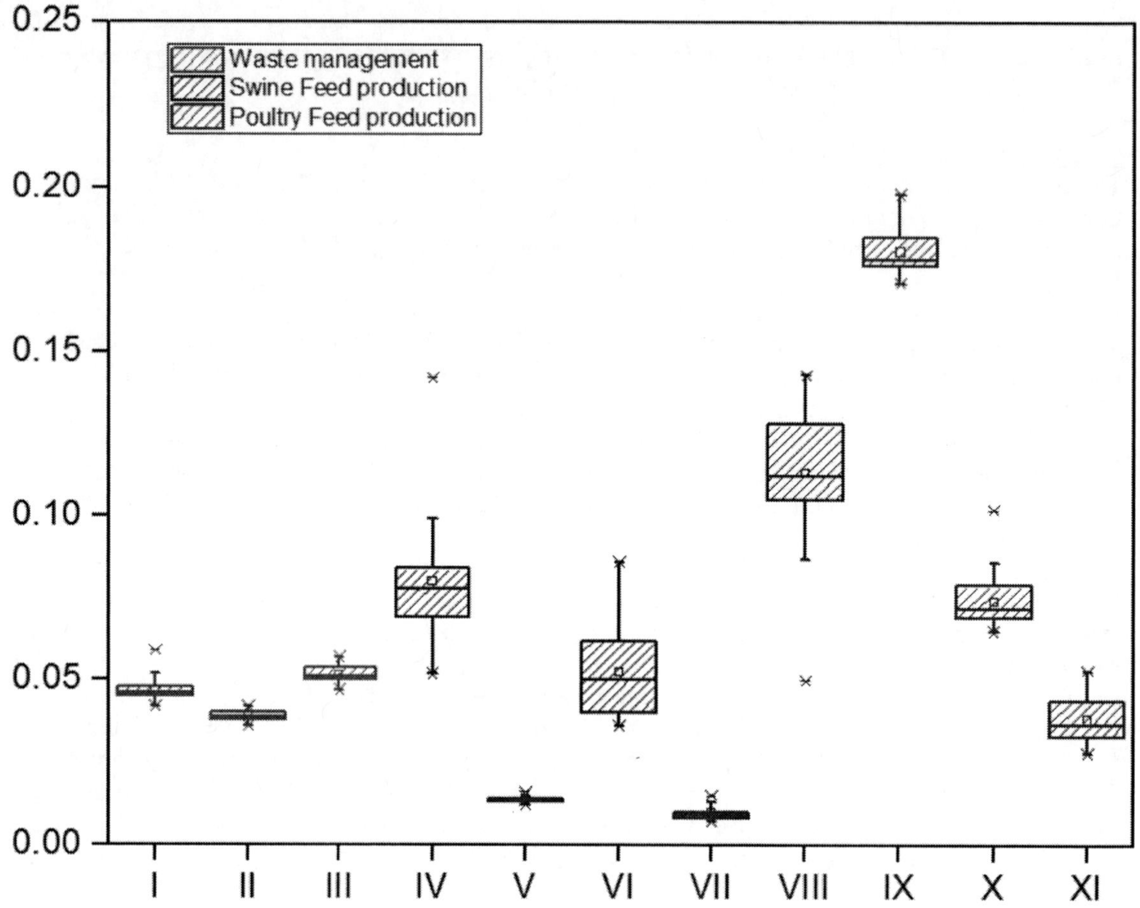

Figure 1 - Distribution of PM2.5 values in each occupational setting

Tu-Po-21

French interregional variability of exposure to sunscreen products

Laëtitia Noël, *Laboratory of risk assessment for consumer, Brest, France*
Audrey Bernard, *Laboratory of risk assessment for consumer, Brest, France*
Anne-Sophie Ficheux, *Laboratory of risk assessment for consumer, Brest, France*
Nicolas Dornic, *Laboratory of risk assessment for consumer, Brest, France*
Gregoire Chevillotte, *Laboratory of risk assessment for consumer, Brest, France*
Alain-Claude Roudot, *Laboratory of risk assessment for consumer, Brest, France*

Sunscreen products are used by a significant part of the French population, between 14 and 46 % for adult depending on the type of product. Sunscreen use depends partly on the need to protect the skin from sun adverse effects and can thus differ according to the region of France the user live. The aim of this work was to assess the variability in the quantity and frequency of sunscreen products use between the Ile-de-France, the North and the South of France and to evaluate the corresponding cutaneous exposure.
A web questionnaire was conducted among 5657 French adults, on the usage patterns of 141 cosmetic products, including sunscreen products. This survey allowed an assessment of the frequency of use for all products used during the past 12 months by each participant. As geographical area of residence was inquired, frequency of use has been assessed for the Ile-de-France, the North and the South of France. Then a face-to-face survey was conducted on 1078 French people in order to assess the amount of cosmetic products consumed by the French population. It was performed in four towns in order to represent the different French region and for solar products; data were also collected during the summer on the beach. The frequency and quantity of sunscreen products use for the different region were then compared by running Mann-Whitney statistical analyses with the XLSTAT software in order to assess the interregional variability. Probabilistic exposure assessment to sunscreen products was performed using Monte Carlo random simulations with @Risk 6 software.
Results obtained for the frequency and the quantity of use of sunscreen in spray indicated significant statistical differences between regions of France. A mean frequency of 86.34 uses per year and a mean amount of 3.46 g/use were obtained for the North. The mean frequency was of 100.84 uses per year and the mean amount was of 5.49 g/use in the South. For Ile-de-France, the mean frequency was of 82.71 uses per year and the mean amount was of 5.55 g/use. The exposure assessment to this product gave a mean cutaneous exposure of 17.49 mg/cm2/year in the North, 26.22 mg/cm2/year in the Ile-de-France and 30.69 mg/cm2/year in the South. Similar results were obtained for other sunscreen products.
Results of this study showed the importance of taking into account the variability between subpopulations in order to ensure the consumer safety correctly.

Tu-Po-23

TTC for botanicals - data analysis to substantiate and extend the TTC approach to botanicals

Kirstin Kosemund, *Procter & Gamble, Schwalbach, Germany*

The presence of botanicals in consumer products has become an important consumer criterion for the purchase of specific product formulations. Botanicals are complex mixture of mostly unknown natural chemicals and classical toxicological data are often lacking. While some botanicals are used broadly for human and animal survival as food, flavours/spices, or medicines, some botanicals are also known to cause various toxic effects. The combination of these uncertainties makes risk assessment for botanicals a challenging task. Establishing safe botanical exposure levels in absence of carcinogenicity and genotoxicity data, which address the most sensitive endpoint becomes a critical and often rate limiting step in botanical risk assessment. To support the presence of single chemical substances at low exposure levels in consumer products the TTC based approach can be used in absence of full toxicological characterization. This approach can also serve as a basis for deriving exposure limits for botanical substances. The TTC decision tree approach starts with the identification and evaluation of possible structural alerts for genotoxicity and high potency carcinogenicity. This step applies an exposure threshold of 0.15 µg/person/day (Munro, 1996, Kroes, 2004). We propose to extend the TTC approach to botanicals, relying on this first TTC exposure limit of 0.15 µg/day (0.0025 µg/kg bw/day) and adjusting it based on the concentration of natural chemical constituents of concern that are found in plants. An evaluation of genotoxic/DNA reactive substances found in plants has been made and the concentration data in plants compiled. Based on the analysis of these data in many hundred plant species, we have derived upper confidence levels for the concentration of substances of concern in plants that can in turn be used to extend the TTC for crude botanical mixtures.

Tu-Po-24

Characterization of indoor air quality by canister sampling and TD-GC-MS analysis

Gwendolyn Beckmann, *Radboud University Nijmegen Medical Centre, Nijmegen, Netherlands*
Rob B.M. Anzion, *Radboud University Nijmegen Medical Centre, Nijmegen, Netherlands*
Paul T.J. Scheepers, *Radboud University Nijmegen Medical Centre, Nijmegen, Netherlands*

Background
A canister is a vacuumed container designed for air sampling. This technology is available for a range of air sampling applications, including personal air sampling. Canisters have a number of interesting features but are still not much used in Europe. We have explored the value of canisters in our field of interest: indoor air quality (IAQ).
Objective
Evaluation of the measurement of volatile organic compounds (VOCs)in indoor air by canister sampling followed by analysis on a thermal desorption gas chromatograph mass spectrometer (TD-GC-MS) system. We performed an exploratory study on the impact of emissions from emergency helicopters and testing diesel-fueled emergency power supplies as potential sources of contamination of our hospitals' IAQ.
Method
For the VOC measurements air samples were collected using Entech Silonite canisters equipped with CS1200 Silonite coated samplers or grab samplers (Interscience). For calibration, we used TO14 and TO15 calibration standard VOC mixtures that were provided in gas cylinders. For calibration the cylinder was connected to an air server (Series 2, Markes). The analytical instrument consisted of a thermal desorber (Unity 2, Markes) and a gas chromatograph mass spectrometer (Focus/ISQ , Thermo). We collected samples of 100 - 250 mL per canister.
Results
The limit of quantification for VOCs was 0.1 µg/m3. Healthcare workers were asked to collect a grab sample when they picked up a kerosene odor. Simultaneously a sample was collected at a service building near the helicopter platform. In a second campaign we collected indoor air samples at two operation rooms before and during the test run of an emergency power supply. These measurements were compared to the results of a reference sample that was collected at the exhaust outlet of the power supply. Relating to the indoor air quality in the hospital, the concentrations of acetone, isopropanol and ethanol were somewhat elevated due to cleaning and disinfection practices. We did not detect VOCs in the indoor air that were characteristic for helicopter emissions or diesel exhaust.
Conclusion
Canisters can be used for self-assessment grab sampling in addition to the conventional use of air sampling pumps.

Tu-Po-25

Recovery rates in the measurements of the concentrations of organophosphorus pesticide metabolites in urine extracted from children's diaper

Naoko Oya, *Nagoya City University Graduate School of Medical Sciences, Nagoya, Japan*
Yuki Ito, *Nagoya City University Graduate School of Medical Sciences, Nagoya, Japan*
Keisuke Hioki, *Nagoya City University Graduate School of Medical Sciences, Nagoya, Japan*
Arisa Aoi, *Nagoya University Graduate School of Medicine, Nagoya, Japan*
Yuka Sugiura, *Nagoya University Graduate School of Medicine, Nagoya, Japan*
Jun Ueyama, *Nagoya University Graduate School of Medicine, Nagoya, Japan*
Tomoko Oguri, *Nagoya City University Graduate School of Medical Sciences, Nagoya, Japan*
Sayaka Kato, *Nagoya City University Graduate School of Medical Sciences, Nagoya, Japan*
Takeshi Ebara, *Nagoya City University Graduate School of Medical Sciences, Nagoya, Japan*
Michihiro Kamijima, *Nagoya City University Graduate School of Medical Sciences, Nagoya, Japan*

Background: Recently epidemiological studies to examine the relationship between pesticide exposure and neurodevelopmental effects have attracted more attention. Although central nervous system rapidly develops in early childhood, no basic information about pesticide exposure of children who cannot control urination has been available in the world.
Objectives: This study aimed to increase the sensitivity of the measurements of concentrations of organophosphorus pesticide (OP) metabolites in urine extracted from used diaper and determine urinary concentrations of OP metabolites in 1.5-year-old children.
Methods: Urine samples were extracted from the diapers by acetone, dried up with a gentle nitrogen stream, and were stored at -80 °C until analyses. Urinary dimethylphosphate (DMP), diethylphosphate (DEP), which passed through the solid-phase extraction (SPE) column, and dimethylthiophosphate (DMTP), diethylthiophosphate (DETP), dimethyldithiophosphate (DMDTP), and diethyldithiophosphate (DEDTP) which were extracted from SPE column using 2.5% NH3 water including 50% acetonitrile, were analyzed by ultra-performance liquid chromatography with tandem mass spectrometry (LC-MS/MS). Both recovery rate throughout SPE procedure and whole procedure including urine extraction from a diaper were calculated. In order to improve recovery rate during SPE procedure, elution conditions were examined. Furthermore, we recruited 18-month-old children participating in Japan Environment and Children's Study (JECS) at the Aichi Regional Center of JECS as an adjunct study, and we collected used diapers from 104 children (18-21 months of age, 53 males and 51 females) from June 25 to July 31 in 2015 (participation rate was 82%) and determined urinary concentrations of OP metabolites. Respective deuterium-labeled dialkylphosphates (DAPs) were used as internal standards.
Results: The condition using 2 ml solution of 2.5% NH3 including 50% acetonitrile at 30°C yielded the highest recovery. Recovery rates of SPE procedure and extracted procedure from diaper of DMP, DEP, DMTP, DETP, DMDTP, and DEDTP at the low (DEDTP 0.22~DMP 26.3 µg/L) and high (DEDTP 20.09~DMP 313.43 µg/L) concentrations were 110.3% and 69.7%, 107.4% and 23.7%, 85.5% and 23.0%, 110.7% and 47.7%, 99.5% and 61.7%, 75.1% and 61.7%, respectively. The geometric means of the urinary DMP, DMTP, DMDTP, DEP, DETP, DEDTP, and total DAPs (ΣDAPs) in 104 children were 6.6 (0.74-98.2), 1.9 (ND-78.3), 1.6

(ND-117.1), 0.2 (ND-88.2), 0.2 (ND-5.9), 0.1 (ND-1.5) μg/L and 116 (8.8-1389.0) nmol/L, respectively. Between-individual variability of ΣDAP was about 160 times. In conclusion, a highly sensitive method for the simultaneous quantitation of urinary OP metabolites from used diaper was developed, which could be applied to determine urinary concentrations of OPs metabolites in 1.5-year-old children.

Tu-Po-26

A study design to assess exposure levels of insecticides in 1.5-year-old children in Aichi Regional Subcohort of the Japan Environment and Children's Study

Yuki Ito, *Nagoya City University Graduate School of Medical Sciences, Nagoya, Japan*
Takeshi Ebara, *Nagoya City University Graduate School of Medical Sciences, Nagoya, Japan*
Naoko Oya, *Nagoya City University Gradute School of Medical Sciences, Nagoya, Japan*
Jun Ueyama, *Nagoya University Graduate School of Medicine, Nagoya, Japan*
Tomoko Oguri, *Nagoya City University Graduate School of Medical Sciences, Nagoya, Japan*
Sayaka Kato, *Nagoya City University Graduate School of Medical Sciences, Nagoya, Japan*
Michihiro Kamijima, *Nagoya City University Graduate School of Medical Sciences, Nagoya, Japan*

Background: A nationwide and government-funded birth cohort study in Japan named the Japan Environment and Children's Study (JECS) started from January 2011 to elucidate environmental factors that affect children's health and development. After 3-year recruitment period of pregnant women living in designated study areas including Aichi, Japan, the children to whom they gave birth are now followed until 13 years old.
Objectives: This presentation aims to describe a study protocol of an adjunct study of JECS to assess exposure levels of insecticides in 1.5-year-old children in Aichi.
Methods: For this adjunct study, we have recruited 1.5-year-old JECS participants since June 2015, which will be completed in June 2016. Their guardians are asked to take part in the study at health-care centers on the occasions of municipal checkups for children at the age, and informed consents for this study were obtained. Food intake questionnaire to assess exposure amount of insecticides from diets was developed based on data about permitted application of pesticides for registered crops in Japan. The intake of each crop is counted as a number of child-size spoons by the guardians at home for each of three meals during the day, and is recorded on the questionnaire. The questionnaire also asks them whether or not spraying insecticides during the preceding week. To collect urine, the children wear designated disposable diapers during the night following the three meals recorded, and the questionnaire and the used diapers are sent to our laboratory as refrigerated cargoes.
Results: We collected the questionnaire and used diapers from 618 children from June 25 through December 31 in 2015 (consent rate was 79.1%). Their months of age were between 17 and 24 and the mode was 19 (54.9%). Three hundred and eighteen male and 300 female children were enrolled. One thousand children in total are expected to take part in this adjunct study by the end of the recruitment period. Age of the mothers at the delivery and the birth weights of the participating children are distributed from 18 to 44 (median, 33; interquartile range, 30-36) and from 815 to 4,400 (median, 3,044; interquartile range, 2,748-3,282), respectively. The rate of low birth weight (<2500 g) was 9.5%. These demographic data suggest that the subjects in this survey represent the population in Aichi regional subcohort of JECS and entire population in Aichi. Thus, it is expected that this study reveals exposure levels of insecticides in the area.

Tu-Po-27

Indoor air quality in French hospitals: large scale sampling campaigns and first physical-chemical results

Arnaud Florentin, *Lorraine University, Nancy, France*
Olivier Blanchard, *EHESP, Rennes, France*
Alexandre Rivier, *CHRU de Nancy, Nancy, France*
Pierre-Yves Donnio, *CHU de Rennes, Rennes, France*
Monique Guillaso, *CHRU de Nancy, Nancy, France*
Pierre Le Cann, *EHESP, Rennes, France*
Amandine Luc, *CHRU de Nancy, Nancy, France*
Sorya Belaz, *CHU de Rennes, Rennes, France*
Fabien Mercier, *EHESP, Rennes, France*
Emilie Surget, *EHESP, Rennes, France*
Jean-pierre Gangneux, *CHU de Rennes, Rennes, France*
Estelle Baures*EHESP, Rennes, France*

Aim
Indoor air quality in hospital is an important issue. Currently, the biological indoor air quality monitoring and control in hospital are necessary and integrated in prevention strategy of hospital-acquired infections. However, the chemical contamination of indoor air in hospitals, although it is established, is little-known and rarely studied. This chemical contamination may be associated with a wide range of specific compounds emitted from various used products and materials but also influenced by the outdoor environment. Finally some activities in link with practical in hospital may also lead to human exposure.
Methods
This study was conducted in June 2014 and February 2015 ("summer" and "winter" sampling campaigns) during twice four consecutive days in two hospitals in Rennes (Brittany, West part of France) and in Nancy (Lorraine, East part of France). For each hospital, air samples were collected in seven rooms (the reception hall, a patient room, a nursing care, the parasitology mycology laboratory, a post-anesthesia care unit, a plaster cutting room and the flexible endoscope disinfection unit) in order to estimate the spatial (related to the healthcare activities and between 2 hospitals) and temporal (daily, weekly and seasonal) variability. During these both campaigns, 34 volatile organic compounds (VOCs), 7 aldehydes and 13 semi-volatile organic compounds (SVOCs) were measured. PM10 and PM2.5 samples were collected and gravimetric analysis was used to determine the particle mass on the filter. In parallel, microbial agents (culture and PCR: bacteria, fungi and viruses), ambient parameters (temperature, relative humidity, pressure and carbon dioxide) and the particles number (from 0.3 to 25 µm) were measured.
Results
The results showed that the main chemical compounds found are in the same order in the both hospital. Mean concentrations were for alcohols (334 and 23 µg/m3 respectively for ethanol and isopropanol) but also several aromatic and halogenated hydrocarbons, aldehydes (4.5 µg/m3 for formaldehyde), ketones (17 µg/m3 for acetone), ethers (9.5 µg/m3 for ether) and terpenes (2.7 µg/m3 for limonene). The SVOCs were quantified in all the sampling rooms (mainly phthalates: 0.26 µg/m3 for diisobutylphtalate). Mean concentrations were for fungi (226CFU/m3) and bacteria (352CFU/m3). The results are in the same order in two hospitals.
Conclusion

Our study showed a low chemical pollution in the two hospitals. We found a spatial and temporal variability of fungal and particulate contamination that seemed correlated with the activity and ventilation.

Tu-Po-28

Testing a procedure for the identification of emerging chemical risks in the food chain

Caroline Gabrielle Merten, *European Food Safety Authority, Parma, Italy*
Jan Oltmanns, *Forschungs- und Beratungsinstitut Gefahrstoffe GmbH (FoBiG), Freiburg, Germany*
Marie-Léonie Bohlen, *Forschungs- und Beratungsinstitut Gefahrstoffe GmbH (FoBiG), Freiburg, Germany*
Sylvia Escher, *Fraunhofer Institute for Toxicology and Experimental Medicine ITEM, Hannovre, Germany*
Oliver Licht, *Fraunhofer Institute for Toxicology and Experimental Medicine ITEM, Hannovre, Germany*
Matthew Macleod, *Stockholm University, Stockholm, Sweden*
Vittorio Silano, *European Food Safety Authority, Parma, Italy*
Nikolaos Georgiadis, *European Food Safety Authority (EFSA), Parma, Italy*
Georges Kass, *European Food Safety Authority (EFSA), Parma, Italy*

The aim of this study was to test whether substance-specific data generated and made available in electronic form under REACH can be used for the identification of substances of potential concern in the food chain. For this purpose, 100 data-rich substances registered under the REACH regulation together with four substances selected as positive controls were evaluated. The procedure consisted of a multi-step selection process following a sequence of selection criteria. The evaluation criteria took into account parameters related to exposure (tonnage information, environmental release, biodegradation, potential for bioaccumulation) and toxicity endpoints (repeated dose toxicity, reproductive and developmental toxicity and genotoxicity). All substances were scored for each parameter grouped into six blocks of parameters. The ACC-HUMAN steady software was used to evaluate the potential for bioaccumulation. The extraction of experimental data generated under REACH was successful in principle, but encountered several problems, both in relation to the extraction process itself and the subsequent evaluation steps. Overall, the approach developed showed a good level of differentiation between the percentage of high and low scores (potential exposure or potential toxicity). Several weighting scenarios were developed to aggregate the parameters related to exposure with those related to toxicity endpoints and to enable a quantitative ranking of the 100 data-rich substances. These scenarios identified substances produced in high tonnage, subject to environmental release with limited biodegradation and high potential for accumulation in food as substances of potential concern in the food chain due to their toxicity profiles. The four positive control substances received a high score. Additional analyses compared the scores derived from experimental data with those derived from predicted (in silico) data for the same set of 100 substances, allowing to adapt our approach to data-poor substances of potential concern. The critical point in the scoring system is that all individual scores are kept separately allowing further differentiation in the possible scenarios. In conclusion a procedure was developed that consisted in a multi-step selection process allowing the identification of chemical substances of potential concern for the food chain and that could be adapted using a tiered approach to screen the entire dataset currently registered under the REACH regulation.

Tu-Po-29

Assessment of exposure to Aflatoxin M1 Oaxaca cheese in the population of Veracruz City, Mexico

Estela Hernandez, *Université Bretagne Occidentale, Brest, France*
Alejandra Ramirez-Martinez, *Université Bretagne Occidentale, Brest, France*
Nathalie Wesolek, *Université Bretagne Occidentale, Brest, France*
Anne-Sophie Ficheux, *Université Bretagne Occidentale, Brest, France*
Nicolas Dornic, *Université Bretagne Occidentale, Brest, France*
Marco Antonio Salgado, *Instituto Tecnológico de Veracruz, Veracruz, Mexico*
Guadalupe del Carmen Rodriguez, *Instituto Tecnológico de Veracruz, Veracruz, Mexico*
Alain-Claude Roudot, *Université Bretagne Occidentale, Brest, France*
Víctor José Robles, *Instituto Tecnológico de Veracruz, Veracruz, France*

In Mexico, several grain producing states have climatic conditions that promote the presence and growth of mycotoxin-producing fungi, such as Veracruz. Veracruz is an important producer of milk for dairy products (such as cheese) and grains for cattle consumption. Taking in consideration that some grains produced on the state may be contaminated by Aflatoxin B1 and may be present in milk products as Aflatoxin M1 it is fundamental to assess the exposure of population to this toxin. The present study focused on the exposure of the population of Veracruz City, one of the biggest urban areas in the state via the consumption of Oaxaca cheese. Oaxaca cheese is one of the most consumed cheeses in the region; its consumption has not been assessed until now. To achieve this goal the intake of Oaxaca cheese, as well as the levels of Aflatoxin M1 in Oaxaca cheese were evaluated. Dietary intake of Oaxaca cheese was assessed via a 7-day food dairy questionnaire. A total of 1100 people were interviewed during 2014 and 2015. 25 samples of Oaxaca cheese from different small markets in the city were randomly sampled during 2014 and 2015. The obtained samples were analyzed to determine the concentration levels of Aflatoxin M1 following validated methods including immunoaffinity chromatography and high performance liquid chromatography with fluorescence detection. The exposure of the population of the city of Veracruz to Aflatoxin M1 was assessed through the combination of the Probabilistic Density Functions (PDF) for Aflatoxin M1 levels and the consumption data of Oaxaca cheese (Probabilistic Methodology). Results showed that 92% of the samples concentration level was below 0.02 µg Aflatoxin M1/kg. This concentration levels are below the safety levels set by the European Commission. Evaluation studies showed that exposure of children was higher than that found in adults (1.1 x 10-4 µg AM1/kg bw/day vs 4.56 x 10-5 µg AM1/kg bw/day, respectively). These values revealed that population of Veracruz City does is not at risk to Aflatoxin M1 via the consumption of Oaxaca Cheese. The present study is the first investigation on the exposure of Mexican population to Aflatoxin M1.

Tu-Po-30

Associations between plasma concentrations of PCB 28 and possible indoor exposure sources in Danish school children and mothers

Lisbeth E. Knudsen, *University of Copenhagen, Copenhagen, Denmark*

Polychlorinated biphenyls (PCBs) are ubiquitously present in the environment and are suspected of carcinogenic, neurotoxic and immunotoxic effects. Significantly higher plasma concentrations of the congener PCB 28 occur in children compared to adults. Exposure in schools may contribute to this difference.

PCB 28 was analyzed in plasma samples from 116 children aged 6-11 years and 143 mothers living in an urban and a rural area in Denmark and participating in the European pilot project DEMOCOPHES (Demonstration of a study to COordinate and Perform Human Biomonitoring on a European Scale). In Denmark, PCBs were used in construction in the period 1950-1977, and year of construction or renovation of the homes and schools was used as a proxy for indoor PCB exposure. Linear regression models were used to assess the association between potential PCB exposure from building materials and lipid adjusted concentrations of PCB 28 in plasma, with and without adjustment for potential confounders.

Amongst the 116 children and 143 mothers, we were able to specify home construction period in all but 4 children and 5 mothers leaving 111 children and 138 mothers for our analyses. The median lipid adjusted plasma PCB 28 concentration was 3 (range: 1-28) ng/g lipid in the children and 2 (range: 1-8) ng/g lipid in the mothers.

Children living in homes built in the PCB period had significantly higher lipid adjusted plasma PCB 28 concentrations compared to children living in homes built before or after the PCB period. Following adjustment for covariates, PCB 28 concentrations in children were 40 (95% CI: 13; 68) percent higher than concentrations of children living in homes constructed at other times.

Our results suggest that PCB exposure in the indoor environment in schools and homes constructed during the PCB period may contribute significantly to children's plasma PCB 28 concentration. Efforts to minimize PCB exposure in indoor environments should be considered.

Egsmose EL, Bräuner EV, Frederiksen M, Mørck TA, Siersma VD, Hansen PW, Nielsen F, Grandjean P, Knudsen LE. (2016) Associations between plasma concentrations of PCB 28 and possible indoor exposure sources in Danish school children and mothers. Environment International 87: 13-19.

Tu-Po-31

General rules for a unified Hazard banding in compliance with the new Globally Harmonised System (GHS) for use in control banding tools

Mario Arnone, *IFA of the DGUV, Sankt Augustin, Germany*

Many control banding tools use hazard banding in risk assessments for the classification of hazards arising from occupational handling of hazardous substances. Hazard bands can be assigned using the hazard communication elements from the label or the safety data sheet (SDS) of chemical products.
The Globally Harmonised System (GHS) has resulted in a change in the hazard communication elements, i.e. hazard (H) statements instead of risk-phrases. In addition to the new H statements, GHS leads also to changes in the dilution rules for the classification of mixtures. Due to this fact hazard banding schemes that depend on the old form of safety information have to be adapted to the new legislation. The simple translation from R-phrases to H statements is problematic and not always possible. Therefore general rules for the assignment of hazards to hazard bands are proposed.
As basis for the assignment of hazard bands, the H statements and the dilution rules from GHS are used. All H statements were assigned to six hazard bands with respect to the severity of the underlying hazard. These hazard bands range from n.a. = "not applicable" for e. g. highly self-diluted products or non-health related statements through A = "low hazard", B = "moderate hazard", C = "high hazard" and D = "very high hazard" to E = "extremely high hazard". To support the important principle of substitution, the last one is reserved for proven carcinogens or mutagens, while all less severe hazards are assigned at worst to hazard band D. Hazard bands are assigned specifically to the route of exposure concerned in the respective H statement. This enables the user to adjust risk management measures to the specific uptake route.
The rules for assigning hazard bands with respect to the severity of the hazard to encourage substitution and the exposure route are in line with the requirements for qualitative risk characterisation described e.g. in the REACH guidance. The SDS of the products as source of the required hazard information and the GHS regulation as basis for the assignment of hazard bands ensures that the new hazard banding is in conformity with the new legislation on classification and labelling of chemicals.
This presentation will demonstrate that the implementation of this GHS hazard banding in various control banding tools can result in a unified classification of health hazards from the handling of hazardous substances.

Tu-Po-32

Identification and Treatment Options for Waste Streams of Certain Bromine Containing Flame Retardants (WAFER)

Mohamed Abdallah, *University of Birmingham, Birmingham, United Kingdom*
Martin Sharkey, *National University of Ireland Galway, Galway, Ireland*
Harald Berresheim, *National University of Ireland Galway, Galway, Ireland*
Stuart Harrad, *University of Birmingham, Birmingham, United Kingdom*

The WAFER project (http://www.wafer-research.com/) aims to evaluate the use of a portable XRF instrument in waste sites across Ireland for the purposes of screening items treated with brominated flame retardants (BFRs) classified as persistent organic pollutants (POPs). A database of the concentrations of polybrominated diphenyl ethers (PBDEs) and hexabromocyclododecane (HBCDD) in various waste categories is also being established to streamline the screening process.
Approximately 1000 samples are being taken from a range of waste items, which maight have been treated with PDBEs and/or HBCDDs including: casings from waste electrical and electronic equipment (WEEE); foam and upholstery from soft furnishings; foam and upholstery from end-of-life vehicles (ELVs); as well as expanded and extruded polystyrene building insulation foams.
Preliminary results obtained for polystyrene foam items have shown a strong correlation ($n=71$, $P<0.05$) between the XRF-measured total bromine content and that of GC/LC-MS measured POP-BFR content. In other subgroups of WEEE plastics from IT waste items, large and small domestic appliances, soft furnishing foams and fabrics and ELVs; not all samples with high Br counts (XRF measured) corresponded to high levels of POP-BFRs. This may be explained by the use of other BFRs in these samples (e.g. TBBP-A) or the use of the deca-BDE commercial formulation, which is not yet listed as a POP.
Overall, XRF readings show considerable variations in bromine content across the entire range of sampled items with notable highs of several tens of thousands of ppm in foam and upholstery items and concentrations in excess of 100,000 ppm in certain waste items, potentially well above the European Commission lower POP concentration limits of 0.1%.

Tu-Po-33

Consumer behaviour - compilation and evaluation of primary data

Astrid Heiland, *Federal Institute for Risk Assessment (BfR), Berlin, Germany*
Jan Oltmanns, *Forschungs- und Beratungsinstitut Gefahrstoffe GmbH (FoBiG), Freiburg, Germany*
Marie-Leonie Bohlen, *Forschungs- und Beratungsinstitut Gefahrstoffe GmbH (FoBiG), Freiburg, Germany*
Eva Kaiser, *Forschungs- und Beratungsinstitut Gefahrstoffe GmbH (FoBiG), Freiburg, Germany*
Stefanie Klenow, *Federal Institute for Risk Assessment (BfR), Berlin, Germany*
Klaus Schneider, *Forschungs- und Beratungsinstitut Gefahrstoffe GmbH (FoBiG), Freiburg, Germany*

Patterns of product use by consumers are essential information in consumer exposure scenarios. Frequency, duration, amount of products used, and product use location are parameters of interest. Fact sheets (e.g. from RIVM) and specific consumer exposure determinants (SCEDs, developed by industry associations) provide defaults derived from studies and expert judgement. These defaults are implemented in exposure calculation tools to support e.g. registrants to fulfil their information requirements under REACH. In this process, the origin and the designated purpose of the exposure parameters are often unclear and poorly documented. The objective of the project was to establish an inventory of the parameter values on the amount, duration, frequency, and location of consumer product use. The compilation was restricted to mixtures regulated under REACH. An extensive literature research was carried out to identify relevant studies that investigated exposure parameters. These parameters along with additional information regarding study design as well as a scoring that considers quality, validity, and applicability were recorded in an MS Excel® file. The products were categorized according to the REACH use descriptor system, which includes e.g. cleaning agents, adhesives, paints and lacquers. Overall, 37 relevant studies were identified that allowed extraction of parameter values, resulting in 822 datasets of which 43% related to use frequency, 26% to product amount, and 23% to use duration. The location of consumer product use is documented in relatively few datasets (9%). Looking at the product types, most datasets (57%) addressed cleaning activities followed by the use of air care products (10%) and paints (8%). To evaluate the quality of parameter values, a scoring system based on features which characterise empirical studies (e.g. study design, number of subjects, level of detail) was developed. The majority of datasets yielded a score of more than 60 (out of 120) for almost all parameters, but only a small fraction (about 10%) received a score of 90 or higher. However, none of the datasets achieved a score of 100 or higher. In conclusion, the knowledge base of published product use is weak. Our findings strongly call for further surveys focusing on consumer behaviour. The project was carried out by the Forschungs- und Beratungsinstitut Gefahrstoffe GmbH (FoBiG) in cooperation and on behalf of the German Federal Institute for Risk Assessment (BfR).

Tu-Po-34

ETS personal exposure levels of Japanese people measured by using a passive nicotine sampler

Satoshi Nakai, *Yokohama National University, Yokohama, Japan*
Miyuki Noguchi, *Seikei University, Tokyo, Japan*
Yoshihiro Suzuki, *Sibata Scientific Technology, Ltd., Souka, Japan*
Keiichi Arashidani, *University of Occupational and Environmental Health, Japan, Kitakyushu, Japan*
Yukio Akiyama, *University of Occupational and Environmental Health, Japan, Kitakyushu, Japan*
Takashi Amagai, *University of Shizuoka, Shizuoka, Japan*
Yoshika Sekine, *Tokai University, Hiratsuka, Japan*
Hiroshi Satoh, *Nagasaki International University, Sasebo, Japan*
Hideaki Matsuki, *Tokai Universi t y, Isehara, Japan*
Yukio Yanagisawa, *Kaisei Academy, Tokyo, Japan*

Background: ETS (environmental tobacco smoke) is one of the public concern. It is necessary to assess ETS personal exposure levels to evaluate the health effects and health risk of ETS. However, there have not been so many quantitative assessments about ETS exposure for epidemiologic study. In order to investigate the ETS exposure levels and the relationship between ETS exposure and factors including smoke perception, we conducted a nicotine exposure study. This study was conducted as a part of "comprehensive study on the social acceptance of ETS evaluated from the personal exposure and perception". In this paper, the distribution of ETS exposure levels was described.
Methods: Twenty-four hour ETS exposure levels for adult and child volunteers were measured by using improved MoNIC (monitor of nicotine) passive sampler at six areas in Japan. They were also asked to record the number of active and passive smoking, time activity pattern, etc. by using a questionnaire. For children, guardians were requested to fill the questionnaire.
Results: One-hundred sixty subjects (age 1 to 79) were included in this study. Median ETS exposure level for adults (>12 years old, passive smoker) was 1.1 µg/m3 (IQR (interquartile range): 0.2-4.6 µg/m3) and for children was 0.3 µg/m3 (IQR: LOT-2.9 µg/m3). A clear relationship between measured exposure levels and the number of passive smoking (self-reported) was observed. ROC (receiver operating characteristic) curve analysis showed that adult non-smokers, passive smokers and active smokers could be well classified based on nicotine exposure. Median Ceq (Cigarette equivalent) exposure estimated based on nicotine was 14 cigarettes per year for adults. Personal cigarette acceptance was not related to ETS exposure.

Tu-Po-35

Features of the air pollution from the pig farm in view of targeted chemical pollutants

Eugene Slautenko, *State Institution "Public Health Institute" National Academy of Medical Sciences of Ukraine, Kiev, Kiev, Ukraine*

Features of the air pollution from the pig farm in view of targeted chemical pollutants
Slautenko E., Turos O., Mykhina L.
State Institution "Public Health Institute" National Academy of Medical Sciences of Ukraine, Kiev

As part of research of air hygiene laboratory field measurements of pollutant emissions have been produced in the air. Measurements were carried out in the summer of 2011, the functioning of the area pig farm with capacity of 30 thousands heads.
Measurements have been carried taking into account the definition of targeted pollutants, are part of the pig farm emissions (hydrogen sulfide, ammonia, total suspended particles (TSP).
Comparison of the results was carried out with the existing Ukrainian criterion for evaluating a time averaging 30 minutes.
After the results are compared with the current criteria, data on concentrations of pollutants were obtained in the air: pollution due to ammonia emissions and TSP are missing, but there are excess hydrogen sulfide (at a distance of 200 meters - 0.038 mg / m^3 - 2.5 times, at a distance of 500 meters - 0.0043 mg / m^3 - 1.1 times, and at a distance of 1000 meters - 0.04 mg / m^3 - 5.0 times).
Thus, the measurement results indicate a lack of contamination of the sanitary protection zone of the air emissions of ammonia and TSP and presence of hydrogen sulfide in concentrations exceeding the statutory criteria, depending on the distance (200, 500 and 1000 meters).
Based on the results, it has been suggested on the summation of the concentrations of hydrogen sulfide emissions from pig and hydrogen sulfide emissions from livestock farms belonging to the local population.
The results make it possible to reduce the size of regulatory sanitary protection zones of livestock farms subject to the application of modern production technologies and the content of fattening pigs, the organization of production of ventilation systems and technological content of the process of cattle that reduce hydrogen sulphide emissions to the level of regulatory criteria.

Tu-Po-36

How to reach harmonised exposure assessment under REACH (on behalf of REACH Exposure Expert Group - REEG)

Astrid Heiland, *Federal Institute for Risk Assessment (BfR), Berlin, Germany*
Morgane Thierry-Mieg, *French Agency for Food, Environmental and Occupational Health and Safety, Maisons-Alfort Cedex, France*

The adoption of the REACH regulation (EC) No 1907/2006 has changed the chemical policy of the European Union (EU). Not only industry, but also exposure assessors of the authorities have to deal with the challenges of the new obligations under REACH and associated topics. It turned out that the primary assumption that an exposure assessment on a very generic base is sufficient in cases where Article 14(4) of the REACH regulation is fulfilled causes several difficulties from a scientific point of view. Thus the question arose, how exposure assessors of authorities can exchange their experience outside of the policy making processes and reach a common understanding for a harmonised exposure assessment. The objective of the REACH Exposure Expert Group (REEG) is to provide a forum where exposure issues regarding worker, consumer, and the environment can be discussed on a scientific basis to reach a common understanding and to identify needs for further developments in order to tackle these issues. Exposure assessors from Member States who are authorised to deal with REACH registration dossiers have organised themselves into an informal network since 2013. Important tools for communication are annual meetings, telephone conferences between the members of fluctuating working groups, and the information exchange via an online exchange platform provided by the French Agency for Food, Environmental and Occupational Health and Safety (Anses). The group currently counts about 50 members from 17 countries. To define the role of REEG, a written mandate has been worked out: REEG "is an informal, permanent group of experts from Authorities. It deals with human and environmental exposure to chemicals in the context of REACH. (...) The REEG aims at maintaining informal exchanges between exposure experts on the technical level. Therefore, no "REEG opinion" can be expected. (...) The condition to join the REEG is to be authorised to deal with registration dossiers." Several topics are being discussed and/or have been presented at the annual meetings and in between since the beginning in 2013: information exchange on daily experience (substance evaluation, national strategies on screening of substances to prioritise), intermittent exposure, combined exposure, children`s exposure, tool developments/evaluation (ConsExpo, Chesar, IUCLID-OHTs on exposure, E-TEAM), guidance documents (national and ECHA), recycling, exposure factors, etc. Information on national exposure projects are compiled within REEG to stimulate synergism between the members for improving the exposure assessment in the context of REACH.

Tu-Po-37

Aggregate exposures to indoor semivolatile organic compounds in France.

Maud Pelletier, *INSERM-U1085, Irset-Research Institute for Environmental and Occupational Health, Rennes, France*
Nathalie Bonvallot, *INSERM-U1085, Irset-Research Institute for Environmental and Occupational Health, Rennes, France*
Corinne Mandin, *INSERM-U1085, Irset-Research Institute for Environmental and Occupational Health, Rennes, France, Marne la Vallée, France*
Olivier Ramalho, *University of Paris-Est, Scientific and Technical Center for Building (CSTB), Marne la Vallée, France*
Fabien Mercier, *INSERM-U1085, Irset-Research Institute for Environmental and Occupational Health, Rennes, France*
Wenjuan Wei, *University of Paris-Est, Scientific and Technical Center for Building (CSTB), Marne la Vallée, France*
Jean-Paul Lucas, *University of South Brittany, Vannes, France*
Barabara Le Bot, *INSERM-U1085, Irset-Research Institute for Environmental and Occupational Health, Rennes, France*
Olivier Blanchard, *INSERM-U1085, Irset-Research Institute for Environmental and Occupational Health, Rennes, France*
Philippe Glorennec, *INSERM-U1085, Irset-Research Institute for Environmental and Occupational Health, Rennes, France*

Aim: Many chemicals emitted indoors - such as plasticizers, flame retardants, pesticides - are semivolatile organic compounds (SVOCs) and partition between gas phase and indoor surfaces, including airborne particles, settled dust and body surface. Indoor exposures thus may occur via inhalation, dust ingestion and dermal contact.
Methods: Two French nationwide surveys were conducted with sampling of airborne particulate (n=285, 2003-2005) and settled dust (n=145, 2008-2009), with analysis of respectively 66 and 48 SVOCs from different chemical families: phthalates, bisphenols, polycyclic aromatic hydrocarbons, pyrethroids, organophosphorus, organochlorines, synthetic musks, polychlorinated biphenyls and polybromodiphenylethers. Gas phase concentrations were modelled from other media according to partitioning theory. Human exposure factors and indoor air transdermal permeability coefficient were used to calculate, with Monte-Carlo simulations, doses through inhalation, dust ingestion and air-skin contact, for children under 6 years old.
Results: Results will be presented as ingestion-equivalent doses, and relative contribution of each pathway to total indoor exposure dose will be displayed.
Conclusion: These aggregate exposures will enable in a next step characterizing cumulative exposures to indoor SVOCs with common toxic effects.

Tu-Po-38

Exposure to particulate matter in temple

Ching-Chun Lin, *National Taiwan University/Institute of Occupational Medicine and Industrial Hygiene, Taipei, China, Republic of (Taiwan)*
Meng-Shan Tsai, *National Taiwan University/Institute of Occupational Medicine and Industrial Hygiene, Taipei, China, Republic of (Taiwan)*
Pau-Chung Chen, *National Taiwan University/Institute of Occupational Medicine and Industrial Hygiene, Taipei, China, Republic of (Taiwan)*

Backgrounds: Many evidence revealed that exposure to ambient particulate matter (PM) is associated with adverse health effects. To pray in temples is a regular activity in Buddhism and Taoism societies; however, the incense-burning in temples produced numerous PM affecting human health.
Methods: We monitored the temple in Taipei when people at high activity period, usual activity period, and weekend. PM concentrations were measured by MOUDI(micro-orifice uniform deposit impactor) and TSI 3321 Aerodynamic Particle Sizers(APS) for long term, and DUSTTRAK II 8530for short term. We adjusted considered relative factors and compared the measurement ten years ago which had 7 incense burners and nowadays with 3 incense burners. We also compared the measured concentrations in the study, Taiwan air quality regulations, and U.S.EPA air quality standards for further decision.
Results: Above 90% of incense-burning particle sizes were less than 2.5μm. Furthermore, measured PM concentrations were higher than Taiwan air quality regulations and U.S.EPA air quality standards.
Conclusions:
The air quality and the pollutants in the temple related to the public health impact. The policy should have the solution to reduce the environmental exposure and the high concentrations of pollutants emitted from incense burning should be more concern.

Tu-Po-39

Measurement of cortisol in human hair as a biological marker of chronic stress

Ching-Chun Lin, *National Taiwan University /Institute of Occupational Medicine and Industrial Hygiene, Taipei, China, Republic of (Taiwan)*
Meng-Shan Tsai, *National Taiwan University/Institute of Occupational Medicine and Industrial Hygiene, Taipei, China, Republic of (Taiwan)*
Pau-Chung Chen, *National Taiwan University /Institute of Occupational Medicine and Industrial Hygiene, Taipei, China, Republic of (Taiwan)*

Background: Cortisol was considered as a “stress hormone” that gets released in order to show “fight or flight response” at stressful conditions. Hair , is an emerging biological specimen , in addition to non-invasive, easy to save , you can also back the case of long-term exposure scenario
Moderate storing condition and explore specific duration of exposure make hair species getting notice in recent years. Stress hormone levels in hair became a novel biomarker to evaluate chronic stress.
Aim:This study aimed to develop an analytical method by LC-MS/MS to quantify the concentration of cortisol, cortisone in hair, serum, urine and saliva. Structured questionnaire were used to evaluate stress level to investigate the correlation between concentrations of biomarkers and stress.
Methods: There were 12 male and 19 female adult volunteers were recruited in this study. Serum, urine and saliva were collected once a month at the morning within three months. Hair samples were collected at the end of the third month. We developed analytical method by liquid chromatography-tandem mass spectrometry to quantify the concentration of hair, serum, urine and saliva. Type A personality, consciously stress level, Pittsburgh Sleep Quality Index (PSQI) and Hospital Anxiety and Depression Scale (HADS) were used to evaluate the stress level of the study subjects. Pearson’s correlation coefficients were applied to evaluate the correlation.

Results: The limit of detection and limit of quantitation were 1.47 and 4.47 (pg/mg), respectively. The detection rate which above limit of quantitation of cortisone levels in hair was 100.00%. Median of hair cortisol levels in our study population was 8.04 (pg/mg). We found that levels of cortisone in serum, urine, and saliva had significant correlation. We did not find significant correlation between levels of cortisol and cortisone in hair and stress-related outcomes.
Conclusions: The cortisone in hair and stress-related biomarker have well correlation in our study population. IN future, cortisone could be used as the indicator of stress level.

Tu-Po-43

Magnitude and spatial patterns of ultrafine particulate matter associated with aircraft arrivals near Boston Logan Airport.

Chloe Kim, *Boston University School of Public Health, Boston, MA, United States*
Yorghos Tripodis, *Boston University School of Public Health, Boston, MA, United States*
Jonathan Levy, *Boston University School of Public Health, Boston, MA, United States*

Background: Aircraft emissions can influence exposures to multiple air pollutants at local and regional scales, but their magnitude and spatial extent have not been well characterized. For ultrafine particulate matter (UFP), studies have shown highly elevated concentrations immediately downwind of departing aircraft, with rapid declines as a function of distance, but the spatial patterns associated with aircraft arrivals are more uncertain. Some studies have shown that traffic-related UFP far exceeds aircraft contributions, while others have shown aircraft contributions exceeding the contribution from traffic over large areas. However, few studies have had the necessary air pollutant and flight activity data to definitively quantify the influence of aircraft arrivals on UFP concentrations.
Objective: The goal of our study was to investigate the effect of aircraft arrivals on ambient UFP concentrations, as well as to determine what characteristics of aircraft have a more significant impact on UFP concentrations.
Methods: Real-time UFP concentrations were collected using TSI 8775 Condensation Particle Counter between March 9, 2011 and May 31, 2011 at a fixed site underneath a flight arrival path but upwind of major roadways in Boston, Massachusetts. We gathered real-time flight activity data from the Federal Aviation Administration's Performance Data Analysis and Reporting System that included the location of each individual aircraft in three dimensions and the aircraft attributes, along with timely meteorological data. We constructed a time series regression model predicting UFP concentrations as a function of flight activity and meteorology, incorporating methods to account for temporality.
Results: Our regression model was able to isolate the contribution of aircraft arrivals to UFP concentrations at a fixed site underneath a flight path, capturing the influence of aircraft location, attributes, and meteorology on measured concentrations.
Conclusions: Using novel real-time data on UFP concentrations and aircraft source attributes, we were able to ascertain the contribution of a defined source to measured concentrations and develop a statistical approach that could be applied to other near-airport settings.

Tu-Po-44

Personal Exposure to Polycyclic Aromatic Hydrocarbons (PAHs), Fine Particulate Matter (PM2.5), and Carbon Monoxide (CO) During Cookstove Use in Rwandan Households

Zoe Frolking, *University of North Carolina at Chapel Hill, Chapel Hill, NC, United States*
Joseph A Pedit, *University of North Carolina at Chapel Hill, Chapel Hill, NC, United States*
Karin Yeatts, *University of North Carolina at Chapel Hill, Chapel Hill, NC, United States*
Pamela Jagger, *University of North Carolina at Chapel Hill, Chapel Hill, NC, United States*
Sara Abdoulayi, *University of North Carolina at Chapel Hill, Chapel Hill, NC, United States*
Sudhanshu Handa, *University of North Carolina at Chapel Hill, Chapel Hill, NC, United States*
Leena A Nylander-French, *University of North Carolina at Chapel Hill, Chapel Hill, NC, United States*

The World Health Organization has estimated that exposure to toxins and smoke emitted from cookstoves leads to 4.3 million premature deaths annually. Nearly three billion people, mostly in developing countries, rely on solid-fuel-burning cookstoves to prepare food and heat their homes. The relationship between personal exposure to indoor air pollutants such as carbon monoxide (CO) and fine particulate matter (PM2.5) and adverse health effects is well established. Little attention has been given to polycyclic aromatic hydrocarbons (PAHs); one of the primary sources of PAH exposure is indoor cookstove use. The goal of our study is to quantify and characterize the exposure to PAHs, PM2.5, and CO produced by cookstove smoke in peri-urban Rwandan households. Comprehensive exposure assessments were carried out in 180 households during July-August 2015. In each household, the primary cook's exposure to PAHs was measured using a PUF/XAD2 sampler (SKC, Eighty Four, PA) over a 24-h period. CO exposure was determined using a CO monitor (EL-USB-CO; Lascar, Erie, PA). Finally, PM2.5 exposure was measured using gravimetric analysis of Teflon filters. We also recorded kitchen ventilation and size as well as fuel type, quantity, and moisture content. Of the households surveyed, 74% cooked primarily inside during the seven days prior to sample collection. The majority used charcoal stoves as their primary cooking method; 70% used portable charcoal stoves, 18% used fixed charcoal stoves, and the remaining 12% used a clay stove, three-stone fire, or other stove. Charcoal was used as the primary fuel source in 94% of the households. Preliminary GC-MS analysis shows that the most prominent PAHs emitted from the charcoal-burning cookstoves were chrysene and indeno(1,2,3-cd)pyrene (IP). Other PAHs detected included naphthalene, benzo(ghi)perylene (BghiP), fluorene, and pyrene. According to relevant literature, the experimental ratio of 0.66 for IP:IP+BghiP concentrations indicates that the observed PAHs originate from the burning of biomass, rather than from liquid fossil fuels. Spikes in CO measurements align well with expected cooking times, as did stove temperature increases measured by stove-use monitors (Digit-TL, LabJack Corporation, Lakewood, CO, USA). Some CO spikes were unaccompanied by increases in stove temperature, indicating the presence of other exposures (trash burning, transportation, unmonitored stoves, etc,). The results of this research will aid in both understanding the personal exposure from cookstove emissions and the development of new, safer methods for food preparation and home heating. This has the potential to improve the health outcomes of millions of people globally.

Tu-Po-45

Viability of Cultured Primary Human Skin Cells Treated with HDI Monomer and HDI Isocyanurate

Leena A Nylander-French, *University of North Carolina at Chapel Hill, Chapel Hill, North Carolina, United States*
Jayne C Boyer, *University of North Carolina at Chapel Hill, Chapel Hill, North Carolina, United States*
Kaitlyn Phillips, *University of North Carolina at Chapel Hill, Chapel Hill, North Carolina, United States*
Laura Taylor, *University of North Carolina at Chapel Hill, Chapel Hill, North Carolina, United States*
Rachel Johnson, *University of North Carolina at Chapel Hill, Chapel Hill, North Carolina, United States*

The monomer and oligomer, 1,6 hexamethylene diisocyanate (HDI) and HDI isocyanurate, respectively, are components in sprayed polyurethane coatings. Exposure via the lungs and skin can lead to sensitization and chemically induced asthma. Much research has focused on effects of exposure on respiratory and immune cells. Using the luminescent ATP-viability assay (CellTiter-Glo®, Promega, Madison, WI), we have studied the effect of a 4-h exposure to HDI monomer or HDI isocyanurate on the viability of three types of cultured primary human skins cells (fibroblasts, keratinocytes, and melanocytes) from several individuals to determine inter-individual variation and cell-type specific toxicity. Preliminary LD50's (50% lethal dose) of HDI- and isocyanurate-treated cells in unsupplemented culture medium range from 30-2000 µM for the HDI monomer and 0.7 µM for the oligomer. Similarly, published IC20 (20% inhibitory concentration) data using HDI-treated respiratory and immune cancer cells lines range 40-500 µM. Aerosolized paints typically contain 1% HDI monomer, which is the equivalent of 60 mM, The lethal doses in cultured cells are well below observed exposure concentrations in occupational settings and, thus, the in vitro data may predict dermatologic health issues with occupational exposures to the monomer. HDI isocyanurate, which can make up to 96% of sprayed polyurethane coatings, may be more toxic than the HDI monomer due to its extra reactive NCO group, and the much greater potential for exposure may make it a more significant health problem. HDI isocyanurate constitutes the largest inhalation and skin exposure and has been shown to possess a greater sensitizing capacity than HDI monomer. Further, HDI isocyanurate penetrates the skin faster than HDI monomer. Our dose/response data obtained with normal human cell cultures indicate that skin cell sensitivity to death by HDI varies among individuals but not between cell types from the same individual. We have also observed a hormesis effect at very low doses in some individuals. Our data will aid understanding of individual sensitivity to diisocyanate exposure as well as the relative risk associated with different diisocyanate forms.

Tu-Po-47

Estimation of the daily soil/dust (SD) ingestion rate of children via hand-to-mouth contact using tracer elements

Jin Ma, *Chinese Research Academy of Environmental Sciences, Beijing, China, People's Republic of*

Soil/dust (SD) ingestion by children is a recognized potential route of exposure to soil pollutants. Estimates of SD ingestion play an important role in risk assessment. Recently, some values of SD ingestion have been reported by USEPA, JPNEPA, AUEPA. However, there is very limited information available regarding the exact amounts of SD ingestion, especially from children's hands, but hand-to-mouth contact has been proven to be an important pathway of SD ingestion for children. Information regarding the rates of SD ingestion for children in China is very limited. We therefore measured the amount of SD on children's hands using hand wipes in Gansu Province, northwest China, and used the results to estimate potential SD ingestion from hand-to-mouth contact.
In this study, a total of 60 children (31 males and 29 females) between the ages of 3 and 12 years were randomly selected from Lanzhou City in Gansu Province, northwest China. Hand (soil/dust) SD samples from these children were collected using hand wipes. We determined the approximate amounts of hand SD and the concentrations of three tracer soil elements (Ce, Y, and V) in these samples.
The approximate amounts of hand SD ranged from 42.28 to 173.76 mg, with a median value of 85.42 mg. In addition, the mean amounts of hand SD estimated using the concentrations of Ce, Y, and V in the samples were 4.63, 3.43, and 3.42 mg, respectively. The amount of hand SD varied greatly among the age groups: primary school children had more hand SD than kindergarten children, males had more hand SD than females, and children from rural areas had more hand SD than those from urban areas. The rates of daily ingestion of hand SD for kindergarten and primary school children were estimated to be 7.73 and 6.61 mg/d, respectively.
SD ingestion rates in this study were much lower than those of previous studies on dust ingestion via the hands, indicating that the values of SD ingestion rates were overestimated. These results will further facilitate the assessment of children's exposure to SD and our understanding of the potential health effects.

Tu-Po-48

Aeroallergenic Monitoring of Ambrosia in Kyiv

Irina Kovtunenko, *SI „O.M. Marzeyev Institute for Public Health, National Academy of Medical Sciences of Ukraine", Kyiv, Ukraine*

Introduction. The prevalence of allergic diseases has increased in recent decades in the industrialized world. On allergy suffers from 30 to 40% of the world's population. The research is devoted to the problem of biological ambient air pollution caused with the pollen of allergenic plants which the past few years has become one of the leading environmental pollutants.

Materials and methods. Duration of investigation was from January 2014 to August 2015. Ambient air are collected for visual identification and enumeration of pollen grains on the building roof at the height of 25 m above ground using 7 day recording volumetric spore trap Burkard. The device is calibrated to collect 10 L/min of air. Annual sampling period was from January 1 to December 31 (at least covering starts of pollination of Alnus and Corylus until the end of Artemisia and Ambrosia period). Daily recording period - 00:00 - 24:00 hours. We are conducting compound light microspore with magnification 400x and resolution (numerical aperture of the objective) 0,65 for examining samples.

Results. It was observed that during 2014 concentration of Ambrosia pollen grains increased at the end of August - at the beginning of September. It was detected sufficiently highest levels of pollen grains - 1245 gr/m3 during August peak and 980 gr/m3 during September peak. In summer and winter months, Ambrosia pollen grains in samples were not found. It was noticed seasonal movements of the concentrations of Ambrosia pollen grain from sporadic in April (7 gr/m3) to some outbreak in May (38 gr/m3) and to some increasing in October (237 gr/m3).

Similar results were observed in 2015. In March, concentration of Ambrosia pollen grains was 3 gr/m3, in May it increased to 262 gr/m3 and in August it increased abruptly to 1828 gr/m3. However, in contrast to 2014, in summer months, 2015 Ambrosia pollen grains were found - during June concentration of Ambrosia pollen grains was 45 gr/m3 and during July, it was 22 gr/m3.

Conclusion. This study, based on time series analysis adjusting for meteorological factors and air pollution variables, assessed the short-term effects of allergenic Ambrosia pollen of Kyiv, Ukraine. It affords us additional opportunities for informing the population about the terms of the highest risk of allergic diseases' outbreaks and help to support the medical treatment of the population.

Tu-Po-49

Health based policy advice on consumption of home-produced eggs to achieve exposure reduction for POPs

Ann Colles, *Flemish Institute for Technological Research (VITO), Mol, Belgium*
Christa Cornelis, *Flemish Institute for Technological Research (VITO), Mol, Belgium*
Mirja Van Holderbeke, *Flemish Institute for Technological Research (VITO), Mol, Belgium*
Els Van de Mieroop, *Provincial Institute of Hygiene, Antwerp, Belgium*
Melissa Paulussen, *Provincial Institute of Hygiene, Antwerp, Belgium*
Greet Schoeters, *Flemish Institute for Technological Research (VITO), Mol, Belgium*
Marjory Desmedt, *Flemish Environment Agency, Antwerp, Belgium*
Maja Mampaey, *Environment & Health, Flemish Government, Brussels, Belgium*

Introduction
The community of Menen, Flanders, hosts a municipal waste incinerator, active until 2005, and an active metal recycling plant. Elevated levels of polychlorinated biphenyls (PCBs) and dioxins have been repeatedly measured in air samples and in locally grown food. In 2002-2006, measurements of persistent organic compounds (POPs) in cord blood and in peripheral blood of adolescents showed higher internal exposure in Menen compared to other regions in Flanders. These higher exposures were associated with consumption of locally grown food. After local authorities discouraged consumption of local eggs since 2003, lower POPs blood levels were observed in 2010.
Aim
To verify whether home-produced eggs in the region of Menen were again safe to consume.
Method
In 2013, at 14 locations in the region of Menen, home-produced chicken eggs were collected. The levels of dioxins, furans and PCBs were determined in the yolks by gas chromatography-high resolution mass spectrometry. The levels of dichloro-diphenyl-trichloro-ethane or DDT compounds were determined by gas chromatography - electron capture detector.
Results
Median concentrations in eggs were 11.64 pg TEQWHO98/g lipids for sum of dioxins, furans and PCBs, 21.30 ng/g lipids for marker-PCBs and 229.45 ng/g lipds for sum of DDT compounds.
The European maximum levels for food do not apply to home-produced chicken eggs. Therefore, reference values for a safe consumption were calculated from our results, based on available toxicological guidance values. Comparing POPs levels in the eggs with these reference values enabled to establish a health based and age dependent consumption advice concerning home-produced eggs. For areas close to the industrial site the previously introduced consumption ban was remained. For parts of Menen outside this precautionary area, a custom-made consumption advice was formulated. Afterwards, this advice was customized for different pollution levels, allowing implementation all over Flanders.
Conclusion
Combining data from multiple environmental compartments enables to identify ways to achieve exposure reduction and establishing health based consumption advice for locally grown food.

This page intentionally left blank.

UTRECHT
THE NETHERLANDS
OCTOBER 9-13
ISES 2016

Tu-Po-51

Personal particulate matter exposure assessment of rural Malawian children and device wearability considerations

Ryan Chartier, *RTI International, Research Triangle Park, North Carolina, United States*
Mukesh Dherani, *The University of Liverpool, Liverpool, United Kingdom*

Aim: Three billion of the world's poorest people rely on solid fuels (wood, crop wastes, dung, charcoal and coal) and simple open fires and stoves for their everyday cooking needs. This results in high levels of exposure to smoke pollution among family members, including young children. This study aims to: (1) quantify the personal particulate matter (PM) exposures of children <5 years of age, (2) examine the child/mother exposure relationship, (3) assess the comfort and wearability of personal exposure platforms for children, and (4) provide wearing platform recommendations to aid future study design.
Methods: This study is being carried out at the Karonga Prevention Study site of the CAPS Trial in conjunction with another effort focusing on the impact of improved stoves on reductions in PM2.5 exposure and nasopharyngeal carriage of streptococcus pneumoniae in 6 month old Malawian children. Newborns are recruited through study site surveillance in homes recruited to CAPS, and for which the intervention group already has the Philips stove. A subset of these children, along with older siblings (< 5 years) and their mothers, are recruited to take part in the current study. The 48-hour PM2.5 personal exposure levels and patterns of the children are measured using the RTI Enhanced Children's MicroPEM (ECM) monitor while the mother wears a v3.2 MicroPEM.
Results: Preliminary data for 6 month and 6 week olds from this cohort indicate that children residing in homes using an improved Philips stove have slightly lower exposures 53.9 [38.1, 69.6] µg/m3 than those in homes using a 3-stone fire 66.8 [47.7, 85.9] µg/m3. Additionally, we observed that PM2.5 exposures are significantly higher ($p=0.0476$) for 6 month old children 72.1 [42.7, 101.4] µg/m3 when compared to 6 week old children 48.7 [42.0, 55.5] µg/m3. The waking hour wearing compliance (% sample time device is moving) was calculated using the monitor's onboard accelerometer and determined to be 44.6% [41.6, 47.6]. These data indicate the device is being properly deployed for the youngest and most at-risk children.
Conclusions: While this work is ongoing, the preliminary results demonstrate that it's possible to confidently and accurately measure the personal PM exposures of small children. Accurate assessment of children's exposure to PM from solid fuel cooking is necessary to define exposure/health relationships and to inform research funding and policy decisions. Additional pilot efforts, with similar goals to this study, will be underway in Peru and Ghana in the spring of 2016.

Tu-Po-53

An Integrative Oxidative Potential Assay for Data Sharing and Validation Across Laboratories

Michelle North, *University of Toronto, Toronto, Ontario, Canada*
Angela Huang, *University of Toronto, Toronto, Canada*
Alison Traub, *University of Toronto, Toronto, Canada*
Jianhuai Ye, *University of Toronto, Toronto, Canada*
Arthur Chan, *University of Toronto, Toronto, Canada*
Krystal Pollitt, *University of Massachusetts, Amherst, Massachusetts, United States*
Greg Evans, *University of Toronto, Toronto, Canada*

Aims: Oxidative potential (OP) provides a biologically-relevant indicator of the potential toxicity of complex multi-component environmental pollutants. However, several methods are employed for the assessment of OP, making sharing and validation of data across laboratories difficult. The overarching goal of this research is to integrate the ascorbate, glutathione and 1,4-dithiothreitol assays of OP and assess the activity of standard reference materials (SRMs) to establish an assay with enhanced transferability and potential applications in multicenter studies. This project focused on enhancing the between-lab standardizability of the glutathione component of the integrated assay. Glutathione depletion can be measured by incubating particulate suspensions in synthetic respiratory tract lining fluid (RTLF), which models the antioxidants in the human lung, including 2mM each of ascorbate, glutathione and urate. The aim of this study was to establish chemical and particulate positive controls, with linear concentration-depletion characteristics against glutathione in RTLF, to aid in inter-assay and inter-laboratory data integration.
Methods: H2O2 was tested as a chemical control at concentrations ranging from 0.25-30%, and the following SRMs from 25-200µg/ml; Steel Flue Dust (Ministry of the Environment (MOE), SIFD-700-1985), non-ferrous dust (Canadian Reference Material PD-1), ultrafine dust (Powder Technology, ISO-12103), and incinerator fly ash (MOE, IFA-100-1987). Incubations in RTLF were carried out at 37C for 4h. The glutatione concentration remaining was calculated by measuring total glutathione and glutathione disulphide (GSSG) using GSSG-reductase-5,5'-dithio-bis(2-nitrobenzoic acid). Additionally, an experimentally-produced naphthalene secondary organic aerosol (SOA) was assessed to pilot the transferability of the assay.
Results: H2O2 was found to be a suitable chemical positive control, with linear depletion at concentrations of 0.25-2.5%. Steel flue dust and incinerator fly ash did not deplete glutatione at study concentrations. Ultrafine and non-ferrous dust demonstrated linear depletion of glutathione, equivalent to 1% and 4.7% depletion per 10µg/ml increase in particulate, respectively. As non-ferrous dust generated the strongest response of the SRMs, it was chosen as the particulate-based positive control for standardization. SOA demonstrated linear depletion of glutathione equivalent to 34% of the activity of non-ferrous dust, confirming the ability to standardize unknown samples against an easily transportable SRM.
Conclusions: The addition of H2O2 and non-ferrous dust standard curves as chemical-based and particulate-based positive controls will aid in enhanced internal validation and multicenter studies. The chemical control demonstrates the standardization of reagents, while the particulate control ensures a similar matrix to experimental samples, and thus will be used to normalize particulate suspension OP activities.

Tu-Po-54

Legacy and Emerging Flame Retardants in Fire Station Dust

Beverly Shen, *University of California, Berkeley, Berkeley, California, United States*
Ranjit Gill, *California Department of Toxic Substances Control, Berkeley, California, United States*
Joginder Dhaliwal, *California Department of Toxic Substances Control, Berkeley, California, United States*
F. Reber Brown, *California Department of Toxic Substances Control, Berkeley, California, United States*
Myrto Petreas, *California Department of Toxic Substances Control, Berkeley, California, United States*
S. Katharine Hammond, *University of California, Berkeley, Berkeley, California, United States*
Sharyle Patton, *Commonweal, Bolinas, California, United States*

Aim
Firefighters are exposed to a unique profile of potentially fire-related chemical contaminants, including dioxins, antimony, and other metals. In 2011, we measured chemical contaminants, including polybrominated biphenyl ethers (PBDEs), in the dust of 20 California fire stations and found elevated levels of PBDEs (median level of BDE-209: 47,000 ng/g) as compared to California residences and other occupational settings from around the world. PBDEs have been phased out of commercial use, but other flame retardants are replacing them. In this study, we measured concentrations of PBDEs and organophosphate flame retardants (OPFRs) in dust from 25 fire stations, 5 from each of 5 states: California, Minnesota, New Hampshire, New York, and Texas.
Methods
Sampling packets were mailed to each fire station for collection of vacuum cleaner bags. Sampling packets included a sample collection protocol, a building characteristics and work practices questionnaire, and a re-sealable polyethylene bag. Dust samples were requested from vacuum cleaner bags used in the firefighter living quarters. At the lab, the dust samples were sieved to remove fibers and debris larger than 150µm. The resulting dust fraction was aliquoted (~50 mg), spiked with carbon-13 labeled PBDE and deuterated OPFR internal standards (9 PBDEs and 5 OPFRs) and extracted by sonication in a 3:1 hexane:acetone solution. The extracts were cleaned using a Florisil gel column and then solvent-exchanged into isooctane and spiked with labeled recovery standards (PCB-209 and TPP). OPFR extracts were further diluted (1:5) before analysis by EI-GC/MS/MS. PBDEs were analyzed using high resolution EI-GC-MS.
Results
Eighteen PBDEs were detected in the fire station dust samples and with concentrations ranging over 5 orders of magnitude; major congeners including BDE-47 and BDE-99 had some of the highest concentrations. Five OPFRs were detected with concentrations ranging over 2 orders of magnitude; the major congener TDCIPP had some of the highest measured concentrations. The median levels of flame retardants were as much as 2-fold greater than those measured in residential settings.
Conclusions
Flame retardants are ubiquitous. Firefighters may potentially be exposed to higher levels of flame retardants, both legacy and emerging, than the general population.

Tu-Po-55

Developing a new Indoor Stationary and Personal Passive Air Sampler from PDMS and XAD-coated PDMS

Joseph Okeme, *University of Toronto, Toronto, Ontario, Canada*
Atousa Abdollahi, *University of Toronto, Toronto, Ontario, Canada*
Suman Dhal, *University of Toronto, Toronto, Ontario, Canada*
Liisa Jantunen, *Environment Canada, Egbert, Ontario, Canada*
Mahiba Shoeib, *Environment Canada, Toronto, Ontario, Canada*
Cassandra Rauert, *Environment Canada, Toronto, Ontario, Canada*
Jeff Brook, *Environment Canada, Toronto, Ontario, Canada*
Miriam Diamond, *University of Toronto, Toronto, Ontario, Canada*

Passive air samplers (PAS), particularly personal passive air samplers (PPAS) have long been used to monitor exposures in occupational settings but are less well characterized for non-occupational use. Our goal was to develop a reliable, versatile, and relatively inexpensive and easily used PAS for stationary measurements and PPAS for personal measurements of a range of semi-volatile organic compounds (SOVs). We report on PAS and PPAS consisting of polydimethylsiloxane (PDMS) and a newly designed PDMS coated with styrene divinyl benzene co-polymer (PDMS-XAD). The target analytes included phthalate esters, halogenated and organophosphate ester flame retardants, PAH and perflourinated compounds. We conducted several calibration studies. First, two indoor air calibration studies were conducted to determine passive sampling rates by deploying stationary PDMS PAS and comparing with air concentrations measured using two low-volume active air samplers analyzed for gas and particle phases separately. Over the study 50 day study, surface-area normalized uptake rates of PDMS were comparable to the more commonly used polyurethane foam (PUF) stationary samplers housed in a single bowl shelter, with rates ranging between 0.6 to 1.5 m3 day-1 dm-2 for brominated flame retardants and phthalates. In the second calibration study we expanded the number of target chemicals with the PDMS-XAD design because of the increased sorptive capacity of XAD. As a PPAS, we again tested the PDMS worn pinned to the lapel (close to the breathing zone) and a co-deployed low-volume active sampler. Sufficient masses of some flame retardants and phthalates were detected after 3 to 4 days of wearing the PPAS for ~7 hours per day. Finally, we tested the PDMS for use as a "mail out" sampler to question whether the mailing process would contaminate the sampler. In general, we found low levels of flame retardants and phthalates in the mailed PDMS (which were in the mail from 2 days to a week), except for DEHP which is ubiquitous in the environment. These results show promise for using PDMS as a stationary and personal passive air sampler with the possibility of distribution and return through the mail system.

Tu-Po-56

Personal, indoor and outdoor PM2.5 exposure characterization for household air pollution related to cooking in Lampang, Thailand

Ilse Ottenbros, *Utrecht University, Nijmegen, Netherlands*
George Downward, *Utrecht University, Utrecht, Netherlands*
Roel Vermeulen, *Utrecht University, Utrecht, Netherlands*
Suleeporn Sangrajrang, *National Cancer Institute Thailand, Bangkok, Thailand*

Aim - Household Air Pollution (HAP) is a large contributor to lung cancer incidences in low- & middle-income countries due to the combustion of solid fuels. This study characterizes personal, indoor and outdoor PM2.5 concentrations related to cooking activity in the province of Lampang, Thailand.
Methods - This study design provides a characterization of air pollution levels in the area of Lampang, Thailand (n=52), to find patterns between cooking characteristics (e.g. type of fuel, volume of fuel used), individual-level characteristics (e.g. age) and household characteristics (e.g. ventilation), since they may contribute to variations in personal exposure between and within individuals. 2 times 24-hour filter measurements were carried out at females cooking on gas, wood or charcoal; after which PM2.5 and PM2.5-absorbance concentrations were derived. In addition, personal real-time aerosol mass concentrations (mg/m3) were collected and temperatures of the cooking stoves were monitored continuously. Stationary indoor (living environment) and outdoor filter measurements were carried out to characterize the local PM¬2.5 concentrations. Household characteristics, meteorological characteristics and activity questionnaires were collected to investigate possible contributing variables to the exposure.
Results - The found GM personal PM¬2.5 concentrations for cooking were 14.0(±1.7), 35.8(±2.0), 38.3(±3.0) and 45.1(±1.8) µg/m3 for respectively gas, wood, charcoal and wood+charcoal. Personal absorbance values were respectively 1.1(±1.8), 4.5(±2.1), 3.7(±2.7) and 5.7(±2.0). Spearman correlation analysis showed high correlations between personal and indoor measurements ($r=0.8$, $p<0.001$) and between indoor and outdoor measurements ($r=0.9$, $p<0.001$). The found peaks in the temperature of the cooking stove could be mainly related to peaks in the real-time aerosol mass concentrations. Per individual aerosol mass concentration peaks not related to cooking matched additional exposure sources, which can be used to correct for the total exposure concentration due to cooking. Linear mixed effect modelling showed that the type and volume of fuel used, the minutes the stove was heated and the age of the subject were the main contributors to personal PM2.5 exposure concentrations.
Conclusion - This study characterizes exposure variables and their interrelationships from both 24-hour and real-time measurements, which contributes to future HAP exposure modelling and prevention of related health effects.

Tu-Po-58

Development of a new microextraction method and on-line derivatization coupled with GC-MS for analyzing of five metabolites of synthetic pyrethroids in urine samples

Anna Klimowska, *Medical University of Gdańsk, Gdańsk, Poland*
Bartosz Wielgomas, *Medical University of Gdańsk, Gdańsk, Poland*

Sample preparation step is usually the most expensive and time consuming part of laboratory work during the assessment of exposure to environmental chemicals. In order to reduce expenses and/or increase throughput of analytical methods much efforts are made for miniaturization and automatization. The main goal of this work was to simplify and automate the sample preparation technique before GC-MS analysis of urinary biomarkers of pyrethroid exposure. A novel microextraction by packed sorbent (MEPS) method coupled with gas chromatography-mass spectrometry (GC-MS) was developed for determination of five urinary metabolites of synthetic pyrethroids: cis-2,2-dimethyl-3-(2-chloro-3,3,3-trifluoro-1-propenyl)-cyclopropanecarboxylic acid, cis/trans-3-(2,2-dichlorovinyl)-2,2-dimethylcyclopropane-1-carboxylic acids, cis-(2,2-dibromovinyl)-2,2-dimethylcyclopropane-1-carboxylic acid and 3-phenoxybenzoic acid.
MEPS is a miniaturized solid phase extraction (SPE) utilizing a manually operated semiautomatic syringe equipped with a needle combined with SPE sorbent. Several significant factors affecting MEPS performance like: sample pH, type of extraction packing, type and volume of washing solvent, number of draw-eject cycles, volume and type of elution solvent were optimized.
MEPS was performed using C18 solid phase, which was conditioned with methanol (4×50μL) and 2% formic acid in water (3×20μL). Subsequently the enzymatically hydrolyzed urine sample (3×100μL) was loaded and the bed was washed with 30% methanol in water (3×50μL). Then the sorbent was dried under vacuum and finally analytes were simultaneously derivatized and eluted with the mixture of 1% 1,1,1,3,3,3-hexafluoroisopropanol and 2% diisopropylcarbodiimide in n-hexane (2×40μL). Forty microliters of the extract were injected into GC-MS system using large volume injection mode (LVI). Optimized method was then validated and LODs in the range of 0.06 - 0.42 ng mL-1, correlation coefficient above 0.990 and precision below 17% were obtained for all analytes.
The proposed method is very fast, simple and environment friendly. In comparison with reference extraction method (liquid-liquid extraction) the volume of sample and organic solvents and time of extraction procedure were significantly decreased. The new approach can be used to routine monitoring studies of evaluation of human exposure to synthetic pyrethroids.

Tu-Po-59

Risk Assessment Guidance for Enzyme-containing Products

Francis Kruszewski, *American Cleaning Institute, Washington, District of Columbia, United States*

Aim:
The purpose of this guidance, as provided by the American Cleaning Institute (Washington, D.C.), is to describe the potential health hazards of enzymes present in consumer products and provide a framework for manufacturers of these products to conduct risk assessments to help ensure the safety of new products containing enzymes.
Methods:
Enzymes generally have good safety profiles. However, enzymes like many other proteins can act as allergens and induce the production of allergen-specific IgE antibody upon repeated inhalation or exposure to mucous membranes that may lead to allergy symptoms, including asthma. The primary challenge associated with enzyme use is preventing the generation of allergen-specific antibody and the development of symptoms of Type 1 hypersensitivity. This hazard is the primary focus for the risk assessment for enzymes and must be managed carefully. Another hazard that also should be addressed is primary irritation of the eye and skin. However, most uses of enzymes in consumer products do not pose a likelihood of causing irritation. If the risks posed by enzymes are not managed appropriately, the consequences may spread beyond a single product or company. This could lead to unwarranted limitations on the use of enzyme technology in other consumer applications. Therefore, it is recommended that companies using enzymes responsibly consider how they are managing enzyme safety including the conduct of appropriate risk assessments and risk management programs. The preferred approach is for product manufacturers to develop comprehensive programs to assess and manage the risks of using enzymes in consumer products. Such programs will include measures to manage exposures to enzymes. The program design should be developed on a case-by-case basis to address parameters specific to the type of product and its applications.
Results:
Experience in the cleaning products industry demonstrates that the potential risk of adverse effects can be successfully managed by identifying the hazards, carefully assessing exposure, characterizing the risk and then applying appropriate risk management.
Conclusions:
Good understanding of the hazards and exposures for enzymes as used in consumer products will lead to informed decisions about the potential risks and the development of sound approaches to manage these risks. This guidance document outlines strategies and methods that have been used successfully by the cleaning products industry.

Tu-Po-60

Characterization of E-cigarette Users: a Descriptive Analysis of Participants Exposed to E-cigarettes in Maryland.

Stephanie Jarmul, *Johns Hopkins University, Baltimore, Maryland, United States*
Angela Aherrera, *Johns Hopkins University, Baltimore, Maryland, United States*
Pablo Olmedo-Palma, *Johns Hopkins University, Baltimore, Maryland, United States*
Rui Chen, *Johns Hopkins University, Baltimore, Maryland, United States*
Ana M. Rule, *Johns Hopkins University, Baltimore, Maryland, United States*
Ana Navas-Acien, *Johns Hopkins University, Baltimore, Maryland, United States*

Background: Electronic-cigarette devices (e-cigarettes) or vaporizers, allow users to inhale an aerosol (usually containing nicotine) into their lungs. These devices are often regarded and marketed as a safe alternative to tobacco cigarettes and their widespread use is transforming the pathway to nicotine addiction, as well as creating new potential toxic exposures.
Objectives: The study objectives were to analyze e-cigarettes as a possible exposure pathway for toxic and carcinogenic metals based on preliminary data of metals found in the vaped aerosols. This poster presents the demography and vaping behaviors of the participants in order to better understand the exposed population and design specific and targeted interventions.
Methods: Seventy participants were recruited during a 7-month period through social media outlets, flyers, and in-person recruitment in Baltimore, Maryland. Participants included never smokers, dual users of both cigarettes and e-cigarettes, and sole e-cigarette users. A questionnaire was administered with 65 questions addressing overall health status, smoking habits, demographic characteristics, and beliefs/perceptions of electronic cigarette safety. Biospecimen samples of hair, urine, saliva, and exhaled breath were also collected from each participant, as well as e-cigarette juice and vaped condensate.
Results: Participants in the study had a mean age of 32 years, and were predominately white males. Ninety percent of sole e-cigarette users were previous smokers. Of those who had never smoked cigarettes prior to vaping, 4 out of 5 (80 %) were under the age of 25. E-cigarette juice consumption varied greatly among participants, with a range of 5-240 ml/week reported by the subjects. A majority of the participants use e-juices with nicotine concentrations below those of tobacco cigarettes, yet less than 50% intend to quit or lower their nicotine levels further. As our preliminary data show metals such as tin, lead, and copper in vaped e-cigarette condensate, these devices may be exposing e-cigarette users to toxic and carcinogenic metals.
Conclusion: Little is known about the vaping behaviors and demography of e-cigarette users, nor the possible exposures to toxic metals from e-cigarette use. This study seeks to provide a descriptive analysis of e-cigarette use in Maryland in order to better understand metals exposure among the study subjects and overall perceptions of e-cigarette users. As a majority of e-cigarette users in this study claim to use these devices as a safe alternative to tobacco cigarettes, more research is needed to define the short and long-term health effects of e-cigarette vapors.

Tu-Po-61

Generation of omics data using 'challenging samples'

Kirsten Hertoghs, *GenomeScan, Leiden, Netherlands*

The effect of external exposures on the individual health status is often measured by -omics analyses. However, sample collection is challenging. Extraction of enough and sufficient quality DNA and/or RNA to generate a biologically relevant dataset, often more so.
In this technical poster, GenomeScan gives a short overview of the most state-of- the-art techniques to measure the whole exposome. Together, methylation assessment, gene-expression analysis and (targeted) SNP profiling give a comprehensive overview of the individual and its response to external stressors.
Large amounts of DNA/RNA or high quality is not a prerequisite anymore. The techniques have been refined so that robust datasets (under ISO 17025) can be obtained. DNA sequencing protocols were adapted so that the input material could be lowered. DNAseq results are generated starting from solely 125 pg DNA (~20 cells). For mRNA, reliable transcriptomics results are now routinely generated using only 5 ng DNA, by either rRNA depletion or poly-A selection.
Furthermore, FFPE samples of which the DNA is partially degraded by e.g. fragmentation and de-amination, can be reliably measured using restoration techniques. Methylation assays were performed using FFPE DNA, fragmented into 100-300 bp lengths. A high concordance of >0.98 was observed between methylated C-residues after restoration. Even for diagnostic purposes, degraded FFPE material can also be used as starting material. A whole exome sequencing (WES) validation test, performed on a selection 8 FFPE DNA samples, yielded the correct diagnosis of all 8 patients.
The latest DNA-, RNA-, and Methylseq techniques can robustly measure genomics, transcriptomics and methylomics. In conclusion, current technology offers sufficient tools to perform genomic cohort analyses , even the most challenging samples.

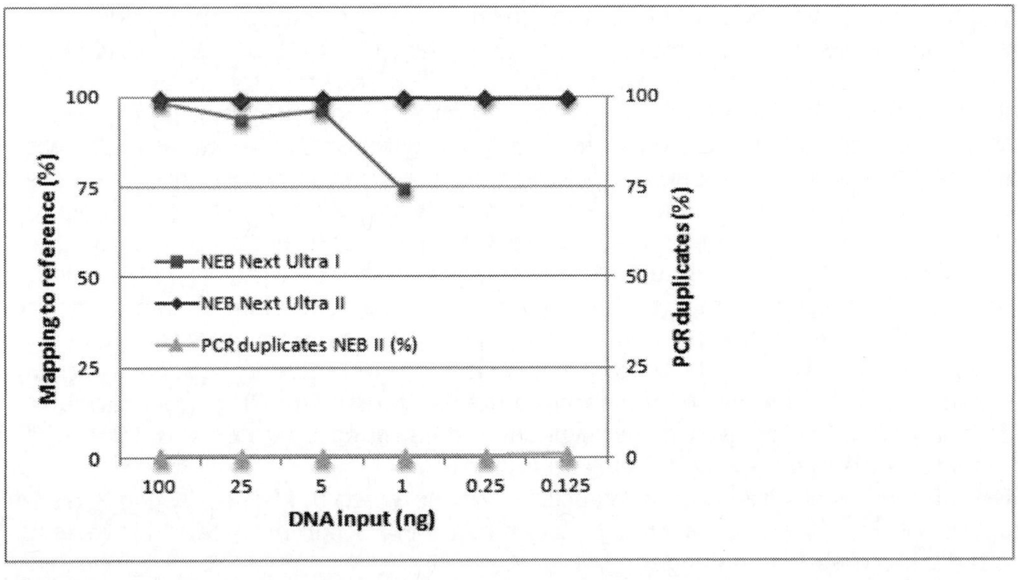

DNAseq on ultra-low input amounts

Late Breaking Abstracts

Tu-LBA-11

Human in vitro skin permeation rates for polycyclic aromatic hydrocarbons (PAHs) are altered with co-exposures to solar ultraviolet radiation (UVs)

Nancy B. Hopf, *Institute for Work and Health (IST), Epalinges-Lausanne, Switzerland*
Philipp Spring, *Centre hospitalier universitaire vaudois CHUV, Lausanne, Switzerland*
Nathalie Hirt-Burri, *Centre hospitalier universitaire vaudois CHUV, Lausanne, Switzerland*
Benjamin Sutter, *Institut National de Recherche et de Securite INRS, Nancy, France*
Silvia Jimenez, *Institute for Work and Health (IST), Epalinges-Lausanne, Switzerland*
David Vernez, *Institute for Work and Health (IST), Epalinges-Lausanne, Switzerland*
Aurélie Berthet, *Institute for Work and Health (IST), Epalinges-Lausanne, Switzerland*

Road construction workers are simultaneously exposed to two carcinogens; solar ultraviolet (UV) radiation and polycyclic aromatic hydrocarbons (PAHs) in bitumen emissions. The combined impact of the two carcinogens may contribute to an increased risk for skin cancer due to photogenotoxicity and enhanced PAH skin permeation rates.
AIM: Our aims were to compare skin permeation rates for selected PAHs with and without simultaneous UV exposures, and to explore two possible photogenotoxicity biomarkers in vitro: p53 for DNA damage and matrix metalloproteinase-1 (MMP1) for degeneration of dermal extracellular matrix.
METHOD: We used flow-through diffusion cells mounted with human viable skin (not previously frozen) to measure permeation rates over 24 hours for five PAHs; naphthalene, chrysene, anthracene, pyrene, and benzo(a)pyrene (BaP) in a mixture (5 mg/ml of each) with co-exposure to solar UV radiation (equivalent to a day of sun exposure ~ 600 J/m2) generated by a solar simulator (Solar Light LS1000). Human skin was obtained from abdominoplasty surgery patients (N=3) after informed consent (DAL biobank). MMP1 was determined by real time PCR and p53 by histology.
RESULTS: Naphthalene, pyrene, and BaP in the PAH mixture permeated human skin greater without compared to with simultaneous exposures to UVs (13.3 vs 6.7 ng/cm2 for naphthalene, 3.32 vs 2.77 ng/cm2 pyrene, and 0.94 and 0.34 ng/cm2 for BaP, respectively). Time until breakthrough (Tlags) were similar for naphthalene (4-5h) and pyrene (7-8h) co-exposed or not to UVs; while rapid (1h) without UVs and longer (3h) with UVs for BaP. Anthracene and chrysene permeated skin to a greater extent with simultaneous exposures to UVs compared to without (1.76 vs 2.89 ng/cm2 for anthracene and 0.49 vs 0.73 ng/cm2 for chrysene, respectively). UV co-exposure did not change the Tlags for anthracene (7h) and chrysene (1h). Permeation rates increased for anthracene and chrysene; and decreased for naphthalene, pyrene, and BaP. Possible explanations are that i) skin metabolizes PAHs thus the sum of the parent compound and metabolites should be measured or ii) UVs reacts with PAHs which undergo radical reactions producing reaction products. MMP1 could not be determined due to insufficient RNA. Qualitative interpretation of p53 indicated greater damage after simultaneous exposure to PAHs and solar UV compared to either exposures separately.
CONCLUSION: All PAHs measured had permeation rates between 0.34 and 13.3 ng/cm2 and permeated skin within 1-7h. Co-exposures to UVs altered PAHs skin permeation rates and could potentially increase DNA damage.

Tu-LBA-12

AirSensEUR: Open platform and open access air quality monitoring

Laurent Spinelle, *European Comission - Joint Research Center, Ispra, Varese, Italy*
Michel Gerboles, *European Commission - Joint Research Center, Ispra, Varese, Italy*
Alexander Kotsev, *European Commission -Joint Research Center, Ispra, Varese, Italy*
Marco Signorini, *Liberaintentio SRL, Malnate, Varese, Italy*

AirSensEUR is an open platform that facilitates the use of low-costs gaseous sensors for the monitoring of air pollution at low concentration levels. The platform has been developed by the Joint Research Centre of the European Commission (JRC) and LiberaIntentio, an Italian SME specialized in IoT. All development aspects are made freely available through the use of public licenses. AirSensEUR is a multi-sensor platform that has the capacity to behave as a node within a network of multi sensors. It has been developed assuring compliance with the INSPIRE Directive (Infrastructure for Spatial Information in the European Community), ensuring interoperability and easy access to the observation data being collected.
This platform is composed of a sensor shield, a host board and web server. The sensor shield (Figure Part A1) was designed to measure the low currents of electrochemical sensors. It is a high precision 4-channel sensor board with temperature/humidity and pressure sensors mounted on an ancillary board. Details of its electronic and operability are given in Gerboles et al., AirSensEUR, Part A: Sensor shield (ISSN 1831-9424).
The host board (Figure Part A2) is based on a low cost Arietta G25 module from ACMESystem. It also accommodates a micro-SD card, a GPS, a GPRS and a Wi-Fi access point. The whole system can be powered by a high capacity battery, through USB or power line. The host board gathers data form the sensor shield and GPS into a local sqlite3 database, stored on the SD card. These data are then pushed via GPRS or Wi-Fi to an external server through a standard-based transactional Sensor Observation Service (SOS-T). Details are given in Gerboles et al., AirSensEUR, Part B: Host platform, influx datapush and assembling of AirSensEUR (in press).
Observations are stored on the AirSensEUR server (Figure Part B) in a PostgreSQL/PostGIS database together with additional metadata. The use of the SOS ensures compliance with the requirements of the INSPIRE Directive. Apart from legal compliance SOS facilitates data interoperability, as it can be retrieved and directly re-used by standard clients. Furthermore, the use of PostgreSQL database makes it easy to interface sensor observations with open source GIS applications. Data correction can be easily done using the “R” statistical package, e.g. allowing the development of on-the-fly calibration of sensors against existing monitoring stations.
In conclusion, AirSensEUR is an easy to configure platform which is sensitive enough to measure ambient air pollution in the range expected at background and traffic sites. It is a promising technology which provides new opportunities for the monitoring of population exposure in mobile context and fixed measurements.

Architecture of AirSensEUR

Tu-LBA-13

Exposure to Flame Retardant Chemicals in the Home and Increased Risk for Papillary Thyroid Cancer

Kate Hoffman, *Duke University, Durham, North Carolina, United States*
Julie Ann Sosa, *DUKE CANCER INSTITUTE, DURHAM, NC, United States*
Amelia Lorenzo, *Duke University, Durham, NC, United States*
Craig Butt, *Duke University, Durham, NC, United States*
Albert Chen, *Duke University, Durham, NC, United States*
Heather Stapleton, *Duke University, Durham, NC, United States*

Aims: Thyroid cancer is the fastest increasing cancer in the US. While increased exposure to radiation has been hypothesized to play a role, recent studies suggest other environmental factors are likely responsible. Exposure to flame retardant chemicals (FRs) also is increasing, raising concerns about potential health impacts, as animal studies indicate that some FRs can disrupt thyroid function and homeostasis. Furthermore, some FRs are classified as probable carcinogens.
Methods: We are conducting a case controlled study investigating the impact of FR exposures on papillary thyroid cancer (PTC) occurrence and severity. We have recruited 52 participants with PTC and 52 matched controls. Because levels of FRs in household dust are strongly correlated with personal exposure, we visited participants' homes and collected dust samples. Participants also provided a blood sample. Several classes of FRs were measured in dust and a few PBDE congeners in serum. Demographic, lifestyle, and environment information were collected via questionnaire, and tumor histology data were abstracted from medical records.
Results: Study participants ranged from 21 to 80 years of age, and the majority were female (83%), reflecting a known gender difference in PTC risk. One third of cases had nodal metastases (36%), and 42% were positive for the BRAFV600E mutation. Our results suggest that higher levels of some FRs, particularly BDE-209 and tris(2-chloroethyl)phosphate (TCEP) in dust are associated with an increased PTC odds. Those with BDE-209 concentrations in household dust above the median were 2.73 times as likely to have PTC (95% confidence interval (95% CI): 1.14, 6.58) relative those with low BDE-209. Associations differed by the presence of BRAFV600E mutation; those with the highest levels of BDE-209 were 10 times as likely to have PTC and be negative for the BRAFV600E mutation (95% CI: 1.10, 92.50). However, BDE-209 tended to be most strongly associated with less aggressive tumors (e.g. smaller tumors confined to the thyroid). Conversely, TCEP was associated with larger, more aggressive tumors. For example, those with high TCEP levels in dust were 6.33 times more likely to have PTC with nodal metastases (95% CI: 1.38, 28.96). Although not related to case status, data suggest that higher serum BDE-153 levels may be associated with more aggressive tumors (i.e. nodal metastases, OR=2.63; 95% CI: 0.68, 10.27), while serum BDE-47 was associated with increased odds of PTC without BRAFV600E (OR=4.83; 95% CI: 0.99, 23.59).
Conclusions: Taken together, our results suggest exposure to FRs in the home environment may well be associated with the occurrence and severity of PTC. More research is needed to verify these results in a large population, and if validated, steps should be taken to mitigate exposures.

Tu-LBA-14

Disposition of Silver Nanoparticles and C60 in Non-pregnant and Pregnant Rats After Intravenous or Oral Exposure and the Effect on the Biochemical Profile in Urine

Ninell Mortensen, *RTI Internatinoal, RTP, NC, United States*
Rodney Snyder, *RTI International, RTP, NC, United States*
Sherry Black, *RTI International, RTP, NC, United States*
Wimal Pathmasiri, *RTI International, RTP, NC, United States*
James Harrington, *RTI International, RTP, NC, United States*
Keith Levine, *RTI International, RTP, NC, United States*
Anita Lewin, *RTI International, RTP, NC, United States*
Timothy Fennell, *RTI International, RTP, NC, United States*
Susan Sumner, *RTI International, RTP, NC, United States*

Aim: To determine the changes due to the physiologic state of pregnancy, in tissue distribution, internal dose, elimination, and metabolic perturbations, due to Engineered Nanomaterials (ENMs) C60 and silver nanoparticles (AgNPs) with different coating and different sizes.
Methods: Single doses of 5% polyvinylpyrrolidone (PVP)-saline vehicle or uniformly carbon-14-labeled C60 [14C(U)]C60 (~0.2mg [14C(U)]C60 kg-1 body weight) in 5% PVP were administered to non-pregnant or pregnant Sprague-Dawley rats intravenously (i.v.). For AgNP, two coatings (PVP and citrate) and two sizes (20 and 110 nm) were administered either i.v. (1 mg kg-1 body weight) or by oral gavage (p.o., 10 mg kg-1 body weight) to investigate the role of AgNP coating, size, and route of administration. To compare the distribution of AgNP and free silver ions, groups of rats were administered silver acetate (AgAc) either i.v. or p.o., at the same nominal silver concentration as AgNP dosed rats. Pregnant rats were exposed to AgNP on gestation day 18 and euthanized at 24 or 48h post-exposure. [14C(U)]C60 was administered to pregnant rats at different stages of pregnancy. The concentration of ENMs in tissues was measured using inductively-coupled plasma mass spectrometry (silver), or liquid scintillation counting (C60). Broad-spectrum NMR metabolomics analysis of urine was conducted as a discovery tool to determine metabolites and metabolic pathways that were perturbed as a result of exposure.
Results: The physiological state was found to have influenced tissue distribution and internal dose of both ENMs. ENMs were detected in both the placenta and fetus for all exposure groups. The distribution of ([14C(U)]C60) in pregnant rats was influenced by both the state of pregnancy and time of termination post exposure. For AgNP the route of administration, nanoparticle size, and coating had a profound impact on tissue concentration of silver. The tissue distribution and internal dose differed between AgNP and AgAc. Metabolomics analysis of urine demonstrated that AgNP exposure in both non-pregnant and pregnant rats impacted carbohydrate, lipid, and amino acid metabolism and transport. In female non-pregnant rats ([14C(U)]C60) impacted pathways related to nicotine signaling and tricarbonic acid cycle. For pregnant rats nicotine signaling pathway and N-acylethanolamine, HSRL5-transacylation pathway were perturbed 24 h post exposure, while vitamin B and regulation of lipid metabolism were perturbed 8 days after exposure.
Conclusions: Our work demonstrated that pregnancy impacts the tissue distribution of [14C(U)]C60 and AgNP in rats, and that both ENMs cross the placenta and reach the developing fetus.
[U19ES019525, Fennell; 1U24DK097193-01, Sumner]

Tu-LBA-15

Assessing the effect of using exposure imputation approaches on the association between nitrate concentrations in public drinking water and birth outcomes in Ohio, 2006-2013

Larissa Pardo, *United States Environmental Protection Agency (U.S. EPA), Washington, DC, United States*
Michael Wright, *United States Environmental Protection Agency (U.S. EPA), Cincinnati, OH, United States*

Background: Epidemiologic evidence suggests that in utero nitrate exposures are associated with adverse reproductive outcomes. Study limitations have included limited sampling frequency and exposure misclassification, which may have diminished the ability to quantify exposure-response relationships.
Objective: This retrospective cohort study examines the effects of using various exposure imputation methods on the association between nitrate exposures in public drinking water and low and very low birth weight (LBW, VLBW) in infants born in Ohio from 2006-2013.
Methods: Sampling frequency of nitrates from 1,045 Ohio public water systems (PWSs) ranged from 1-92 times per year. Most PWSs sampled annually, wherein an imputation method based on the exposure distribution for each year by quarter was used to estimate quarterly exposures. Birth data included all term singleton births (n=1,085,948). Nitrate data were matched to maternal zip code at time of birth. Nitrate exposure was calculated as an average concentration of exposure during pregnancy. Logistic regression was used to estimate the association between nitrate exposure and both LBW and VLBW. Potential confounders included maternal age, number of prenatal care visits, change in maternal weight (during pregnancy), pre-pregnancy body mass index, marital status, maternal education, smoking, race/ethnicity, gestational age, and parity.
Results: Non-imputed and imputed nitrate exposure values were highly correlated ($r_s=0.87; p<0.05$) with an overall range of 0.003 to 57.9 mg/L. When comparing non-imputed versus imputed exposure quartile classifications, the 72% of non-imputed quartile 1 scores that were reclassified were all now in the intermediate quartiles. Quartile 4 results were largely unchanged, with only 7% of non-imputed scores reclassified as quartile 3. When comparing non-imputed versus imputed exposures with VLBW (aOR=0.89; 95%CI: 0.63, 1.26 vs. aOR=1.15; 95%CI: 0.90, 1.48), preliminary results appear to be attenuated. We saw null results for LBW irrespective of the exposure metric used.
Discussion: Although our preliminary results showed little evidence of associations between in utero nitrate exposures and fetal growth retardation; we saw changes in VLBW when comparing the use of non-imputed versus imputed exposures. We also saw some evidence of exposure misclassification based on preliminary imputations. Given the potential for seasonal fluctuations in nitrate concentrations, the available monitoring data does not allow for consideration of more narrow exposure assessment windows (e.g., trimesters). Future analyses will consider integration of private water system data, more advanced imputation techniques to address un-captured temporal variability, and consideration of the cumulative impacts of joint exposure to co-occurring pesticides.
Disclaimer: The views expressed in this abstract are those of the authors and do not necessarily reflect the views or policies of the U.S. EPA or the ORISE.

This page intentionally left blank.

Tu-LBA-16

Biomarkers to assess exposure to nickel and chromium from e-cigarette use

Angela Aherrera, *Johns Hopkins University Bloomberg School of Public Health, Baltimore, Maryland, United States*
Pablo Olmedo-Palma, *Johns Hopkins University Bloomberg School of Public Health, Baltimore, Maryland, United States*
Stefan Tanda, *University of Graz, Graz, Austria*
Walter Goessler, *University of Graz, Graz, Austria*
Maria Grau-Perez, *Johns Hopkins University Bloomberg School of Public Health, Baltimore, Maryland, United States*
Stephanie Jarmul, *Johns Hopkins University Bloomberg School of Public Health, Baltimore, Maryland, United States*
Rui Chen, *Johns Hopkins University Bloomberg School of Public Health, Baltimore, Maryland, United States*
Joanna Cohen, *Johns Hopkins University, Baltimore, Maryland, United States*
Ana Rule, *Johns Hopkins University Bloomberg School of Public Health, Baltimore, Maryland, United States*
Ana Navas-Acien, *Johns Hopkins University Bloomberg School of Public Health, Baltimore, Maryland, United States*

Background
Electronic cigarette (e-cigarette) use is increasing worldwide, yet little is known about the chemical components of e-cigarette devices and e-liquids, including their potential as an exposure pathway to metals. Nickel (Ni) and chromium (Cr) are frequent components of e-cigarette heating coils and have been found at high levels in e-liquid and aerosol. We assessed the association of e-cigarette use patterns and e-liquid metal concentrations with Ni and Cr biomarker concentrations in e-cigarette users from Maryland.
Methods
We recruited 64 e-cigarette users from December 2015 to March 2016. Urine, saliva, and exhaled breath condensate (EBC) samples were collected along with e-liquid from the dispenser (no contact with the coil), the condensed aerosol, and the remaining e-liquid in the tank (in contact with coil). Ni and Cr concentrations were measured using ICP-MS.
Results
The median Ni and Cr concentrations were 0.73 and 0.39 µg/g of creatinine in urine, 3.11 and 1.71 µg/L in saliva, and 1.25 and 0.29 µg/L in EBC. Urine Ni concentrations were positively associated with increased Ni concentrations in condensed aerosol (p for trend 0.03). Saliva Cr concentrations were positively associated with Cr concentrations in condensed aerosol (p for trend 0.02) and with the e-liquid in the tank after contact with the heating coil (p for trend 0.02). We found no association between e-liquid in the dispenser and metal biomarkers. Regarding e-cigarette use patterns, increased urine Ni concentrations were observed in participants who had an earlier time to first vape in the morning ($\leq$ 15 minutes)(p for trend 0.014). Participants who consumed more e-liquid per week (35 to 60 ml) and who also vaped at a higher voltage (4.1 to 4.5 volts) had higher saliva Ni concentrations than those who consumed less e-liquid (5 to 30 ml) and vaped at a lower voltage (2.1 to 4 volts).
Conclusion
Nickel in urine and chromium in saliva were positively associated with the concentrations of these metals in condensed aerosol and/or e-liquid in the tank, but not with the e-liquid

before coil contact. Higher e-cigarette use and potentially higher voltage were associated with higher Ni biomarkers. Additional studies with a larger sample size and comparison to a reference group are needed to confirm that e-cigarette use increases chromium and nickel exposure.

Metal levels in urine (μg/g of creatinine) by participant characteristics. Horizontal lines, interquartile ranges; squares, medians; dotted vertical line, the geometric mean for the overall study sample.

Tu-LBA-17

Metal concentrations in e-cigarette liquid and aerosol samples: the contribution of the metallic coils

Pablo Olmedo-Palma, *Johns Hopkins Bloomberg School of Public Health, Baltimore, Maryland, United States*
Stefan Tanda, *University of Graz, Graz, Austria*
Walter Goessler, *University of Graz, Graz, Austria*
Maria Grau-Perez, *Johns Hopkins Bloomberg School of Public Health, Baltimore, Maryland, United States*
Stephanie Jarmul, *Johns Hopkins Bloomberg School of Public Health, Baltimore, Maryland, United States*
Angela Aherrera, *Johns Hopkins Bloomberg School of Public Health, Baltimore, Maryland, United States*
Rui Chen, *Johns Hopkins Bloomberg School of Public Health, Baltimore, Maryland, United States*
Ana Navas-Acien, *Columbia University Mailman School of Public Health, New York, New York, United States*
Ana Rule, *Johns Hopkins Bloomberg School of Public Health, Baltimore, Maryland, United States*

Aim: In recent years, e-cigarette use has markedly increased worldwide. Potentially healthier than tobacco, these devices contain a metallic coil (made of Ni, Cr, Al and other metals) which heats the e-liquid to produce the aerosol. Our objective was to investigate the possible transfer of metals from the heating coil to the e-liquid and generated aerosol.
Methods: We sampled 57 modified devices (called MODs) and 37 disposable cartridges (22 used for aerosol samples 15 used for liquid samples) for 5 different brands of "cig-a-like" devices available in Maryland (United States). MOD liquid samples were collected before they were added to the device and after the device had been used to generate the aerosol. Cig-a-like liquid samples were centrifuged from the cartomizers before being vaped. Aerosol samples from MODs and cig-a-like devices were collected using a method that condenses the aerosol back into liquid form. Samples were diluted in acid and analyzed by ICP-MS.
Results: In modified devices (MODs), the median (IQR) in e-liquid from the dispenser (no contact with the coil), the condensed aerosol, and the remaining e-liquid (in contact with coil) was, respectively, 11.6 (7.2-21.0), 16.2 (12.1-20.8), 31.4 (17.5-115) µg/kg for Al; 0.35 (0.35-2.25), 10.9 (0.4-43.8), 57.9 (18.0-222) µg/kg for Cr; 2.08 (0.71-41.04), 63.6 (6.2-286.3), 246.0 (72.1-762.3) µg/kg for Ni; 13.2 (6.9-23.6), 496 (228-806), 431 (154-1505) µg/kg for Zn and 0.47 (0.24-1.04), 14.4 (3.3-34.8), 40.1 (13.7-188.2) µg/kg for Pb. In cig-a-like devices, the median (IQR) in the condensed aerosol and e-liquid from the cartomizer was, respectively, 112 (105-123), 193 (152-226) µg/kg for Al; 1.25 (0.4-11.35), 62.2 (18.5-192.9) µg/kg for Cr; 62.2 (17.8-238), 338 (62-1026) µg/kg for Ni; 681 (324-1283), 1215 (664-14060) µg/kg for Zn and 1.49 (0.56-7.98), 24.1 (2.7-747.2) µg/kg for Pb. We also found detectable and potentially high concentrations of other metals such as Mn, Fe, Cu, and Sn in MODs and cig-a-like devices, as well as Sb and W in the cig-a-likes.
Conclusions: E-cigarettes are a source of toxic metals such as Cr, Ni and Pb. For MOD devices, metal concentrations were low in the original dispenser and higher in the condensed aerosol and the e-liquid remaining in the tank, supporting that coil contact induced e-liquid contamination. Metal concentrations were generally higher in cig-a-likes

versus modified devices. In MODs, metal concentrations in the aerosol increased with voltage.

Tu-LBA-18

Metal concentrations in processed meat samples.

Pablo Olmedo-Palma, *Johns Hopkins Bloomberg School of Public Health, Baltimore, Maryland, United States*
Rui Chen, *Johns Hopkins Bloomberg School of Public Health, Baltimore, Maryland, United States*
Maria Grau-Perez, *Johns Hopkins Bloomberg School of Public Health, Baltimore, Maryland, United States*
Keeve Nachman, *Johns Hopkins Bloomberg School of Public Health, Baltimore, Maryland, United States*
Fernando Gil, *University of Granada, Granada, Spain*
Ana Rule, *Johns Hopkins Bloomberg School of Public Health, Baltimore, Maryland, United States*
Ana Navas-Acien, *Columbia University Mailman School of Public Health, New York, New York, United States*

Aim: Processed meat products could constitute a relevant source of cadmium (Cd) exposure as they contain heavily processed animal tissues, potentially including organ meats (e.g. kidneys, liver) which are known to accumulate this metal. A recent study found that processed meat consumption is associated with higher urinary Cd levels in American Indian populations. We analyzed Cd together with zinc, arsenic, tin and lead in heavily processed meat products commonly found in U.S. supermarkets.
Methods: 75 samples of processed meat products were acquired in Baltimore, MD from 3 types of supermarkets. First, we acquired 5 different products (beef franks, wieners, pork cheese sausages, bologna and spam) at a low-cost supermarket (5 samples of each product, n=25). Then, we bought the same products from the same brands in a medium-cost supermarket (n=25). Finally, we purchased 5 different similar products (beef hotdogs, chicken hotdogs, turkey hotdogs, cheese franks and turkey bologna) from a high-cost supermarket (5 samples of each product, n=25). Samples were acid digested in a microwave-assisted system and analyzed by ICP-MS.
Results: The median (IQR) Cd for the products with the highest Cd levels was 7.95 (7.83-8.67) µg/kg for beef franks (from low-cost supermarket), 7.81 (7.34-8.00) µg/kg for pork cheese sausages (from low-cost supermarket), and 6.55 (3.56-6.59) µg/kg for bologna (from medium-cost supermarket). The product with the lowest Cd concentrations was spam (from medium-cost and low-cost supermarkets) having all the samples under the limit of detection. The median (IQR) Zn for the products with the highest Zn levels was 25.8 (24.2-26.3) mg/kg for cheese franks (from high-cost supermarket) and 25.1 (23.3-26.0) mg/kg for beef hotdogs (from high-cost supermarket) while bologna (from medium-cost supermarket) with 8.1 (7.9-10.0) mg/kg showed the lowest Zn levels. All samples tested displayed undetectable As, Sn, and Pb concentrations except one sample of beef franks with detectable Pb levels.
Conclusions: Processed meat products analyzed generally presented detectable Cd levels and practically undetectable levels of other potentially toxic elements such as As, Sn and Pb. Zn was widely detectable in all of the samples tested, evidencing that these meat products are a source of this essential metal. More research is needed to understand the source of cadmium in processed meats with detectable concentrations and to assess the importance of these foodstuff, including a wider variety of processed meat products, as a dietary source of Cd and other metals.

Wednesday, October 12, 2016

We-SY-A1: Detection of new and emerging risks of chemicals (NERCs); the need for interdisciplinary cooperation

We-SY-A1.1

SIGNAAL, OSH-vigilance put into practice

Annet Lenderink, *Netherlands Center for Occupational Diseases, Amsterdam, Netherlands*
Lode Godderis, *KULeuven, Center for Environment and Health, Leuven, Belgium*

Background
Changes at work may lead to new occupational health risks. Our goal was to develop, implement and evaluate an online tool called SIGNAAL for the notification and assessment of these new occupational risks in the Netherlands and Belgium. The presentation will outline the possibilities and problems of the system as a first step in Occupational Health and Safety (OSH) vigilance; a possible approach to handle new and emerging health risks.
Methods
An online reporting tool was developed with an online form, a public website and an evaluation procedure. Since July 2013, Dutch and Belgian occupational physicians can report suspected new combinations between health problems, exposure and / or the work situation at www.signaal.info. Each report is reviewed by at least two occupational health experts. For unknown and new cases, targeted research on the etiology is carried out in the scientific literature using a special search string developed for Evidence Based Occupational Medicine. Finally, on the basis of a joint and preferably interdisciplinary consultation, we deduced the occupational nature of the reported diseases as well as their newness.
Results
Between 2 July 2013 and 1 December 2015, 21 reports were registered in SIGNAAL. The cases mentioned relate to various health problems in different sectors. Currently, 16 reports were fully reviewed while the other cases are still under investigation. Of these 16 cases, one case was considered a well-known work-related disease. One case was considered to be a new association of health problems and exposure at work. Then, six cases were known, but rarely reported work-related diseases and in eight cases the known disease was reported in a work situation which was not described before. Some examples will be presented to illustrate how possible new occupational health risks can be assessed, but also which obstacles and pitfalls were encountered.
Conclusion
An online reporting system designed within the occupational health framework can provide valuable data on the possible risks of new and emerging occupational diseases by creating a structured tool for reporting and evaluating new associations of health problems and exposure on the workplace. SIGNAAL is a first step in Occupational Health Vigilance. Further work is needed to develop tools for strengthening and validating these signals through thorough methods of assessment, interdisciplinary discussion between experts and proper dissemination of the results to relevant stakeholders.

We-SY-A1.2

The Dutch approach on handling occupational dermatology

Thomas Rustemeyer, *VUmc university medical centre, Amsterdam, Noord Holland, Netherlands*

Western communities have realised a great improvement on work safety and effects on the health of employees. This is by far related to the reduction of acute intoxications and incidences. On a long term perspective, a diversity of chemical exposures and working conditions may have a negative effect on the health siutaion of employees. In the Netherlands a national system warning system has been established to detect new risks and to alarm on the presence of known existing risks. This network includes medical specialists seeing patients with potential occupational related dermatoses. On a regular emerging risks asre discussed and the management of existing risks are improved in order to lower occupational related dermatological and allergological related diseases. Furthermore, a strong emphazise has been put on post-academical education of professionals to improve early detection of known diseases and warning of potentially new risks.

We-SY-A1.3

Talc dust: Food for thought

Jos Rooijackers, *NECORD, Utrecht, Netherlands*
Remko Houba, *NECORD, Utrecht, Netherlands*

Introduction: Chronic health effects of occupational exposure to dust, gases and vapours are not well recognised by health professionals and neglected by public authorities and employers, leading to missed diagnoses and putting employees in danger. In 2012 an employee of a chocolate products plant was diagnosed with talcosis. Talc (magnesium silicate) is often used in the food industry. Although talc is considered to be safe by oral route, inhalation of talc is a well known cause of granulomatous lung disease and fibrosis. The company was not aware of this risk.
Objectives: To identify exposed workers at risk of talcosis and to define control measures.
Methods: Exposure assessment consisted of semi-quantitative evaluation for all job titles with potential talc exposed tasks. In addition, personal respirable talc measurements were performed. Medical evaluation in 111 workers consisted of a questionnaire on occupational history and respiratory symptoms. Cumulative exposure was estimated as the product of the total number of days worked and job title with relevant talc exposure as dummy variable. Based on estimated cumulative exposure workers were referred for clinical investigation including a HRCT scan of the thorax.
Results: Full shift personal respirable talc exposure varied between job titles (range TWA 0.05- 0.54 mg/m3) and were often close to or exceeding the Dutch OEL of 0.25 mg/m3. During some tasks high peak exposures occurred, e.g. up to 9 mg/m3 respirable talc during filling the talc storage box. HRCT scan was performed in 18 highest exposed workers with 8-40 work years. In addition to the index case, for one worker with NSIP on the HRCT talcosis was confirmed by lung biopsy. In another worker HRCT showed a nodular pattern. Several control measures have effectively reduced exposure.
Discussion: Once an index-case has been diagnosed, further actions depend on many players and factors. In the Netherlands, limitations may occur on every level, both at expertise in occupational medicine, multidisciplinary team work, executing and financing a surveillance. Alternatively, tools for early identification of health risks of occupational exposure to agents aimed at primary prevention is even more challenging.
Conclusion: Inhaled talc was an unidentified hazard in this food processing plant. This may apply to other industries as well.
This study showed that comprehensive surveillance programmes including exposure assessment and structured medical evaluation are the keystone of prevention and contribute to a safe and healthy workplace.

We-SY-A1.4

The NIOSH Health Hazard Evaluation Program and Investigations of New and Emerging Hazards

Teresa Seitz, *U.S. Centers for Disease Control and Prevention / National Institute for Occupational Safety and Health, Cincinnati, Ohio, United States*

The National Institute for Occupational Safety and Health (NIOSH) Health Hazard Evaluation (HHE) Program conducts evaluations at U.S. workplaces to learn whether workers are exposed to hazardous materials or harmful conditions, including new and emerging chemicals. The Program is legislatively mandated to conduct these workplace evaluations when requested by the employer, a union representative, or three or more employees. This presentation will highlight the Program's authority to conduct these workplace evaluations and discuss its operating procedures and products. This public health practice program has received over 10,000 requests for assistance since it began in 1971, following the passage of the U.S. Occupational Safety and Health Act of 1970 that created NIOSH and the Occupational Safety and Health Administration (OSHA). The Program currently receives about 250 requests for assistance each year. A triage process is used to prioritize requests that will receive on site evaluations. An interdisciplinary team of industrial hygienists, physicians, and other subject matter experts from NIOSH (health communicator, psychologist, epidemiologist, ergonomist) conduct the workplace evaluations. An important component of this work is the inclusion of employee and employer representatives during all phases of the evaluation. Most evaluations have both an exposure and health component. The exposure assessments can include personal and area air monitoring, surface sampling (including dermal assessments), record reviews, workplace observations, and assessment of engineering and administrative controls and personal protective equipment use. The health component can include confidential employee interviews, medical record reviews, exams, questionnaires, biological monitoring, and medical testing. The final product of the evaluations is a report that is provided to employer and employee representatives and is posted in the workplace. Final reports provide recommendations to correct identified hazards and contain a plain language summary that is posted in the workplace along with the final report. Final reports are also available to the general public on the NIOSH website. To further disseminate important findings from these evaluations, results are often shared in peer-reviewed and trade publications, blogs and other social media outlets, and at technical and trade conferences. Results from these investigations have been used in developing NIOSH, OSHA, and ACGIH occupational exposure limits, and other national guidance documents. Case studies from two HHEs involving new and emerging chemical hazards will be discussed.

We-SY-A1.5

Panel Discussion

Nicole Palmen, *RIVM, Bilthoven, Netherlands*

General Discussion
This symposium will give an overview of several existing examples of organizations that identify, prioritize, evaluate and establish potential NERCs. Once a potential NERC is identified, the causal relationship between exposure (or work) and the health effect needs to be established by an interdisciplinary expert group. They should study whether the signal is real and whether additional research is necessary to confirm the signal.
During a general discussion, the speakers and attendants will discuss:

- Minimal conditions needed to identify, prioritize and evaluate the causal relationship between exposure to the NERC and the health effect;
- Effect of national policies;
- Bottlenecks in identification, prioritization and evaluation of NERCs;
- How can these bottlenecks be solved;

This will be organized by putting forward some propositions at which both public and speakers can reflect.

We-SY-B1: The Worker Health and Efficiency (WE) Program: Understanding and mitigating the risks of Chronic Kidney Disease in El Salvadorian Sugarcane Cutters.

We-SY-B1.1

Mesoamerican Nephropathy - A Primer

Ilana Weiss, *La Isla Foundation, Chicago, Illinois, United States*

Over the course of the last two decades a form of Chronic Kidney Disease not related to the traditional risk factors of diabetes and hypertension has emerged throughout Central America and Mexico. This new form of kidney disease, called Mesoamerican Nephropathy (MeN) or Chronic Kidney Disease of unknown cause (CKDu), predominantly affects sugarcane cutters. Prevalence in several sugarcane communities in Nicaragua is as high as 41%, with high numbers in similar populations also documented in Costa Rica, Guatemala, and El Salvador.

There is a general consensus within the scientific community that MeN is multi-factorial and that the primary driver is occupational—heavy labor in extreme heat. Sugarcane cutting is repetitive high-intensity work carried out in conditions of formidable heat stress. Exposure to pesticides and silica are other hypotheses that need to be further explored.

Those who cut cane come from impoverished, vulnerable communities living in precarious conditions, without access to adequate healthcare, education, food or housing. In addition to grinding poverty, many contend with some of the worst gang violence in the world. Workers are usually paid per ton cut—a rate that varies from less than $1 USD/ton in Nicaragua, to about $2.40/ton in El Salvador. Some places in El Salvador pay workers per tarea, or predetermined area. Salaries are generally about $5/day.

Work cutting sugarcane is often the only option for poor populations with little or no schooling. Sugarcane is a monoculture that has grown rapidly in the last 40 years displacing traditional subsistence crops and uprooting traditional farming populations. As land under cane has quadrupled, so has the likelihood of CKD for men in the cane producing areas of the region.

We-SY-B1.2

Pesticide and silica exposure in sugarcane cutters in El Salvador

Brian Curwin, *National Institute for Occupational Safety and Health, Cincinnati, OH, United States*
Ilana Weiss, *La Isla Foundation, Ada, MI, United States*
Cynthia Striley, *National Institute for Occupational Safety and Health, Cincinnati, OH, United States*
Marissa Alexander-Scott, *National Institute for Occupational Safety and Health, Cincinnati, OH, United States*

One occupational risk being considered as a potential etiologic factor in chronic kidney disease of unknown origin (CKDu) in Central America is pesticide exposure. There have been a few studies in El Salvador, Sri Lanka, and the US that have suggested pesticides may be linked to CKD and end stage renal disease (ESRD). Silica exposure has also been shown to be associated with nephrotoxicity and CKD in several occupations. The purpose of this study was to assess occupational exposures among sugarcane workers in El Salvador to 2,4-D and glyphosate, 2 commonly used herbicides in sugarcane production, and to a lesser extent to assess exposure to silica. The study took place in El Salvador during the sugarcane harvest (zafra) in March 2016. Forty sugarcane cutters were surveyed, 20 from a coastal sugarcane field and 20 from an inland field at higher elevation. Each worker was sampled for three consecutive days. Each sampling day, hand wipe and urine samples were collected from each worker and analyzed for glyphosate, and 2,4 D; area air samples were collected near the sugarcane being harvested and analyzed for glyphosate, 2,4-D, and silica. Additionally, drinking water samples for each cutter was collected and analyzed for glyphosate, and 2,4-D.
The fieldwork was ongoing at the time of abstract submission. A total of 236 urine samples, 118 hand wipe samples, 30 air samples, and 40 water samples were collected. The relationship between pesticide and silica exposure and various factors and practices such as field location (inland versus coastal), personal protective equipment (PPE), pesticide application schedule, and harvest production, was evaluated. The relationship between pesticide and silica exposure and markers of kidney function including urinary albumin and serum creatinine was also evaluated by combining the pesticide exposure results with kidney function results obtained from the Worker Health and Efficiency (WE) study.

We-SY-B1.3

Assessing Heat and Dehydration in Sugarcane Harvesters

Rebekah Lucas, *University of Birmingham, Birmingham, United Kingdom*
Ilana Weiss, *La Isla Foundation, Chicago, Illinois, United States*
Sandra Peraza, *Universidad de El Salvador, San Salvador, El Salvador*
Emmanuel Ricardo Jarquin Romero, *Agency for Agricultural Health and Development (AGDYSA), San Salvador, El Salvador*
Theo Bodin, *Karolinska Institutet, Stockholm, Sweden*

Introduction: Sugarcane cutters work in difficult, hot and strenuous work conditions. These work conditions are believed to be key causal factors in the chronic kidney disease of unknown origin (CKDu) epidemic occurring in Mesoamerica. The aim of this study was to assess the level of heat stress and dehydration in sugarcane cutters. This study formed part of a larger Worker Health and Efficiency (WE) Program, an intervention implementing OSHA's 'Water.Rest.Shade' recommendations.
Methods: Data were collected in a cohort of two groups of sugarcane cutters (totaling 60 individuals) during 2015 and 2016 harvests. Outdoor Wet Bulb Globe Temperatures (WBGT) was calculated (WBGT (outdoor) = 0.7WB + 0.2G + 0.1DB) via the QuesTemp °34. Heart rate (HR, Polar) was recorded in 10-11 workers per day, for seven workdays in 2015. An algorithm using sequential HR data was used to estimate body core temperature (Tcore) in these workers. Heart rate, gastro-intestinal temperatures (TGI, Equivital, vitalsense), heat and dehydration symptoms, and water consumption were recorded in 11 workers during three workdays in 2016.
Preliminary Results: WBGT reached 32.5°C (95% confidence interval [CI]: 33.9 to 31.1°C), with 77% (95% CI: 83 to 72%) of the day spent working at a WBGT above 26°C (threshold limit for continuous harvesting at 100%). Heart rates averaged 54%HRmax (95% CI: 56 to 53%HRmax) across all workdays, with workers spending on average 36% (95% CI: 42 to 31%HRmax) of their workshift (including rest breaks) at and above 50%HRmax. This corresponded to an average estimated Tcore of 37.5°C (95% CI: 37.6 to 37.4°C) and a maximum Tcore of 38.0°C (95% CI: 38.1 to 37.9°C) across all workdays. On average, workers' estimated Tcore exceeded 37.9°C for 14% of their workshift (95% CI: 19 to 9%). TGI averaged 37.5°C (95% CI: 37.7 to 37.4°C) across a workshift and reached a maximum of 38.4°C (95% CI: 38.7 to 38.1°C). Two out of 11 workers TGI exceeded 39.0°C during their workshift.
Conclusions: The hot and strenuous nature of sugarcane cutting determines that internal body temperature for the majority of workers exceeds 38°C and can on occasion reach very high temperatures. This is unsurprising given that the large majority of the workshift is carried out in hot conditions exceeding international threshold limits for continuous work. The long term health impact of such chronic heat exposure is still to be elucidated.

We-SY-B1.4

Assessing Heat Stress Symptoms using Repeated Symptom Questionnaires in Sugarcane Cutters

Catharina Wesseling, *Karolinska Institutet, Stockholm, Sweden*
Theo Bodin, *Karolinska Institutet, Stockholm, Sweden*
Ilana Weiss, *La Isla Foundation, Chicago, Illinois, United States*
Dorien Faber, *La Isla Foundation, León, Nicaragua*
David Wegman, *University of Massachusetts Lowell, Lowell, Massachusetts, United States*

Aim. Heat stress from extremely heavy work in a hot environment is thought to be the main driver of the Mesoamerican nephropathy epidemic in Central America. The Worker Health and Efficiency (WE) Program assessed the feasibility of implementing an intervention providing water, rest and shade in cane cutters in El Salvador during the harvest 2014-2015, and has extended the effort during the harvest 2015-2016. Repeated surveys are an important method to assess symptoms of heat stress and dehydration.
Methods. A group of 60 cane cutters participated in the first year of the intervention that was implemented two months into the harvest and effected January-April 2015. Questionnaires were administered at baseline (November 2014), pre-intervention (January 2015), and post-intervention (at the end of harvest in April 2015); in addition, a short questionnaire was completed biweekly in-between the main data collection events. 41 cutters completed the 3 main questionnaires. Baseline questionnaires addressed work history, general health, fluid intake and symptoms of heat stress and dehydration. Follow-up questionnaires addressed fluid intake and recent symptom occurrence. Similar methods are currently used during the second year of the intervention, harvest 2015-2016, with a larger study population of 250 cutters in three locations.
Results: During the first year of the WE-Program, water intake increased 25% on the group level post- compared to pre-intervention. Little difference in symptoms occurred between baseline and pre-intervention but a decrease occurred in most symptoms post-intervention compared to pre-intervention. Reduced symptom reports occurred for very little urine, feeling feverish, exhaustion, heart racing, cramps, nausea, stomachache, diarrhea, dizziness and disorientation while there was no change in symptoms of dysuria and dark urine. On the individual level, associations between symptoms and water intake and workload over time were unclear during the first year of the study. These relationships are further assessed with the larger group of the 2015-2016 harvest, as well as potential associations between symptoms and biomarkers of kidney function and hydration.
Conclusions. Overall, the decrease in heat stress symptoms post-intervention indicate improved working conditions at the group level post-intervention. With the experience gained by the WE-Program in year-1, the questionnaire was improved, both with regard to symptom questions and questions about fluid intake. The possibility to use symptoms not only as outcomes of heat exposure but also as proxies of heat exposure with biomarkers of kidney function and hydration as outcomes is being explored.

We-SY-B1.5

From Intervention to Policy

Jason Glaser, *La Isla Foundation, Chicago, Illinois, United States*

Authors: Jason Glaser (La Isla Foundation, London School of Hygiene and Tropical Medicine)
Title: From Intervention to Policy
Background: In Mesoamerica CKDu (Chronic Kidney Disease of unknown cause) is epidemic among sugarcane workers and present in other workers. Excessive heat stress and workload are believed to contribute to onset and acceleration of CKDu. The Worker Health and Efficiency (WE) program is the first intervention evaluated that addresses excessive heat stress and workload in sugarcane workers.

Aims:
- Develop a demonstration program to illustrate features and challenges associated with the WE program implementation.
- Demonstrate need for governments and industry to address CKDu and excessive heat stress in sugarcane and other populations.
- Use the resulting press, political and industry attention to push for a wider agenda of worker protections.

Methods: The WE Program is an open lab that evolved into an observatory. Results were used to inform industry and governments. Coordination with media outlets supported good actors while isolating less noble actors while providing a way forward.
Results: Health and productivity indicators from program are encouraging. Results drove policy discussions and measurable change in companies and governments. This has led to private and public policy changes that benefit the health of sugarcane and other at-risk workers. Advances include: leading mills in the region sharing best practices, a presidential decree in Costa Rica on heat stress and mitigating risks of CKDu, and new changes ensuring protections against heat stress and CKDu in sustainable certification programs and industry giants like Nestle. A CKDu specific pilot project also started for The US Department of Labor. The speed of these exchanges and policy changes has exceeded expectation.
Conclusion: An evidenced-based dialog between sugar industry farmers, millers, buyers, and governments was created. This has led to private and public policy changes that benefit the health of sugarcane and other at-risk workers.
Qualifications:
I conceptualized WE Program, led international effort for improved working conditions, hired the research team, and was directly involved with the US Governments and Costa Rican Governments prioritizing this issue as part of their labor and occupational health agendas.

We-SY-C1: What are the requirements for nanomaterial exposure models? - I

We-SY-C1.1

caLIBRAte - establishment of the next generation nano-risk risk governance framework

Keld Alstrup Jesen, *National Research Centre for the Working Environment, Copenhagen, Denmark*

Background: caLIBRAte is a new project addressing the EU Horizon NMP30 call for a next generation risk governance framework to support the European nanotechnology innovation. The general uncertainty about the potential human and environmental risks of manufactured nanomaterials (MN) and new nanoproducts may be significantly reduced if safety is taken into consideration during innovation and development. Due to lack of exposure data and incomplete documentation of MN hazards, such risk assessment need to strongly rely on precautionary or predictive model estimates. Before true benefits of such a paradigm can be reached, confidence must be established in the results generated by such prospective models. This can be achieved if predictive models are thoroughly tested, calibrated, and demonstrated in relevant use scenarios.
Objective: The key objective of the caLIBRAte project is to establish a Systems-of-Systems (SoS) framework for nano-risk governance, which consists of calibrated qualitative to quantitative predictive models for assessment and management of both human and environmental risks of MN and MN-enabled products as well as scanning tools for identification of emerging risks. The SoS framework will specifically support safety in innovation by aligning suitable framework models to support the different decision steps in a "Cooper Stage-Gate®"-type of Idea-to-Launch innovation model as well as the conventional ISO 31000 risk governance framework.
Methods: The caLIBRAte nano-risk governance framework will link different models for: 1) screening of apparent and perceived risks and trends in nanotechnology; 2) control banding, qualitative and fully integrated predictive quantitative risk assessment operational at different information levels; 3) safety-by-design and multi-criteria decision support methods; 4) risk surveillance, -management and -guidance. The SoS Nano-Risk governance framework:
- shall enable systematic risk analysis with clear understanding of data gaps and uncertainties in the assessments for further risk mitigation and management.
- models will all be refined, tested, calibrated and documented to the greatest extent possible.
- enable high reliability and quality of risk assessment and management of MN and MN-enabled products to increase the trust and confidence in all steps of the industrial and regulatory risk governance and communication and risk transfer between stakeholders.
Results: The establishment of a specific and scientifically documented risk governance system for MN and MN innovation is expected to be a major step forward. The outcome of this approach should be safer and state-of-the-art-assessed MN and MN-based products with the chance for faster implementation and better competitiveness and profit of MN as a key-enabling technology.

We-SY-C1.2

Development of ConsExpo nano: a tool to investigate potential consumer exposure to nanomaterials in consumer spray products

Christiaan Delmaar, *RIVM, Bilthoven, Netherlands*
Benjamin Makkes van der Deijl, *RIVM, Bilthoven, Netherlands*
Roel Schreurs, *RIVM, Bilthoven, Netherlands*
Susan Wijnhoven, *RIVM, Bilthoven, Netherlands*

The potential exposure of consumers to nanoparticles raises concern about possible adverse health effects. As the dermal and oral absorption of nanoparticles form non-food consumer products seems to be limited, these concerns are mainly on the potential inhalation of nanoparticles as aerosol from consumer products in applications as spray cans, pump sprays or powders.
To assess consumer exposure to nanoparticles from sprays and powders, RIVM has developed ConsExpo nano (see Figure), an adaptation of the 'exposure to spray' model from the ConsExpo tool for nanomaterial spray scenarios. In this new tool, apart from mass, alternative dose metrics such as total number or total surface area of the nanoparticles inhaled have been suggested, allowing the exposure assessor to evaluate various alternatives.
Because the most relevant effect after inhalation exposure to nanomaterials is the induction of inflammation in the alveoli (Braakhuis et al. (2014)), and one of the most critical determinants of this effect is both the magnitude and duration of the alveolar load of a nanomaterial, ConsExpo nano combines models that estimate the external aerosol concentration in indoor air, with models that estimate the deposition in and clearance of inhaled aerosol from the alveolar region. The tool is currently online, but additional comments are highly appreciated via consexpo@rivm.nl.
Reference
Hedwig M Braakhuis, Margriet VDZ Park, Ilse Gosens, Wim H De Jong and Flemming R Cassee
Physicochemical characteristics of nanomaterials that affect pulmonary inflammation. Particle and Fibre Toxicology 2014, 11:18

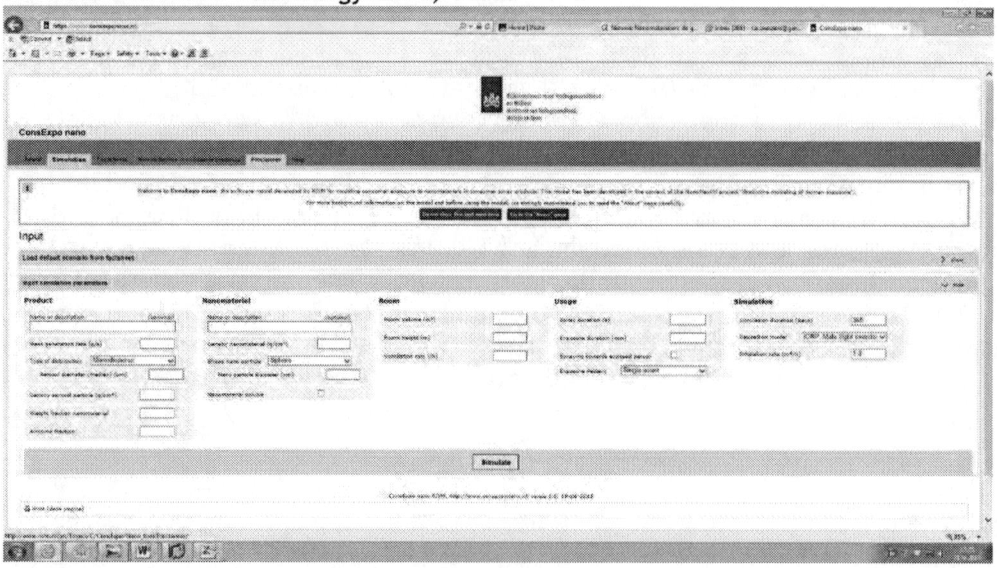

ConsExpo nano home screen

We-SY-C1.3

The SUN 3-Tier modelling-based consumer and worker exposure assessment models

Antti Joonas Koivisto, *National Research Centre for the Working Environment, Copenhagen, Denmark*
Aiga Mackevica, *Technical University of Denmark, Kgs. Lyngby, Denmark*
Martie Van Tongeren, *Institute of Occupational Medicine, Edinburgh, United Kingdom*
Araceli Sanchez, *Institute of Occupational Medicine, Edinburgh, United Kingdom*
Laura MacCalman, *Institute of Occupational Medicine, Edinburgh, United Kingdom*
Wouter Fransman, *TNO, Zeist, Netherlands*
Ilse Tuinman, *TNO, Zeist, Netherlands*
Steffen Foss Hansen, *Technical University of Denmark, Kgs. Lyngby, Denmark*
Keld A Jensen, *National Research Centre for the Working Environment, Copenhagen, Denmark*

Exposure assessment models are critical when the impact of nanomaterials on human health is to be estimated. Human exposure may occur in occupational environment or during use of the product. The SUN project develops a modelling-based 3-tier model for the assessment of inhalation, dermal and inadvertent oral consumer and occupational exposure to nanomaterials. In tier-approach the potential level of exposure decreases when model tier-level (accuracy) is increased. The Tier-1 assessment is based on risk categorization or control-banding procedures, while Tier-2 is based on first-order semiquantitative exposure prediction and Tier-3 is a quantitative exposure prediction. All three tiers will be developed for inhalation exposure, whereas Tier-2 or Tier-3 models will, to the extent possible, be established for dermal and inadvertent oral exposure assessment. A key-criterion for the higher-tier SUN models is the access to quantitative data on the source term. This is exemplified by the Danish NanoSafer Control Banding and quantitative occupational exposure assessment model.
The occupational exposure is usually well controlled where the processes and environmental conditions are known. Thus, the exposure assessment can be made by using Tier-2 or Tier-3 models where conceptual information is needed. In consumer exposure, the exposure scenarios conceptual information is usually no as well known and nanomaterial release from the product may change during the use. Thus, in consumer exposure assessment, all exposure pathways (inhalation, dermal, oral) needs to be estimated using the highest potential nanomaterial release.
Here is presented the parameterization principle of the exposure models. Models performance was tested by comparing predicted inhalation exposure potentials with measured occupational exposure levels in paint industry (Figure 1) and nanodiamond handling in laboratory and to laboratory measurements during electrostatic spray deposition process and powder pouring. Dermal exposure potentials and inadvertent oral exposure levels were predicted for the occupational exposure scenarios and compared with the measured levels in laboratory experiments.

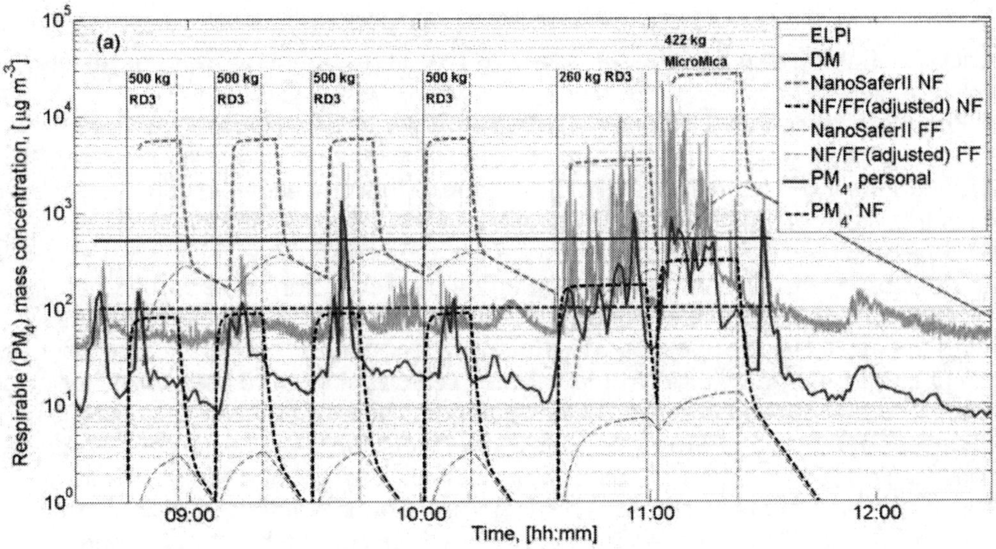

Figure 1. Measured and modelled (NF/FF model and NanoSaferII) near field (NF) and far field (FF) respirable mass concentration time series during powder pouring in a paint factory Gravimetrical personal and NF samplers show the mean mass concentration lev

UTRECHT
THE NETHERLANDS
OCTOBER 9-13
ISES 2016

We-SY-C1.4

Guidance for linking exposure assessment to risk assessment of nanomaterials

Albert Koelmans, *Wageningen University, Wageningen, Netherlands*

Our understanding of the environmental fate and effects of engineered nanomaterials (ENMs) is in a state of fast transition. Recent scientific developments open new and powerful perspectives to define a framework for the prospective risk assessment of ENMs in aquatic ecosystems. This requires abandoning the reductionist's approach of mechanistic analysis on particle or cellular scales and calls for engineering solutions that deal with uncertainties by applying assessment factors and probabilistic approaches. An ecological risk assessment (ERA) framework for ENMs is similar to that for other classes of substances, in that it requires clear protection goals based on ecosystem services, evidence-based concepts that link exposure to effects, and a transparent tiered effect assessment. This presentation discusses approaches to assess and link exposure and effects of ENMs in the natural environment. This includes recent developments in validated spatially resolved ENP fate modeling (i.e. NanoDUFLOW), which greatly expands the potential of retrospective as well as prospective exposure assessments. For the effect assessment, we advise a cost-effective screening based on principles of read-across as a conservative first tier. The feasibility of using species sensitivity distributions (SSDs) as a higher tier option is discussed. Controlled model ecosystem field experiments are proposed as a highest experimental tier, and are required for calibration of the lower tiers. An outlook to unify information from various tiers by experimental work, fate modeling, and effect modeling as cost-effective prospective tools for the ERA of ENMs is provided.

We-SY-C1.5

Modeling the fate of nano- and microplastic in freshwater systems

Ellen Besseling, *Wageningen University, Wageningen, Netherlands*
Joris Quik, *RIVM, Bilthoven, Netherlands*
Albert Koelmans, *Wageningen University, Wageningen, Netherlands*

Riverine transport to the marine environment is an important pathway for microplastic. However, fate and transport models for nano-, and microplastic are lacking. Here we present a spatiotemporally resolved hydrological model that accounts for advective transport, homo- and heteroaggregation, sedimentation-resuspension, polymer degradation, presence of biofilm, and burial of nano- to millimetre sized microplastic (100 nm - 10 mm). Literature data were used to parameterize the model, except for the attachment efficiency for heteroaggregation, which was determined experimentally. The attachment efficiency ranged from 0.004 to 0.2 for 70 nm and 1050 nm polystyrene particles aggregating with kaolin or bentonite clays. Modelled effects of polymer density (1 - 1.5 kg/L) and biofilm formation were not large, due to the fact that variations are largely overwhelmed by excess mass of suspended solids that form heteroaggregates with pristine microplastic. Particle size had a dramatic effect on the modelled fate and retention of microplastic and on the positioning of the accumulation hot spots in the sediment along the river. Remarkably, retention was lowest (18-25%) for intermediate sized particles of about 5μm, which implies that the smaller submicron particles as well as larger microplastic are preferentially retained. Our results suggest that not all microplastic reaches the sea, and that river hydrodynamics affects microplastic size distributions with profound implications for emissions to marine systems.

We-SY-D1: Biomonitoring: The Genie is out of the Bottle: Challenges in Data Quality and Interpretation

We-SY-D1.1

1. Study quality: Biomonitoring, Environmental Epidemiology and Short-lived Chemicals (BEES-C) Instrument

Judy LaKind, *LaKind Associates, LLC, Catonsville, MD, United States*

The quality of exposure assessment is a major determinant of the overall quality of any environmental epidemiology study. The use of biomonitoring as a tool for assessing exposure to ubiquitous chemicals with short physiologic half-lives began relatively recently. These chemicals present several challenges, including their presence in analytical laboratories and sampling equipment, difficulty in establishing temporal order in cross-sectional studies, short- and long-term variability in exposures and biomarker concentrations, and a paucity of information on the number of measurements required for proper exposure classification. To date, the scientific community has not developed a set of systematic guidelines for designing, implementing and interpreting studies of short-lived chemicals that use biomonitoring as the exposure metric or for evaluating the quality of this type of research for WOE assessments or for peer review of grants or publications. We describe key issues that affect epidemiology studies using biomonitoring data on short-lived chemicals and propose a systematic instrument the Biomonitoring, Environmental Epidemiology, and Short-lived Chemicals (BEES-C) instrument for evaluating the quality of research proposals and studies that incorporate biomonitoring data on short-lived chemicals. Quality criteria for three areas considered fundamental to the evaluation of epidemiology studies that include biological measurements of short-lived chemicals are described: 1) biomarker selection and measurement, 2) study design and execution, and 3) general epidemiological study design considerations.

STUDY ASSESSMENT COMPONENTS	TIER 1	TIER 2	TIER 3
Biomarker Selection and Measurement			
Biological relevance Exposure biomarker			
Effect biomarker			
Specificity			
Method sensitivity			
Biomarker stability			
Sample contamination			
Method requirements			
Matrix adjustment			
Study Design and Implementation			
Epidemiology study design			
Exposure variability and misclassification			
General Epidemiological Study Design Considerations			
Study rationale			
Study participants			
Reporting			
Data analysis			

STUDY ASSESSMENT COMPONENTS	TIER 1	TIER 2	TIER 3
Biomarker Selection and Measurement			
Biological relevance Exposure biomarker			
Effect biomarker			
Specificity			
Method sensitivity			
Biomarker stability			
Sample contamination			
Method requirements			
Matrix adjustment			
Study Design and Implementation			
Epidemiology study design			
Exposure variability and misclassification			
General Epidemiological Study Design Considerations			
Study rationale			
Study participants			
Reporting			
Data analysis			

Figure 1. Biomonitoring, Environmental Epidemiology and Short-lived Chemicals (BEES-C) Instrument

We-SY-D1.3

Urinary dilution—do we know what we are doing? Correction methods and controversies

Ana Navas-Acien, *Columbia University, New York, NY, United States*
Chin-Chi Kuo, *China Medical University Hospital, Taichung, China, Republic of (Taiwan)*
Pablo Olmedo-Palma, *Johns Hopkins Bloomberg School of Public Health, Baltimore, MD, United States*

Urine biomonitoring is critical in exposure sciences and environmental epidemiology. 24-hour urine collection, however, is challenging and both biomonitoring and environmental epidemiology generally rely on spot urine samples. A major limitation of spot urine samples is measurement error introduced by differences in urine dilution, which can be quite substantial. While for many environmental chemicals concentrations in spot urine samples are generally considered valid surrogates of exposure and internal dose, there are concerns regarding the best method to control for urine dilution. The most common approach includes dividing or adjusting for urinary creatinine. Urine creatinine is a useful molecule to adjust for urine dilution because it is constantly produced and secreted in the urine throughout the day. While urinary creatinine is a standard method clinically used to adjust for urine dilution, it can itself induce confounding and/or measurement error as creatinine production is related to muscle mass, nutritional status and other factors that vary across individuals. Additional methods to account for urine dilution include specific gravity and osmolality. Those methods have also limitations as they can be influenced by factors that affect urine density such as glucosuria or proteinuria. In this presentation, different methods for correcting urine dilution and their impact in biomonitoring and environmental epidemiology will be presented, including case studies of the association of arsenic and cadmium with chronic health outcomes.

We-SY-D1.4

Capturing temporal variability and transient exposures—is biomonitoring the right tool?

John Meeker, *University of MIchigan, Ann Arbor, MI, United States*
Susan L. Teitelbaum, *Icahn School of Medicine at Mount Sinai, New York, NY, United States*

The use of biomonitoring in epidemiology studies to investigate the potential role of exposure to environmental contaminants in the development of many diseases and disorders has rapidly increased in recent years. Research efforts to date have varied greatly in study design as well as in their application of exposure biomarkers, and, consequently, in their contribution to our current knowledge of these exposure-disease relationships and utility in risk assessment policy setting efforts. Exposure measurement error and misclassification stemming from temporal variability in exposure can be immense and have detrimental impacts on the quality of an epidemiology study and interpretation of study results. While the potential for measurement error can be substantially reduced in many cases through careful considerations in study design and selection of the appropriate exposure measures, these details are often overlooked which can result in poor or inefficient use of research resources given the current high measurement cost for many exposure biomarkers. Evidence from recent studies on temporal stability metrics such as intraclass correlation, sensitivity, specificity, and positive or negative predictive value will be reviewed for biomarkers of exposure for a range of legacy and emerging chemicals of concern. Consequences of these and other metrics on the potential for the different types of exposure measurement error, and how that information should be used to inform study design and translation, will be discussed. Potential enhancements or alternatives to these approaches aimed at improving exposure assessment for epidemiology in the future will be proposed, as will future research needs in this area.

We-SY-D1.5

Challenges in Interpreting Biomonitoring Data: Special Considerations in Childhood and Pregnant Women

Mary Mortensen, *U.S. Centers for Disease Control, Atlanta, GA, United States*

Pharmacokinetics encompasses the absorption, distribution, metabolism, and elimination of a chemical from the body and can be influenced by life stages. Biomonitoring, the measurement of a chemical or metabolite, usually in blood or urine, is a tool to assess human exposure. This presentation will focus on how biomonitoring measurements and their interpretation can be affected by physiologic differences in childhood relative to adulthood and changes that occur during pregnancy. Examples will illustrate the following situations:

- Children have a smaller muscle mass than adults, so the typical child produces less creatinine than the adult. When used to "adjust" a urinary biomarker measurement, the child's lower urinary creatinine concentration can lead to confusing results, with a considerably higher "adjusted" concentration than the original (volumetric) measurement. Numerous other factors including time of day, age, sex, race/ethnicity, and certain disease states also affect urine creatinine, limiting its value as a urinary dilution adjustor. Alternatives to creatinine adjustment have been proposed, including use of urine creatinine as a covariate in regression models and calculation of the biomarker excretion rate.
- Compared to adults, children may demonstrate behaviors (e.g., mouthing), dietary (e.g., reliance on dairy and less dietary variety) and other differences (e.g., higher respiratory rate, greater ventilation ratio relative to adults) that influence exposure likelihood to ingest or inhale chemicals. The child's resulting exposure may be greater than the adult, on a bodyweight basis.
- During the first trimester of pregnancy, renal clearance and total body water increase dramatically. Non-persistent chemicals that are primarily eliminated in urine may have a shortened elimination time. Depending on collection timing, urine concentrations may be exceptionally high if maximum elimination occurred during the sampling interval, or very low or non-detectable if most of the chemical was eliminated prior to sampling.
- Placental transfer and breast milk can be significant elimination routes for low molecular weight or lipid soluble chemicals. Particularly for lipid soluble chemicals that are stored in fat, successive pregnancies may reduce serum chemical concentrations, so parity is an important variable to ascertain in studies of women exposed to these chemicals.

Epidemiologists should be attentive to physiologic and behavioral differences that may influence sample collection design and the interpretation of biomonitoring measurements.

We-SY-E1: Exposure to SVOCs in the Indoor Environment - Products, Emissions, Exposure, Pharmacokinetics and Biomarkers - I

We-SY-E1.1

Exposure to SVOCs in the Indoor Environment - Products, Emissions, Exposure, Pharmacokinetics and Biomarkers

John Little, *Virginia Tech, Blacksburg, VA, United States*

A systematic and efficient strategy is needed to assess and manage potential risks to human health that arise from the manufacture and use of thousands of chemicals. Among available tools for rapid assessment of large numbers of chemicals, significant gaps are associated with the capability to evaluate exposures that occur indoors. For semi-volatile organic compounds (SVOCs), an important class of indoor pollutants which includes plasticizers, flame retardants, and pesticides, exposure is strongly influenced by the types of products in which the SVOCs are present (for example, are they additives or are they sprayed or applied to interior surfaces), the characteristics of the indoor environment in which the emissions occur (for example, the air exchange rate and the concentration and type of airborne particles), the behavior of the occupants who are present in the environment (for example, the food they eat, the cosmetics they apply and the clothes that they wear) and their physiological characteristics (for example, their breathing rate and their metabolic rate). In this symposium, we will begin with products and emissions and work our way through exposure, pharmacokinetics, and biomarkers, illustrating and integrating the complex interactions that govern the entire exposure pathway.

We-SY-E1.2

Comparison of available methods to measure the source/sink characteristic parameters important for estimating indoor exposure to SVOCs

Yinping Zhang, *Tsinghua University, Beijing, China, People's Republic of*
Jianping Cao, *Tsinghua University, Beijing, China, People's Republic of*

Aim: Widely used in various indoor materials and products, semi-volatile organic compounds (SVOCs) are ubiquitous in indoor environments. Due to extremely low vapour pressure, SVOCs tend to redistribute from their source materials to indoor air, interior surfaces (e.g., the ceiling, walls, indoor material surfaces, human skin, and clothing), dust, and suspended particles. Associations between human exposure to several indoor SVOCs and adverse health effects (e.g., asthma, birth defects, obesity, and cancer) has been made. To quantify exposure to indoor SVOCs, characterizing the emission behaviour of SVOC source materials and adsorption behaviour of indoor sink materials is a prerequisite. Through mass transfer analysis, key parameters that characterize the SVOC source or sink behaviours have been identified. Various measuring methods have been developed and used to determine these characteristic parameters and examine the relevant mass transfer models. The aim of this paper is to review existing methods for measuring the SVOC source/sink characteristic parameters, focusing on their principles, precisions and time durations. Areas for further research are also identified.
Methods: The papers are found using the key words “SVOC” (or “phthalate”, “flame retardant”, and “polychlorinated biphenyl”), “measure”, “emission” or “sorption” in Google Scholar, as well as other papers that cite these papers or have been cited by these papers.
Results: The methods were reviewed in the order of years they were proposed. In the early studies, the methods widely used for VOCs measurement were directly employed to measure the SVOC source/sink parameters. However, several features of SVOCs, including the low gas-phase concentration, strong sorption onto surfaces (including the interior surfaces of test chamber and the sampling lines), and ubiquity in laboratory environments, reduce the measurement precision and accuracy, prolong the experimental duration and complicate the analysis of resulting data. Lately, the methods specially designed for SVOCs were developed based on deeper mass transfer analysis, shortening the experimental duration and increasing the measurement accuracy.
Conclusions: Further research may focus on developing methods that can simultaneously, rapidly and accurately measure SVOC source and sink parameters in a single test run. In addition, to better identify the accuracy of determined parameters, inter-laboratory study using either the same method or different methods to test the same SVOC source/sink parameters will be valuable.

We-SY-E1.3

Predicting SVOC Emissions into Air and Foods in Support of High-Throughput Exposure Assessment

Kristin Isaacs, *US EPA, Research Triangle Park, North Carolina, United States*
Chantel Nicolas, *Oak Ridge Institute for Science and Education, Research Triangle Park, North Carolina, United States*
Derya Biryol, *Oak Ridge Institute for Science and Education, Research Triangle Park, North Carolina, United States*
John Wambaugh, *US EPA, Research Triangle Park, North Carolina, United States*

The release of semi-volatile organic compounds (SVOCs) from consumer articles may be a critical human exposure pathway. In addition, the migration of SVOCs from food packaging materials into foods may also be a dominant source of exposure for some chemicals. Here we describe recent efforts to characterize emission-related parameters for these exposure pathways to support prediction of aggregate exposures for thousands of chemicals For chemicals in consumer articles, Little et al. (2012) developed a screening-level indoor exposure prediction model which, for a given SVOC, principally depends on steady-state gas-phase concentrations (y0). We have developed a model that predicts y0 for SVOCs in consumer articles, allowing exposure predictions for 274 ToxCast chemicals. Published emissions data for 31 SVOCs found in flooring materials, provided a training set where both chemical-specific physicochemical properties, article specific formulation properties, and experimental design aspects were available as modeling descriptors. A linear regression yielded R2- and p- values of approximately 0.62 and 3.9E-05, respectively. A similar model was developed based upon physicochemical properties alone, since article information is often not available for a given SVOC or product. This latter model yielded R2 - and p- values of approximately 0.47 and 1.2E-10, respectively. Many SVOCs are also used as additives (e.g. plasticizers, antioxidants, lubricants) in plastic food packaging. Migration of these chemicals into foods is a complex kinetic process; the speed of migration and ultimate partitioning of chemical depends on the properties of the food, the chemical migrant, and the packaging material. A linear regression model was developed for equilibrium chemical migration (mass/area) from publically-available data collected in a variety of foods and food simulants for different conditions (e.g. temperatures; polymer formulations). The regression yielded an R2=0.71; significant predictors included the initial concentration of the migrant in the packaging, chemical properties (logP; vapor pressure), temperature, and food type (e.g., fatty, aqueous). Migration predictions were combined with food intakes from the National Health and Nutrition Examination Survey to estimate screening-level exposures to over 1500 additives and contaminants potentially present in food packaging. This abstract does not necessarily reflect U.S. EPA policy.

We-SY-E1.4

A quantitative visual dashboard to explore exposures to consumer product ingredients

Daniel Vallero, *National Exposure Research Laboratory, Research Triangle Park, North Carolina, United States*
Peter Egeghy, *National Exposure Research Laboratory, Research Triangle Park, NC, United States*
Heidi Hubbard, *ICF International, Durham, NC, United States*
Tao Hong, *ICF International, Durham, NC, United States*
Cara Henning, *ICF International, Durham, NC, United States*

The Exposure Prioritization (Ex Priori) model features a simplified, quantitative visual dashboard to explore exposures across chemical space. Diverse data streams are integrated within the interface such that different exposure scenarios for "individual," "population," or "professional" time-use profiles can be interchanged to tailor exposure and quantitatively explore multi-chemical signatures of exposure, internalized dose (uptake), body burden, and elimination. Ex Priori will quantitatively extrapolate single-point estimates of both exposure and internal dose for multiple exposure scenarios, factors, products, and pathways. Currently, EPA is investigating its usefulness in life cycle analysis, insofar as its ability to enhance exposure factors used in calculating characterization factors for human health.

We-SY-E1.5

PFASs and PFRs as SVOCs: Measurements and Modeling

Thomas Webster, *Boston University School of Public Health, Boston, MA, United States*

Aim: Per- and poly-fluoro alkyl substances (PFASs) and organophosphate flame retardants (PFRs) are compounds of emerging concern as indoor contaminants. Recent research suggests that indoor exposure may make substantial contributions to body burdens. Indoor air concentrations of fluorotelomer alcohols (FTOHs) have been linked to body burdens of stable PFAS (following metabolic conversion). Organophosphate flame retardants/plasticizers such as triphenyl phosphate (TPHP), tris(2-chloroisopropyl) phosphate (TCIPP), tris(2-chloroethyl) phosphate (TCEP) and tris(1,3-dichloro-isopropyl) phosphate (TDCIPP) in indoor air and/or dust also lead to exposure and their metabolites can be measured in urine. However, the routes of indoor exposure to these compounds are not well understood. The aim of this paper is to estimate the relative importance of three indoor pathways—inhalation, dust ingestion, and vapor-to skin transfer followed by dermal absorption.
Methods: Physical-chemical properties were estimated using SPARC for the following compounds: 6:2 FTOH, 8:2 FTOH, 10:2 FTOH, TPHP, TCIPP, TCEP, TDCIPP. Measurements of these compounds in indoor air or dust were abstracted from the research of our group, our collaborators or the literature. The gas phase concentrations of the compounds were estimated from the air or dust concentrations. The screening-level model of Little et al (2012) was then used to estimate indoor exposure and relative contributions from the three pathways for adults and children (dermal absorption following contact with surfaces is not in the model).
Results and Conclusions:
FTOHs are on the volatile end of SVOCs and the screening model estimates that indoor exposure to the three FTOHs is nearly 100% via inhalation. Inhalation is also the dominant pathway for two of the PFRs: TCIPP and TCEP. In contrast, dermal exposure (following absorption of vapor to skin) equals or exceeds inhalation, as does inadvertent dust ingestion—for the other two PFRs: TPHP and TDCIPP. These latter results (combined with the absence in the model of dermal absorption following contact with surfaces) suggest application of dermal absorption models to skin wipe measurements, followed by comparison with urinary biomarkers. Important caveats of these conclusions will be discussed, including the poor information on dust ingestion rates, particularly for adults.

We-SY-F1: Exposure science informing policy decision-making - I

We-SY-F1.1

Improving chemical exposure scenarios for informed regulatory risk management

Jean-Christophe Dewart, *Cefic aisbl, Brussels, Belgium*
Tanya Dudzina, *Exxon Mobil Petroleum and Chemical B.V.B.A., Machelen, Belgium*
Frank Schnoeder, *DuPont de Nemours (Deutschland) GmbH, Neu-Isenburg, Germany*
Donna Seid, *Ashland Inc., Barendrecht, Netherlands*
Jan Urbanus, *Shell (c/o Belgian Shell NV), Brussels, Belgium*

Under EU REACH, Exposure Scenarios (ESs) are a set of Operational Conditions (OCs) and Risk Management Measures (RMMs) that describe how a substance that is hazardous to human health or the environment, as such, or in a mixture or an article, can be safely used at each stage of its lifecycle. When such a substance is manufactured or imported in quantities greater than 10 tonnes per year, the ESs are developed and become part of the substance's chemical safety report (CSR) submitted to ECHA and forwarded to downstream users (DUs) via Safety Data Sheets (extended SDSs).
The derivation of an ES is an iterative process involving communication up and down the supply chain aiming to arrive at an accurate description of use and use conditions (OCs and RMMs). The registrants of substances can then update their CSRs and ESs accordingly. The ultimate goal is to ensure ESs are meaningful, comprehensible, realistic and up-to-date. . The CSR/ES Roadmap* was launched as a cross-stakeholder action plan involving registrants, their customers, industry sector organisations and Competent Authorities to address these challenges. It builds on the experience drawn from these actors towards achieving the goals of the REACH Regulation for the safe use of chemicals. Keeping up-to-date information on the actual conditions of use of chemicals communicated through supply chains is crucial to ensure safe use downstream and to allow for appropriate prioritisation of substances for further regulatory actions.
The speaker will present key deliverables generated under the CSR/ES Roadmap. The talk will introduce a package of standard phrases (ESCom) developed to make the supply chain communication easy, efficient and transparent. In addition, the presenter will cover sector use maps and related specific exposure assessment inputs (i.e. SpERCs, SWEDs and SCEDs) developed by DU sector associations. Use maps are intended to be an efficient and effective means to inform registrants about realistic OCs and RMMs implemented at a DU level for generic uses of substances. The presenter will also showcase tools developed to support DUs in derivation of safe use information for mixtures and verifying conformity with the advice communicated by substance suppliers in the extended SDSs.
*The CSR/ES Roadmap, a cross-stakeholder plan of actions to 2018, ECHA, July 2013

We-SY-F1.2

Approaches for feeding use and exposure information into prioritisation of substances for regulatory action under REACH

Andreas Ahrens, *European Chemicals Agency, Helsinki, Finland*

The presentation will provide an overview on which kind of information on use and exposure ECHA receives with the registration dossiers and how this information can help in priority setting for post-registration action under REACH. The presentation will cover current practice (and its limitations) and future opportunities.
Post registration action under REACH includes evaluation of registration dossiers (compliance check), substance evaluation, harmonised classification and labelling, identification of candidate substances for authorisation, and restrictions. Together with the information on the substance properties, information on the types of uses (and the related exposure potential) form the basis for setting priorities on substances that matter. The information on uses includes: Indication on whether the substance is used at industrial sites, by professionals outside industrial sites, in consumer products and/or in articles; characteristics of worker tasks carried out with the substance, types of consumer products, environmental release patterns; tonnage of substance per use; indication on whether the substance is used in a rigorously contained manner. Substances with wide dispersive uses are given priority over substances with limited number of users and low exposure potential. The challenge is however to assess the information provided with the registration dossiers against these criteria in practice. The presentation will explain how ECHA has operationalised the generic criteria, and how IUCLID 6 and Chesar 3, ECHA's tools for registrants, facilitate the relevant information to be reported to ECHA. Finally the presentation will point out the common interest between industry and authorities in obtaining correct and relevant use and exposure information for reducing the rate of false positives and false negatives in priority setting.

We-SY-F1.3

Potential policy impact of REACH restrictions (Article 68.2) on CMR substances present in construction articles in the EU and the related consumer exposure

Katleen De Brouwere, *VITO, Mol, Belgium*
Lieve Geerts, *VITO, Mol, Belgium*
Marc Lor, *VITO, Mol, Belgium*

Aim
Exposure of consumers to substances that are carcinogenic, mutagenic or toxic for reproduction (CMR), is being addressed in the REACH legislation (EC No 1907/2006) of which restriction is one potential management option. The aim of the project was to screen for substances classified as CMR 1A/1B (CLP Regulation EC No 1272/2008) which are likely to be present in construction materials and for which consumer exposure is possible.
Methods and Results
In a first part of the study, we used REACH Article 33 requests, the publically accessible part of the ECHA database and other information sources on release of substances from construction articles (e.g. BUMA database) to identify the CMR 1A/1B substances that are likely to be present in construction articles present on the nowadays EU market. In total, 31 CMR 1A/1B substances with at least one use in a construction article (CA) have been identified. An overview of the identified CMRs, in relation to specific CA categories will be presented.
For substances and mixtures, merely the presence of a substance is enough to assume exposure, based on the precautionary principle. However, for articles (REACH definition Art.3.3), it is the release rather than the presence of a substance in the article that drives potential consumer exposure.
In the second part of the study, it was investigated whether the presence of the 31 indentified CMR 1A/1B substances in CAs actually resulted in release and consumer exposure.
In order to assess the likelihood of exposure arising from CAs, 5 criteria have been applied: 1) presence of CMR 1A/1B substances in CA; 2) emission of CMR 1A/1B substances from CA (from test chamber results); 3) estimated exposure (modelling); 4); measured indoor exposure (monitoring Indoor Air Quality in buildings); 5) source apportioning supporting that CA are the main source of indoor exposure. Moving from criterion 1 to 5 gives increasing evidence for exposure to CMR 1A/1B substances in relation to the presence of CA.
Conclusion
For some of the 31 identified CMR 1A/1B substances, consumer exposure arising from construction articles has been demonstrated. For other substances, monitoring data suggest that they do not occur in indoor environments, and thus the articles do not lead to consumer exposure. However, for the majority of the substances, the evidence is less clear and more monitoring data are needed to make a sound assessment.
Acknowledgements: this study was commissioned by the European Commission, DG Environment.

We-SY-F1.4

Approaches for refining the assessment of short-term infrequent consumer exposures in support of risk management decision making

Tatsiana Dudzina, *ExxonMobil Biomedical Science Inc., Brussels, Belgium*
Hua Qian, *ExxonMobil Biomedical Science Inc., Annandale, New Jersey, United States*
Rosemary Zaleski, *ExxonMobil Biomedical Science Inc., Annandale, New Jersey, United States*
Jennifer Foreman, *ExxonMobil Biomedical Science Inc., Annandale, New Jersey, United States*
Carlos Rodriguez, *Procter and Gamble, Brussels, Belgium*

Consumer exposure to substances in products may last from seconds to hours per use event and occur on a regular or sporadic basis (e.g. every day vs. few times a year). The default approach in consumer exposure assessment is to assume that the products are used daily taking the event exposure as a starting point, for which the control of risk should be demonstrated. However, for realistic risk assessment it is essential to consider product application time patterns to match the actual exposure duration and frequency with the corresponding DNEL. For the purpose of comparison to a long-term daily DNEL, ECETOC has developed a use frequency banding approach and included it as a refinement option in the latest version of the ECETOC TRA tool. The four frequency bands allow adjusting the daily event exposure up to a factor of 100 aligning the exposure scenario frequency with the long-term DNEL that assumes daily exposure. The approach follows common practices from other well established higher tier exposure tools, and yields more conservative exposure estimates than a straight linear averaging approach.
In the context of the third update of the Guidance on Information Requirements and Chemical Safety Assessment, the European Chemical Agency and interested stakeholders discussed different approaches to the assessment of risk for long-term effects from short-term infrequent exposures. The concept was reviewed and adapted several times during the Partner Expert Group (PEG) consultation bringing together experts from industry and national competent authorities. By comparing the results of the differing approaches, an assessment of the relative impact on exposure and risk modeled output will be made. When developing risk assessment approaches, integrated engagement from both the toxicology and exposure fields is ideal.

We-SY-F1.5

Development of an Ontology for Occupational Exposure

Heidi Hubbard, *ICF International, Durham, North Carolina, United States*
Arun Varghese, *ICF International, Durham, North Carolina, United States*
Peter Egeghy, *US EPA, Research Triangle Park, North Carolina, United States*
Daniel Vallero, *US EPA, Research Triangle Park, North Carolina, United States*

When discussing a scientific domain, the use of a common language is required, particularly when communicating across disciplines. This common language, or ontology, is a prescribed vocabulary and a web of contextual relationships within the vocabulary that describe the given domain with a view to organizing information. This presentation describes a methodology to ontology development that uses machine learning and natural language processing algorithms, including vector space language models, lexical relation extraction, and topic discovery algorithms, to define an ontology for describing occupational exposures. By applying these automated processes to support expert judgment, a much larger body of literature can be considered than if an individual was required to evaluate each document. Additionally, computer-generated synonym lists can work as an aid to researchers by suggesting keywords that may not otherwise be considered.
In order to use the automated tools, publicly available scientific abstracts from PubMed were gathered using keywords related to "occupational exposure". The titles and abstracts from each study were combined into a single text field. This textual vector was analyzed using ICF's DoCTER (Document Classification and Topic Extraction Resource) tool to determine clusters and inter-cluster distances that were used to suggest taxonomies. Then, ICF's L-Rex (Lexical Relationship Extractor) and ToxSyn (Toxicologic Semantic Similarity Discovery) tools were used to propose ontology rules, by finding synonyms, antonyms, hyponyms and hypernyms, discovering range-domain relationships, and assessing term similarity queries. These results were used to develop a semantic model or visual representation of the ontology pattern suggested for describing occupational exposures.
Disclaimer: The views of the authors of this presentation are those of the authors and do not represent Agency policy or endorsement.

We-SY-G1: Pesticide Exposure: Developing Monitoring, Methods and Modeling in Human Health Risk Assessments (Consumer and Worker Risk) - I

We-SY-G1.1

The use of small-scale human volunteer studies in pesticide exposure assessment

Kate Jones, *Health & Safety Laboratory, Buxton, United Kingdom*

Background
Although there are many pesticides approved for use globally and the approvals process is quite rigorous in the amount of data required, there is often a lack of human data. Small human volunteer studies can often provide valuable information about metabolism, biomarker output and toxicokinetics that can help inform sampling strategies for pesticide exposure assessment, particularly in the absence of other human exposure data. The UK Health and Safety Laboratory has conducted a number of such studies for different pesticides.
Methods
Controlled human volunteer studies for pesticides have generally looked at oral or dermal exposures, inhalation studies are rarely done. Oral dosing usually complies with the Acceptable Daily Intake (ADI) for the particular active substance; dermal dosing is usually calculated from the ADI assuming a likely 'worst-case' penetration rate. Multiple biological samples are then taken for several days following the dose. Blood samples may be analysed for the active substance or for effect markers (such as cholinesterase activity for organophosphate pesticides). A complete urine collection over a defined time period is usually accomplished using timed voids. Urine samples are analysed for the parent compound or relevant metabolites as well as creatinine concentration, and volume is also measured.
Results
Volunteer studies have been used to determine whether metabolites identified in animal studies are relevant to humans - this allows biomonitoring to be considered as part of exposure assessment. Examining the toxicokinetics of the absorption and elimination and calculating the half-life of excretion can inform the sampling strategy, determining the timing of samples to best capture the potential exposure. Dermal dosing studies can provide more realistic estimates of uptake, which can be used to better inform models and to determine whether control measures are needed specifically to prevent skin exposure. The metabolite concentrations measured from a fixed known dose are also useful for putting results from actual exposure scenarios into context and can provide a view on whether a benchmarks such as ADIs (for the general population) or Acceptable Operator Exposure Levels (for workers) have been exceeded. Although volunteer studies are, by necessity, limited in scope and power they are nevertheless a valuable resource in contextualising exposure data.
[Examples for at least three different pesticides will be discussed]

We-SY-G1.2

Exposure assessment using biological monitoring for pesticide users in amenity horticulture.

Alison Connolly, *National University of Ireland, Galway, Galway, Ireland*
Kate Jones, *Health and Safety Laboratory, Buxton, United Kingdom*
Karen Galea, *Centre for Human Exposure Science (CHES), Edinburgh, United Kingdom*
Laura MacCalman, *Centre for Human Exposure Science (CHES), Edinburgh, United Kingdom*
Anne Sleeuwenhoek, *Centre for Human Exposure Science (CHES), Edinburgh, United Kingdom*
Laura Kenny, *Health and Safety Laboratory, Buxton, United Kingdom*
Marie Coggins, *National University of Ireland, Galway., Galway, Ireland*

It is estimated that approximately 2.5 billion tonnes of pesticides are used annually worldwide. Occupational exposures to pesticides products have been linked to a number of adverse health effects including development of cancers, respiratory diseases, detrimental reproductive health, neurological diseases and mental illnesses. A significant proportion of the research on occupational exposures to pesticides has focused on agriculture use and little data is available on exposure levels during pesticide use in horticultural and amenity gardening.
This research project involved using biological monitoring to assess exposure to pesticides in horticulture and amenity gardening. The study was conducted from June to October 2015 among horticultural workers involved in the maintenance of historic properties and ornamental gardens. Two active ingredients, fluroxypyr and glyphosate, were chosen as they were the pesticides of highest volume used by the horticulture group. Pesticide metabolites were measured in urine samples, collected before work tasks and within one hour of completing the work task. Four similar exposure groups were defined: spraying glyphosate with a manual knapsack, spraying glyphosate with a pressured applicator, spraying glyphosate with a large droplet applicator and spraying fluroxypyr with a tractor mounted boom sprayer. A total of 80 exposure measurements were collected (40 paired samples). Contextual information regarding the worker, the task and the environment was recorded for each task sampled.
Qualitative dermal exposure assessments will be performed using the GuLF DREAM assessment tool, (the GUlf Long-term Follow-up DeRmal Exposure Assessment Method). Model predictions will be correlated against actual exposure data. The biological monitoring results will be presented at the conference.

We-SY-G1.3

Developing an operator exposure database in Brazil

Daniele Lautenschalaeger, *PROHUMA, Sao Paulo, Sao Paulo, Brazil*

Regulatory safety decisions should consider both the hazard(s) and the exposure(s) associated with a pesticide and its uses. An extensive dataset of toxicological studies to identify pesticide hazards and characterize its levels of concern for registration purposes is often available. In contrast, data to characterize operator exposures from mixing, loading and applying pesticides in Brazil are fewer and less robust. To overcome this limitation, PROHUMA was established. The main objective is to develop a generic pesticide handler exposure database with data that are considered to be representative of Brazilian scenarios in order to support operator risk assessment in the country.
The database will ultimately combine existing handler exposure data with new data from studies that will be conducted locally if necessary. In 2013, a preliminary comparison between Brazilian scenarios and scenarios from generic databases to assess their representativeness was done and estimated high similarity between Brazilian and North American practices. Next step is a deeper analysis of these and other exiting data for possible inclusion into the Brazilian database. The result will then be used to identify gaps in the Brazilian exposure data to cover handler scenarios of interest. PROHUMA will address these gaps by conducting, state-of-the-art operator exposure studies in Brazil, using Brazilian resources and expertise.
The systematic knowledge generated by the whole process will enable the development of a science based operator risk assessment that reflects Brazilian agriculture. Also, the deeper understanding of the current agricultural practices will allow the revision and improvement of risk management measures currently in place.

We-SY-G1.4

Residential Exposure Assessment to Direct Spray Drift in the United States: A Review of the Environmental Protection Agency Approach and Comparison to the European Approach

Curt Lunchick, *Bayer, Durham, North Carolina, United States*

In the United States the U.S. Environmental Protection Agency (EPA) evaluates the potential non-occupational, residential exposure to conventional pesticides that results from the direct deposition of spray drift to residential lawns and from the volatilization of conventional pesticides after completion of the application. Exposure to direct drift is not assessed in the United States because this is a label violation that is addressed through enforcement. This presentation will focus on the direct deposition of spray drift and compare the EPA approach to the EU approach which involves estimating direct drift exposure. In November 2013, the EPA published its proposed approach to assessing residential exposure resulting from spray drift. Although the proposed approach has not been formally finalized it has been used by both the EPA and registrants since 2013. The approach taken in the US combines the drift modeling from the AgDRIFT model with the residential exposure assessment methodology presented in the 2012 Residential Exposure Assessment Standard Operating Procedures. AgDRIFT is a model based on drift data developed by the Spray Drift Task Force and permits a first tier assessment of drift deposition resulting from aerial, groundboom, and orchard airblast applications. The Residential SOPs permits a first tier assessment of adult and child exposure resulting from contact with pesticide residues deposited on residential turf. An assessment of adult and toddler exposure to residential turf adjacent to an agricultural groundboom pesticide application will be presented and compared to the EU approach of estimating the exposure to direct drift from a similar groundboom pesticide application.

We-SY-G1.5

Residential Exposure Assessment to Direct Spray Drift in Europe: A Critical View on EFSA's Default Values for Groundboom Applications

Christian Kuester, *Bayer, Monheim, Germany*

In the last years three approaches have been developed to assess residential exposure to Plant Protection Products during application via direct spray drift in Europe:
The EUROPOEM II module and the UK model are based on constant exposure figures in ml spray/person while the approach by Martin et al. (2008) is based on drift percentage values.
On 1st of January 2016 a new EFSA guidance on assessing exposure to operator, workers, bystanders and resident has entered into force. New direct drift figure values have been introduced leading to exposure values for groundboom applications which are up to 125 times higher than before. These new drift values were derived from the "Bystander and Residential Exposure Assessment Model (BREAM)", a model developed by the SILSOE spray drift application unit under the commission of UK's Chemical Regulatory Directorate (CRD). Unfortunately, no access to the model was granted to stakeholders during the reviewing process of the EFSA guidance, making the whole model developing process highly opaque and thus incomprehensible. After several enquiries industry got finally access to the model by mid of 2015, but still without having enough information about the underlying data and the applied algorithms to understand the mechanism behind. It remained a black box and parts of the data-set were not disclosed. After intensive investigative work some of the main mechanisms behind the model have become a bit clearer. These mechanisms will be critically reviewed in the presentation and considerations for a more realistic approach for exposure calculations will be proposed.

1)

Current approach?!:

Correlation model output	**95th**			**X**
	75th		**X**	
	50th	**X**		
VS.		**50th**	**75th**	**95th**
		Airborne spray result		

2)

Possible refinement:

Correlation model output	**95th**			
	75th			
	50th	**X**	**X**	**X**
VS.		**50th**	**75th**	**95th**
		Airborne spray result		

Worst Case x Worst Case Assumptions vs. Realistic Conservatism - A proposal for a Reasonable Refinement.

We-SY-H1: Tool and methods for an exposure driven safe by design approach for nanomaterials - I

We-SY-H1.1

Safe by Design for nanomaterials, products and processes: the role of exposure science.

Martie Van Tongeren, *Institute of Occupational Medicine, Edinburgh, United Kingdom*
Jerome Rose, *CEREGE, Aix-en-Provence, France*
Armand Masion, *CEREGE, Aix-en-Provence, France*
Paul Westerhoff, *Arizona State University, Temp, Arizona, United States*

Nanotechnology is a fast growing sector with ever increasing variety and complexity of new materials. Currently, risk assessment is struggling to keep up with the innovation and hence more emphasis will need to be placed on Safe by Design, whereby health, safety and environment are taken into account early in the innovation chain to ensure that risks are managed properly. The SbD concept for nanomaterials is currently being developed and tested in a number of EU FP7 and H2020 projects (NANOREG, PROSAFE, NANOREG2). Exposure plays an important role in this area, as Safe by Design refers not just to reducing the hazard potential of the nanomaterials but also to reducing the release and exposure potential of products and processes. This paper will summarise the results of a two day workshop organised by the EU-US Community of Research working group on Exposure to nanomaterials through its life cycle, on 26-28 April 2016 in Aix-en-Provence, France. This workshop brought together leading exposure scientists in the field of nanotechnology from Europe and the US and aimed to develop exposure-driven risk assessment approaches and how such approaches can be used within the safe by design concept. It will also provide an introduction to the symposium on “Safe by Design for nanomaterials, products and processes: the role of exposure science”.

We-SY-H1.2

Framework on Release of nanomaterials: Concept, test procedures and link to exposure assessments

Thomas Kuhlbusch, *IUTA e.V., Duisburg, Germany*
Heinz Kaminski, *IUTA, Duisburg, Germany*
Araceli Sanchez Jiminez, *IOM, Edinburg, United Kingdom*
Yaobo Ding, *Universities of Lausanne and Geneva, Lausanne, Switzerland*
Martie van Tongeren, *IOM, Edinburg, United Kingdom*
Michael Riediker, *IOM, Singapore, Singapore*
Simon Clavaguera, *CEA, Grenoble, France*
Henk Goede, *TNO, Rijswijk, Netherlands*
Burkhard Stahlmecke, *IUTA, Duisburg, Germany*

Release is the prerequisite for exposure and may also significantly alter the hazard potential of a given nanoobject. Hence the understanding of release as well as its linkage to emissions and exposures is of basic importance to be able to predict potential effects on humans or the environment.

The presentation will introduce a „Framework on Release“ which is based on material properties and release processes enabling the linkage to exposure scenarios via the development of release scenarios. One release scenario may consist of several release processes such as mechanical, thermal or chemical stress. The combined effect of the release processes are subsequently combined to specific release scenarios representing real world work places in companies but also during construction or other activities. Figure 1 also depicts some other uses of the "Framework on Release".

To facilitate the assessment of the effect of release processes on possible emissions, test stands have to be developed, evaluated and linked to real work processes. Based on these laboratory results as well as the construction of release scenarios the combined information may be used to e.g. form release classes which by itself can be linked to exposure and hence to e.g. tier 1 of the tiered exposure assessment approach.

Furthermore, by ranking the release classes, based on their dependence on nanomaterial and release process, a safer-by-design is facilitated.

The presentation will give an overview on the conceptual approach, the state-of-the art in release testing and will open up the discussion on the linkage of the "Framework on Release" to existing regulation.

The research leading to these results has received funding from the European Research Council under the European Union's FP7 (FP/2007-2013)/ERC Grant Agreement n.263215 (MARINA project) and Grant Agreement n.604602 (FNN project).

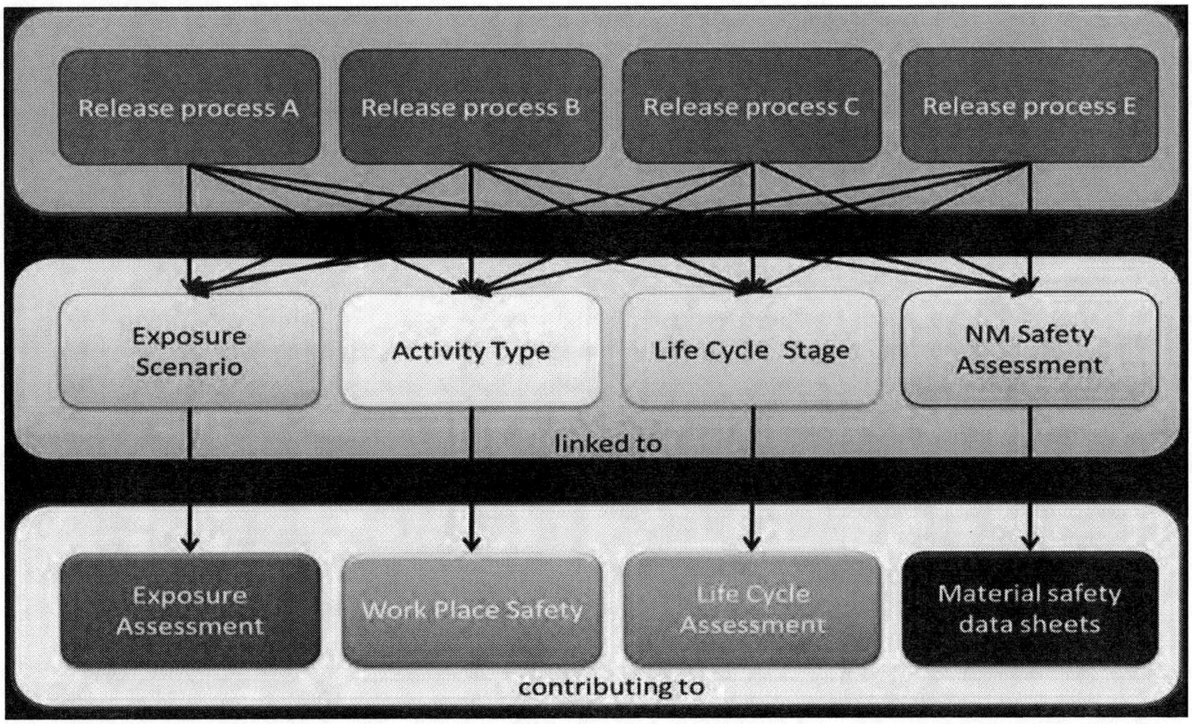

Figure 1. Linkage of release processes to regulatory and safety needs in nanotechnology.

We-SY-H1.3

Safety by design using dustiness and release rate data in modelling of potential exposure

Keld Alstrup Jesen, *National Research Centre for the Working Environment, Copenhagen, Denmark*

Background
The ability of a powder to liberate dust during processing and handling is an important indicator of exposure. Standards in development for dustiness testing include determination of respirable mass concentrations along with the airborne particle size-distributions and the dust particle generation rates during testing. Dustiness data can therefore also be applied for safety-by-design assessment: a lower dustiness index, a coarser dust particle size-distribution and slow particle generation rates would indicate the lowest risk of exposure.
Objective
The objective of this study is to illustrate how application of powder dustiness data can be used to select and produce materials in a safer-by-design approach.
Methods
Dustiness data are generated using a rotating drum dustiness tester, which enables gravimetric data (mg/kg powder) on respirable dustiness levels (D50=4μm) and real-time data on particle number concentrations and particle size-distributions in the dust cloud covering the entire size-range of respirable dust. These data enables calculation of a time-resolved particle generation rate. The dustiness data are further used for model estimates of exposure in given exposure scenarios using the NanoSafer v 1.1 predictive first order exposure model.
Results
An already published study demonstrated the potential use of dustiness data considering dustiness levels and dustiness kinetics of four different molecular pharmaceutical active powder ingredients (Levin, Koponen, and Jensen. Journal of Occupational and Environmental Hygiene 11(3):165-77, 2014). Investigations into different material compounds illustrate that powder dustiness may be affected in different ways as a consequence of physicochemical modifications. For example, chemical surface modifications of a series of Ca-carbonates showed increased dustiness and finer particle sizes after chemical modification. Surface modification of bentonite and conversion into different organoclays on the other hand appears to reduce the dustiness levels and release kinetics. Consequently, dustiness tests may be used to assess whether different material modifications may also be beneficial in a safety-by-design approach.

We-SY-H1.4

Mesocosms: an approach for a realistic assessment of environmental release of nanomaterials

Armand Masion, *CNRS-CEREGE, Aix en Provence, France*
Catherine Mouneyrac, *UCO, Angers, France*
Carlos Fito, *ITENE, Barcelona, Spain*
Melanie Auffan, *CNRS-CEREGE, Aix en Provence, France*

Nanomaterials are released to the environment at various stages of their life cycle and thus various stages of their transformations. Standard aging procedures give valuable information that is relevant for both consumer and environmental exposure but do address biological factors. Mesocosm are an elegant approach of assessing the effects of nanomaterials in a somewhat controlled setting while preserving the complexity of a "miniature" ecosystem. The present presentation describes the implementation of indoor mesocosm set-ups to the monitoring of the fate of nano enabled products in fresh- and saltwater environments. Careful equilibration of the mesocosm before the contamination phase is a pre-requisite for a meaningful experiment. Exposure scenarios can be simulated to be acute (one dosing of the nanoresidue) or chronic (several increments of nanomaterials added over time). Aggregation behavior, and consequently bioavailability of the nanomaterial, depend on the contamination mode. Commercial available CeO_2 and Ag0 based products at several stages of alteration were introduced in these mesocosm. The distribution of the nanomaterials / nanoresidue in the mesocosms (water column, sediment, biota) is controlled by the surface chemistry of the introduced material. The results can used for biovavilability and transfer prediction purposes.

We-SY-H1.5

GUIDEnano safe by design strategies to reduce release of NM from nano-enabled products

Alejandro Vilchez, *LEITAT Technological Centre, Barcelona, Spain*
Camilla Delpivo, *Leitat Technological Center, Barcelona, Spain*
Olga Chibová, *INOTEX spol, Dvůr Králové n.L., Czech Republic*
Cristiano Citterio, *LATI Industria Termoplastici, Vedano Olona VA, Italy*
Rinaldo Zanasca, *LATI Industria Termoplastici, Vedano Olona, Italy*
Alexei Antipov, *PlasmaChem GmbH, Berlin, Germany*
Socorro Vázquez-Campos, *LEITAT Technological Center, Barcelona, Spain*

GUIDEnano is an EU project aimed at developing a web-based Tool to evaluate and manage human and environmental health risks posed by nano-enabled products, considering the whole product life cycle. The Tool will be validated by different real case studies in collaboration with industrial partners. The final version of GUIDEnano Tool will guide the nanotech-industries to apply the most suitable risk assessment and risk mitigation strategies for their nano-enabled products.
This presentation will be focused on GUIDEnano Safer-by-design (SbD) strategies proposed to reduce NM release into the human and environmental compartments, NM that are more compatible with the matrices in which they are incorporated and consequently reducing their release during the use of nano-enabled products. The results will be presented for the following two case studies:
1) An antibacterial treatment of textiles employing nano-silver (Ag NM). During the process, Ag NM adsorbed on the textile fibres are easily washed off, ending in laundering waters. To reduce the Ag release and improve the duration of the textile antibacterial property, AgNPs were modified in their surface. To validate the SbD strategies, textiles were subjected to a common house-hold washing (simulated by laboratory washing machines). Then, the collected waters were analyzed to determine NM concentration and form.
2) TiO_2 NM used in polymeric nanocomposite as UV filtering agent, however the TiO_2 photocatalytic activity cause polymer degradation, that promote NM release from the substrate. To reduce release of NM, surface chemistry on the TiO_2 was performed. Polymer samples treated with TiO_2 and with the SbD NM were exposed to different ageing cycles, including UV-light irradiation, to simulate their use. UV filtering properties were monitored for aged samples and the run-off waters

We-PL-I1: Waterborne Contaminants

We-PL-I1.1

Circadian Exposomics and Diurnal Variation of Urinary Trihalomethanes

Pantelis Charisiadis, *Cyprus University of Technology, Limassol, Cyprus*
Stephanie Gaengler, *Cyprus University of Technology, Limassol, Cyprus*
Xanthi Andrianou, *Cyprus University of Technology, Limassol, Cyprus*
Ratanesh Seth, *University of South Carolina, Columbia, United States*
Saurabh Chatterjee, *University of South Carolina, Columbia, United States*
Konstantinos Makris, *Cyprus University of Technology, Limassol, Limassol, Cyprus*

Aim: We applied external and internal exposure metrics (metabolomics) to evaluate the exposure-effect association between diurnal exposures to common disinfection by-products (trihalomethanes, THM) and 4-hydroxynonenal histidine adducts (HNE-His), a surrogate of hepatic CYP2E1 enzyme activity and lipid peroxidation product.
Methods: Seven young healthy adults were recruited. An activity day was designed where each participant conducted four well-controlled activities that could generate THM in the surrounding indoor environment (showering, hand dishwashing, mopping, and bathroom cleaning); another sampling day was included with none of the aforementioned activities (control). Each participant collected spot urine samples at predetermined intervals (before and after each activity) on both days. Urinary THM and HNE-His levels were measured using gas chromatography tandem mass spectrometry (GC-MS/MS) and immunoassay test, respectively. Urine samples were prepared for the metabolomics protocol, followed by urea depletion, extraction and derivatization prior to obtaining full mass spectra. Spectral data preprocessing, i.e. deconvolution, and compound identification techniques were implemented and followed by chemometrics analysis.
Results: During the activity day, a discrete diurnal pattern was consistently observed with higher creatinine-adjusted urinary THM levels during the late afternoon towards evening than those during the morning. Temporal trends of within-subject variation between urinary HNE-His and urinary THM levels were observed. Chemometric analyses explored the differential expression of metabolite profiles under the circadian influence and accounted for the effect of external THM exposures on hepatic biological pathways.
Conclusions: The inclusion of diurnal measurements in exposome studies is warranted to better describe the dynamic changes in exposure patterns between and within subjects that are prospectively followed during critical life stages. Research is needed to understand whether such diurnal fluctuations of biomarkers of exposure and effect could be linked with alterations in an individual's circadian rhythm.

We-PL-I1.2

Disinfection by-product exposures and the risk of specific cardiac birth defects

Wright J. Michael, *US EPA/ORD, Cincinnati, OH, United States*
Amanda Evans, *Campbell University, Lillington, NC, United States*
John Kaufman, *Association of Schools and Programs of Public Health, cincinnati, OH, United States*
Zorimar Rivera-Núñez, *Washington University, St. Louis, MO, United States*
Michael Narotsky, *US EPA/ORD, Research Triangle Park, NC, United States*

Background: Epidemiological studies suggest that women exposed to disinfection by-products (DBPs) in treated water have an increased risk of delivering babies with cardiovascular defects (CVDs), though evidence for specific DBP-birth defect associations is limited. It also remains unclear which DBP metrics of the complex mixtures found in drinking water are the best surrogates for assessing potential risk related to reproductive toxicants.
Methods: We conducted a case-control study of all birth defects in Massachusetts from 2000-2004 with complete trihalomethane (THM) and haloacetic acid (HAA) data. We randomly matched 904 CVD cases to 10 controls (n=9040) based on week of conception. We used weight-averaged aggregate first trimester DBP exposures across all quarterly monitoring sample locations linked to individuals based on residence at birth. Adjusted odds ratios (aORs) were calculated for nine CVDs in relation to categorical DBP exposures including bromoform, chloroform, dibromochloromethane (DBCM), bromodichloromethane (BDCM), monobromoacetic acid (MBAA), dichloroacetic acid (DCAA), trichloroacetic acid (TCAA), and summary DBP measures (HAA5, THMBr, THM4 and DBP9).
Results: We detected strong associations for Tetralogy of Fallot and the upper exposure categories for TCAA, DCAA, and HAA5 (aOR Range: 3.34-6.51) including positive exposure-response relationships for DCAA and HAA5. aORs consistent in magnitude were detected for atrial septal defects and bromoform (aOR=1.56; 95%CI: 1.01, 2.43), as well as DBCM, chloroform, and THM4 (aOR Range: 1.26-1.67). With the exception of chloroform, TCAA, and HAA5, consistently elevated aORs were detected for ventricular septal defects (VSDs) and every DBP metric including bromoform (aOR=1.85; 95%CI: 1.20, 2.83), MBAA (aOR=1.81; 95%CI: 0.85, 3.84), and DBCM (aOR=1.54; 95%CI: 1.00, 2.37).
Conclusions: Overall, we saw limited evidence of risk of CVDs based on DBP surrogate mixture measures such as THM4 and DBP9; however, several associations were detected between individual DBP species and specific types of CVDs. For example, bromoform was consistently associated with elevated aORs for all the individual and group CVDs that were examined. To our knowledge, this is the first epidemiological study of birth defects to develop multi-DBP adjusted regression models and is only the second study to evaluate brominated THMs or HAAs. Our findings, therefore, inform exposure specificity for the consistent associations previously reported between THM4 and CVDs including the VSDs.
The views expressed in this abstract are those of the authors and do not necessarily reflect the views or policies of the US Environmental Protection Agency.

We-PL-I1.3

Associations Between Musculoskeletal Birth Defects and Disinfection By-Product Exposures in Massachusetts, USA

John Kaufman, *Association of Schools and Programs of Public Health, Cincinnati, United States*
Michael Wright, *US EPA, Cincinnati, United States*
Amanda Evans, *Campbell University, Lillington, North Carolina, United States*
Zorimar Rivera-Nunez, *Washington University, Saint Louis, Missouri, United States*

Epidemiological studies have shown that in utero exposures to disinfection by-products (DBPs) in treated water are associated with increased risks for some birth defects, though evidence for musculoskeletal defects (MSDs) is limited. Most studies of DBPs and birth defects have been limited by the use of surrogate metrics which may not accurately represent the complex mixture of reproductive toxicants.
We used a case-control design of MSDs in Massachusetts from 2000-2004 with complete trihalomethane (THM) and haloacetic acid (HAA) data. We randomly matched each of the 187 cases to 10 controls based on week of conception. We used weighted averages of first trimester DBP exposures across quarterly water sampling locations linked to individuals based on residence at birth. We calculated adjusted odds ratios (aORs) between five MSD variables and 13 DBP metrics categorized based on distributions of the available data, including bromoform, chloroform, bromodichloromethane (BDCM), dibromochloromethane (DBCM), THMBr (sum of bromoform, BDCM, DBCM), THM4 (sum of chloroform and THMBr), trichloroacetic acid (TCAA), dichloroacetic acid (DCAA), monochloroacetic acid (MCAA), dibromoacetic acid (DBAA), monobromoacetic acid (MBAA), HAA5 (sum of TCAA, DCAA, MCAA, DBAA, and MBAA), and DBP9 (sum of THM4 and HAA5).
Compared to the lowest exposure categories, we observed elevated aORs for the combined MSD group for all quintiles of DBP9 (aOR range: 1.73-2.80) and for the highest quartiles of THM4 (aOR=3.75; 95%CI: 1.33-10.56) and chloroform (aOR=2.82; 95%CI: 0.98-8.10). We observed elevated aORs between upper limb reduction (n=53) and the highest bromoform decile (aOR=1.83; 95%CI: 0.42-7.89), and for all quartiles of THM4 (aOR range: 2.41-7.59), chloroform (aOR range: 3.32-6.59), and THMBr (aOR range: 1.34-2.03), with positive exposure-response relationships detected for THM4 and chloroform. We observed elevated aORs for diaphragmatic hernia (n=41) for the upper decile of dibromoacetic acid (aOR=2.18; 95%CI: 0.53-8.98), as well as positive exposure-response relationships for tertiles of THM4 (aOR range: 3.36-5.81), chloroform (aOR range: 1.94-3.51), and BDCM (aOR range: 1.34-1.58). We observed elevated aORs for gastroschisis or omphalocele (n=66) for all HAA5 quintiles (aOR range: 1.55-1.90) and all TCAA quartiles (aOR range: 1.83-2.19).
Although these rare outcomes resulted in limited statistical power, we saw elevated risks associated with different surrogate DBP metrics as well as with several individual DBP species. Though our analyses were limited by aggregate exposure data which may result in measurement error, this is the first study of MSDs and DBPs to examine HAAs, and the first MSD study to develop multi-DBP-adjusted regression models.

We-PL-I1.4

Spatial-temporal Indoor Exposures in Homes Affected by Trichloroethylene (TCE) - contaminated Soil and Groundwater - Preliminary Findings

Leonid Turczynowicz, *University of Adelaide, Adelaide, South Australia, Australia*
Dino Pisaniello, *University of Adelaide, Adelaide, South Australia, Australia*
Terence Wiliamson, *University of Adelaide, Adelaide, South Australia, Australia*

Aim
The evaluation of indoor exposures to trichloroethylene (TCE) arising from vapour intrusion has not yet resulted in structured measurement methodologies which ensure confidence in risk assessment. A study was designed to examine indoor spatial and temporal TCE concentration changes and influencing variables for the purposes of improved exposure assessment. This is considered an area requiring exploration, as the toxico-kinetics following TCE inhalation is relatively rapid.
Methods
An abandoned house in a TCE-affected area was monitored over a 16-month period for indoor TCE concentrations plus a range of variables known to influence such measurements. Indoor spatial TCE distribution was assessed using five locations with passive sampling undertaken at a consistent height of 1.5 metres. Two outdoor locations were used on the north and south-western sides of the house. A solar powered meteorological station was established on the northern side of the property. A combination of experiments concurrently measured indoor air TCE concentrations over 4-h, 6-h, 24-h and 7-day periods, outdoor TCE concentrations, real-time meteorological variables and indoor temperatures; air exchange rates; soil vapour and sub-slab TCE concentrations, and indoor and outdoor TCE flux concentrations. Real-time changes in indoor volatiles were also examined during detailed observation periods using a suitably calibrated photo-ionisation detector with logging capabilities.
Results
Seasonal and spatial differences in indoor air TCE concentrations were observed. Outdoor TCE concentrations were a minor contributor to indoor concentrations. Winter indoor air concentrations based on 24-h averages were up to an order of magnitude greater than those in summer. Spatial differences were observed between the front and rear of the house with the rear consistently higher across all seasons. Concurrent short-term (4-h and 6-h) and longer term (24-h and 7-d) averaged monitoring suggested shorter period peak concentrations were occurring which were consistent with elevations observed during periods with real-time PID logging.
Conclusions
Measurement methods need to account for spatial and seasonal indoor air concentration changes. The magnitude of the observed differences may in some cases shift the evidenced-based decision-making process. Shorter-term peak concentrations were contributing to exposures but were not obvious if longer term averages were the only data examined. The significance associated with short-term peak inhalation exposures warrants closer examination in terms of adverse pathology considering the relatively rapid toxico-kinetics of TCE inhalation. Further exploration of the data using multivariate methods will examine influencing variables.

We-PL-I1.5

New Approaches to Legionella Detection in Environmental Samples - The Way to Better Risk Assessment in Outbreak Situations

Bernadett Bartha, *Bavarian Health and Food Safety Authority, Munich, Germany*
Sandra M. Walser, *Bavarian Health and Food Safety Authority, Munich, Germany*
Bernhard Brenner, *Bavarian Health and Food Safety Authority, Munich, Germany*
Stefanie Huber, *Bavarian Health and Food Safety Authority, Munich, Germany*
Christiane Höller, *Bavarian Health and Food Safety Authority, Munich, Germany*
Michael Seidel, *Technical University of Munich, Munich, Germany*
Caroline E. W. Herr, *Bavarian Health and Food Safety Authority, Munich, Germany*

Background
Legionnaires´ disease is a severe acute pneumonia caused by bacteria of the genus Legionella. Outbreaks occur throughout the world, and the fast identification of the infection sources is essential for a successful outbreak management. The culture-based standard method used for Legionella quantification is impeded by uncertainties, the concentrations are often underestimated and analysis takes 3 to 12 days. In most cases culture-independent methods, like quantitative polymerase chain reaction (qPCR), give a better estimation of exposure levels and provide results within a few hours. However, conventional qPCR does not distinguish between living and dead cells or remaining DNA fragments and could lead to an overestimation of the risk.
Objectives
For a better quantification of viable Legionella spp. and Legionella pneumophila in water and aerosol samples the live/dead qPCR method was tested in a BMBF framework project (LegioTyper). The method is based on the selective removal of DNA from dead cells using the intercalating agent propidium monoazide (PMA).
Our aim is to investigate the usability of culture-independent methods, like the PMA-qPCR for routine application in the laboratory praxis or during an outbreak situation. Its use in combination with other analysis methods (e.g. antibody microarray) may lead to more rapid identification of potential sources.
Methods
Legionella pneumophila ATCC33152 strain was cultivated on GVPC agar plates. After five days of incubation, a part of a single colony was suspended in sterilized tap water and serial dilution was prepared in order to find the suitable cell concentration for qPCR measurements. The bacteria suspension was divided into two aliquots. To prepare dead cells, one of the aliquots was heat-inactivated and used as control. PMA treatment conditions were optimized by exposing the samples to different PMA concentrations followed by different light-exposure times.
Results
The cell concentrations were adjusted between 103 and 104 GU/100 µl in the spiked water probes according to the quantification range of the qPCR method. PMA treatments of spiked water samples containing L. pneumophila in this range showed just a moderate signal reduction in case of heat-killed cells compared to live cells. Based on the results, there is an urgent need for further optimization of the method in relation to environmental samples to obtain an effective and reliable method for a better risk assessment.

UTRECHT
THE NETHERLANDS
OCTOBER 9-13
ISES 2016

We-SY-A2: New Data Streams for 21st Century Exposure Science

We-SY-A2.1

Merging methods, measurements and models to estimate metabolism rates in fish and select mammal species

Jon Arnot, *ARC Arnot Research and Consulting, Toronto, Canada*
Karen Foster, *Karen Foster Environmental Research, Peterborough, Canada*
Alexandra Looky, *ARC Arnot Research and Consulting, Toronto, Canada*
Trevor Brown, *ARC Arnot Research and Consulting, Halifax, Canada*
James Armitage, *AES Armitage Environmental Sciences, Toronto, Canada*
Ester Papa, *University of Insubria, Varese, Italy*
John Nichols, *US EPA, Duluth, United States*

Chemical concentrations in humans and ecological receptors are required for the exposure and risk assessment of thousands of chemicals; however, there are few or no measured concentration data available for the vast majority of chemicals. In the absence of measured concentrations, models are often used to predict exposures and concentrations in receptors. A key parameter required to calculate concentrations is the chemical half-life in a receptor. Half-life data are also required for reconstructing exposures and interpreting biomonitoring data, i.e., relating external exposure estimates (e.g., intake rates in mg/kg/d) with internal concentrations (blood, tissues) and biomarkers of exposure. For hydrophobic, low volatility chemicals the chemical half-life is largely determined by the biotransformation (metabolism) rate constant. Chemical biotransformation rates are also required for in vitro to in vivo extrapolation and reverse toxicokinetics. Despite the fundamental value of biotransformation rate information, relatively few measured in vivo data are available compared to the thousands of commercial chemicals requiring evaluation. The objectives of this research are to compile, evaluate and compare existing in vivo, in vitro and in silico data streams for estimating biotransformation rates for organic chemicals in fish and select mammalian species. The literature and existing publicly available databases of in vitro (S9, hepatocytes, microsomal assays) and in vivo biotransformation rate estimates in mammals (humans and rodents) and fish are collected and evaluated. In vitro to in vivo extrapolation models are developed and applied to the in vitro data to obtain estimates of hepatic clearance, and as applicable, whole body biotransformation rate (clearance, or half-life) estimates. In vitro biotransformation rate estimates and in silico predictions from existing screening-level quantitative structure-activity relationships (QSARs) are compared to in vivo biotransformation rate estimates. The data compilation includes: (1) whole body biotransformation rate constant estimates for approximately 940 chemicals in humans and 700 organic chemicals in fish and (2) in vitro biotransformation rate constants measured for 8,000 chemicals in humans and 130 chemicals in fish. Key uncertainties and challenges comparing the datasets are described and a strategy to address data gaps and uncertainty for estimating biotransformation rates is discussed.

We-SY-A2.2

Leveraging Publically Available Chemical Functional Use Data in Support of Exposure Prediction

Kristin Isaacs, *US EPA, Research Triangle Park, North Carolina, United States*
Katherine Phillips, *Oak Ridge Institute for Science and Education, Research Triangle Park, North Carolina, United States*
Peter Egeghy, *US EPA, Research Triangle Park, North Carolina, United States*
Kathie Dionisio, *US EPA, Research Triangle Park, North Carolina, United States*
John Wambaugh, *US EPA, Research Triangle Park, North Carolina, United States*

The U.S. EPA Exposure Forecasting (ExpoCast) project aims to provide rapid screening-level exposure predictions for thousands of chemicals, most of which lack detailed exposure data. Chemical functional use - the role a chemical plays in processes or products (e.g. solvent, antimicrobial, plasticizer) - may be a useful heuristic for predicting exposure potential in that it reflects both the compound's likely physical properties as well as the product formulations, consumer articles, or industrial processes in which it may be used. Functional use information is also critical in alternatives assessment, in which safer chemicals that can perform a particular role in products are identified. Here, data on chemical functional use for more than 14,000 chemicals were collected from publically available government, manufacturer, and industry sources. A new standardized Functional Use (FUse) database was created by harmonizing 240 function categories across sources. The FUse database was used to build machine-learning classifier models for function and consumer product weight fraction using descriptor sets of either chemical structure or a combination of predicted physical-chemical properties and chemical structure. Statistically robust models (i.e., those passing a y-randomization test and having 5-fold cross-validation error of <25%) were built for 44 functions and weight fractions. The final models were applied to a library of 8,500 mostly data-poor chemicals, including those being tested using high-throughput methods in the U.S. interagency Toxicology in the 21st Century (Tox21) program. In addition, the predictions generated by the classification models were used to screen the chemical library for potential alternatives on the basis of an average in-vitro bioactivity metric generated from a suite of 16 Tox21 assays. Functional role could be predicted with high probability (>80%) for 2,332 chemicals; of these chemicals, 1,034 had a lower bioactivity metric than at least one known chemical with that function. The presented functional use database and models have wide applications both in alternatives assessment and in the refinement and parameterization of heuristic-based and mechanistic models of human exposures in ExpoCast. This abstract does not necessarily reflect U.S. EPA policy.

We-SY-A2.3

New Data from EPA's Exposure Forecasting (ExpoCast) Project

John Wambaugh, *U.S. E.P.A., Research Triangle Park, NC, United States*
Alice Yau, *Southwest Research Institute, San Antonio, TX, United States*
Anne Marie Gregg, *Battelle Memorial Institute, Columbus, OH, United States*
Ann Louise Sumner, *Battelle Memorial Institute, Columbus, OH, United States*
Kristin Favela, *Southwest Research Institute, San Antonio, TX, United States*
Derya Biryol, *Oak Ridge Institute for Science and Education (ORISE), Research Triangle Park, NC, United States*
Peter Egeghy, *U.S. Environmental Protection Agency, Research Triangle Park, NC, United States*
Chris Grulke, *U.S. Environmental Protection Agency, Research Triangle Park, NC, United States*
Kristin Isaacs, *U.S. Environmental Protection Agency, Research Triangle Park, NC, United States*
Xiaoyu Liu, *U.S. Environmental Protection Agency, Research Triangle Park, NC, United States*
Kamel Mansouri, *Oak Ridge Institute for Science and Education (ORISE), Research Triangle Park, NC, United States*
Chantel Nicolas*North Carolina Central University, Durham, NC, United States*
Katherine Phillips, *U.S. Environmental Protection Agency, Research Triangle Park, NC, United States*
Ann Richard, *U.S. Environmental Protection Agency, Research Triangle Park, NC, United States*
Caroline Ring, *Oak Ridge Institute for Science and Education (ORISE), Research Triangle Park, NC, United States*
Woodrow Setzer, *U.S. Environmental Protection Agency, Research Triangle Park, NC, United States*
Jon Sobus, *U.S. Environmental Protection Agency, Research Triangle Park, NC, United States*
Antony Williams, *U.S. Environmental Protection Agency, Research Triangle Park, NC, United States*
Russell Thomas, *U.S. Environmental Protection Agency, Research Triangle Park, NC, United States*

The health risks posed by the chemicals in our environment depends on both chemical hazard and exposure. However, relatively few chemicals have estimates of exposure intake, hampering risk estimations for thousands of chemicals. The U.S. EPA Exposure Forecasting (ExpoCast) project aims to provide rapid, provisional exposure predictions for all commercially used chemicals. In order to provide rapid predictions of human and ecological exposure, the EPA is developing mathematical models, organizing and analyzing extant data, and using new tools such as screening-mode mass spectrometry (MS) to collect new data on chemical properties, use, and occurrence. The pilot phase of the ExpoCast data collection has focused on four activities: 1) high throughput physicochemical property measurements, 2) new biomonitoring data, 3) chemical emissivity data for articles of commerce, and 4) chemical deformulation of consumer products and articles of commerce. As an example of ExpoCast data collection, a selection of 100 objects that might be found in the home was screened using gas chromatography

(GC) x GC time of flight MS, and 3803 unique chemical signatures were observed in test objects. 1608 of the signatures could be confirmed or tentatively identified. Only 184 of the 1608 chemicals had previously been known to have potential proximate or “near field” sources of exposure. The new data streams will be used to expand the domain of applicability, and to refine, and validate existing ExpoCast models. This abstract does not necessarily reflect U.S. EPA policy.

We-SY-A2.4

Rapid methods to estimate exposure to VOCs and SVOCs in the indoor environment

John Little, *Virginia Tech, Blacksburg, VA, United States*

A systematic and efficient strategy is needed to assess and manage the potential risks to human health that arise from the manufacture and use of thousands of chemicals. Among available tools for rapid assessment of large numbers of chemicals, significant gaps are associated with the capability to evaluate exposures that occur indoors. For both volatile organic compounds (VOCs) and semi-volatile organic compounds (SVOCs), exposure is strongly influenced by the types of materials and products in which the chemicals occur. For VOC emissions, important parameters governing exposure are the diffusion coefficient (D), and initial material-phase concentration (C0). A simple method for determining D and C0 using data from ventilated chamber tests and dimensionless analysis is developed and validated using VOC emission data from a material emissions database. With these parameters in hand, screening-level estimates of inhalation exposure to VOCs can be made. For SVOCs, there are two primary SVOC source classes: additives in products used indoors and ingredients in products sprayed or applied to interior surfaces. In both cases, important parameters governing exposure are the gas-phase concentration in equilibrium with the material or product (y0) and the partition coefficient between airborne particles and air (Kp). We have developed simple methods to measure these two parameters. Then, accounting for product use, emission characteristics, and the properties of the SVOCs, we estimate exposure via inhalation of SVOCs in the gas-phase, inhalation of SVOCs sorbed to airborne particles, ingestion of SVOCs sorbed to dust, and dermal sorption of SVOCs from the air into the blood. Further development of a comprehensive set of models for estimating exposure to volatile chemicals in materials and products is needed. When combined with rapid toxicity estimates, screening-level exposure estimates for both VOCs and SVOCs can be used for health-risk-based prioritization of a wide range of chemicals of concern.

We-SY-A2.5

Targeted and Non-Targeted Approaches to Measuring SVOC Exposure Using Handwipes and Indoor Dust

Heather Stapleton, *Duke University, Durham, NC, United States*
Lee Ferguson, *Duke University, Durham, United States*
Bernadette Vogler, *Duke University, Durham, United States*
Stephanie Hammel, *Duke University, Durham, United States*
Kate Hoffman, *Duke University, Durham, United States*
Craig Butt, *Duke University, Durham, United States*
Thomas Webster, *Boston University, Boston, United States*

Several classes of semivolatile organic compounds (SVOCs) are used as additives in plastics and textiles found in a number of consumer products and building materials. As a result, SVOCs are ubiquitously detected in the indoor environment, and a number of these compounds are suspected of being linked with health effects including neurodevelopmental effects and metabolic disorders. Human exposure can occur through contact with consumer products, indoor furnishings, inhalation, and inadvertent ingestion of indoor dust. Many of these contaminants are endocrine disruptors, including phthalates, flame retardants, and pesticides. Using mass spectrometry based approaches, we have previously shown that a range of brominated and organophosphate SVOCs are ubiquitous and abundant in indoor dust samples and are frequently detected in handwipes collected from people. For several of these chemicals, levels measured on handwipes are predictive of levels measured in the body. However, there are likely other SVOCs present in these samples that are "missed" using our targeted analyses. Therefore, we developed a non-targeted, HPLC-HRMS analytical method to identify mixtures of SVOCs present in 10 paired samples of hand-wipes and household dust. We utilized a stepwise workflow based on use of accurate molecular mass, high-fidelity isotope measurements, and data-directed HRMS/MS spectra for querying public molecular databases. An LTQ-Orbitrap Velos mass spectrometer operated in either positive or negative ESI mode was programmed to acquire continuous high-resolution (R>100,000) full-scan (m/z 150-2000) data as well as data-dependent CID spectra concurrently. Chromatograms were then processed for molecular feature detection and annotation with molecular formula and structure based on accurate mass, isotope abundance, intensity, MS/MS spectrum, and retention time filtering. Tentative identifications of compounds in individual samples, by accurate mass (80%) and MS/MS library match scores (>50%), were used to compile a list of putatively identified compounds. Using this approach, 13 chemicals were tentatively identified and demonstrated significant associations between handwipes and dust extracts. These SVOCs included tricresyl phosphate (plasticizer), triclocarbon (antimicrobial agent), Sudan 3 (diazo dye) and imazalil (fungicide). These results demonstrate that non-targeted analytical approaches can enable a much more comprehensive exposure data sets for SVOCs than targeted approaches alone, and can provide unique "leads" for further investigation of health effects.

UTRECHT
THE NETHERLANDS
OCTOBER 9-13
ISES 2016

This page intentionally left blank.

We-PL-B2: Exposure Factors

We-PL-B2.1

Update to the U.S. EPA's Guidelines for Human Exposure Assessment

Nicolle Tulve, *U.S. Environmental Protection Agency, Research Triangle Park, North Carolina, United States*
Marian Olsen, *U.S. Environmental Protection Agency, New York, New York, United States*
Michael Broder, *U.S. Environmental Protection Agency, Washington, District of Columbia, United States*

The mission of the U.S. EPA is to protect human health and the environment by understanding, characterizing, and reducing risks associated with exposure to environmental contaminants. Exposure science characterizes, estimates, and predicts exposures and provides information for developing exposure and risk assessments as well as effective strategies for reducing exposure and risk. When conducting a risk assessment, an assessor needs to understand whether an agent may cause an adverse health effect and how exposure to that agent may be reduced. Advances in the field of exposure science require updated resources for conducting exposure and risk assessments. The Guidelines for Human Exposure Assessment has been prepared to provide an updated resource for exposure and risk assessors and managers both within and outside the Agency. This document builds on the 1992 Guidelines, incorporating advances in the field that have occurred since the Guidelines were originally published. It reflects the best science currently conducted across the Agency. This updated document describes the principles of exposure assessment, provides guidance on approaches to conduct an exposure assessment, presents references for more detailed information, and supplies hyperlinks to exposure assessment tools and technical documents. The Guidelines are arranged into chapters, each of which explores a component of the exposure assessment process, including: basic concepts and principles in exposure science; planning and scoping; incorporating lifestages, vulnerable groups, and populations of concern into an assessment; collecting and using data; using models; planning an observational exposure measurement study; evaluating uncertainty and variability; and presenting and communicating results. This presentation will highlight and showcase many of the updates in the document and provide an update on the review and publication status of the document.

We-PL-B2.2

Total Exposure Health - A Revolutionary Way to Think of Exposure and Primary Prevention

Richard Hartman, *United State Air Force, Falls Church, Virginia, United States*
Col Kirk Phillips, *US Air Force, Fall Church, Virginia, United States*

Background: Private and public sector organizations in the US have made strides toward achieving the President's Precision Medicine Initiative (PMI), an approach for disease prevention and treatment that takes into account an individual's unique genes, environment, and lifestyles to provide personalized healthcare. However, they have overlooked one key factor that influences individual health risks, which drives policy as well as protective and clinical interventions: the exposure.
In response to the PMI, we created Total Exposure Health (TEH) which associates exposures to the individual's DNA enriching primary prevention with forward vision using advancements in medicine, science, technology, and informatics.
Objectives: To present TEH as it's being operationalized in the US Air Force and demonstrate how: 1) TEH advances epidemiology, bioinformatics, and "Big Data" by aggregating and analyzing large amounts of specific group and individual exposure data using advanced informatics to provide individual and population health risk analysis; 2) TEH incorporates environment, workplace, and lifestyle exposures by accounting for all exposures and their ties to genetics/genomics and a person's predispositions to disease; and 3) TEH fosters research and technology by supporting sensor development and applied toxicology models for rapid identification of unknown threats and low-level exposure biomarkers in human genomics. Collectively, TEH provides a pathway to personalized health to maximize human performance and ultimately overall well-being.
Methods: Currently, we have the ability to collect refined information on the individual based on their exposure and genetics which allows us to focus on unique interventions. We will discuss various operational models to show how TEH takes our existing knowledge of "exposures" and connects them to the individual's organ systems, cellular function, and DNA, along with how classic exposure modeling advances science, technology, medicine, and informatics. We will also show how TEH will position "exposure scientists" to improve the patient/provider experience with a focus on individual exposures (unique and targeted).
Results: Exposure means different things to different people, so we conveniently packaged TEH into a simple brand. We revealed TEH as a catalyst to move exposure health away from animal data and population models to individual personalized effects of exposure. We also found TEH fosters innovations in research and technology development and can promote economic development, particularly in science, technology, engineering and mathematics (STEM) career fields. Lastly, we found TEH to be a system integrator between programs, policies, and disciplines that creates collaborations to bridge the divide between "Precision Medicine" and "Population health".

We-PL-B2.4

Emerging exposure and policy interventions: A vulnerability analysis for urban population to air-borne particulate matter

Audil Rashid, *PMAS Arid Agriculture University, Rawalpindi, Punjab, Pakistan*
Faiza Naseem, *University of Science and Technology of China, Hefei, China, People's Republic of*
Muhammad Ubaid Ali, *PMAS Arid Agriculture University, Rawalpindi, Pakistan*

Human exposure to environmental contaminants is unprecedented. World Health Organization has highlighted environmental disease burden a major health risk for developing countries. In the successive years (2012-14), two major hospitals in Rawalpindi city have witnessed rapid increase in air borne health anomalies among population such as upper respiratory tract infections, chest congestion, allergic response etc. Human exposure to air pollutants in urban environment cause considerable disease burden however, a scientific assessment is necessary to attribute health risks. Based on this premise, this study was conducted in urban area of Rawalpindi city with an aim to analyze population exposure to dust and particulate matter (PM) in relation to their vulnerability for disease susceptibility. Data from hospital records as well as through questionnaire-based survey was collected about exposure estimates, work environment, disease history, socio-demographic aspects and health risk type. The observed population (males=452 and females=128) of adult ages had numerous exposure durations ranging from 4 to 12 h day-1. Chi square test revealed 'age' and 'occupation' significant but 'gender' inconsistent with respiratory symptoms. A Cronbach's alpha value of 0.74 was maintained for reliability of health variables. Logistic regression analysis showed 'shortness of breath' (β= 2.62; odds ratio=13.8; 95% CI = 3.63-52.41) has highest risk factor followed by 'eye redness' (β= 1.14; odds ratio=3.1; 95% CI =1.55-6.30). Overall a direct relationship between exposure to dust and PM with population illness was observed especially during construction of Rawalpindi Metro Bus Project in 2014. We conclude that degradation of environmental health has strong negative impact on general urban population that has not only lowered their functional capacity but also placed them at higher risk category. To overcome this challenge, prevention of human exposure to air pollution can be an effective intervention measure especially for people who are more vulnerable.

We-PL-B2.5

Human Exposure Factors as a Potential Determinant of Heterogeneity in City-Specific Associations between PM2.5 and Mortality

Lisa Baxter, *United States Environmental Protection Agency, Research Triangle Park, North Carolina, United States*
Kathie Dionisio, *United States Environmental Protection Agency, Research Triangle Park, North Carolina, United States*
Lucas Neas, *United States Environmental Protection Agency, Chapel Hill, North Carolina, United States*

The inability to explain the city-to-city heterogeneity in PM2.5 mortality risk estimates observed in multi-city studies remains a key uncertainty in the examination of the relationship between short-term PM2.5 exposures and mortality. Measurements from fixed-site monitors are often used as surrogates of exposure but may introduce bias into the observed risk estimates if the relationship between fixed-site monitor measurements and personal exposures varies by city. Factors that can affect personal exposures include housing characteristics (e.g. home age and size), commuting patterns (e.g. commuting distance and times), and climatic factors (e.g. cooling and heating days).

The objective of this analysis is to determine whether human exposure factors can help explain the observed heterogeneity in the associations between PM2.5 measured at fixed-site monitors and total non-accidental mortality. Rate ratios were generated for 313 core-based statistical areas and their metropolitan divisions across the continental United States for 1999-2005. Log rate ratios (betas) were derived from a model adjusting for time, an interaction with age-group, day of week, and natural splines of current temperature, current dew point, and unconstrained temperature at lags 1, 2, and 3. The heterogeneity in the betas was assessed by linear regression with inverse variance weights. The human exposure factors examined included housing type (e.g. detached vs. attached home), number of rooms in residence, commuting time and modes, type of heating fuel used, and annual cooling and heating degree days.

Overall a 1.02% (95% CI 0.86-1.18) increase in non-accidental mortality per 10 micrograms per cubic meter increase in 24-hour average PM2.5 concentrations at lag 1 was observed. Factors related to home size, fraction of duplex homes and median number of rooms, were associated with a 0.13% and 0.26% increase in mortality, respectively. A positive association of 0.32% was also observed with annual number heating degree days. Therefore larger homes and colder temperature were associated with increases in mortality. For heating fuel type, results depended on the type of fuel used. A larger fraction of homes heated using utility gas was negatively associated with mortality (-0.41%) while having a larger fraction of homes heated with oil was positively associated (0.14%) with the percent increase in mortality.

Multi-city population-based epidemiological studies have observed heterogeneity between city-specific PM2.5-mortality effect estimates. One possible reason for the differences observed between cities may be differences in human exposure factors.

We-SY-C2: What are the requirements for nanomaterial exposure models? - II

We-SY-C2.1

A Multimedia Model For Nanoparticle Fate And Biotic Update In The Environment

Stephen Lofts, *NERC Centre for Ecology and Hydrology, Lancaster, United Kingdom*
Egon Dumont, *NERC Centre for Ecology and Hydrology, Wallingford, United Kingdom*
Virginie Keller, *NERC Centre for Ecology and Hydrology, Wallingford, United Kingdom*
Richard Williams, *NERC Centre for Ecology and Hydrology, Wallingford, United Kingdom*
Antonia Praetorius, *University of Vienna, Vienna, Austria*
Frank von der Kammer, *University of Vienna, Vienna, Austria*
Geert Cornelis, *Swedish Agricultural University, Uppsala, Sweden*
Susana Loureiro, *University of Aveiro, Aveiro, Portugal*
Nico van den Brink, *Wageningen University, Wageningen, Netherlands*

Predicting the behaviour of manufactured nanomaterials (MNMs) following release into the environment is key to assessing their ultimate fate and potential risks. Developing models of MNM transfer through environmental compartments following release is therefore an essential component of assessing their environmental sustainability.
MNMs present a specific challenge to models due to their chemical and physical diversity, for example metals, metal oxides, carbon nanotubes, quantum dots. Furthermore, MNMs are frequently manufactured with capping agents or other types of coatings, or may be enclosed within product matrices (for example, antifouling paints containing zinc or copper-based MNMs). Such modifications increase the physicochemical complexity of the particles and may further modify their environmental behaviour. Additionally, MNMs may transform prior to environmental release (for example, in sewage treatment works) or following environmental release (for example, by dissolution or heteroaggregation). Environmental models of MNMs thus require the capability to simulate a wide range of physicochemical forms and behaviours, within a single framework. Such a framework ideally needs to be readily updatable, to permit simulation of MNMs not yet in commercial use.
The NanoFASE Horizon2020 project (2015-2019) aims to tackle the particular issues of multimedia nanoparticle modelling by developing a spatially-explicit, gridded dynamic model of nanoparticle transport, transformation and biouptake (the NanoFASE model) using object-oriented programming (OOP) concepts. The OOP approach allows for the system to be considered as a linked set of 'objects' representing entities such as a layer of soil or sediment, a population of organisms, or a population of nanoparticles. The complete model system will comprise terrestrial and aquatic compartments, with a link to atmospheric deposition modelling, to allow holistic simulation of nanoparticle fate and biouptake. It is intended for application at scales up to that of a large European river catchment.
The model will be divided into 'transport' and 'transformation' components. The transport component will handle the bulk movement of MNMs, for example in river flow, settling sediments, soil porewater and surface soil erosion to waters. The transformation component will handle key MNM transformations and reactions within environmental compartments, such as heteroaggregation and adsorption of environmental species such as dissolved organic matter. MNMs will be categorised by 'type', where all MNMs within a

single 'type' will share common algorithms for transformations within each environmental compartment. The number and nature of the types simulated will be readily extensible, thus allowing for the efficient addition of new nanoparticle types to the model.

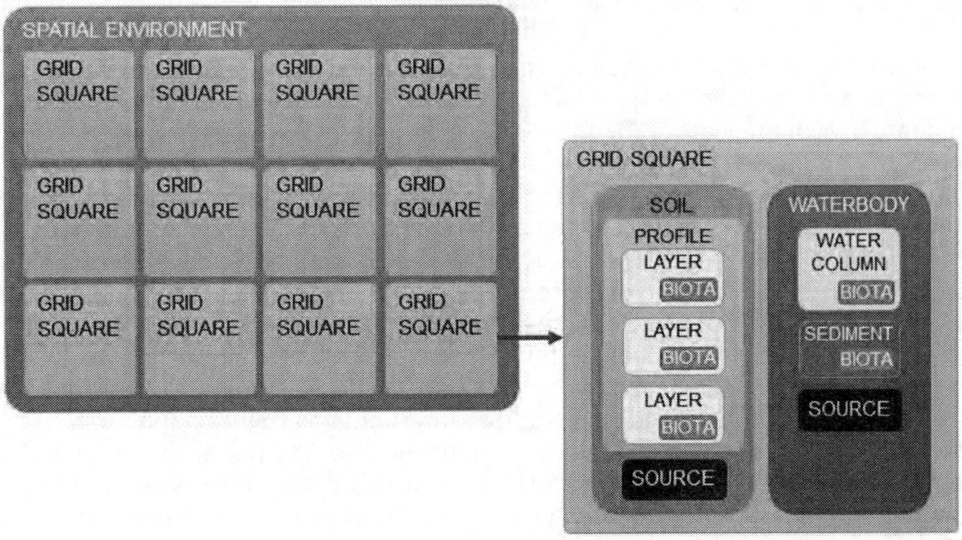

Object arrangement for the transport component of the NanoFASE model

We-SY-C2.2

Modeling Environmental Interactions of Nanomaterials in Aquatic Ecosystems

Mark Wiesner, *Duke University, Center for the Environmental Implications of NanoTechnology, Durham, North Carolina, United States*

Environmental transformations and exposure are key elements in determining the environmental and health effects of nanomaterials. The study of nanomaterial impacts on environment, health and safety (nanoEHS) has been largely predicated on the assumption that exposure and hazard can be predicted from physical-chemical properties of nanomaterials. This approach is rooted in the view that nanoöbjects essentially resemble chemicals with additional particle-based attributes that must be included among their intrinsic physical-chemical descriptors. The complexity of a system's effects on nanoparticle properties tends to obscure predictive links between intrinsic properties of nanomaterials and their impacts. An expedient alternative to directly linking ENM properties to impacts is to focus on standardized reproducible measurements of relevant nanomaterial behaviors in relevant systems that can inform near-term decision-making. We refer to procedures that produce such measurements as functional assays.
Tools for predicting the environmental behaviour include functional assays that can be used to evaluate nanomaterial properties in complex or reference systems. Simulations show that nanoparticles introduced in a complex, albeit greatly simplified environment exhibit a wide range of behaviors depending on their affinities for each other and their concentrations. The complexity of these interactions appears to be governed by the relative affinity of nanoparticles for each other (autoaggregation) and with background particles (heteroaggregation) and other native surfaces. A functional assay for determining the affinity of nanoparticles for complex mixtures of native particles will be presented. This talk will summarize a series of laboratory and mesocosm studies designed to evaluate such interactions. Cases of bioüptake, trophic transfer, material transfer and cycling of nanomaterials observed in these experiments will be presented in concert with analysis of the relative toxicity of engineered nanomaterials observed to date compared with more conventional contaminants. Principles for conducting studies using nano-scale phases in toxicity studies will be discussed in the context of transport and transformations of nanomaterials to be considered.

We-SY-C2.3

Environmental Exposure Modeling Of Engineered Nanoparticles

Johannes Meesters, *Radboud University Nijmegen, Nijmegen, Netherlands*
Joris Quik, *RIVM, Bilthoven, Netherlands*
Dik Van de Meent, *Radboud University Nijmegen, Nijmegen, Netherlands*
Jan Hendriks, *Radboud University Nijmegen, Nijmegen, Netherlands*
Bart Koelmans, *Wageningen University, Wageningen, Netherlands*

The novel developed tool SimpleBox4nano (SB4N) covers the necessary adjustments to make the multimedia fate model SimpleBox (SB) used for environmental risk assessment of chemicals fit for nano. SB4N enables simulation of the environmental fate of engineered nanoparticles (ENPs) attached to natural particles and the chemical distribution within compartments to be calculated with rates instead of partitioning equilibria, including dissolution as a removal or transformation process. SB4N is a deterministic model requiring single values for physical and chemical properties of the ENPs and emissions as default inputs. In reality the model parameters reflecting the environment are subjected to natural variability, whereas the input parameters are subjected to uncertainty. Therefore, data have been collected for all of SB4N's input and model parameters reflecting realistic distributions of variability and uncertainty. These distributions have been inserted in Monte Carlo (MC) simulations of the environmental fate of the three mostly used metal oxide ENPs in Europe nano-TiO2, nano-ZnO and nano-CeO2. From this evaluation of the confidence in the predicted environmental concentrations (PECs) calculated with SB4N, it is concluded that screening level multimedia fate models are appropriate for conservative estimations of environmental exposure to ENPs. Uncertainties in ENP emissions, physicochemical properties and natural variability of the environmental system only leads to a variation in total PECs that is comparable to that of conventional chemicals: a factor 10 for air and water, 10,000 for sediments, and 100 for soil. Species concentrations of ENPs as free pristine , hetero-aggregated with natural colloid particles, or attached to coarse particles is less feasible to extrapolate to other nanomaterials, because they strongly depend on their physicochemical properties. Evaluation of the influence of ENP properties on their environmental fate indicates that the most environmentally persistent ENPs are determined to be insoluble (< 10-10 s-1), small (persistency decreases with ENP size), and have attachment efficiencies high enough to accumulate in soil (> 10-6) and sediments (>10-6), whereas ENPs that are persistent in a free pristine state are large, insoluble (<10-10 s-1)and have low attachment efficiencies (<10-6). Ultimately, the goal of environmental exposure estimation is to predict whether exposure concentrations exceed predicted no-effect concentrations. A case study on nano-TiO2 proves that accounting for exposure to hetero-aggregates is crucial in environmental risk assessment.

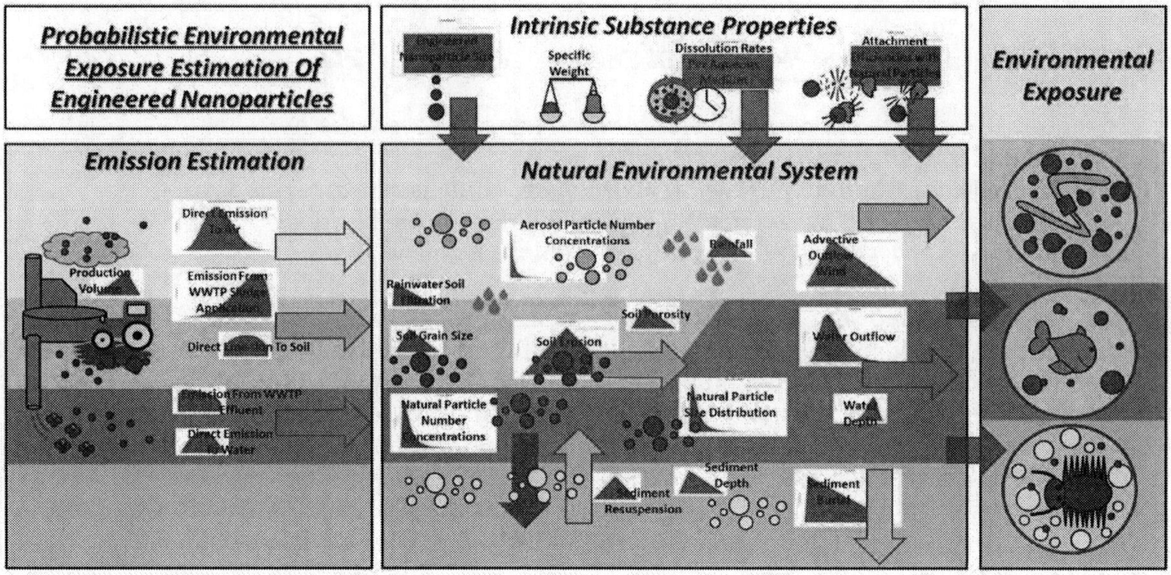

Natural variability and uncertainty in probabilistic environmental exposure estimation.

We-SY-C2.4

Panel discussion on the requirements for nanomaterial exposure models

Joris Quik, *RIVM, Bilthoven, Netherlands*

Exposure models for nanomaterials are currently available and under active development. However, the relevance of these models for application by industry or to support policy is not always clear. What is the relevant output of such exposure models for nanomaterials in order to support safe innovation? How does that output relate to the regulatory definition of a nanomaterial? What are relevant or acceptable input data requirements, considering users in policy and industry? The previous presentations will form the basis for a panel discussion of these topics.

We-SY-D2: Biomonitoring: Uses in Policy and Regulations and Enhancements as a Result of Collaborative Efforts

We-SY-D2.1

The US National Health and Nutrition Examination Survey (NHANES)

Antonia Calafat, *US Centers for Disease Control and Prevention, Atlanta, United States*

People in modern societies are potentially exposed to thousands of environmental chemicals. Some of these chemicals are toxic in animal studies and replacement chemicals are entering consumer markets. Understanding the extent of exposures to both original and replacement chemicals is of public health interest. Biomonitoring measurements (i.e., amounts of a given chemical present in the body) are used more and more to quantify exposures within populations. Biomonitoring programs are particularly useful for assessing human exposures to environmental chemicals. In the United States, one of these programs, the National Health and Nutrition Examination Survey (NHANES) is conducted annually since 1999 by the Centers for Disease Control and Prevention. NHANES participants undergo a physical examination, answer comprehensive questionnaires on demographics and health behaviors (including diet), and provide detailed medical history, as well as biological specimens (i.e., blood and urine)—some of which are used to assess exposure to select chemicals. NHANES biomonitoring data have important uses in public health. NHANES data showed that exposure to some chemicals is prevalent and may reflect lifestyle differences. NHANES biomonitoring data have also been used to establish reference ranges, to provide exposure information for risk assessment (e.g., set intervention and research priorities, evaluate effectiveness of public health measures), and to monitor exposure trends. For example, NHANES data suggest that reformulation of commercial products and regulations limiting phthalate plasticizer content in certain applications in the United States during the last decade may have had important implications for exposures to phthalates and their commercial replacements.

We-SY-D2.2

Health Canada's human biomonitoring initiatives and their use in public policy

Douglas Haines, *Health Canada, Ottawa, Ontario, Canada*
Ellen Lye, *Healthy Environments and Consumer Safety Branch, Ottawa, Ontario, Canada*
Tye E. Arbuckle, *Healthy Environments and Consumer Safety Branch, Ottawa, Ontario, Canada*
Shawn Donaldson, *Healthy Environments and Consumer Safety Branch, Ottawa, Ontario, Canada*

Human biomonitoring (HBM) is used as an indicator and quantitative measure of exposure by measuring environmental chemicals, their metabolites or reaction products in biological specimens. We describe Health Canada's HBM initiatives in the general population, pregnant women, indigenous peoples, and northerners and highlight the use of their results in public policy.
The Canadian Health Measures Survey (CHMS) launched in 2007 is a nationally-representative cross-sectional direct health measures survey. National HBM blood and urine data are available for ages 6-79 years in CHMS cycle 1 (2007-2009), and for 3-79 years in cycles 2 (2009-2011) and 3 (2012-2013). Field collection has been completed for cycle 4 (2014-2015), with cycle 5 (2016-2017) in progress and planning for cycle 6 (2018-2019) being finalized. HBM results for about 270 chemicals are expected over these cycles.
The Maternal-Infant Research on Environmental Chemicals Study was established to obtain Canadian HBM data for pregnant women and their infants and to examine potential adverse effects on pregnancy and infant health. About 2000 pregnant women from 10 sites across Canada were recruited between 2008 and 2011. Maternal blood, urine, hair and breast milk, cord blood and infant meconium were analyzed for a range of environmental biomarkers.
The First Nations Biomonitoring Initiative is a representative survey of First Nations peoples 20 years and older, living on reserve south of the 60th parallel. It was conducted in 2011 and has provided HBM blood and urine data for 97 chemicals.
The Northern Contaminants Program, established in 1991, has undertaken targeted HBM studies in Canadian Arctic communities with a focus on metals and persistent organic pollutants (POPs).
The chemicals measured in these initiatives include metals, trace elements, PCBs, organochlorines, dioxins, furans, flame retardants, perfluoroalkyl substances, environmental phenols, triclocarban, acrylamide, chlorophenols, pesticides, phthalates, volatile organic compounds, PAHs, and tobacco biomarkers.
Results from these initiatives have established baseline HBM concentrations in Canadians. They have been used in federal regulatory risk assessment and management of chemicals (e.g. cobalt, lead , perfluorooctanoic acid, selenium, triclosan, phthalates). HBM studies have informed public health advisories on the consumption of traditional diets and exposure to contaminants. HBM data are part of the Canadian Environmental Sustainability Indicators national reporting and fulfill international reporting under the Stockholm Convention on POPs, the Minamata Mercury Convention and the Arctic Monitoring and Assessment Programme. Concurrent efforts are underway to develop statistically- and risk-based concepts and tools to interpret HBM data.

We-SY-D2.3

The German Human Biomonitoring Program

Marike Kolossa-Gehring, *German Environment Agency (UBA), Berlin, Germany*
André Conrad, *German Environment Agency (UBA), Berlin, Germany*
Christine Schulz, *German Environment Agency (UBA), Berlin, Germany*
Ulrike Fiddicke, *German Environment Agency (UBA), Berlin, Germany*
Petra Apel, *German Environment Agency (UBA), Berlin, Germany*
Gerda Schwedler, *German Environment Agency (UBA), Berlin, Germany*

The German population is still substantially exposed to chemicals. Some of which have already been restricted and cannot be reasonably further regulated, others just recently raised concern. Exposure levels as well as their sources and development over time are well documented by the German human biomonitoring (HBM) program consisting of the population-representative German Environmental Survey (GerES) and the German Environmental Specimen Bank (ESB). GerES and ESB data are used to derive reference values for selected chemicals, analyse trends, identify sources, derive exposure reduction measures, and identify highly exposed sub-groups.
The information is transferred specifically to the federal government and other policy makers, the scientific community and the general population including sub-groups with special needs for risk-communication.
Policy makers are supplied with vital information via specific reports including proposed mitigation measures. Peer reviewed papers, congress participations and organisation of workshops are used to put our scientific approaches and results under scientific scrutiny and stimulate further research. Results are communicated to the general public via brochures, TV and radio interviews, respective websites, as well as personal counseling. For ethical reasons, participants of GerES and USB are informed as soon as possible about the concentrations of analysed substances in their blood, urine, drinking water and indoor air samples or whether a concentration was elevated according to either the statistical derived reference value or the toxicologically derived HBM-value, both stipulated by the German HBM Commission. The reference values are defined according to the 95th percentiles of the measured concentrations of selected substances in human samples (e. g. blood or urine). Hence, they allow for identifying unusually high body burden of individuals, they cannot serve to evaluate health risks. Later can be achieved by mean of the health-related biological exposure assessment values (HBM values). The derivation of HBM-values is another key activity of the HBM Commission. The HBM-I value represents the concentration of a substance in blood, urine or serum at or below which, according to current knowledge and assessment by the HBM Commission, there is no risk of adverse health effects. Above HBM-I health effects cannot be ruled out any more with sufficient certainty. The HBM-II value defines the concentration above which adverse health effects are to be expected. Consequently, exposure reduction measures and the provision of biomedical advice are strongly recommended when HBM-II is exceeded.

We-SY-D2.4

The Flemish Environment and Health studies, a participative approach with impact on policies

Greet Schoeters, *VITO, Mol, Belgium*
Ilse Loots, *Uantwerpen, Antwerp, Belgium*
Karen Van Campenhout, *Flemish Governement, Brussels, Belgium*
Hana Chovanova, *Flemish governement, Brussels, Belgium*
Baeyens Willy, *Vrije Universiteit Brussel, Brussels, Belgium*
Van Larebeke Nick, *Vrije Universiteit Brussel, Brussels, Belgium*
Elly Den Hond, *Scientific Institute of Public Health, Brussels, Belgium*
Tim Narwot, *University Hasselt, Hasselt, Belgium*
Isabelle Sioen, *Ghent University, Ghent, Belgium*
Bert Morrens, *University of Antwerp, Antwerp, Belgium*
Ann Colles, *VITO, Mol, Belgium*
Vera Nelen*PIH, Antwerp, Belgium*

Background: Flanders, the Northern part of Belgium, is heavily industrialised, very densily populated and has a dense network of traffic roads. Since 2002, chemical exposure of the population is monitored by human biomonitoring. This HBM program is the core activity of the Center of Expertise on Health and Environment which combines surveillance with an interdisciplinary research program that is engrafted on the biomonitoring framework.
Objectives: "Monitoring for action" has been the first slogan and primary goal of the Flemish Environment and Health Surveillance (FLEHS) program. This implies that a scientific approach has been developed to make sure that the HBM data are fit for purpose and used to inform policy options.
Methods: Each HBM cycle has been framed in a code of conduct that has been agreed upon by the different stakeholders. Transparency of the process, open communication of the results, the willingness to take into account different perspectives and subsidiarity are the corner stones of our "playing rules ." Multicriteria decision frameworks have been developed to select priority pollutants for biomonitoring, to select hot spot areas for biomonitoring and to prioritise the outcome of human biomonitoring programs for further actions. These consultations involve representatives from environment and health agencies, experts from inside and outside the FLEHS consortium and representatives from the civil society.
Results:
In different successive HBM campaigns we obtained exposure data of more than 50 prioritised environmental chemicals in three age groups (newborn - mothers, adolescents and adults) in a geographically representative sample of the Flemish population. We have evaluated in three hot spot areas whether exposure in adolescents was higher than in the reference adolescent population. Exposure markers were evaluated in relation to effect biomarkers and to specific health parameters such as performance in neurobehavioural tests, reporting of doctors' diagnosed asthmas and allergies, health records on puberty development.
Conclusion:
The participative evaluation of the results has resulted in actions including sensibilisation of the public, regulatory measures and optimization of the environmental monitoring networks.

We-SY-D2.5

European Human Biomonitoring Initiative: objectives and strategy

Catherine Ganzleben, *Euroepan Environment Agency, Copenhagen, Copenhagen City, Denmark*
Greet Schoeters, *VITO, Antwerp, Antwerp, Belgium*
Argelia Castaño Calvo, *Institute of Health Carlos III, Madrid, Madrid, Spain*
Denis Sarigiannis, *Aristotle University of Thessaloniki, Thessaloniki, Greece*
Marike Kolossa-Gehring, *Umweltbundesamt, Berlin, Greece*
Ovnair Sepai, *Public Health England, London, United Kingdom*
Jean Philippe ANTIGNAC, *L'Ecole Nationale Vétérinaire, Agroalimentaire et de l'Alimentation Nantes-Atlantique, Nantes, France*
Jana Klánová, *Masaryk University, Brno, Czech Republic*
Erik Lebret, *National Institute for Public Health and the Environment (RIVM), Bilthoven, Netherlands*
Robert BAROUKI, *French National Institute of Health and Medical Research (INSERM), Paris, France*

The proposal for a European Human Biomonitoring initiative was developed by a consortium of representatives from 26 countries, with input from the European Environment Agency, in response to a Horizon 2020 call, under the Work Programme on Health, demographics changes and well-being. The proposal was submitted in April 2016 and, if accepted, the initiative will launch early 2017. This presentation will outline the objectives and strategy of the initiative.
The overarching goal of the initiative is to generate knowledge to inform the safe management of chemicals and so protect human health in Europe. We will use human biomonitoring to understand human exposure to chemicals and resulting health impacts and will communicate with policy makers to ensure that our results are exploited in the design of new chemicals policies and the evaluation of existing measures.
Key objectives include:

- Harmonizing procedures for human biomonitoring across 26 countries, to provide policy makers with comparable data on human internal exposure to chemicals and mixtures of chemicals at EU level;
- Linking data on internal exposure to chemicals to aggregate external exposure and identifying exposure pathways and upstream sources. Information on exposure pathways, including environmental, occupational, consumer and dietary exposure, is critical to the design of targeted policy measures to reduce exposure;
- Generating scientific evidence on the causal links between human exposure to chemicals and negative health outcomes; and
- Adapting chemical risk assessment methodologies to use human biomonitoring data and account for the contribution of multiple external exposure pathways to the total chemical body burden.

We will achieve these objectives by harmonizing human biomonitoring initiatives in 26 countries, drawing on existing expertise and building new capacities. To this end, we will create a robust Human Biomonitoring Platform at European level, supported by National Hubs in each country. The National Hubs will consolidate expertise and experience at national level, feed priorities up to EU level, and coordinate activities between the national and EU level.

The initiative will contribute directly to the improvement of health and well-being for all age groups, by investigating how exposure to chemicals affects the health of different groups, such as children, pregnant women, foetuses and workers. We will also investigate how factor such as behavior, lifestyle and socio-economic status influence internal exposure to chemicals across the EU population. This knowledge will support policy action at EU and national levels to reduce chemical exposure and protect health.

We-SY-E2: Exposure to SVOCs in the Indoor Environment - Products, Emissions, Exposure, Pharmacokinetics and Biomarkers - II

We-SY-E2.1

Investigating Associations Between Flame Retardant Application in Televisions and Furniture with Indoor House Dust Levels

Stephanie Hammel, *Duke University, Durham, North Carolina, United States*
Kate Hoffman, *Duke University, Durham, North Carolina, United States*
Amelia Lorenzo, *Duke University, Durham, North Carolina, United States*
Heather Stapleton, *Duke University, Durham, North Carolina, United States*

Flammability regulations such as California Technical Bulletin 117 have historically influenced flame retardant (FR) chemical use in consumer products that are common to indoor environments. The use of FRs in consumer products has led to their near ubiquitous presence in indoor air and dust and to exposure among the general population. Studies have demonstrated that FR levels in house dust are predictive of human serum levels, suggesting most exposure occurs in the home. However, it is unclear which products contribute most to FR levels detected in house dust. With recent global phase-out of polybrominated diphenyl ethers, organophosphate FRs and other brominated compounds have been increasingly used in residential furniture. Decabromodiphenyl ether (BDE-209) has historically been applied to electronics such as televisions but are slowly being replaced by other compounds such as decabromodiphenylethane (DBDPE) and 2,4,6-tris(2,4,6-tribromophenoxy)-1,3,5-triazine (TTBP-TAZ). We collected paired samples of polyurethane foam from residential furniture (n=97), TV wipes (n=111), and house dust (n=103) from participants' living areas to determine if the presence and levels of FRs in furniture and TVs were predictive of dust levels. Higher levels of pentaBDEs, tris(1-chloro-2-isopropyl)phosphate (TCIPP), and brominated components associated with Firemaster® 550 (FM550) in dust were significantly associated with their respective detections in furniture (Z=2.4-3.0, p=0.01). BDE-209 and DBDPE were detected in over 80% of TV wipes with TTBP-TAZ detected in 20%. A weak positive association between BDE-209 on TV surfaces and in dust was observed in paired samples (rs=0.2, p=0.1). Levels of BDE-209 on TV wipes were significantly greater in cathode ray tube (CRT) TVs compared to flatscreen TVs (Z=5.4, p=0.0001). A comparison of foam and TVs suggests that TV wipes may experience more confounding with other variables (e.g. time TV is on, cleaning, etc), which should be explored further. In dust samples, flooring type (carpet vs. wood) was not found to be associated with FR dust levels; however, it was associated with dust mass loading within the designated living space.

We-SY-E2.2

Fate and Transport of Phthalates in Indoor Environments and the Influence of Temperature: A Case Study in a Test House

Ying Xu, *The University of Texas at Austin, Ausitn, United States*
Chenyang Bi, *The University of Texas at Austin, Austin, United States*

A case study in a test house was conducted to investigate the fate and transport of benzyl butyl phthalate (BBzP) and di-2-ethylhexyl phthalate (DEHP) in residential indoor environments and the influence of temperature. Total airborne concentrations of phthalates were sensitive to indoor temperatures, and their steady-state concentration levels increased by a factor of three with an increase in temperature from 21 to 30 °C. Strong sorption of phthalates was observed on interior surfaces, including dust, dish plates, windows, mirrors, fabric cloth, and wood. Equilibrium partitioning coefficients for phthalates adsorbed to these surfaces were determined, and their values decreased with increasing temperature. For impervious surfaces, dimensionless partitioning coefficients were calculated and found to be comparable to reported values of the octanol-air partition coefficients of phthalates, Koa, suggesting that an organic film may develop on these surfaces. In addition, sorption kinetics was studied experimentally, and the equilibration time scale for impervious surfaces was found to be faster than that of fabric cloth. Finally, using an indoor fate model to interpret the measurement results, there was good agreement between model predictions and the observed indoor air concentrations of BBzP in the test house.

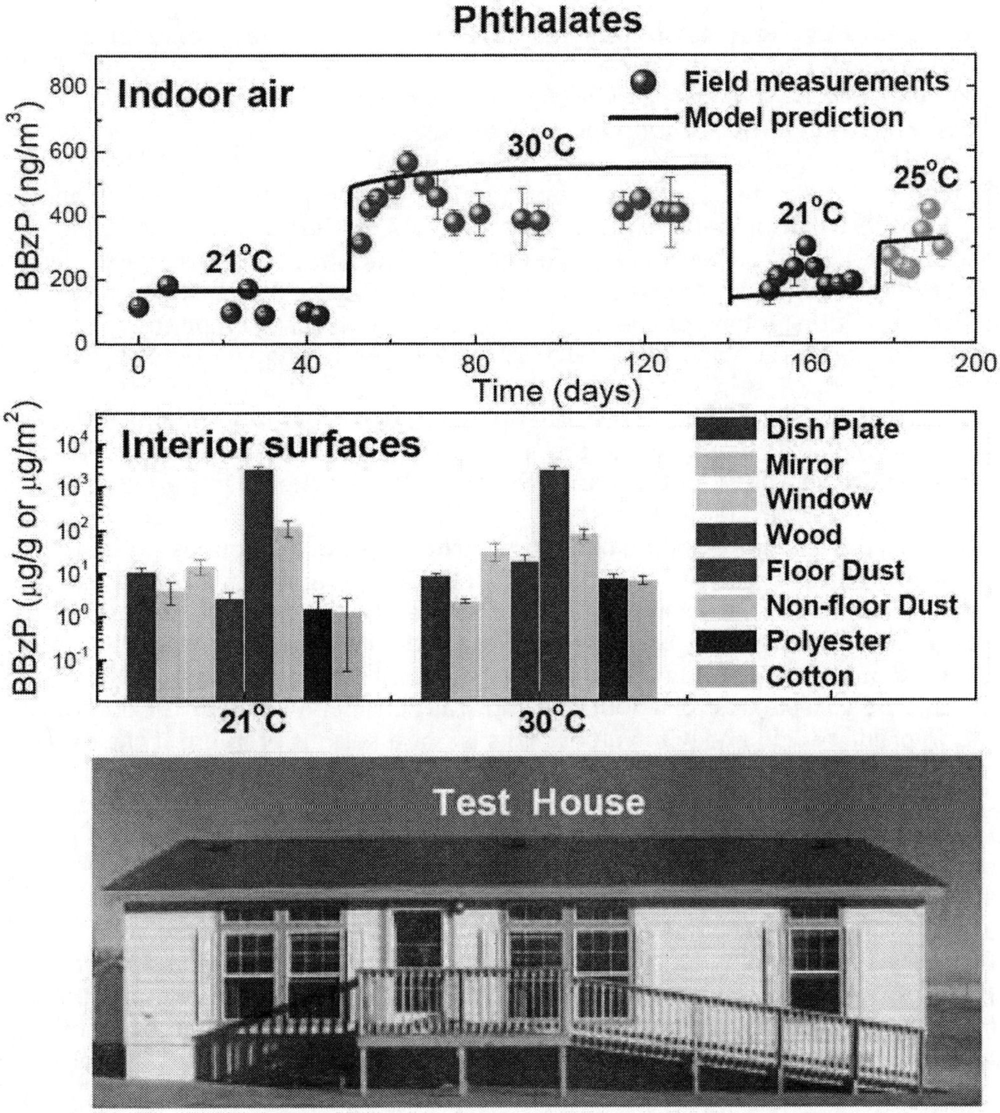

Abstract Figure

We-SY-E2.3

Distribution of SVOCs between gas phase, particle phase and settled house dust

Tunga Salthammer, *Fraunhofer WKI, Braunschweig, Germany*
Tobias Schripp, *Fraunhofer WKI, Braunschweig, Germany*

BACKGROUND: The physical and chemical properties of compounds are frequently applied for modeling their distribution between different media. A similar approach is used in indoor sciences for estimating their distribution and dynamics between gas phase, particle phase and settled dust for estimating human exposure (see Figure 1). This, however, requires a detailed understanding of the environmentally important compound parameters, their interrelation and of the algorithms for calculating kinetic and partitioning coefficients.
OBJECTIVES: Parameter uncertainties and variations of indoor conditions might influence the distribution behavior of compounds in the indoor environment. A problem occurs for compounds of medium volatility. In this case the description of their gas/particle distribution behavior is due to large errors.
METHODS: The gas/particle partitioning of semi volatile organic compounds (SVOCs) is usually estimated on basis of physico-chemical models. The parameters of major concern for the determination of the partitioning constant Kp are the saturation vapor pressure (p0), the Henry's law constant (H), the octanol/water partition coefficient (KOW), the octanol/air partition coefficient (KOA) and the air/water partition coefficient (KAW).
RESULTS: Calculated gas/particle distributions and fractions can widely differ due to the uncertainties in predicted p0 and KOA values. This is not a serious problem if the target com-pound is of low or high volatility, but in the intermediate region even small changes in p0 or KOA will have a strong impact on the expected partition behavior. The precision in prediction is also affected by a superposition of uncertainties in models and the physical parameters. Gas/particle partitioning is usually based on adsorption or absorption theory and does not consider the physical and chemical structure of the particle surface. The particle concentration [TSP] also has a strong influence on the particle associated fraction. The KOA value can only be used for particle absorption from the gas phase if the organic portion of the particle is high. The same is true for absorption in settled house dust. For most SVOCs, reliable experimental vapor pressures are not available. Estimation methods often lead to partially significant deviations in the predicted values. Moreover many algorithms do not distinguish between structural isomers. KOA is commonly derived from KOW and KAW coefficients by KOA=KOW/KAW. Moreover, KAW and p0 are fundamentally related to H. Experimental KOA and H values are not available for most SVOCs and have to be calculated from QSPR approaches. This means that KOW can be calculated from p0 and vice versa.

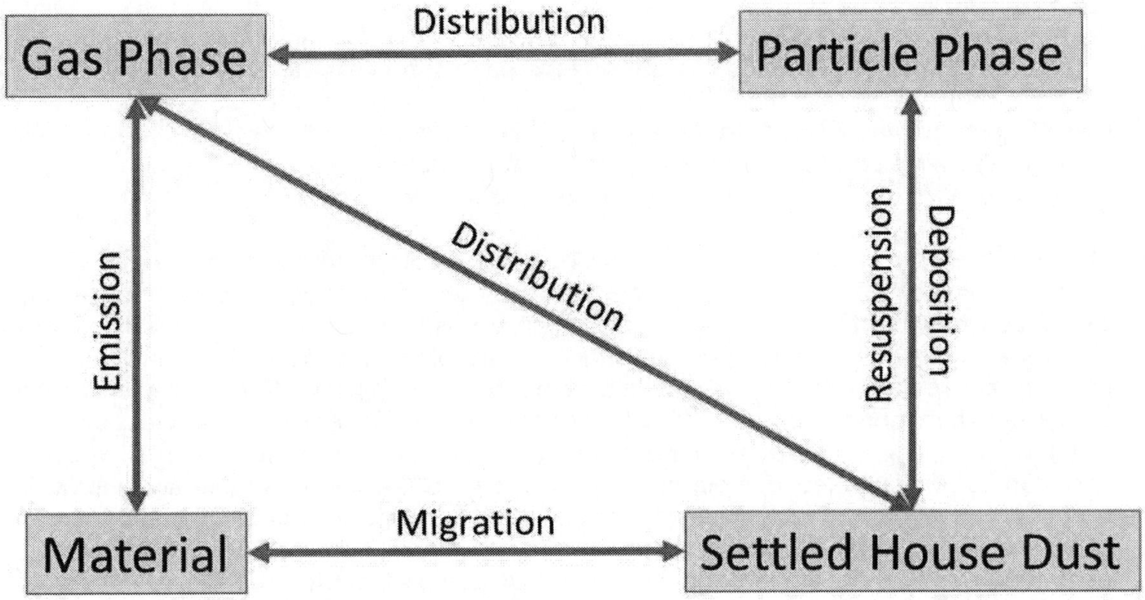

Figure 1: Distribution of SVOCs between gas phase, particle phase and settled house dust in the indoor environment

We-SY-E2.4

A rapid method for measuring the air/surface partition coefficient of SVOCs

Clara Eichler, *Virginia Tech, Blacksburg, VA, United States*
Yaoxing Wu, *Virginia Tech, Blacksburg, VA, United States*
John Little, *Virginia Tech, Blacksburg, VA, United States*

The potential health risks of human exposure to semi-volatile organic compounds (SVOCs) emitted from consumer products and materials in the indoor environment are of public health concern, but the assessment of this exposure is still difficult. Models estimating the exposure rely on key emission parameters which are often not available. Therefore, methods are required to measure these parameters. One important key parameter is the air/surface partition coefficient, KS, because it controls the temporal dynamics and distribution among many states in indoor environments. An estimation of KS using the partition coefficients between octanol and air, Koa, and between octanol and water, Kow, is not always sufficient, as KS changes for different combinations of SVOC compounds, sources and sorption materials. A simple and rapid method based on passive sampling technique was thus developed to measure air/surface partitioning coefficients of SVOCs. The method uses disks made of the targeted material as the receiving phase. Phthalates in two types of polyvinyl chloride flooring (VF) were selected to test the method as the emission source, and aluminum was chosen as the test material surface. Solvent extraction and chromatographic technique were used for sample analysis. A diffusion model has been developed to predict the uptake rate of the passive sampler that collects phthalates emitted from the VF surface. The values of KS and an additional parameter y0, the gas-phase SVOC concentration immediately adjacent to the material surface in a consumer product, were obtained by fitting the diffusion model to the sampling data. Fitting the experimental data for the diffusion of DEHP from one type of VF to aluminum to a diffusion model shows a KS of 320 m. The results agree well with those measured in previous tests. The method proved useful and could be easily expanded to other combinations of SVOCs, sources and sorption materials. As increasing numbers of such measurements are completed, the method would make a great contribution to the assessment of the potential exposure to SVOCs in indoor environments and can help with the exposure based prioritization of chemicals and products.

We-SY-F2: Exposure science informing policy decision-making - II

We-SY-F2.1

Analyzing short-term benzene exposure data to assess the effectiveness of control measures in the refining sector

Daan Huizer, *Caesar Consult, Nijmegen, Netherlands*
Joost van Rooij, *Caesar Consult, Nijmegen, gld, Netherlands*

Background
Recent research has suggested that regular, short-term exposures to benzene may be associated with an increased risk of developing myelodysplastic syndrome (MDS); haematological (blood-related) medical conditions with ineffective production of the myeloid class of blood cells.
It is generally believed that exposures to benzene and benzene-containing products are well controlled within industry. However, given the recent findings of a possible association of MDS at exposure levels that approach some Short Term Exposure Limits (STELs), it would seem prudent to summarize benzene exposures in various operations to evaluate the effectiveness of available control measures for short-term benzene exposures.
Objectives
In this presentation the gathering, compiling and evaluation of monitoring data on regular short-term peak exposures to benzene at workplaces in the supply chain for petroleum products are discussed. Information on existing industry practices and control strategies in place are reviewed to allow the formulation of guidance in this context.
Methods
A baseline review of short-term benzene exposure data in peer-reviewed literature and CONCAWE sector-reports was performed. Next, a standardized collection format was developed to collect short-term benzene exposure data from individual companies in the sector. Available benzene data were extracted for identified work area, job groups and tasks, in accordance with the REACH task descriptors (CONCAWE) as used within the sector. A task-exposure matrix was built in which summary statistics of the collected data are presented for each task, when data availability allowed this.
Further analysis was performed to identify tasks for which existing industry practices appeared either sufficient or insufficient to control short-term exposure levels of benzene. Additional gap analysis identified tasks for which no or limited data were available.
Results
Over 2000 short-term benzene exposure measurements were collected from 8 relevant work areas in the refining sector. A task-exposure matrix covering up to 25 job groups and 40 unique tasks/activities was built. Although the data collection has been finalized, data analysis is currently ongoing.
In the presentation the final results will be presented, covering exposure levels based on the aggregated monitoring data and identified efficiencies of available risk management measures (RMM) and/or operational conditions (OC).

We-SY-F2.2

Practical workplace specific risk communication including exposure assessment data

Koen Verbist, *Cosanta B.V., Amstelveen, Netherlands*

Background
To comply with legislations like REACH and the CAD much effort and money is invested by companies in the exposure and risk assessment process and in determining control of risks. Where needed additional risk management measures are implemented to reduce exposure. However, the step of adequately informing and instructing employees to work safely with the substances should not be neglected or forgotten.
Objectives
To make sure that employees work safely with hazardous materials a project was organized to develop ready-to-use risk based workplace instruction cards that include information on the exposure assessment. The cards will be generated with the already available and widely used internet tool Stoffenmanager®.
Methods
Stoffenmanager® Premium clients were invited to participate in the project and give their input and comments on the draft versions of the instruction cards. Generic model exposure parameters were translated into understandable language for employees. A visual language expert reviewed all texts and further refined the instructions.
Results
Specific risk driven instruction cards were developed that can be generated for each performed risk assessment within Stoffenmanager®. This can be for a single substance of for a mixture. The instruction card is a translation of the exposure assessment parameters into understandable language. The cards has a visual look-and-feel. If personal protective equipment is required, this is directly visible. Two categories of instructions are included. The first category can be influenced by the worker (e.g. applying LEV). This category is in the imperative mood. The second category describes these parameters that are more related to the process and much less likely to be influenced by the worker (e.g. the room volume, of the task descriptor). These parameters are descriptive in language.
Discussion
Stoffenmanager® is being used both under REACH as higher tier tool and under the CAD legislation. For REACH mostly ECETOC TRA will be used as primary tool for the risk assessment. As a result exposure scenarios (for mixtures) coming from REACH and translated into eg. SUMI's will mainly include the ECETOC TRA parameters and will be generic. Downstream users applying Stoffenmanager® can easily compare the REACH exposure scenarios with their own specific risk driven instruction cards to see if they comply to the REACH regulation. If not, they can adjust their operational conditions and risk management measure or choose to forward their risk assessment report.

We-SY-F2.3

Interval testing: A new validation method for models in occupational safety and health

Dorothea Koppisch, *Institute of Occupational Safety an Health of the German Social Accident Insurance (IFA), 53757 Sankt Augustin, Germany*
Mario Arnone, *Institute for Occupational Safety and Health of the German Social Accident Insurance (IFA), 53757 Sankt Augustin, Germany*
Rainer Van Gelder, *Institute for Occupational Safety and Health of the German Social Accident Insurance (IFA), 53757 SanktAugustin, Germany*
Stefan Gabriel, *Institute for Occupational Safety and Health of the German Social Accident Insurance (IFA), 53757 SanktAugustin, Germany*

Aim
Model developers in the field of occupational safety and health often see themselves confronted with the demand that the exposure estimation should be conservative. This means that if a working place is classified as safe, it should really be safe. This demand is often translated into: overestimation by a model is better than underestimation. On the other hand model results should correlate with measured exposure.
To meet both requirements exposure modelling in the field of occupational safety and health should include not only a point estimate for the exposure level. In addition to providing estimates for the mean or median of the exposure distribution it would be better to provide also an estimate for the variation so that higher and lower percentiles can be modelled. If these considerations are met and the variation of exposure is also modelled, it is important that this is also reflected by the validation method.
This presentation will therefore introduce a new validation method called interval testing. In order to show the usefulness of this method, it will be demonstrated using the Stoffenmanager® model and data from the German exposure database MEGA.
Methods
For interval testing a model that estimates not only the mean exposure, but also percentiles, is needed. It is then possible to define percentile intervals. For this study we chose the following set of percentile intervals: (0, 50], [51, 60], [61, 70], [71, 75], [76, 80], [81, 90], [91, 95], [95, 100). For every data point of the validation data set the measured value is compared with the modelled percentiles in order to sort it into the appropriate percentile interval. The number of data points in every percentile interval is then compared with the expected number and Chi^2 is calculated.
Two model algorithms from Stoffenmanager® are used as examples.
Results
The interval testing method can be used to validate Stoffenmanager®. For algorithm one (handling of powders and granules) a Chi^2 of 13.90 ($p>0.05$) is found and for the second algorithms (abrasive processing of wood and stone) a Chi^2 of 122.98 ($p<0.001$).
Discussion/Conclusion
The variability of exposures over time, between and within workers should be reflected in modelling and model validation. If a model gives not only a point estimation of the exposure height, but also percentiles, the new validation method presented in this talk - interval testing - fulfils these requirements.

We-SY-F2.4

Integrated exposure assessment to PAHs arising from the use of petroleum substances

Katleen De Brouwere, *VITO, Mol, Belgium*
Gudrun Koppen, *VITO, Mol, Belgium*
Chris Money, *Cynara Consulting, Brockenhurst, United Kingdom*
Hans Ketelslegers, *Concawe, Brussels, Belgium*

Aim
Exposures to polycyclic aromatic hydrocarbons (PAHs) are ubiquitous. There are many sources and many routes by which human exposure to these substances occurs. However, the contribution of petroleum substances (PS) to PAH exposures for the general population (i.e. non-professional) has not been widely characterised and therefore the impact is not fully established. In view of the potential for petroleum substances to be included in the different REACH processes (notably Evaluation and Authorisation), the aim of the project was to identify integrated multi-source, multi-route (MSMR) exposure model(s) suitable for characterising exposure to PAHs including those arising from the direct consumer use of petroleum substances, as well as those occurring indirectly (such as those arising as the result of the combustion of fuels).
Methods and results
A list of 24 'available integrated exposure tools' was compiled based on models screened within 2 recent projects on integrated exposure modelling, namely the project 4-Fun (http://4funproject.eu/: "The FUture of FUlly integrated human exposure assessment of chemicals") and the the CEFIC LRI project TAGS (http://cefic-lri.org/projects/b5-certh-realistic-estimation-of-exposure-to-substances-from-multiple-sources-tags/). After a first screening of model relevance for PAHs, a systematic inventory of various model aspects of relevant models was made, such as 1) model purpose, 2) exposure pathways considered in the models, 3) model applicability domain, 4) model parameterization, 5) exposure pathways aggregation method, 6) ease of use of model, The result of this inventory will be presented.
A workshop was organized to discuss with model developers the application options of the models for predicting exposures to PAHs from the use of petroleum substances, and the possibility to verify the outcome with suitable validation data.
As an outcome, 2 models (MerlinExpo and INTEGRA) were selected as the most promising MSMR tools to assess PAH exposures arising from the use of petroleum substances, and to compare the predicted exposure with validation data such as biomonitoring data.
Way forward
Exposure to PAHs in 5 distinct consumer use and environmental exposure scenarios of petroleum substances will be modelled using MerlinExpo and INTEGRA, and compared with biomonitoring data. Preliminary results and experiences with the models will be presented.

We-SY-F2.5

A study preparing for a strategy for a non-toxic environment, according to the 7th Environmental Action Programme

Urban Boije af Gennäs, *DG Environment of the European Commission, Sweden*

This presentation will provide a background and present an ongoing study preparing for the strategy for a non-toxic environment, which according to the EU 7th Environment Action Program (7th EAP) is to be presented by 2018. The study focuses on seven topics being substitution, substances in articles and non-toxic material cycles, protection of children and vulnerable groups, very persistent chemicals, innovation, development of green chemicals and early warning of approaching chemicals threats.

Aim

The aim is to study a selection of the topics highlighted in the 7th EAP in the context of a future non-toxic environment strategy. The study will identify gaps, deficits and options for improvement in current policies as well as gaps regarding knowledge and methodologies. A final report is due in the first half of 2017 and will form part of the basis of a non-toxic environment strategy, which is to be presented by the Commission in 2018.

Methods

For each topic, the study includes a literature review, describing the health and environmental issues, the current state of play of policy, gaps and deficits as well as improvement opportunities and best practices. Further, a workshop was held in Brussels on June 8-9 to collect input from experts and stakeholders, in particular on improvement opportunities and experiences from past and ongoing activities. Interviews and questionnaires are also used in the study.

Results

The results of literature review, workshop and other collection of facts and views is to be presented in an interim study report during the summer 2016. Apart from descriptions of status quo for the different sub-study topics, the interim report will include preliminary listings of improvement opportunities to address the identified gaps and deficits as well as best practices. These findings will be further processed during the remaining part of the study and presented in a final report during the first half of 2017.

Conclusions

A preliminary conclusion is that there is a large interest in the topics included in the study in academia, among different kinds of stakeholders both in Europe and beyond. The level of activity in looking for and trying out different kinds of solutions to the problem involved is also considerable. There also seems to be strong interconnections and possible also synergies between different measures to address gaps and deficits identified.

Keywords: EU, 7th Environment Action Program, strategy for a non-toxic environment, substitution, substances in articles, non-toxic material cycles, protection of children, vulnerable groups, very persistent chemicals, innovation, development of green chemicals, early warnings, approaching chemicals threats.

This page intentionally left blank.

We-SY-G2: Pesticide Exposure: Developing Monitoring, Methods and Modeling in Human Health Risk Assessments (Consumer and Worker Risk) - II

We-SY-G2.1

Pesticide Exposure: Developing Monitoring, Methods and Modeling in Human Health Risk Assessments (Consumer and Worker Risk)

Muriel Ramirez-Santana, *Universidad Catolica del Norte, Coquimbo, Chile*
Floria Pancetti, *Universidad Católica del Norte, Coquimbo, Chile*
Paul Scheepers, *Radboud UMC, Nijmegen, Netherlands*
Koos Van der Velden, *Radboun UMC, Nijmegen, Netherlands*
Nel Roelevelt, *Radboud UMC, Nijmegen, Netherlands*
Liliana Zuniga, *Universidad Catolica del Norte, Coquimbo, Chile*
Rodrigo Sandoval, *Universidad Catolica del Norte, Coquimbo, Chile*
Sebastian Corral-Zavala, *Universidad de Chile, Santiago, Chile*

Introduction: Chile has experienced increase of agricultural activity in recent years, leading to enhanced use of pesticides. Organophosphates pesticides (OPP) cause 39% of occupational acute pesticide intoxications. While acute poisonings are easily diagnosed, chronic exposure often goes unnoticed. Biomarkers available today only serve to diagnose acute poisoning. There is no biomarker available to monitor chronic exposure.
Aim: We studied the activity of the erythrocyte enzyme acyl peptide hydrolase (ACPH) as a potential new biomarker for chronic exposure to OPP, relating its activity to cognitive performance and comparing ACPH performance with established biomarkers.
Methods: A total of 268 study participants were recruited: 81 environmentally exposed (EE), 87 occupationally exposed (OE), and 100 in a reference group (RG). The population was homogeneous in age, smoking habits, alcohol and drugs consumption. Blood was collected and analyzed for erythrocyte acetylcholinesterase (AChE), plasma cholinesterase (BChE) and erythrocyte ACPH exopeptidase activity. The neuropsychological assessment included general mental state, memory, language, attention, praxis, executive function, psychomotricity and mood.
Results: During fumigation, the biological tolerance value (BTV) based on 70 % of individual baseline enzyme activity was exceeded for AChE in 28.2% of EE and 24.4% of OE and in 29.5 % and 16.7% for BChE, respectively and 33.3% and 11.5% for ACPH, respectively (for RG these measurements were not performed). For cognitive performance a fair performance in RG (2% low scores) was observed, whereas both EE and OE showed a significantly lower performance in nearly all tests. The most affected endpoints were memory, executive function and psychomotricity. A predictive model was constructed to relate enzyme activities to cognitive outcomes. In this model the most influential variable was an exposure index based on the number of years of employment and information on OPP exposure from a questionnaire; whereas the biomarkers did not contribute significantly to predict cognitive outcome.
Conclusions: Both residents and workers in an agricultural Chilean setting showed cognitive impairment. Biomarker levels indicated higher frequencies of impaired enzyme activities in residents than in workers but did not predict neuropsychological outcome. An index of exposure based on information provided by questionnaire (years living in agricultural area or working in contact with pesticides) was more informative in this respect.

We-SY-G2.2

The Cumulative Aggregate Risk Evaluation System - Next Generation (CARES NG) Model: Progress and Next Steps

Bruce Young, *Bayer CropScience, RTP, North Carolina, United States*
Jennifer Lantz, *Bayer CropScience, RTP, North Carolina, United States*

The Cumulative Aggregate Risk Evaluation System - Next Generation (CARES NG) model is an updated version of the previous CARES model developed in 2001. The CARES NG model is designed to estimate consumer exposures to pesticides in the United States from food, drinking water, and residential use. The software has undergone extensive upgrades to a cloud-based application built upon public data, updated exposure algorithms, and improved user functionality. The dietary (food and water) module utilizes US consumption data (National Health and Nutrition Examination Survey 2005 to 2010) translated to raw agricultural commodities (RAC) based on EPA's recipe file. The dietary user interface allows input of residue values, processing factors, and percent crop treated with built in rules to create the appropriate distribution of values according to EPA SOP. Multiple approaches to estimating dietary exposure are available: acute non-temporal, multi day temporal (repeating diet and match diet), chronic, and within day (event or minute based). The residential module consolidates the EPA 2012 residential SOPs for both handler and post-application exposure calculations into a simple decision tree containing the default parameter inputs. The residential user interface allows input of product use scenarios for a given residential use pattern with product-specific parameters. Approaches to estimating residential exposure depend on the level of refinement required and the data available and are accommodated in this module: non-temporal (deterministic), temporal by day (probabilistic), and temporal within day (event or minute based). The temporal approach will incorporate probability of product use and the human behavioral data (Consolidated Human Activity Database; CHAD) to estimate exposure within day (event based). Using the temporal based dietary and residential approach, the CARES NG model has the ability to aggregate exposure (single chemical) and cumulative exposure (multiple chemicals) by all possible exposure routes: oral (dietary), dermal, inhalation, and oral (incidental). The within day exposure approach provides minute by minute exposure estimates for all routes of exposure which can then be used in physiologically based pharmacokinetic (PBPK) models. This presentation will highlight the features of this "state of the science" model that can accommodate both low to high tier assessments.

We-SY-G2.3

Approaches to Assessing Longitudinal Dietary Exposure in the CARES NG Software

Giulia Vilone, *Creme Global Ltd, Dublin, Ireland, Ireland*
Cian O'Mahony, *Creme Global ltd, Dublin, Ireland, Ireland*

The Cumulative and Aggregate Risk Evaluation System - New Generation (CARES NG®) is a web-based platform for conducting exposure and risk assessments for pesticide residues on foods consumed over several time frames, from the single-day acute assessments up to chronic intakes. In particular, the CARES NG platform was designed to assess exposure levels over 365 days for the US population, using different averaging periods within the 365 day period. CARES NG provides two models to carry out this type of analysis, both based on the food intake data recorded in the NHANES/FCID 2005-2010 database. These models require information on subjects' characteristics such as their diet or the possible change in their bodyweight over the time period considered. The NHANES surveys contain a large amount of data about each participant which provides a detailed picture of their health and physical situation at the time of interview and up to two days of food consumption, but no long-term data are present in the database. Since no data exists covering this period of time, these long-term data must be modelled. The first exposure model uses the short-term data in NHANES to cover the 365 days; the body-weight is kept constant and the two days of food consumption data are randomly repeated throughout the year. The second model attempts to refine the longitudinal exposure levels by 1) using body-weight growth models to assess the changes in the body-weight of children, adolescents and pregnant women over one year and 2) mixing the subject's food consumption 2-day diaries with the consumption data of other NHANES subjects having similar characteristics.
The processes followed to create the CARES NG long-term exposure models will be presented, as well as comparisons between the summary statistics of the outputs obtained from the two models to show the impact that the underlying assumptions have on the results.

We-SY-G2.4

Case Study Comparison of Acute and 21-Day Rolling Average Dietary Exposure Assessments Conducted with DEEM-FCID and CARES NG

Jason Johnston, *The Acta Group, Washington, DC, United States*

Estimating dietary exposures to pesticide residues in treated agricultural commodities, livestock commodities and water plays an important part in the regulation of the safe use of pesticide products. Software tools to conduct such assessments have been developed over time. The U.S. Environmental Protection Agency has used DEEM-FCID (the Dietary Exposure Evaluation Model - Food Commodity Intake Database) successfully for such purposes for many years. However, as regulatory requirements have evolved, increasingly complex assessments are now sometimes required. In an effort to meet those needs, a task force was created to update and expand the Cumulative and Aggregate Risk Evaluation System - Next Generation (CARES NG). For DEEM-FCID n-day average exposure calculations are typically conducted using 2-day average food consumption estimates and n-day water residues. CARES NG includes a feature that permits calculation of n-day rolling average exposures using time-series drinking water residues with one of two approaches to assessing food consumption, i.e., alternate use of food consumption data from two-day surveys or creation of food consumption time-series profiles using data from similar individuals. The aim of this case study is to compare acute and 21-day rolling average dietary exposure estimates for a chemical generated using DEEM-FCID and CARES NG. Similarities and differences in model-generated exposure estimates will be discussed in terms of underlying modeling differences.

UTRECHT
THE NETHERLANDS
OCTOBER 9-13
ISES 2016

We-SY-G2.5

Quantification of Dermal Pesticide Absorption from Dried Foliar Residues

James Clarke, *University of Bath, Bath, United Kingdom*
Sarah Cordery, *University of Bath, Bath, United Kingdom*
Neil Morgan, *Syngenta, Bracknell, United Kingdom*
Peter Knowles, *Syngenta, Bracknell, United Kingdom*
Richard Guy, *University of Bath, Bath, United Kingdom*

Pesticides go through rigorous assessments to ensure that their use does not represent an unacceptable risk to human health. For re-entry workers who may come into contact with treated surfaces after application during tasks such as crop inspection or harvest, dermal exposure is estimated using simple predictive models. Determination of a systemic dose from this predicted exposure relies on applying a factor for dermal absorption. Currently, dermal absorption studies involve the concentrated product and one or more representative spray dilutions. There is no recognised protocol for measuring dermal absorption for foliar residues and in the EU risk assessments adopt the highest measured value from these studies or a more precautionary default, although previous work (Belsey et al. 2011) showed that absorption from dried residues was different than from aqueous solutions.

A key aim of this study was to develop a novel method for assessing the dermal absorption of pesticides from dried foliar residues, with the ultimate aim of using this method to obtain more realistic absorption values for risk assessment.

It is important that this method is as close to a real field scenario as possible, yet is simple and easily reproducible. To this end, a laboratory technique was developed (Clarke et al., 2015) based on applying pesticides to an inert platform to create uniform dried deposits of pesticide mimicking foliar residues, which could be transferred by a standardised process to skin membranes and absorption measured in vitro in conventional Franz diffusion cells. Absorption values from a range of pesticides as dried residues were measured and compared to those from spray dilutions applied to the skin at an equivalent dose level. This demonstrated that the percentage absorption from the dried residue was consistently lower than from the spray dilution.

Further work using this method is investigating the effects of dose and formulation type on absorption from residues. The work is providing valuable insight into a poorly documented area of exposure science and has the potential to allow more realistic risk assessments than those which may currently overestimate exposure and prevent the registration of safe and effective products.

References

Belsey, N.A., Cordery, S.F., Bunge, A.L. & Guy, R.H., 2011. Assessment of dermal exposure to pesticide residues during re-entry. Environmental Science & Technology, 45: 4609-4615.

Clarke J.F., Cordery S.F., Morgan N.A., Knowles P.K. & Guy RH. In vitro method to quantify dermal absorption of pesticide residues. Chem Res Toxicol. in press, 2015

We-SY-H2: Tool and methods for an exposure driven safe by design approach for nanomaterials - II

We-SY-H2.1

Harmonisation of exposure assessment strategies and data storage to support data-driven safe by design approaches

Wouter Fransman, *TNO, Zeist, Netherlands*

Nanotechnology is a fast growing sector with ever increasing variety and complexity of new materials. Currently, risk assessment is struggling to keep up with the innovation and hence more emphasis will need to be placed on Safe by Design, whereby health, safety and environment are taken into account early in the innovation chain to ensure that risks are managed properly. Exposure plays an important role in this area, as Safe by Design refers not just to reducing the hazard potential of the nanomaterials but also to reducing the release and exposure potential of products and processes. This presentation will emphasize the need for an exposure database, collating measurement data of airborne nanoparticles to permit data-driven safe by design approaches by exposure modelling, exposure scenario building, risk management and development of occupational exposure limits.

In the past decade a series of international workshops have been organized to discuss nano-specific issues in exposure assessment research related to the three identified topics: (i) measurement strategies; (ii) analyzing, evaluating, and reporting of exposure data; and (iii) core information for (exposure) data storage. Preliminary recommendations were achieved with respect to (i) a multimetric approach to exposure assessment, a minimal set of data to be collected, and basic data analysis and reporting as well as (ii) a minimum set of contextual information to be collected and reported. To make progress in the process of harmonization, it was concluded that for research in studying exposure to nanoparticles, there is a need for an occupational exposure database to permit data-driven safe by design approaches by exposure modelling, exposure scenario building, risk management and development of occupational exposure limits. Amongst a working group of PEROSH institutes a database structure called NECID (Nano Exposure and Contextual Information Database) was developed, which include exposure data and contextual information. The database facilitates the comparing and sharing of nano exposure data, because the exposure data of different institutes are collected and stored in a harmonized way. The database is based on the characteristics of existing databases (ART, MEGA) and the NANOSH dataset. As nanomaterials have distinctive characteristics and the measurement strategy is based on a multimetric approach, additional variables have been introduced.

We-SY-H2.2

Occupational exposure during the production, simulated use and end-of-life stages of nanoenabled products for energy harversting and energy storage

Simon Clavaguera, *CEA, Grenoble, France*
Sébastien Artous, *CEA, Grenoble, France*
Cécile Philippot, *CEA, Grenoble, France*
Dominique Locatelli, *CEA, Grenoble, France*
Sébastien Jacquinot, *CEA, Grenoble, France*
Bryony Ross, *IOM, Edinburgh, United Kingdom*
Laura MacCalman, *IOM, Edinburgh, United Kingdom*
Martie Van Tongeren, *IOM, Edinburgh, United Kingdom*

Nanotechnology is a fast growing sector with ever increasing variety and complexity of new materials. Currently, risk assessment is struggling to keep up with the innovation and hence more emphasis will need to be placed on Safe by Design, whereby health, safety and environment are taken into account early in the innovation chain to ensure that risks are managed properly. Exposure plays an important role in this area, as Safe by Design refers not just to reducing the hazard potential of the nanomaterials but also to reducing the release and exposure potential of products and processes.

This work presents the results of comprehensive experimental campaigns focused on the assessment of occupational exposure to next-generation nanomaterials covering two case studies along the life cycle of nanoenabled products for energy harversting and energy storage.

The first value chain investigated was the production and simulated used of nanoenabled thermoelectric generators. The following scenarios were monitored: mechanosynthesis, sintering, grinding, diamond sawing, thermopressing and simulated use. The second value chain was the production and the end-of-life of nanoenabled electrodes for Li-ion batteries. The following scenarios were monitored: powder handling, weighing, mixing, transferring and mechanical recycling (shredding) of electrodes.

Measurements were performed complementary in both lab and pilot line facilities in order to reproduce an industrial environment. The boundaries of the system relevant for the present study are the indoor air and indoor surfaces which are considered potential occupational endpoints for inhalation and dermal exposure. The methodology used for the exposure evaluation followed the "French approach" (Durand et al. 2012) and the nanoGEM / OECD Tiered approach. Thanks to several granulometers and counters measurements include concentration of particles, either background or activity, size, size distribution, state of agglomeration / aggregation, specific surface area, morphology and chemical composition. Exposure data and contextual information were gathered according to NECID requirements to facilitate the comparing and sharing of nano exposure data.

The results will be presented in terms of exposure and control bands (ISO/TS12901) to identify and document the best operative conditions on health, safety and environment for each scenario studied.

The research leading to these results has received funding from the European Union's FP7 Grant Agreement n.604602 (FutureNanoNeeds project) and 310584 (NanoReg project).

C. Durand, et al. (2012) NanoSafe 2012 French approach for characterizing potential emissions and exposure to aerosols released from nanomaterials in workplace operations.
C. Asbach, et al. (2012) nanoGEM Standard Operation Procedures.
OECD report ENV/JM/MONO(2015)19.

We-SY-H2.3

Prevention through design (PtD): selection of proven risk management measures (RMMs) to control the exposure to ENMs

Carlos Fito, *ITENE, Paterna, Spain*
Maida Domat, *ITENE, Paterna, Spain*

The use of engineered nanomaterials (ENMs) is growing continuously due to the increasing number of applications, promoting the development of a new generation of innovative products that have created tremendous growth potential for a large number of sectors. However, along with the benefits, there is an on-going debate about their potential effects on the human health or the environment.

For a comprehensive risk assessment of ENMs, information is needed with respect to the intrinsic harmfulness of the particle, likelihood of exposure, and efficacy of workplace controls. Major investments have been done so far on the characterization of the toxicological profile. However, research aiming to improve our understanding of the possible exposure, as well as on the effectiveness of common risk management measures is far less advanced.

This work presents experimental data on the effectiveness of respiratory and dermal protection equipment, and local exhaustive ventilation (LEV) systems to control the exposure to ENMs in occupational settings. New experimental data on the protection factors achieved under representative exposure scenarios, as well as recommendations for the design of PPE and ECs will be presented. The testing activities were conducted after the validation of a set of standardized procedures, including the evaluation of the permeation to ENMs for dermal protective equipment, total inward leakage (TIL) inward leakage for respirators and filters, and capture efficiency for ventilation systems. The experimental work was conducted in a dedicated nano-aerosol exposure chamber where several exposure scenarios can be reproduced.

The results from the test suggest that the control of exposure via inhalation is a key priority. Respirators provided medium performance levels of filtration efficiency against NMs. The performance levels determined suggest that face seal leakage, and not filter penetration, is a key parameter to be considered when working with NMs. The evaluation of dermal protective equipment showed very low permeation levels, meaning that common measures are effective. The capture efficiency of the LEV systems was demonstrated to be adequate.

The data are compiled in a library of nano-specific RMMs developed using Microsoft Excel®. The library helps stakeholders to select proper measures depending of the type of ENM and process, guiding the user in the selection of proven risk management measures.
The research leading to these results has received funding from the European Union's FP7 Grant Agreement n. 310584 (NanoReg project), and the LIFE project NanoRISK (LIFE ENV/ES/000178)

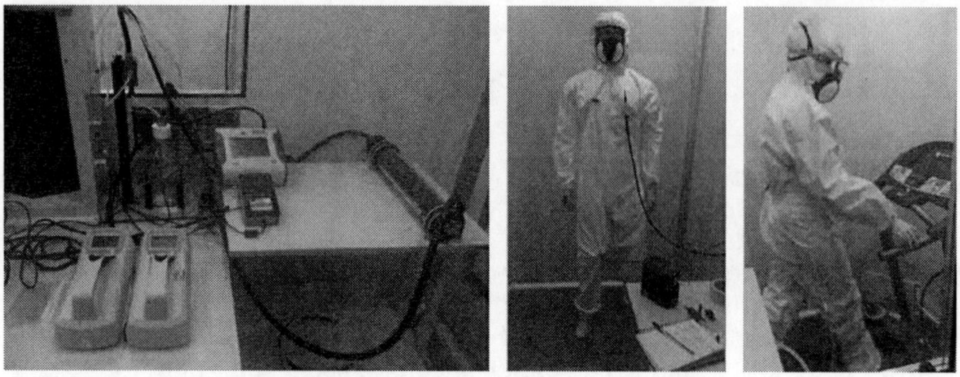

Figure 1. Experimental set up for effectiveness testing studies

We-SY-H2.4

The use of quantitative exposure models within the safe by design concepts

Laura MacCalman, *Institute of Occupational Medicine, Edinburgh, United Kingdom*
Araceli Sanchez-Jiménez, *Institute of Occupational Medicine, Edinburgh, United Kingdom*
Emmanuel Belut, *INRS, Nancy, France*
John W Cherrie, *Institute of Occupational Medicine, Edinburgh, United Kingdom*
Lang Tran, *Institute of Occupational Medicine, Edinburgh, United Kingdom*
Martie Van Tongeren, *Institute of Occupational Medicine, Edinburgh, United Kingdom*

Measurement of engineered nanomaterials (ENM) is not straightforward, particularly because of the limitations of current equipment and the lack of commercially available personal monitors. For this reason exposure models are being increasingly used to estimate potential exposure to ENM in the workplace and for consumer exposure. Current tools available for risk assessment of ENM are qualitative, or semi-quantitative and typically aim to provide a category of exposure potential (along with hazard and risk) rather than a quantitative air concentration.

As part of the NANoREG project a quantitative two-box source-receptor exposure model has been developed. The model predicts size-resolved aerosol number concentration over time given information on the emission rate and pattern of emission, along with the dimensions of the room and ventilation rate. The model accounts for: deposition of particles to walls, floors and other surfaces, dilution due to ventilation and agglomeration of particles over time, using particle-size specific equations. Validation of the model is being undertaken to compare the estimates to measurements obtained during a large-scale experiment.

The model can be used to evaluate the impact of changing the exposure scenario (e.g. increasing/decreasing ventilation rate, changing the local controls used, amending the size of the room, adding more emission sources). This enables an assessment of the impact of changes and by exploring the different aspects of the exposure scenario allow for the process, and associated controls, to be implemented in such a way to sufficiently control exposure, thus promoting the "safer by design" paradigm. The model is currently being developed into a user-friendly tool which will enable users to easily make such evaluations and assessments.

We-SY-H2.5

Is the environment the great "post release equaliser" for nanomaterials, and can we design to help it?

Claus Svendsen, *CEH, Oxford, United Kingdom*

As many nanomaterials are specifically designed to be highly reactive or have special properties to enable reactive their functions.. It should be considered if materials could be designed to "lose their nano properties" once they are released from the product or place where the function is required. Using the NanoFASE (http://nanofase.eu/) project's exposure assessment framework, this talk will look methods to identify what properties could be targeted as features of materials to help drive such post release "accelerated degradation" and properties that help "targeted fate properties determination" to encourage only less reactive forms of NMs are released and that these where possible NMs to end up in the least hazardous form in the least hazardous place in the environment.

We-PL-I2: Indoor Environment

We-PL-I2.1

Flammability Standards Impact Flame Retardant Concentrations in Dust

Robin Dodson, *Silent Spring Institute, Newton, MA, United States*
Kathryn Rodgers, *Silent Spring Institute, Newton, MA, United States*
Joseph Allen, *Harvard T.H. Chan School of Public Health, Boston, MA, United States*
Jose Guillermo Cedeno Laurent, *Harvard T.H. Chan School of Public Health, Boston, MA, United States*
Adrian Covaci, *University of Antwerp, Wilrijk-Antwerp, Belgium*
Giulia Poma, *University of Antwerp, Wilrijk-Antwerp, Belgium*
Govindan Malarvannan, *University of Antwerp, Wilrijk-Antwerp, Belgium*
Ruthann Rudel, *Silent Spring Institute, Newton, MA, United States*
Gale Carey, *University of New Hampshire, Durham, NH, United States*

Aim: Furniture flammability standards are typically met with chemical flame retardants (FRs), and the nature and amount of FRs used depends on the properties of the material and on how the performance-based test is conducted. In the U.S., most furniture purchased by colleges and universities meets one of two flammability standards: Technical Bulletin (TB) 117 or TB 133. In the absence of national flammability standards, California's TB 117 and TB 133 have become the de facto national standards. Because TB 133 requires furniture to withstand a much larger test flame than TB 117, we hypothesize that TB 133 furniture have different FR profiles and potentially higher levels of FRs compared to TB 117 furniture, and that these FRs will migrate out of furniture and into dust.
Methods: We collected 96 vacuum dust samples from residential spaces on 2 northeastern U.S. college campuses adhering to either TB 117 or TB 133. Chemical analysis targeted 54 FRs, including 12 polybrominated diphenyl ether (PBDE) congeners, 20 other brominated FRs, 2 Dechlorane Plus isomers, 3 hexabromocyclododecane isomers, 12 organophosphate flame retardants (OPFRs), and 5 polybrominated biphenyls.
Results: PBDEs and OPFRs were found in the majority of dust samples, and OPFRs tended to have the highest median dust concentrations. Median levels were comparable to our previous measurements in California house dust; however, maxima were up to 100x higher than previous residential measurements. The maximum TDCIPP (chlorinated "tris") concentration was 170,000 ng/g, higher than levels previously reported in U.S. dust, even office dust, which tends to be higher than house dust. Dust concentrations of several FRs, including BDE 209, decabromodiphenylethane (DBDPE), anti-Dechlorane Plus, and tri-(2ethylhexyl) phosphate (TEHP), were significantly higher on the TB 133 campus compared to the TB 117 campus. BDE 209, and its replacement DBDPE, are used in textile back-coatings to meet stricter upholstered furniture flammability standards, like TB 133. Dust concentrations in samples collected from student dorm rooms (i.e. sleeping spaces) were generally higher and more variable than concentrations in samples collected from common spaces in residence halls. This is likely a result of additional furnishings and electronics brought in by students.
Conclusions: FR concentrations varied by flammability standard used. The high density of FR-treated products, including furniture, furnishings, and electronics, in student dorm

rooms likely led to elevated levels for some FRs. Results will be used to inform institutional purchasing policies and local flammability standards.

We-PL-I2.2

Individual-Level Home Environmental Exposures are Associated with Respiratory Outcomes in the Kingston Allergy Birth Cohort (KABC)

Michelle North, *University of Toronto, Toronto, Ontario, Canada*
Elizabeth Lee, *Queen's University, Kingston, Canada*
Vanessa Omana, *Queen's University, Kingston, Canada*
Jenny Thiele, *Queen's University, Kingston, Canada*
Yuchao Wan, *University of Toronto, Toronto, Canada*
Miriam Diamond, *University of Toronto, Toronto, Canada*
Jeff Brook, *University of Toronto and Environment Canada, Toronto, Canada*
Anne K. Ellis, *Queen's University, Kingston, Canada*

Aims
The KABC was instigated to study environmental influences on the developmental origins of allergic disease. Kingston General Hospital was chosen as the collection site, as it serves a mixture of rural and urban residents with diverse socioeconomic status (SES) and a high prevalence of maternal smoking. The aim of this study was to evaluate prenatal and early life associations between various indoor air quality/home environment factors and parental reports of wheeze or cough without a cold in cohort children to age 2.
Methods
Pregnant women gave informed consent and completed a health/environmental survey (n=557). Umbilical cord blood was collected from 413 deliveries, and follow-up surveys to age 2 yielded data on 232 children. We examined home-environment characteristics by urban/rural residence and SES. We employed multivariate Cox proportional hazard models to examine factors associated with the development of parentally-reported respiratory symptoms (wheeze or cough without a cold) to age 2 years.
Results
The KABC encompassed a high proportion of rural residents (42.4%), and prenatal exposure to smoke (25.9%). Rural participants exhibited higher SES income measures, but lower education, compared to urban dwellers. Urban and low-SES families were more likely to report living near traffic-related air pollution sources, and low-SES families reported living in older homes. The incidence rate of parental reports of respiratory symptoms was 0.19 cases/person-year. Breastfeeding for the first 6 months of life was significantly associated with a lower rate of respiratory symptom development, while low-SES was associated with a higher risk of symptoms. Important indoor environmental factors were identified. The regular use of air fresheners in the home and the self-reported presence of mold in the home was associated with a higher incidence of respiratory symptoms in the children.
Conclusions
We found that both residing in rural/urban and high/low-SES areas affected characteristics of the home and potential environmental exposures. Early exposure to mold, the regular use of air fresheners in the home and tobacco smoke were associated with a higher incidence of parental reports of wheeze and cough, while breastfeeding was negatively associated with those symptoms. Home visits have been carried out for the collection of dust samples for chemical analyses, including phthalates, flame retardants, and polycyclic aromatic hydrocarbons. Analyses of dust samples in tandem with epigenetic analysis of

umbilical cord and peripheral blood are ongoing to reveal environmental biomarkers that may underlie the increased risk for respiratory symptoms observed in this study.

We-PL-I2.3

The application of the DYLOS to assess indoor residential PM2.5 aerosols in the HEALS pilot study

Anjoeka Pronk, *TNO, Zeist, Utrecht, Netherlands*
Remy Franken, *TNO, Zeist, Netherlands*
Eelco Kuijpers, *TNO, Zeist, Netherlands*
John Bartzis, *IPTA, Greece, Greece*
Thomas Maggos, *IPTA, Greece, Greenland*
Mina Stamatelopoulou, *IPTA, Greece, Greece*
Denis Sargiannis, *Aristotle University of Thessaloniki, Thessaloniki, Greece*
Susanne Steinle, *IOM, Edinburgh, United Kingdom*
Miranda Loh, *IOM, Edinburgh, United Kingdom*
John Cherrie, *IOM, Edinburgh, United Kingdom*

Background
Particulate matter (PM) is a major source of indoor air pollution. Low cost real time particle counters are becoming available. As part of the HEALS pilot protocol, the DYLOS is being used in 150 homes of families across Europe to estimate exposure to PM. The aim of this study was 1) to assess the validity of the DYLOS and 2) to calibrate and validate a method for the conversion of the particle number concentrations (PNC) obtained by DYLOS to PM2.5 mass.
Methods
Side by side measurements were collected for 3-5 days in the homes of four volunteers with the Dylos DC1700™ (0.5 - 20µm, 2 size bins) and the APS™ Aerodynamic Particle Sizer® (0.5 - 20µm, 52 size bins). For two volunteers a total of ten 24h gravimetric PM2.5 samples (Harvard impactor or Harvard PEM) were collected simultaneously. Correlations between the obtained PNCs were calculated. DYLOS-PNC0.5-2.5 was converted to PM2.5 mass concentration as follows. First, the APS-PNC0.5-2.5 was converted to the PM2.5 mass based on the mean particle density (1.6 g/cm3) and mean aerodynamic diameter by size bin (n=22). The gravimetric data were used for calibrating this conversion. Then a second order polynomial equation was used to fit the DYLOS-PNC0.5-2.5 on the calculated PM2.5 mass for a random sample of 33% of the data. The derived model was validated internally with the remaining 67%. In addition, for external validation, thirty 24h gravimetric PM2.5 samples were collected in parallel with the DYLOS data among HEALS pilot study participants.
Results
Preliminary results indicate that the PNC obtained by Dylos correlated well with the PNC obtained by APS ($R^2>0.92$). It was observed that the Dylos slightly underestimated particle counts at particle number concentrations above 30 particles/cm3. The internal validation of the model for converting DYLOS-PNC0.5-2.5 to PM2.5 mass demonstrated a high correlation ($R^2=0.76$). The external validation of the model with the gravimetric samples is ongoing.
Discussion

This study in a real home setting indicated a high correlation between PNC0.5-2.5 obtained by DYLOS and APS. The fitted model for converting the Dylos PNC0.5-2.5 to PM2.5 mass gave results that are comparable to applying a previously published model based on an experimental setting. The model will be used to obtain PM2.5 mass concentrations in the HEALS pilot study. Preliminary conversions for the 40 Dutch homes demonstrate a median modelled PM2.5 mass of 6.7 µg/m3 (P5-P95: 6-8 µg/m3).

We-PL-I2.4

Naturally ventilated schools located near traffic hotspots in developing countries: Risks and exposure to carcinogenic pollutants

Darpa Jyethi, *Indian Statistical Institute, Tezpur, New York, India*
P S Khillare, *School of Environmental Sciences, Jawaharlal Nehru University, New Delhi, Delhi, India*
S. Sarkar, *NUS Environmental Research Institute, National University of Singapore, Singapore, Singapore*

Polycyclic aromatic hydrocarbons (PAHs) associated with the inhalable fraction of particulate matter were determined for 1 year at a school site located in proximity of industrial and heavy traffic roads in Delhi, India.
PM10 (aerodynamic diameter ≤10 μm) levels were ~11.6 times the World Health Organization standard. Vehicular (59.5 %) and coal combustion (40.5 %) sources accounted for the high levels of PAHs (range 38.1-217.3 ng m−3) with four- and five-ring PAHs having ~80 % contribution. Total PAHs were dominated by carcinogenic species (~75 %) and B[a]P equivalent concentrations indicated highest exposure risks during winter. Extremely high daily inhalation exposure of PAHs was observed during winter (439.43 ng day−1) followed by monsoon (232.59 ng day−1) and summer (171.08 ng day−1). Daily inhalation exposure of PAHs to school children during a day exhibited the trend school hours>commuting to school>resting period in all the seasons. Vehicular source contributions to daily PAH levels were significantly correlated with the daily inhalation exposure level of school children. A conservative estimate of ~11 excess cancer cases in children during childhood due to inhalation exposure of PAHs has been made for Delhi.

We-PL-12.5

Indoor Exposure to Particulate Matter - What do we know about exposures and their health consequences?

David Butler, *National Academies of Sciences, Engineering, and Medicine, Washington, DC, United States*
Guru Madhavan, *National Academies of Sciences, Engineering, and Medicine, Washington, DC, United States*

Aim
The National Academies of Sciences, Engineering, and Medicine---in response to a request from the US Environmental Protection Agency---conducted a February 2016 workshop intended to benchmark the state of the science regarding indoor exposure to particulate matter, with a focus on PM2.5 and ultrafine particles. The workshop addressed the indoor and outdoor sources of indoor PM, particulate dynamics and chemistry, the determinants of exposure levels, characterization of the nature of exposures, exposure mitigation, identified and emerging health concerns, interventions and risk communication. Special attention was paid to attributes of the exposures that are of greatest concern for occupant health, exposure modifiers, vulnerable populations, risk management, and gaps in the science. The workshop brought together engineers, epidemiologists, building professionals, clinicians, risk communications specialists and other researchers interested in the interface between the indoor environment and occupant health. Participants included in-person attendees and over 400 people from 12 countries who connected to the event via a live webcast.
Methods and Results
Workshop speakers identified some of the key drives of variation in indoor levels; new and relatively underappreciated sources of PM, including e-cigarettes and desktop 3-D printers; and building characteristics that influence the penetration of outdoor sources. They discussed the role of particle resuspension in personal exposure as well the utility and limitations of present-day exposure measurement devices. Speakers considered the ways that occupants influence the composition of indoor PM, the socioeconomic determinants of exposures, the effectiveness of filtration in exposure mitigation, and how weatherization and other energy-conservation measures may have unintended health consequences. New research on indoor PM and cardiovascular health, birth outcomes, and neurological and psychiatric disorders was shared. And the challenges of communicating indoor PM risks and exposure mitigation strategies were discussed.
Conclusions
The conference presentation will summarize the major issues identified in the workshop and the participants' suggestions for addressing knowledge gaps. These included the needs for better research on how exposures are influenced by the chemical breakdown of building materials, more integration between epidemiologists and exposure scientists to address exposure misclassification and improve health effect estimates, and to draw lessons from decision science to inform how best to communicate indoor PM risks and the means to mitigate them.

We-SY-A3: New Frontiers in Toxicology Create New Challenges for Risk Assessment: What must Exposure Scientists do to Meet the Challenge?

We-SY-A3.2

PBPK Modelling for Environmental Chemicals: Linking to In Vitro Data

Harvey Clewell, *ScitoVation, Durham, North Carolina, United States*
Miyoung Yoon, *ScitoVation, Durham, North Carolina, United States*

The field of toxicology is currently undergoing a global paradigm shift to use of in vitro approaches for assessing the risks of chemicals and drugs, yielding results more rapidly and more mechanistically based than current approaches relying primarily on live animal testing. However, the use of in vitro data in risk assessment entails a number of new challenges associated with translating the in vitro data on bioactive concentrations into estimates of safe in vivo exposures.

When used within a Mode of Action / Adverse Outcome Pathway framework, physiologically based pharmacokinetic (PBPK) models provide an effective tool for conducting quantitative in vitro to in vivo extrapolation (IVIVE). Their physiological structure facilitates the incorporation of in silico- and in vitro-derived chemical-specific parameters in order to predict in vivo absorption, distribution, metabolism and excretion. In particular, the combination of in silico and in vitro parameter estimation with PBPK modeling can be used to predict the in vivo exposure conditions that would produce chemical concentrations in the target tissue equivalent to the concentrations at which effects were observed with in vitro assays of tissue/organ toxicity. They can also support the identification of potentially susceptible populations associated with age-dependent pharmacokinetics or metabolic polymorphisms.

This presentation will describe the key elements of IVIVE and the critical issues that must be addressed to move forward. Two examples of PBPK-based IVIVE will be described: the use of in vitro assays and PBPK modeling to estimate a margin of exposure for endocrine active compounds and the use of in vitro metabolism data and PBPK modeling to evaluate early life sensitivity to pesticides.

We-SY-A3.3

Multi-route Temporal Exposure Models for Pesticides in CARES NG and Linking to PBPK Modelling

Cian O' Mahony, *Creme Global, Dublin, Ireland*
Giulia Vilone, *Creme Global, Dublin, Ireland*

The Cumulative and Aggregate Risk Evaluation System (CARES) for is a model and software for determining pesticide exposure in US consumers from food, drinking water, and residential exposure due to the use of products containing pesticides as active ingredients. The software has been migrated to a cloud based system accessible via the web, and the models and databases updated. This is with a view to developing a suite of fast and robust exposure models, from simple point estimate calculations of exposure to high-tier, temporal cumulative and aggregate exposure models covering the dermal, oral and inhalation routes. In the case of the latter, one of the key requirements of the system is to be able to produce output that can be linked with PBPK models, in order to develop more refined estimates of consumer exposure based on the time-course of a chemical within the body. This is with a view to developing a more refined estimate of consumer risk in a population, rather than just estimating the total daily externally applied dose of pesticide per route of exposure.
The technical details of the multi-route, subject-based temporal exposure model will be presented, as well as the how a generic output structure can be generated with that can in turn be linked to a given PBPK model. Considerations that will be discussed will include subject-based anthropometric data, time steps and dose metrics used for different routes and sources of exposure.

We-SY-A3.4

Going from in vitro ("hazard") data to final assessment and the need for refined exposure estimates in the assessment of genotoxicity risk

Christina Hickey, *Firmenich, Plainsboro, New Jersey, United States*
Tetyana Kobets, *New York Medical College, Valhalla, New York, United States*
Matthew Tate, *Gentronix, Alderley Edge, United Kingdom*
Gary M. Williams, *New York Medical College, Valhalla, New York, United States*
Benjamin Smith, *Firmenich, Plainsboro, New Jersey, United States*

For genotoxic events the traditional assumption is that there may not be a threshold dose,and that some degree of risk exists at any level of exposure; leading to recommendations that exposure should be as low as reasonably achievable (ALARA). Such recommendations are of limited value, especially for materials that are present naturally in the diet, as it does not allow comparison with estimates of human intake nor carcinogenic potency. While negative results in an in vitro tier one test indicate that DNA damage is unlikely to be induced, any chemicals that produce positive results are then evaluated through higher tier testing in animal models. These second tier tests are designed to reduce false positive results, yet they still maintain a certain level of uncertainty. False positive results in tier one tests can often result from high doses of chemicals being used that are not relevant to actual human exposures. Further, it has been demonstrated that some chemicals which generate DNA damaging effects at high doses do not act the same at low doses, suggesting that thresholds for DNA damaging chemicals do exist. To expand on this concept we use Point of Departure (PoD) estimates from an in vitro screening assay (Bluescreen™HC), the Turkey Egg Genotoxicity Assay (TEGA), and available in vivo data, in an attempt to identify if tier one in vitro thresholds can provide sufficient information for human health risk assessments and how these compare to estimated dietary intakes. Identifying thresholds estimated to be well above potential human exposures, estimated from current dietary intake values, would suggest that these chemicals are not a risk to human health. Furthermore, modifying the testing approach to consider relevant human exposures, could allow for exposure-based data waiving, thereby reducing the number of animal tests conducted and the uncertainty associated with them.

We-SY-A3.5

A computational framework for incorporating dermal penetration and elimination pathway predictions into provisional PBPK models: A practical tool in high throughput chemical risk assessment

John Troutman, *The Procter & Gamble Company, Cincinnati, Ohio, United States*

To further the use of non-animal hazard data into emerging risk assessment frameworks, a combination of pharmacokinetic and toxicity information is needed to calculate internal and external dose metrics. Internal dose can provide a linkage to in vitro effect data, allow for a combination of external doses to facilitate aggregate exposure assessment, maximize the ability to compare studies on related chemicals done by different dose routes and species. In this presentation, first we demonstrate the direct linkage between a dermally applied dose and the resulting internal dose and then explore methods to understand the resulting internal dosimetry when measured data to parameterize a PBPK model are missing or incomplete. We have developed an approach for rapid parameterization of dermal PBPK model based solely on in silico QSAR-derived chemical inputs. A computational model for the a priori prediction of renal and metabolic clearance mechanisms was employed in the development of this screening level model. It is anticipated that this screening level information can be used to assess the need for additional data generation when greater accuracy is required (based on projected worst case margins of safety). Importantly, being able to predict whether a compound will be renally eliminated without biotransformation will decrease the number of chemicals for which hepatic clearance will need to be measured experimentally. These concepts can be evaluated and used to help drive decisions and efficiency in safety testing of cosmetics and personal care products. A comparison of model simulations to experimental data will illustrate the approach.

We-SY-B3: Occupational Exposure Models - Development and/or Evaluation of REACH and other European and US models and tools (including tool for nanomaterials) - I

We-SY-B3.1

The challenge of model building

John Cherrie, *Institute of Occupational Medicine, Edinburgh, United Kingdom*

Exposure models and associated software tools are increasingly being used for regulatory purposes and in epidemiological studies. This presentation discusses some of the fundamental issues involved in constructing exposure models and tools.

Models should have a clear theoretical conceptual framework that is ideally articulated in advance of more detailed model building. The conceptual model is a simplification of reality that embodies enough detail of the real situation to enable predictions that are sufficiently reliable for a specific need. The clear advantage of models based on a conceptual framework is the ability to more explicitly define the model applicability domain. Surprisingly, there is no real consensus on the theoretical basis for human exposure, and this is a major impediment to reliable and consistent model building.

Regardless of the form of the final model it needs to have some mathematical expression to facilitate prediction of exposure in new circumstances, e.g. deterministic or probabilistic. The parameters for the model can be derived from prior data or can be assigned a priori. For inhalation exposure models there is a considerable amount of data available, although not always with the necessary metadata to define all model parameters. For dermal and inadvertent ingestion exposure good quality exposure data is almost completely lacking.

The key to judging the utility of any model or tool is its ability to predict an outcome in realistic scenarios. Assessment of the validity of a model across the whole applicability domain is an essential perquisite for trusting the reliability of the results. To do this properly is expensive and time-consuming and so most exposure models are inadequately validated. Limited availability of exposure data again hampers the validation of reliable models.

Models may be implemented in the form of a software tool or otherwise to predict exposure in practice. There should be some appropriate guidance and exemplars for individuals using the tool and appropriate quality assurance procedures to ensure that ongoing use continues to produce reliable predictions. The latter point is almost completely ignored in exposure science where there has been little attempt to evaluate the within and between assessor variability in modelled exposure, or provide quality assurance systems for tool use.

UTRECHT
THE NETHERLANDS
OCTOBER 9-13
ISES 2016

We-SY-B3.2

ETEAM: Overview of the project background

Martin Tischer, *Federal Institute for Occupational Safety and Health (BAuA), Dortmund, NRW, Germany*

Several 1st tier exposure models such as ECETOC TRA, MEASE, EMKG-EXPO-TOOL, STOFFENMANAGER and RISKOFDERM are recommended by the European Chemicals Agency (ECHA) for estimating occupational exposure. The risk assessment under REACH follows a tiered approach in which the first tier should provide a conservative (i.e., protective) system that can discriminate between substances in scenarios of some concern and those which are considered save. Although the tier 1 models claim to have a broad range of applicability, none of these models has been extensively validated during their development. The German Federal Institute for Occupational Safety and Health, (BAuA) has therefore initiated and sponsored a comprehensive Evaluation of the Tier 1 Exposure Assessment Models (ETEAM). Carried out by the Institute of Occupational Medicine (IOM Edinburgh) and the Fraunhofer Institute for Toxicology (ITEM Hannover) the ETEAM project was intended to compare and contrast the different REACH Tier 1 exposure assessment models using an integrated approach (s. fig1) that includes a conceptual evaluation, an external validation exercise and a between user reliability (BURE) / user-friendliness study. An international Advisory Board has provided objective scientific advice to the project and made available workplace exposure data for use in the external validation process.
The results of the ETEAM project will assist industry and registrants to choose the most appropriate model for a given exposure situation. In addition, its results will help authorities to assess whether or not an exposure scenario presented by a registrant is safe and to estimate how conservative the exposure estimates are. Finally the results are intended to identify areas of concern where model developers are encouraged to revise and improve the models.
Fig.1: Integrated approach of the ETEAM study

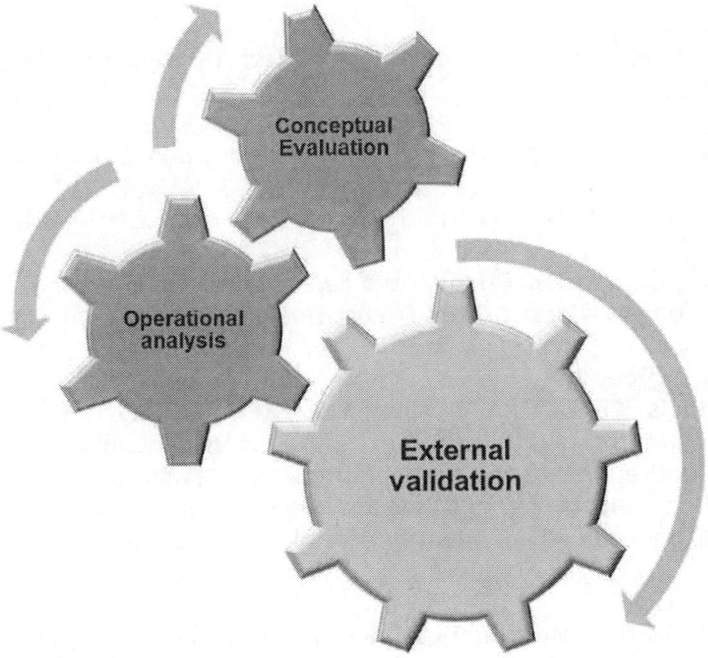

Integrated approach of the ETEAM study

We-SY-B3.3

Conceptual Evaluation and Uncertainty of Tier 1 Exposure Assessment Models Used Under REACH

Susanne Hesse, *Fraunhofer Institute for Toxicology and Experimental Medicine, Hannover, Germany*
Stefan Hahn, *Fraunhofer Institute for Toxicology and Experimental Medicine, Hannover, Germany*
Judith Lamb, *Occupational Health and Safety Services, Edinburgh, United Kingdom*
Martie van Tongeren, *Institute of Occupational Medicine, Edinburgh, United Kingdom*

Aims and background:
The tier 1 tools ECETOC TRA, MEASE, EMKG-EXPO-TOOL, STOFFENMANAGER© and RISKOFDERM are frequently used in exposure assessments under REACH. In this context the ETEAM project represents the first comprehensive evaluation of these tools. The conceptual and uncertainty analyses are two parts of this project, which aim to describe the models' background and identify possible reasons for a deviation between estimate and reality.
Methods and tasks:
In the course of the conceptual evaluation a general evaluation of the models' concepts was done, including a description of the tools, their design and use as well as their historical background. The algorithms, underlying principles and data were described. An applicability matrix was developed, which summarises the models' scope and can be used to identify appropriate models for different exposure situations. A so-called usemap was created that facilitates the conversion of the different use categorisation systems into each other (e.g. DEO units, PROCs).
In the course of the uncertainty analysis different aspects were evaluated which may lead to an uncertainty of the model estimate, i.e. a difference between estimate and experimental exposure value. These sources of uncertainty include assumptions within the model algorithm, but also omitted influences. Sources of uncertainty can also be the model's input parameters, i.e. their definition within the model or their model inherent reflection, e.g. efficiency of ventilation.
All identified sources of uncertainty were categorised as far as possible according to transparency, knowledge base, input parameter quality and their effect on the exposure estimate. Results were collected in an evaluation matrix and discussed in a qualitative way.
Results and conclusion:
It can be summarised that based on the models' concept alone no recommendation of a "best" model can be made as the models show very different scopes and designs. All models are uncertain depending on situation and substance assessed. Many input parameters show a high vagueness which may lead to high variability concerning their assignment (e.g. use category, intrinsic dustiness).

We-SY-B3.4

Validation and between-user variability of tier 1 exposure models

Martie Van Tongeren, *Institute of Occupational Medicine, Edinburgh, United Kingdom*
Judith Lamb, *Institute of Occupational Medicine, Edinburgh, United Kingdom*
Karen Galea, *Institute of Occupational medicine, Edinburgh, United Kingdom*
Laura MacCalman, *Institute of Occupational Medicine, Edinburgh, United Kingdom*

Risk assessment under REACH follows a tiered approach in which the first tier should provide a conservative system that can discriminate between substances in scenarios of concern and those which are considered safe. Several 1st tier assessment tools such as ECETOC TRA, MEASE, EMKG-EXPO-TOOL, STOFFENMANAGER and RISKOFDERM are recommended by the European Chemicals Agency (ECHA) for estimating occupational exposure. In this paper we present results of comparison of model estimates with measured exposure levels and the between-user reliability of the different Tier 1 tools. Measurement data with descriptions of exposure situations and were obtained from providers in Europe and the US. Information on the exposure situation was used to generate exposure estimates using different tools. The level of conservatism was determined by the fraction of the measurement that exceeded the tool estimates (high: ≤10%; medium 11≤25% and low >25%). The impacts of various exposure determinants as implemented in the tools were investigated using linear mixed effects statistical models. Differences in the level of conservatism for all of the tools were observed between exposure category, PROC codes, data providers and the presence/ absence of local exhaust ventilation. Correlations between the measurement results and tool predictions were generally stronger for powders and non-volatile liquids than for the other exposure categories.
The between-user reliability for the Tier 1 tools was investigated using a remote-completion exercise and focus group. Tool parameters and other factors potentially associated with between-user variability, for example user demographics and previous exposure assessment and tool-use experience, were identified and evaluated. In the remote-completion exercise, participants (N=146) generated dermal and inhalation exposure estimates (N=4066) from a defined set of exposure situation descriptions/Tier 1 tool combinations over a fixed time period. Qualitative information on decision-making processes associated with tool use was collected during the focus group. Significant variation was observed between users when selecting task/ activity, dustiness and risk management measures within the tools. Considerable variability in the resultant user-generated exposure estimates for the same situation was observed, which appeared to be unrelated to user characteristics.
The results of these analyses show that tools are generally conservative, although some tools may not be sufficiently conservative for all types of exposures and exposure situations. More importantly, the high level of between-user differences suggests that quality control / assurance procedures are important when using these exposure tools.

We-SY-B3.5

Implications of the eteam project results

Martin Tischer, *Federal Institute for Occupational Safety and Health (BAuA), Dortmund, NRW, Germany*

Overall the comparison of the tool estimates with measurement data within ETEAM suggests that whilst the tools tend overall to be conservative, they may not be sufficiently conservative in all situations. In addition the statistical analyses of the between user reliability study (BURE) results suggest that when presented with brief, identical descriptions of exposure situations, user variation in the choice of input parameters can lead to very different results. Both underestimation of exposure and the impact of user variation could have serious consequences. Workers' health might be at risk, if an exposure scenario is incorrectly diagnosed as 'safe'. The economic situation of organizations could be unnecessarily burdened if an exposure scenario is incorrectly diagnosed as 'unsafe', which could lead to costly over-engineering.
As a consequence more confidence in the level of conservatism and accuracy of the model may be necessary. The registrant can help to reduce the uncertainty within risk characterisation by comparing the estimates from a range of sources, including other tools and measured data. For competent authorities REACH offers a regulatory basis to request such independent measurement data by way of the substance evaluation process. It should be highlighted therefore, that member states competent authorities should be aware of exposure estimates that may underestimate exposure, whereby the need for further investigation increases if the risk characterisation ratio approaches 1. In consequence the risk assessment is always a trade-off between uncertainty and level of required conservatism that should be considered in the substance evaluation process. Further tool developments and improvements should consider user friendliness implications, the ability of users to choose the correct input parameters and the level of detail that the tool provides. The BURE has shown that there are some parameters which are prone to induce a high level of variability due to their vague definition. In particular these are: the use categorisation for all tools, the intrinsic dustiness which is defined qualitatively, the type of setting (professional/ industrial) and the definition of risk management measures. The resulting variability can potentially be decreased in different ways. Obviously, the definition of the corresponding parameters should be as precise as possible to reduce the need for subjective interpretation. However, the knowledge of the user about their tool is also of high relevance, therefore to decrease the total level of uncertainty, it is crucial that they are well informed about both the models and the situations that will be assessed.

We-SY-B3.6

External validation of exposure assessment tools used under REACH

Eun Gyung (Emily) Lee, *National Institute for Occupational Safety and Health, Morgantown, West Virginia, United States*
Judith Lamb, *Institute of Occupational Medicine, Edinburgh, United Kingdom*
Nenad Savic, *Institute for Work and Health (IST), Epalingen-Lausanne, Switzerland*
Bojan Gasic, *Swiss State Secretariat for Economic Affairs (SECO), Zurich, Switzerland*
Christian Jung, *Federal Institute for Occupational Safety and Health (BAuA), Dortmund, Germany*
Martin Tischer, *Federal Institute for Occupational Safety and Health (BAuA), Dortmund, Germany*
Jongwoon Kim, *KIST Europe, Korea Institute of Science and Technology, Saarbruecken, Germany*
Martin Harper, *National Institute for Occupational Safety and Health, Morgantown, West Virginia, United States*

Exposure assessment tools used under REACH were evaluated using measurement data. Five tier 1 models: the ECETOC TRAv2, TRAv3, EMKG-EXPO-TOOL, MEASEv1.02.01, Stoffenmanager®v4.5, and one tier 2 model, the Advanced REACH Tool (ART), were included. Sixty-seven Exposure Situations (ES) based on tasks/chemicals were developed from NIOSH field surveys to collect workers' exposure measurements, and circulated to seven organizations (from the US and Europe) for coding of the situations into the tool parameters. Input parameters for each model were then agreed and each model was used to generate estimates of exposure for each ES. The exposure data and model estimates were compared in six categories: Aqueous solutions (n=4; for MEASE, Stoffenmanager, and ART), Liquids with a vapor pressure (VP) ≤ 10 Pa at room temperature (n=5; for all but MEASE), Liquids with a VP > 10 Pa at room temperature (n=419; for all but MEASE), Metal processing (n=15; only for MEASE), Powder handling (n=20; for all models), and Solid objects (n=20; for all models). The level of conservatism of the model estimates were defined as high, medium, and low if the proportion of exposure measurements (%M) exceeding the model estimates (T) was ≤10%, 11≤25%, and >25%, respectively. The comparison was made using T derived from the point estimates for the TRAv2, TRAv3 and MEASE, the upper range value for the EMKG-EXPO-TOOL, and 90th percentile value for the Stoffenmanager and ART. Overall, the level of conservatism was in the order of the EMKG-EXPO-TOOL (%M>T=2%; highest), Stoffenmanager (%M>T=8%), TRAv2/MEASE (%M>T=9%), TRAv3 (%M>T=26%), and ART (%M>T=59%; lowest). All tier 1 models exhibited high levels of conservatism except for the following categories: Liquids with a VP > 10 Pa for TRAv3 (%M>T=29%) and Solid objects for MEASE (%M>T=25%) and Stoffenmanager (%M>T=100%). The TRAv3 was less conservative than the TRAv2. Stoffenmanager resulted in 100% of %M>T for the solid objects because of an assumption of zero emissions from solid objects. The ART tool resulted in low levels of conservatism for all exposure categories showing a range of %M>T from 30% to 80%. The study findings clearly suggest needs of improvements for each model. Although the present study covers a broader range of exposure situations, still further validation studies are necessary, especially for those categories with insufficient data.

We-SY-C3: Wristband Samplers Advancing Chemical Exposure Science - I

We-SY-C3.1

The wristband sampler saga

Kim A. Anderson, *Oregon State University, Corvallis, Oregon, United States*
Glenn R. Wilson, *Oregon State University, Corvallis, Oregon, United States*
Richard P. Scott, *Oregon State University, Corvallis, Oregon, United States*
Steven G. O'Connell, *Oregon State University, Corvallis, Oregon, United States*

The nature of the silicone wristband sampler will be explored. Fundamentals, uptakes and stability testing for over 100 organic contaminants with the wristband sampler will be described. Organic chemicals that will be discussed include flame retardants, polycyclic aromatic hydrocarbons, oxygenated polycyclic aromatic hydrocarbons, BTEX, alkanes, polychlorinated biphenyls, fragrance and other consumer products, pesticides and phthalates. The volatile and semi-volatile organic chemicals (boiling point <450□C) for a series of stability studies were quantified from the wristband for multiple times (e.g. 3 days, 1 week, 4 weeks, 3 months) and temperatures (e.g. -20, 4, and +30C) and will be reported. Most of the chemicals, over 95%, were within 25% of the original starting concentrations. Comparisons with other approaches, and limitations of the technology will be present.

We-SY-C3.2

Moving Forward: Personal Exposure Monitoring, Citizen Science, and Disaster Research

Aubrey Miller, *National Institutes of Health, Bethesda, Maryland, United States*
Liam O'Fallon, *NIEHS, Dur, N, United States*
Joseph Hu, *N, Durham, N, United States*
April Bennett, *N, Bethesda, Maryland, United States*
Richard Kwok, *NIEHS, Du, No, United States*
Leslie Reinlib, *NIE, Du, No, United States*
David Balshaw, *NI, Du, N, United States*

Background
Responses to various disasters and emerging threats including the World Trade Center attack, Gulf Oil Spill, Superstorm Sandy, and the Ebola outbreak have revealed the dire need for improved ability to perform rapid data collection and research for such events. As such, the National Institutes of Health (NIH) Disaster Research Response Program (DR2) was created to build and promote tools, processes, and relationships to collect vital exposure and health information in response to environmental disasters. Additionally, the advent of new personal exposure monitoring devices such as "wristband samplers" is opening new frontiers for research, including community-engagement and citizen science in response to disasters.
Objectives
Facilitate understanding of the current gaps in critical human health and exposure data needed to inform risk assessment and and applied public for disasters and other emerging threats. Provide information regarding newly evolving programs, tools, and exposure assessment strategies to enhance rapid data collection for time-critical responses. Strengthen awareness of the efforts and challenges associated with the implementation worker and community data collection and environmental data management.
Methods
The National Institute of Environmental Health Sciences (NIEHS) intramural and extramural research responses to situations such as the Gulf Oil Spill, hydraulic fracturing, ebola epidemic, and other events will be used to highlight various efforts to implement timely health and exposure data collection, including the use of mobile devices, portable samplers, and the inclusion of citizen science. Additionally, NIEHS reviewed hundreds of articles and websites related to disasters to identify and make publicly available data collection tools for use by the research community. For identified tools, metadata was also developed to help researchers review and understand the utility of the various tools for differing situations.
Results
The NIEHS program has created a publicly accessible repository of over 165 questionnaires and data tools used in past disasters for use in future situations. Additionally, a novel human subject reviewed protocol that can be rapidly used for future disaster situations has been developed. This protocol also includes the ability to perform medical testing and the collection of exposure data, including, biospecimens. Large-scale tabletop exercises and new networks linking academia, public health officials, and impacted communities have also been created to test initiatives and to implement enhanced environmental health and exposure research in response to emergencies. As such, the research community is now poised to begin moving into a new era of “strategic science” as part of disaster response efforts.

We-SY-C3.3

Using Simple Wristband Samplers to Detect Chemical Exposures, Engage Citizen Scientists, and Inform Policy

Lindsay McCormick, *Environmental Defense Fund, Washington, DC, United States*
Jennifer McPartland, *Environmental Defense Fund, Washington, DC, United States*
Christopher Portier, *Environmental Defense Fund, Thun, Switzerland*
Sarah Vogel, *Environmental Defense Fund, Washington, DC, United States*

Aim: Environmental Defense Fund (EDF) works to better understand health impacts from environmental exposures in order to drive health protective chemicals policies. EDF is interested in innovative approaches to improve the real-world exposure knowledgebase. Here we discuss a pilot project that explored the use of passive, silicone wristband samplers to characterize individual chemical exposures and our future planned activities. The objective of the pilot was to explore the functionality and utility of the wristbands and develop effective risk communication approaches. EDF's broader goals are to 1) help fill knowledge gaps on individual chemical exposures, 2) raise awareness through active engagement with citizen scientists, and 3) use the generated data to inform policy.
Methods: EDF deployed passive, silicone wristband samplers to 28 non-random volunteers (worn continuously for one week). Oregon State University performed a qualitative analysis for the presence of 1,418 chemicals and a quantitative analysis of a panel of 40 flame retardants. EDF developed individualized reports that were electronically delivered and verbally reviewed during individual in-person or phone meetings.
Results: In total, 57 chemicals were detected in the qualitative analysis across all wristbands (average 15 chemicals/wristband). Detected chemicals included polycyclic aromatic hydrocarbons, flame retardants, synthetic and natural fragrances, pesticides, preservatives and plasticizers. There was considerable overlap in chemical exposures across participant wristbands; for example, the synthetic fragrance galaxolide was detected in every wristband. The quantitative flame retardant analysis identified 12 distinct compounds across all the wristbands, including PBDEs and other halogenated flame retardants. Exposure levels varied greatly; among wristbands with detects, PBDE 49 had the narrowest range of detection (3x concentration difference) and PBDE 99 had the largest range (255x concentration difference).
Conclusions: While this was not a random sample of individuals, we found the wristbands to be a highly effective engagement tool. Participants reported high compliance, increased awareness, and a desire to learn more. We experienced several challenges in risk communication owing to data gaps on exposure sources, current inability to derive external to internal dose estimations, and lack of safe/regulatory action levels for many of the detected chemicals. We found comparing individual results to the group to be an effective communication method. We are exploring future projects that would engage a geographically-diverse network of "citizen scientists" to collect and share chemical exposure data. Possible policy uses of these projects include informing prioritization of chemicals for targeted assessment and evaluating the effectiveness of chemical exposure mitigation strategies.

We-SY-C3.4

Assessing preschool children's exposure to flame retardants, using silicone wristbands, and links with teacher-rated social behaviors

Molly Kile, *Oregon State University, Corvallis, OR, United States*
Shannon Lipscomb, *Oregon State University, Bend, OR, United States*
Megan MacDonald, *Oregon State University, Corvallis, OR, United States*
Megan McClelland, *Oregon State University, Corvallis, OR, United States*
Richard Scott, *Oregon State University, Corvallis, OR, United States*
Steven O'Connell, *Oregon State University, Corvallis, OR, United States*
Kim Anderson, *Oregon State University, Corvallis, OR, United States*

Background: Young children are exposed to a mixture of flame retardants. Silicone wristbands can be used as passive sampling tools for measuring personal environmental exposure to organic compounds. There is also concern that some flame retardants negatively impact neurocognitive development.
Methods: We recruited a cohort of 92 preschool aged children (3-5 years) in the state of Oregon to wear a silicone wristband passive sampling device for one week. The wristbands were analyzed on an analytical method that could detect 41 different flame retardant compounds including brominated diphenyl ethers (BDEs) and organophosphate flame retardants (OPFRs). Children's social behaviors were rated by their preschool teachers using the Social Skills Improvement System Rating Scale. Covariates were measured through a family survey and included child age, gender, family context (parent education, employment, income, and home learning environment), and adverse experience (e.g. lived with family member with substance abuse or mental illness, experienced violence or trauma, neglect, or witnessed domestic violence).
Results: Seventy-seven caregivers returned the wristbands for analysis of 35 PBDEs, 4 OPFRs, and 2 other brominated flame retardants. A total of 20 compounds were detected above the limit of quantitation during the 7 day exposure assessment period. Multiple regression analyses (controlling for child age, gender, family context, and adverse experience) indicated that total polybrominated diphenyl ether exposure was linked to children's poorer assertion skills on the teacher-rated scale. Total organophosphate flame retardant exposure was linked to children's lower responsibility and higher externalizing behaviors based on teacher ratings.
Conclusions: This descriptive cross sectional study showed that the vast majority of preschool children tolerated the silicone wristband sampler and that they were exposed to a mixture of volatilized PBDEs and OPFRs. A dose-response relationship was observed between total exposure to flame retardants and poorer social skills (e.g., lower assertion, responsibility and higher externalizing behaviors) observed in preschool. Further studies are warranted that would identify sources of exposure and further explore a potential causal relationship between flame retardant mixtures and social behaviors in children.

We-SY-C3.5

Quantifying Exposure to Flame Retardants and Polyfluorinated Compounds using Silicone Wristbands and Handwipes

Stephanie Hammel, *Duke University, Durham, North Carolina, United States*
Kate Hoffman, *Duke University, Durham, North Carolina, United States*
Craig Butt, *Duke University, Durham, North Carolina, United States*
Rebecca Siebenaler, *Duke University, Durham, North Carolina, United States*
Rochelle Cameron, *Duke University, Durham, North Carolina, United States*
Thomas Webster, *Boston University, Boston, Massachusetts, United States*
Kim Anderson, *Oregon State University, Corvallis, Oregon, United States*
Heather Stapleton, *Duke University, Durham, North Carolina, United States*

Flame retardants (FRs) and polyfluoroalkyl substances (PFASs) are semi-volatile compounds that partition in varying degrees between gas phase and particles. Applied to consumer products, they are emitted over time and are exposure sources in indoor environments. Recently, we experimented with handwipes as metrics of exposure to FRs and PFASs. Silicone wristbands have also been used to characterize exposure to volatile and semi-volatile contaminants, including FRs. Here, we compared the utility of handwipes and wristbands for predicting internal dose and exposure to FRs and PFASs. Two cohorts of 40 participants each were recruited in 2015 to examine exposure to FRs and PFASs, separately. We collected urine to evaluate FR metabolites and serum samples to examine PFASs. Two FRs measured on wristbands, tris(1,3-dichloroisopropyl)phosphate (TDCIPP) and tris(1-chloro-2-propyl)phosphate (TCIPP), were more highly correlated with their corresponding urinary metabolites (rs=0.6, p=0.001) compared to handwipes. This provides a strong indication that FR concentrations captured by wristbands are representative of internal dose. Three of four FR compounds (TDCIPP, TCIPP, mono-substituted isopropyl triaryl phosphate) analyzed on wristbands were also associated with handwipe levels (rs=0.3-0.7, p=0.05). Decabromodiphenyl ether (BDE-209) was measured on all wristbands (geometric mean (GM)=47.9 ng/band) despite having low vapor pressure (estimated 9.02E-13 Pa), suggesting that wristbands capture particle-associated chemicals. 6:2 and 8:2 flurotelomer alcohols (FTOHs) and fluorotelomer diphosphate esters (diPAPs) were commonly detected on wristbands (GM=1.1-190.4 ng/band) and handwipes (GM=0.9-19.3 ng/wipe). We did not observe significant correlations with serum levels of perfluorinated carboxylic acids (e.g. PFOA) or sulfonic acids (e.g. PFOS). However, 6:2 FTOH, 6:2 diPAP, and 8:2 diPAP on wristbands were highly correlated with handwipe levels (rs=0.4-0.8, p=0.05). Our results demonstrate that both handwipes and wristbands are useful in assessing inhalation and dermal exposure to FRs and PFASs; however, wristbands may be more useful in predicting internal dose than handwipes.

We-SY-D3: UBA HBM Colloquium I - Human Biomonitoring in International Population Studies Improving our Knowledge of Environmental Public Health

We-SY-D3.1

Describing exposures to pesticides in French pregnant women: results from the perinatal component of the French HBM program based on the Elfe cohort

Clémentine Dereumeaux, *Santé publique France, Saint-Maurice, France*
Vérène Wagner, *Santé publique France, Saint-Maurice, France*
Sarah Goria, *Santé publique France, Saint-Maurice, France*
Perrine de Crouy-Chanel, *Santé publique France, Saint-Maurice, France*
Clémence Fillol, *Santé publique France, Saint-Maurice, France*
Sébastien Denys, *Santé publique France, Saint-Maurice, France*

Pesticides are extensively used in France for both agricultural and residential uses. Although results of epidemiological studies are still controversial, exposures of pregnant women to environmental pesticides is suspected to have adverse effects on pregnancy outcomes and infant development. However, little is known about impregnation levels by pesticides among French pregnant women and the potential sources of exposure, including the residential proximity to crops.
In this context, Santé publique France, the French national public health agency, has implemented a perinatal component as part of the French human biomonitoring (HBM) program. The aim of this study was to describe impregnation levels by various chemicals (metals, bisphenol A, phthalates, pesticides and persistent organic pollutants) and to identify their determinants, in French pregnant women. The presentation will focus on outcomes concerning pesticides.
The study population was based on a random selection of 1 077 mothers who have been enrolled in the Elfe cohort (the French Longitudinal Study since Childhood) in 2011. Exposure biomarkers of pesticides (metabolites of atrazine, glyphosate, propoxur, chlorophenols, dialkylphosphates and pyrethroids) were measured in spot urine samples collected from pregnant woman just after her admission to the maternity unit for delivery. Simultaneously, data about potential sources of exposure to pesticides during pregnancy related to food intakes and life style characteristics were collected, as well as sociodemographic and anthropometric characteristics. The presence of crops in the vicinity of pregnant woman's municipality of residence was also used to identify the determinants of exposure to pesticides.
In this study, metabolites of pyrethroids were quantified in all French pregnant women, with the exception of 4-F-3-PBA. One out of two pregnant women had quantified levels of dialkylphosphates, however percentages of quantification were lower for propoxur (and its metabolite, 2-IPP), chlorophenols and herbicides (atrazine and metabolites, glyphosate and AMPA). The results of the study have shown that pyrethroids levels increased with the domestic use of pesticides during pregnancy (insecticides, anti-lice and anti-mite), the consumption of tobacco and alcohol. The possible presence of crops close to the place of residence was also related to higher pyrethroids levels. However interpretation of these findings warrants caution because of potential misclassification of exposure due to the short half-life of pesticides in the human body, glyphosate in particular.

For the first time in France, this study provides a national representative description of impregnation levels by pesticides among French pregnant women and their determinants. These results will provide relevant information for Public Health actors.

We-SY-D3.2

The German Human Biomonitoring Program: a Powerful Tool for Accomplishing Public Health Tasks

André Conrad, *German Environment Agency (UBA), Berlin, Germany*
Christine Schulz, *German Environment Agency (UBA), Berlin, Germany*
Maria Rüther, *German Environment Agency (UBA), Berlin, Germany*
Hans-Wolfgang Hoppe, *Medical Laboratory Bremen, Bremen, Germany*
Steffen Uhlig, *QuoData, Dresden, Germany*
Marike Kolossa-Gehring, *German Environment Agency (UBA), Berlin, Germany*

Introduction:
As surveillance tool for human exposure to chemicals and other substances, human biomonitoring (HBM) serves important public health tasks: HBM can be applied for elucidating associations between exposure and health issues, gaining early warnings on public health problems, providing the scientific basis for public health strategies, evaluating the impact of policy actions, monitoring the degree of achievement of public health objectives, and facilitating priority-setting. This is demonstrated by results of the German HBM Program of the German Environment Agency (UBA).
Methods:
The German HBM program consists of the German Environmental Survey (GerES) and the Environmental Specimen Bank (ESB). GerES is a cross-sectional population study carried out repeatedly since 1985. In addition to HBM GerES comprises indoor and drinking water monitoring as well as extensive interviews. Since 1985, the ESB collects human samples from 20-29 years old participants on a yearly basis which are cryo-archived and (retrospectively) analyzed for various pollutants.
Results:
ESB data documents an increase of the glyphosate background exposure in Germany from 2001 to 2012: The fraction of quantifiable concentrations in 24 h-urine increased from 10 % to almost 60 %. The subsequent decrease to 40 % in 2015 might indicate the impact of changes of glyphosate application on the human exposure. This, however, needs to be confirmed by ongoing ESB and GerES monitoring.
GerES results on the association between urinary Hg concentrations and the number of amalgam fillings triggered a public health recommendation of the German Federal Health Office in 1992 to consider dental amalgam cautiously for children. Also against this background, the fraction of 6 to 14 years old German children exceeding health-based HBM assessment values for Hg in urine decreased from approx. 2 % in 1990/92 to almost 0 % in 2003/06. Statistical analysis of ESB data reveals regional differences in the overall decrease in amalgam fillings in Germany. As dental amalgam became less relevant, food consumption gains relatively more importance for the internal Hg exposure.
Time-trends of perfluorinated compounds in ESB blood plasma samples collected from 1982 to 2010 document changes in human exposure and confirm i. a. the effect of voluntary and regulatory action. However, as current GerES participants still exceed the health-based HBM assessment values for PFOS and PFOA, a further reduction of the exposure in Germany is necessary.
Conclusions:
By way of various examples, GerES and ESB demonstrate how HBM serves key public health tasks. The German HBM program underlines numerous health gains due to environmental regulation. On the other hand, both studies timely reveal needs for increased attention

and/or further action in environmental policy-making for maintaining and improving health and wellbeing in Germany.

References:

WHO 2016, Public health surveillance. http://www.who.int/topics/public_health_surveillance/en/
UBA 2016, German Environmental Specimen Bank. https://www.umweltprobenbank.de/en/
UBA 2016, German Environmental Survey. http://www.uba.de/geres/

We-SY-D3.3

NHANES: Biomonitoring experience and results

Antonia Calafat, *US Centers for Disease Control and Prevention, Atlanta, United States*

Biomonitoring provides a quantitative measure of the amount of a given chemical present in the human body. Biomonitoring measures, which integrate all sources and routes of exposure, are increasingly used to estimate human exposures. In particular, biomonitoring programs are useful for investigating human exposure to environmental chemicals. One of these programs, the National Health and Nutrition Examination Survey (NHANES) is conducted annually by the Centers for Disease Control and Prevention. NHANES collects data on the health and nutritional status of the general U.S. population, as well as biological specimens which can be used to assess exposure to select chemicals. NHANES biomonitoring data have shown that exposure to some chemicals is widespread. NHANES data also suggest variability in exposure by sex, age, and race/ethnicity, likely as a result of lifestyle differences. This presentation will provide an overview on the use of NHANES biomonitoring data to establish reference ranges, provide exposure information for risk assessment (e.g., set intervention and research priorities, evaluate effectiveness of public health measures), and monitor exposure trends.

We-SY-D3.4

Biomonitoring as part of exposome measurement in Japan Environment and Children's Study

Shoji F. Nakayama, *National Institute for Environmental Studies, Tsukuba, Japan*

Japan Environment and Children's Study (JECS) is a national birth cohort study launched in 2011 by the Ministry of the Environment. JECS is designed to evaluate the effect of the environment on children's health and development. A total of 103,000 mother-child pairs were registered. Biological samples, such as blood, urine, cord blood, breast milk and hair were collected during pregnancy, at birth and a month after birth from mothers, children and fathers. Questionnaires have been administered to obtain the exposure and health outcome information. Numerical models are used to estimate air pollutants and physical environment (e.g. noise, radiation).
JECS considers the concept of 'exposome' seriously. Every exposure during pregnancy and childhood could affect children's health and development. While JECS considers the environment broadly including chemical, physical and biological factors as well as socio-economic status, behavioural environment and community environment, chemical exposure is one of the main focuses of the study. In JECS, exposures are measured/estimated by a variety of methods including biomarkers, questionnaires, interviews, direct observations, personal sampling, sensing and simulation models. Biomonitoring is a major tool for the chemical exposure measurements.
Biomonitoring for epidemiological studies needs to be very well designed. It is not population based but individual based. The sampling scheme is important but options are limited for large-scale studies. In most cases, only spot samples can be collected. Thus, biological and statistical consideration plays an important role. In the session, JECS approach to the specific chemicals as well as untargeted ones will be presented.

We-SY-D3.5

Canadian Health Measures Survey: Derivation of human biomonitoring reference values for the general population

Douglas Haines, *Health Canada, Ottawa, Ontario, Canada*
Cheryl Khoury, *Health Canada, Ottawa, Ontario, Canada*
Gurusankar Saravanabhavan, *Health Canada, Ottawa, Ontario, Canada*
Kate Werry, *Health Canada, Ottawa, Ontario, Canada*

The interpretation of human biomonitoring (HBM) data and its incorporation in the development of policy decisions is valuable for protecting public health. HBM reference values (RV95), defined as the 95th percentile of the measured chemical concentration of the reference population, within the 95% confidence interval, is an instrument enabling such interpretation. RV95s indicate the upper bound of background exposure to a given chemical at a given time. The nationally representative Canadian Health Measures Survey (CHMS) is the most comprehensive direct health measures survey conducted for the general population in Canada. The CHMS is ongoing and, to date, HBM data for 176 chemicals in blood and urine, including metals and trace elements, persistent organic pollutants (POPs) and non-persistent chemicals and corresponding questionnaire and health biomarker information are available from three cycles of the CHMS (2007-2013). We report the RV95s developed for a range of environmental chemicals measured as part of the CHMS.

We used a systematic approach based on the reference interval concept proposed by the International Federation of Clinical Chemistry and Laboratory Medicine and the International Union of Pure and Applied Chemistry to derive RV95s for chemicals using the latest CHMS biomonitoring data. Biomarkers were chosen based on specific selection criteria, including widespread detection in Canadians (≥ 66% detection rate). For each chemical, a reference population was constructed based on an a posteriori selection approach with specific criteria for exclusion (e.g. smoking, seafood consumption, fasting) and partitioning (age, sex) of the data. Separate RV95s were derived for sub-populations in cases where partitioning was deemed necessary. RV95s were computed for 12 metals and trace elements in blood and 14 in urine, for 21 POPs in blood plasma, and for 41 non-persistent chemicals including six in blood, 33 in urine and two haemoglobin adducts. RV95s ranged as follows: metals and trace elements in blood from 0.18 µg/L (cadmium) to 7900 µg/L (zinc), in urine from 0.17 µg/L (antimony) to 1400 mg/L (fluoride); POPs in plasma from 0.018 µg/L (PCB 201) to 21 µg/L (perfluorooctane sulfonate); non-persistent chemicals in blood from 0.072 µg/L (o-xylene) to 0.21 µg/L (toluene), in urine from 0.063 µg/L (4-hydroxyphenanthrene) to 790 µg/L (triclosan), and haemoglobin adducts between 100 pmol/g Hb (acrylamide) and 130 pmol/g Hb (glycidamide).

These RV95s are the first reference values derived for the general Canadian population. RV95s are statistical values used strictly as indicators of exposure. Because toxicological information of the environmental chemicals is not incorporated in their derivation, RV95s cannot be used directly to predict any adverse health outcomes in the population. Nonetheless, they provide a reference point against which individual and population HBM results from other surveys and studies can be compared. RV95s are not fixed but can be updated using HBM data from future cycles of the CHMS.

We-SY-E3: Exposure to SVOCs in the Indoor Environment - Products, Emissions, Exposure, Pharmacokinetics and Biomarkers - III

We-SY-E3.1

Using Ultrafine Particles as a Metric for Characterizing SVOC Contamination of Surfaces

Lance Wallace, *None, Santa Rosa, CA, United States*
Charles Weschler, *Rutgers Univ, New Brunswick, New Jersey, United States*
Wayne Ott, *Stanford Univ, Palo Alto, Ca, United States*
Alvin Lai, *City University, Hong Kong, China, People's Republic of*

Background: A 2014 study (1) suggested that ultrafine particles (UFP) from heated metal surfaces may be created by desorption of SVOCs from the surfaces, followed by nucleation as the air diffusing from the heat source cools and becomes supersaturated with vapor-phase SVOC molecules.
Objectives:
1. Test this theory by applying to surfaces other than metal, including porcelain and glass
2. Estimate the buildup over time of SVOCs encountered in a residence
3. Consider transfer of skin oil to cooking pans as determined by UFP counts
Methods: Use electric burner or laboratory hot plate to heat cooking pans, Petri dishes, and aluminum foil, measuring UFP by a condensation particle counter (CPC) and a Scanning Mobility Particle Sizer (SMPS) to provide size-resolved emissions as a function of temperature and time exposed to indoor air.
Results: Most of the surfaces tested could be driven to near-zero particle production following repeated heating to temperatures in the range of 150-300 degrees C. Aluminum foil from inner portions of a newly purchased roll appeared to be free of SVOC contamination. Newly purchased Petri dishes had varying amounts of contamination, sometimes near-zero. These "clean" surfaces were then exposed to indoor air for increasing periods of time up to 150 days. Total mass produced ranged from 500 µg for longer exposures. Longer exposures shifted the UFP size distribution to the right (from modes of 5 nm to >50 nm). Total particle concentrations in a 25.8 m3 room ranged from a few thousand to more than a million per cubic centimeter. Washing pans with detergent produced no particles if sterile gloves were employed, but copious particles if bare hands were employed.A single thumbprint on a previously cleaned (by repeated heating) pan could produce one million particles, although most were greater than10 nm in diameter and thus had negligible (greater than 0.1 µg). Multiple thumbprints were capable of producing one hundred million particles and greater than 100 µg mass. (1) Wallace, L.A., Ott, W.R., and Weschler, C.J. (2014) Ultrafine particles from electric appliances and cooking pans: experiments suggesting desorption/nucleation of sorbed organics as the primary source. Indoor Air 2015:536-546.

CPC, SMPS, and Petri dishes set out for exposure to SVOCs

UTRECHT
THE NETHERLANDS
OCTOBER 9-13
ISES 2016

We-SY-E3.2

Contribution of Dermal Absorption to Body Burdens of SVOCs: Absorption from Air vs. Absorption from Skin Surface Lipids

Charles Weschler, *Rutgers University, Piscataway, New Jersey, United States*

Aim: This presentation will contrast SVOC uptake directly from room air and SVOC uptake from skin surface lipids following contact transfer with contaminated surfaces.

Methods: kp_b is the permeability coefficient that applies to the transport of an SVOC from air at the skin's surface through the stratum corneum/viable epidermis composite (sc/ve) to dermal capillaries, while kp_l is the permeability coefficient that applies to the transport from skin surface lipids through the sc/ve to dermal capillaries. Partitioning between surface lipids and interfacial air is describe by the coefficient Klg and occurs relatively quickly. Klg can be used to relate kp_b and kp_l:

$$kp\_b = kp\_l \times Klg$$

kp_g is the permeability coefficient that applies to the transport of an SVOC from room air through the layer of air adjacent to skin and then through the sc/ve to dermal capillaries. The coefficient for mass transport through the layer of air adjacent to the skin is denoted as hm (typically ~ 6 m/h). kp_g can be estimated with a resistor in series approach:

$$1/kp\_g = 1/hm + 1/kp\_b$$

Using methods described in Weschler & Nazaroff (Atmos Environ 2008), kp_b, kp_l, kp_g and Klg have been calculated for SVOCs commonly found indoors.

Results: SVOCs can be divided into three categories with respect to kp_b:

i) kp_b > 500 m/h. In this category resistance across the air adjacent to skin is much larger than across the sc/ve. Hence kp_g ~ hm, and, for equivalent SVOC activities in room air and skin surface lipids, uptake from air occurs at a much slower rate than from surface lipids.

ii) 500 m/h > kp_b > 0.2 m/h. In this category kp_g is impacted by both resistance across the air layer and resistance across the sc/ve. For equivalent SVOC activities, uptake from room air occurs at a somewhat slower rate than from surface lipids.

iii) kp_b < 0.2 m/h. In this category resistance across the air layer is much smaller than resistance across the sc/ve. Hence kp_g ~ kp_b, and, for equivalent SVOC activities, uptake from air occurs at the same rate as from surface lipids.

Conclusions: For SVOCs with large permeability coefficients, if their activities are equivalent in room air and surface lipids, dermal uptake from surface lipids is much larger. Given kinetic constraints, contact transfer from indoor surface films may be necessary to achieve activities in skin surface lipids approaching those in the room air.

We-SY-E3.3

Determination of SVOC Volatilization from Porcine Skin for Assessing Inhalation Exposure Following the Use of Cosmetics: Experimental Study for Decamethylcyclopentasiloxane (D5)

Tatsiana Dudzina,
Elena Garcia Hidalgo, *ETH Zurich, Zurich, Switzerland*
Natalie von Goetz, *ETH Zurich, Zurich, Switzerland*

Background
Leave-on cosmetic ingredients of low or moderate volatility are SVOCs. Consumer exposure to these substances is often assumed to occur primarily via dermal absorption. In reality they may volatilize from skin and represent a significant source for inhalation exposure. Often, evaporation rates of pure substances from inert surfaces are used to inform human exposure assessment to product mixtures.
Objectives
We developed a method for measuring chemical evaporation rates of substances in cosmetics under realistic consumer exposure conditions from porcine skin in vitro. For the test substance D5 volatilization was compared between neat substance and two relevant cosmetic formulations from both inert and skin surfaces.
Methods
Series of experiments were carried out in a custom-made ventilated chamber fitted with a vapor trap. Single doses were applied neat and in commercial deodorant and face cream formulations to aluminum foil and porcine skin membranes mounted on static diffusion cells at ambient air (23°C) and skin (32°C) temperature. The condition-specific evaporation rates were determined as the chemical mass loss per unit surface area at 1-1.25 h post dose time intervals. Product weight loss was monitored gravimetrically and the residual D5 concentrations in formulations were analyzed with GC/FID.
Insert attachment
Figure 1: Experimental chamber for determining the evaporation rates
Results
For neat D5, the evaporated mass increased linearly with time. From aluminum surfaces the release of D5 occurred very fast with mean rates of 0.029 mg cm−2 min−1 and 0.060 mg cm−2 min−1 at 23°C and 32°C, respectively. The effect of surface temperature on the evaporation rate was statistically significant. Unlike observed in the experiments with aluminum foil, the mean evaporation rates from porcine skin in vitro were similar for neat and formulated forms of D5 (mean group difference is not statistically significant) and ranged between 0.056 and 0.058 mg cm−2 min−1 (at 32°C). The results for the face cream deviate from the overall trend showing a substantially faster evaporation from skin surface compared to aluminum.
Statistical analysis of experimental data confirmed a significant effect of cosmetic formulations on the evaporation of D5 with the largest effect (twofold decrease of the evaporation rate) observed for the neat/face cream-pair at 32 °C.
The developed method can be used to assess more accurate volatilization rates for dermally applied SVOCs. These rates are important for achieving an appropriate route-apportionment (between inhalation and the dermal route) in exposure modeling for dermally applied SVOCs.

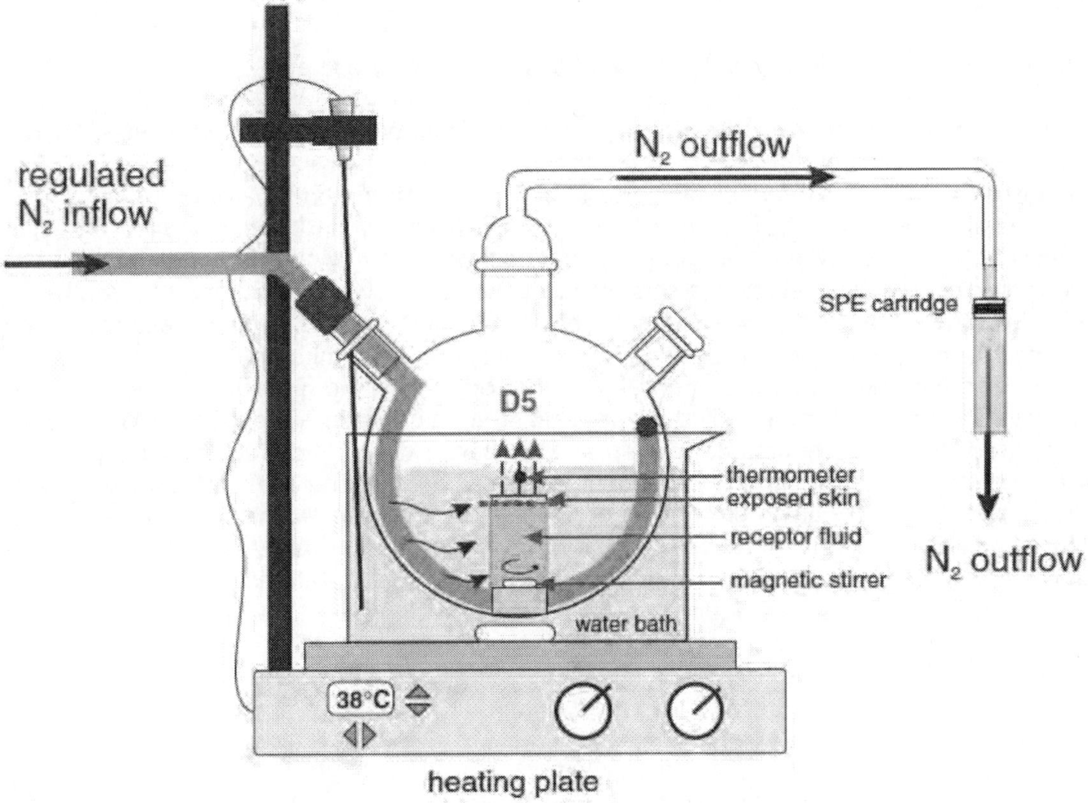

Figure 1: Experimental chamber for determining the evaporation rates

We-SY-E3.5

The role of clothing in dermal uptake of SVOCs from indoor air

Glenn Morrison, *Missouri University of Science & Technology, Rolla, MO, United States*

Analyses of exposure to indoor air pollutants focus primarily on inhalation of gases and particles. However, dermal uptake directly from air has recently been shown to be significant for some semi-volatile organic compounds (SVOCs). We hypothesized that wearing clothing should enhance this effect if the fabric was first allowed to equilibrate with SVOCs present in indoor air. Simple mass transport models demonstrate that close-fitting clothing would reduce external mass-transport resistance and increase uptake relative to bare skin. To experimentally assess the effect, we measured uptake of selected airborne phthalates for an individual wearing clean clothes or air-exposed clothes and compared these results with dermal uptake for bare-skinned individuals under otherwise identical experimental conditions. When compared against the average results for bare-skinned participants, clean clothes were protective, whereas clothes exposed to phthalates dramatically increased dermal uptake of DEP and DnBP. An advanced model of dermal uptake that includes clothing predicts that dermal uptake is most sensitive to a very human variable: the frequency of laundering clothing.

We-SY-E3.6

Towards understanding the role of clothing in human exposure to SVOCs

Miriam Diamond, *University of Toronto, Toronto, Ontario, Canada*
Amandeep Saini, *University of Toronto, Toronto, Ontario, Canada*
Clara Thaysen, *University of Toronto, Toronto, Ontario, Canada*
Rachel McQueen, *University of Alberta, Edmonton, Alberta, Canada*
Liisa Jantunen, *Environment Canada, Egbert, Ontario, Canada*
Joseph Okeme, *University of Toronto, Toronto, Canada*

Estimates have shown that clothing, followed by household textiles, have the largest surface area of all materials indoors. Clothing is unique amongst indoor materials in its intimate role in everyday life. As with other materials indoors, clothing can act as a sink and source with respect to indoor SVOC concentrations. Our aim was to investigate the role of clothing as a sink and source of SVOCs. A series of experiments were conducted to quantify the uptake of selected SVOCs from air to fabrics and then SVOC release during laundering. The results showed the high sorptive capacity of cotton, polyester and rayon for gas- and particle-phase halogenated flame retardants (HFRs) and phthalate esters. Measured fabric-gas phase distribution coefficients were 6.5-7 (log units) after 56 days with modeled equilibrium partition coefficients of 8-12 (log units). Concentrations of Σ5phthalates, Σ10HFRs and Σ8OPEs were 200, 70 and 500 ng/dm2 after 30 days of uptake. Uptake prior to reaching equilibrium appeared to be air-side controlled for gas-phase compounds, but equilibrium is estimated to be reached after >10 years for PBDEs. Uptake rates of 0.4-0.9 m3 air equivalent/day.dm2 fabric translate into the accumulation of SVOCs in 100 m3 of equivalent air per day by 2 m2 of clothing typically worn by a person. Cotton, polyester and rayon accumulated similar masses of halogenated flame retardants (not including organophosphate esters or OPEs) and higher molecular weight phthalates when expressed on a planar basis but cotton had lower concentrations when expressed according to specific surface area, which was high for cotton. Cotton accumulated more OPEs and lower molecular weight phthalates than polyester for which several explanations are offered. Cotton and polyester showed similar release of SVOCs to laundry water where the percentage release was a function of water solubility and KOW. In controlled laundry experiments, release to laundry water was > 80% of OPEs and low molecular weight phthalates, ~50% for OPEs with aromatic structures, and <20% of high molecular weight HFRs and phthalates. In conclusion, these results point to the role of clothing and other textiles in the fate of SVOCs indoors; results suggest that clothing is a continual sink for non-polar SVOCs (e.g., BFRs) and a transient sink for polar SVOCs (e.g., OPEs). The results also suggest different implications for dermal exposure from clothing to HFRs vs OPEs, with phthalates being intermediate.

We-SY-F3: Measuring marijuana exposure in a changing legal landscape

We-SY-F3.1

Exposure to THC in Dutch suspected impaired drivers

Beitske Smink, *Netherlands Forensic Institute, The Hague, Netherlands*

Aim of this presentation is first to describe the legal procedure, sampling and analytical methods that are used to detect exposure to THC in suspected impaired drivers and second, to explain the scientific base of the proposed legal limits for THC.
According to the Dutch Road Traffic Act, it is forbidden to drive under the influence of a substance of which a driver ought to know that it can affect the driving performance negatively. For alcohol, legal limits, related to accident risk, have been laid down. For drugs, no legal limits have been laid down and driving impairment has to be evaluated case-by-case, based on the results of blood analysis. In 2015, the Dutch Road Traffic Act was changed; it is expected to come into force in 2017. The new law comprises limits for the nine most frequently detected drugs in blood as well as the use of oral fluid tests as a screening for drugs. The results of this oral fluid test will have to be confirmed in blood.
After alcohol, the next most frequently detected drug in drivers is cannabis. Based on scientific literature, the proposed legal limit for single cannabis use is 3.0 ng THC/ml whole blood (impairment limit). In case of multi-drug use, the proposed legal limit for THC is 1.0 ng/ml in whole blood (analytical limit). These limits take into consideration that THC may come from passive inhalation of cannabis smoke. The prevalence of THC in blood of suspected impaired drivers was investigated by reviewing the results of the Netherlands Forensic Institute (NFI) during the years 2009-2012. The identification and quantification of THC and metabolites was performed by using a validated UPLC-MS/MS (ultra performance liquid chromatography-tandem mass spectrometry) method. The limit of quantification was 1 μg/L.
THC was demonstrated in 36% (1085/3038) of the blood samples of suspected impaired drivers. In 69% (748/1085) of the cases, no other illicit drug was detected. The most frequently detected combinations of THC and other illicit drugs in drivers were THC and amphetamines 14% (153/1085), THC and cocaine 5.7% (62/1085), THC and GHB 2.7% (29/1085). Alcohol was not included in the review of the cases because the results of the alcohol breath test (if performed) were unknown.
Introduction of threshold values in the law is expected to make prosecution of an impaired driver more efficient, because it will obviate discussions on e.g. circumstances, tolerance and passive inhalation of cannabis.

We-SY-F3.2

Exposures Related to Marijuana use by Smoking, Vaping, and Ingesting

Benjamin Blount, *Centers for Disease Control and Prevention, Chamblee, Georgia, United States*

Abstract: Ten marijuana users were studied in a metabolic ward following controlled smoking, vaping, and ingesting of marijuana products. Study participants were restricted to non-tobacco smokers so that tobacco exposures did not confound marijuana exposures. Urine samples collected before and at multiple times after each exposure will be assayed for cannabinoids (5 metabolites), nicotine (7 metabolites), volatiles (6 metabolites), PAHs (9 metabolites), heterocyclic amines (7 metabolites), and creatinine. Exposure doses will be calculated for cannabinoids, and exposure to harmful combustion products will be discussed in the context of potentially harmful exposures to active users. Potential exposure from secondhand and thirdhand marijuana smoke exposure will also be discussed.

We-SY-F3.3

Marijuana smoke exposure among hospitalized children exposed to tobacco smoke

Karen Wilson, *University of Colorado School of Medicine, Aurora, CO, United States*
Binnian Wei, *Centers for Disease Control and Prevention, Atlanta, GA, United States*
Michelle Torok, *University of Colorado School of Medicine, Aurora, CO, United States*
Langing Wang, *Centers for Disease Control and Prevention, Atlanta, GA, United States*
Michelle Robinson, *Children's Hospital, Colorado, Aurora, CO, United States*
Benjamin Blount, *Centers for Disease Control and Prevention, Atlanta, GA, United States*

Background: Marijuana smoking is becoming increasingly legal in the United States. Most places with legal recreational use do not restrict use in the presence of children, but the prevalence and effects of exposure on children are not well known. Parents who smoke tobacco are also more likely to also use marijuana, increasing the risk for children's exposure to multiple types of smoke.
Objectives: To determine the prevalence of marijuana smoke exposure among children with a parent who smokes tobacco.
Methods: As part of a randomized controlled trial of an inpatient parent smoking cessation intervention, we have recruited 123 children with at least one parent who smokes tobacco. Parents completed a survey about tobacco and marijuana use in the home, as well as collecting the child's health information and demographic characteristics. 101 children provided a urine sample, and there was sufficient remaining for this study from 58; samples were tested for cotinine (LC/MS) and NNAL (LC/MS) at the laboratory at UCSF. Samples from children whose parents consented for future research (N=43) will be anonymized and shipped on dry ice to the laboratory at the Centers for Disease Control and Prevention, and will be analyzed for Δ9-tetrahydrocannabinol (THC) and 11-nor-9-carboxy-THC (COOH-THC) using UHPLC-MS/MS, with limits of detection (LOD) of 0.005 and 0.015 ng/mL, respectively. Chi-square tests will be done in SAS to assess bivariable differences in exposure by demographics, cotinine level, and clinical status.
Results: We expect to find a high prevalence of marijuana smoke exposure in this cohort, and that marijuana levels will correlate strongly with cotinine and NNAL. This information will help us to understand the relationship between tobacco smoke and marijuana smoke exposure in children.

We-SY-F3.4

Exposures Resulting From Active Use of Marijuana via Smoking

Peggy Van der Pol, *Trimbos-instituut, Utrecht, Netherlands*
Raymond Niesink, *Trimbos-instituut, Utrecht, Netherlands*

Background and aims: In the Netherlands, Δ9-tetrahydrocannabinol (THC) concentrations increased from approximately 5-7% in the 1970s and an average 8% in the 1990s up to 20% in 2004 and decreased thereafter (Pijlman et al., 2005; Niessink at al., 2015). Although increased THC concentrations in cannabis may lead to higher THC-exposure, cannabis dependence and treatment need, users self-report to adapt the actual intake of THC through reduced inhalation of THC containing smoke (titration). This is supported by protocolled laboratory studies (Hunault 2008), but users' ability to estimate the dose and potency of their cannabis remains unknown and titration has not been assessed in a naturalistic setting.
Methods:
In a naturalistic experiment, heavy cannabis users (n = 98) brought their own cannabis, rolled a joint and smoked it ad libitum. First, THC concentration of their cannabis and the total amount of cannabis in their joint were objectively measured and compared with self-reported estimates of dose and potency. Second, it was assessed whether those using stronger cannabis used lower doses or inhaled less smoke.
Results: Objective estimates of doses per joint (0.07-0.88 g/joint) and cannabis potency (1.1-24.7%) varied widely. Self-reported measures of dose were imprecise, but at group level, average dose per joint was estimated accurately with the number of joints made from 1 g, whereas a photo card resulted in serious underestimation. THC concentration in cannabis was associated with subjective potency and with cannabis price, but not with level of intoxication.
THC concentration in cannabis was correlated positively with cannabis dose per joint, but the resulting THC concentration per joint (range 0.24-15.72%) was associated negatively with inhalation volume.
Conclusions: Self-report measures relating to cannabis use appear at best to be associated weakly with objective measures. Of the self-report measures, number of joints per gram, cannabis price and subjective potency have at least some validity.
Although more potent cannabis was used in larger doses per joint, cannabis users titrate their delta-9-tetrahydrocannabinol intake by inhaling lower volumes of smoke when smoking strong joints. However, this does not fully compensate for the higher cannabis doses per joint when using strong cannabis. Thus, users of more potent cannabis are generally exposed to more delta-9-tetrahydrocannabinol. (Van der Pol et al., 2013, Addiction, 108, 1801-1808; 109, 1101-1109)

We-SY-F3.5

Diversity of Modes of Exposure to Marijuana in a Sample of US Adult Co-Users of Marijuana and Tobacco

Arno Hazekamp, *Bedrocan Beheer BV, Veendam, Netherlands*
Erica Peters, *Battelle Public Health Center for Tobacco Research, Baltimore, MD, USA*

Aim
Vaporizing has become a popular alternative to smoking for recreational and medicinal consumption of cannabis. Currently available vaporizing devices range from large tabletop units connected to electric outlets, to small battery operated handheld vape-pens. Some vaporizers need to be loaded by the user with herbal cannabis, while others rely on preloaded cartridges filled with so-called e-liquids containing cannabis extracts. Besides the psychoactive cannabis compound THC, increasingly the non-psychoactive cannabinoid CBD is also consumed in this form. Although it is claimed that vaporizing is healthier than smoking, virtually none of the currently available vaporizers, or the cartridges that are used with them, have been tested for any form of safety or efficacy. This presentation will present an overview of scientific data currently available about vaporizing cannabis and cannabinoids.
Methods
This presentation provides an overview of analytical data on the safety, reproducibility and efficacy of vaporizing cannabis for medicinal purpose. This data was obtained by the author during his scientific work over the last 15 years on medicinal cannabis as part of the Dutch Medicinal Cannabis Program. Vaporizers covered include the Volcano Digit and Medic by Storz&Bickel, the MiniVap vaporizer by Hermes Medical Engineering, and various others. Special attention will be given to the chemical composition of the vapor released by vaporizers, and the risks of environmental exposure to secondhand vapor. Parameters studied include type and dose of cannabis used, airflow, temperature setting, and type of heating source.
Results
Based on the lessons learned on vaporizing cannabis and cannabinoids, an optimized strategy is presented for the application of vaporizing for medical use. By a systematic evaluation of tested parameters, evaporation of active cannabis components is maximized, while avoiding the creation of harmful degradation products. Based on the obtained results, a prototype of an optimized vaporizer is currently entering clinical trials, and is being approved as a medical device. These data are important for use of vaporizers in public spaces, as well as for the set-up of clinical trials, where occupational exposure to cannabis compounds is often considered a serious risk by medical staff.
Conclusions
Vaporizing has the potential to replace smoking of cannabis, specifically for medical use. Lessons learned in this arena can be applied to consumption of recreational cannabis use, in order to reduce health risks for cannabis users, and limit secondhand exposure to THC and other harmful compounds.

We-SY-G3: Exposure Science and 21st century oil and gas development - I

We-SY-G3.1

Unconventional Natural Gas Waste Injection and Public Health

Erin Haynes, *University of Cincinnati, Cincinnati, Ohio, United States*

Natural gas extraction has boomed in the last decade due to the new technique of high volume horizontal hydraulic fracturing or unconventional natural gas (UNG) extraction. Approximately five to six million gallons of water are used to fracture each shale gas well, and it's estimated that 30% to 70% returns to the surface. This flow back water or brine contains the complex chemical mixture used for fracturing and also naturally occurring toxicants from underground including metals, volatile organics, and radioactive compounds. These fluids contain toxics that may result in reproductive or developmental toxic exposures if found in drinking water. The primary option for management of UNG extraction waste is underground injection. In Ohio, these well are referred to as Class II wells, which were originally drilled and designed for conventional gas or oil extraction and are being repurposed UNG extraction waste injection. In Ohio there are 214 active Class II injection wells that receive UNG extraction waste In 2014, Ohio injected 924 million gallons of UNG extraction waste fluid into these wells, about half from neighboring states. Very little research has focused on the environmental public health hazards of UNG extraction waste management. This talk will include a summary of the current peer-reviewed literature on UNG extraction waste and the potential environmental public health issues identified in Ohio.

We-SY-G3.2

A summary of fires, explosions, and pollutant releases at Oil and Gas Operations in Colorado from 2000 to 2014.

Benjamin Blair, *University of Colorado, Aurora, CO, United States*
Lisa McKenzie, *University of Colorado Denver, Aurora, CO, United States*
Benjamin Allshouse, *University of Colorado Denver, Aurora, CO, United States*
Lindsay Taylor, *University of Colorado Denver, Aurora, CO, United States*
John Adgate, *University of Colorado Denver, Aurora, CO, United States*

Oil and gas operations have the potential for catastrophic incidents and accidents, such as fires, explosions, and pollutant releases. However, the prevalence and causes of these events are unclear. Using publicly available documents, we estimated the number of fires, explosions, and pollutant releases at oil and gas development sites in Colorado from 2000 to 2014 and investigated the cause of each event. A total of 124 fires and explosions were reported between 2000 and 2014. There was an average annual rate of 1 fire or explosion for every 4,490 active wells with a highest total rate was 1 fire or explosion for every 2,660 active wells in 2008. The proximity of homes to fires and explosions was much closer in the more populated Denver Julesburg Basin than the less populated Piceance Basin. We also evaluate 4,673 pollutant release reports between 2000 and 2013 and we find an average of 1.0% and up to 1.5% of active wells had a reported release per year. The ignition sources of the fires and explosions and the cause of pollutant releases will also be discussed. To our knowledge, this is the first summary of fire, explosions, and notable releases at oil and gas sites at a state level and provides a template to determine rates and assess catastrophic risks from these operations.

We-SY-G3.3

Spatial and Temporal Patterns of Methane, Carbon Dioxide, Carbon Monoxide, and Ozone in the Denver-Julesburg Basin Across Regions of Various Oil and Gas Production Densities

Joanna Casey, *University of Colorado at Boulder, Boulder, Colorado, United States*
Ashley Collier, *University of Colorado at Boulder, Boulder, Colorado, United States*
Ricardo Piedrahita, *University of Colorado at Boulder, Boulder, Colorado, United States*
Michael Hannigan, *University of Colorado at Boulder, Boulder, Colorado, United States*

In recent years along the Colorado Front Range, there has been an influx of people in the midst of an expanding oil and gas production industry. These trends have resulted in more people living in areas of increasingly dense oil and gas production activities. To explore how the density of oil and gas production activities can influence air quality, a set of atmospheric trace gases were measured via a network of ten air quality monitors over the course of 3 months in 2015 and 2016. The attached figure includes a map of the field sites and a picture of an air quality monitor. The measured gases included methane, carbon dioxide, carbon monoxide, and ozone. Relevant environmental variables including temperature, humidity, wind speed, and wind direction were also measured. Through analysis of the relative abundance and correlations among the measured gases at each sampling site, we gained information about the regional distribution of emissions and mixing patterns throughout the basin. Patterns of short-term, dynamic enhancements of methane, carbon monoxide, and carbon dioxide were observed at some of the sampling sites, which indicate the presence of nearby sources. Wind speed and direction were used to determine likely sources of these plumes. Carbon monoxide was used as a combustion tracer. Ratios among these gases pointed to the nature of emission sources. Measured concentrations of carbon monoxide and ozone, both known to be harmful to human health, are presented in context with World Health Organization exposure limits. Spatial variability in methane, carbon dioxide, carbon monoxide, and ozone were observed across the basin. In general, higher levels of methane were observed in areas of higher oil and gas production density, particularly at night when stable atmospheric conditions and a low planetary boundary layer generally settled into place.

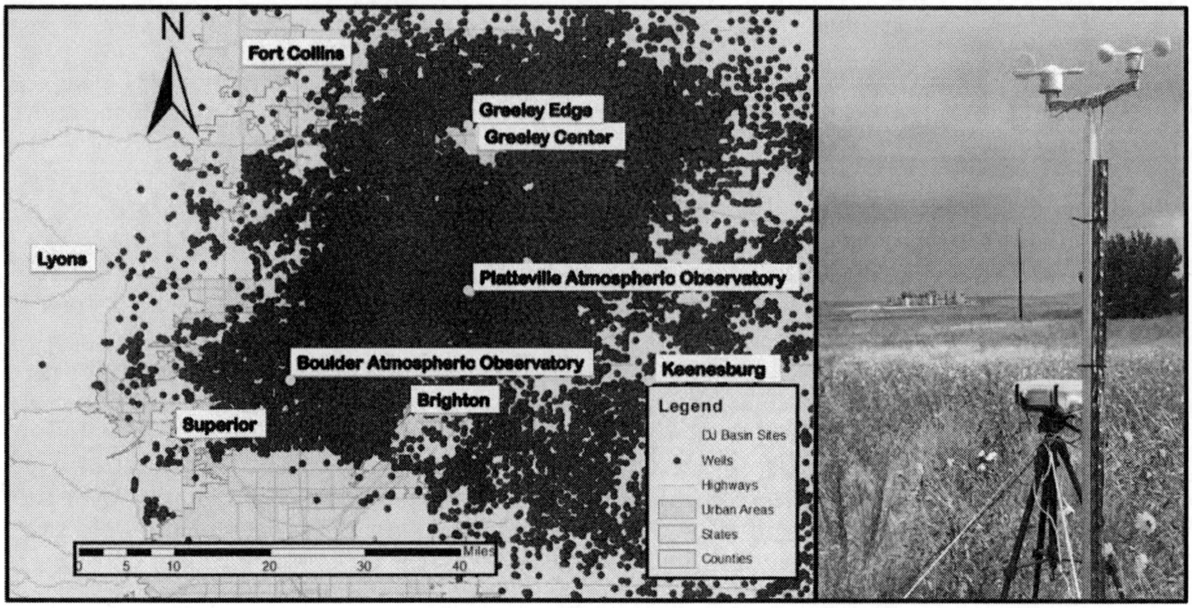

Left: Map of the Denver-Julesburg Basin with air quality monitoring sites indicated by yellow circles and oil and gas wells indicated by green circles. Right: Air quality monitor at the Platteville Atmospheric Observatory field site with oil and gas pr

We-SY-G3.4

Noise Levels from a Producing Oil Well Pad and their Potential Impacts on the Surrounding Community

William Allshouse, *Colorado School of Public Health, Aurora, Colorado, United States*
Benjamin Blair, *Colorado School of Public Health, Aurora, CO, United States*
Madeline Newman, *Colorado School of Public Health, Aurora, CO, United States*
Stephen Brindley, *Colorado School of Public Health, Aurora, CO, United States*
Lisa McKenzie, *Colorado School of Public Health, Aurora, CO, United States*
John Adgate, *Colorado School of Public Health, Aurora, CO, United States*

With the rise of unconventional oil and gas (O&G) development, including hydraulic fracturing and horizontal drilling, public concern about potential health effects for those living in proximity of these operations has also increased. Research on exposures to date has mainly focused on chemical releases into the air and water from O&G development, with little documentation about the noise produced. Noise is generated during all stages of O&G well development and operation life cycle. In order to document the noise experienced by a community living close to a well pad, we measured noise levels near an oil production site at approximately 350, 500, and 1000 feet in multiple directions and at different times of day over the course of several months. We removed measurements that were influenced by documented persistent community noise not originating from the well pad (n=48), and analyzed the remaining data points (n=192). Noise levels of the remaining space/time data points ranged between 35.3 and 62.6 A-weighted decibels (dBA), with a mean of 48.5 dBA (SD=5.4 dBA). The noise levels recorded at 500 feet from the well pad (the current setback between a well and residence in Colorado), exceeded 50 dBA in 34.4% of measurements. Additionally, noise was highest in the evening compared to morning and afternoon ($p<0.01$), when people were most likely to be home. Since noise levels exceeding 50 dBA have been associated with adverse health effects in other studies, we conclude that noise from O&G operations warrants further investigation.

We-SY-G3.5

Movement of PAHs emitted from natural gas extraction wells

L. Blair Paulik, *Oregon State University, Corvallis, Oregon, United States*
Kevin A. Hobbie, *Oregon State University, Corvallis, Oregon, United States*
Diana Rohlman, *Oregon State University, Corvallis, Oregon, United States*
Richard P. Scott, *Oregon State University, Corvallis, Oregon, United States*
Laurel Kincl, *Oregon State University, Corvallis, Oregon, United States*
Erin N. Haynes, *University of Cincinnati, Cincinnati, Ohio, United States*
Kim A. Anderson, *Oregon State University, Corvallis, Oregon, United States*

Natural gas extraction (NGE) has expanded rapidly in the United States in the last 15 years. Air emissions are a major pathway through which NGE may impact the health of nearby communities and workers. However, few studies have directly measured emissions coming from NGE. Recent research has suggested that NGE emits polycyclic aromatic hydrocarbons (PAHs). This study used passive air samplers to measure PAHs in two concentric rings, around active NGE wells (n=3) and sites permitted to host future well pads (n=2). At each site, an inner ring of three samplers was placed approximately 60 meters from the well pad and an outer ring of three samplers was placed approximately 120 meters from the well pad. At sites without wells, samplers were placed in these two rings around the proposed well pad location. The study was conducted in a rural Ohio community with a high density of NGE activity. Volunteer landowners were identified through collaboration with a local concerned citizens group, and were engaged as citizen scientists. Samplers were deployed for 20-28 days in spring 2014. Citizen scientists returned samplers with 100% compliance. Samples were analyzed for 62 PAHs using GC-MS/MS, and total levels were summed (ΣPAH). Benzo[a]pyrene equivalent values, BaPeq, were calculated using the EPA's 2010 relative potency factors to compare carcinogenic potency of PAH mixtures. Isomer ratios were used to identify sources of PAH mixtures. ΣPAH levels were significantly higher at sites with active NGE wells than at sites without wells (Wilcoxon rank sum test, $p < 0.005$). Median ΣPAH levels were two-fold higher at sites with active NGE wells than at sites without wells. Isomer ratios indicated that PAH mixtures at sites with active NGE wells had more petrogenic signatures, while sites without wells had more pyrogenic signatures. This is consistent with NGE well sites being more heavily affected by emissions from within the earth. At sites with NGE wells, there was an increasing trend in BaPeq measured in inner rings (closer to NGE well) than in outer rings. This suggests that carcinogenic potency of PAH mixtures may increase closer to NGE wells.

We-PL-H3: Spatio-Temporal Measures - I

We-PL-H3.1

Measuring and assessing individual exposures of external radiation doses in the evacuation zone in Fukushima

Wataru Naito, *National Institute of Advanced Industrial Science and Technology, Tsukuba, Japan*
Motoki Uesaka, *National Institute of Advanced Industrial Science and Technology, Tsukuba, Japan*
Chie Yamada, *National Institute of Advanced Industrial Science and Technology, Tsukuba, Japan*

Background: The accident at Fukushima Daiichi Nuclear Power Plant on March 11, 2011, released radioactive material into the atmosphere and contaminated the land in Fukushima and several neighboring prefectures. Five years after the Fukushima accident, the radiation levels have greatly decreased due to physical decay, weathering, and decontamination operations in Fukushima. The populations of 12 communities were forced to evacuate after the accident; as of March 2016, the evacuation order has been lifted in only a limited area, and permanent habitation is still prohibited in most of the areas. In order for the government to lift the evacuation order and for individuals to return to their original residential areas, it is important to assess current and future realistic individual doses in the evacuation areas.
Aim: We used personal dosimeters, called "D-shuttle", along with the Global Positioning System and Geographic Information System to elucidate realistic individual external doses and to relate individual doses, ambient doses, and activity-patterns of individuals in the evacuation zones in Fukushima.
Results: The results showed that the measured individual doses were well correlated to the ambient doses based on the airborne monitoring survey, and the results of linear regression analysis suggested that the additional individual doses were on average about one-fifth that of the additional ambient doses. The reduction factors, which are the ratios of the individual doses to the ambient doses, were calculated to be on average about 0.15 and 0.3 for time spent at home and outdoors, respectively. Analysis of the contribution of various activity patterns to the total personal dose demonstrated good agreement with the average fraction of time spent daily in each activity, but the contribution due to being outdoors varied widely.
Conclusions: Our results are a valuable contribution to understanding realistic personal doses and the corresponding airborne monitoring-based ambient doses and time-activity patterns of individuals in the affected areas. Furthermore, the results provide important information for predicting future cumulative doses after the return of residents to evacuation zine in Fukushima.

We-PL-H3.2

Bridging Exposure Science and Randomized Controlled Clinical Trials (RCCT's

Ellen Kinnee, *University of Pittsburgh, Pittsburgh, Pennsylvania, United States*
Jessie Shmool, *University of Pittsburgh, Pittsburgh, Pennsylvania, United States*
Fernando Holguin, *University of Pittsburgh School of Medicine, Pittsburgh, Pennsylvania, United States*
Jane Clougherty, *University of Pittsburgh, Pittsburgh, Pennsylvania, United States*

New therapeutic interventions are normally evaluated using randomized controlled clinical trials (RCCT), wherein randomization is assumed to balance both measured and unmeasured confounders across treatment allocation arms. Social and environmental exposures, however, vary by location, and geographically-clustered or population-stratified RCCT's may fail to fully account for their effects. In the case of asthma, accounting for these exposures may be critical to assessing treatment efficacy.
We developed a database of residential addresses for participants of three clinical trials being conducted as part of the AsthmaNet research network at 29 sites across the United States. Participants consented to sharing their primary residential address and length of residence at that location with our team. Addresses were uploaded by the clinical trial coordinators at each site to a secure web-based platform. Baseline measurements were uploaded by the AsthmaNet Data Coordinating Center and addresses and medical information stored under separate logins, linked only by Patient ID. This separation throughout the geographic analyses ensures patient confidentiality.
Residential addresses were geocoded in GIS using a composite locator to determine the most accurate point locations and then linked to baseline measures to obtain lung function indicators. 583 residential points across 19 cities were successfully uploaded, geocoded, and matched to baseline data. We found significant variation in mean baseline PFEV1 from 77-101 (+/- 6.8%) and other measures of lung function by city. We also found a substantial variation across the cohort in Census Tract SES and distance to nearest major road, both within and between cities. Patients were disproportionately recruited from areas of lower Census Tract SES, compared to city averages. Currently, we are evaluating correlations among lung function, distance to the nearest major roadway, and other environmental exposure indicators.
Incorporating GIS-based techniques and strategies for addressing spatial exposures into RCCTs is a novel solution towards better understanding which chronic exposures may meaningfully affect response to a clinical intervention. This innovation has broad implications for both clinical networks and multicenter studies.

We-PL-H3.3

Long-Duration Spatiotemporal Modeling of Nitrogen Oxides for Exposure Estimation at High Resolutions

Lianfa Li, *University of Southern California, Los Angeles, California, United States*
Fred Lurmann, *Sonoma Technology, Inc., Petaluma, California, United States*
Rima Habre, *University of Southern California, Los Angeles, California, United States*
Robert Urman, *University of Southern California, Los Angeles, California, United States*
Beate Ritz, *University of California, Los Angeles, Los Angeles, California, United States*
Frank Gilliland, *University of Southern California, Los Angeles, California, United States*
Jun Wu, *University of California, Irvine, Irvine, California, United States*

Background: Spatiotemporal models of air pollutants, especially those of high spatial and temporal resolution, are often needed in the epidemiological studies on health effects of chronic and acute exposure to air pollution. However, for the estimation of long-duration exposures, model predictions are often limited by spatial and temporal availability of the field measurements and the value range of the covariates from the training sample.

Aim: To develop a multi-level spatiotemporal model for southern California that can reliably estimate nitrogen oxide concentrations over a long time span (22 years, 1992-2013) with high spatiotemporal resolution by combining long-term monitoring data with high temporal definition from SCAQMD and data from sporadic measurement campaigns conducted by UCI, UCLA and USC that are short-term but highly spatially defined.

Methods: Continuous biweekly measurement (44 sites; 16,168 biweekly measures) from the US EPA's routine monitoring stations and sporadic measurements from intensive field campaigns (1320 sites; 3,014 biweekly measures) were collected from 1992 to 2013. Temporal basis functions were used to extract the long-duration temporal trends from continuous routine monitoring data. Thiessen polygons were constructed around the selected sampling sites to capture influence of spatial effects from the neighboring areas. Seasonal and local variation of pollutant concentrations were modeled in a non-linear fashion using spatiotemporal covariates such as meteorological parameters, traffic density, and estimated local traffic emissions from the CALINE4 dispersion model. In order to minimize the influence of uncertainties of the covariates, constrained optimization was used to adjust the estimates of concentrations at the target locations. Leave-one-community-out cross validation were conducted for model evaluation.

Results: The most important predictors (variances explained>=10%) included the first temporal basis trend, traffic density, population, county-level means of concentration and spatial effects. Meteorological factors accounted for a smaller portion (approximately 9%) of the variance, likely due to the first temporal basis function having also captured the majority of the seasonal fluctuations. Spatial effects indicated considerable influence of spatial autocorrelation on the target communities. The final R2 was about 0.90, comparing observed biweekly trends of concentrations to predicted values at individual sampling site. Leave-one-community-out cross validation produced an R2 of 0.71-0.79 (root mean square error: 5.40 ppb for NO_2; 10.43 ppb for NO_x).

Conclusions: Our model can reliably estimate spatiotemporal concentrations of nitrogen oxides over a large metropolitan area for a long overall period of 22 years. This model will be useful in studies of acute or short-term health effects of air pollution.

We-PL-H3.4

Measurement and Geospatial Modelling of Ultrafine Particle Concentrations

Greg Evans, *Southern Ontario Centre for Atmospheri Aerosol Research, University of Toronto, Toronto, Ontario, Canada*
Kerolyn Shairsingh, *Southern Ontario Centre for Atmosperic Aerosol Research, University of Toronto, Toronto, Ontario, Canada*
Kelly Sabaliauskas, *Southern Ontario Centre for Atmospheric Aerosol Research, University of Toronto, Toronto, Ontario, Canada*
Cheol-Heon Jeong, *Southern Ontario Centre for Atmospheric Aerosol Research, University of Toronto, Toronto, Ontario, Canada*
Jonathan Wang, *Southern Ontario Centre for Atmospheric Aerosol Research, University of Toronto, Toronto, Ontario, Canada*
Nathan Hilker, *Southern Ontario Centre for Atmospheric Aerosol Research, University of Toronto, Toronto, Ontario, Canada*
Jeff Brook, *Environment and Climate Change Canada, Toronto, Canada*

Aims: Ambient ultrafine particle concentrations in cities can provide a valuable metric of exposure to traffic related air pollution. However, these concentrations can be influenced by multiple sources and processes. Data mining can enable deconvolution of signals to help isolate these different sources and processes.
Methods: Particles smaller than 100 nm have been measured in and around Toronto, Canada for almost ten years. The applied methodologies have included continuous measurements at fixed sites, multi-site intensive sampling campaigns, and mobile sampling on foot, in cars, and in a mobile laboratory. The instruments used have included high time resolution nanoparticle sizers and portable particle counters. Intensive sampling campaigns were conducted in winter and summer and included measurements at 22 sites across the city.
Results: Mining of these data allowed resolution of a range of temporal and spatial patterns. Factors governing the high frequency, diurnal, seasonal and multi-year patterns thereby identified will be discussed. Different strategies were applied to isolate temporal signals within the ultrafine particle concentration time series data. These included isolation of the regional background using signal deconvolution. This method of estimating regional background was found to yield concentrations that were comparable to coincident measurements made at background monitoring sites. Short lived spikes (<1 min) caused by plumes from passing vehicles were found to be useful for estimating emission factors for individual vehicles. These data revealed that a wide range of emission factors exist across Toronto's vehicle fleet, and that a large fraction of the emissions originate from the smaller portion of high emitting vehicles. Finally, isolation of a local urban signal was found to improve geospatial modeling through land use regression. Creation of geospatial surfaces based on data collected from fixed sites at different times of the day allowed concentration surfaces to be produced illustrating the combined temporal and spatial patterns.
Conclusions: Ultrafine particles were found to exhibit a high level of spatial and temporal variability across Toronto, Canada. Evaluation of these patterns required collection of large quantities of data using a range of methodologies. Statistical mining was found to enhance the types of knowledge that could be extracted from these data.

This page intentionally left blank.

We-PL-H3.5

Seasonal difference of exposure factors of personal care products in Korea

Miyoung Lim, *Graduate School of Public Health, Seoul National University, Seoul, Korea, South*
Ji Young Park, *Graduate School of Public Health, Seoul National University, Seoul, Korea, South*
Wonho Yang, *Daegu Catholic University, Daegu, Korea, South*
Seungyul Park, *World Research Inc., Seoul, Korea, South*
Kiyoung Lee, *Graduate School of Public Health, Seoul National University, Seoul, Korea, South*

Personal care products (PCPs) are widely used in the world. PCPs contain a variety of chemicals that can enter the body and cause potential adverse health effects. Exposure to these chemical can be affected by PCPs usage pattern. Since PCPs usage pattern may be different by season, understanding seasonal variation is critical for accurate measurement of consumer product exposure. The aim of this study was to determine seasonal difference of PCPs usage and exposure factors. For consumer product exposure assessment, we developed a database of national representative exposure factors of consumer products. The exposure factors of 5 PCPs including toothpaste, shampoo, hair conditioner, face cleanser and body cleanser, were determined in winter of 2013 and summer of 2014. Trained interviewers visited each household and conducted face-to-face interviews using the questionnaire. This questionnaire contained detailed information of PCPs usage. We obtained survey data from 10,000 people (5,010 men and 4,990 women) in winter of 2013 and 3,000 people (1,282 men and 1,718 women) in summer of 2014. The study population was selected by regional, gender and age prorated square root extraction in Korean population.

For toothpaste and shampoo, no difference was observed between two seasons. Hair conditioner had 10.1% use rate difference between winter and summer. For hair conditioner, face cleaner and body cleanser, the use rate was higher in summer than in winter. The use rates also differed by gender, age, and income groups. Most PCPs were used more frequently by female, the young (15-34years) and middle (35-49 years) aged groups, medium (monthly household income $2,000-$4,000) and high (monthly household income >$4,000) income groups. The five PCPs investigated in this study were used on a daily basis. Frequency of every 5 PCPs usage was higher in summer than in winter. Otherwise, seasonal variation of use time and amount per application was different by products. Use time and amount of toothpaste, shampoo, and hair conditioner per application was higher in summer than in winter. Face cleanser and body cleanser showed opposite trend.

The national representative exposure factor showed seasonal difference of the 5 PCPs usage. With higher exposure factors in summer, more conservative exposure and risk assessment should utilize summer exposure factors of the 5 PCPs Accurate exposure factors are a critical component of exposure and risk assessment to ensure the safe use of PCPs.

We-PL-I3: Sensor Technology

We-PL-I3.1

A Trial of Low-Cost Sensors to Observe Variations and Sources of Airborne Particle Levels in Homes in a Wood-Burning Community

Ian Longley, *National Institute of Water & Atmospheric Research, Auckland, New Zealand*
Gustavo Olivares, *National Institute of Water & Atmospheric Research, Auckland, New Zealand*
Sam Edwards, *National Institute of Water & Atmospheric Research, Christchurch, New Zealand*
Erin Riley, *University of Washington, Seattle, Washington, United States*

Substantial exposure to air pollutants, even of outdoor origin, occurs within the home. Exposure may occur due to infiltration, direct exposure to indoor sources (including combustion and non-combustion sources), and from the faulty or unintended use or performance of heating appliances. However very little data exists describing the variation in exposure between homes and across populations, and the relative contributions of each source-impact pathway.
A sensor package has been developed ("PACMAN") with the aim of characterising indoor concentrations of particulate matter and the contributing sources by using rapid direct-reading dust and gas sensors and analysing the temporal structure of the data collected to infer source activities. 10 PACMAN devices were deployed in private homes in the New Zealand town of Rangiora in late winter 2015 for a period of approximately one month as part of the "Community Observation Networks for Air" initiative.
Data quality was variable with some sensors suffering from unexplained drift. Nevertheless, "emission events" were able to be extracted from the time series of most of the dust sensors. In most homes these events were not regular and were observed on an average of two events per week or less. In one home an average 19 emission events were detected per week. Events were classified on whether they were associated with a simultaneous increase in CO2 (presumed combustion sources). These events were found to decrease in frequency during the study whereas dust events not associated with CO2 remained consistently frequent. The provision of precise timing of these events has provided the basis of preliminary source identification. The high resolution data also provides substantial data on air mixing rates and relative exposures. Once these emission events were removed from the dataset, evidence of infiltration could be observed by comparison with outdoor PM10 and dust levels.

We-PL-I3.2

Metabolomic and inflammatory responses to in-vehicle traffic pollution in a panel of car commuters

Chandresh Ladva, *Emory University, Atlanta, GA, United States*
Rachel Golan, *Ben Gurion University of the Negev, Beersheba, Israel*
Roby Greenwald, *Georgia State University, Atlanta, GA, United States*
Karan Uppal, *Emory University, Atlanta, GA, United States*
Donghai Liang, *Emory University, Atlanta, GA, United States*
ViLinh Tran, *Emory University, Atlanta, GA, United States*
Tianwei Yu, *Emory University, Atlanta, GA, United States*
Dean Jones, *Emory University, Atlanta, GA, United States*
Jeremy Sarnat, *Emory University, Atlanta, GA, United States*

Aim: Advances in high-resolution liquid chromatography-mass spectrometry (LC-MS) have enabled metabolomics to emerge as a sensitive tool for measuring environmental exposures and corresponding biological response. Using measurements collected as part of a large, panel-based study of car commuters, the current analysis examines in-vehicle air pollution concentrations, targeted inflammatory biomarker levels, and metabolomic profiles to trace potential metabolic perturbations associated with on-road traffic exposures.
Methods: A 60-person panel of adults participated in a crossover study, where each participant conducted a highway commute and randomized to either a side-street commute or clinic exposure session. In addition to in-vehicle exposure characterizations, the participants contributed multiple dried blood spots over the course of each day for targeted pro-inflammatory and vascular injury biomarkers and plasma twice each day for high-resolution metabolomics. Samples were analyzed on a Thermo QExactive MS system in positive electrospray ionization (ESI) mode and resolution of 70,000 with C18 chromatography. Data were processed using apLCMS and xMSanalyzer on the R statistical platform.
Results: Four of 7 targeted inflammatory cytokines, including hs-CRP, IL-1β, and TNF-α, were significantly higher after highway exposure but not following clinic exposures in stratified analyses by commute type. Metabolomic analyses across all plasma samples yielded 4,177 robust features with coefficients of variation across triplicates below 30% in the positive mode. Within highway commute days (N = 35), 2 features differed significantly (False Discovery Rate < 0.05) between morning and evening plasma collections.
Conclusions: Initial results indicated that the protocol yielded rich targeted and untargeted biological information, with some indication of exposure response. Our results demonstrate that detecting metabolic perturbations in a panel study are feasible, but only in a subset of highway commutes without controlling for subject effects. Multivariate analysis with metabolomic profiles and targeted biomarkers is currently being conducted to further trace potential metabolic perturbations associated with corresponding changes in the targeted inflammatory biomarkers.

We-PL-I3.3

Using Wearable Cameras to Identify Microenvironments Relevant to Particulate Exposure in India

Maëlle SALMON, *ISGlobal, Center for Research in Environmental Epidemiology (CREAL), Barcelona, Spain*
Santhi Bhogadi, *Public Health Foundation of India, New Delhi, India*
Srivalli Addanki, *Public Health Foundation of India, New Delhi, India*
Pavitra Madhira, *Public Health Foundation of India, New Delhi, India*
Margaux Sanchez, *ISGlobal, Center for Research in Environmental Epidemiology (CREAL), Barcelona, Spain*
Julian D. Marshall, *University of Washington, Seattle, United States*
Cathryn Tonne, *ISGlobal, Center for Research in Environmental Epidemiology (CREAL), Barcelona, Spain*

Aim: Time-activity data can help identify activities with high personal exposure to particles such as cooking with solid fuel. We compared two methods for obtaining detailed information on exposure-relevant activities: hands-free wearable cameras and self-reported time-activity diaries.
Methods: In a peri-urban/rural area near Hyderabad, India, we recruited 40 adult participants to wear a hands-free Autographer camera on 6 occasions for non-sleeping periods within 24 hours. Images were annotated by trained coders, who applied codes from 5 non-exclusive categories: travel, occupation, cooking, indoor/outdoor location and presence of other combustion. Coders underwent training ensuring inter-rater agreement. We analyzed codes using an open-source package in the R programming language developed as part of the project. Participants also completed an hourly time-activity questionnaire including 12 exclusive possible activities and 8 exclusive possible locations.
Results: Photos from 218 participant-days were collected; annotation by coders (4 hours of work per each participant-day) is underway. Preliminary annotation results (n=10 participant-days) indicate photos covered on average 12 hours per day, with on average 1483 pictures. The median number of pictures per hour was 100. Time spent in the kitchen had a median of four minutes (IQR: 2-28min). Average travelling time was one hour (range: 0 to 2.5 hours). Nearly all (99%) pictures were codable as indoor/outdoor. Participants spent on average (sd) 7 hours (4) indoors and 5 hours (2) outdoors. Out of time spent outdoor according to self-reported questionnaire, 18% was classified as indoor by coded pictures. Conversely, out of the time spent indoor according to the questionnaire, 32% was classified as outdoor by coded pictures. Pictures seem to provide more accurate and detailed information than self-reported time-activity diaries about microenvironments and duration spent in them. Microenvironments potentially associated with high air pollution exposure such as "visible flame or smoke" were identified from the wearable camera data (two events identified in two participants, respectively one hour and 5 minutes), which were not reported in the questionnaire.
Conclusions: Wearable cameras are a relatively recent tool. Compared to self-reported hourly time-activity diaries, they offer considerably higher temporal resolution and a more objective means to identify activities and locations relevant to air pollution exposure. Future analyses will investigate how activity codes derived from wearable cameras correspond with personal particulate exposure measured by a collocated RTI MicroPEM.

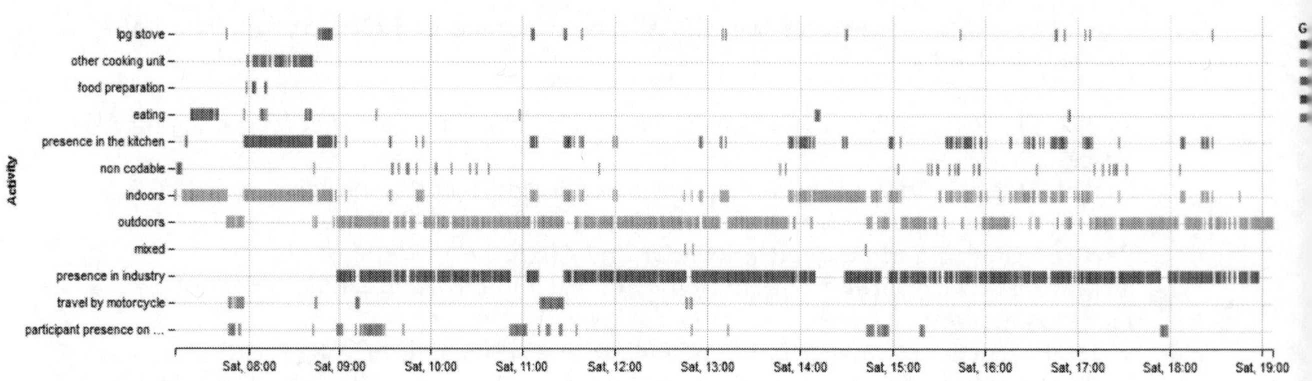

Example of annotated participant-day: each tick corresponds to one picture. Codes are grouped in 5 categories.

We-PL-I3.4

Individual variation in temporal relationships between exposure to radiofrequency electromagnetic fields and non-specific physical symptoms: A new approach in studying 'electrosensitivity'

Rik Bogers, *National Institute for Public Health and the Environment (RIVM), Bilthoven, Netherlands*
Anne Van Gils, *University Medical Center Groningen, Groningen, Netherlands*
Sander Clahsen, *National Institute for Public Health and the Environment (RIVM), Bilthoven, Netherlands*
Wendy Vercruisse, *National Institute for Public Health and the Environment (RIVM), Bilthoven, Netherlands*
Irene Van Kamp, *National Institute for Public Health and the Environment (RIVM), Bilthoven, Netherlands*
Christos Baliatsas, *Netherlands institute for health services research, Utrecht, Netherlands*
Judith Rosmalen, *University Medical Center Groningen, Groningen, Netherlands*
John Bolte, *National Institute for Public Health and the Environment of the Netherlands (RIVM), Bilthoven, Netherlands*

Background
Everyday exposure to radiofrequency electromagnetic fields (RF-EMF) emitted from wireless devices such as mobile phones and base stations, radio and television transmitters is ubiquitous. Some people attribute non-specific physical symptoms (NSPS) such as headache and fatigue to exposure to RF-EMF. Previous laboratory studies or studies that analyzed populations at a group level did not find evidence of an association between RF-EMF exposure and NSPS.
Objectives
We explored the association between exposure to RF-EMF in daily life and the occurrence of NSPS in individual self-declared electrohypersensitive persons using body worn exposimeters and electronic diaries.
Methods
We selected seven individuals who attributed their NSPS to RF-EMF exposure. The level of and variability in personal RF-EMF exposure and NSPS were determined during a three-week period. Data were analyzed using time series analysis in which events in exposure were correlated with NSPS.
Results
We found statistically significant correlations between (perceived and actual) exposure to wireless internet (WiFi) and base stations for mobile telecommunications (GSM+UMTS downlink) and NSPS scores in four of the seven participants. In two persons a higher EMF exposure was associated with higher symptom scores, and in two other persons it was associated with lower scores.
Conclusions
RF-EMF exposure was associated with NSPS in some but not all of the selected self-declared electrohypersensitive persons.

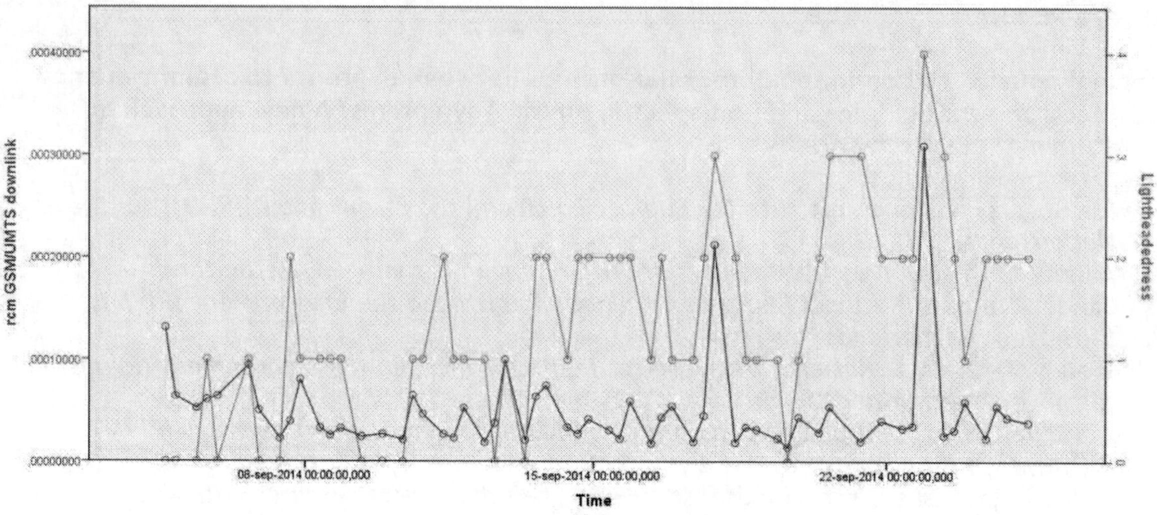

Self-rated score for lightheadedness and measured downlink exposure (rate of change metric) in participant #8

We-PL-I3.5

Low-cost sensors to estimate long-term exposure to household air pollution

Ariadna Curto, *Centre for Research in Environmental Epidemiology (CREAL), Barcelona, Spain*
David Donaire-Gonzalez, *Centre for Research in Environmental Epidemiology (CREAL), Barcelona, Spain*
Gregory A Wellenius, *Brown University School of Public Health, Providence, RI, United States*
Julian D Marshall, *University of Washington, Seattle, WA, United States*
Mark J Nieuwenhuijsen, *Centre for Research in Environmental Epidemiology (CREAL), Barcelona, Spain*
Cathryn Tonne, *Centre for Research in Environmental Epidemiology (CREAL), Barcelona, Spain*

Aim: Household air pollution (HAP) is a major public health threat in developing countries, where people are highly exposed to smoke from combustion of solid fuels. However, there is no low-cost gold-standard sensor for assessing long-term exposure to HAP. This study compares three low-cost air pollution sensors against equivalent benchmark sensors in a controlled wood-combustion experiment.
Methods: Sampling was conducted in a single-family house in Spain during five non-consecutive days. Sensors were co-located 1 m away from an indoor fireplace and 0.6 m above the ground. Fire was set once per day with hardwood logs and kept burning for 12 hours including a minimum of 2 hours with an opened window. To measure particles less than 2.5 µm in diameter (PM2.5), two low-cost light-scattering particle counters, the HAPEX (HAPEX Nano, Climate Solutions Consulting) and the TZOA-R (Model RD02, MyTZOA), were compared against a DustTrak (Model 8534, TSI Inc.). A low-cost electrochemical data-logger, the EL-USB-CO (Lascar Electronics Ltd.), was compared against a Q-Trak (Model 7575, TSI Inc.) for carbon monoxide (CO) measurements. To study within-device variability, multiple examples of the low-cost devices were employed each day. PM2.5 values from DustTrak were corrected for relative humidity. Statistical analyses were conducted using Spearman correlation and Concordance Correlation Coefficient (CCC).
Results: An average of 50.7 kg of wood per day was consumed. Temperature and relative humidity reached during fire hours ranged from 18 to 44.8°C and from 4.4 to 42.5%, respectively. One out of the two units of HAPEX and two out of the three units of TZOA-R failed during sampling; reasons are unknown. The three units of EL-USB-CO failed two out of the five sampling days, also owing to unknown reasons. Correlations were: 0.63 for HAPEX/DustTrak, 0.78 for TZOA-R/DustTrak, and 0.54 for EL-USB-CO/Q-Trak (all $P<0.001$). Agreement was moderate for HAPEX/DustTrak (CCC (95%CI) = 0.68 (0.67-0.69)) and TZOA-R/DustTrak (0.46 (0.43-0.49)), and high for EL-USB-CO/Q-Trak (0.79 (0.78-0.80)). Agreement between the three EL-USB-CO units among the two successful sampling days was 0.69 (0.53-0.86).
Conclusions: Our tests reveal moderate to high correlations and agreement among low-cost devices, but also a higher than expected device-failure rate. All of the measurements require post-processing (e.g., moving average, outlier detection), rather than straightforward use of devices' raw output. Low-cost sensors for measuring PM2.5 and CO are desirable, but may not yet be ready to replace more established exposure assessment methods.

We-SY-A4: Exposure Sciences with Stakeholders in Contested Societal Debates About the Risk of Toxic Substances

We-SY-A4.1

Including stakeholders in exposure science. An introduction to tailor-made approaches in The Netherlands

Jeroen Devilee, *RIVM, Bilthoven, Urecht, Netherlands*

In the exposure science during a contested societal debate, we react differently than in the old days. In a part of the approaches, we follow the advice of the proponents of post normal science to start a dialogue with an 'extended peer community' consisting of all those affected. This is specifically meaningful, when facts are uncertain, values in dispute, stakes high and decisions urgent.
The inclusion of such an 'extended peer community' in our scientific practice is a type of risk governance. Risk governance is a process in which the different stakeholders collectively try to minimize the exposure to a hazard and to keep health risks to an acceptable level. Besides a traditional risk assessment, this process includes a characterization of the type of risk (simple, uncertain, disputed), a concern assessment and a certain level of stakeholder engagement. In the events that a governance process failed, stakeholders collectively try to repair it. Exposure science is a part of this process. In the design of exposure science studies, the advice of advocates of this post normal science is implemented in very different ways. In fact, the governance approaches in the aftermath of an exposure to toxic substances in the Netherlands are tailor-made. The approaches chosen are dependent upon the given micro contexts of risk governance. One of the elements of this context are the ideas of experts, the public and other stakeholders about the relation between science, policy and society. Dependent upon the dominant ideas in a particular case, an approach is developed. To illustrate: Weiss (2003) categorizes experts in: environmental absolutists, cautious environmentalists, environmental centrist, technological optimists and scientific absolutists.
Another element of the micro context of risk governance, is the effect of media attention. Public turmoil is often the result of the social amplification of risk. Social amplification is sometimes the primary reason why a topic gets on the policy agenda and is a recurrent phenomenon. It makes that the discussion about a subject can go up and down for a significant period. Because of revitalized discussion, additional governance arrangements are sometimes needed.
Other relevant elements of micro contexts include policy mandate, level of risk and concern, budgets etc. In this symposium, different governance approaches that suit the different micro contexts of the exposure to toxic substances will be presented. All cases will present the roles for exposure science, the expectations about the outcomes and relevant agenda building mechanisms.

We-SY-A4.2

Exposure to Hexavalent Chromium and Health Effects among Military Personnel: Designing the Study Using Input from Stakeholders

Rob Beelen, *National Institute of Public Health and the Environment (RIVM), Bilthoven, Netherlands*

A group of former employees of the Dutch Ministry of Defense, who linked their health effects to occupational chromium exposure, contacted the Dutch media. Due to this media attention and the increasing number of (former) employees who were worried (>2,000 individuals), the Dutch Ministry of Defense asked the National Institute of Public Health and the Environment (RIVM) to conduct a study on the health effects of occupational chromium exposure.

The Ministry asked the RIVM to:

1. Define research questions by consulting the stakeholders
2. Conduct a study to answer these research questions

A committee consisting of members of the Ministry and trade unions supervises the study. All stakeholders (e.g. employees, trade unions, Ministry of Defense) were invited to send in their questions for the study. This has resulted in more than 1,400 questions. These questions were summarized in 34 research questions. The committee decided which questions had to be investigated.

The study was divided in sub studies based on the research questions, and for each sub study a study proposal was developed. These proposals were evaluated by a review group of experts nominated by the different stakeholders.

The research questions were divided in:

1. General questions about hexavalent chromium
2. What was the situation at work during the time of occupational exposure, including the amount of occupational exposure and comparison with occupational standards (e.g. appropriate preventive measures and medical surveillance)?
3. Which health risks were/are associated with occupational exposure to hexavalent chromium?
4. Is the Ministry of Defense liable for health damage?

For answering the second research question the stakeholders are important. Especially the (former) employees themselves have useful information on exposure assessment. Therefore a participatory approach of research is used.

While the study is unfolding it is interesting to see that this study started off with a suggested link between occupational exposure to hexavalent chromium and health effects and, by consultation of different stakeholders, the scope of the study has widened to include many more related aspects.

We-SY-A4.3

Public risk perception of exposure to poly-urethane used in house insulation - The Dutch approach of developing an evidence-based diagnostic guideline

Thomas Rustemeyer, *VUmc university medical centre, Amsterdam, Noord Holland, Netherlands*

The insulation of private homes with poly-urethane foam is popular in the Netherlands since few years. Soon thereafter, occupants started to report on a diversity of health complaints. Among these are headaches, tiredness, loss of concentration, weakness and shortness of breath. The affected occupants experienced severe impact on their own and their family lives. For several occupants leaving their own homes seemed to be the only solution to reduce symptoms. Public media took up these reports, which soon thereafter resulted in a political discussion. An early conclusion was the lack of knowledge on the appropriate medical diagnostic procedures in occupants with suspected health problems related to exposure to polyurethane insulation foam. The Dutch ministry of social affairs and employment stimulated the formation of a scientific working group on PUR insulation foams. The primary goal is the development of a multidisciplinary guideline for the appropriate diagnostic work-up procedures. Hereto, several medical specialists and public health experts were contacted about their willingness to participate. The secondary goal will be broad implementation of the guideline in daily practice and monitoring of the effectivity. Finally, these scientific data will be published and made publically available.

We-SY-A4.4

Exposure science, what else? Pesticides and residents.

Mark Montforts, *National Institute for Public Health and the Environment (RIVM), Bilthoven, Netherlands*

Possible health effects of pesticides used in bulb fields in the vicinity of homes raised public concern following broad media coverage in 2011. Consequently, the Ministry of Environment asked the Dutch Health Council for advice. In the complete absence of data, the Health Council recommended to initiate an exposure study among residents. The ministry commissioned RIVM to undertake this research.

We will present how RIVM used laymen experts and stakeholders to both formulate research aims and make choices in strategic issues during the research. An advisory board of stakeholders (residents, farmers, industry, local administrators, health professionals, and environmental organizations) formulated a shared objective for this exposure research. Accepting that there may be diverging opinions on the final recommendations, they agreed that knowledge on the exposure was lacking and that this should be the primary concern of the study. How much are people, living close to agricultural fields, exposed to pesticides?

By design, the research bridges disciplines of epidemiology, toxicology, experimental field research and environmental fate modelling, working together in a consortium. The research proposal was well received in an international peer review process, as well as by a national science advisory committee and the stakeholders. The researchers will gather samples of outdoor and indoor air, dust, soil and garden plants, as well as the urine of participating residents. Residents will record dietary and behavioral patterns. This information will be connected to data and measurements of pesticide spraying in, and emission from, nearby fields.

In the first phase (2015-2018) only spray applications in bulb fields are examined. The experiences and results of the first phase will be used to optimize the design of a second phase, in stone fruits. In 2016 the first growers and residents were included for the field experimental work. Monitoring, research into toxicokinetics, and preparations for analytics and modelling are ongoing. Therefore no study results will be presented.

The process made way, given the complexity of the issue (intensive agriculture in a densely populated landscape, variable climatic and geographic conditions), for a research that not just collects anecdotal information in a few subjects. It aims to capture the driving factors so that, no matter the situation (crop, weather, substance, location), predictions can be made. Exposure science is expected to enable an evidence based assessment and appraisal of the issue, resulting in (if deemed relevant) targeted research into potential health effects, validation of authorization procedures, or other measures.

We-SY-B4: Occupational Exposure Models - Development and/or Evaluation of REACH and other European and US models and tools (including tool for nanomaterials) - II

We-SY-B4.1

TREXMO: a Translation Tool to Support the Use of Regulatory Occupational Exposure Models

Nenad Savic, *Institute for Work and Health (IST), Lausanne, Switzerland*
Bojan Gasic, *State Secretariat for Economic Affairs (SECO), Bern, Switzerland*
David Vernez, *Institute for Work and Health (IST), Lausanne, Switzerland*

Background: Occupational exposure models vary significantly in their complexity, purpose and the level of expertise required from the user. The available supporting documentation does not prevent a correct use of the models. This may result incorrect or different choices made by the users and further affect risk characterization.
Objective: The goal was to implement the most often used exposure models in a single platform and to provide a support to improve their use. Decreasing the number of choices required from the users seemed to be a good strategy to limit possible implementation errors and, ultimately, between-user variability.
Methods: Briefly reviewed published documentation for six models: Advanced REACH Tool (ART), STOFFENMANAGER®, ECETOC TRA (v.3), MEASE, EMKG-EXPO-TOOL and EASE (v.2.0), were used a systematic comparison and establishing of the translations between them. Defined parameters options in these models together with their workflows were "cloned" into a unique platform. The established translation rules were programmed to connect every pair of the six models. Furthermore, a high number (~1000 per exposure type) of exposure situations (theoretically possible) was systematically generated and used to statistically examine capabilities of the developed tool.
Results: The new tool developed, TREXMO or Translation of Exposure Models enables semi-automatic translations between the six models. A set of the options selected in one model can be partly or entirely converted into the corresponding options in the other five models describing the same exposure situation. The users, therefore, may or may not be required to make additional choices. Depending on the number of additional choices required and the model selected for a translation from the ART, TREXMO can reduce the number of possible outcomes by 1-4 orders of magnitude. Guided by these translations, a higher consistency, between the users, is expected. Reduced number of choices and therefore lower possibility for an erroneous option selected should support a better use of the models. Furthermore, as less expertise is required, the users may become more motivated to use several models for a same exposure situation and therefore make risk characterization more robust in TREXMO.

We-SY-B4.2

Confirmation of the NIOSH Occupational Exposure Banding Process

Lauralynn McKernan, *National Institute for Occupational Safety and Health, Cincinnati, OH, United States*

NIOSH recognizes that chemicals are being introduced at a rate that significantly outpaces occupational exposure limit (OEL) development. While NIOSH develops new OELs and updates existing OELs, guidance is needed for the thousands of chemicals that lack exposure limits. To that end, NIOSH has developed a draft occupational exposure banding protocol to address the myriad unregulated chemicals in commerce. The protocol would sort chemicals into five air concentration bands based on toxicity. Chemicals with the lowest toxicity would be grouped in band A, while band E would include the most toxic chemicals. These band assignments are known as occupational exposure bands (OEBs). The proposed protocol uses a three-tiered evaluation system and gathers available toxicological data from preselected sources to select the appropriate band or range of chemical concentrations. Important questions include the reliability of the protocol over a variety of chemical types and families and the reproducibility of the system across users. In preliminary testing of the draft occupational exposure banding protocol, the concentration range corresponding to each band was compared with published OELs for 600 chemicals banded in Tier 1 of the protocol. Overall, the resulting bands were as or more protective than the published OEL >90% of the time. Preliminary evaluation of novel user experiences with the Tier 2 banding protocol indicated some inconsistencies in application of the protocol across users. Further analysis indicated that one of the primary reasons for the inconsistencies was the user not following the instructions. NIOSH staff are working to simplify and clarify the instructions to ameliorate that issue.

We-SY-B4.3

Evaluation of available dermal exposure models.

Jody Schinkel, *TNO, Zeist, Netherlands*
Henk Goede, *TNO, Zeist, Netherlands*
Wouter Fransman, *TNO, Zeist, Netherlands*
Anjoeka Pronk, *TNO, Zeist, Netherlands*
Suzanne Spaan, *TNO, Zeist, Netherlands*
Rianda Gerritsen, *TNO, Zeist, Netherlands*

Aims and background
Many occupational exposure models have been developed to estimate occupational exposure levels (as a point estimate or ranges) to hazardous chemicals via inhalation and/or the skin as part of the risk assessment process. In spite of the great acceptance and use of these tools in Europe and the US, these models have not been comprehensively validated. In this part of the symposium the available generic dermal exposure models or tools will be presented and evaluated.
Methods
Generally available dermal exposure models were evaluated based on their structure, their methodology, their applicability domain, and in the way they express dermal exposure. Furthermore, an overview of the results of available validation studies will be provided.
Results and conclusions
Although dermal exposure is a highly relevant exposure route for workers, the efforts undertaken with regard to dermal exposure modeling are limited compared to inhalation exposure modelling. Less (sophisticated) models are available and the validation of these models is very limited.

We-SY-B4.4

Evaluation of tools for estimating (occupational) exposure to nanomaterials

Laura MacCalman, *Institute of Occupational Medicine, Edinburgh, United Kingdom*
Sami Goekce, *IST - Institute for Work and Health, Lausanne, Switzerland*
Lucie Sikorova, *VSB - Technical University Ostrava, Ostrava, Czech Republic*
Radka Prichystalova, *VSB - Technical University Ostrava, Ostrava, Czech Republic*
Carla Alexander, *Institute of Occupational Medicine, Edinburgh, United Kingdom*
Martie Van Tongeren, *Institute of Occupational Medicine, Edinburgh, United Kingdom*
Araceli Sanchez Jimenez, *Institute of Occupational Medicine, Edinburgh, United Kingdom*

A number of tools have been developed to estimate the exposure, hazard and risk of nanomaterials. These tools are generally focussed on occupational exposure, although some do exist which consider consumer exposure. These tools are being increasingly used to make some assessment of the potential risk associated with the use of nanomaterials so it is important to understand how reliable these tools are. As part of the NANoREG project a number of these tools have been evaluated by comparing the results of the various tools to each other and to measurements, where applicable. As part of this work a small-scale inter-user study has also been undertaken. The inter-user study focussed on 3 commonly used occupational nano control banding tools (Stoffenmanager Nano, Nanosafer and Control Banding Nanotool), one consumer tool (Consexpo Nano) and the Advanced REACH Tool (ART), which could be applicable for nanomaterials in certain circumstances. Over 40 people completed exercises where they were asked to use a specific tool for a specific exposure scenario and provide information on both the answers and the inputs. The analysis will focus on evaluation of the variability of these inputs, and the answers obtained. The analysis will look at how different groups of users compare and at whether there are key parameters within each tool which drive the variability between these groups. The results will provide valuable information to users and tool developers on how variable the tools are and the potential difficulties with using such tools, particularly for exposure scenarios which are not fully characterised and all required information in not available.

We-SY-B4.5

ChemSTEER: A Computerized Tool for Assessing Workplace Releases and Exposures of Chemicals

Nhan Nguyen, *US EPA, Washington DC, Virginia, United States*

The Office of Pollution Prevention and Toxics (OPPT) of the United States Environmental Protection Agency (US EPA) has developed several exposure assessment tools, databases and models to evaluate the fate of chemicals when they are used and released to the environment and how workers, the general public, consumers and aquatic ecosystems may be exposed to the chemicals.
This presentation will provide an overview of a computerized tool for estimating workplace releases and exposures of chemicals during various activities including manufacture, processing (e.g. formulation) and industrial and commercial uses. The tool consists of peer reviewed mathematical models and industry- specific release and exposure scenarios that can be used to estimate exposures and releases in the absence of data and/or to augment available data. The tool is used routinely as part of the risk assessment of new chemicals in the Office of Pollution Prevention and Toxics (OPPT) of the US EPA. It can also be used to develop estimates of workplace exposures and releases of existing chemicals.
This tool, Chemical Screening Tool for Exposures and Environmental Releases (ChemSTEER), was originally designed for internal use by staff in EPA/OPPT for estimating workplace exposures and releases of new chemicals. The latest version of the tool, ChemSTEER version 3.1, has significant improvements over earlier versions: (1) it is more user-friendly, (2) it includes a new mass balance model for estimating vapor concentrations using a near-field/far-field approach, (3) it includes pre-defined scenarios that minimize the effort required to prepare assessments, and (4) it allows easier access to release and exposure models.
Note: The views in this abstract are those of the authors and do not represent Agency policy or endorsement.

We-SY-C4: Wristband Samplers Advancing Chemical Exposure Science - II

We-SY-C4.1

Silicone Wristbands as a Complementary PAH Exposure Assessment Tool

Holly Dixon, *Oregon State University, Corvallis, OR, United States*
Laurel Kincl, *Oregon State University, Corvallis, OR, United States*
David Camann, *Southwest Research Institute, San Antonio, TX, United States*
Antonia Calafat, *Centers for Disease Control and Prevention, Atlanta, GA, United States*
Julie Herbstman, *Columbia University Mailman School of Public Health, New York, NY, United States*
Kim Anderson, *Oregon State University, Corvallis, OR, United States*

Assessment of an individual's chemical exposures is vital in determining if such chemical exposures are associated with adverse health effects. However, there is currently a lack of inexpensive, easy-to-use technology to accurately assess human exposure to environmental chemicals. Current exposure science studies primarily rely on questionnaires, biological samples, and active air sampling devices to evaluate human toxicant exposure. In this study, silicone wristbands were used within an established Columbia Center for Children's Environmental Health birth cohort and compared to conventional exposure assessment methodologies. Wristbands and air monitoring backpacks were deployed together on 20 pregnant women for 48 hours. At the end of the 48 hours, a single spot urine sample was also collected. The wristband extracts were analyzed for 62 PAHs at Oregon State University, the polyurethane foam (PUF) and filter extracts from the backpacks were analyzed separately for 20 PAHs at Southwest Research Institute, and the urine samples were analyzed for 8 hydroxy-PAHs (OH-PAHs) at the Centers for Disease Control and Prevention. PAH concentrations in the backpack (PUF and filter combined) and the OH-PAHs in the urine samples are not correlated for seven of the eight PAH comparisons; however, naphthalene and 1-OH-naphthalene, are moderately correlated ($R_s=0.50$ and $p=0.02$). The backpack and wristband comparisons result in significant correlations for naphthalene ($R_s=0.72$, $p=0.0003$) and fluorene ($R_s=0.68$, $p=0.001$). The wristband was also correlated with the 1-OH-metabolites in urine, including 1-OH-naphthalene ($R_s=0.48$, $p=0.03$) and 1-OH-phenanthrene ($R_s=0.74$, $p<0.01$). Overall, the PAH/OH-PAH concentrations in backpacks and urine samples are more positively and significantly associated with PAH concentrations in wristbands than the associations between concentrations in backpacks and urine samples. We acknowledge the limitations of a small sample size, yet these results suggest that wristbands are more predictive of predominant OH-metabolites in urine than any component in the backpack, either the PUF cartridge or filter.

We-SY-C4.2

Passive wristband samplers assess PAH exposure of individuals living near natural gas extraction

L. Blair Paulik, *Oregon State University, Corvallis, Oregon, United States*
Erin N. Haynes, *University of Cincinnati, Cincinnati, Ohio, United States*
Diana Rohlman, *Oregon State University, Corvallis, Oregon, United States*
Kevin A. Hobbie, *Oregon State University, Corvallis, Oregon, United States*
Laurel Kincl, *Oregon State University, Corvallis, Oregon, United States*
Kim A. Anderson, *Oregon State University, Corvallis, Oregon, United States*

Natural gas extraction (NGE) has expanded rapidly in the United States in the last 15 years. Air emissions are a major pathway through which NGE may impact the health of nearby communities and workers. However, no study has directly measured the individual exposures of people living or working near NGE. Recent research has suggested that NGE emits polycyclic aromatic hydrocarbons (PAHs). This study used passive wristband samplers (PWS) to measure individual PAH exposures in a rural Ohio community with a high density of NGE activity. Volunteers were identified through collaboration with a local concerned citizens group, and the study engaged participants as citizen scientists. Participants (n=23) each wore one PWS for 20-22 days in May 2014. Participants were asked to complete daily exposure and health logs. We had over 91% participant compliance with the PWS. Each PWS was analyzed for 62 PAHs using GC-MS/MS and the total levels were summed (ΣPAH). Results were divided into three groups: active NGE well reported on participant's property (n=3), active NGE well reported on neighbor's property (n=4), and no active NGE well reported on property (n=14). These groups corresponded to participants living less than 0.75 km, between 0.75 and 2.0 km, and farther than 2.0 km from an active NGE well, respectively. ΣPAH levels were significantly higher in PWS from participants who had NGE wells on their property than in PWS from participants without NGE wells on their property (Wilcoxon Rank Sum Test, $p < 0.05$). Median ΣPAH was more than four-fold higher in PWS of participants living within 0.75 km of active NGE wells than in PWS of participants living farther than 2.0 km from active NGE wells. Additionally, ΣPAH in PWS was not correlated with primary or secondary exposure to cigarette smoking, as reported in the daily health logs (r2 = 0.01). These findings suggest that having an NGE well on your home property may increase personal exposure to PAHs.

We-SY-C4.3

Passive Sampler Devices (PSDs) Adapted for Use in Horses -Sentinels for Human Health Risks

Lane G. Tidwell, *Oregon State University, Corvallis, Oregpm, United States*
Kathleen R. Mullen, *College of Veterinary Medicine - Cornell University, Ithaca, New York, United States*
Dorothy Ainsworth, *College of Veterinary Medicine - Cornell University, Ithaca, New York, United States*
Kim Anderson, *Oregon State University, Corvallis, Oregon, United States*

Passive sampling devices (PSDs) have been established as a robust technology for detecting persistent organic pollutants in the aqueous and atmospheric compartments. Many different persistent organic pollutants have been associated with the potential for increased frequency of adverse health outcomes in humans. We adapted this methodology to investigate a Pennsylvania (PA) farm that had experienced a high number of neonatal (unrelated) foals born with neurodevelopmental disorders (dysphagia, altered mentation). The farm was situated in a region active in unconventional natural gas development (UNGD, hydraulic fracturing). Neonatal foals, born on a control farm situated 400 km east in New York (NY) and owned by the same individual, were free of the same neurological signs. We hypothesized that environmental chemical exposure was associated with the development of the neurological disorders. PSDs, affixed to the halters of 23 horses residing on the affected and control farms, were deployed and retrieved at 6-week intervals over a 13-month period (2014-2015). In addition to the halter PSDs, samplers were also deployed in the ambient atmosphere and in the water sources at both farms. PSDs were qualitatively and quantitatively analyzed using gas chromatography-mass spectrometry for 62 polycyclic aromatic hydrocarbons, and for 1182 chemicals of concern respectively. During the study period, a total of 161 halter samplers were deployed and 95% were retrieved. In preliminary data, 5 compounds—diisobutyl phthalate, phenanthrene, pyrene, cyhalothrin and PCB 96—were detected either solely, or in considerably higher concentrations at the PA farm compared to the control farm. Current efforts are underway to determine if environmental chemicals are detectable in equine tissues and if their presence is linked to neurodevelopmental defects and/or endocrine dysfunction in the horses. To our knowledge, this is the first instance of PSDs being deployed on horses for the purpose of assessing organic contaminant exposure. Easy adaption of PSD methodology to the horse enhances investigations of environmental chemical exposure and the associated adverse health risks in domestic animals and in humans. (Funding NIEHS; Zweig Memorial Fund for Equine Research).

We-SY-C4.4

Silicone wristbands detect individuals' pesticide exposures in West Africa

Carey Donald, *Oregon State University, Corvallis, OR, United States*
Richard Scott, *Oregon State University, Corvallis, OR, United States*
Kathy Blaustein, *Oregon State University, Corvallis, OR, United States*
Mary Halbleib, *Oregon State University, Corvallis, OR, United States*
Makhfousse Sarr, *United Nations, Dakar, Senegal*
Paul Jepson, *Oregon State University, Corvallis, OR, United States*
Kim Anderson, *Oregon State University, Corvallis, OR, United States*

We detected between 2 and 10 pesticides per person with novel sampling devices worn by thirty-five participants who were actively engaged in farming in Diender, Senegal. Participants were recruited to wear silicone wristbands for each of two separate periods of up to 5 days. Individual pesticide exposure profiles were highly individualized with only limited associations to demographic data. Using a 63-pesticide dual-column gas chromatography-electron capture detector (GC-ECD) method, we detected pyrethoid insecticides most frequently, followed by organophosphate pesticides which have been linked to adverse health outcomes. This work provided the first report of individualized exposure profiles among smallholder farmers in West Africa, where logistical and practical constraints have prevented use of more traditional approaches to exposure assessment in the past. The wristbands and associated analytical method enabled detection of a broad range of agricultural, domestic, legacy and current-use pesticides, including esfenvalerate, cypermethrin, lindane, DDT, and chlorpyrifos. The method is a candidate for more widespread use in pesticide exposure and health monitoring, and in the development of evidence-based policies for human health protection in an area where food security concerns are likely to intensify agricultural production and pesticide use in the near future.

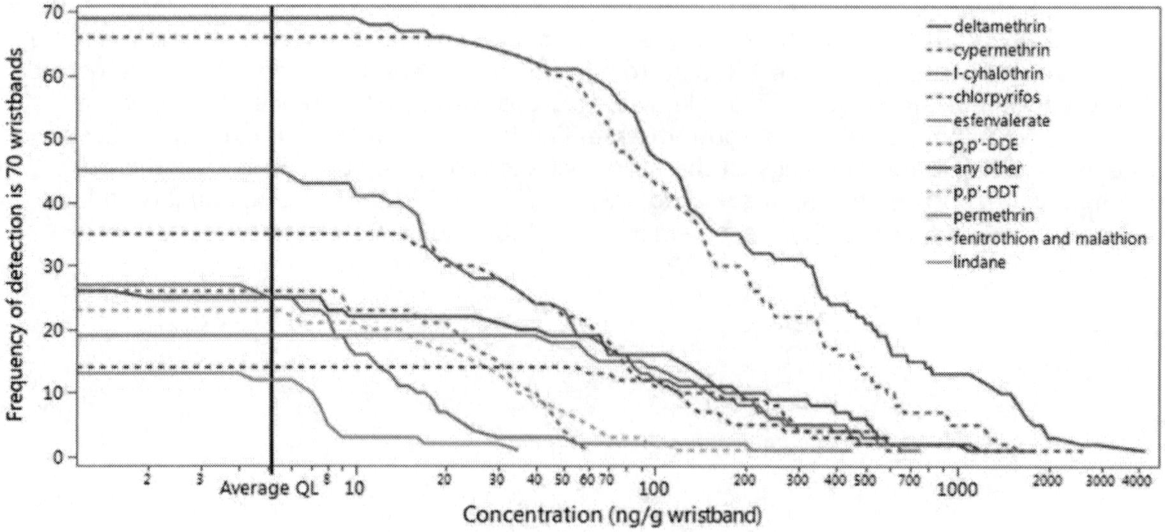

Frequencies of detected pesticides by concentration. Each line represents the frequency that met or exceed a given concentration threshold

We-SY-C4.5

Personal Passive Sampling in Peru: Magnitude and Sources of Diverse Chemicals Measured with Silicone Wristbands

Alan Bergmann, *Oregon State University, Corvallis, Oregon, United States*
Luis Vasquez, *Yantalo Peru Foundation, Yantalo, Peru*
Paula North, *Medical College of Wisconsin, Milwaukee, Wisconsin, United States*
Kim Anderson, *Oregon State University, Corvallis, Oregon, United States*

Silicone wristbands are a novel passive sampling device (PSD) for measuring lipophilic chemicals in the personal environment of people, bridging the gap between environmental and internal concentrations. Inexpensive and noninvasive, wristband PSDs are efficient at evaluating personal exposures in distant populations. We demonstrate this technology in a developing region of Peru, the Alto Mayo. Sixty-nine volunteers from four communities of the Alto Mayo wore wristbands for 31-34 days. We analyzed the wristbands for 63 pesticides by GC-ECD, 62 polycyclic aromatic hydrocarbons (PAHs) by GC-MS/MS, and screened for 1408 compounds with GC-MS coupled with automated spectral deconvolution and reporting software. We compared the wristband results between demographic groups in the Alto Mayo and to water and air (ambient) samples collected with low density polyethylene PSDs. For all deployments, trip and field blanks accounted for artifacts of sampling methods and surrogate compounds accounted for loss during extraction. Chlorpyrifos, DDT and its metabolites, and cypermethrin were the most commonly detected pesticides in both the ambient and personal environment of Alto Mayo residents. Chlorpyrifos was detected up to 2 ng/L in water and from 17 to 9000 ng/g in wristbands. Summed PAHs ranged three orders of magnitude, from 315 to 172,000 ng/g wristband highlighting a large discrepancy among individual exposures, but there were no differences between groups. PAH isomer ratios indicate a pre-dominantly pyrogenic signature of PAHs in wristband samples. An additional 65 compounds were detected with the GC-MS screening method. Differences in the presence and magnitude of pesticides and personal care products suggest chemical use patterns that influence exposure between regions. Silicone wristbands account for many sources of contaminants that contribute to human exposure including ambient water and air. These results demonstrate the utility of wristbands to evaluate individual and population exposures to many contaminants and link environmental contamination to human exposure.

We-SY-D4: UBA HBM Colloquium II - Human Biomonitoring in Europe Harmonising Instruments and Data for Science and Policy-making

We-SY-D4.2

New biomonitoring methods for HBCDD, BHT and 4-MBC

Wolfgang-Julius Gries, *Currenta GmbH & Co. OHG, Leverkusen, Germany*
Gabriele Leng, *Currenta GmbH & Co. OHG, Leverkusen, Germany*

1. Analytical determination of hexabromocyclododecane in plasma
HBCDD is a brominated flame retardant which is used in polystyrene plastics, electronic devices and especially in insulation boards' in house buildings. Technical HBCDD is a mixture of the three major diastereomeres α- HBCDD, β-HBCDD and γ-HBCDD. HBCDD can be brought into the environment during the whole life time cycle of an HBCDD containing end product and can be detected worldwide in soil, sediment and water as well as in fat tissues of organisms. As HBCCD was found in mother milk a sensitive analytical method for the determination of HBCDD in the population was required. The developed analytical method allows a LOD of 0.03 µg/L and a LOQ of 0.1 µg/L by applying LC-MS/MS and threefold lower LOD and LOQ could be achieved with UPLC-MS/MS.
2. Rapid and selective UPLC-MS/MS determination of 2,6-Di-tert-butyl-4-hydroxytoluene (BHT) residues in human urine
BHT is used as an antioxidant in food, animal feedstuff, cosmetic products, lacquers and drugs, i.e. an exposure of population is probable. The object of this method development was the sensitive and selective determination of the BHT metabolite 3,5-Di-tert-butyl-4-hydroxybenzoic acid (BHT-acid) in human urine. Analytical determination was done after enzymatic hydrolysis by LC-LC/MS-MS technique in which the first column was used for sample clean up and the second one was used for chromatographic separation. BHT-acid could be measured in ESI + mode reaching a LOQ of 0.2 µg/l urine.
3. Rapid and sensitive UPLC-MS/MS determination of specific metabolites of 3-(4-Methylbenzyliden)camphor in human urine
3-(4-Methylbenzylidene) camphor is used as an UV filter in sunscreen lotions up to a concentration of 4 %. Human studies have shown that the dermal absorption is in the range of approximately 0.5%. The identified metabolites 3- (4-carboxybenzylidene) camphor (cx-MBC) and 3-(4-carboxybenzylidene)-6-hydroxycamphor (cx-MBC-OH) were selected as specific marker in urine. Analytical determination was done after enzymatic hydrolysis by LC-LC/MS-MS technique in which the first column was used for sample clean up and the second one was used for chromatographic separation. The obtained LOQ was 0.15 µg/l for cx-MBC and 0.3 µg/l for cx-MBC-OH in ESI - mode.

We-SY-D4.3

Urinary metabolites of the UV filter Ethylhexyl Salicylate to be used as biomarkers of exposure in human biomonitoring

Daniel Bury, *Institute for Prevention and Occupational Medicine of the German Social Accident Insurance (IPA), Bochum, Germany*
Peter Griem, *Symrise AG, Holzminden, Germany*
Frank-Hinrich Köster, *Symrise AG, Holzminden, Germany*
Thomas Brüning, *Institute for Prevention and Occupational Medicine of the German Social Accident Insurance (IPA), Bochum, Germany*
Holger Martin Koch, *Institute for Prevention and Occupational Medicine of the German Social Accident Insurance (IPA), Bochum, Germany*

2-Ethylhexyl salicylate (syn. EHS, octyl salicylate, octisalate) is a UV filter substance used in one third of sun screen formulations and in other personal care products in concentrations up to 5% (maximum authorized concentration within the USA and the EC)[1,2]. Fish model experiments indicated endocrine disrupting activity of EHS [3]. Because of the likely exposure of the general population, EHS was selected as a substance of interest by the cooperation project between the German Federal Ministry for Environment (BMUB) and the German Chemical Industry Association (VCI), which has the aim to provide biomarker based exposure data for fifty emerging substances of concern.
In a human metabolism study (one male volunteer) three alkyl chain oxidized EHS metabolites (hydroxy EHS, oxo EHS, and EHS carboxylic acid) were tentatively identified as possible biomarkers of exposure. After custom synthesis of analytical standards (unlabeled and stable isotope labeled) and development of an analytical method post dose urine samples (5 mg EHS) of three male volunteers were analyzed. We report the elimination kinetics for these EHS metabolites. The results also allowed the calculation of urinary conversion factors.
In further studies the occurrence of the identified EHS metabolites in samples from the general population and their suitability as biomarkers of exposure will be investigated.
The study has been approved by the ethical review board of the Ruhr-University Bochum (Reg. No.: 4288-12).

UTRECHT
THE NETHERLANDS
OCTOBER 9-13
ISES 2016

We-SY-D4.4

Human biomonitoring of the exposure to the flavorant 2-(4-tert-butylbenzyl)propionaldehyde (lysmeral)

Gerhard Scherer, *ABF Analytisch-Biologisches Forschungslabor GmbH, Munich, Germany*
Max Scherer, *ABF Analytisch-Biologisches Forschungslabor GmbH, Munich, Germany*
Holger Koch, *Institute for Prevention and Occupational Medicine of the German Social Accident Insurance, Institute of the Ruhr-Universität Bochum (IPA), Bochum, Germany*
Andre Schütze, *Institute for Prevention and Occupational Medicine of the German Social Accident Insurance, Institute of the Ruhr-Universität Bochum (IPA), Bochum, Germany*
Nikola Pluym, *ABF Analytisch-Biologisches Forschungslabor GmbH, Munich, Germany*
Dusan Krnac, *ABF Analytisch-Biologisches Forschungslabor GmbH, Munich, Germany*
Gerhard Gilch, *ABF Analytisch-Biologisches Forschungslabor GmbH, Munich, Germany*
Edgar Leibold, *BASF SE, Ludwigshafen, Germany*

Aim: 2-(4-tert-Butylbenzyl)propionaldehyde, also known as lysmeral, lilial or lily-aldehyde (CAS No 80-54-6) is a synthetic fragrance used in a variety of consumer products like perfumes, after shave lotions and cosmetics. Due to its broad application, lysmeral was selected for the development of a biomonitoring method for the general population within the frame of the cooperation project of the German Federal Ministry for the Environment (BMUB) and the German Chemical Industry Association (VCI).
Methods: A method based on UPLC-MS/MS was developed for the simultaneous determination of potential biomarkers of lysmeral in human urine samples. Sample clean-up was performed by liquid-liquid extraction (LLE) after enzymatic hydrolysis of the conjugates. Quantification was achieved by standard addition using stable-isotope labeled, authentic reference standards. The method was applied to urine samples collected in an ethicall-approved study with 5 subjects, who received one oral dose of lysmeral as well as to spot urine samples provided by 40 volunteers from the general population.
Results: The method is characterized by its robustness, reliability and excellent sensitivity as proven during method validation according to FDA and DFG guidelines. The following four lysmeral metabolites were identified as suitable biomarkers of exposure for lysmeral in human urine samples: lysmerol, lysmerylic acid, hydroxylated lysmerylic acid and tert-butylbenzoic acid (TBBA). After oral administration, the urinary excretion reached its maximum (Tmax) after 2 - 5 h, showing faster kinetics for the primary (lysmerol, lysmerylic acid) as compared to the secondary metabolites (hydroxylated lysmerylic acid, TBBA). More than 90 % of all measured lysmeral metabolites were excreted after 12 h. After 48 h, when the renal excretion is virtually complete, TBBA, lysmerol, lysmerylic acid and hydroxyl-lysmerylic acid represent on average 14.3, 1.82, 0.20 and 0.16 %, respectively, of the dose administered. In total, the 4 metabolites represent about 16.5 % of the dose. With the conversion factors derived from the human study, we estimated median exposure doses for lysmeral in a group of 40 human volunteers of approximately 140 - 220 µg per day.
Conclusion: The lysmeral metabolites lysmerol, lysmerylic acid, hydroxyl-lysmerylic acid and TBBA in urine are suitable biomarkers of exposure and can be applied, either single or in any combination, for biomonitoring of the general population

We-SY-D4.5

Human metabolism of the biocidal compounds methylisothiazolinone and chloromethylisothiazolinone: excretion kinetics of N-methylmalonamic acid in urine

Thomas Schettgen, *Medical Faculty, RWTH Aachen University, Aachen, Germany*
Jens Bertram, *RWTH Aachen University, Aachen, Germany*
Thomas Kraus, *RWTH Aachen University, Aachen, Germany*

Introduction: Methylisothiazolinone (MIT) and the mixture of Chloromethylisothiazolinone (CIT) and MIT (CIT/MIT (3:1)) are important biocidal active substances (ASs) and are used widespread as preservative agents in industrial applications, cosmetic and household products. Both substances have skin-sensitizing properties and are well absorbed orally and dermally. An increasing frequency of contact allergies to both MIT and CIT/MIT (3:1) in recent years has raised concerns about the use of these biocides. Human biomonitoring might help to evaluate the extent of exposure to MIT and CIT/MIT (3:1) in the general population and to elucidate main exposure pathways. The main urinary metabolite of MIT and CIT in rats was reported to be N-methylmalonamic acid (NMMA). However, data on human metabolism of both substances and kinetics of urinary excretion of NMMA are lacking. Therefore, it was the aim of the present study to fill this gap by the investigation of the urinary excretion of NMMA after oral dosage of isotopically labelled MIT and CIT to human volunteers.
Methods: Four volunteers (2 m/2 f) received one dosage (2 mg) of 13C3-MIT and d3-CIT separately and at least 2 weeks apart. Consecutive urine samples were collected over 48 h. For the determination of urinary NMMA, a previously developed GC/MS/MS-method was applied. The study has been reviewed and approved by the institutional review board of the RWTH Aachen University (EK 336/14).
Results: Both substances are rapidly metabolized in humans. The mean creatinine-corrected peak excretions of urinary NMMA from four dosings were 2.5 h and 2.9 h after oral dosage of 13C3-MIT and d3-CIT, respectively. The mean urinary half-lives of excretion of labelled NMMA were determined to be 6.1 h and 7.6 h for 13C3-MIT and d3-CIT, respectively. With respect to MIT, mean excretion of NMMA within 48 h accounts for 23.7 % (18 - 30.9 %) of the dose. Concerning CIT, urinary NMMA accounts for 13.3 % (10.9 - 15.9 %) of the dose with more than 90 % excreted within the first 24 h. No gender differences in toxikokinetics were observed.
Conclusion: The present study is the first to investigate human metabolism of the biocidal compounds MIT and CIT. The results of this study confirm rapid metabolism of both substances and the urinary excretion of NMMA as major human metabolite and potential biomarker of exposure. The kinetic data obtained in this study might be useful for exposure assessment in the general population.

We-SY-D4.6

Human Biomonitoring of the fragrant compound Geraniol - The challenge of exposure assessment of natural products and nature-identical chemicals

Thomas Jaeger, *BASF SE, Ludwigshafen, Germany*
Sandra Baecker, *BASF SE, Ludwigshafen, Germany*
Edgar Leibold, *BASF SE, Ludwigshafen, Germany*
Holger Koch, *Institute for Prevention and Occupational Medicine of the German Social Accident Insurance, Institute of the Ruhr University Bochum, Bochum, Germany*
Michael Bader, *BASF SE, Ludwigshafen, Germany*

Geraniol is a fragrant compound with a rose-like smell, and is versatilely utilized in cosmetic and hygiene products as well as in household cleaners. The annual worldwide use exceeds 1,000 metric tons. For this purpose, geraniol is either extracted from essential oils (natural product) or is chemically synthesized (natural-identical flavouring) on an industrial scale. Within a collaboration project between the German Chemical Industry Association (VCI) and the German Federal Ministry for the Environment, Nature Conservation, Building and Nuclear Safety (BMUB), an analytical method for the determination of geraniol metabolites in human urine was developed. In addition, a volunteer study was carried out in order to investigate the human metabolism of geraniol.

A method was developed and validated for the analysis of the main urinary metabolites of geraniol, namely Hildebrandt acid, 8-carboxygeraniol, geranic acid and 3-hydroxycitronellic acid. Sample preparation involves enzymatic hydrolysis and liquid-liquid extraction. The analyses are carried out by liquid chromatography with tandem mass spectrometry (LC-MS/MS) which offers a sensitive and specific quantification of the four metabolites. The volunteer study showed that a large percentage of the administered geraniol is metabolized to urinary eliminated Hildebrandt acid, geranic acid and 3-hydroxycitronellic acid. 8-Carboxygeraniol is only a minor metabolite (<2 % of the dose). However, it seems to be a specific biomarker for geraniol while the other metabolites may have different precursors, e.g. citral. The elimination of all metabolites was fast, with peak concentrations at 1.5 - 5 hours after oral ingestion and subsequent elimination half-lives of 2 - 5 hours. In the course of the method validation, background values of all metabolites were detected in spot urine samples of persons without occupational exposure to geraniol.
In conclusion, the novel analytical method enables the determination of 8-carboxygeraniol, Hildebrandt acid, geranic acid and 3-hydroxycitronellic acid in human urine. 8-Carboxygeraniol seems to be a specific and thus promising biomarker for geraniol exposure. The application of the method to samples of the general population, together with the results of the metabolism study may assist in the exposure assessment and evaluation of the daily uptake of geraniol by food and/or consumer products.

We-SY-E4: Exposure to SVOCs in the Indoor Environment - Products, Emissions, Exposure, Pharmacokinetics and Biomarkers - IV

We-SY-E4.1

Probabilistic prediction of indoor aggregate SVOC exposure

John Kissel, *University of Washington, Seattle, Washington, United States*

EPA's CTEPP study collected extensive environmental measurements and urinary biomarkers relevant to SVOC exposures in probabilistically selected populations totaling roughly 250 children in Ohio and North Carolina. Environmental measurements were focused on indoor contaminant levels in residences and daycare facilities and included duplicate dietary samples. For a very limited subset of the target SVOCs a mass balance between exposures predicted from environmental measurements and those estimated from observed biomarker excretion can feasibly be attempted. Requisite data exist for the parent/metabolite pair of chloropyrifos (CPS) and trichloropyridinol (TCPy), and for pentachlorophenol (PCP) and 2,4-dichlorophenoxyacetic acid (2,4-D), which are excreted as conjugates of the parent compound. Initial evaluations were limited to consideration of dietary and non-dietary ingestion and inhalation exposure. In all three cases, application of commonly applied exposure assumptions and measured environmental contamination levels leads to under-prediction of observed urinary biomarker excretion (assuming pseudo-steady state conditions). This gap remains if alternative sources of urinary metabolites are considered. Barring fundamental sampling or analytical error, non-trivial contributions from additional exposure pathways are a possible explanation. Candidate pathways include dermal absorption and ingestion by means other than dust/soil or dietary inputs. Weschler and Nazaroff have proposed that dermal absorption of vapor phase SVOCs can exceed inhalation exposure under some circumstances and have presented a mechanism for evaluation of the relative contributions of those pathways to aggregate exposure. However, given the apparently very minor contribution of inhalation to predicted aggregate exposures of the CTEPP subjects to the three feasibly evaluated SVOCs, plausible multiples of inhalation exposures also appear inadequate to close the mass balances. Examination of surface harvesting by dermal contact and subsequent dermal absorption or of hand-to-mouth or object-to-mouth contact with surface films (as opposed to dust) permits back-calculation of requisite exposure factors. However, evaluation of the plausibility of those results suffers from a lack of prior empirical investigations of key underlying phenomena (e.g., hand washing efficiency or surface contamination replenishment rates). The CTEPP database remains a suitable test bed for aggregate exposure models.

We-SY-E4.2

Characterizing Cumulative Uptake of Indoor SVOCs Based on Physicochemical Interactions Between Humans and Their Residential Environments

Tom McKone, *University of California, Berkeley, California, United States*
Srinandini Parthasarathy, *University of California, Berkeley, Berkeley, California, United States*

Humans spend a significant proportion of their daily schedule inside of buildings. In modern buildings there are concerns about human exposure to a broad range of semi-volatile organic compounds (SVOCs) that are transferred from building materials, furniture, consumer products and personal-care products to indoor surfaces and indoor dust. The relative contributions from dermal uptake, inhalation, and non-dietary ingestion to cumulative uptake have often been separated into studies that consider these as separate pathways. The cumulative dermal uptake (DU) along with inhalation and non-dietary ingestion (NDI) intakes of SVOCs in residential environments are complex activity-based processes that cannot be fully understood without linking human pharmacokinetics with building system chemistry and physics. In this paper we will, through theory and examples, explore an alternative approach to finding simple relationships among the complexity of this multi-route exposure problem. The field of environmental chemistry has set out a process for considering competing pathways of active and passive mass transport in environmental systems that include indoor environments. This process depends on an explicit treatment of chemical-solution thermodynamics and mass balance, with a clear emphasis on chemical potential and mass transport via diffusion- and advection-driven processes. The sources and fate of SVOCs in the indoor environment are altered not only by human activity and the materials of the building, but also by the chemistry of clothing, skin, saliva, lungs, liver etc. Once we capture the complex interactions of humans within the various media of the indoor environment, we find parsimonious patterns that arise to help see our way through some of the complexity. But the common modeling approach of distinguishing for DU and NDI between active and passive exposure fails to explain this pattern. Using an alternative approach that captures multiple and more complex interactions provides key insight for both ongoing modeling and experimental efforts. Among the insights gained we illustrate how this approach reveals the importance of considering characteristic system (residential, human) time scales for assessing the potential of human intake of chemicals released to residential environments.

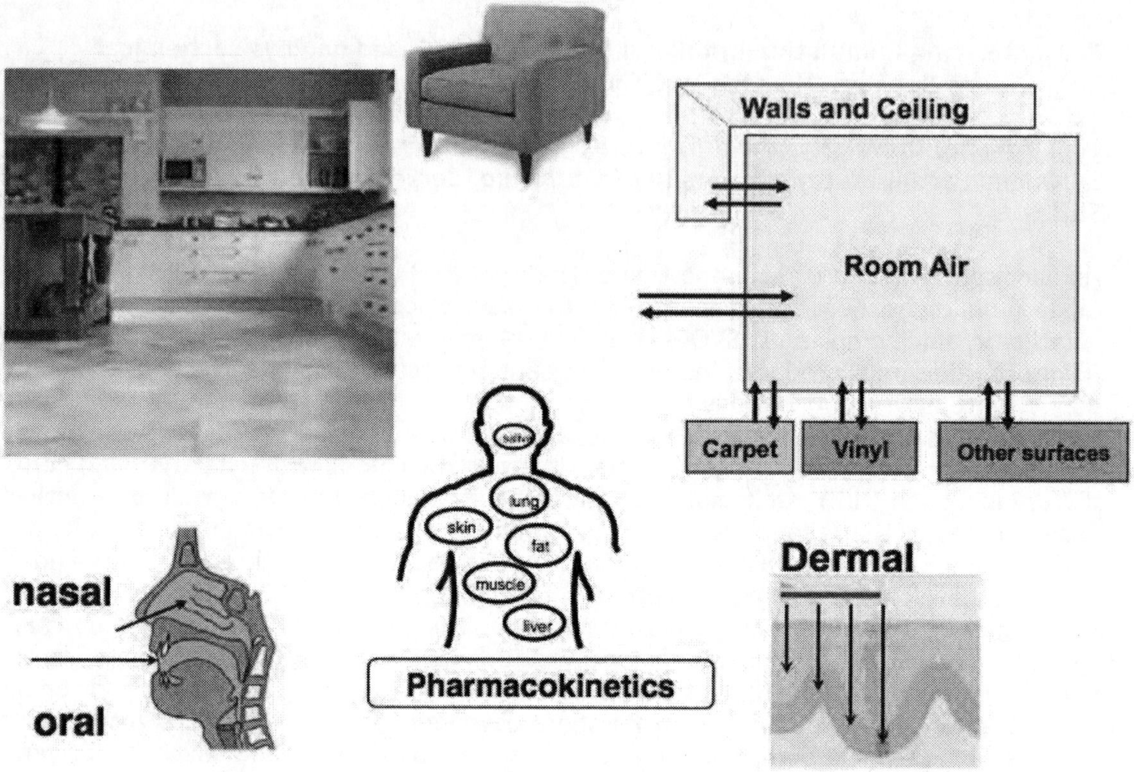

Multiple routes of exposure in residential environments

We-SY-E4.3

A modelling framework to link aggregate exposure pathways with internal exposures

Jon Arnot, *ARC Arnot Research and Consulting, Toronto, Canada*
Lauren Hughes, *ARC Arnot Research and Consulting, Toronto, Canada*
Xianming Zhang, *Harvard University, Cambridge, United States*
John Westgate, *ARC Arnot Research and Consulting, Toronto, Canada*
James Armitage, *AES Armitage Environmental Sciences, Toronto, Canada*
Peter Egeghy, *US EPA, Research Triangle Park, United States*

Humans are exposed to numerous chemicals from various exposure pathways. Exposure pathways include direct ingestion of food and beverages, inhalation of air, indirect (passive) dermal, direct dermal, hand-to-mouth contact and dermal contact with indoor materials and surfaces. Exposure can be quantified based on multimedia contact rates and with absorbed intake rates. Chemical concentrations in the body (internal exposures) are a function of chemical contact rates, absorption efficiencies, internal distribution, biotransformation rates and elimination rates (passive and active). Measured exposure data are generally quite limited compared to the numbers of chemicals in use and requiring exposure and risk evaluations. Empirical and mechanistic models are necessary to address data gaps and to improve understanding of chemical exposure pathways corresponding to concentrations associated with biological activity and possible adverse effects. The objectives of this research are to develop and evaluate a screening-level model for estimating potential risks to exposures of organic chemicals from multimedia, multipathway exposures. The Risk Assessment IDentification And Ranking-Indoor and Consumer Exposure (RAIDAR-ICE) model is an evaluative, steady-state mass balance model that links far-field exposure pathways (i.e., ingestion of food and water) with indirect and direct near-field exposure pathways (i.e., dermal, inhalation and non-dietary ingestion) to predict human intake rates and concentrations (i.e., skin surface, whole body). Effects (toxicity hazard) data can be included as model input for risk estimation and risk-based prioritization. RAIDAR-ICE is an extension of the Indoor Chemical Exposure Classification/Ranking Model (ICECRM, 2014) that links an indoor fate mass balance model with a three compartment human toxicokinetic model and indirect near-field exposure pathways. The model is described and applied to evaluate chemicals for exposure and risk potential including preliminary model evaluations with select semi-volatile organic chemicals (SVOCs).

We-SY-E4.5

Biomarkers of human exposure to organophosphate flame retardants

Adrian Covaci, *University of Antwerp, Wilrijk-Antwerp, Belgium*
Nele Van den Eede, *University of Antwerp, Wilrijk-Antwerp, Belgium*

While the demand for flame retardants, such as organophosphate esters (PFRs), has increased following the banning of several brominated flame retardants, there are still major uncertainties regarding the human exposure to these compounds. Biomonitoring is considered the best approach to monitor internal exposure, yet little is known regarding the metabolic processes PFRs undergo in the human body and the identity of biomarkers of exposure. In the present study, we aim at presenting recommendations on the selection of the target biomarkers to be used when assessing the human exposure to PFRs through biomonitoring. These recommendations are largely based on our recently published in vivo and in vitro data completed with in vivo literature data. More precisely, we evaluate the usefulness of oxidative metabolites specific for one parent PFR as compared to non-specific or infrequently detected diesters formed by hydrolytic processes. Exposure biomonitoring studies for rapidly cleared chemicals require identification of stable (and specific) major metabolites for parent chemicals of interest.
We have reported the in vitro metabolism of several PFRs, including tris(1-chloro-2-propyl) phosphate (TCIPP), tris(1,3-dichloro-2-propyl) phosphate (TDCIPP), tris(2-butoxyethyl) phosphate (TBOEP), ethylhexyl diphenyl phosphate (EHDPHP), triphenyl phosphate (TPHP), resorcinoldiphenyl phosphate (RDP) and V6, by human liver enzymes. Several metabolites, such as diesters and oxidative metabolites, have been suggested as potential useful for biomonitoring in urine. For the selection of biomarkers for PFR exposure, we have based our conclusions on findings from biomonitoring (in vivo) studies in urine and on the identification of PFR metabolites in in vitro liver preparations.
Together with the corresponding diesters (BDCIPP, BDCIPP, BBOEP, and DPHP), specific metabolites formed by oxidative metabolism, such as bis(1-chloro-2-propyl) 1-hydroxy-2-propyl phosphate (BCIPHIPP), (bis(2-butoxyethyl) 3′-hydroxy-2-butoxyethyl phosphate (HO-TBOEP), bis(2-butoxyethyl) 2-hydroxyethyl phosphate, and (hydroxyphenyl diphenyl phosphate (HO-TPHP), are suggested to be used in biomonitoring studies. The in vivo formation of some PFR metabolites has not been confirmed, and as a result, their suitability for human biomonitoring is still unknown.

We-SY-F4: E-Cigarettes, Exposures, and (Health) Effects

We-SY-F4.1

An Overview of Centers for Disease Control and Prevention (CDC)'s E-cigarette work

Patrick Breysse, *Centers for Disease Control and Prevention, Atlanta, Georgia, United States*

Electronic cigarettes (e-cigarettes) or electronic nicotine delivery systems (ENDS) are rapidly growing in popularity, especially among youth and young adults. E-cigarettes are battery powered aerosol generating devices that use a resistive heating coil to vaporize a solution containing propylene glycol, glycerin, flavors, frequently nicotine, and sometimes ethanol and water. The solution, also known as e-liquid, is contained in a disposable or refillable cartridge depending on the design of the e-cigarette. This presentation will open the e-cigarette symposium with Aan overview of the current state of the science, including recent work this presentation will discuss the e-cigarette work from the CDC's Division of Laboratory Sciences Tobacco Lab characterizing potentially harmful chemicals and components of the devices. Specifically, the presentation will include recent information about flavors, nicotine, and pH in the e-liquid, as well as the potential for exposure to metals (from sources such as the heating coil or solder joints), and volatile organics (from sources such as combustion at the heated coil. The factors affecting nicotine delivery in e-cigarettes and the select toxicity profile of e-cigarettes will be presented.

We-SY-F4.2

Properties of E-Cigarette Emissions that Promote Secondhand Exposure

Jonathan Thornburg, *RTI International, Research Triangle Park, NC, United States*
Seung-Hyun Cho, *RTI International, Research Triangle Park, NC, United States*

Use of electronic nicotine delivery systems (ENDS), such as electronic cigarettes (e-cigarettes), is increasing in the United States and globally. Despite the growing use of e-cigarettes, little is known about the physical and chemical properties of e-cigarette emissions exhaled by the user that determine the extent of secondhand exposure. The objective was to produce a physiologically relevant e-cigarette aerosol for assessing secondhand exposure by mimicking the temperature and humidity found in a user's respiratory system. This approach produced an exhaled aerosol with more representative physical and chemical properties than sampling directly from the e-cigarette. The output from the system corresponded to the expected aerosol size distribution and chemical composition in the user's lungs. We used the multi-path particle dosimetry (MPPD) model to predict the deposited and exhaled fractions of the e-cigarette aerosol. Our experiments evaluated the emissions produced by two e-liquids from one device. The aerosol size distribution produced by both liquids under dry and humid conditions were different. We found that elevated humidity and residence time inside the simulated lung activated the growth of condensation nuclei. The resulting aerosol size distribution inside the simulated lung had a smaller median diameter (184 nm versus 220 nm) but had a broader range (GSD of 3.4 versus 2.8). The aerosol chemical composition changed under humid conditions because the humectants promoted absorption of nicotine, flavorings, and preservatives into the liquid droplets. Nicotine and propylene glycol were the only compounds found in both the gas and aerosol phase. The measured e-cigarette emission aerosol size distribution was the input into the MPPD model. The dosimetry model predicted that 47% of mass of inhaled emission were deposited in the lung, with 40% in the alveolar region, and 53% was exhaled. These initial data provide evidence that secondhand exposure to e-cigarette emissions can be significant. The size distribution and chemical composition of the exhaled emissions promote a stable aerosol that can travel significant distances within an environment.

We-SY-F4.4

The Health Risks of Using E-cigarettes

Wouter Visser, *Dutch National Institute for Public Health and the Environment (RIVM), Bilthoven, Netherlands*
Liesbeth Geraets, *Dutch National Institute for Public Health and the Environment (RIVM), Bilthoven, Netherlands*
Walther Klerx, *Dutch National Institute for Public Health and the Environment (RIVM), Bilthoven, Netherlands*
Ed Stephens, *University of St. Anders, St Andrews, United Kingdom*
Lya Soeteman-Hernández, *Dutch National Institute for Public Health and the Environment (RIVM), Bilthoven, Netherlands*
Esther Croes, *Trimbos Instituut, Utrecht, Netherlands*
Paul Schwillens, *Dutch National Institute for Public Health and the Environment (RIVM), Bilthoven, Netherlands*
Hans Cremers, *Dutch National Institute for Public Health and the Environment (RIVM), Bilthoven, Netherlands*
Peter Bos, *Dutch National Institute for Public Health and the Environment (RIVM), Bilthoven, Netherlands*
Reinskje Talhout, *Dutch National Institute for Public Health and the Environment (RIVM), Bilthoven, Netherlands*

E-cigarette use has increased rapidly in the past few years. Accordingly, concerns regarding the health effects associated with the use of these devices has grown. We have performed an assessment of the health risk to e-cigarette users in three steps: Firstly, given the enormous variety of products on the market and differences in vaping behaviour, a market survey was conducted to identify exactly which products (devices and e-liquids) were popular and to obtain data regarding the vaping behaviour of users. Secondly, we established the concentrations of harmful components in the aerosol and in e-liquids by chemical analysis. Finally, we used the data from the market survey and the chemical analysis to perform an assesment of the health risks e-cigarettes present to users, taking into consideration different user profiles.
Several impurities were found in e-liquids and aerosol, including di- and triethylene glycol, formaldehyde, acetaldehyde, acrolein, metals and nitrosamines. The observed aldehydes were highly variable and did not originate from the e-liquid but were generated during the heating process. Similarly, metals in the aerosol originated from the vaporiser rather than the e-liquid.
The levels of several components in the aerosol present a risk to human health. Inhalation of the aerosol can lead to irritation and damage of the respiratory tract, palpitations, a decreased lymphocyte count and an increased risk of developing cancer. However, the risks are highly dependent on the behavior of the e-cigarette user and are not as severe as the risk associated with smoking tobacco sigarettes.

We-SY-F4.5

Electronic Cigarette Social Gatherings: Attendees and Exposures

Jona Johnson, *UGA, Athens, Georgia, United States*

E-cigarette use is so popular that conventions designed to bring e-cigarette users together in a social setting are held nearly monthly across the United States. These gatherings can range from being held at a local e-cigarette shop with a crowd of 100 attendees to being as large as a regional convention that draws thousands of people We developed a 25-question, multiple-choice survey to disseminate at a Southeastern e-cigarette convention. Of particular interest were questions designed to characterize potential for secondhand exposure in public places and health effects experienced by both users and those around them during e-cigarette use. 125 respondents completed this survey. Of these respondents, 71.2% were males (n=89) and 28.8 % were female (n=36). The average age of males was 31.2 years old (95% CI: 28.9-33.5); the average age of females was 28.8 years old (95% CI: 24.8-32.8). 50.4% (n=66) of respondents reported using their e-cigarette more than 30 times a day. 53.4% (n=70) of respondents have used an e-cigarette for 1-5 years, though 41% (n=54) of respondents began using an e-cig in the past year, indicating that e-cigarette use is growing. 85.5% (n=112) of respondents were prior users of traditional tobacco cigarettes. Most respondents (86%, n=112) use a second generation, or "mod", e-cigarette device. These devices are known to increase the amount of nicotine and VOCs delivered to and exhaled by the user. 40% (n=44) of respondents use their e-cigarette in public places (i.e. churches and shopping malls), demonstrating the opportunity for public secondhand exposures. Over 20% (n=28) of respondents reported experiencing adverse health effects while using an e-cigarette, including throat and eye irritation, headaches, nausea, and coughing. These effects are likely a result of the VOCs in the e-cigarette vapor. Roughly a quarter of respondents reported that while using an e-cigarette, someone around them has complained of skin or eye irritation and headaches, though some (n=14) reported complaints about the vapor or smell produced by the e-cigarette. Our data indicate that e-cigarette use is increasing and use in public places is common. It is noteworthy that the health effects reported are correlated with the VOCs identified in literature to be in e-cigarette vapor. Additionally, we measured salivary cotinine in non-vaping volunteers who attended an e-cigarette convention to characterize secondhand exposures. Data were analyzed by API LC/MS/MS. Data will be compared with current secondhand exposure among the U.S. population measured by serum cotinine in the NHANES database.

We-SY-G4: Exposure Science and 21st century oil and gas development - II

We-SY-G4.1

Exposure Assessment in Unconventional Natural Gas and Health Studies

Sara Rasmussen, *Johns Hopkins Bloomberg School of Public Health, Baltimore, MD, United States*
Hugh Ellis, *Johns Hopkins University, Baltimore, MD, United States*
Kirsten Koehler, *Johns Hopkins Bloomberg School of Public Health, Baltimore, MD, United States*
Brian S. Schwartz, *Johns Hopkins Bloomberg School of Public Health, Johns Hopkins School of Medicine, and Geisinger Health System, Baltimore, MD, United States*

Background: Unconventional natural gas development (UNGD) has environmental and community impacts. Psychosocial stress, changes in socioeconomic status, and exposure to noise, vibration, light, and air and water pollution are biologically plausible pathways for UNGD to affect health. Epidemiology studies have used geographic information system (GIS)-based metrics as proxies for UNGD activity and have found associations between these metrics and health outcomes. However, studies have used different metrics, making comparison across studies and understanding what each metric captures difficult, and these metrics have only incorporated wells, whereas other components of UNGD, namely compressors and impoundments, are also have potential environmental impacts.
Objective: Our goal was to characterize impoundments and compressors related to UNGD in Pennsylvania, explore the relationships among metrics capturing different components of UNGD, and compare GIS-based UNGD metrics used in health studies to date.
Methods: To characterize compressors, we visited Pennsylvania Department of Environmental Protection offices and abstracted data on compressor stations' locations, engines, dates, and emissions. For impoundments, we used aerial imagery to identify locations, sizes, and dates. To evaluate metrics capturing different components of UNGD, we created a fishnet grid across 38 counties in Pennsylvania and assigned grid points the metrics for impoundments, compressors, and four phases of well development (pad preparation, drilling, stimulation, and production) on January 1 and July 1 of 2005-13. We used principal component analysis (PCA) to understand the relationships among these six metrics. To compare different UNGD metrics used in health studies to date, we identified three types of objective asthma exacerbations (oral corticosteroid medication orders, emergency department visits, and hospitalizations) among asthma patients (n= 35,508) in the Geisinger Heath System in Pennsylvania using electronic health records from 2005-2012. We assigned patients different GIS-based UNGD metrics (inverse distance, inverse distance squared, and in buffers) that have been used in health studies and compared associations.
Results: We identified 457 compressor stations and 1,218 impoundments related to UNGD in Pennsylvania. The metrics for impoundments, compressors, and four phases of well development were highly correlated. The first PCA component, which explained 58 to 94% of the variance on the days evaluated, was strongly positively correlated with all six metrics. The regressions comparing GIS-based UNGD metrics found different associations between the UNGD metrics and asthma exacerbations.

We-SY-G4.2

Study Design and Implementation Approaches for Conducting Population-Based Studies Near Oil and Natural Gas Development Sites: A Case Study from the Denver Julesburg Basin

John Adgate, *University of Colorado Denver, Aurora, CO, United States*
Lisa McKenzie, *University of Colorado Denver, Aurora, CO, United States*
William Allshouse, *University of Colorado Denver, Aurora, CO, United States*
Benjamin Blair, *University of Colorado, Aurora, CO, United States*
Stephen Brindley, *University of Colorado Denver, Aurora, CO, United States*
Jennifer Peel, *Colorado State University, Fort Collins, CO, United States*
Stephanie Malin, *Department of Colorado State University, Fort Collins, CO, United States*

The process of estimating exposure and health risk from the wide range of stressors present near oil and gas development (OGD) sites in communities is complex. This is partially due to the recent changes in production practices, recent shifts in the intensity of development, and the challenge of recruiting a representative sample of subjects from the population at risk. In this presentation, we discuss the study design deployed to evaluate differences in exposure and response to chemical and nonchemical stressors from OGD in two Colorado communities in the Denver Julesburg Basin. Our multifaceted study design used community level surveys, measurement of biomarkers of subclinical effects, and community noise and air monitoring to obtain data on chemical and non-chemical stressor exposure and response in both communities. Challenges in implementing the study included uncertainty introduced by the recent reduction in the pace OGD and varying levels of subject willingness to participate in the two communities. Using examples from our work we describe strategies to address these design and implementation challenges, discuss results, and present our approach to communicating our results to community partners and other stakeholders.

We-SY-G4.3

Assessing the potential link between chemical exposures from unconventional oil and gas development and risk of childhood leukemia

Elise Elliott, *Yale School of Public Health, New Haven, CT, United States*
Pauline Trinh, *Yale School of Public Health, New Haven, CT, United States*
Xiaomei Ma, *Yale School of Public Health, New Haven, CT, United States*
Brian Leaderer, *Yale School of Public Health, New Haven, CT, United States*
Mary Ward, *National Cancer Institute, Rockville, MD, United States*
Nicole Deziel, *Yale School of Public Health, New Haven, CT, United States*

Introduction: Knowledge of health risks of unconventional oil and gas development is sparse, though epidemiologic studies are emerging. Childhood leukemia can be an important outcome to study, as it may be an early indicator of environmental hazards due to the short disease latency and vulnerable exposed population. The objective of this analysis was to evaluate evidence for potential carcinogenic and leukemogenic chemical contaminants in air and water associated with unconventional oil and gas development to inform exposure and health studies. Methods: We obtained a list of 1178 chemicals detected in hydraulic fracturing fluids and wastewater from the US Environmental Protection Agency and constructed a list of 135 air pollutants potentially associated with unconventional oil and gas development based on a comprehensive literature review. We systematically assessed the carcinogenicity and leukemogenicity of these chemicals by searching International Agency for Research on Cancer (IARC) monographs and evaluating whether the monographs cited findings related to leukemia and/or lymphoma. Results: Most chemicals were not evaluated by IARC (91% and 79% of potential water and air pollutants, respectively). Of the 119 unique compounds evaluated (111 and 29 potential water and air pollutants, respectively), 55 (49 and 21 potential water and air pollutants, respectively) were known, probable, or possible human carcinogens. Of these 55 compounds, 21 (17 and 7 potential water and air pollutants, respectively) were associated with increased risk of leukemia and/or lymphoma. Examples include 1,3-butadiene, benzene, cadmium, diesel exhaust, and polycyclic aromatic hydrocarbons. Conclusions: Our assessment underscores the need for air, water, and/or biological monitoring in communities near oil and gas activity and provides support for the investigation of possible associations between unconventional oil and gas development and risk of cancer, particularly leukemia/lymphoma.

We-SY-G4.4

Childhood Leukemia and Residential Proximity to Oil and Gas Development

Lisa McKenzie, *Colorado School of Public Health, University of Colorado Anschutz Campus, Aurora, Colorado, United States*
William Allshouse, *Colorado School of Public Health, University of Colorado Anschutz Campus, Aurora, Colorado, United States*
Tim Byers, *Colorado School of Public Health, University of Colorado Anschutz Campus, Aurora, Coloraod, United States*
Edward Bedrick, *University of Arizona, Tucson, Arizona, United States*
Berrin Serdar, *Colorado School of Public Health, University of Colorado Anschutz Campus, Aurora, Colorado, United States*
John Adgate, *Colorado School of Public Health, University of Colorado Anschutz Campus, Aurora, Colorado, United States*

Background: In the United States, oil and gas development (O&GD) has grown rapidly over the past 15 years because of technological advances in hydraulic fracturing, horizontal drilling, and 3D-seismicity. These technological advances have facilitated extraction of petroleum reserves from shale and other tight formations, resulting in extensive de-centralized dispersion of oil and gas wells and associated facilities across populated areas. OG&D has the potential to emit known carcinogens into the air and water.
Objective: We examined associations between childhood acute lymphocytic leukemia (ALL) and non-Hodgkin lymphoma (NHL) and residential proximity to O&GD using a population-based case-control study design.
Methods: Children were aged 0-24 years, living in rural Colorado, and diagnosed with cancer between 2001-2013. We calculated inverse distance weighted oil and gas well counts within a 10-mile radius of residence at cancer diagnosis for each year in a 10 year latency period to estimate residential exposure. Logistic regression, adjusted for known risk factors, such as gender, race, and elevation of residence, was used to estimate associations across exposure tertiles for 87 ALL cases and 50 NHL cases, compared to 528 controls with non-hematologic cancers
Results: Overall, ALL cases among 0-24 years old were more likely to live in the highest exposure tertiles compared to controls (trend p-value = 0.046), but findings differed substantially by age. For ages 5-24, ALL cases were 4 times as likely to live in the highest exposure tertile, compared to controls, with a monotonic increase in risk across exposure tertiles (trend p-value = 0.013); while ALL cases ages 0-4 years were no more likely to live near O&GD than controls. No association was found between NHL and exposure to O&GD.
Conclusions: We observed an association between ALL at ages 5-24 and exposure to O&GD. Future studies should incorporate information on O&GD activities and production levels near homes, schools, day care centers, provide age-specific residential histories, and address other potential confounders, and environmental stressors.

We-PL-H4: Spatio-Temporal Measures - II

We-PL-H4.1

Assessing the Impact of the El Niño Southern Oscillation Phenomenon upon Extreme Weather/Climate Events at the Local and Regional Level Across the Contiguous United States

Narges Khanjani, *Kerman Medical University, Kerman, Iran*
Chengsheng Jiang, *University of Maryland, College Park, College Park, Maryland, United States*
Sutyajeet Soneja, *University of Maryland, College Park, College Park, Maryland, United States*
Ghassem Asrar, *University of Maryland, College Park, College Park, Maryland, United States*
Amir Sapkota, *University of Maryland, College Park, College Park, Maryland, United States*

Aim
Increasing body of literature suggests that some characteristics of extreme events (e.g., frequency, duration, and intensity) are affected by a changing climate. Large-scale weather phenomenon, such as the El Nino Southern Oscillation (ENSO), is known to affect weather events at regional and local level, globally. No studies to date have investigated how ENSO may modulate the frequency of extreme heat and precipitation events at a local level. In this study, we utilized a historical weather/climate dataset to provide quantitative estimates on how ENSO has influenced extreme heat and precipitation events at the local and regional level across the contiguous United States. This information is of particular importance for understanding the impact and risk of these events on human health, for preparedness, recovery, and long-term adaptation measures.
Methods
We used a 30-year baseline (1960-1989) dataset to derive extreme heat and precipitation events for counties within the contiguous U.S. from 1960 to 2010. We obtained information regarding the phases of ENSO (El Niño, La Niña, and Neutral condition) from the National Weather Service Climate Prediction Center. Stratifying across the ENSO phases, we computed descriptive statistics of the spatial (Census division) and temporal (seasonal, inter-annual, as well as longer time periods) characteristics of extreme heat and precipitation events. We also determined the difference of extreme heat and precipitation events for the three phases of ENSO for each climate region within the U.S.
Results
We found the Northwest, Upper Midwest, Central, and Northeast climate regions of the U.S. to have less extreme precipitation events during El Niño, relative to the ENSO Neutral phase. In contrast, the West, Southwest, Southeast, and South climate regions showed more extreme precipitation events during El Niño relative to the ENSO Neutral phase. During La Niña events, the Northwest, Central Northwest, and Upper Midwest climate regions showed more extreme precipitation during La Niña relative to the ENSO Neutral phase. However, the West, Southwest, South, Southeast, and Northeast climate regions showed less extreme precipitations events during La Niña in comparison to the ENSO Neutral phase. ENSO's impact upon extreme heat events also varied considerably by climate region.

Conclusion

Our study demonstrates that the frequency of extreme weather events varies considerably during ENSO phases, which has a strong but uneven influence on weather across the globe. Studies investigating the link between climate change and mitigation/adaptation measures, including those focused on assessing impact to human health, need to account for this phenomenon.

We-PL-H4.2

Spatial Variability of Air Quality Data from Extensive Mobile Monitoring with Google Street View Cars

Kyle P Messier, *University of Texas at Austin; Environmental Defense Fund, Austin, Texas, United States*
Shahzad Gani, *University of Texas at Austin, Austin, Texas, United States*
Steven P Hamburg, *Environmental Defense Fund, Boston, Massachusetts, United States*
Roel Vermeulen, *Utrecht University, Utrecht, Netherlands*
Joshua S Apte, *University of Texas at Austin, Austin, Texas, United States*

Aim: The distribution of urban concentrations of air pollutants such as ultrafine particles and black carbon can vary sharply over fine spatial scales (~ 1 - 300 m). This spatial variation is not usually well-represented by routine fixed-site observations. Data from mobile air quality monitoring platforms can contribute to a richer understanding of intraurban variation in air pollution. However, previous mobile monitoring campaigns have deployed few repeat measurements. Goals of this analysis are to (1) investigate the stability of spatial trends detected by mobile monitoring and (2) quantify the spatial variability of mobile monitoring data using geostatistical techniques.
Methods: We collected an extensive dataset of routine mobile air quality measurements in the San Francisco Bay Area using a fleet of Google Street View cars with fast-response (1 Hz) particle instrumentation. These vehicles collect 40-60 h per week of daytime air quality measurements, including black carbon (BC) and nitrogen oxides (NO and NO2). The dataset presented here incorporates >1200 h and >28,000 km of on-road data collected from May-Dec 2015. Within our core sampling region, we made ≥ 20-40 sampling trips along every public street, yielding ~20,000 repeatedly sampled 30 m road segments. We systematically sub-sample this dataset to investigate how many samples are required to develop stable estimates of long-term spatial trends. We estimate the spatial autocorrelation of temporally reduced concentration fields for BC, NO, and NO2.
Results: Preliminary results indicate BC, NO, and NO2 medians for 30 m road segments have high correlation (R-Squared ≥ 0.75 for BC and NO; R-Squared ≥ 0.60 for NO2) with their long-term medians after about 15 drives. Over all driving, median BC, NO, and NO2 concentrations were roughly twice as high on highways as on major arterial roads, and twice as high on arterials as on residential streets. Geostatistical covariance analyses indicate distinct near-source and regional-scale processes that govern the spatial variability of pollution.
Conclusions: Our results on the stability of long-term medians will help inform future plans for mobile monitoring air quality campaigns. Nitrogen oxides and BC are substantially higher on highways compared to major arterial roads, which are higher than residential roads. Spatial auto-correlation ranges can help inform monitoring density needs for both mobile and fixed-site monitors.

We-PL-H4.3

Temporal Trends in Exposure to C4-C8 Perfluoroalkyl Substances among U.S. Adults

Sherry (Xiaoyun) Ye, *Centers for Disease Control and Prevention, Atlanta, GA, United States*
Kayoko Kato, *Centers for Disease Control and prevention, Atlanta, GA, United States*
Janice Ma, *Centers for Disease Control and Prevention, Atlanta, GA, United States*
Akil Kalathil, *Centers for Disease Control and Prevention, Atlanta, GA, United States*
Lily Jia, *Centers for Disease Control and Prevention, Atlanta, GA, United States*
Antonia Calafat, *Centers for Disease Control and Prevention, Atlanta, GA, United States*

Aim: Polyfluoroalkyl and perfluoroalkyl substances (PFASs) may persist in people and the environment. Since early 2000s, manufacturing practices of some PFASs, such as perfluorooctane sulfonate (PFOS) and perfluorooctanoate (PFOA), have changed, and shorter alkyl-chain PFASs are increasingly used in commerce. Because short-chain PFASs have been detected in tap and surface water, interest exists in evaluating human exposure to these chemicals. Unlike PFOA and PFOS, short-chain PFASs have relatively short elimination half-lives and likely eliminate in urine.
Methods: In 2001, 2009, 2012, and 2015, we collected 431 spot urine samples anonymously from convenience groups of demographically diverse adults in the Southeastern United States. We assessed exposure to C4-C8 PFASs from the urinary concentrations of PFASs quantified by mass spectrometry; limits of detection (LODs) were at or below 0.1 ng/mL for all PFASs.
Results: We did not detect perfluorobutane sulfonate, perfluorohexane sulfonate, or PFOA. By contrast, we detected perfluorobutanoic acid (PFBuA), perfluoropentanoic acid (PFPeA), and perfluorohexanoic acid (PFHxA) in samples collected in 2015. PFBuA was most frequently detected (>50%) with 90th percentile concentration at 0.35 ng/mL; detection frequency and 90th percentile concentrations for the other compounds were 28%, 0.14 ng/mL (PFPeA) and 5%, <LOD (PFHxA). PFOS and perfluoroheptanoic acid (PFHpA) were only detected in urines collected in 2001; detection frequency and 90th percentile concentration were relatively low: 29%, 0.20 ng/mL (PFOS) and 19%, 0.20 ng/mL (PFHpA).
Conclusions: PFOS and PFHpA were only detected in urine samples collected in 2001 when PFOS was still being manufactured in the USA. However, we detected C4-C6 perfluoroalkyl acids in urine collected in 2015 suggesting potential human exposure to short-chain PFASs in recent years. Urine could be an appropriate matrix for assessing human exposure to emerging short-chain PFASs.

We-PL-H4.4

Spatial and temporal variation of traffic-related air pollution in two urban neighborhoods in the Boston metropolitan area (MA, USA)

John Durant, *Tufts University, Medford, Massachusetts, United States*
Matthew Simon, *Tufts University, Medford, Massachusetts, United States*
Doug Brugge, *Tufts University, Boston, Massachusetts, United States*

Aim: Accurate quantification of exposures to traffic-related air pollution in near-roadway neighborhoods is challenging due to the high degree of spatial and temporal variation of pollutant concentrations. The objective of this study was to measure air pollutant concentrations in two urban areas over a wide range of traffic and meteorological conditions using a mobile monitoring platform.
Methods: We studied a 40-km2 area in Boston, and a 6.4-km2 area in Chelsea. Interstate highways 93 and 90 border the Boston study area and carry 150,000 and 110,000 vpd, respectively. In Chelsea, U.S. Route 1 (77,000 vpd) dissects the city diagonally. A mobile platform equipped with rapid-response instruments was driven repeatedly along a 40-km route in Boston on 49 days (~120 hours) between Dec. 2011 and Nov. 2013, and along a 21-km route in Chelsea on 46 days (~180 hours) between Dec. 2013 and May 2015. Monitoring was performed in 3-6-hour shifts in the morning, afternoon and evening on weekdays and weekends in winter, spring, summer and fall. Measurements were made of particle number concentration (PNC; 4-3,000 nm), PM2.5, PAH, BC, CO, NO and NOx.
Results: The highest concentrations were measured 0-50 m from major roadways (>20,000 vehicles/day) with distance-decay gradients varying depending on pollutant, traffic and meteorology. The most pronounced variations were observed for PNC. Median PNC 0-50 m from major roads were ~45% higher compared to areas 400-800 m from major roads. PNC were highest in winter (37,000 #/cc in Boston; 35,000 #/cc in Chelsea) and lowest in summer (18,000 #/cc in Boston; 14,000 #/cc in Chelsea), higher on weekdays compared to weekends, and higher during morning rush hour compared to later in the day. Spatial variations in PNC distance-decay gradients were non-uniform largely due to contributions from local street traffic. Similar spatial and temporal patterns were observed for the other pollutants; however, the near-roadway gradients were less pronounced compared to PNC.
Conclusions: Datasets containing fine-scale temporal and spatial variation of air pollution concentrations near busy urban roadways and highways may be useful for informing exposure assessment efforts.

We-PL-H4.5

Relationships of Indoor, Outdoor, and Personal Exposure to Fine Particles Concentrations in Hong Kong

Kin Fai Ho, *The Chinese University of Hong Kong, Hong Kong, Hong Kong*
Xiao-Cui Chen, *The Chinese University of Hong Kong, Hong Kong, Hong Kong*
Steven Sai Hang Ho, *Institute of Earth Environment, CAS, Xi'an, China, People's Republic of*
Chi Sing Chan, *The Chinese University of Hong Kong, Hong Kong, Hong Kong*
Junji Cao, *Institute of Earth Environment, CAS, Xi'an, China, People's Republic of*

Aim: The aim of this study is to investigate the relationships of indoor, outdoor and personal exposure to PM2.5 and their chemical components in Hong Kong.
Method: A random sampling strategy was applied to recruit forty non-smoker adults (age > 18 years of age) resided in different areas of Hong Kong for non-occupational personal exposure measurement purpose. Twenty-four hours integrated personal exposure samples of PM2.5 were collected for two consecutive days during summer and winter of 2014. 10 out of 40 subjects have been selected to conduct simultaneous indoor sample, outdoor ambient sample, and personal exposure sample collection during summer and winter, respectively. Gravimetric PM2.5 mass analyses were determined in all samples.
Results: Personal PM2.5 exposures demonstrated the significant seasonal difference ($p < 0.01$) with a higher average concentration in winter (35.2 ± 16.7 μg m-3) than in summer (18.1 ± 10.3 μg m-3). Personal exposures were lower than corresponding indoor (27.0 ± 17.6 μg m-3) and outdoor (28.7 ± 14.8 μg m-3) concentrations during the study period. Generally, there are high statistically significant correlation coefficients of personal exposure associated with both indoor ($r = 0.647$, $p < 0.01$) and outdoor concentrations ($r = 0.640$, $p < 0.01$), which suggesting that personal exposures were influenced by both ambient and indoor particle sources.
Discussion: Average indoor, outdoor and personal PM2.5 levels were higher in winter when compared to summer, these findings are consistent with the results reported previously in the literature. Although less indoor activities was observed during the measurement period, P/O and I/O ratios higher than unity were observed during the sampling days. Personal exposure levels were more strongly associated with indoor air pollution levels than with outdoor concentrations. Although PM2.5 mass was the primary focus of analysis, future plans included the investigation of chemical and toxicological characteristic of the PM2.5 samples.

We-PL-I4: Continuous/Real Time Measures

We-PL-I4.1

Wearable and Stationary IoT Chemical and Location Sensor Devices with Emergency Communication.

Kenneth Brown, *NIOSH, Cincinnati, Ohio, United States*
Kenneth Mead, *NIOSH, Cincinnati, Ohio, United States*
Peter Shaw, *NIOSH, Cincinnati, Ohio, United States*
Ronald Kovein, *NIOSH, Cincinnati, Ohio, United States*
Robert Voorhees, *University of Cincinnati, Cincinnati, Ohio, United States*
Adam Brandes, *MeasureNet Technology, Cincinnati, Ohio, United States*
Samuel Glover, *NIOSH, Cincinnati, Ohio, United States*
John Snawder, *NIOSH, Cincinnati, Ohio, Uruguay*
Michael Breitenstein, *NIOSH, Cincinnati, Ohio, United States*

Real-time knowledge of both a worker's location and hazardous exposure improves vital emergency response time and protects emergency responders by providing critical environmental hazard data prior to the responders entering the accident scene. NIOSH recently published research on a wearable device called the Chemical Exposure Monitor with Indoor Positioning (CEMWIP). (J Occup Environ Hyg. 2016 Jan 19:1-37. EPUB). The project adopted an interdisciplinary team approach that included a chemist, industrial hygienist, statistician, electrical engineer, and a software developer to produce a real-time direct reading exposure assessment method. The CEMWIP system was laboratory tested for chemical sensor and real-time location system (RTLS) accuracy and precision. The method combined the RTLS and a wireless direct-reading method (DRM), and provided simultaneous exposure alerts to both the exposed worker and a remote monitor with location and exposure data at set exposure levels. Data were wirelessly and simultaneously collected from sensors every second, for volatile organic compounds (VOCs) concentration, location, temperature, humidity, and time. The streaming data were collected and graphically displayed in real-time onto digital floorplans and could subsequently be evaluated as a three-dimensional hazard maps showing peak exposure with location. While CEMWIP used a PID sensor to measure VOCs for proof of concept, different sensors could be exchanged depending upon the expected types of exposure. A new proposal will be also presented that applies the CEMWIP method for research into laboratory safety by adding manual and automated emergency communication for times of incapacitating events to notify emergency medical services, when response time and location are critical for life. The laboratory sensor would continuously monitor and remotely display worker hazardous exposure information. An incapacitating event would be detected by an accelerometer if the worker is down triggering automatic emergency communication. Pressing the alert button would contact monitoring personnel and provide emergency responders with worker location and hazard information. Research will be extended from indoors to outdoors for field workers venturing into lone-worker situations. New research will also develop, and test stationary sensors with a triggered sampling device with further enhancements two-way communication and the remote alarm capabilities of the CEMWIP system. Finally, a pilot study with a cohort of laboratory personnel will provide data on wearability, worker compliance, dependability, durability, accuracy, precision, sensitivity, and effectiveness for improving safety both in the laboratory and field.

3-D VOC-Location Hazard Map

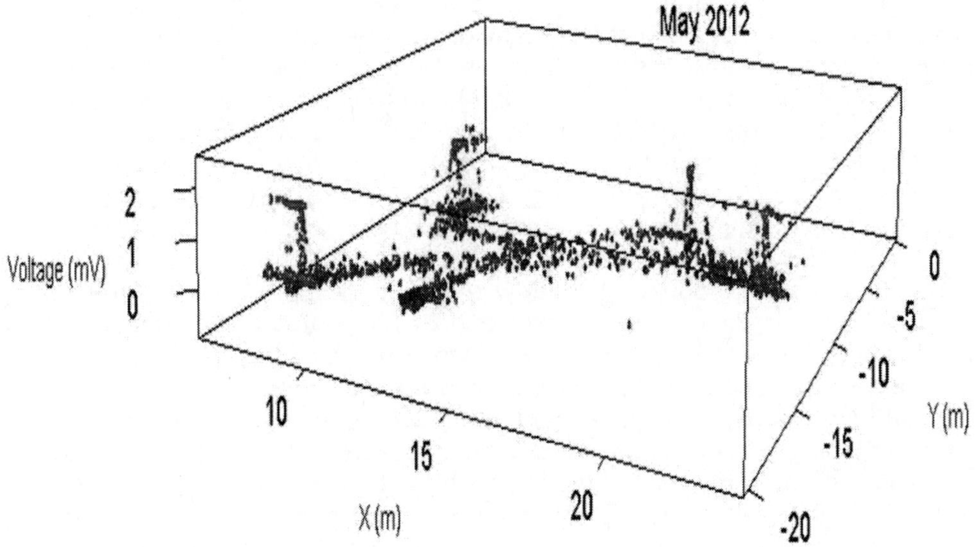

CEMWIP method provided indoor VOC concentration and location data for a 3-D hazard map.

We-PL-I4.2

Exposure Assessment Using Long Term Sampling with Evacuated Canisters in both Occupational and Non-Occupational Indoor Environments.

Alan Rossner, *Clarkson University, Potsdam, NY, United States*
Ryan Ryan LeBouf, *National Institute for Occupational Safety and Health,, Morgantown, West Virgina, United States*
Marley Carroll, *Clarkson University, Potsdam, NY, United States*
Emily Sullivan, *Clarkson University, Potsdam, NY, United States*

There is an essential need for tools that result in more effective indoor air quality and vapor intrusion (VI) exposure assessments that provide representative data of long-term human exposures. Evacuated canisters have been used for many years to evaluate ambient and indoor air environments. Recently, capillary flow controllers have been used to sample at very low flow rates (0.1 to 0.5 mL/min). Capillary flow controllers have a well-defined drop in flow rate over the sampling period as the pressure in the canister rises. Under extreme circumstances, this will give rise to a positive or negative bias in sampling results when peak exposures are present. This study was conducted to assess the performance of the capillary flow controller coupled with evacuated canisters during exposures to volatile organic compounds. Six flow controllers coupled with 400-mL evacuated canisters were tested in a small chamber (32 L) to evaluate sampling bias with respect to grab samples using canisters. All samples were analyzed by a gas chromatograph/flame ionization detector. A 2ppm concentration of toluene was generated in a chamber as a background concentration and peaks of 200 ppm (100x) were generated at the beginning of the test period to assess positive sampling bias and also at the end of the period to assess negative sampling bias. A series of experiments were run for 4 and 8 hours, as well as several for up to 3 week sampling periods with six replicate canisters per experiment. The reference concentration was established using a series of canisters drawn directly from the chamber. Comparison of the reference values to the concentrations collected by the capillary flow controllers allowed for an assessment of the capillary flow controller sampling bias. The bias for all experiment trials ranged from 0.01% to -25% as compared to the reference concentrations. Relative standard deviations ranged from 1.0% to 16.3% for the trials. Reducing the sampling period from 8 hours to 4 hours caused a decrease in sampling bias from -25% to -16% for a peak at the end of the sampling period. Samples collected at low flow rates with the capillary flow controller were found to provide results comparable to the reference method. Sampling bias can be reduced by filling the canister to ~ 35% capacity during sampling. The new canister method captures the advantages of both canisters and sorbent samplers without their limitations by allowing for long term (hours to weeks) exposure assessments.

We-PL-I4.3

Innovative Sensors and Models for City-Level Air Pollution Exposure Monitoring

Nicholas Hamm, *University of Twente, Enschede, Netherlands*
E. Alfredo Vásquez Gómez, *University of Twente, Enschede, Netherlands*
Ipsit Dash, *University of Twente, Enschede, Netherlands*
Rene Otjes, *Energy Centre of the Netherlands (ECN), Petten, Netherlands*
Sandra van der Sterren, *Gemeente Eindhoven, Eindhoven, Netherlands*
Vera van Zoest, *University of Twente, Enschede, Netherlands*
Andre van der Wiel, *Scapeler.com, Eindhoven, Netherlands*
Gerard Hoek, *Utrecht University, Utrecht, Netherlands*
Alfred Stein, *University of Twente, Enschede, Netherlands*

Aims: Space-time resolved estimates of air pollution exposure require information on the movement and location of study subjects together with predictions of air pollution concentration; however, the possibilities are typically limited by the available data. The Dutch city of Eindhoven (90 km2, 225,000 inhabitants) has two monitoring stations in the LML (Dutch national air quality monitoring system), yielding hourly data. This has been augmented (2013) by 35 "Airboxes" equipped with low-cost sensors for particulate matter (PM), NO2, O3 and ultra-fine particles delivering measurements every 10 minutes, as part of the ILM AiREAS initiative. These data pose new opportunities but raise new challenges, specifically: How should the data be organized? What are their quality and how can this be evaluated automatically? What are the appropriate models to provide spatial-temporal measures of exposure? What spatial and temporal resolutions can be achieved?
Methods: The variability in the PM data was evaluated using descriptive statistics and the variogram (to explore spatial autocorrelation), allowing exploration of different spatial and temporal resolutions. The ILM PM data were validated against the LML data in 2013 and 2015. For analysis we used Bayesian maximum entropy (BME) and space-time dynamic models. Both allowed additional information, for example dispersion model output and weather data, to be incorporated. We used a service oriented architecture (SOA), based on standards from the Open Geospatial Consortium (OGC), to organize the data.
Results: The exploratory analysis showed that the variability in the ILM PM data increased from 10% to 16% (coefficient of variation) between 2013 and 2015 The ILM data tended to record lower values than the LML data, although the peaks and troughs were still observed. The 10-minute data were typically too noisy to allow the identification of spatial correlation, although this was clear when averaged to 1-hour averages and for longer time periods. Using BME we were able to integrate the dispersion model and ILM data, yielding an RMSE of 1 μg m-3 for daily PM2.5. The SOA was effective to combine different datasets (e.g., ILM and LML) and to implement simple geostatistical models in an automated fashion.
Conclusions: Using low-cost sensors allowed us to identify spatial and temporal patterns that are valuable for spatial prediction at sub-daily time resolutions. These patterns can be identified and modelled using space-time geostatistics e.g., BME. Standards based methods should be used for organizing, archiving and disseminating data and are essential for future automated analysis.

We-PL-I4.4

Real time detection and characterization of bioaerosols from environmental sources

Sean Tyrrel, *Cranfield University, Cranfield, United Kingdom*
Zaheer Nasir, *Cranfield University, Cranfield, United Kingdom*
Mohanad Jawad Jawad, *Cranfield University, Cranfield, United Kingdom*
Gill Drew, *Cranfield University, Cranfield, United Kingdom*
Alan Nelson, *Cranfield University, Cranfield, United Kingdom*

Aim:
Biowaste and intensive agriculture industries emit bioaerosol of significance to human health. Whilst progress has been made in characterising emissions from these industries relatively little headway has been made in: understanding exposure of the general public to bioaerosol from these sources; putting process-based exposures into the context of background exposure to natural bioaerosol; or in quantifying health risk and setting health-based standards. A critical limiting factor is the lack of advanced microbiological methods to characterise and quantify bioaerosol emissions and dispersion. Our current evidence base is almost entirely reliant on short duration "snapshot" sampling and culture-dependent microbiology.
Methods:
Among emerging techniques, laser induced fluorescence has shown promise in exploring bioaerosol properties with high time and size resolution (Pan et al. 2015). Continuous real time measurements were carried out to monitor the number concentration of different particulate categories detectable using a novel Spectral Intensity Bioaerosol Sensor (SIBS)(Droplet Measurement Technologies, USA) which is a development of the WIBS 4-A technology described by Toprak and Schnaiter (2012). Measurements were made at distance of approximately 50 m from the operational area of a green waste composting facility in eastern England during a 4 hour period encompassing periods of active composting as well as a period of relatively low activity.
Results:
The SIBS device generated real time information on the number concentration of total and fluorescent particles which are assumed to represent the bioaerosol fraction. There was an association between periods of composting activity and higher particulate concentrations. There were distinct differences in fluorescence emission characteristics from particles detected at the composting and a background site (data not shown) demonstrating the possibility of classifying bioaerosols.
Conclusions:
Fluorescence based real time measurement of bioaerosols will contribute significantly to advancing the existing state of knowledge on bioaerosol detection and emission characteristics from different environmental sources which will in turn inform enhanced exposure assessment and modelling.
Reference:
Pan, Y. L. (2015). J. Quant. Spectrosc. Radiat. Transf. 150, 12-35
Toprak and Schnaiter (2012) Atmos. Chem Phys. 12, 17607-17656
Acknowledgement:
This work was supported by the Natural Environment Research Council (NERC) and Defence Science and Technology Laboratory (Dstl). This award is made under the auspices of the Environmental Microbiology and Human Health programme.

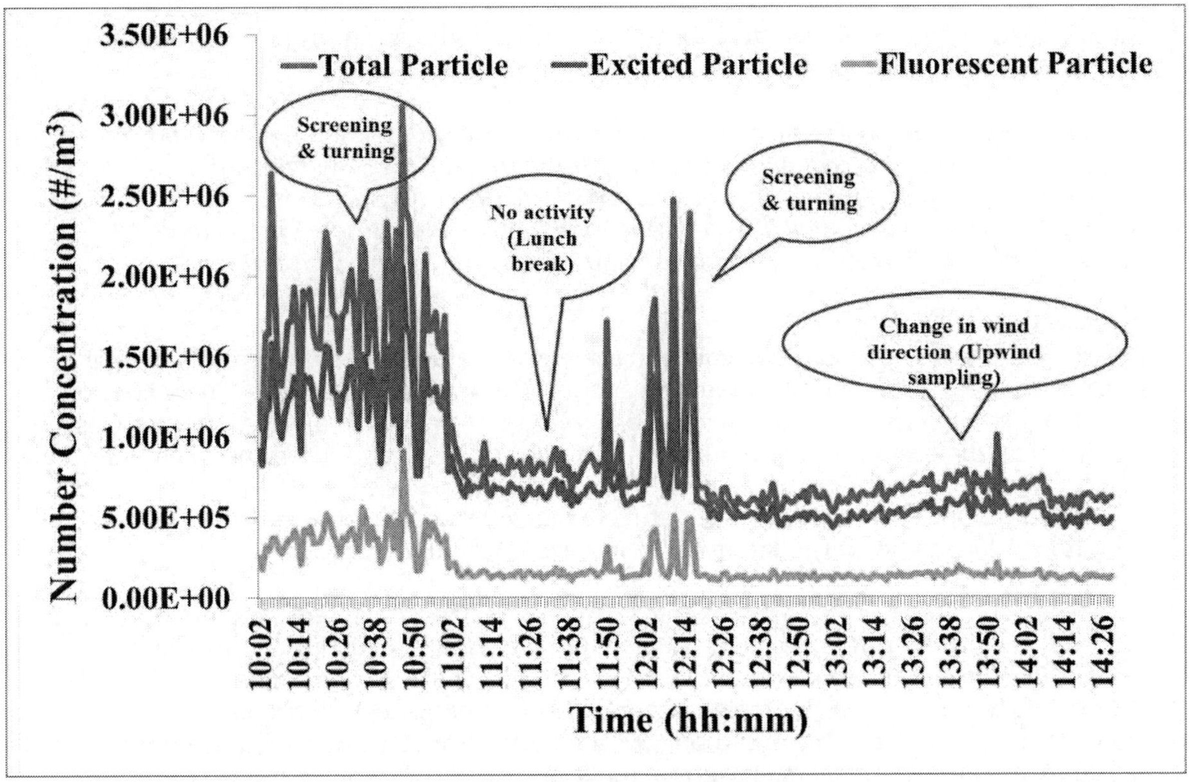

Representative number size concentration of particles in different categories generated by the SIBS at an operational composting site during a 4 hour sample of the working day

We-PL-I4.5

Characterizing real-time vertical air pollution gradients in an urban environment - Vegas (vertical gradient study)

Marloes Eeftens, *Swiss Tropical and Public Health Institute, Basel, Switzerland*
Mark Davey, *Swiss Tropical and Public Health Institute, Basel, Switzerland*
Alex Ineichen, *Swiss Tropical and Public Health Institute, Basel, Switzerland*
Danyal Odabasi, *Swiss Tropical and Public Health Institute, Basel, Switzerland*
Ming-Yi Tsai, *Swiss Tropical and Public Health Institute, Basel, Switzerland*

Background
Spatial contrasts in air pollution concentration are characterized at increasingly high resolution. Yet, air pollution monitoring and modelling studies commonly only consider proximity to sources in the horizontal plane while differences in air pollution exposure in the vertical dimension; eg, depending on the floor of residence, are rarely characterized.
Objectives
We aimed to understand how the real-time vertical gradients for three different air pollutants depend on traffic intensity, street configuration and seasonality.
Methods
Measurement sites were selected along 11 streets in Basel, Switzerland on two quiet (50-100 vehicles/30min), four medium-traffic (100-250 vehicles/30min) and five busy streets (over 300 vehicles/30min), with different street configurations. We measured Particle Number Concentration (PNC, a measure for ultrafine particles), particulate matter smaller than 2.5µm (PM2.5) and black carbon (BC, a measure for soot) using real-time instruments at up to six different heights above ground (1.5, 4, 7, 11, 17 and 25 meters) simultaneously, during the winter and (upcoming) summer seasons of 2016. Baskets containing the instruments were hung from a bucket truck, which remained at each sampling site for 30 minutes during off-peak hours. Average 30-minute concentrations were calculated for each street and for each height.
Results & discussion
Concentrations were highly correlated between the different sampling heights (e.g. for 1.5m and 7m) for UFP (R2=0.98), BC (R2=0.74), and (R2=96). Wintertime PNC concentrations were typically 19% lower at 7m than at 1.5m, and 50% lower at 25m than at 1.5m. The decrease in concentration was consistently larger for busy streets than for medium-traffic streets and for quiet streets (e.g. for a height difference 1.5m and 7m: 29%, 22% and 14%, respectively). We found no significant vertical gradient for PM2.5, even between 1.5m and 25m. PM2.5 is known to be a secondary aerosol showing less spatial variability. For BC, we found no substantial differences between the heights of 1.5m and 7m, whereas we saw a median reduction of 25% in BC concentration with increasing height from 1.5m to 25m. The absence of a BC gradient between 1.5m and 7m may be due to a major source of BC: the tailpipes of diesel trucks, which are commonly higher than the tailpipes of cars.
Conclusion
If we measure and model air pollutants only at ground level, we may substantially overestimate exposure at higher floors for some pollutants. This is especially true for pollutants whose levels decrease sharply within close proximity of the source, such as PNC.

Poster sessions Wednesday, October 12, 2016

Biomonitoring

We-Po-01

Relationship between the external exposure and biomarker of 1-bromopropane in workplace

Yu-Wen Lin, *Fu Jen Catholic University, New Taipei City, China, Republic of (Taiwan)*
Jyun-De Wu, *Chang Jung Christian University, Tainan City, China, Republic of (Taiwan)*
Cheng-Ping Chang, *Chang Jung Christian University, Tainan City, China, Republic of (Taiwan)*
Hui-Ling Lee, *Fu Jen Catholic University, New Taipei City, China, Republic of (Taiwan)*
Chung-Han Ko, *Fu Jen Catholic University, New Taipei City, China, Republic of (Taiwan)*

Aims: 1-Bromopropane (1-BP) has been applied as the metal cleaning agents in manufacture industries since the ozone-depletion substances were banned. Exposure of 1-bromopropane has been found in related with the peripheral neuropathy. The ACGIH classified 1-BP as an A3 chemical. Several occupational disease cases due to the 1-BP exposure were confirmed in Japan, USA, China and Taiwan. This study investigated the exposure profiles of workplaces using 1-BP. Also, the 1-BP metabolite, n-acetyl-S-(n-propyl)-L-cysteine (AcPrCys) in urine was quantified to establish the relationship between the external dose and internal dose of 1-BP.
Methods: Three factories that using 1-BP for metal cleaning were investigated. The 1-BP sampling protocol was modified from OSHA Methods No. 1017 and NIOSH No. 1025. The sampling flowrate was set at 200 mL/min. The sampling durations were 6 hours. All samples were analyzed by GC/MS with a method detection limit of 0.84 ppb. The urine samples were collected on the day that the personal air samples collected. n-Acetyl-S-(n-propyl)-L-cysteine (AcPrCys), the 1-BP metabolite, was selected as the biological exposure index (BEI) and quantified by HPLC-MS/MS. The limit of quantitation was 0.023 ng/mL. A total of 100 area and personal air samples and 76 urine samples (before and after the work shift) were collected from three plants. The study was proofed by the Fu Jen Catholic University IRB.
Results: The 1-BP concentrations of the 95th percentiles of exposure group (cleaning operation) were ranged from 31.44 to 41.96 ppm for personal samples. The 95th percentile1-BP concentrations of area samples air were ranged from 20.43 to 41.84 ppm. The AcPrCys concentrations in urines were between 11.58 mg/g cre and 4,945.71 mg/g cre before shifts, and the AcPrCys concentrations in urines were from 3.72 mg/g cre to 7,818.26 mg/g cre after shift. The correlation between the after-shift urine AcPrCys concentrations and the 1-BP concentrations of personal air samples was significant ($r = 0.679$, $p = 0.05$). This implied that "after-shift" is a better specimen collecting time for 1-BP BEI, AcPrCys.

Conclusions: The personal air sampling results exceeded Cal OSHA PEL 5 ppm and ACGIH TLV®-TWA 0.1 ppm. The after-shift urine concentration of AcPrCys was a good biomarker to represent the internal dose of 1-BP exposure. Meanwhile, from the occupational hygiene concern,effective engineering control, respiratory protection program and dermal protection program should be implanted to limit the exposures.

We-Po-02

Human Urinary Biomarkers of the UV Filter Ethylhexyl Salicylate

Daniel Bury, *Institute for Prevention and Occupational Medicine of the German Social Accident Insurance (IPA), Bochum, Germany*
Peter Griem, *Symrise AG, Holzminden, Germany*
Frank-Hinrich Köster, *Symrise AG, Holzminden, Germany*
Thomas Brüning, *Institute for Prevention and Occupational Medicine of the German Social Accident Insurance (IPA), Bochum, Germany*
Holger Martin Koch, *IPA, Bochum, Germany*

2-Ethylhexyl salicylate (syn. EHS, octyl salicylate, octisalate) is a UV filter substance regularly used in sun screen formulations and in other personal care products [1,2] in concentrations up to 5% (maximum authorized concentration within the USA and the EC). Fish model experiments showed endocrine activity of EHS under the conditions of the test [3]. Due to its widespread use, internal exposure of the general population towards EHS is not unlikely. Thus, EHS was selected as a substance of interest by the cooperation project between the German Federal Ministry for the Environment (BMUB) and the German Chemical Industry Association (VCI), which has the aim to provide biomarker based exposure data for fifty emerging substances of concern.
We investigated metabolism and renal excretion of EHS after oral dosage (5 mg) in three male individuals. Consecutive urine samples were collected for a period of 48 h after dosage. Urine samples were analyzed with online-SPE-LC-MS/MS after enzymatic deconjugation.
In this manner, we tentatively identified three predicted alkyl chain oxidized metabolites of EHS (hydroxy EHS, oxo EHS, and EHS carboxylic acid). Analytical standards as well as stable isotope labeled internal standards for these metabolites were obtained, to allow quantitative analysis of urinary metabolite concentrations. We here report the presence of the three metabolites in urine after oral exposure, as well as their elimination kinetics and urinary conversion factors.
Further studies will investigate the occurrence of the identified EHS metabolites after dermal exposure, as well as their occurrence in samples from the general population. The suitability of the metabolites as biomarkers of exposure will be evaluated.
The study has been approved by the ethical review board of the Ruhr-University Bochum (Reg. No.: 4288-12).
References
[1] Kerr et al. Clinical and experimental dermatology 2011;36:541-3.
[2] Manová et al. The British journal of dermatology 2014;171:1368-74.
[3] Kunz et Fent Aquatic Toxicology 2006;79:305-24.

This page intentionally left blank.

We-Po-03

Detection of tetrahydroxylated metabolites in hair as biomarkers of human exposure to Polycyclic Aromatic Hydrocarbons

Nathalie Grova, *Luxembourg Institute of Health, Human Biomonitoring Research Unit, Luxembourg*
Emilie, Marie Hardy, *Luxembourg Institute of Health, Esch-sur-Alzette, Luxembourg*
Brice M.R. Appenzeller, *Luxembourg Institute of Health, Esch-sur-Alzette, Luxembourg*

This study aimed at developing a method for the determination of tetrahydroxylated metabolites (tetra-OH-PAHs) in hair as new biomarkers of exposure to Polycyclic Aromatic Hydrocarbons. This method based on gas chromatography coupled with tandem mass spectrometry allowed the quantitative analysis of 10 tetrahydroxylated metabolites representative of 4 parent PAH's (phenanthrene (Phe), chrysene (Chry), benz[a]anthracene (B[a]A) and benzo[a]pyrene (B[a]P)). Negative chemical ionization was selected for the analysis of tetra-OH-PAHs in hair. The calibration curve performed on 10 concentration levels was linear from the LOQ up to 40 pg/mg for all the isomers of tetra-OH-Phe, tetra-OH-Chry, tetra-OH-B[a]A and tetra-OH-B[a]P tested. The coefficients of determination were above 0.970 and the recoveries established for each compound were evaluated between 55.0 % and 82.6 %. This method allows reaching LOQs ranging from 0.05 to 1 pg/mg in hair depending on compound.
The applicability of tetra-OH-PAH analysis in hair as biomarkers of PAH exposure was evaluated in a dose-response study conducted on 64 rats (Long Evans females / n= 8 per groups) under repeated exposure (3 times per week) to a mixture of 16 PAHs at low doses (0.01 - 0.8 mg/kg) for 90 days. The analysis of rats' hairs demonstrated the presence of 1 isomer of tetra-OH-Phe, 4 isomers of tetra-OH-B[a]A, 2 isomers of tetra-OH-Chry and 4 isomers of tetra-OH-B[a]P. With the exception of B[a]A-r-7,t-8,c-9,t-10-tetrahydrotetrol, B[a]A-r-7,c-8,t-9,t-10-tetrahydrotetrol and B[a]P-r-7,t-9,t9,c-10-tetrahydrotetrol, which were only measured at the two highest level of exposure, all tetra-OH-PAHs were detected in hairs of rats, whatever the dose of exposure. Strong linear relationship (R2 ranging between 0.805 and 0.964, $p<0.001$) was observed between the administered dose and the tetra-OH-PAH concentration in hairs for 7 out of the 10 analytes.
To confirm whether the method was sufficiently sensitive to monitor environmental levels of exposure in humans, 34 hair specimens collected from the general population were analyzed. The results demonstrated the presence of 10 different tetra-OH-PAHs in the hair of volunteers, the most common being Phe-r1,t-2,t-3,t-8-tetrahydrotetrol and B[a]P-r-7,t-8,t-9,c-10-tetrahydrotetrol. By widening the range of PAH metabolites used as biomarkers of exposure so as to include the analysis of PAH tetrahydroxylated forms (especially those exhibiting more than 5 aromatic rings), the method presented here will enable multi-exposure assessments which are more accurately representative of actual situations of exposure to these compounds.

We-Po-04

Trisaminohexyl Isocyanurate, a Biomarker for HDI Isocyanurate Exposure

Zachary Robbins, *University of North Carolina at Chapel Hill, Chapel Hill, North Carolina, United States*
Wanda M Bodnar, *University of North Carolina at Chapel Hill, Chapel Hill, North Carolina, United States*
Zhenfa Zhang, *University of North Carolina at Chapel Hill, Chapel Hill, North Carolina, United States*
Avram Gold, *University of North Carolina at Chapel Hill, Chapel Hill, North Carolina, United States*
Leena A Nylander-French, *University of North Carolina at Chapel Hill, Chapel Hill, North Carolina, United States*

Exposures to monomeric and polymeric 1,6-hexamethylene diisocyanate (HDI) in the automotive refinishing industry have been well characterized. However, biological monitoring is limited to a hydrolysis product of HDI monomer, 1,6-hexamethylene diamine (HDA). Further, inhalation and skin exposures to HDI monomer during painting operations are low compared to its oligomers. HDI isocyanurate constitutes the largest fraction of inhalation and skin exposures in automotive spray paints and has been shown to possess a greater sensitizing capacity than HDI monomer. Additionally, HDI isocyanurate penetrates the skin at a faster rate than HDI monomer, emphasizing the need to shift research focus towards the fate of HDI isocyanurate after exposure. We have developed a sensitive and specific method to quantify trisaminohexyl isocyanurate (TAHI), a hydrolysis product of HDI isocyanurate, in urine. Two end of day urine samples were collected from two spray painters exposed to HDI isocyanurate during automotive spray-painting operations. Urine samples were hydrolyzed with sulfuric acid, made basic with sodium hydroxide, and extracted with dichloromethane. The extracts were derivatized with acetic anhydride and excess reagent was removed with phosphate buffer and sodium sulfate before sample was dried and reconstituted with water for analysis. A calibration curve was created by spiking urine from non-exposed persons with the synthesized standard TAHI with a concentration range of 0.04-5.00 µg/L and a synthesized internal standard trisaminoheptyl isocyanurate (TAHpl; 2.5 µg/L). Samples were analyzed with nanoUPLC-ESI-MS/MS for the precursor ions m/z 553.3 (trisacetamidohexyl isocyanurate, TAAHI) and m/z 595.3 (trisacetamidoheptyl isocyanurate, TAAHpl) using selected reaction monitoring. Urine samples collected from two workers with significant breathing-zone and skin exposure to HDI monomer (breathing zone: 216 and 79.5 µg/m3; skin: 8.3 and 1.3 µg/mm3) and HDI isocyanurate (breathing zone: 65432.3 and 20926.6 µg/m3; skin: 3949.1 and 366.0 µg/mm3) had TAHI levels of 0.43 and 2.29 µg/L, respectively, while HDA concentrations in the same samples were 0.21 and 0.37 µg/L, respectively. The results indicate that this method can be used to quantitate HDI isocyanurate biomarker TAHI in urine of exposed workers. Quantitation of HDI isocyanurate biomarker in urine in conjunction with HDI monomer biomarker from workers exposed to HDI-containing spray paints will aid in the investigation of the adverse effects of inhalation and skin exposure and individual susceptibility in exposed workers.

Environmental/Human Health

We-Po-05

Indoor environmental quality in multi storey office buildings and implication on the health and safety of workers. Evaluation of Lagos State Government buildings in Nigeria

Shamusideen Kadiri, *Zub Chord Tech Ventures, Lagos, Lagos, Nigeria*

INTRODUCTION:
Indoor environmental quality (IEQ) refers to the quality of a building's environment in relation to the health and wellbeing of those who occupy space within it. IEQ is determined by many factors, including lighting, air quality, and damp conditions. Workers are often concerned that they have symptoms or health conditions from exposures to contaminants in the buildings where they work. One reason for this concern is that their symptoms often get better when they are not in the building. An office building should satisfy occupants' needs and promote efficiency of indoor environmental quality. The success or failure of a building depends on the implementation and sustainability of the IEQ. The building should be designed with the aim of producing a high-quality interior environment, so that the health and safety (H&S) of the occupants or employees are not compromised. In this paper we describe health risks associated with indoor environments, illuminate barriers to overcoming these risks, and provide policy recommendations to achieve healthier indoor environments.
METHODS. The overall purpose of the study was to determine the level of satisfaction of building occupants' in terms of Indoor Environmental Quality (IEQ) and how it affect their health and safety vis-a-vis their productivity at work. The questionnaire was in five sections (A-E). The designs of the questionnaire envisage a maximum of 15 minutes for its completion.
RESULTS.Observations from the data led to the view that the satisfactory level of IEQ awareness is low among the employees It was found out most these employees faces a multitude of Hazards in their offices which include biological and chemical contaminants, as well as poor ergonomics, lighting, and physical design. These hazards cause and exacerbate a variety of adverse health effects in them, ranging from asthma to sick building syndrome to cancer.
CONCLUSIONS Organisational structure needs to be formed that will enlighten occupants about factors that contribute to poor indoor air quality (IAQ). Employees must be well informed of such risks in order to make useful health decisions; they must also understand both the health consequences of poor indoor environmental quality, and some simple and feasible interventions to improve IEQ. Indeed, policy changes at multiple levels are needed to achieve healthy workplace indoor environments.

We-Po-06

Long-term exposure to ambient air pollution and ischemic heart disease among elderly residents of Tokyo metropolitan area, Japan

Haruya SAKAI, *Japan Automobile Research Institute, Tsukuba, Japan*
Tazuko Morikawa, *Japan Automobile Research Institute, Tsukuba, Japan*
Yukika Toda, *Japan Automobile Research Institute, Tsukuba, Japan*
Tsuyoshi Ito, *Japan Automobile Research Institute, Tsukuba, Japan*
Akiyoshi Ito, *Japan Automobile Research Institute, Tsukuba, Japan*
Hiroshi Koike, *Japan Automobile Research Institute, Tsukuba, Japan*
Hiroki Kishikawa, *Mukogawa Women's University, Nishinomiya, Japan*
Masaji Ono, *National Institute for Environmental Studies, Tsukuba, Japan*
Kenichi Azuma, *Kinki University Faculty of Medicine, Osakasayama, Japan*
Satoshi Nakai, *Yokohama National University, Yokohama, Japan*
Iwao Uchiyama, *Louis Pasteur Center for Medical Research, Kyoto, Japan*

Numerous epidemiological studies have demonstrated an association between traffic-related air pollution and ischemic heart disease (IHD) in the United States and Europe. However, there are few studies describing the association between air pollution and IHD in Japan. The objective of this study was to examine the association between traffic-related air pollution and IHD in elderly people living in Tokyo metropolitan area, Japan. The subjects included 6,000 elderly people (≥ 65 years old in April 2014) who lived in roadside (< 50 m from highway) and non-roadside (> 500 m from highway) areas. IHD was assessed using self-reported doctor diagnosis and history of medication for myocardial infarction and/or angina pectoris collected by questionnaire. The questionnaire comprised 52 items, including body height, body weight, smoking status, drinking habits, and medical history of both the study participants and their parents. To assess the individual levels of exposure to traffic-related air pollution, the annual concentrations of nitrogen oxides (NOx) and elemental carbon (EC) in fine particles at participants' residential addresses in 2009 were estimated using two plume dispersion models: the National Institute of Advanced Industrial Science and Technology - Atmospheric Dispersion Model for Exposure and Risk Assessment (AIST-ADMER) and the Ministry of Economy, Trade and Industry - Low Rise Industrial Source dispersion model (METI-LIS). A total of 3,190 participants answered the questionnaire (1,589 from roadside areas and 1,601 from non-roadside areas). The estimated annual exposure levels of NOx and EC for each participant varied from 11.5 to 110 ppb, and 0.217 to 3.29 μg/m3, respectively. We stratified all participants into four groups by exposure levels for each pollutant. After adjusting for confounders (sex, age, body mass index, smoking status, and drinking habits), multiple logistic regression analyses revealed that compared to participants with the lowest exposure levels of NOx (11.5-27.9 ppb) and EC (0.217-0.577 μg/m3), odds ratios for participants with the highest NOx exposure (45.7-110 ppb) and participants with the highest EC exposure (1.14-3.29 μg/m3) were 1.47 (95% confidence interval [CI] 0.96-2.27) and 1.41 (95% CI 0.92-2.20), respectively. These results suggest that there is a positive association between long-term exposure to traffic-related air pollution and IHD among elderly residents of Tokyo metropolitan area, Japan.

We-Po-08

Life-stage specific windows of susceptibility to lead and manganese exposure and children's behavior

Megan Horton, *Icahn School of Medicine at Mount Sinai, New York, New York, United States*
Paul Curtin, *Icahn School of Medicine at Mount Sinai, New York, New York, United States*
Sandra Martinez, *Instituto Nacional de Perinatología, Mexico City, Mexico*
Chris Gennings, *Icahn School of Medicine at Mount Sinai, New York, New York, United States*
Lourdes Schnaas, *Instituto Nacional de Perinatología, Mexico City, Mexico*
Martha Maria Téllez Rojo, *Instituto Nacional de Salud Pública, Cuernavaca, Mexico*
Manish Arora, *Icahn School of Medicine at Mount Sinai, New York, New York, United States*
Robert Wright, *Icahn School of Medicine at Mount Sinai, New York, New York, United States*

Background: One challenge in children's environmental health is to identify windows of susceptibility for environmental toxicants such as neuroactive metals. We recently developed cutting-edge methods to objectively identify windows of susceptibility to environmental toxicant exposure throughout gestation and early childhood. Using our novel tooth biomarker with distributed lag models (DLMs), we identified sensitive windows for the associations between perinatal exposure to lead (Pb) and manganese (Mn) with children's behavior.
Methods: We collected deciduous teeth from 85 subjects enrolled in the ELEMENT cohort and used DLM regression to examine the time dependent associations between perinatal exposure to Pb and Mn with performance on the composite scores of the Behavior Assessment System for Children (BASC-2) at 4-5 years of age. These models regress tooth Pb and Mn along a moving window of weeks starting in the 2nd trimester (-4m) through the first 12 months of life (12m) to predict behavior at 4-5 years of age. Susceptibility windows are identified by time points where Bonferroni-adjusted 95% confidence intervals on the association between exposure and outcome do not include 0. Analyses controlled for child sex and socioeconomic status (SES).
Results: Exposure to higher Pb or Mn near 12m was associated with significantly worse BASC internalizing scores ($p < 0.05$). Mn had 2 windows of vulnerability; higher Mn exposure in the 2nd trimester was associated with significantly better BASC-2 performance (i.e., lower scores). This is biologically consistent with Mn acting as both a nutrient and toxicant at different developmental windows.
Discussion: Using novel tooth-matrix biomarkers that provide both prenatal and postnatal measures of exposure, we observed that higher Pb and Mn exposure at 12-months of age was associated with lower performance on behavior measures at 4-5 years of age. Prenatal Mn exposure, however, was associated with better behavioral outcomes.

We-Po-11

Short-term effects of exposure to air pollution and mortality: are those previously diagnosed with cancer at greater risk?

Paul Villeneuve, *Carleton University, Ottawa, Canada*
Yi Li, *Carleton University, Ottawa, Ontario, Canada*
Shirley Mills, *Carleton University, Ottawa, Ontario, Canada*
Keith Van Ryswyk, *Carleton University, Ottawa, Ontario, Canada*
Eric Lavigne, *Health Canada, Ottawa, Canada*

Aim: It is well recognized that short-term, or daily increases, in air pollution concentrations increases the risk of mortality. A number of studies have also found that those with pre-existing disease such as diabetes, or a history of heart disease, may be particularly susceptible to these effects. To date, there have been few attempts to examine whether those previously diagnosed with cancer may be more vulnerable due to their comprised health status.
Methods: We applied a time-stratified case-crossover study to investigate whether increases in NO2 and PM2.5 were associated with increased risks of mortality. These methods were applied to approximately 200,000 deaths that occurred within the Ontario Tax cohort. These deaths occurred in a follow-up of approximately 660,000 adults who lived in the province of Ontario and who were followed up between 1981 and 2004. Diagnoses of cancer were determined through record linkage to national cancer incidence data. Daily estimates of air pollution were assigned based on the reported place of death obtained from death certificates. Conditional logistic regression methods were applied to characterize the risk of death, and adjustment was made for meteorological variables including temperature, and relative humidity.
Results: We found that daily increases in NO2 were associated with increased mortality during the warm season, but not during the cool season. In particular, a 5 ppb increase in NO2 was found to increase mortality by 0.9% (95% CI=0.2%-1.6%). Stratified analyses of individuals based on whether they had previously diagnosed with cancer or not revealed no substantial differences in risk.
Conclusions: This study confirmed previous findings that short-term increases in air pollution are associated with increases in mortality. Importantly, our findings suggest that those diagnosed with cancer are not a susceptible population

We-Po-13

Factors Affecting Occupational Health Among Mushroom Farmers Dond Poo Daeng Village Huai Po Sub - Distric, Muang Distric,Kalasin Province

Kallaya Harnpicharnchai, *Faculty of Public Health, Mahasarakham, Thailand*

Factors affecting occupational health were examined on mushroom farmers in Dond Poo Daeng Village Huai Po Sub - Distric, Muang Distric, Kalasin Province. The study was devided into two parts. The first part was collecting the data of mushroom farmers using interview forms. The first part included general information, working history, smoking habit, exercise habit, and habit of wearing personal protective equipment and past health history symptoms. The second part was collecting atmospheric bioaerosal (n=120) by the area sampling technique, in accordance to NIOSH Manual of Analytical Method number 0800. The samples were the 41 mushroom farmers working the day shift.

The results showed that most mushroom farmers were female (94.6%). Their average age was 44.3±7.57 years old. Most education level was primary school (88.9%). Approximately, 82.50%, 36.7% and 60.0% of them had non-smoking, drinking alcohol and exercising, respectively. For the results of the air sampling analysis showed that the highest average concentrations of bioaerosal, fungi and bacteria , were 1182 ± 546 CFU/m3 and 2601 ± 462 CFU/m3 respectively.

The occupation (Working history), congenital disease, skin dermatitis and allergy (past health history) and habit of wearing personal protective equipment was significantly associated with respiratory symptoms (p-value<0.05).

KEY WORDS: OCCUPATIONAL HAELTH / MUSHROOM FARMERS / BIOAEROSAL

We-Po-15

Mercury exposure and it's health effect on children in six cities, China

Daeseon Kim, *National Institute of Environmental Research, Korea, Inchon, Inchon, Korea, South*
Si Eun YOO, *National Institute of Environmental Research, Inchon, Inchon, Korea, South*
Sang Hoon Nam, *National Institute of Environmental Research, Inchon, Inchon, Korea, South*
Seung Do Yu, *National Institute of Environmental Research, Inchon, Inchon, Kyrgyzstan*
Kyunghee CHOI, *National Institute of Environmental Research, Inchon, Inchon, Korea, South*
Xiaochuan Pan, *School of Public Helath, Peking university, China, Beijing, Beijing, China, People's Republic of*

Aims: "Mercury exposure and it's health effect on children in China" has been conducted from 2012 to enhance the infrastructure for international joint research by building the Korea-China cooperation network and to protect children's health from environmental pollution between the two countries.
Methods: 1,008 students of fourth grade from twelve elementary schools were recruited in six areas in six provinces which were ranked in high mercury emission on 2005, Suning in Hebei, Lanzhou in Gansu, Tieling in Liaoning, Nanning in Guangxi, Xinxiang in Henan Province, and Shizuishan Ningxia Hui Autonomous Region. Urine and hair samples were collected from twelve areas as biomarker samples, but blood samples were collected from only one area, Suning, along with a questionnaire survey, neurobehavioral test and balance test.
Results: In the geometric mean(GM), blood mercury was 0.44 µg/L, urine mercury 0.68 µg/g-cr and hair mercury 0.28 µg/g in total samples. The GM of urine mercury were 0.36 in Suning, 0.28 in Lanzhou, 0.82 in Tieling, 0.44 in Nanning, 0.44 in Xinxiang and 0.44 µg/g-cr in Shizuishan. The GM of hair mercury were 0.14 in Suning, 0.18 in Lanzhou, 0.35 in Tieling, 0.50 in Nanning, 0.12 in Xinxiang and 0.15 µg/g in Shizuishan. Samples were divided into four groups by mercury concentration, the 1st group was the highest 25% samples and 4th group was the lowest 25% samples in mercury concentration. In the neurobehavioral test, the 1st group in urine showed relatively higher score than the 4th group in memory scanning(MS) test ($p<0.05$). Hair mercury was found that higher score in the 1st group in visual retention, MS, aim tracing test ($p<0.01$). In postural reaction, the 1st group in hair mercury showed higher value in tremor test and body sway test, the 1st group in urine showed the relation with sway area in body sway test.
Conclusion: Tieling, the largest Hg-emitting area among 32 provinces in China, showed the highest Hg concentrations in urine among 6 survey areas and other 5 areas showed no significant differences. Nanning showed the highest mercury in hair, 2-3 times higher that other areas. In the NES results, the high group (25%) with high Hg concentrations showed high neurobehavioral performance. In balance test, the 1st high Hg group showed statistically high significance in reaction time and body sway. In exposure factors, hair mercury showed relation with family income, adjacency with car road, frequency of fish intake, frequency of marine product intake.

We-Po-16

Toward a Comprehensive Assessment of the Health Effects of Chronic VOC Releases from Gas Stations

M. Hilpert, *Johns Hopkins Bloomberg School of Public Health, Baltimore, Maryland, United States*
B. Adria-Mora, *Johns Hopkins Bloomberg School of Public Health, Baltimore, MD, United States*
M. Grau-Perez, *Johns Hopkins Bloomberg School of Public Health, Baltimore, MD, United States*
K.E. Nachman, *Johns Hopkins Bloomberg School of Public Health, Baltimore, MD, United States*
J. Ni, *Johns Hopkins University, Baltimore, MD, United States*
A.M. Rule, *Johns Hopkins Bloomberg School of Public Health, Baltimore, MD, United States*

Aims: The aim of this study is to provide a better understanding of the health effects of hydrocarbon releases, which chronically occur at gas stations. Such releases are of concern, because fuel contains toxic chemicals such as BTEX. We are specifically interested in quantifying the amounts released to the aqueous, atmospheric, and subsurface environments, and in assessing associated environmental exposures.
Methods: We reviewed the literature in order to assess vapor releases that occur during vehicle refueling and that occur through the vent pipe of storage tanks. We also reviewed policies with regard to the implementation of pollution prevention technology. Furthermore, we performed laboratory experiments in order to determine the fate of spilled fuel.
Results: Employees at service stations are among those with greatest exposure to benzene originating from gas stations. However, also occupying residences, businesses, and other structures neighboring gas stations can be exposed to fuel vapors originating at the gas station, though typically at lower concentrations. While it is clear that populations living near gas stations are exposed to released chemicals, health effects are not well understood.
Our experiments with spilled fuel droplets have shown that evaporation is greater for gasoline, while infiltration is greater for diesel spills. Diesel has therefore a higher potential for soil contamination because of the higher infiltrated mass.
Discussion: Despite the poor understanding of the health effects of living near gas stations, policies are not uniformly implemented that would minimize chronic vapor releases by adoption of available pollution prevention technology. This is in part due to the fact that cost-benefit analyses typically only account for the cost of implementation and maintenance of such technology. We believe that policy makers should also account for public health burdens due to released pollutants and energy-saving benefits due to valuable hydrocarbons not wastefully released to the environment.

UTRECHT
THE NETHERLANDS
OCTOBER 9-13
ISES 2016

We-Po-17

Municipal Solid Waste Burning: Discoloring the Taj Mahal and Human Health Impacts in Agra

Raj Lal, *Georgia Institute of Technology, Atlanta, GA, United States*
Ajay Nagpure, *University of Minnesota, Minneapolis, United States*
Lina Luo, *Georgia Institute of Technology, Atlanta, Georgia, United States*
Sachi Tripathi, *Indian Institute of Technology - Kanpur, Kanpur, India*
Anu Ramaswami, *Minnesota, Minneapolis, United States*
Michael Bergin, *Duke University, Durham, United States*
Armistead Russell, *Georgia Tech, Atlanta, United States*

The Taj Mahal - an iconic World Heritage monument built of white marble - has become discolored with time, due, in part, to high levels of particulate matter (PM) soiling its surface. Such discoloration has required extensive and costly treatment and despite previous interventions to reduce pollution in its vicinity, the haze and darkening persists. PM responsible for the soiling has been attributed to a variety of sources including industrial emissions, vehicular exhaust and biomass burning, but the contribution of the emissions from the burning of open municipal solid waste (MSW) may also play an important role. A recent source apportionment study at the Taj Mahal showed biomass burning emissions, which would include MSW emissions, accounted for nearly 40% of organic matter (OM) - a component of PM - deposition to its surface; dung cake burning, used extensively for cooking in the region, was the suggested culprit and banned within the city limits, although the burning of MSW, a ubiquitous practice in the area, may play a more important role in local air quality. Using spatially detailed emission estimates and air quality modeling, we find that open MSW burning leads to about 150 (+-130) mg m-2 yr-1 of PM being deposited to the surface of the Taj Mahal compared to about 12 (+-3.2) mg m-2 yr-1 from dung cake burning. Those two sources, combined, also lead to an estimated 713 (377-1050) premature mortalities in Agra each year, dominated by waste burning in socioeconomically lower status neighborhoods. An effective waste management strategy would reduce soiling of the Taj Mahal, improve human health, and have additional aesthetic benefits.

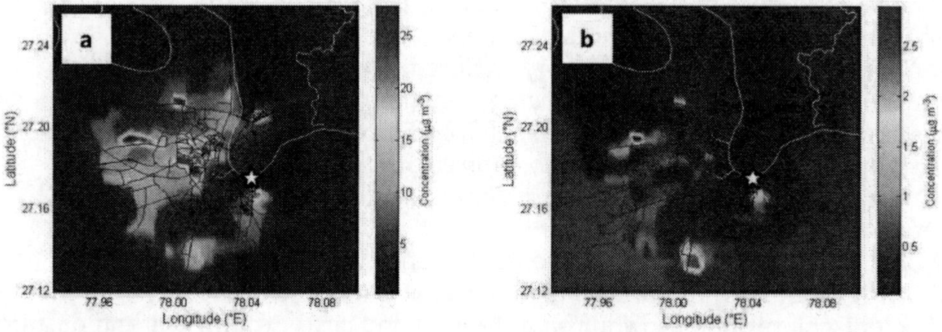

Fig 1. Annual average particulate matter (PM) concentrations in Agra from: a, open MSW burning **b,** dung cake burning. Modeled [PM] at the Taj Mahal (depicted by the white star) was 4.31 ($\pm$ 3.87) $\mu g\ m^{-3}$ from MSW emissions and 0.335 ($\pm$ 9.15 x 10^{-2}) $\mu g\ m^{-3}$ from dung cake burning emissions. These concentration profiles generated in AERMOD showed higher pollution from both forms of biomass burning concentrated in areas of lower SES. Organic matter (OM) and black carbon (BC), the PM components modeled, concentration profiles show the same spatial variation, but OM concentrations contribute more than BC to ambient PM (SI Fig. 5).

Modeled PM concentrations from MSW and dung cake burning in Agra

Measuring/monitoring/strategy

We-Po-18

Associations among personal care product use patterns and exogenous hormone use in the NIEHS Sister Study

Kyla Taylor, *NIEHS, Durham, United States*
Donna Baird, *NIEHS, DURHAM, United States*
Amy Herring, *UNC, Chapel Hill, United States*
Hazel Nichols, *UNC, Chapel Hill, United States*
Lawerence Engel, *UNC, Chapel Hill, United States*
Dale Sandler, *NIEHS, Durham, United States*
Melissa Troester, *UNC, Chapel Hill, United States*

Background: It is hypothesized that certain chemicals in personal care products may alter risk of adverse health outcomes. The primary aim of this study was to use a data-centered approach to classify complex patterns of exposure to personal care products and to understand how these patterns vary according to use of exogenous hormone exposures, oral contraceptives (OCs) and postmenopausal hormone therapy (HT).
Methods: The NIEHS Sister Study is a prospective cohort study of 50,884 U.S. women. Limiting the sample to non-Hispanic blacks and whites (N=47,019), latent class analysis (LCA) was used to identify groups of individuals with similar patterns of personal care product use based on responses to survey questions. Personal care products were categorized into three product types (beauty, hair, and skincare products) and separate latent classes were constructed for each type. Prevalence differences (PD) were calculated to estimate the association between exogenous hormone use, as measured by ever/never OC or HT use, and patterns of personal care product use.
Results: Three latent classes were identified for both the beauty and hair product groups; the skincare product group had four classes. There were strong differences in latent class distribution by race, particularly for hair care products. Irrespective of race, exogenous hormone exposures were associated with higher levels of product use, especially beauty and skincare products.
Discussion: Personal care product usage patterns differed by race and were associated with ever OC and HT use. Future studies should consider personal care product exposures with other exogenous exposures when modeling health risks.

Figure 1. Item-response probability conditional on class membership

1a. Beauty Products

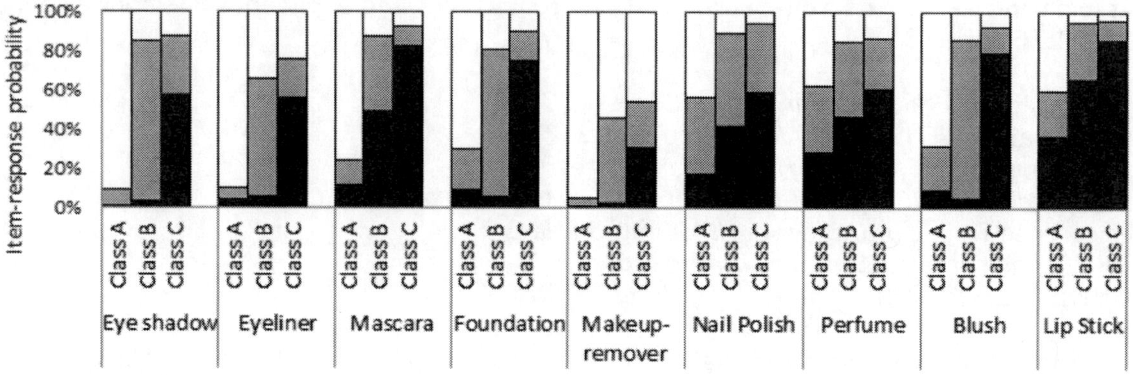

1b. Hair Products

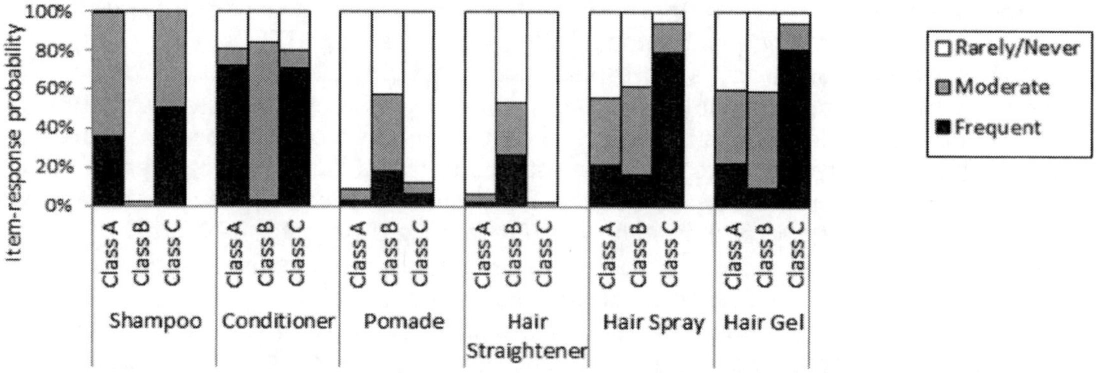

1c. Skincare Products

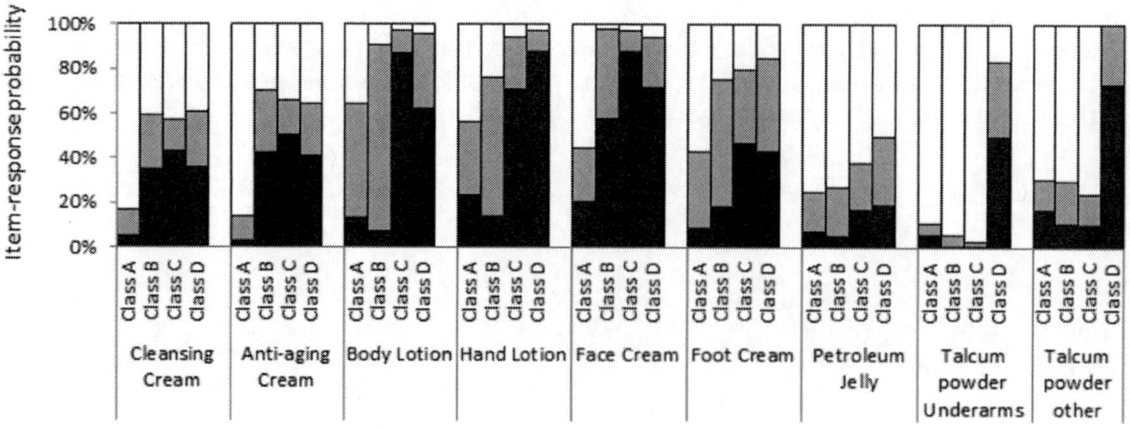

This page intentionally left blank.

We-Po-19

Factors determining the variability of exposure to contact allergens from topical aromatherapy

Olivia YAOUANC, *LERCCO, Brest, France*
Nicolas DORNIC, *LERCCO (laboratoire d'évaluation du rsique chimique pour le consommateur - Laboratory of chemical risk assessement for the consumer), brest, France*
Anne-Sophie FICHEUX, *LERCCO, Brest, France*
Audrey BERNARD, *LERCCO, Brest, France*
Grégoire CHEVILLOTTE, *LERCCO, Brest, France*
Alain-Claude ROUDOT, *LERCCO, Brest, France*

Essential oils are products of complex composition, which can be used in different ways by the consumer, i.e. per oral, inhalation or dermal route. For the latter, it appears that the consumption of such products entails exposure to contact allergens in varying amounts, depending on the type of oil applied. The aim of this work was to study the factors that may influence inter-individual variability of exposure to these allergens.
A study was conducted on the Lavender, an essential oil particularly consumed, as demonstrated our previous work on essential oils consumption by the French population. To do this, the qualitative and quantitative composition of the different types of Lavender oils has been sought in the available publications. Different oils sold on the market were then bought in different locations, e.g. pharmacies or shops, and then weighed to determine the exact mass of the essential oils drops. The properties of the containers, i.e. size of the bottles, diameters and length of the dropper were also studied. Different scenarios were then considered in order to assess the exposure to contact allergens from lavender by a probabilistic method, using Monte Carlo random simulations with @Risk 6 software.
Our results show that among the "classic" determining factors such as the amount applied on the skin and the frequency of use, types of hybrid, geographical origin of the original plant but also the manufacturing process are factors that may influence the exposure. Surprisingly, for an equivalent consumption with two references of oils, an a priori innocuous parameter such as the size of the dropper can vary the exposure in µg / cm2 in the ratio of one to two for an individual.
With so many different factors, the exact exposure of the consumer resulting from essential oil consumption is difficult to ascertain. Besides this, our study shows the value of setting standards at a Community level for essential oils, in order to protect consumers against skin sensitization.

We-Po-20

Task-based approach used on surfaces sampling strategy definition - The case of antineoplastic occupational exposure

Susana Viegas, *ESTeSL-IPL, Lisbon, Portugal*
Ana Cebola Oliveira, *Environment and Health Research Group/ Lisbon School of Health Technology, Lisbon, Portugal*
Mário Pádua, *Environment and Health Research Research Group/Lisbon School of Health Technology, Lisbon, Portugal*

Task-based approach implicates identifying all the tasks developed in each workplace aiming to refine the exposure characterization. The starting point of this approach is the recognition that only through a more detailed and comprehensive understanding of tasks is possible to understand, in more detail, the exposure scenario. In addition allows also the most suitable risk management measures identification. This approach can be also used when there is a need of identifying the workplace surfaces for sampling chemicals that have the dermal exposure route as the most important. In this case is possible to identify, through detail observation of tasks performance, the surfaces that involves higher contact (frequency) by the workers and can be contaminated.
A study was developed in one Portuguese Hospital aiming to identify the surfaces to sample when performing occupational exposure assessment to antineoplastic agents. The selected sampling spots were judged to be the surfaces potentially contaminated and, simultaneously, more frequently handled/touched by the workers in each task. 5-fluorouracil (5FU) was used as surrogate marker for surfaces contamination by all antineoplastic drugs. Samples were collected by wipe-sampling method and analyzed by HPLC-DAD.
45 samples were analyzed from different surfaces (antineoplastic preparation and administration services). Results ranged from < LOD to 75.24 ng/cm2. The higher value was obtained in a support table of the antineoplastic preparation room. The following two values (42.57 and 57.70 ng/cm2) were obtained in surfaces of the administration room handle/touch normally without protection gloves: chair used by the nurses for doing therapeutic registers and the phone, also used only by the nurses of the administration room.
Results point out for the importance of task-based approach for surfaces sampling strategy definition and for the improving of the cleaning protocols (surfaces to clean, cleaning products and cleaning frequency). Additionally, this approach allowed the identification of the tasks (surfaces) that represent higher risk for workers.
Keywords: Task-based approach; occupational exposure assessment; surfaces sampling; antineoplastic agents

We-Po-22

The MAPEC_LIFE Study: indoor/outdoor air pollution exposure and lifestyles of the prospective cohort

Antonella De Donno, *University of Salento, Lecce, Italy*
Francesco Bagordo, *University of Salento, Lecce, Italy*
Claudia Zani, *University of Brescia, Brescia, Italy*
Elisabetta Ceretti, *University of Brescia, Brescia, Italy*
Milena Villarini, *University of Perugia, Perugia, Italy*
Samuele Vannini, *Univeristy of Perugia, Perugia, Italy*
Annalaura Carducci, *University of Pisa, Pisa, Italy*
Beatrice Casini, *University of Pisa, Pisa, Italy*
Sara Bonetta, *University of Torino, Torino, Italy*
Tiziana Schilirò, *University of Torino, Torino, Italy*
Silvia Bonizzoni, *Comune di Brescia, Brescia, Italy*
Alberto Bonetti*C.S.M.T. Gestione S.c.a.r.l., Brescia, Italy*
Tiziana Grassi, *University of Salento, Lecce, Italy*
Marcello Guido, *University of Salento, Lecce, Italy*
Gabriele Devoti, *University of Salento, Lecce, Italy*
Umberto Gelatti,University of Brescia, Brescia, Italy
MAPEC_LIFE Study Group, University of Brescia, Brescia, Italy

Background. The MAPEC-LIFE project (Monitoring Air Pollution Effects on Children for Supporting Public Health Policy) is a multicentre study that seeks to assess the association between concentrations of certain atmospheric pollutants and early biological effects in children aged 6-8. The study protocol envisages: recruitment of 1000 primary schoolchildren in five Italian cities (Brescia, Lecce, Perugia, Pisa and Torino); sampling in two seasons of exfoliated buccal mucosa cells and salivary leukocytes; assessment of genotoxic damage in the sampled cells respectively by micronucleus cytome assay and comet assay; atmospheric monitoring near the schools involved, evaluation of the concentration of genotoxic contaminants and the in vitro toxicity of PM 0.5.
In order to evaluate the confounding role of other factors to which the subject may be exposed, the parents of the children participating in the study were asked to fill in an ad hoc questionnaire preliminarily subjected to feasibility and realibility tests. The results of the investigation of the indoor/outdoor exposure and some aspects of the lifestyles of the children enrolled are presented.
Methods. The questionnaire was subdivided into different sections: criteria for inclusion in the study, personal information, and information about the parents, children's homes, lifestyles, indoor/outdoor exposure and diet. The questionaire was filled in twice, during each biological sampling (winter 2014-2015 and spring 2015).
Results. 1356 valid questionnaires were collected in the first season and 1164 (50.9% males, 94.4% born in Italy) in the second, with a fall of 14.2%. The analysis of the data on specific exposures highlights differences between the various cities and between the two seasons.
Conclusions. Information on outdoor and indoor environmental exposure and the lifestyles of participating children can be integrated with the results of environmental and biological monitoring in order to construct a global model of genotoxic risk that can be used to support environmental policies.

Components of the MAPEC_LIFE Study Group:
Serio F., Idolo A., De Giorgi M., Tumolo M.R., Verri T., Covolo L., Donato F., Feretti D., Festa A., Limina R.M., Viola G.C,V., Zerbini I., Fatigoni C., Levorato S., Monarca S., Moretti M., Salvatori T., Donzelli G., Verani M., Bruni B., Bonetta Si., Carraro E., Gilli G., Pignata C., Romanazzi V., Furia C., Codenotti R., Colombi P., Crottini S., Gaffurini L., Zagni L..

We-Po-23

Interpolation in between Road Measurements in Radiofrequency Electromagnetic Field Exposure Assessment

Sam Aerts, *Ghent University, Ghent, Belgium*
Wout Joseph, *Ghent University, Ghent, Belgium*
Loek Colussi, *Radiocommunications Agency, Amersfoort, Netherlands*
Jos Kamer, *Radiocommunications Agency, Amersfoort, Netherlands*
Luc Martens, *Ghent University, Ghent, Belgium*
John Bolte, *National Institute for Public Health and the Environment of the Netherlands (RIVM), Bilthoven, Netherlands*

In some European countries, radio communication agencies carry out large-scale radiofrequency (RF) electromagnetic field (EMF) measurements for ether regulation. In this study, we assess the possibility of using this existing database for the assessment of RF exposure over large areas. Using a car-mounted frequency-selective measurement system, signals from mobile-phone base stations in the 900 and 1800 MHz bands were measured within (I) and around (R) a residential area. Then we interpolated the data on the edge (along both a closed and an open loop) complemented with increasing amounts of inner data to achieve progressively accurate exposure models. Through analysis of a 50-point validation, we found that 80 inner data points per km2 could be sufficient to obtain an accurate interpolation model

Three regions in Amersfoort, The Netherlands with car measurements along ring (red) and inner (blue) roads. "Inner" measurements outside the area demarcated by the ring road were removed from the inner data set (I).

We-Po-24

Exploring determinants of exposure to formaldehyde in a hospital pathology laboratory

Esther Beckers, *Radboudumc, Nijmegen, Netherlands*
Cornelis van Loon, *Radboudumc, Nijmegen, Netherlands*

Exploring determinants of exposure to formaldehyde in a hospital pathology laboratory
Esther AB Beckers1, Cornelis J van Loon1, Coos Diepenbroek2, Paul TJ Scheepers3

1. Master students Biomedical Sciences, Radboud university medical centre/Radboud University, Nijmegen, the Netherlands;
2. Department of Pathology, Radboud university medical centre, Nijmegen, the Netherlands;
3. Department for Health Evidence, Radboud university medical centre, Nijmegen, the Netherlands

Introduction: Formaldehyde is used for tissue fixation in pathologic-histological practice of an academic hospital in the Netherlands. Due to its high vapor pressure, formaldehyde evaporates quickly at room temperature, resulting in inhalation exposure in hospital workers. In 2012, the IARC has classified formaldehyde as a group 1 human carcinogen. Therefore, additional exposure monitoring is enforced in the Netherlands since January, 2015. Currently, the Dutch occupational exposure limit (OEL) is 0.15 mg/m3 (0.1 ppm) for 8 hours (8-h OEL) and 0.5 mg/m3 (0.4 ppm) for 15 minutes (15-min OEL).
Aim: The aim of the study was to identify which determinants are contributing to formaldehyde exposure and could be used in further mitigation steps.
Methods: Background levels, 8-h and 15-min exposure concentrations of formaldehyde were measured as time-weighted averages (TWA), using 2,4-dinitrophenylhydrazine (DNPH) impregnated adsorption tubes. These were used for active sampling: both in the breathing zone of workers and on fixed locations. In addition, local exhaust, room ventilation, room air exchange rate and air flow patterns were determined.
Results: The 8-h and 15-min OELs were not exceeded. All TWA concentrations were below 0.15 mg/m3. Inter-individual differences in personal exposure levels indicated that the distance to emission sources is a significant determinant, e.g. body height was identified as a contributing factor to higher exposure levels, at least in one worker. Waste bins were identified as an important contributing source: a TWA concentration of 6.5 mg/m3 over a period of 15 min was found.
Conclusion: Although no exceedances of current OELs were observed, some determinants of exposure need further mitigation. Future exposure assessments will be required in order to confirm the effectiveness of interventions reducing formaldehyde exposure.

We-Po-25

Pesticide residues in bayberry (Myrica rubra) and cumulative exposure assessment for consumers in Zhejiang, China

Guiling Yang, *Zhejiang Academy of Agricultural Sciences, Institute of Quality and Standard of Agricultural Products, Hangzhou, China, People's Republic of*

As the presence of pesticides in bayberry has raised serious concern from the public in China, 44 pesticides in 157 bayberry samples were determined from 2013-2014. 99 samples had at least one pesticide. 77 samples contained more than two pesticide residues. Probabilistic exposure assessments were performed for single detected pesticides and for cumulative assessment groups: organophosphate, benzimidazole, triazole, pyrethroids and pesticides with anti-androgenic effects. The respective mean hazard quotient (HQ) of all the pesticides ranged from 0.005 - 0.16 below 1. EDI of cyhalothrin at P97.5th was 1.11 of acceptable daily intake (ADI) for children; the estimated short-term intake (ESTI) at P97.5th is 1.9 and 1.78 of acute reference dose (ARfD) for adult and children, respectively. EDI of the pesticides with anti-androgenic effects ranged from 0.15-2.46 of ADI, the probability of exposure exceeding the ADI was 7.1% and 31.1% for adults and children, respectively. EDI of pyrethroids pesticides ranged from 0.01-1.11 of ADI, and the probability of exposure exceeding the ADI was 3.8% for children. Exposures for other pesticide groups were below 1.0. Actually the occurrence of frequency was 9.55% for the combination of cyhalothrin and cypermethrin and the combination of pesticides with anti-androgenic effects has not been found. Therefore, the unacceptable risk should be concerned from pythroids.

We-Po-26

Combination of food monitoring and total diet studies in a combined food safety approach - Results from the TDS-Exposure Project

Anna Elena Kolbaum, *Federal Institute for risk assessment (BfR), Berlin, Berlin, Germany*
Pia Pfennigwerth-Pelka, *Federal Institute for Risk Assessment (BfR), Berlin, Berlin, Germany*
Katrin Uhlig, *Federal Institute for Risk Assessment (BfR), Berlin, Berlin, Germany*
Oliver Lindtner, *Federal Institute for Risk Assessment (BfR), Berlin, Berlin, Germany*

Total diet studies (TDS) as well as food monitoring (FM) aim to collect data about contaminants in foods. However, although both approaches pursue the same objective, they are different in ways of food sampling, preparation and analysis. Consequently, the results are of different quality depending on the research question. One objective of the TDS-Exposure project was to clarify for which questions TDS or FM data are the suitable choice and how both approaches can optimally be combined to achieve the best possible results in terms of exposure assessment and risk management decisions. A literature review on the objectives, benefits and limitations of both approaches was carried out in the databases PubMed, Web of Science, Scopus and LitDok (internal database, BfR). Titles and abstracts from 663 publications from the last five years were screened for relevance, and 153 finally classified as relevant. Advantages and disadvantages of TDS and FM data were worked out and described. A decision tree and one flow chart for combining TDS and FM data were drafted and then filled and adapted with information extracted from the literature. The decision tree describes the application of TDS and FM data in exposure assessment. It clearly documents that FM has its strength mainly where knowledge about variability is required, like acute exposure assessment, checking compliances of maximum levels, or chronic exposure assessment based on high percentiles of concentrations. In contrast, TDS data is preferred as cost-effective approach in questions where total food intake has to be considered for a large number of substances. Further TDS can be used to refine exposure by considering foods as consumed and addresses substances like process contaminants that can't be analysed in unprocessed foods. These findings are further used to develop a food safety concept where TDS and FM complement each other in an integrated approach. The concept proposes to start with a "screening TDS" to draw a first landscape of substance distribution in the food supply. These data are then extended in a "refined TDS" and data gaps are complemented by advising the FM respectively. Subsequently, benefits from each method are used for a cost-effective organization of follow up monitoring activities; e.g. results from TDS can set priorities for FM towards a more targeted approach, whereas information about variability from FM can help to support adequate planning of following TDS projects.

We-Po-27

Analysis of Toolkit and Strategy Developments for the Exposure Assessment of Nanomaterials in Consumer Products

Yasmin Sommer, *Federal Institute for Risk Assessment (BfR), Berlin, Germany*

Aim: Consumer products involving engineered nanoparticles and nanomaterials are available on the market with a wide variety of applications and uses in different regulatory fields. Considerable scientific effort has been made to ensure their safe use, identifying nano-specific requirements and adapt toolkits and strategies for a robust realistic exposure assessment. This study aims to state the current status for assessments in a regulatory context.
Methods: Based on a literature research characteristic and boundaries of models employed for exposure assessments were compiled and trends in model development and additional information sources as well as remaining challenges and uncertainties in the strategies analysed.
Results: Several nano-specific tools are available and developments of additional tools envisaged. Many nano-specific tools currently available only support initial, low tier assessments. Since 2015 ConsExponano enables nano-specific higher tier assessments for inhalative exposures in seven dose measure metrics. Validation of tools for quantitative assessments is generally missing. In the context of the REACH regulation field measurements are regarded as a current alternative to model based assessments. Publications on release measurements from consumer products are increasing especially for particles associated to a high prevalence. For regulatory purposes their value depends on a thorough characterization and documentation of analytical set-up and results. Employment of different protocols may impact the results obtained.
Conclusion: The best strategy for an exposure assessment needs to be decided on a case by case basis. Some assessments will involve (non)nano-specific tools or nano-specific tools for workers. Field measurements are a mean to address uncertainties in terms of realistic exposure prediction. However, both ways of quantitative assessment incorporate challenges for realistic assessments. Efforts for improvement would profit from a better knowledge of the use of nanomaterials in consumer products.

We-Po-28

Development of an on-line analytical method for the quantification of carbamate pesticides and metabolites in human matrices

Pim Leonards, *VU University, Amsterdam, Netherlands*

Worldwide, serious concern has arisen about the increased incidence of learning and developmental disorders in children. From a scientific point of view, there is no doubt that exposure to neurotoxic chemicals during early brain development can adversely affect learning and development. Various recent epidemiological studies have indicated that exposure to low doses of environmental biologically active contaminants during human development can have deleterious effects on cognitive development in childhood. The European commission-funded project DENAMIC "Developmental Neurotoxicity Assessment of Mixtures in Children" investigates neurotoxic effects (e.g. learning and developmental disorders) of low-concentration mixtures of pesticides and a number of common environmental pollutants in children. Because of recent concerns of cognitive and neurobehavioral effects related to pesticide exposure and if proven necessary the possibility to protect future generations by regulatory measures, DENAMIC will primarily focus on possible neurotoxic effects of pesticides (e.g. organophosphate, carbamates, pyrethroids, organochlorine). One of the aims is to study perinatal and early-childhood exposure in maternal urine and cord blood, as well as breast milk and urine of the child. Data and samples from existing European cohorts with different exposure profiles, both pre-and postnatal, (Norway, Netherlands, Slovakia and Spain) are studied, which offers the possibility to distinguish between pre-and postnatal exposure effects and identify the susceptible period. Associations between chemical exposure and learning (cognitive) and neurobehavioural (ADHD, ASD, anxiety) development or disorders will be studied. The aim of this paper is to present the development of an on-line LC-MS/MS for the detection and quantification of carbamates and metabolites in urine, serum and breast milk. The developed sample prepartion method consist of a deconjugation step, and an on-line extraction, clean-up, and separation method combined with LC-MS/MS. The compounds of interested were fractionated using a Restricted Access Material (RAM) from the matrix, and then transferred to the analytical column before detection with LC-MS/MS. Different parameters for the RAM material were optimised to provide acceptable recoveries.

We-Po-29

Mercury exposure in small and artisanal gold mining in Suriname

Romilda Boerleider, *Radboud Institute for Health Sciences, Radboudumc, Nijmegen, Gelderland, Netherlands*
Daphne van den Oetelaar, *Radboud Institute for Health Sciences, Radboudumc, Nijmegen, Gelderland, Netherlands*
Lisa Nijzink, *Radboud Institute for Health Sciences, Radboudumc, Nijmegen, Gelderland, Netherlands*
Bianca Jubitana, *Medische Zending, Paramaribo, Netherlands*
Jan Quik, *Bureau voor Openbare Gezondheidszorg, Paramaribo, Suriname*
Paul T. Scheepers, *Radboud Institute for Health Sciences, Radboudumc, Nijmegen, Gelderland, Netherlands*

In Suriname small and artisanal gold mining (ASGM) is a source of livelihood for more than 20,000 people. Gold miners use mercury during the extraction of gold from ore and may be exposed to mercury vapors by inhalation. Both miners and inhabitants of villages in the mining areas may be exposed to methyl mercury through dietary intake. Both short-term and long-term inhalation exposure to high concentrations of mercury vapor and methyl mercury can lead to serious health effects in the miners population and their offspring. The primary aim of our study was to evaluate the level of awareness concerning the potential health effects due to mercury exposure. Secondly, we assessed the feasibility of introducing a program to monitor mercury exposures by use of biological monitoring.
We recruited residents from villages and miners in the small scale gold mining regions in the inland of Suriname. The study protocol was submitted to the ethical committee in Suriname. Information on the awareness and knowledge of health implications of mercury exposures was collected by interviews using semi-structured questionnaires by researchers who were familiar with the cultural background and could speak the local language. An information program was introduced involving individual consultations and group information sessions. Two months later the questionnaire study was repeated in the same participants to find out about the effectiveness of the information campaign. In a subgroup of the population additional interviews were conducted to find out about the feasibility of collecting biological materials in adults and children.
Conclusion: The acquired data can be used to evaluate the need for information regarding the health risk of mercury. The effectiveness of an information campaign was assessed and resulted in improvements of this campaign. It is considered feasible to initiate a pilot study for a monitoring program involving the collection of biological media. Cultural aspects need to be addressed and the least invasive methods of sample collection are preferred, especially in children.

We-Po-30

Research on Fugitive Formaldehyde and TVOC in Public Exhibition Area—A Measurement and Control Study

Chane-Yu Lai, *Chung Shan Medical University, Taichung, China, Republic of (Taiwan)*
Chuan-Chen Yu, *Chung-Shan Medical University, Taichung, China, Republic of (Taiwan)*
Wen-Hsin Su, *Chung-Shan Medical University, Taichung, China, Republic of (Taiwan)*

INTRODUCTION
Many exhibitions usually take place in large exhibition centers, for instance, art, painting, sculpture, or popular science. Due to discrepancies of interior design following by exhibition topics or exhibition duration, the exhibition partition changes couple times every year. Labors who rebuild the exhibition partition often enclose the construction field or process to avoid produce nuisance. For this manner, the labors in the enclosure construction area will expose to high concentration of pollution. During the reconstruction of exhibition partition, lots of plywood is used for exhibition decoration. However, the plywood contains a lot of formaldehyde base resins, and will slowly and repeatedly release the formaldehyde and volatile organic compound species (VOCs) which is great harmful to people who work in exhibition places.
MATERIALS/METHODS
In this study, formaldehyde and TVOC evaporated from exhibition places in a public museum were taken for analysis to estimate the exposure of workers in the exhibition partitions, and the using or not using green building materials (GBM, class F1 plywood) were compared and used as suggestions for controlling the hazard from exhibition partition. The MiniRAE 2000 VOC direct reading instrument and Taiwan Environmental Protection Agency (EPA) NIEA (A715.15B) TVOC standard sampling methods was utilized to monitor and evaluate the TVOC concentration at the exhibition partitions.
RESULTS
The result showed that the museum used energy-saving, recirculation system, positive pressure, and few exhaust air and the mean of ACH was only 2.3. As for the STER A1, using the GBM, reused plywood and under charcoal filtered recirculation air condition, the concentration of formaldehyde was between 0.01 and 0.02 ppm; however, while not using the GBM, during the decoration and floor waxing, the highest concentration was around 0.09 ppm.
DISCUSSION
As figure 1(a), after a sampling cycle, the concentration of formaldehyde using the GBM was low. As figure 1(b), concentrations of TVOC during floor waxing was high, the standard analysis methods show that was caused by 2 - (2-Ethoxyethoxy) ethanol; without consider alcohol, the TVOC were under the standard value 0.56 ppm.
CONCLUSIONS
The local exhaust ventilation devices and the GBM, or other air filters were suggested to control the above hazardous pollutants in this study. The ACH should increase to 8 ~12, and the recirculation air volume should reduce to lower the accumulation of formaldehyde.

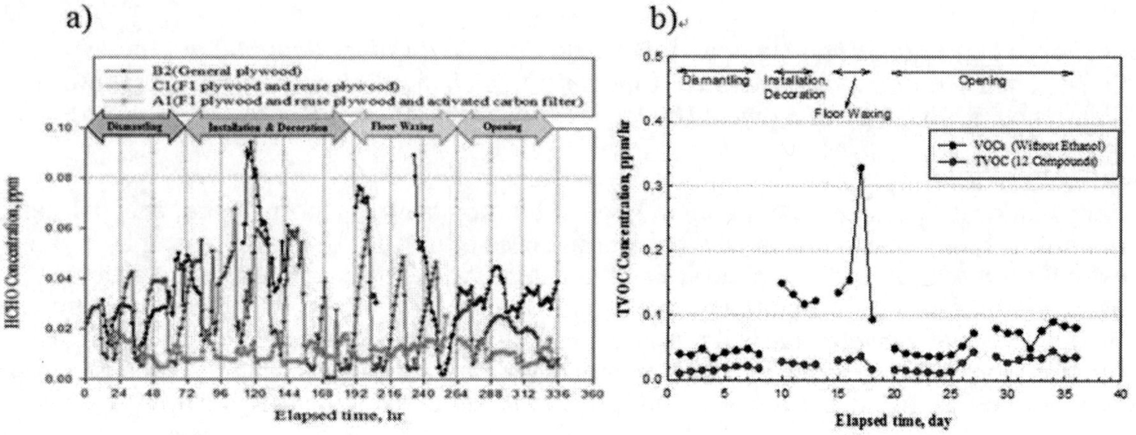

Figure 1. Concentration in the STER. a) Formaldehyde, b) TVOC.

We-Po-31

Study on Bioaerosol Characteristics in Semi-indoor Wood Processing Workplace

Chane-Yu Lai, *Chung Shan Medical University, Taichung, China, Republic of (Taiwan)*
Hsiang-Yu Huang, , *Chung-Shan Medical University, Taichung, China, Republic of (Taiwan)*
Kai-Ling Huang, *Chung-Shan Medical University, Taichung, China, Republic of (Taiwan)*

INTRODUCTION
From literature review, wood processing workers were found to expose to occupational hazards and induced to have occupational asthma. For the reason, the study investigated the biological exposures, included concentration of wood dust, bioaerosol, identification of bioaerosols, endotoxin, and mycotoxin in wood processing workplaces.
MATERIALS/METHODS
An Anderson 6 stage bioaerosol sampler, Biosampler, AGI-30, and SKC Aluminum cyclone with polycarbonate (PC) filter were used to sample bioaerosols. Moreover, SKC Aluminum cyclone was used to collect respirable sawmill wood dust.
RESULTS
The results as showed in Fig.1 a and b: the highest concentration of contained endotoxin was found in cultured Pantoea agglomerans, and was about 0.8 EU/m3. The range of endotoxin concentration sampled by using SKC Aluminum cyclone with PC filter was about 2.2E-6~5.2E-6 EU/m3.
DISCUSSION
The concentration of Gram-negative bacteria appeared statistically associated with the ambient temperature and humidity ($P<0.05$). The concentration of endotoxin measured in the study did not exceed the DECOS health exposure recommendation 50 EU/m3 criteria.
CONCLUSIONS
At present, Taiwan have no complete specification for content restrictions and exposure limit in respirable wood dust, endotoxin and mycotoxin in workplace environment. The study recommended the related regulations should be established to prevent the increase of the disease in the future.

a)

b)

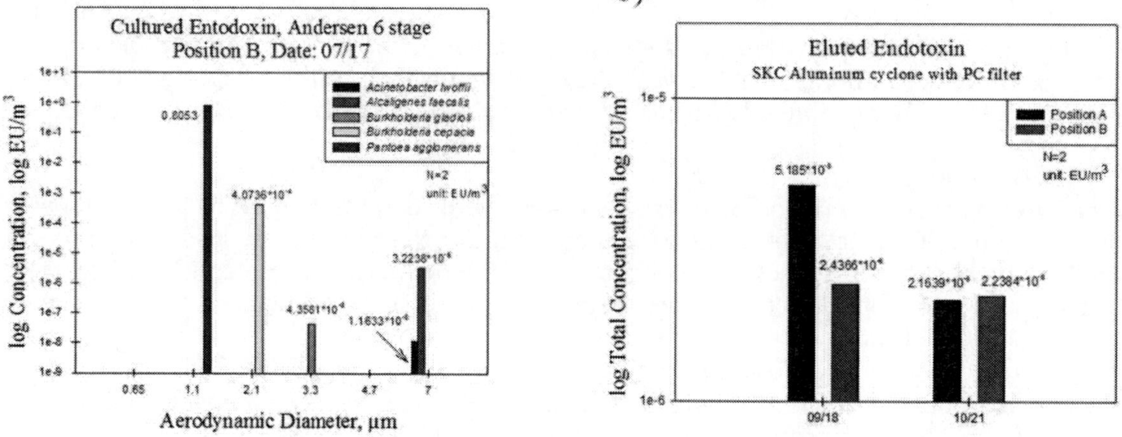

Fig.1. a) Concentration of endotoxin distribution within the culture of Gram-negative bacteria using Anderson 6 stage sampler; b) Concentration of endotoxin eluted from PC filter using SKC Aluminum cyclone.

We-Po-32

Sampling Evaluation of Bioaerosol and Antibiotic-Resistant Characteristics in Intensive Care Unit

Chane-Yu Lai, *Chung Shan Medical University, Taichung, China, Republic of (Taiwan)*
Nien-Hsin Wu, *Chung Shan Medical University, Taichung, China, Republic of (Taiwan)*
Ya-Hue Lin, *Chung Shan Medical University, Taichung, China, Republic of (Taiwan)*
An-Ru Kuo, *OSH co., Ltd., Taichung, China, Republic of (Taiwan)*

Objective: Our research was based in a medical center's Internal Medicine Intensive Care Unit (MICU) and Surgery Intensive Care Unit (SICU) located in central Taiwan. The research objective focus on the bioaerosols and their antibiotic-resistant characteristics in both MICU and SICU.
Methods: Three bioaerosol samplers were utilized (Anderson six-stage, AGI-30, and BioSampler) for sampling before and during patient visiting. Upon acquisition of samples, they were inoculated and cultured on BBL ™ Trypticase ™ Soy Agar (with 5% Sheep Blood) medium for growth. The bacterial colonies were later identified and analyzed for antibiotic-resistant characteristics via BD Phoenix ™medium ted microbial identification and susceptibility test analyzer.
Results : Research results have showed from the bioaerosol samples acquired within the MICU that dominant concentration of bacteria and fungi were below cut off size of 3.3 μm, and they had high possibility to enter human lung's alveolar regions of the body, thereby causing opportunistic infections. The factor of season and air change rate per hour did not statistically associate with bioaerosol concentration ($P>0.05$) ; However, factor of patient visiting and temperature, relative humidity during sampling showed statistically agreement with bioaerosol concentration ($P<0.001$) . In terms of bacterial strain identification, Gram-positive bacteria were mainly isolated with risk group (RG) of II. As for antibiotic-resistant bacteria analysis of MICU, strains were identified 63.5 % that were resistant to National Health Insurance Administration (NHIA) designated first (17 types) and second (18 types) line antibiotics. This phenomenon could very likely affect the medical staffs working within the hospital environment.
Conclusions: As a result, recommendations for MICU ventilation designs should be carefully evaluated for the effectiveness of controlling nosocomial infections as well as proper implementation of personal protective equipment in order to reduce bioaerosol opportunistic infections and harmful exposure effects.

We-Po-33

POLYDIMETHYLSILOXANE AS A PERSONAL PASSIVE AIR SAMPLER (PPAS) FOR MEASURING SEMI-VOLATILE ORGANIC COMPOUNDS

Joseph Okeme, *University of Toronto, Toronto, Canada*
Miriam Diamond, *University of Toronto, Toronto, Canada*
Yuchao Wan, *University of Toronto, Toronto, Canada*
Michelle North, *University of Toronto, Toronto, Canada*
Arthur Chan, *University of Toronto, Toronto, Canada*
Jeff Brook, *University of Toronto, Toronto, Canada*
Greg Evans, *University of Toronto, Toronto, Canada*
Paul Demers, *University of Toronto, Toronto, Canada*
Roel Vermeulen, *Utrecht University, Utrecht, Netherlands*
Jelle Vlaanderen, *Utrecht University, Utrecht, Netherlands*
Lutzen Portengen, *Utrecht University, Utrecht, Netherlands*
Anke Huss*Utrecht University, Utrecht, Netherlands*
Gerard Hoek, *Utrecht University, Utrecht, Netherlands*
Shelly Tse, *Chinese University of Hong Kong, Hong Kong, Hong Kong*
XianQiang Lao, *Chinese University of Hong Kong, Hong Kong, Hong Kong*
Kin-Fai Ho,*Chinese University of Hong Kong, Hong Kong, Hong Kong*
Feng Wang, *Chinese University of Hong Kong, Hong Kong, Hong Kong*

Exposure to semi-volatile organic compounds (SVOCs) that are released from a variety indoor sources has been linked to a range of potential health effects. Limiting human and ecosystem exposure to SVOCs can be intiated from knowing their abundance. This objective is usually met by measuring stationary SVOC levels in micro-enviroment. Personal levels, due to personal activity and proximity to SVOC sources, are usually higher and more reflective of an individual's actual exposure circumstance than are stationary levels. Personal Passive air samplers (PPAS) have been used to monitor exposures in occupational settings but are less well characterized for non-occupational use. This project was aimed at developing an easily used PPAS for measuring levels of SVOCs in a non-industrial indoor environment. We have used commonly available rubber, polydimethylsiloxane (PDMS), as PPAS. PDMS collects and retains a wide range of chemicals and is easy to use.

PDMS wristbands were worn by participants at all times for a duration of a week while leading their normal lives. Post-deployment samples were extracted and analysed for phthalate esters, halogenated and organophosphate ester flame retardants (BFRs and OPFRs) and polycyclic aromatic hydrocarbons (PAH) using gas chromatography mass spectrometry (GC-MS). Surrogate recoveries raged from 80 to 120%. Preliminary results show that BFRs levels were generally < MDL. Σ5phthalates ranged from < 0.10 to 62 µg/wristband. PAH and OPFRs were measured in the range of <MDL to 0.25 µg/wristband and <MDL to 4.7 ng/wristband, respectively. This study shows the promise of using PDMS as a passive air sampler for measuring personal levels of SVOCs indoors.

We-Po-34

Non-Euclidean distance based kriging, water quality monitoring, and remote sensing data to predict Vibrio parahaemolyticus in the Chesapeake Bay

Benjamin Davis, *Johns Hopkins School of Public Health, Baltimore, Maryland, United States*
Benjamin Zaitchik, *Johns Hopkins University, Baltimore, Maryland, United States*
John Jacobs, *National Oceanic and Atmospheric Association, Oxford, Maryland, United States*
Frank Curriero, *Johns Hopkins School of Public Health, Baltimore, Maryland, United States*

Background: Vibrio parahaemolyticus bacteria are a major cause of seafood-borne illnesses. The number of vibriosis cases has steadily increased in recent years, despite reductions in other foodborne illnesses. These bacteria are naturally prevalent in brackish waters and can concentrate in shellfish before being harvested. The bacteria's survival and growth are highly influenced by its surrounding environment, particularly by water temperature and salinity. The geostatistical method known as kriging can be used for predicting the distribution of these bacteria. However, spatial correlation structures used in these techniques are limited to proximity measures based on Euclidean (i.e. straight-line) distances, while V. parahaemolyticus are often found in the winding tributaries of estuaries.
Objectives: The goal of this study was to characterize the concentrations of Vibrio parahaemolyticus bacteria in the Chesapeake Bay. A new geostatistical technique was developed to accommodate predictions using non-Euclidean (e.g. water) distances. This technique was used with both direct and remotely sensed water quality data to create an accurate and precise prediction model for the concentrations V. parahaemolyticus bacteria that included spatially varying prediction uncertainty.
Methods: The new kriging method is based on a multi-dimensional scaling approach applied to the spatial correlation structure, as characterized by the semivariogram, which ensures validity when used with non-Euclidean distances. Water samples collected by the National Oceanic Atmospheric Association (NOAA) were analyzed using qPCR to determine concentrations of Vibrio parahaemolyticus genetic material. Direct measurements from the Chesapeake Bay Program's Water Quality Database, and Ocean Color data from the MODIS sensor about the Aqua satellite were obtained and used as environmental covariates. These data and the developed non-Euclidean kriging method were applied to spatially predict V. parahaemolyticus throughout the Chesapeake Bay.
Results: Results from cross validation evaluations support improved prediction performance of the developed non-Euclidean kriging method when compared to current competing methods and naive use of Euclidean distance. The Vibrio parahaemolyticus prediction model was able to include satellite data with missing values imputed via the new method, and performed well in preliminary analysis. Future work will continue to improve the accuracy and reliability of this modeling approach so that it can be incorporated into an ecological forecasting system for enhanced management on the openings and closings of shellfish harvesting beds.

We-Po-35

Integration of Alternative Methods for an Ab Initio Chemical Safety Assessment

Alicia Paini, *Joint Research Centre, Ispra, Italy*
Elisabet Berggren, *Joint Research Centre, Ispra, Italy*
Frederic Y. Bois, *INERIS, Verneuil-en-Halatte, France*
Catherine Mahony, *Procter & Gamble, Egham, United Kingdom*
Gladys Ouedraogo, *L'OREAL R&I, Aulnay-sous-bois, France*
Andrea-N Richarz, *Joint Research Centre, Ispra, Italy*
Andrew White, *Unilever PLC, Bedford, United Kingdom*

Assessing chemical safety using only non-animal methods is a major challenge within the European Union, especially considering the ban to market cosmetics with ingredients tested on animals (Regulation 1223/2009). One of the outcomes of the SEURAT-1 initiative (http://www.seurat-1.eu/), co-sponsored by the European Commission 7thFP and Cosmetics Europe, was to develop case studies addressing that issue. Besides a read-across case-study, where "new approach" data was used to strengthen the confidence in reading across data from an already assessed chemical to a structural analogue, an ab initio case study was carried out and is presented here.
The ab initio case study was an attempt to structure knowledge and data in a logic workflow which could be used for decision making to predict whether the intended exposure could be considered safe based on data solely from alternative methods. The workflow is includes the following steps: i) identification of the exposure/use scenario; ii) data collection on the chemical; iii) toxicokinetic and toxicodynamic modeling; iv) evaluation of other alternative methods (in vitro or in silico) which are available and could provide evidence for a hypothesis on the chemicals' mode of action; v) confirmation of the hypothesis with targeted testing using selected in vitro or in silico methods; vi) propose a risk valuation, as well as estimating uncertainties and identifying data gaps. The exposure scenario is an initial and essential step in this workflow. Whether exposure is intentional or not it should be considered, and in both cases estimates of the doses, expected routes of exposure, frequency and length of exposure should be made.
The ab initio workflow was applied to the piperonyl butoxide (PBO), as a hypothetical case, for which the mode-of-action was assumed to be unknown. The only data on the compound was the molecular structure and a concentration of PBO in a consumer product formulation. An exposure scenario was set up and data were generated for simulation of internal exposure for a "healthy human" using PBPK modeling. The concentration in blood and liver predicted by the model were then used to design in vitro tests (performed by SEURAT-1 partners). The test results were then evaluated and used to estimate whether PBO would be safe to use for the selected exposure scenario.
To our knowledge, the SEURAT-1 ab initio case study is a first attempt to shift our minds to a fully non-animal based chemical risk assessment relying only on alternative methods.

We-Po-36

Toxicokinetics Strategy highlighting In vitro to In vivo Extrapolation

Alicia Paini, *Joint Research Centre, Ispra, Italy*
Maurice Whelan, *Joint Research Centre, Ispra, Italy*
Andrew Worth, *Joint Research Centre, Ispra, Italy*

The EURL ECVAM Toxicokinetics (TK) Strategy aims to promote a better use of TK data in the framework of the 3Rs to reduce, refine, and ultimately replace animal testing in the assessment of systemic toxicity. The strategy comprises four key objectives: 1. Development of standards for human in vitro adsorption, distribution, metabolism and excretion (ADME) methods. For example, characterisation of in vitro metabolism methods to identify key metabolic routes or to estimate clearance as a surrogate for in vivo metabolism. This also provides a basis to develop in vitro to in vivo extrapolations (IVIVE) and to predict potential TK interactions for hazard characterisation of chemicals. 2. Kinetic Modelling and good modelling practice: So far a number of biokinetic models have enabled potential TK interactions to be investigated and used for IVIVE. This objective is expected to facilitate the acceptance and use of biokinetic models (such as physiologically based kinetic models) in the risk assessment process; for instance to improve extrapolation from in vitro effects on target cells to an actual exposure dose by applying reverse dosimetry. 3. Analytical data collection and database development to serve kinetic modelling. This includes information on ADME and TK properties, Mode-of-Action, dose-response and toxicity of chemicals in a harmonized one-stop-web based portal. 4. Regulatory anchoring: expresses the need to develop guidance on how to generate, interpret and use ADME/TK data in a regulatory setting, for a better risk assessment processes.
The TK Strategy will have a significant short to mid-term 3Rs impact, and at the same time will lay the foundation for a risk assessment approach that is increasingly based on human ADME/TK data.

We-Po-37

Examining The Association Between Natural Gas Compressor Stations and Residential Noise In West Virginia, USA

Sutyajeet Soneja, *University of Maryland, College Park, College Park, Maryland, United States*
Meleah Boyle, *University of Maryland, College Park, College Park, Maryland, United States*
Lesliam Quiros-Alcala, *University of Maryland, College Park, College Park, Maryland, United States*
Amy Sapkota, *University of Maryland, College Park, College Park, Maryland, United States*
Laura Dalemarre, *University of Maryland, College Park, College Park, Maryland, United States*
Thurka Sangaramoorthy, *University of Maryland, College Park, College Park, Maryland, United States*
Sacoby Wilson, *University of Maryland, College Park, College Park, Maryland, United States*
Donald Milton, *University of Maryland, College Park, College Park, Maryland, United States*
Amir Sapkota, *University of Maryland, College Park, College Park, Maryland, United States*

Aim
From 2000 to 2014, natural gas production increased 35%, from 19.7 quadrillion British Thermal Units (BTU) to 26.6 quadrillion BTU across the United States, much of which is due to technological advances in horizontal drilling and high-volume hydraulic fracturing. Exposure to chemical, physical, and psychosocial hazards has become a growing concern as a result of this development. Noise associated with natural gas compressor stations has been identified as a major concern for nearby residents and communities, yet limited studies exist assessing this exposure. Our pilot study investigated how residential noise levels near a natural gas compressor station varied based upon time of day (daytime versus nighttime), location (indoors versus outdoors), and varying distances.
Methods
We collected 24-hour noise measurements indoors and outdoors concurrently at 8 homes across varying distances within 750 meters of a natural gas compressor station and 3 homes located more than 1 kilometer away, considered control homes, from April 11-17, 2014 in Doddridge County, West Virginia. We used generalized estimation equation regression models to assess how A-weighted decibel exposure changed based upon factors previously mentioned.
Results
Indoor noise levels for control homes (mean: 42.5 dBA; standard error (SE): 2.6) were 10.9 dBA ($p<0.01$) less than indoor levels for homes within 750 meters of the station. No statistically significant difference was found for outdoor noise levels for control vs. measurements <750m (average of 51.9 and 55.3 dBA, respectively). Focusing on measurements 50dBA. In contrast, measurements >1km were found to be on average 9.4 dBA (SE=3.9) higher outdoors relative to indoors ($p=0.02$). Examining noise levels across multiple distance categories, all measurements within 300 meters of the station were deemed to be 10 dBA (SE= 3.6) higher (p1km). Lastly, daytime noise levels for all locations

within 750m of the station were found to be 3.4 dBA (SE=1.2) higher relative to nighttime levels (p<0.01).

Conclusions

Our findings suggest that living near a natural gas compressor station results in high environmental noise exposures for nearby residents. Future studies on a larger scale are needed to confirm these findings and evaluate potential health impacts.

We-Po-38

Determining Exfiltration Estimates for Particulate Matter from the Use of Alternative Cookstoves in a Village-Like Household in Rural Nepal

Sutyajeet Soneja, *University of Maryland, College Park, College Park, Maryland, United States*
James Tielsch, *George Washington University, Washington DC, United States*
Subarna Khatry, *Nepal Nutrition Intervention Project Sarlahi, Kathmandu, Nepal*
Benjamin Zaitchik, *Johns Hopkins University, Baltimore, Maryland, United States*
Frank Curriero, *Johns Hopkins University, Baltimore, Maryland, United States*
Patrick Breysse, *Johns Hopkins University, Baltimore, Maryland, United States*

Aim

Half of the world's population utilizes solid fuels for cooking (in developing countries), with resultant pollution levels well over WHO air quality guidelines. Alternative stoves, an intervention option to reduce household air pollution, are designed to burn solid fuels more efficiently and in many cases redirect pollution outdoors via the use of a chimney. Particulate matter (PM) generated from cooking fires settles and deposits on surfaces within the home, as well as exits the home thereby entering ambient air. The amount of air pollution exiting homes when alternative stoves with chimneys are utilized is not known. In this study, PM exfiltration estimates are presented for four types of alternative stoves within a village-like home in rural Nepal. Developing an understanding of these exfiltration estimates allows for a more comprehensive assessment to be provided on the impact that alternative cookstoves utilizing chimneys have upon outdoor air quality and subsequent effects including health and climate change.

Methods

A test house representing a village kitchen was built in rural Nepal, with a water-boiling test conducted for all trials to simulate a typical cooking session. Four alternative stoves with chimneys were examined, including an alternative mud brick stove, original Envirofit G3355 model, manufacture altered Envirofit G3355, and locally altered Envirofit G3355. Multiple linear regression was utilized to determine estimates of PM exfiltration.

Results

Overall exfiltration fraction average (converted to a percent) for the four stoves were: Alternative mud brick stove with chimney 56%, original Envirofit G3355 model with chimney 87%, manufacture altered Envirofit G3355 model with chimney 69%, and locally altered Envirofit G3355 model with chimney 69%. This is in contrast to a previous study conducted by our group where we determined overall exfiltration from a traditional, mud-based stove with no chimney to be 23%.

Conclusion

Alternative cookstoves resulted in higher overall average exfiltration due to direct (via a chimney) and indirect (windows, doors, and cracks in the wall) ventilation relative to traditional, mud-based stoves. This contrast emphasizes the need for an improved understanding of the climate and health implications that are believed to come from implementing alternative stoves on a large scale and the resultant shift of exposure burden from indoors to outdoors.

UTRECHT
THE NETHERLANDS
OCTOBER 9-13
ISES 2016

We-Po-39

Indoor exposure to outdoor air pollutants controlled by different urban design strategies

Zhiwen Luo, *University of Reading, Reading, United Kingdom*
Jian Hang, *Sun Yat-sen University, Guangzhou, China, People's Republic of*
Man Lin, *Sun Yat-sen University, Guangzhou, China, People's Republic of*

Background:
In compact mega-cities such as London and Beijing, motor vehicle emissions have been a major source of local urban pollutions especially from the urban viaduct. However, a significant part of exposure to outdoor pollution occurs indoors as people spend over 80% of their time living indoors and the outdoor pollutants can penetrate indoors. Urban design can affect the outdoor air pollution as well as indoor-outdoor exchange, and therefore indoor exposure, however, little research has quantified such impact.
Aim:
The aim of this paper is to investigate the impact of urban design strategies on the indoor exposure to outdoor traffic pollutants for different age-groups of population.
Methods:
Coupling indoor and outdoor Computational fluid dynamic (CFD) model is used to simulate the airflow and pollution dispersion between the street canyon and building interior space. Building intake fraction was introduced to evaluate the accumulated intake fraction indoors from the outdoor mobile vehicle emission.
Results:
Our studies provide new insights on how to reduce indoor exposure to outdoor origins by improving outdoor urban design rather than focusing on building design itself. Detailed results drawn from this study show: 1) When the aspect ratio equals 1, indoor personal exposure to street pollution will be reduced if the single pollutant source is elevated by the viaduct; 2) High upstream velocity can depress the pollutant concentration. Pollutant dispersion can be stronger by great ground heating intensity under same upstream velocity; 3) The indoor exposure to street pollution can be lessened when broadening the street under the condition of same building height and source strength; 4) Personal intake fraction ranges from 1500 to 6500 ppm depending on the different urban canyon and building design.

We-Po-40

Associations Between Lifestyle and Air Pollution Exposure

Maciej Strak, *Institute for Risk Assessment Sciences (IRAS), Utrecht, Netherlands*
Nicole Janssen, *National Institute for Public Health and the Environment (RIVM), Bilthoven, Netherlands*
Rob Beelen, *National Institute for Public Health and the Environment (RIVM), Bilthoven, Netherlands*
Oliver Schmitz, *Utrecht University, Utrecht, Netherlands*
Derek Karssenberg, *Utrecht University, Utrecht, Netherlands*
Danny Houthuijs, *National Institute for Public Health and the Environment (RIVM), Bilthoven, Netherlands*
Martin Dijst, *Utrecht University, Utrecht, Netherlands*
Bert Brunekreef, *Institute for Risk Assessment Sciences (IRAS), Utrecht University, Utrecht, Netherlands*
Gerard Hoek, *Institute for Risk Assessment Sciences (IRAS), Utrecht University, Utrecht, Netherlands*

Aim
Administrative data cohorts have size advantages over individual cohorts in investigating air pollution risks, but often lack information on individual risk factors related to lifestyle. If there is a correlation between lifestyle and air pollution then omitted lifestyle variables result in biased risk estimates.
Methods
We used a recent Dutch national health survey of 387,195 adults to investigate the associations of PM10, PM2.5, PM2.5-10, absorbance, OPDTT, OPESR and NO2 annual average concentrations from ESCAPE land use regression models with smoking habit, alcohol consumption, physical activity and BMI. We assessed the associations with and without adjustment for neighborhood- and individual background characteristics typically available in administrative cohorts.
Results
Current smoking and alcohol consumption were generally positively associated with air pollution, e.g., smoking 10 cigarettes/day was associated with 1% increased annual average NO2 concentration. Physical activity and overweight were negatively associated with air pollution. The associations were small but significant and remained after adjusting for other potential confounders. Direction and magnitude of the associations depended on the pollutant (PM2.5 the least), use of continuous vs categorical scale of the confounder and level of adjustment.
Conclusions
In a recent Dutch national health survey, individual lifestyle-related risk factors were weakly associated with long-term exposure to air pollution. However, this association could potentially lead to bias in risk estimates from administrative cohorts.

We-Po-44

IPCheM: the reference platform for chemical monitoring data in Europe

Silvia Dalla Costa, *European Commission, Joint Research Centre, Ispra, Italy*
Stylianos Kephalopoulos, *European Commission, Joint Research Centre, Ispra, Italy*
Alberto Cusinato, *European Commission, Joint Research Centre, Ispra, Italy*
Paolo Leva, *European Commission, Joint Research Centre, Ispra, Italy*
Otmar Geiss, *European Commission, Joint Research Centre, Ispra, Italy*
Anita Radovnikovic, *European Commission, Joint Research Centre, Ispra, Italy*
Vittorio Reina, *European Commission, Joint Research Centre, Ispra, Italy*
Salvatore Tirendi, *European Commission, Joint Research Centre, Ispra, Italy*
Alexandre Zenie, *European Commission, Joint Research Centre, Ispra, Italy*

One of the major gaps in our knowledge base supporting chemical policies is the lack of information on the burden of chemical exposure to humans and the environment. Occurrence data on chemicals in the environment, food and human body are produced by many European governmental and research institutions and industrial partners, through monitoring programs and various funded projects at both EU and national/regional levels. IPCheM (The Information Platform for Chemical Monitoring Data) was recently developed on initiative by the European Commission (DG ENV and DG JRC) in close co-operation with the European Environment Agency and the European Food Safety Authority. It represents a single access point for searching, retrieving, pulling and analyzing chemical occurrence data across various media (environment, human biomonitoring, food/feed, indoor air and products) from multiple underlying databases hosted by several collaborating entities both in Commission Services, EU Agencies, EU Member States and other international organisations. Through its tailored access, it supports the needs of several categories of end-users including academia, industry and policy making.
IPCheM is handling in a transparent way data retrieved from heterogeneous data sources serving a multitude of chemical policies and provides different level of data accessibility to different users' typology in accordance with the conditions of data access and use defined by the data providers.
The long-term ambition of IPCheM is to help enhancing and boosting comparability, quality and standardisation of data and metadata information, thus increasing the knowledge base for sound exposure and risk assessment, management and communication.
This paper will outline the policy context and present the IPCheM's architecture, capabilities, functionalities and tools for selecting, visualizing, viewing spatial and temporal trends, filtering, comparing and performing statistical analyses of actually more than 20 million of occurrence data on chemicals from 19 studies at pan-European and Member States levels.

We-Po-45

Development of a Source-Exposure Matrix for Occupational Exposure Assessment of Electromagnetic Fields in the INTEROCC Study

Javier Vila, *ISGlobal, Centre for Research in Environmental Epidemiology (CREAL), Barcelona, Spain*
Joseph D. Bowman, *National Institute for Occupational Safety and Health (NIOSH), Cincinnati, OH, United States*
Jordi Figuerola, *ISGlobal, Centre for Research in Environmental Epidemiology (CREAL), Barcelona, Spain*
David Morina, *1ISGlobal, Center for Research in Environmental Epidemiology (CREAL), Barcelona, Spain*
Laurel Kincl, *5Oregon State University (OSU), , Oregon, USA, Corvallis, Oregon, United States*
Lesley Richardson, *University of Montreal Hospital Research Centre (CRCHUM), Montreal, Quebeq, Canada*
Elisabeth Cardis, *ISGlobal, Center for Research in Environmental Epidemiology (CREAL), Barcelona, Spain*

Introduction: As part of the INTEROCC study, we collected information on work with sources of exposure to electric and magnetic fields (EMF) by using a detailed source-based questionnaire, covering the entire EMF frequency range (from 0 Hz to 300 GHz). To estimate occupational exposure of study subjects to the EMF sources identified, we constructed a database of source-based measurements from published and unpublished literature. The aim of the current work was to summarize these measurements into a source-exposure matrix (SEM), accounting for their quality and relevance. Methods: We developed methods for combining available measurements based on the assumptions of log-normality and symmetric quantiles of EMF data. Pooled estimates were weighted by our confidence in the combined measurement data. Arithmetic and geometric means, as well as estimates of variability and maximum exposure were calculated by EMF source, frequency band and dosimetry type. Results: The SEM contains confidence-weighted exposure estimates for the electric (E-field) and magnetic (H- and B-field) fields for 312 EMF exposure sources. Operator position geometric mean electric field levels for RF sources ranged between 0.8 V/m (plasma etcher) and 320 V/m (RF sealer), while magnetic fields ranged from 0.02 A/m (speed radar) to 0.6 A/m (microwave heating). For ELF sources, electric fields ranged between 0.2 V/m (electric forklift) and 11,700 V/m (HVTL-hotsticks), while magnetic fields ranged between 0.14 μT (visual display terminals) and 17 μT (TIG welding). Conclusion: The methodology developed allowed the construction of an EMF-SEM from measurements collected from the literature and may be used to summarize similar exposure data for other physical or chemical agents. The SEM will be used together with detailed information on distance to the source, automation, and other determinants of exposure reported by the study subjects, to calculate indices of cumulative exposure to EMF for their use in the analysis of brain tumours risk associated with these exposures. The SEM will also be offered publicly for its use by other researchers.

We-Po-46

Development of a generic PBPK model for pyrethroids to assess the cumulative exposure of populations

Paul QUINDROIT, *French National Institute for Industrial Environment and Risks (INERIS), Verneuil-en-Halatte, France*
Celine BROCHOT, *French National Institute for Industrial Environment and Risks (INERIS), Verneuil-en-Halatte, France*

Context and objective: Pyrethroids are ubiquitous insecticides used in many areas such as agriculture, housing maintenance, and human or veterinary medicine. Over the last decade, biomonitoring studies have shown the wide exposure of human populations in many countries. The human biomarkers are usually the concentrations of pyrethroid metabolites in urine. The interpretation of these biomarkers to assess the environmental exposure of populations to a specific pyrethroid can be difficult since pyrethroids share metabolic pathways and common metabolites. In this work, we propose to develop a model that will link the exposure to three pyrethroids (permethrin, cypermethrin and cyfluthrin) to the urinary concentrations of their common metabolites (DCCA, 3-PBA and F-BPA).

Methods: The model is based on toxicokinetic models that describe the fate of the compounds in the human body: a generic and gender-dependent physiologically based pharmacokinetic (PBPK) model was adapted to the toxicokinetics of permethrin, cypermethrin and cyfluthrin, and simple compartmental models were developed to describe the levels of the metabolites in blood and urine. The PBPK model for the parent compounds and the compartmental models for metabolites were then connected together at the level of the metabolic sites. The whole model is therefore able to link the exposure of humans to three pyrethroids to the levels of urinary metabolites. A literature review was conducted for the parameterization. Human specific data were available for the absorption and excretion rates. In silico QSAR models were used to predict the other unknown parameters such as the partitioning into the tissues, the unbound fractions in tissues and blood, and the metabolic rates.

Results and conclusion: Human toxicokinetic data obtained under controlled exposure studies or after poisoning were used to evaluate the model predictions. Several scenarios were tested: exposure to one parent compound, different pathways of exposure (oral, dermal and inhalation) and different biological matrices (blood, urine). On average, our results showed good agreement between the predicted and measured levels. Then we predicted the urinary DCCA concentration following a cumulative exposure to environmental concentrations of permethrin, cypermethrin and cyfluthrin and identified the key determinants of the DCCA concentration in urine (i.e., metabolic rates in liver, liver:blood partition coefficients, urinary flow...). This model will be extended to other pyrethroids and metabolites, and could then be used to assess the exposure of human population using individual biological measurements collected in biomonitoring campaigns.

UTRECHT
THE NETHERLANDS
OCTOBER 9-13
ISES 2016

We-Po-47

Computer simulation of particulate matter formation during heating commercial cooking oils

Mehdi Amouei Torkmahalleh, *Nazarbayev University, Astana, 53 Kabanbay Batyr Ave., Kazakhstan*
Aida Kadyrbayeva, *Nazarbayev University, Astana, Kazakhstan*
Ulmeken Kaibaldiyeva, *Nazarbayev University, Astana, Kazakhstan*

Aim: Developing a new simulation tool to predict PM formation during cooking activities People mostly spend their time indoors which increases the importance of indoor particles. Cooking was found to be one of the main sources of indoor particulate matter (PM). Cooking emissions may have significant effects on human health, and thus, understanding the contribution of cooking elements such as oil to PM emissions is critical. Cooking oils compose of triglycerides such as palmitin and linolein. The primary mechanisms for producing PM by cooking could be the supersaturation of the generated vapour organic compounds due to rapid cooling after mixing with indoor air followed by homogenous or heterogeneous nucleation.
Method: The current study simulated the heating process of safflower and olive oils to predict the supersaturation of the produced organic vapour and resulting particulate matter (PM) mass. Heating of the oils was simulated for the temperature range of 50 to 197C (Figure 1). Heated oils were mixed with air flow at 23C and RH=40%, simulating dilution of the produced oil vapour in normal residential kitchens. NRTL activity coefficient model and ideal gas law were employed to predict the phase equilibria of the heated oils. Saturation level of air-oil vapour mixture was estimated by the following equation.
$S=(P\sum Yi)/P\_dew$
where P_dew is dew pressure of the PM phase at the mixture temperature and PΣYi is sum of the partial pressure of the condensable components after mixing with air. The PM formation was considered to occur when S values exceeded 1.
Results: When S exceeded 1 at 197 C, PM formed. Increased Relative humidity (RH) or moisture content of air flow was found to increase PM mass which could be attributed to the condensation of water vapour onto the PM phase. Further simulation studies were performed to investigate the impact of table salt on supersaturation of the produced vapour chemicals. It was found that addition of table salt to the heated oil reduced the supersaturation level and the PM mass due to the reduced vapour pressure of the organic compounds compared to the heated pure oil at the same temperature. This observation is in good agreement with the experimental finding of Amouei Torkmahalleh et al. (2013).
Conclusion: Computer simulation showed a good agreements with experimental finding and can help us to better understand the PM formation and resulting PM composition.

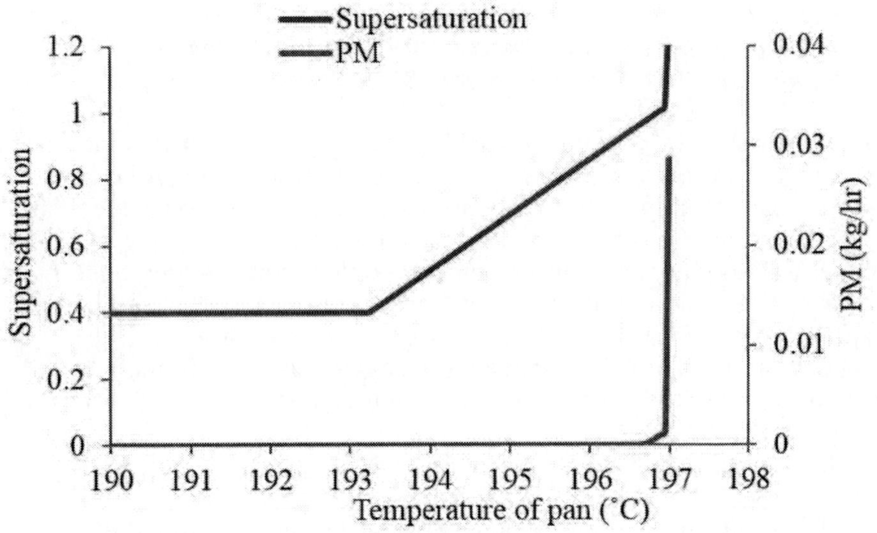

Flowsheet Diagram

UTRECHT
THE NETHERLANDS
OCTOBER 9-13
ISES 2016

We-Po-48

On the importance of developing integrative modelling approaches within the framework of human exposure assessment.

Mouhamadou Sy, *Federal Institute for Risk Assessment (BfR), Berlin, Germany*
Matthias Greiner, *Federal Institute for Risk Assessment (BfR), Berlin, Germany*

Human populations can be exposed to chemical hazards present in the environment and in particular to chemicals present in food. In practice, environmental and human exposure assessments are generally conducted independently on the basis of multiple sources data and relying on separately developed methods, models and scenarios. Connecting the assessments of environmental and human exposure by developing multimedia modelling is of great interest in order to better characterize the different routes across which human populations can be exposed to chemicals. Significant efforts were recently made to improve information by conducting Total Diet Studies (TDS) and collecting human biomonitoring data although these study types have not been considered in combination to study total exposure.
Aim: The background objective of this study is to develop integrative approaches that accommodate available data sources for relating external and internal exposure. Specifically, this study aims at developing methodological tools to account for the different routes while assessing human exposure to chemical contaminants or residuals.
Methods: Approaches will be developed for multi-pathway and multimedia exposure models using food monitoring data, the results of total diet studies and environmental and biomonitoring data. Bayesian modelling will preferably be adopted enabling the combination of a priori knowledge (e.g. expert advices) and multiple data sources (e.g. TDS and human biomonitoring data), and to capture uncertainty/variability. This conceptual approach will be applied focusing on case-studies relevant to German dietary patterns by considering the results of the German total diet study.
Results and discussion: This study will allow better understanding the total chemical exposure of the German population by the identification of sensitive exposure patterns (e.g. high consumers of particular food items). These results, which aim at bridging the gaps between environmental and human exposure assessment models, will improve the characterization of the contribution of environmental processes while predicting on a population level the internal exposure of human to chemical (or radioactive) substances.
Keywords: chemical exposure assessment, external exposure, internal exposure, integrative approaches.

We-Po-51

Conceptual framework describing a child's total (built, natural, social) environment in order to optimize health and well-being

Nicolle Tulve, *U.S. Environmental Protection Agency, Research Triangle Park, North Carolina, United States*
Jazmin Ruiz, *Xerox Corporation, Webster, New York, United States*
Kim Lichtveld, *University of Findlay, Findlay, Ohio, United States*
Sally Darney, *National Institute of Environmental Health Sciences, Research Triangle Park, North Carolina, United States*
James Quackenboss, *U.S. Environmental Protection Agency, Las Vegas, Nevada, United States*

The complexity of the components and their interactions that characterize children's health and well-being are not adequately captured by current public health paradigms. Children are exposed to combinations of chemical and non-chemical stressors from their built, natural, and social environments at each lifestage and throughout their lifecourse. Children's inherent characteristics (e.g., sex, genetics, pre-existing disease) and their activities and behaviors also influence their exposures to chemical and non-chemical stressors from these environments. We describe a conceptual framework that considers the interrelationships between inherent characteristics, activities and behaviors, and stressors (both chemical and non-chemical) from the built, natural, and social environments in influencing children's health and well-being throughout their lifecourse. This framework is comprised of several intersecting circles that represent how stressors from the total environment interact with children's inherent characteristics and their activities and behaviors to influence their health and well-being at each lifestage and throughout their lifecourse. We used this framework to examine the complex interrelationships between chemical and non-chemical stressors for two public health challenges specific to children: childhood obesity and general cognitive ability. One systematic scoping review showed that children's general cognitive ability was influenced not only by chemical exposure (e.g., chlorpyrifos), but by the interrelationships between chemical and non-chemical stressors (e.g., neighborhood-level socioeconomic factors and chlorpyrifos exposure). This systematic scoping review also suggested that non-chemical stressors may modify the response to chemical exposures (for general cognitive ability: e.g., prenatal lead exposure and maternal self-esteem). Another systematic scoping review showed that numerous chemical and non-chemical stressors are linked to childhood obesity. Using this conceptual framework and these systematic scoping reviews, we hypothesize that multiple chemical and non-chemical stressors are interacting to impact childhood obesity and general cognitive ability, suggesting the importance of a conceptual framework describing the interrelationships between inherent characteristics, activities and behaviors, and stressors (both chemical and non-chemical) from the built, natural, and social environments. By better understanding these complex interactions, decision makers can make informed choices for child-specific environments that optimize health and well-being within the home and community.

We-Po-53

A three dimensional land use regression model for NO2 in an urban environment - Vegas (vertical gradient study)

Danyal Odabasi, *Swiss Tropical and Public Health Institute, Basel, Switzerland*
Alex Ineichen, *Swiss Tropical and Public Health Institute, Basel, Switzerland*
Mark Davey, *Swiss Tropical and Public Health Institute, Basel, Switzerland*
Marloes Eeftens, *Swiss Tropical and Public Health Institute, Basel, Switzerland*

Background
Numerous land use regression models exist that explain and predict spatial contrasts of air pollution concentrations at a high spatial resolution. Long-term exposure estimates derived from such models have been linked to substantial differences in the risk of respiratory disease and mortality. However, the exposure assessment for these studies typically does not consider differences in air pollution levels with increasing height above the ground despite the very conceivable likelihood that people living on lower floors (closer to a major source: traffic) are differently exposed than their neighbours on higher floors.

Objectives
We aim to study how traffic intensity, street configuration and season influence the vertical pollution gradient. We further aim to build a 3-dimensional land use regression model for NO2 for Basel, which can explain and predict both horizontal and vertical pollution patterns.

Methods
We selected 25 buildings along residential streets in Basel, Switzerland where we measured NO2 concentrations at 3 different heights per building simultaneously. Small shelters containing passive Passam samplers were hung on the front façade (street-side) of each building. 30-Minute traffic counts were performed during off-peak hours. The aspect ratio (average building height/width of the street) was determined as a measure of street configuration for each location. The samplers remained in place for a 14-day period from 25/26 February until 10/11 March 2016 and were subsequently analyzed in the laboratory. The same locations will be sampled once more in early summer 2016.

Results & discussion
Based on the traffic counts, the locations were grouped into 10 quiet streets (fewer than 100 cars per 30 minutes), 5 medium traffic streets (100-250 cars per 30 minutes) and 10 busy streets (over 250 cars per 30 minutes) based on off-peak traffic counts. Aspect ratios were on average 0.72 (range 0.34 -1.20). As next steps, we will evaluate how these and other factors can explain the measured contrasts in NO2 concentration at ground level, as well as the vertical gradient. It is likely that the vertical gradients are different in summer and winter: in the summer season, traffic emissions from the nearby road are the only major local contributor to NO2, while in winter, both traffic emissions and rooftop-level emissions from residential heating constitute major sources, creating a less predictable vertical gradient.

We-Po-54

Modeling the Health Benefits of Local and Regional Emission Control Policies in the US Aviation Sector

Lindsay Underhill, *Boston University School of Public Health, Boston, MA, United States*
Stefani Penn, *Boston University School of Public Health, Boston, MA, United States*
Scott Boone, *University of North Carolina at Chapel Hill, Chapel Hill, NC, United States*
Sarav Arunachalam, *University of North Carolina at Chapel Hill, Chapel Hill, NC, United States*
Jonathan Levy, *Boston University School of Public Health, Boston, MA, United States*

BACKGROUND. The impact of aircraft emissions on human health is of growing concern due to increasing demands for air transportation and expected decreases in other prominent sources of combustion-related pollutants. Ground level ozone (O3) and fine particulate matter (PM2.5) are believed to drive the majority of health-related impacts, such as premature mortality, caused by air pollutants overall and from aviation sources. While the health implications of national-scale emissions control strategies for aviation and other source sectors have been estimated, few studies have developed and implemented modeling approaches that could examine the differential health implications of an array of geographically targeted policy measures.
METHODS. In this study, we applied the Community Multiscale Air Quality (CMAQ) model with the Decoupled Direct Method (DDM) to predict airport-specific contributions of individual precursor pollutants to ambient concentrations of O3 and PM2.5 for 66 airports across the US, which collectively represent 77% of national aviation fuel burned. We quantified health damage functions, in terms of premature mortality per ton of landing and take-off (LTO) cycle emissions by pollutant and airport, and constructed regression models to explain variability as a function of basic population and meteorological factors. We applied these regression models to estimate health damage functions for 203 additional airports, allowing for health impacts to be calculated for 97.5% of national aviation fuel burned. We developed multiple hypothetical policy scenarios, including regional to national-scale implementation of alternative fuels as well as geographically-focused flight activity measures, and examined the air quality and public health implications.
RESULTS and CONCLUSIONS. Results suggest important regional differences in health damage functions, related to background pollutant concentrations and population patterns, with corresponding differences in health benefits of regional implementation of alternative fuels. Variability in health damage functions is much greater between airports than between regions, given the significance of population density, suggesting the heightened importance of airport-specific health risk estimation for policy measures that can be targeted toward individual airports. Overall, this study demonstrates the utility of health damage function modeling as a low-cost, efficient approach to evaluating the health benefits associated with a variety of emission reduction strategies involving diverse sets of pollutants across a range of populations and geographic regions.

We-Po-56

Combining background and local effects models of ambient ultrafine particle concentration to predict exposure at residences in an urban area

Matthew Simon, *Tufts University, Medford, MA, United States*
Alex Bob, *City of Somerville, Somerville, MA, United States*
Allison Patton, *Rutgers University, Piscataway, NJ, United States*
Doug Brugge, *Tufts University, Boston, MA, United States*
John Durant, *Tufts University, Medford, MA, United States*

Aim: Exposure to traffic-related ultrafine particles (UFP; <100 nm diameter) is associated with markers of cardiovascular disease risk. More accurate exposure estimates are needed to reduce exposure misclassification in health studies; however, because UFP concentrations are highly variable in space and time, accurate estimation is a considerable challenge. Our aim was to develop UFP exposure estimates for individuals in the Boston Puerto Rican Health Study cohort by developing hourly regression models of traffic-related particle number concentration (PNC; a proxy for UFP) that consider both urban background and local-source contributions.
Methods: To build the models, PNC was measured by continuous monitoring at a stationary site in Boston, Massachusetts (USA) and by mobile monitoring along a fixed route in the 40-km^2 study area on 49 days (3-6 hours/day) between December 2011 and November 2013. The background model was generated from log-transformed hourly first percentile of 1-minute PNC measurements from the stationary site. The local effects model was generated from the residuals of the log-transformed mobile PNC measurements relative to the background measurements. Average hourly traffic, meteorology, chemical species indicators of secondary photochemical oxidation pathways, and land-use data (e.g., distance from highways) were used as explanatory variables in the two models. Both models were evaluated by R^2 and root-mean-square-error (RMSE), and validated by 10-fold cross-validation. Additionally, results from the summation of the two models were compared to ambient PNC measured outside 14 homes in the study area. Measurements were made continuously for six consecutive weeks at each home between May 2012 and November 2013.
Results: Measured UFP concentrations at homes ranged from 700-210,000 particles/cc and modeled concentrations ranged from 800-260,000 particles/cc. Model adjusted-R^2 was 0.58 (RMSE = 0.42). Cross validation resulted in an R^2 standard deviation of 0.02. Compared to ln(PNC) at the 14 homes, the model overestimated observations, but predicted ln(PNC) were all within a factor of 1.45 of observed ln(PNC) values.
Conclusions: We found that separating background and local effects improved overall UFP exposure estimates at residences in an urban area. Additionally, accounting for secondary particle formation pathways improved model fit.

We-Po-57

Analyzing participant interactions with personalized report-back: data from DERBI, an online reporting tool

Herbert Susmann, *Silent Spring Institute, Newton, MA, United States*
Ruthann Rudel, *Silent Spring Institute, Newton, Massachusetts, United States*
Katherine Boronow, *Silent Spring Institute, Newton, Massachusetts, United States*
Laurie Havas, *Child Health and Development Studies, Berkeley, California, United States*
Kenneth Arnold, *Harvard University, Cambridge, Massachusetts, United States*
Robin Dodson, *Silent Spring Institute, Newton, Massachusetts, United States*
Krzysztof Gajos, *Harvard University, Cambridge, Massachusetts, United States*
Barbara Cohn, *Child Health and Development Studies, Berkeley, California, United States*
Julia Brody, *Silent Spring Institute, Newton, Massachusetts, United States*

Participants in studies of emerging contaminants have traditionally not received their own results, although, in response to changing ethical norms, studies increasingly provide personal results as print material. Web-based reports provide novel ways to efficiently provide personalized reports and also to research participants' experience receiving results. We developed DERBI (Digital Exposure Report-Back Interface) to generate web-based personalized exposure reports for 295 women in the Child Health and Development Studies monitored for 42 environmental contaminants, including organochlorine pesticides, PCBs, flame retardants, and perfluorinated compounds. Participants accessed their report through a secure online interface. 147 participants were given access to reports that included individual and study-wide results, while the other 148 participants initially received only the study-wide results. The individual reports show personal results, along with contextual information about exposure sources, possible health effects, and what can be done to reduce exposure. Graphs show comparisons to the median for women of similar age in the National Health and Nutrition Examination Study (NHANES). The web reports include individualized summaries that "headline" important aspects of each participant's results. We recorded participant actions on the website, including page loads, mouse movements, and page scrolling. From these low level events we derived analytics like how long each participant spent viewing her results and the most common paths participants take through their reports. We found that participants spent a median of 20 minutes viewing their report, and that 89% of participants viewed the detailed results page of at least one of the chemical groups headlined for them. Furthermore, participants who had access to their individual results spent twice as much time on their report than participants who received only study-wide results. The differences in each participant's use of the DERBI, as evidenced in the analytics data, illustrates how digital report-back allows researchers to support the individual interests, concerns, and learning styles of participants in exposure studies.

We-Po-58

Urban air quality assessments using low-cost mobile sensor ‘AirBeam’

Chris Lim, *NYU School of Medicine, New York, NY, United States*
Ruzmyn Vilcassim, *NYU School of Medicine, New York, United States*
Terry Gordon, *NYU School of Medicine, New York, United States*
George Thurston, *NYU School of Medicine, New York, United States*

Background: Low-cost sensors are increasingly being utilized to assess personal-level exposures and to measure ambient air pollution levels. A citizen-science approach for data collection, especially in urban locations, allows for identification of spatiotemporal hotspots and characterization of small-scale pollutant gradients. The sensors’ capabilities, accuracy, and inter-instrument variability must be tested and characterized in detail before deployment.
Methods: AirBeam fine particulate matter monitors were calibrated against a standard method, DataRAM pDR-1500, which was previously calibrated against integrated gravimetric filters. The instruments concurrently measured CAPs (concentrated ambient air particles) concentration levels in a controlled laboratory setting.
The calibrated AirBeams were used to conduct systematic sampling sessions in Seoul, South Korea, across five routes around fixed-site monitors. Sampling was done daily during morning, afternoon, and nighttime, for approximately three weeks for a total of 180 hours. The collected PM2.5 data were then utilized to construct predictive mobile land use regression (LUR) models and compared with fixed-site LUR models.
Results: Each AirBeam unit was fit with individual calibration curves to adjust for between-instrument variability, which was greatest at higher concentration levels. Concurrent measurement sessions with both AirBeams and pDR-1500 showed high correlations at 1-minute averages, with a non-linear relationship at higher concentrations (>40 µg/m3). Sampling in Seoul revealed high variability in concentration levels depending on time and location, with elevated PM2.5 levels found nearby major roadways and during afternoon rush hours. AirBeam concentration levels also showed high agreements with nearby central monitor 1-hour average values (R2>0.6).
Conclusion: Additional testing will be conducted to better characterize the sensor responses to different particle types and sizes. Overall, AirBeams (and similar sensors) represent a valid and practical technology, with potential for inexpensive personal level exposure assessments and high-resolution mapping of urban air quality, especially in locations with sparse air pollution data.

We-Po-59

NanoSafer version 1.1. Demonstration of a dynamic web-based precautionary risk assessment and management tool for manufactured nanomaterials

Keld Alstrup Jesen, *National Research Centre for the Working Environment, Copenhagen, Denmark*

Background

Industry and workplace inspectors are often challenged in workplace assessments of productions and applications of manufactured nanomaterials. Control banding may aid in making sufficiently precautionary risk assessments and management of nanomaterial exposures as emerging chemicals. NanoSafer 1.1 is an advanced web-based control-banding tool specifically developed to enable safe production and use of manufactured nanomaterials (MNM) in the working environment (http://nanosafer.i-bar.dk/). NanoSafer was originally launched in 2010 and has now been improved as part of a process to develop a 3-Tier NanoSafer modelling-based risk assessment and management framework.

Objectives

NanoSafer 1.1 was developed to aid industry, workplace professionals and administrative inspectors to perform Tier 1 information gathering for exposure assessments as proposed by OECD to manage the potential risks associated with production and use of MNM and MNM-enabled products. NanoSafer 1.1 currently covers assessments of powder handling and episodes of constant release rates from point sources / fugitive sources.

Results

The NanoSafer CB tool bases its assessments on information normally available from the producuct technical data sheet, the material safety sheet and the work situation. These data are used in four modeling sections: 1) A nanomaterial identifier and classifier, which based on physicochemical characteristics and naming ensures discrimination between nanomaterials and non-nanomaterials. 2) A hazard grouping and ranking model considering hazard information on bulk materials and physicochemical properties, if nanospecific hazard data are not available. 3) A two-box instant mixing aerosol dispersion model for assessment of acute and daily near-field and far-field exposure potentials in specific work situations. 4) A combined hazard and exposure-based risk ranking mode resulting in an automated identification of suitable exposure reduction measures.

Testing of NanoSafer demonstrates that the model has a wide dynamic range and ability to perform balanced evaluations of small to large scale production and use of low to potentially highly hazardous MNM. Application of assessments of risks associated with both acute and daily/chronic exposure gives valuable information to improve risk management measures.

The e-learning tool is functional and primarily designed to aid research and development laboratories as well as small and medium size companies who may not have experience in working and managing the potential risks of MNM or can benefit from inspiration exposure management performed in other relevant work situations.

Conclusions

NanoSafer 1.1 provides a versatile modeling-based tool for precautionary risk assessment and management of risk associated with occupational production and use of nanomaterial powders.

We-Po-60

Application of Integrated urban models to simulating health risks

Tor Oiamo, *University of Windsor, Windsor, Canada*

Knowledge translation in the field of environmental health remains an important area in need of advancement. Policy makers, urban and environmental managers are increasingly concerned with sustainable urban development and environmental health is a key aspect in this regard. Therefore, the research presented here aims to facilitate for practitioners who are not familiar with exposure science the application of concentration response functions (CRF) to urban development scenarios. This is achieved through the linkage of a health risk and benefit assessment tool with an integrated urban modelling system (IUM). Traditionally, IUMs have been used to assess how feedback processes that link transportation and land use in urban environments will be affected over a given time horizon within a certain development scenario.
In a medium sized Canadian city, we developed an IUM to represent baseline conditions and simulate the impact of a new rapid transit system on future transportation and land use changes. Estimated changes in network traffic volumes from baseline to scenario conditions were used to assess potential impacts on air emissions. Simulated changes in localized emissions of ozone and fine particulate matter (PM2.5) were consequently linked to CRFs for a number of chronic and acute health outcomes. The simulation modelling system produced information on both urban aggregate and localized impacts of air pollution on morbidity and mortality related to the rapid transit system. Overall, the methodology shows promise with respect to facilitating knowledge translation. Future challenges and opportunities in terms of mobilizing the modelling approach will be discussed.

Late Breaking Abstracts

We-LBA-19

Indoor Air Quality Assessment in an Electronic Cigarette Vaping Convention

Rui Chen, *Johns Hopkins University, Baltimore, Maryland, United States*
Angela Aherrera, *Johns Hopkins University, Baltimore, United States*
Stephanie Jarmul, *Johns Hopkins University, Baltimore, United States*
Chineye Isichei, *Johns Hopkins University, Baltimore, United States*
Pablo Olmedo, *Columbia University, New York, United States*
Ana Navas-Acien, *Columbia University, New York, United States*
Ana Rule, *Johns Hopkins University, Baltimore, United States*

Background: E-cigarette vaping conventions or vape expos are public events that promote sales and provide information on e-cigarette products. These events attract a large number of vaping enthusiasts, vendors, and local residents, and allow the use of e-cigarette inside the indoor venues. The large concentration of vapers and poor air ventilation result in indoor air contamination and potential health impacts especially for people that attend these events regularly such as vendors. Our objective was to evaluate indoor air quality induced by e-cigarette use in a vaping convention.
Methods: Air sampling was conducted during an e-cigarette vaping convention on April 2016, in Baltimore, Maryland. Real-time concentrations of particulate matter (PM)10 were measured with a SidePak; total volatile organic compounds (TVOC), carbon monoxide (CO), carbon dioxide (CO2), and nitrogen dioxide (NO2) were measured with a GrayWolf multi-gas monitor also in real-time during one 7-hour event. Active integrated sampling was performed to measure air nicotine concentrations.
Results: Mean (range) PM10 concentration was 8,699 (9-17,860) μg/m3. The estimated 24-hour time weighted average (TWA) PM10 was 1800 μg/m3, which is 12 fold higher than the EPA 24-hours regulation (150 μg/m3). Mean TVOC, CO2 and CO concentrations were 0.13 ppm, 811 ppm and 0.05 ppm. These were all below their suggested/established air quality guidelines (1.0 ppm, 1,000 ppm, 9 ppm respectively). Mean indoor NO2 concentration was 100 ppb, reaching the US National Ambient Air Quality Standards (NAAQS) of 1-hour daily maximum of 100 ppb. PM10 concentrations were correlated with CO2 concentrations (Pearson correlation coefficient $r=0.76$, $p<0.001$). TVOC concentration was also highly correlated with CO2 ($r=0.81$, $p<0.001$). Indoor CO2 concentration reflected the number of people and the air exchange within the room. More vapers and poor ventilation aggravated indoor air quality. Air nicotine concentration was 125 μg/m3, which is 88 times higher than the average concentration measured in waterpipe cafes in Baltimore (1.42 μg/m3) and equivalent to concentrations measured in bars and nightclub using active sampling.
Conclusion: E-cigarette aerosol is a source of indoor air exposure to of PM10, TVOC, NO2 and air nicotine. Moreover, PM10 and air nicotine during the vaping convention exceeded indoor air guidelines. More robust regulation to ban indoor e-cigarette use is warranted.

We-LBA-20

Evaluation of the Association between Airborne Real-Time Concentrations of Black Carbon (BC) and Fine Particulate Matter (PM2.5) in Urban Hotspots of South Korea

Sungroul Kim, *Soonchunhyang University, Asan, Korea, South*
Sol Yu, *Soonchunhyang University, Asan, Korea, South*
Seonyup Lee, *Soonchunhang University, Asan, Korea, South*
Sujung Park, *Soonchunhyang University, Asan, Korea, South*

According to a recent report by the World Health Organization, daily mortality or hospital admissions due to black carbon (BC) are higher than that due to particulate matter (PM; PM10 and PM2.5) when expressed in μg/m3 as estimates of single-pollutant effects. BC is an indicator of a mixture of particulates from several combustion sources. In urban hotspots, where traffic type and volume vary, spatial or temporal variations in BC concentrations can be different from that of PM2.5. However, information on diurnal distributions of BC concentrations in urban hotspots of South Korea is unavailable. Thus, we measured differences in BC and PM2.5 concentrations for several spots at a fixed distance (spot2: 300m, spot3: 500m) from potential diesel emission sources (spot1) located in urban hotspots. We evaluated the spatial and temporal association between the measurements according to distance.
The measurements for BC and PM2.5 were conducted from July 2014 to December 2014 at entrances of four hotspots including Seoul Express Bus Terminal, Cheonan Express Bus Terminal, Cheonan-Asan Express Train Station, and Namdong industrial complex for more than 4 times per each site in different season using a real time PM or BC monitor.
The correlation coefficient for BC concentrations between spot 2 (300m from spot1) and spot 3 (500m from spot1) was 0.8 and between spot 2 or spot 3 and spot1 was approximately 0.6. The coefficients for PM2.5 concentrations were higher than that of BC concentrations and were consistent (0.9) for all spots. Irrespective of the distances, the unit increase in PM2.5 concentration (μg/m3) displayed the highest slope factor (0.12~0.16) for morning BC measurements (μg/m3) followed by afternoon (0.06~0.07) and night (0.01 ~ 0.02) measurements.
The differences in the association of BC with PM2.5 concentrations between morning and night may possibly affect the risk estimates and source identification. Therefore, a future study may need to be started
Keywords: Black carbon, PM2.5, Urban hotspot

We-LBA-21

Exposure Science and Policy Challenges of the Future: Learning from European Experience

Ekaterina Svyatets, *University of Southern California, Los Angeles, California, United States*

This multidisciplinary research explores the EU practices of bridging exposure science and policy decisions in order to design efficient policies that protect human health and ecosystem health. In particular, these policies are designed to limit human exposure to hazardous chemicals in food and consumer goods. These practices then are compared with those in the United States, and policy recommendations are provided for a possible adaptation of European experience in the U.S. legislative landscape. Using qualitative methods of policy analysis, combined with the latest findings in exposure science, the results of this research show that European standards are more protective of the consumer because of the precautionary principle that requires produces to use the most advance exposure science methods in order to prove the absence of adverse effects in their products. In the United States, however, the burden of proof of harm from hazardous chemicals falls on the consumer through lawsuits. In conclusion, the way to improve the U.S. legal landscape for consumer protection is to expand the exposure science education at universities and high schools, which in turn will bring more visibility to such issues and more weight in the policy landscape.

Various Effects of Toxic Chemicals

We-LBA-22

Operator, worker and bystander tool (OWB) for screening assessment of co-formulants in plant protection products

Matthias Wormuth, *Syngenta Crop Protection AG, Basel, Switzerland*
Volker Mostert, *Extera, Langenfeld, Germany*
Christopher Dobe, *Syngenta Crop Protection AG, Basel, Switzerland*
Sebastien Bonifay, *DuPont de Nemours, Mechelen, Belgium*
Joachim Krass, *BASF SE, Limburgerhof, Germany*
Renate Vosswinkel, *Bayer CropScience AG, Monheim, Germany*

In the European Union, manufacture and import of chemicals is regulated under the REACH Regulation (EC) No 1907/2006 (Registration, Evaluation and Authorisation of Chemicals). An environmental and human exposure and risk assessment has to be carried out under REACH for hazardous substances used in quantities above 10 tonnes per year. This assessment needs to cover all identified uses throughout the whole life cycle of a substance.
The European Crop Protection Association (ECPA) has developed the OWB (operator, worker and bystander) screening tool for the assessment of human exposure to co-formulants in plant protection products in a standardised way. The four generic exposure scenarios developed by ECPA for the spray and granular application of crop protection products by professionals and consumers are integrated in OWB. The tool requires a limited set of input parameters and conducts the assessment for commonly used formulation types, to support manufacturers and importers of substances used as co-formulants as they often have limited knowledge of typical use conditions of plant protection products in the agricultural sector.
OWB supports a REACH-compliant screening assessment of human exposure to co-formulants by using established models and approaches for the risk assessment of active ingredients of plant protection products in the EU. The tool gives relevant output for the communication of use conditions and risk management measures ensuring the safe use of co-formulants. OWB is based on models developed by the European Food Safety Authority (EFSA), which since 2016 are replacing the models and tools formerly used for the assessment of active ingredients of plant protection products in the EU. The presentation highlights the main features of the OWB tool, describes the required input parameters, and gives examples of representative assessments of co-formulants and how the output can be used in the communication of safe conditions of use to downstream users.

We-LBA-23

Approach to correct the MicroPEM shifting baseline issues

Ting Zhang, *Columbia University, New York, New York, United States*
Steve N. Chillrud, *Columbia University, New York, New York, United States*
Masha Pitiranggon, *Columbia University, New York, New York, United States*
James Ross, *Columbia University, New York, New York, United States*
Junfeng Ji, *Nanjing University, Nanjing, China, People's Republic of*
Beizhan Yan, *Columbia University, New York, New York, United States*

Background: Fine particulate matter (PM2.5) is associated with various adverse health outcomes. The miniaturized real-time portable particulate sensors with integrated filters of PM2.5 (MicroPEMs by RTI) have been widely used for exposure assessment to PM2.5. However, an unrealistic or shifting baseline issue has been observed in the nephelometer real-time data of MicroPEMs. Objective: The aim of this study is to find out a method to solve the shifting baseline problem in MicroPEM. Methods: We developed a running baseline correction method based on central site monitored PM2.5 (RBCS). We also compared our corrected results with field based high-efficiency particulate arrestance (HEPA) method fixed data. In addition, gravimetric correction was conducted after baseline correction given the potential highly variable optical reflectivity of local PM sources. 142 personal or residential samples including 7 duplicates generated by MicroPEMs from a cohort of children in New York City and Baltimore were used to validate the baseline correction method. Results: As for 79 of the 142 deployments that have valid start and end HEPA correction, 55.6% of the collected raw data met our validity criteria, while 46.8% and 96.2% of them met the criteria after HEPA and RBCS correction, respectively. As for the remaining 63 deployments, 61.9% of them met the validity criteria before RBCS correction and 100% met the standards after correction. The Pearson correlation coefficient of average PM2.5 concentration for the 7 groups of duplicates increased from 0.87 ($p < 0.005$, slope = 0.60) for raw data to 0.99 ($p < 0.001$, slope = 0.71) for RBCS corrected data. And the slope increased to 1.00 after both RBCS and gravimetric correction. Conclusions: RBCS correction could well fix the baseline shifting issue in MicroPEM monitoring. And it is more effective compared with HEPA correction method. Combination of RBCS and gravimetric correction could help precisely interpret MicroPEM real-time data.

We-LBA-24

Occurrence and exposure to phthalate metabolites and bisphenol analogues in urine from Korean children

Yunsun Jeong, *Hanyang University, Ansan, Korea, South*
Min Kyu Park, *Hanyang University, Ansan, Korea, South*
Sunmi Kim, *Seoul National University, Seoul, Korea, South*
Gowoon Lee, *Seoul National University, Seoul, Korea, South*
Kyungho Choi, *Seoul National University, Seoul, Korea, South*
Hyo-Bang Moon, *Hanyang Univerisity, Ansan, Korea, South*
Jeongim Park, *Soonchunhyang University, Ansan, Korea, South*

Phthalates are a group of chemicals widely used in consumer products such as plasticizers, solvents and additives. Due to the short half-lives of phthalates in human body, urinary phthalate metabolites are used as biomarkers of recent human exposure to phthalates. BPA is primarily used in the production of polycarbonate and epoxy resins. Because of regulations on BPA, bispenol analogues (BPs) such as bisphenol S (BPS) and bisphenol F (BPF) are used as alternatives of BPA. Phthalates and BPs are of concern worldwide because they disrupt endocrine system in human body. In our study, urinary concentrations of 18 phthalate metabolites and 7 bisphenol analogues (BPs) were determined from 133 Korean children (2¬-8 years) in 2015 using LC/MS/MS. To avoid variance of these contaminants according to the sampling point, the first urine samples were collected on Friday. There was no significant difference depends on sex and age. Among analyzed compounds, 10 phthalate metabolites, BPA and BPS were detected over 60% of total urine samples. Total concentrations of phthalate metabolites and BPs ranged from 11.6 to 350 ng/mL (median: 95 ng/mL) and 0.3 to 62 ng/mL (median: 2.5 ng/mL), respectively. In urine samples, metabolites of di(2-ethylhexyl) phthalate (DEHP) such as mono(2-ethyl-5-carboxypentyl) phthalate (MECPP, 52%), mono-[(2-carboxymethyl)hexyl] phthalate (MCMHP, 11%), monoethyl phthalate (MEP, 8%), mono-n-butyl phthalate (MBP, 6%), and monomethyl phthalate (MMP, 6%) showed the highest proportions to total concentrations of phthalate metabolites. Our finding implies the higher consumption of DEHP than other phthalates in Korean industry. Among BPs, BPS (64%) showed highest contribution to the total BP concentrations, BPA (26%) was the next contributor. The dominance of BPS in urine samples could be associated with the increasing demand for BPS in Korean industry due to strong regulation on BPA. The estimated daily intake to MEP (mean: 6.2 μg/day), MBP (7.2 μg/day) and BPA (1.1 μg/day) showed lower than the reference dose values proposed from U. S. Environmental Protection Agency.

We-LBA-25

Perfluoroalkyl acids among Korean children and adolescents: serum levels in 4 to 18 years of age and related exposure sources

Habyeong Kang, *Seoul National University, Seoul, Korea, South*
Hyun-Kyung Lee, *Hanyang University, Ansan, Korea, South*
Jy Eun Lee, *Seoul National University, Seoul, Korea, South*
Sunmi Kim, *Seoul National University, Seoul, Korea, South*
Hyo-Bang Moon, *Hanyang University, Ansan, Korea, South*
Jeongim Park, *Soonchunhyang University, Asan, Korea, South*
Kyungho Choi, *Seoul National University, Seoul, Korea, South*

Humans are ubiquitously exposed to perfluoroalkyl acids (PFAAs) in daily life. Although exposure levels of these compounds in general population have been widely reported, PFAAs exposure among children and adolescents is not well understood. This study was conducted to evaluate levels of PFAAs among Korean children and adolescents. In 2012 and 2014, we collected serum samples from 150 children and adolescents with 4-18-year-old living in Seoul and Gyeonggi, Korea. In the samples, 16 PFAAs were analyzed. To identify potential sources of exposure, dietary and behavioral factors were investigated from the 150 participants. PFOS was detected in all the samples, and the concentration was highest among the analyzed PFAAs with the median concentration of 5.68 ng/mL. PFOA, PFNA, and PFHxS were also detected in all the samples with the median concentrations of 1.88, 0.938, and 0.652 ng/mL, respectively. Detection rates of PFUnDA, PFDA, and PFDS were 98.7, 79.3, and 63.3%, respectively, while detection rates were less than 50% for the other PFAAs. Among 16-18-year-old adolescents, sum of total PFAAs concentration was significantly higher among boys than in girls. Similar trends of increase in boys were observed for other PFAAs in the age groups after 10 years old. However, this sex difference was not significant in the age groups before 10 years old (Figure). Concentrations of several PFAAs, i.e. PFOA, PFNA, and PFOS, decreased as age increased. Using questionnaire, several potential exposure sources of PFAAs, i.e. intake of breastmilk, fish/shellfish consumption, use of frying pan, and use of waterproof coating cloths, were identified. The results of this study will help understand current exposure status of and manage exposure of major PFAAs among children and adolescents. In addition, high detection of long chain PFAAs, i.e. PFUnDA, PFDA, and PFDS, warrants further efforts to investigate exposure source of these emerging compounds.

Geometric mean concentrations of (A) total PFAAs, (B) PFOA, (C) PFNA, (D) PFUnDA, (E) PFHxS, and (F) PFOS in serum according age group and sex. Comparisons between two sexes were conducted with t-test. Symbols #, *, and ** represent p<0.1, p<0.05, and p<0

We-LBA-26

Human exposure assessment of nonylphenol and octylphenol - a review

Gowoon Lee, *Seoul National University, Seoul, Korea, South*
Habyeong Kang, *Seoul National University, Seoul, Korea, South*
Kyungho Choi, *Seoul National University, Seoul, Korea, South*

Alkylphenols are biodegradation products of alkylphenol ethoxylates (APEOs), which are used in various industries such as polymer, plastics, electronics, and fabrics industries. Due to their estrogenicity, alkylphenols have been regulated in many parts of the world. Considering toxicity and possible human exposure to these compounds, identification of exposure sources and management of alkylphenols are necessary. In this review, exposure levels of alkylphenols in general and susceptible populations, and their major exposure route were examined through literature survey. Target compounds of this review were limited to nonylphenol (NP) and octylphenol (OP). For these compounds, information on physico-chemical characteristics, exposure sources, estimated daily intake, and occurrence levels in human biological samples were obtained. In addition, relative contribution of each exposure source and route of NP and OP was estimated.
NP and OP have been detected in food. Fish (mean of 431 µg/kg), meet (262.9 µg/kg), grain (431 µg/kg), and vegetables (131 µg/kg) are determined as major sources of exposure to these compounds for general populations. People can be also exposed by drinking water, and personal care products (PCPs). Up to 39.1 mg/kg of alkylphenols was reported in PCPs. Indoor air and house dust were exposure sources in indoor environment. Among general populations, contributions of food consumption (89%) was estimated to be the greatest, followed by inhalation of indoor air (7%) and use of PCPs (4%). Among the populations with high exposure, contribution of the use of PCPs (72%) was the greatest, followed by food consumption (23%) and inhalation of indoor air (5%).
NP and OP have been detected in human samples, e.g. urine, breast milk, blood, adipocyte, and placenta. When median values of reported alkylphenols concentrations in breast milk and baby food were applied, estimated daily intake (EDI) of NP for infants exceeded tolerable daily intake (TDI) of 5 µg/kg/day. For general populations, EDI did not exceed TDI, even if the maximum detected concentrations were applied. Identification and management of NP exposure of infants warrant further investigation.

We-LBA-27

Occupational Hazards of Standing Work: Work-Related Effects on Musculoskeletal Discomforts

Wei-Cheng Lin, *National Kaohsiung First University of Science and Technology, Kaohsiung, China, Republic of (Taiwan)*
Yu-Li Huang, *National Kaohsiung First University of Science and Technology, Kaohsiung, China, Republic of (Taiwan)*

Prolonged standing at work is a known occupational hazard that leads to various musculoskeletal discomforts (MSD). Poor work environment, improper posture and work overloads are among important risk factors to occupational musculoskeletal hazards, but the influence of prolonged standing at work remain largely unexplored. In this study, literature data were collected from different occupations to examine how prolonged standing at work affects risk of MSD. Thesis and journal publications published between 2005 - 2015 were screened using the online literature databases. Studies on musculoskeletal discomforts among specific groups of workers were identified and included for data analyses. Potential effects of personal and occupational factors related to musculoskeletal discomforts were extracted from each study and combined for further analyses. In addition, work-related characteristics in standing work were compared for their potential influence on prevalence of MSD. In addition, personal and job-related characteristics were further analyzed with individual questionnaire data from a case study to determine whether personal factors affect the risk of MSD form prolonged standing at work.

The results showed that prolonged standing at work, when combined with back bending, weight carrying, walking, kneeling or static burdens, would increase the risk of musculoskeletal discomforts, specifically on the lower limbs. Further analyses were performed with individual data to determine the influence on MSD due to personal characteristics and working conditions. The results showed that women had higher risk of musculoskeletal discomforts than men. On the other hand, by controlling for work conditions, there was no evidence of gender difference in MSD risks. This suggested that work burden may be more important in influencing the risk of MSD than gender... Compared to office workers, prolonged standing work is more likely to cause musculoskeletal discomforts, and work conditions that involve frequent bending, kneeling, prolonged standing, carrying or walking objectives may increase the risk of MSD. Finally, workers who have scheduled breaks at work had about the same risk of MSD as those without breaks, suggesting that current approaches in break arrangement may not be enough to reduce MSD risks.

In summary, prolonged standing is likely to increase risk of musculoskeletal discomfort, especially in lower limbs. Work-related activities such as bending back, bending knees, carrying and walking may increase risk of MSD. The risk of musculoskeletal discomfort should be further assessed with respect to standing work conditions for better hazard prevention strategies.

We-LBA-28

Particulate matter (PM10, PM0,5) and early biological effects in children living in Lecce (Italy) by buccal micronucleus cytome assay

Francesca Serio, *University of the Salento, Lecce, Italy*
Francesco Bagordo, *University of the Salento, Lecce, Italy*
Mattia De Giorgi, *University of the Salento, lecce, Italy*
Tiziana Grassi, *University if the Salento, lecce, Italy*
Marcello Guido, *University if the Salento, lecce, Italy*
Adele Idolo, *University if the Salento, lecce, Italy*
Elisabetta Ceretti, *University of Brescia, Brescia, Italy*
Sara Levorato, *University of Perugia, Perugia, Italy*
Massimo Moretti, *University of Perugia, Perugia, Italy*
Milena Villarini, *University of Perugia,, perugia, Italy*
Marco Verani, *University of Pisa, pisa, Italy*
Sara Bonetta, *University of Torino, Torino, Italy*
Silvia Bonizzoni, *Comune di Brescia, Brescia, Italy*
Alberto Bonetti, *Centro Servizi Multisettoriale e Tecnologico - CSMT Gestione S.c.a.r.l., Brescia, Italy*
Umberto Gelatti, *University of Brescia, Brescia, Italy*
Antonella de Donno,*University if the Salento, lecce, Italy*

Background
Air pollution is one of the most important worldwide health problem. Several studies have found an association between exposure to particulate matter and the incidence of chronic diseases.
The aim of the PJS study (Progetto Jonico Salentino) is to define the exposure levels to atmospheric pollutants of the population living in a macro-area which included in the province of Lecce, Brindisi and Taranto by extending the cohort of 240 6-8 years-old schoolchildren living in Lecce, and enrolled in the MAPEC_LIFE study (Monitoring Air Pollution Effects on Children for Supporting Public Health Policy).
The authors present results of micronucleus cytome assay performed in oral mucosa cells of subjects related to lifestyle and factors associated with exposure to indoor/outdoor including the level of PM10 and PM0,5, context family residential, physical activity and weight status.
Methods
Recruitment is done on a voluntary basis after receiving from children's parents the signed consent form. Parents who accepted to participate at the study were administered, in two seasons (winter and spring) a questionnaire which included the exclusion criteria and some information regarding personal, anthropometric and health status as well as exposure factors related to the home context.
Results on children eligible for the study were carried out sampling exfoliated buccal cells from oral mucosa using a soft-bristled toothbrush, for the tests of the MN.
At the same time, in addition to biological sampling, it was conducted air monitoring by high-volume sampler "Air Flow PM10-HVS" (AMS®Analitica) near the schools attempted by the enrolled children. It was conduced a PM collection for 72h, with membrane replacement every 24h, in each site and in each season.
Results

426 samples collected from 106 (49.8%) males and 107 (50.2%) females were tested. 43% (44.6% in the first season and 41.3% in the second) of the samples tested positive (presence of at least one MN) with an average frequency of MN equal to 0.28 (0.32±0,44 in the first season, 0.24±0.32 in the second) MN/1000 differentiated cells.
Environmental sampling showed a higher concentration of PM10 and PM0.5 in the first season than the second.

Conclusions

The results concerning the frequency of MN seem in line with low particulate levels recorded and related to certain factors regarding family environment and lifestyles. These data will be integrated with those of other areas involved in PJS study.

We-LBA-29

Fruit intake as a source of organophosphate pesticide exposure among pregnant woman in the Netherlands

Michiel Van den Dries, *Erasmus MC, Rotterdam, Netherlands*
Anjoeka Pronk, *TNO, Zeist, Netherlands*
Mònica Guxens, *ISGlobal Alliance, Barcelona, Spain*
Suzanne Spaan, *TNO, Zeist, Netherlands*
Trudy Voortman, *Erasmus MC, Rotterdam, Netherlands*
Todd.A. Jusko, *University of Rochester School of Medicine & Dentistry, Rochester, New York, United States*
Pamela A. Shaw, *Perelman School of Medicine University of Pennsylvania, Philadelphia, United States*
Matthew P. Longnecker, *National Institute of Environmental Health Sciences (NIEHS), Research Triangle Park, North Carolina, United States*
Henning M. Tiemeier, *Erasmus MC, Rotterdam, Netherlands*

Aim: Organophosphate (OP) pesticides are frequently used in agricultural settings for pest control. Exposure to high doses of OP pesticides is neurotoxic to humans and animals. However, concern exists about potential health impacts of low-level OP pesticide exposures among the general population. We aimed to identify sources of prenatal exposure to organophosphate pesticides in an urban population of Dutch women.
Method:
Urinary concentrations of six dialkyl phosphate (DAP) metabolites, the main urinary metabolites of OP pesticides, were determined at 25 weeks of pregnancy in 784 pregnant women participating (between 2004 and 2006) in the Generation R study, a large prospective birth cohort study in The Netherlands. Diet during the preceding 3 months as a potential exposure source was assessed with a questionnaire in the first trimester. Mean food intake in grams per day from completed surveys was estimated using the Dutch food composition database. Linear mixed models were carried out to identify associations between the potential exposure sources and log transformed DAP metabolite concentrations. Results were adjusted for maternal age, height, pre-pregnancy weight, pre-pregnancy energy intake, marital status, maternal smoking, education, income, ethnicity, and parity.
Results: Median concentration total DAP metabolites was 310.34 nmol/g creatinine (Cr) for the first trimester, 316.48 nmol/g Cr for the second trimester and 308.32 nmol/g Cr for the third trimester. A higher intake of fruit, but not other food groups, was significantly associated with an increase in total DAP metabolite concentrations for all three trimesters. A 100 g increase in fruit intake was associated with an increase of 28.3 nmol/g Cr in total DAP metabolites. More specifically, intake of apples ($P<0.001$), lemons ($P<0.05$), peaches ($P<0.05$), and plums ($P<0.05$) was positively associated with total DAP metabolite concentrations.
Discussion: The data indicated that the women participating in this general population study from the Netherlands have been relatively highly exposed to OP pesticides compared to studies from other countries (e.g. 72 nmol/g Cr total DAP metabolites from the US NHANES study, 2001-2002). Our results suggest that fruit intake was the main dietary source of exposure. The extent to which DAP metabolite concentrations reflect exposure to the active parent pesticide rather than to less toxic metabolites remains unclear.

Further research will be undertaken to investigate the possible health effects in offspring of this relatively high low-level OP pesticide exposure.

Thursday, October 13, 2016

Plenary Address 4: Potential of metabolomics in chemical risk analysis

Gaud Dervilly Pinel, LABERCA, Nantes, France

Food safety has become a major issue worldwide and in particular, detecting the presence of toxins, contaminants or residues of chemical substances along the food chain and in fine in food items constitutes a strong consumers demand. Generally all these substances and corresponding metabolites of interest are analysed using efficient targeted methodologies. However, in some cases these targeted approaches do not allow the detection of either those substances or emerging compounds/practices and therefore new approaches and strategies are demanded to efficiently assess the exposure to chemical contaminants. Thereby the study of physiological perturbations induced upon exposure to a given chemical substance has emerged as an interesting alternative approach to be applied in chemical food safety. This presentation is focus to review and describe the most significant applications of metabolomics in the field of chemical food safety. Through various examples, the different risk analysis steps (i.e. assessment, management and communication) will be addressed to illustrate such an approach is fit-for purpose answering the expectations and requirements of chemical risk analysis. It can be considered as an innovative tool to predict the likely occurrence and nature of risks, together with improving detection methods, in the aim of answering global safety issues and anticipating human health problems.

Th-SY-A1: The exposome: a transdisciplinary paradigm for improved environment and health associations - I

Th-SY-A1.1

Multiscale connectivity in HEALS - a high dimension biology approach to unravel the exposome

Dimosthenis Sarigiannis, *Aristotle University of Thessaloniki, Thessaloniki, Greece*

Aim: The exposome represents the totality of exposures from conception onwards. Its exploration means identifying, characterizing and quantifying the exogenous and endogenous exposures and modifiable risk factors that predispose to and predict diseases throughout a person's life span. Using it for improved risk assessment implies that both environmental exposures and genetic variation are reliably measured and linked through mechanistic analysis of toxicity pathways rather than only phenotypically associated. It is thus expected to contribute to the determination of causal associations between environmental factors and human health taking into account genetic susceptibility.
Methods: Making this vision come true poses significant scientific and technological challenges in terms of both untangling the complex biological networks that regulate our body's response to external stressors and processing and analyzing the large datasets generated from the use of multiple high throughput analytical platforms (-omics technologies). To understand the interaction between environmental exposure and disease, we need to: (a) capture the biological perturbations initiated by exposure to environmental stressors; and (b) identify which of these perturbations overcome the homeostasis barrier, resulting in observed alterations of the cell/tissue environment and eventually to pathologic phenotypes. The connectivity approach brings together environmental, socio-economic, exposure, biomarker and health data; in addition, it includes all the procedures and computational sequences necessary for applying advanced bioinformatics coupling advanced data mining, biological and exposure modeling so as to ensure that environmental exposure-health associations are studied comprehensively. Based on preliminary results indicating that major environmental factors defining individual exposome arise from exposure on diet and air pollution, examples of exposome analysis applied on ambient and indoor air pollutants will be given.
Results: With regard to co-exposure to VOC mixtures (e.g. BTEX), transcriptomics analysis confirmed that BTEX co-exposure significantly alters the initiation of early biological effects than single exposure to benzene. Among the most important molecular pathways differentially expressed were the ones regarding inflammation mediated by chemokine and cytokine signaling, apoptosis as well as oxidative stress. With regard to exposure to ambient air PAHs, it was shown that age-dependent differences in internal dose and early exposure are major components of later life health status.
Conclusions: The conclusions of this study contribute to the effective regulation of volatile and semi-volatile compound levels in occupational settings and public spaces. The connectivity approach applied opens the way towards more cost-effective public health protection by prevention of exposure settings that may increase health risk from combined exposure to multiple chemicals.

Th-SY-A1.2

The HEALS environmental data management system - the backbone of the external exposome dataset in Europe

John Bartzis, *University of Western Macedonia, Kozani, Greece*
Evangelos Tolis, *University of Western Macedonia, Kozani, Greece*
Panagiotis Karagiannis, *University of Western Macedonia, Kozani, Greece*
Dimosthenis Sarigiannis, *Aristotle University of Thessaloniki, Thessaloniki, Greece*

One of the key activities in HEALS Project is to develop an integrated external exposure assessment methodology with sufficient resolution in time and space and taken into consideration the whole life-course of individuals in order to be able to estimate external exposures of individuals and population subgroups to multiple stressors via different pathways to help unravel the individual exposome. Such an activity needs, among the others, the optimal utilization of the appropriate environmental data that include emissions of stressors, concentrations of toxic substances in environmental media (outdoor and indoor air, soil, water), in food and in drinking water, various meteorological conditions etc. Such data are collected and stored in a coherent environmental data management system (EDMS) . EDMS besides accommodating HEALS own datasets is able to retrieve existing data from European and Worldwide Organizations as well as various European and national projects and literature published reviews. The data are classified to the following categories: Land use/Land cover, Meteorological Data, Comfort data indoors , Air emission data, Air emission data - consumer products, Pollutant concentration data in air, soil, dust , water, drinking water and food, Population data, Noise data and buildings characteristic data. Each dataset includes location, location characteristics, the time period and the time resolution, units of measurement, measurement methodology, equipment used and the quality assurance and quality control of the related data. EDMS is implemented using MySQL as database management system, in order to grant interoperability in data storage, management and exchange with the HEALS Geo-Database Platform. It is also is in compliance with the INSPIRE Directive and it seeks compatibility with JRC IPCHeM database.

Th-SY-A1.3

Multi-sensor data fusion for location and activity tracking in HEALS

Anjoeka Pronk, *TNO, Zeist, Utrecht, Netherlands*
Miranda Loh, *IOM, Edinburgh, United Kingdom*
John Cherrie, *IOM, Edinburgh, United Kingdom*
Dimitris Chapizanis, *Aristotle University of Thessaloniki, Thessaloniki, Greece*
Spyros Karakitsios, *Aristotle University of Thessaloniki, Thessaloniki, Greece*
Eelco Kuijpers, *TNO, Zeist, Netherlands*
Emmeke Aarts, *TNO, Zeist, Netherlands*
Thomas Maggos, *Demokritos, Athens, Greece*
Mina Stamatelopoulou, *Demokritos, Athens, Greece*
John Bartzis, *Demokritos, Athens, Greece*
Zdravko Spiric, *OIKON, Zagreb, Croatia*
Christian Schierberle*Institute of Energy Economics and the Rational Use of Energy (IER), Stuttgart, Germany*
Denis Sarigiannis, *Aristotle University of Thessaloniki, Thessaloniki, Greece*

Introduction: The determination of presence in micro environments including indoor vs outdoor is critical for modelling personal exposure based on time-location-activity data. The aim of this study was to investigate the potential use of multiple sensors for location and activity tracking.
Methods: As part of the HEALS project time-location-activity data were collected from 28 office workers across 6 European cities for 7 days with the MOVES app on a personal smartphone and the Fitbit Flex. In addition, real time personal air temperature (Elitech RC) was measured for all participants and real time personal UV level (Extech Luxmeter with Semrock 300/80 nm filter) was measured at 4 participants, both devices were attached to the outer clothing. Paper logs were kept by each participants for logging time-activity and indoor and outdoor locations.

Results: The MOVES classification (place(=cluster), walk, cycle, transport) and the paper log correlated well, except for outdoor stationary which was misclassified as 'place' by MOVES. Exploration of the personal activity, air temperature (T) and UV level indicated a correlation between the (variability in) activity, T and UV levels and being indoors or outdoors. The data of the multiple sensors were fused in random forest models for classification of location and activity. Preliminary results indicate a moderate to high accuracy (65-99%) for the different study subjects.
Discussion: The preliminary results indicate that when using MOVES to assess personal time-location-activity information additional sensor data may be used to optimize the classification. Advantages and limitations will be discussed.

Th-SY-A1.4

Personal exposure assessment fusing multi-sensor data and Agent Based Modelling (ABM)

Dimitris Chapizanis, *Aristotle University of Thessaloniki, Thessaloniki, Greece*
Alberto Gotti, *Aristotle University of Thessaloniki, Thessaloniki, Greece*
Spyros Karakitsios, *Aristotle University of Thessaloniki, Thessaloniki, Greece*
Dimosthenis Sarigiannis, *Aristotle University of Thessaloniki, Thessaloniki, Greece*

Aim: The downsizing of monitoring technologies and costs makes it possible for various environmental stressors and exposure factors to be measured more easily. Thus a more reliable "time - geography of exposure" can be generated shifting the current paradigm from a population to an individual level. This study examines the feasibility of using multiple wearable sensors for tracking personal location and activities in order to develop a personal exposure assessment model. This is achieved by coupling sensors data with Agent Based Modelling (ABM), a simulation technique where a system -in this case a city- is modelled as a collection of individual heterogeneous actors, the agents.
Methods: Twenty-five participants in Thessaloniki, Greece, carried a series of devices such as (a) a temperature logger to detect changes between indoor/outdoor conditions, (b) a fitness tracker to capture motion and intensity of activity, (c) a GPS device to track coordinates and speed along with (d) Moves, a smartphone application that enables tracking of location and activity. Additionally, a time activity diary was filled out each day. Location, motion and intensity of activity data was used as input to an Artificial Neural Network (ANN) model, aiming to derive a time-activity model based solely on sensors data. Using Monte Carlo analysis, distributions of participant movement and activities derived from the sensors experiments were extrapolated to a larger population. Using a geographically explicit ABM model, the trajectory of individual participants was modelled and projected on a geo-referenced layer and finally superposed onto high spatial resolution urban air quality modelled maps of PM10 concentration. Personal exposure to air pollutants, expressed as inhalation-adjusted exposure, was then evaluated by assigning pollutant concentrations to a human agent based on his/her coordinates, physical activities and the corresponding inhalation rate.
Results: By estimating the daily time-activity patterns (predicted by the coupled sensors-ANN-ABM platform) of vulnerable subgroups of population, we were able to estimate their personal exposure and intake dose per body weight. On average, personal exposure results were between 10 and 20% more accurate than the equivalent estimate using ambient air concentration of PM as exposure proxy.
Conclusions: An individual exposure model was developed feeding a population-based exposure assessment system without imposing prior bias, but rather basing its estimations onto emerging properties of the agent behaviour. This approach permits the computationally cost-effective identification of refined exposure profiles throughout the day, leading to useful conclusions regarding capping exposure to high pollution levels in cities.

Th-SY-A1.5

The importance of internal dosimetry in unravelling the exposome

Dimosthenis Sarigiannis, *Aristotle University of Thessaloniki, Thessaloniki, Greece*
Spyros Karakitsios, *Aristotle University of Thessaloniki, Thessaloniki, Greece*
Alberto Gotti, *Aristotle University of Thessaloniki, Thessaloniki, Greece*
Evangelos Handakas, *Aristotle University of Thessaloniki, Thessaloniki, Greece*

Aim: Lifelong human exposure (towards exposome analysis) needs to be reckoned based on actual internal dose at the target tissues, for the different life stages. Moreover, in many cases, internal exposure has to be estimated starting from biological indices of exposure (biomarkers).
Methods: A major methodological component of the HEALS project towards the analysis of exposome is the development of a generic Physiology Based BioKinetic (PBBK) model. The model is applicable for a broad variety of chemicals under proper parameterization, through the use of advanced QSAR models. Aiming at capturing in-utero exposure, the model also describes the functional interaction of the mother and the developing fetus through the placenta. The PBBK model is geared with reverse modeling algorithms in order to reconstruct exposure from human biomonitoring (HBM) data. Dynamic flux balance analysis ensures the realistic description of toxicokinetics, under real life exposure scenarios.
Results: The model was parameterized for the assessment of a highly controversial industrial chemical with widespread applicability in consumer goods, namely bisphenol-A (BPA). For the majority of the investigated exposure scenarios, the estimated internal dose was close to 0.002 μg/L and only in the case of bottle fed infants, internal exposure concentrations were up to 0.023 μg/L. This is partially explained by the neonates immaturity of the detoxification pathway, resulting to higher internal doses for the same bodyweight normalized dose compared to children older than 1 year old or adults. The biologically effective dose of the developing fetus during gestation was found to be slightly increased to the one in maternal blood. In addition, exposure to BPA was reconstructed based on real-life HBM data, using an average urine BPA-Glu equal to 2.8 μg/L across Europe, covering different age groups. The results indicated that the overall daily intake is below 1 μg/kg_bw/d and the estimated internal dose was close to 0.002 μg/l, far below any internal dose derived reference value, corresponding to the lower estimates of the already considered exposure scenarios.
Conclusions: Assessment of real life exposure scenarios can be estimated following either a bottom-up (starting from exposure estimates), or a top-down (starting from biomonitoring data). In any case, the assessment is efficiently refined if internal dose metrics are used as reference doses for risk characterization. The latter can be derived by extrapolating from in vivo or in vitro results, taking stock of the wealth of data rapidly produced by modern high-throughput platforms.

Th-SY-B1: Tooth-matrix biomarkers to reconstruct the early life exposome

Th-SY-B1.1

Reconstructing the Early Life Environment Using Micro-Spatial Analysis of Teeth

Manish Arora, *Icahn School of Medicine at Mount Sinai, New York, United States*
Christine Austin, *Icahn School of Medicine at Mount Sinai, New York, United States*
Syam Andra, *Icahn School of Medicine at Mount Sinai, New York, United States*
Birgit Claus Henn, *Boston University, Boston, United States*
Chris Gennings, *Icahn School of Medicine at Mount Sinai, New York, United States*
Brent Coull, *Harvard TH Chan School of Public Health, Boston, United States*
Robert Wright, *Icahn School of Medicine at Mount Sinai, New York, United States*

The exposome concept proposes a comprehensive assessment of environmental exposures, including diet, chemicals and social stressors, from the prenatal period onwards. However, reconstructing these environmental influences during the prenatal and early childhood periods remains a major challenge in environmental epidemiology. This presentation will highlight recent developments in tooth-matrix biomarkers that permit retrospective determination of several important components of the exposome, ranging from metal and organic toxicants, essential dietary elements, homeostatic disruptions and biologic response variables. To support the subsequent presentations in our symposium, this talk will lay the conceptual framework of tooth matrix biomarkers and link the developmental physiology of teeth with their application in environmental epidemiologic studies. Detailed validation studies have been undertaken of the proposed biomarkers in humans and animals for chemical mixtures that include metal toxicants (e.g. lead), essential dietary nutrients that may be harmful at higher exposures (e.g. manganese and zinc) and exposure to stress (physical and social stressors). In a prospective cohort study we compared microspatial analyses of lead (Pb) in teeth with concentrations in maternal blood during the second and third trimesters, umbilical cord blood, and child's serial blood. We also undertook comparisons of tooth lead levels with maternal bone lead (measured using K-X ray fluorescence). The results indicated significant positive correlations between tooth matrix biomarkers and the lead levels in other matrices (r range = 0.4 to 0.69; $p<0.05$). Similar validation has also been undertaken for manganese. To develop and validate a biomarker of early life dietary transitions, in a parallel human and macaque study, we observed that barium signatures corresponded to a move from breast milk to infant formula; introduction of formula was associated with ~1.5 higher barium uptake in dentine than breast milk (Austin presentation in this symposium). We have extended the macaque component of that study to show that external stressors, ranging from illness, injury, separation from mother and introduction to a new group, impart measurable signatures in macaque teeth. This presentation will conclude by providing conceptual links between the development of the biomarker for diet (Austin presentation), organic chemicals (Andra presentation), novel statistical methods to analyze these high dimensional data (Gennings and Coull presentation), and the application of these biomarkers and statistical methods to study children's neurodevelopment (Claus Henn presentation).

Th-SY-B1.2

Novel Biomarkers of Dietary Transitions and Composition

Christine Austin, *Icahn School of Medicine at Mount Sinai, New York, United States*
Syam Andra, *Icahn School of Medicine at Mount Sinai, New York, New York, United States*
Birgit Claus Henn, *Boston University School of Public Health, Boston, United States*
Brent Coull, *Harvard T.H. Chan School of Public Health, Boston, United States*
Chris Gennings, *Icahn School of Medicine at Mount Sinai, New York, United States*
Robert Wright, *Icahn School of Medicine at Mount Sinai, New York, United States*
Manish Arora, *Icahn School of Medicine at Mount Sinai, New York, United States*

Infant and childhood diets are major determinants of lifelong health trajectories with some evidence suggesting that longer duration of breastfeeding is associated with improved immunologic function, respiratory health and neurodevelopment. Risks or benefits of partial breastfeeding with formula supplementation are unclear and complicated by the fact that questionnaires may not accurately capture the relative "dose" of breast milk vs formula, especially when collected retrospectively. A biomarker that can retrospectively quantify the "dose" and composition of breast milk and formula, and its duration of use, would be a major advancement. Furthermore, if collected late in childhood it would facilitate cost-effective case-control studies of infant feeding practices in childhood diseases, which may be years after breastfeeding has ended.
We have developed a novel biomarker that can objectively reconstruct the timing of past infant diet transitions using deciduous teeth. This biomarker combines sophisticated histological and chemical analyses to precisely sample dentine layers corresponding to specific life stages, generating integrated, longitudinal weekly nutrient estimates during early childhood, and prenatally. These dentine assays can be undertaken years after the exposure occurred, are stable over years/decades, and not subject to recall error or bias. We used this biomarker to reconstruct breastfeeding practices in monkey and human teeth, and even a Neanderthal tooth specimen (see Nature doi10.1038/nature121691). Barium (and in some cases strontium) signatures identified the transition from the prenatal period to onset of breastfeeding and the introduction of infant formulas by successive increases in tooth concentration (p for trend <0.05). We investigated the association of Ba, as a proxy for infant diet, with neurodevelopment using a reverse distributed lag model (DLM). There was no significant association of Ba with IQ from birth until about 10 months when a significant negative association was observed. If breast milk is considered the base nutrition required for optimal development, while diet is predominated by breast milk no significant association of Ba and IQ is expected, as development is uninterrupted. As formula use increased with age, a negative association between Ba and IQ was observed. Infant formula is significantly higher in Ba than breast milk and therefore high postnatal Ba levels are expected to indicate a high proportion of formula in the diet. This trend indicates that a greater proportion of formula intake is negatively associated with neurodevelopment, which is consistent with other reports of poorer neurodevelopment outcomes associated with shorter duration of breastfeeding.

Th-SY-B1.3

Prenatal and Early Childhood Exposure to Multi-class Organic Chemicals Using Tooth-Matrix Biomarkers

Syam Andra, *Icahn School of Medicine at Mount Sinai, New York, United States*
Christine Austin, *Icahn School of Medicine at Mount Sinai, New York, United States*
Manish Arora, *Icahn School of Medicine at Mount Sinai, New York, United States*

The exposome concept proposes a comprehensive assessment of environmental exposures from the prenatal period onwards. For decades, teeth have been used to estimate long-term cumulative exposure to metals. Recently developed high-dimensional analytical methods that combine sophisticated histological and chemical analysis to precisely sample tooth layers that correspond to specific life stages have the potential to reconstruct the exposome in the second and third trimesters of prenatal development and during early childhood. We reconstructed the prenatal and early childhood exposure to multiple organic chemical classes using teeth. We performed global screening of small molecules in trimester-specific formed dentine layers from deciduous teeth using liquid chromatography coupled quadrupole time-of-flight mass spectrometry (QTOF-LC/MS) metabolomics approach. QTOF-LC/MS analyses showed unique and differential chemical signatures of environmental exposure that are individual and development-stage dependent. The results of this study (a) revealed more than 12,000 unique chemical signatures in trimester-specific dentine layers, (b) indicate high inter- and intra-child variability in screened chemical profiles, (c) show novel 'known unknowns' and 'suspected unknowns' compounds, (d) demonstrate exposure misclassification error that can cause misleading inferences about causality, and (e) most importantly, the reconstruction of exposure was done 7 to 10 years after prenatal and early childhood exposure. The chemicals detected included phenols (BPA, BPS, BPF), phthalate metabolites, parabens, tobacco markers (cotinine and nicotine). Validation of key signatures is underway. Specifically, phthalate monsters detected in children's deciduous teeth are being compared to levels in maternal urine during pregnancy. Similarly, analysis are also underway for perfluorinated compounds. A retrospective temporal exposomic approach that precisely measures exposure intensity and timing during prenatal and early childhood development would substantially aid epidemiologic investigations, particularly case-control studies of rare health outcomes.

Th-SY-B1.4

Novel Statistical Methods to Uncover Time Varying Critical Developmental Windows to Chemical Mixtures

Chris Gennings, *Icahn School of Medicine at Mount Sinai, New York, NY, United States*
Brent Coull, *Harvard Chan School of Public Health, Boston, MA, United States*
Ander Wilson, *Harvard Chan School of Public Health, Boston, MA, United States*
Syam Andra, *Icahn School of Medicine at Mount Sinai, New York, NY, United States*
Christine Austin, *Icahn School of Medicine at Mount Sinai, New York, NY, United States*
Birgit Claus Henn, *Boston University, Boston, MA, United States*
Robert Wright, *Icahn School of Medicine at Mount Sinai, New York, NY, United States*
Manish Arora, *Icahn School of Medicine at Mount Sinai, New York, NY, United States*

Tooth matrix biomarkers generate complex data structures on multiple chemicals and nutrients over fine scale developmental windows (1 to 2 weekly exposures) that may be linked to a range of health outcomes. The dearth of statistical approaches to study such complex higher-dimensional interactions limits the use of this biomarker in epidemiologic studies. To use this tooth biomarker to identify critical developmental windows associated with chemical mixtures, we embed distributed lag modeling (DLM) techniques, which have proven useful for detecting critical windows associated with a single chemical, within methods for estimating health risks associated with a chemical mixture. Specifically, we integrate DLM methods within weighted quantile regression (WQS), which we term weighted distributed lag models (wDLM) and also consider distributed lag functions within a kernel machine regression framework (DL-KMR). The wDLM approach has the advantage that it yields an interpretable index score for the mixture, and then estimates how the association between this score and outcome varies across the exposure period. The distributed lag kernel has the advantage that it allows for non-additivity of chemical exposures observed at different timepoints. The methods can be applied to the tooth chemical biomarker as well as more broadly to chemical mixture studies. We used time-window specific tooth data from the second trimester to 1 year of age to identify sensitive windows of exposure to each metal in the ELEMENT cohort study.

UTRECHT THE NETHERLANDS OCTOBER 9-13 ISES 2016

Th-SY-B1.5

Critical Windows of Neurodevelopmental Susceptibility to Chemical Mixtures: Case Studies from Across the Globe

Birgit Claus Henn, *Boston University, Boston, MA, United States*
Brent Coull, *Harvard Chan School of Public Health, Boston, MA, United States*
Chris Gennings, *Icahn School of Medicine at Mount Sinai, New York, NY, United States*
Christine Austin, *Icahn School of Medicine at Mount Sinai, New York, NY, United States*
Syam Andra, *Icahn School of Medicine at Mount Sinai, New York, NY, United States*
Robert Wright, *Icahn School of Medicine at Mount Sinai, New York, NY, United States*
Manish Arora, *Icahn School of Medicine at Mount Sinai, New York, NY, United States*

Aim: A major barrier to epidemiologic studies of neurodevelopmental effects of prenatal chemical exposure is the lack of suitable biomarkers to measure fetal uptake. Consequently, prenatal susceptibility windows when chemical exposure may be most strongly linked to neurodevelopmental outcomes have not been identified. Manganese (Mn), lead (Pb), and other metals exposure can now be quantified using teeth, which can determine dose and exposure timing. We apply tooth-matrix biomarkers in two pediatric cohort studies to uncover discrete prenatal and postnatal developmental windows of susceptibility when metals exposure is most strongly associated with neurodevelopmental outcomes.
Methods: We studied mother-child pairs drawn from the longitudinal birth cohort studies in Mexico City that comprise the Early Life Exposures in MExico and NeuroToxicology (ELEMENT) project, which investigates the long-term consequences of prenatal environmental factors on child development. Child neurodevelopment was assessed between 6- and 14-years of age using the Wide Range Assessment of Visual Motor Abilities. We also studied 10-15 year old Italian children living near a ferromanganese plant who are participants of the Public Health Impact of Mixed Element Exposure (PHIME) project. Neurodevelopment was assessed using the Conner's Comprehensive Behavior Rating Scales. Mn, Pb and other metals were measured using the tooth-matrix biomarkers. We applied distributed lag models, multivariable regression, and weighted quantile sum regression to study the association of metals with behavior and visuospatial learning.
Results: Visual-motor scores were significantly associated with Mn exposure in the Mexico City cohort and this relationship varied markedly by children's Pb exposure status. In low-Pb children, prenatal Mn levels were not associated with visual-motor scores, but approximately 100 days after birth, higher Mn levels were correlated with better performance on the visual-motor tests ($p<0.05$). In high-Pb children, however, the postnatal tooth Mn levels were inversely associated with visual-motor scores ($p<0.05$). Fine motor scores also showed two critical windows for Mn exposure, with positive associations prenatally and negative associations postnatally. In the Italian cohort, prenatal tooth Mn levels were associated with fewer self-reported inattentive behaviors ($\beta = -0.09$, $p=0.05$ for highest Mn quintile vs lowest), while tooth Mn levels from early childhood were associated with more inattentive behaviors ($\beta = 0.05$, $p=0.3$ for highest Mn quintile vs lowest).
Conclusions: Using tooth-matrix biomarkers and recently developed statistical methods, we were able to uncover complex associations dependent on the developmental age and level of concurrent exposure to multiple metals, which varied across the neurodevelopmental domains we examined.

Th-SY-C1: Development of personal sampling devices and chemical screening methods for large-scale epidemiology and human biomonitoring studies - I

Th-SY-C1.1

Importance of Human Biomonitoring for Public Health and Chemical Risk Management

Antonia Calafat, *US Centers for Disease Control and Prevention, Atlanta, United States*

Data on the health significance of human exposure to many chemicals used in modern societies are limited and, at times, even contradictory. However, studies to investigate the prevalence of such exposures are of public health importance because of the potential harmful health effects of some of these chemicals, as often shown in animal studies. Advances in analytical techniques in the last few decades have contributed to the increase in biomonitoring research as a tool to determine internal dose, and biomarkers concentrations are increasingly used to quantify exposures within populations. Biomonitoring programs are particularly valuable for investigating human exposure to environmental chemicals. One of these programs, the National Health and Nutrition Examination Survey (NHANES), is conducted annually in the United States by the Centers for Disease Control and Prevention (CDC). NHANES is designed to collect data on the health and nutritional status of the U.S. general population. Since 1999, concentrations of select chemicals in urine and blood of NHANES participants have provided the most comprehensive assessment of Americans' chemical exposures. This presentation will provide an overview on the use of NHANES biomonitoring data to establish reference ranges, to identify determinants of exposure and populations with higher exposures, to provide exposure information for risk assessment (e.g., set intervention and research priorities, evaluate effectiveness of public health measures), and to monitor exposure trends.

Th-SY-C1.2

Recent Developments in Microsampling for Quantitative Bioanalysis

Neil Spooner, *Spooner Bioanalytical Solutions, Hertford, United Kingdom*

In order to accurately determine circulating concentrations of analytes, a high quality sample is required that reflects the physiological situation at the time of collection. Further, for the monitoring of environmental exposure, samples ideally need to be representative of the time and place of exposure and sample collection. Therefore, simple approaches are required so that high quality samples can be obtained in non-hospital settings, potentially facilitating self collection of the sample in the home, or other convenient location.

This presentation will explore the characteristics that are required for such a device to be successful, from both the usability / ergonomic and the bioanalytical point of view. This will be illustrated by considering technologies that are currently widely commercially available and the types of experiments that are required in order for the user to be confident that the concentrations that are determined are valid and reflective of the physiological situation. In addition, the presentation will showcase emerging technologies and future approaches that may better suit the requirement for simple determination of circulating analyte concentrations.

Th-SY-C1.3

hemaPEN, an intuitive device for unassisted and accurate collection of Dried Blood Spot samples

Florian Lapierre, *University of Tasmania, Ringwood, victoria, Australia*
Ricardo Neto, *University of Tasmania, Ringwood, Victoria, Australia*
David Bailey, *Trajan, Ringwood, Victoria, Australia*
Andrew Gooley, *Trajan, Ringwood, Victoria, Australia*

Have you recently tested your blood? Are you a scientist conducting an epidemiological study over a population? How do you usually manage blood sampling?
The bottleneck in the blood analysis process seems to originate right from the blood collection. Going to the clinic, waiting for the nurse to be ready, requiring skilled use of needles to draw blood, and secure transportation and storage of blood samples, are barriers to technology adoption for blood analysis studies.
Dried Blood Spot (DBS) sampling methodology is a commonly used technique to collect a small amount of blood from a heel prick for neonatal studies. However, this technology is generally performed by trained nurses, and attempts to modify this technique for self-collection have not been successful. The reasons for this are the fact that DBS technology has not improved since the 1960s, in terms of usability for both patient and clinician, as well as the level of sample protection.
hemaPEN (Figure 1) redefines blood collection and storage from the fingertip in order to improve access to personal sampling. By assembling existing regulatory approved components - 4 glass capillaries of 3μL and 4 pre-punched PKI226 papers - inside a pen-like device, the hemaPEN enables accurate blood collection without the need for professional medical assistance.
Our results for collected blood volume show that the hemaPEN provides a higher level of precision (CV=1%) than other dispensing methods (i.e. pipette, syringe, CV=3-5%), eliminating human error during collection and processing. The capillary collection mechanism dispenses an accurate volume of blood, leading to superior accuracy during analysis (95% recovery compare to 87% with other dispensing methods). Moreover, the integrated compartments protect samples at every stage of the process, and advanced sample tracking could be enabled by NFC technology.
In conclusion, the hemaPEN provides technical support for epidemiological studies and enables home-based, unassisted blood collection. Adoption into current healthcare systems is expected to save time, cost, and improve accuracy of blood analysis, setting industry standards.

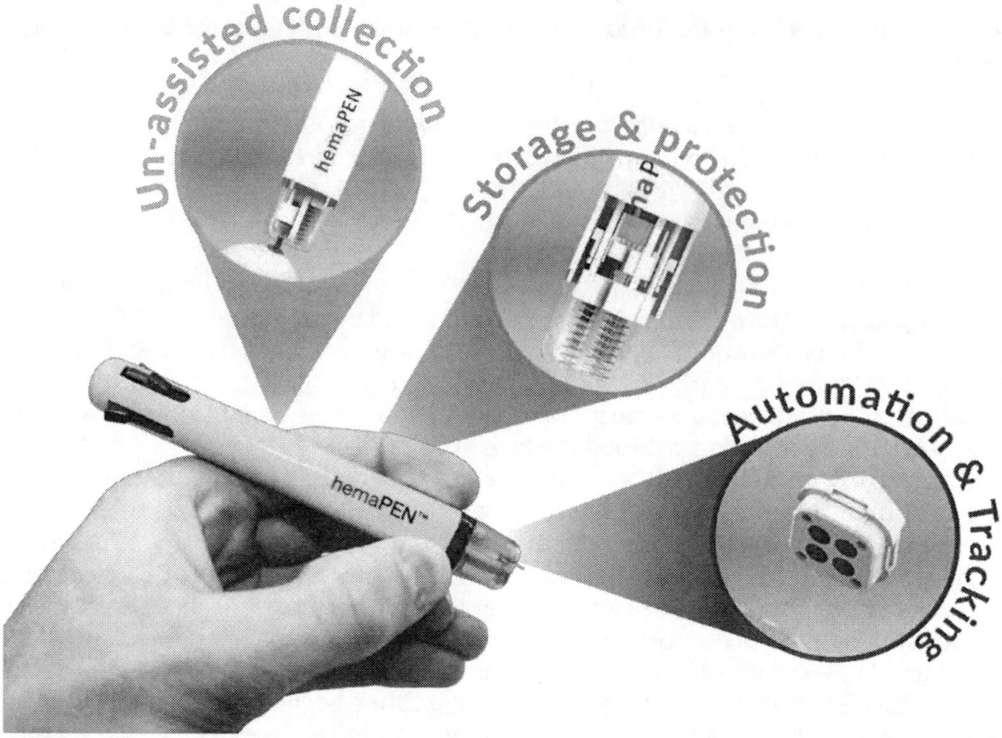

hemaPEN enables unassisted blood collection by anyone, anytime, anywhere. With just a button click, hemaPEN's innovative microfluidic extraction takes very small and precise blood samples from the fingertip, instantly transferring them into an integrated

Th-SY-C1.4

Quality Assurance and Quality Control of Portable Devises using a Standard Reference Material

Jessica Reiner, *NIST, Charleston, South Carolina, United States*
John Bowden, *NIST, Charleston, South Carolina, United States*

Standard Reference Materials (SRMs) are homogeneous, well-characterized materials that are used to validate measurements and improve the quality of analytical data (www.nist.gov/srm). The National Institute of Standards and Technology (NIST) has a wide range of SRMs that have values assigned for clinically important analytes, legacy organic pollutants, and toxic metals in human matrices. Examples of some SRMs include organic contaminants in human serum, human milk, and human urine, lead in caprine blood, arsenic species in human urine, and toxic elements in human urine. These SRMs can serve as materials for quality control when developing methods on a portable devise. Currently NIST has looked at different sampling techniques, including dried blood spot cards (DBS), and different portable devises, including CardioChek and VetScan. SRM 1958 Organic Contaminants in Fortified Human Serum (Freeze-Dried), SRM 955c Toxic Metals in Caprine Blood, and a solution of 25-hydroxyvitamin D3 were spotted onto DBS cards to test the feasibility of using DBS to screen for analytes of interest. Preliminary results for the measurements indicate encouraging data for screening some analytes. SRM 1950 Metabolites in Frozen Human Plasma was used to assess the precision and accuracy of the CardioChek and VetScan devises for field research. Results indicate that precise and accurate measurements for many compounds can be made using both portable devises. This presentation will discuss how the SRMs were used as a control material and the preliminary results. Additionally the benefits of validating measurements with a SRM will be discussed.

Th-PL-D1: Exposure Modeling

Th-PL-D1.1

Effect of Model Choice on Estimates of Ultrafine Particle Number Concentrations near Roadways

Allison Patton, *Rutgers University, Piscataway, New Jersey, United States*
Chad Milando, *University of Michigan School of Public Health, Ann Arbor, Michigan, United States*
Stephen Zemba, *CDM Smith, Boston, MA, United States*
Prashant Kumar, *University of Surrey, Guildford, United Kingdom*
John Durant, *Tufts University, Medford, MA, United States*

Aim: Exposure to ultrafine particles (UFP; less than 100 nm in diameter) is associated with cardiovascular disease risk markers. UFP are emitted in vehicle exhaust, leading to sharply elevated UFP concentrations near busy roadways. The aim of this work was to compare UFP number concentrations predicted by dispersion models and multivariable regression at addresses within 200 m of highways.
Methods: We studied two Boston-area (MA, USA) neighborhoods near interstate highways: a residential area near I-93 (Somerville) and an urban center near I-93 and I-90 (Chinatown). Models used to predict the spatial distribution of UFP included spatial-temporal regression models and four line source dispersion models: CALINE 4, a steady-state Gaussian dispersion model with atmospheric stability classes; R-LINE and AERMOD, dispersion models that parameterize atmospheric stability with Monin-Obukhov length; and QUIC, a Lagrangian model. Traffic emission factors were generated locally; meteorological and traffic data were obtained from state agencies. UFP measurements for model evaluation were made with a mobile laboratory. Models were compared for wind directions parallel and perpendicular to each highway as well as for typical winter and summer temperatures (4 scenarios in Somerville and 6 in Chinatown). Model performance was evaluated by R2, fraction of estimates within a factor of two of measurements (fac2), and fractional bias relative to measurements (FB). UFP concentrations were predicted with each model at 22 Somerville residences and 133 Chinatown residences. Pearson correlations among predictions from the models were calculated.
Results: Measured and modeled UFP were consistently highest near the highways. In both neighborhoods, QUIC predicted the highest UFP concentrations and the regression model predicted lower UFP concentrations close to the highways. Sharper near-highway gradients were measured and modeled in Somerville than in Chinatown. Compared to measurements, the models performed better in Somerville (R2>0.51, fac2>45%, FB between -0.76 and 0.37) than in Chinatown (R2≤0.44, fac2>20%, FB between -0.97 and 0.77), where complex building geometry and lack of local traffic data resulted in poor agreement with measurements. Although Pearson correlations among model predictions for all scenarios combined were >0.8, correlations for individual temperature-wind scenarios varied and were strongest for wind directions across highways in winter.
Conclusions: Five air pollution models predicted higher UFP concentrations closer to highways in two Boston-area neighborhoods, although the magnitude of spatial differences was not consistent among models. The differences in UFP estimated from these models suggest that model choice may substantially affect estimates of UFP concentrations used in near-road exposure assessment.

Th-PL-D1.2

Human exposure to household cleaning products: Application of a two-field model

Gan Wang, *University of Michigan, Ann Arbor, Michigan, United States*
Lei Huang, *University of Michigan, Ann Arbor, Michigan, United States*
Vy Nguyen, *University of Michigan, Ann Arbor, Michigan, United States*
Olivier Jolliet, *University of Michigan, Ann Arbor, Michigan, United States*
Vy Nguyen, *University of Michigan Medical School, Ann Arbor, MI, United States*

Indoor cleaning activity is a major human household activity and cleaning products contain volatile organic compounds (VOCs) that constitute a potential threat to indoor air quality and occupants' health. The present study aims 1) to establish a dynamic modeling framework which accounts for the near-person inhalation and dermal exposure to indoor cleaning products; 2) to determine the evolution of chemical mass and concentration associated with indoor cleaning activity, and to determine the key factors affecting chemical fate and exposure in indoor environment; and 3) to determine the short-term human intakes and product intake fractions (PiFs).
We modify a two-zone model to describe the indoor environment where cleaning products are applied, identifying four transfer compartments, i.e. near-person surface, near-person air, far-person surface, and far-person air. Transfers between compartments are described by first-order transfer rate constants structured in a matrix (K matrix) to describe the mass flows between different compartments. The exposure matrix (XP matrix) relates the mass in a given compartment to the intake by human. Three exposure pathways are considered: inhalation exposure to near-person air, inhalation exposure to far-person air, and dermal exposure during application.
The model is applied to 20 common ingredients in household surface cleaning products, with a focus on two representative chemicals, n-hexane (high volatility) and 2-butoxyethanol (low volatility). The application of cleaning product is assumed to last for 1.5 hours (Phase 1), and a time period of 200 hours after application (Phase 2). For both representative chemicals, mass in each compartment keeps increasing during Phase 1 and reaches its peak at the end of Phase 1. However, for n-hexane the two air compartments have larger chemical mass after the product is applied, whereas the mass in the two surface compartments are higher for 2-butoxyethanol. Dermal intake during application dominates exposure for n-hexane, while inhalation exposure during Phase 2 dominates for the less volatile 2-butoxyethanol. Across the 20 chemicals studied, inhalation intake dominates the total product intake in most cases (Figure 1) and is driven by the chemical's air-water partition coefficient Kaw.
The adapted dynamic two-zone model enables us to describe well and with parsimony the dynamic of chemical mass and near-field intakes during the application of household surface cleaning products. The results will be further compared to exposure estimates for the application of cleaning products in washing machines and dishwashers to elucidate the relative importance of various transfer pathways in the near-field exposures to cleaning products.

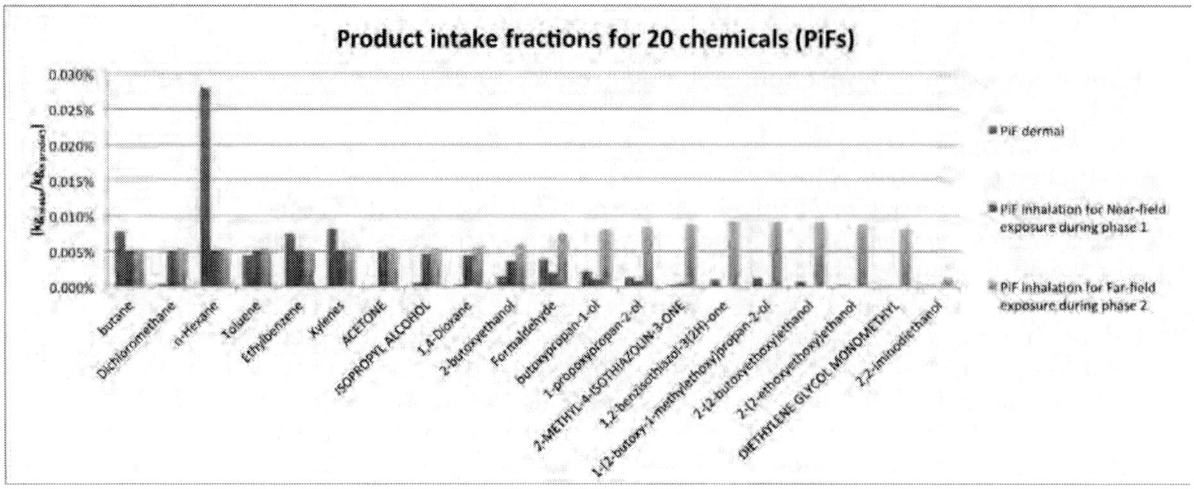

Figure 1. Product intake fractions for 20 chemicals

Th-PL-D1.3

Modelling PM2.5 exposure and health impact from agricultural emissions: dairy farms in three U.S. location

Katerina Stylianou, *University of Michigan, Ann Arbor, MI, United States*
Christopher Tessum, *University of Washington, Seattle, WA, United States*
Julian Marshall, *University of Washington, Seattle, WA, United States*
Joshua Apte, *University of Texas at Austin, Austin, TX, United States*
Olivier Jolliet, *University of Michigan, Ann Arbor, MI, United States*

Agricultural practices are associated with primary and secondary fine particulate matter (PM2.5). Here, we investigate PM2.5 exposures and health impacts attributable to dairy farms in three U.S. locations.
We model PM2.5 intake fractions (iFPM2.5) for emissions of four pollutants (NH3, NOx, SO2, PM2.5) in three locations - Wisconsin (WI), New Jersey/Pennsylvania (NJ/PA), and New York (NY) - using the Intervention Model for Air Pollution (InMAP). These iFs are then coupled with available exposure-response functions to provide region-specific characterization factors (CF) for primary and secondary PM2.5. For each of the three locations, we estimate CFs for two farm sizes (150 versus 1,500 cow) and for two approaches to feeding and manure application: conventional versus a lower-impact "Beneficial Management Practices" (BMPs) approach.
The iFPM2.5,NH3 estimate in NJ/PA is ~10 and 30 times higher than for NY and WI, respectively, reflecting differences in population density and atmospheric chemistry. Of the total PM2.5 population exposure, nearly all (90%) happens within 20 km from the NJ/PA source, 350 km from NY, and 1500 km from WI. Differences in atmospheric chemistry are also reflected by the fact that the iFPM2.5,NH3 is 18 times higher than iFPM2.5,SO2 in NJ/PA (where NOx and SO2 are in abundance), but only 2 times higher in WI (where NH3 is in abundance), suggesting that PM2.5 population exposures resulting from adding 1kg NH3 are higher in NH3-limited regions compared to NH3-abundant regions. Combining the iF values reported here with an exposure-response for rural U.S. regions (170 DALY/kg PM2.5 inhaled) results in CFs (units: 10-5 DALY/kg precursor emitted) of 5-145 (NH3), 2-8 (NOx), 7-200 (PM2.5), and 3-8 (SO2). CFs are applied to a case study of milk production in a 1,500 cow farm located in the three locations to estimate total PM2.5-related health impacts; NH3 emissions contribute ~63% to 73% of total PM2.5-related health impacts, with the highest contribution in NJ/PA where milk production induces the highest total PM2.5-related health impacts (24 x 10-5 DALY/kg milk). These impacts are reduced by ~40% when employing BMPs.
There is substantial spatial variation of iFPM2.5 in the U.S. linked to population density and atmospheric chemistry. Milk production in highly populated NH3-limited regions has substantially higher PM2.5-related health impacts to populations downwind than production in agricultural regions where NH3 is in abundance.This research contributes to spatially-explicit CFs for the agricultural sector.

Th-PL-D1.4

Testing the accuracy of the MERLIN-Expo modelling tool in predicting human Pb biomonitoring data.

Philippe Ciffroy, *Electricité de France (EDF) R&D, Chatou, France*
Johan Bierkens, *Flemish Institute for Technological Research -VITO, Mol, Belgium*
Tine Fierens, *Flemish Institute for Technological Research -VITO, Mol, Belgium*
Mirja Van Holderbeke, *Flemish Institute for Technological Research - VITO, Mol, Belgium*

Aim
Evaluating a model depends on its specific objective(s), its expected decision(s) and the structure of data available for comparison. When guarding specific regulatory thresholds, the model's capability to generate true positives is more important than evaluating accuracy for the whole range of available data. When the model is expected to be conservative, the model results should be 'safe' values rather than 'true' values. Within this context the accuracy of MERLIN-Expo, a newly developed exposure model, has been tested using Pb biomonitoring data via different accuracy approaches and metrics.
Methods
Using 2.25 □g.L-1 Pb in blood of children as the trigger value, a contingency table (or confusion matrix) was constructed, describing the number of occurrences in which measured data and model output are both above the threshold (True Positives - TP), both below the threshold (True Negatives - TN), the number of alarms missed by the model (False Negatives - FN) and that of false alarms (False Positives - FP). Based on this contingency table different accuracy metrics were calculated. The Probability of detection (or true positive rate), equals 1 when all measured positives are detected by the model. The Bias score indicate whether the model has a tendency to underestimate (1) measurement data and to what extent (ranging from 0 to +∞). Finally, in binary classification, the F1 score considers both precision (number of TP divided by the number of all positive results) and sensitivity (i.e. Probability of detection) and can be interpreted as a weighted average of precision and sensitivity. The F2 score weights sensitivity higher than precision (i.e. which gives more importance to FN than to FP).
Results
We calculated a Probability of detection of 0.93, reflecting that almost all blood levels above the trigger value are predicted as such by the model; a Bias of 1.87, i.e. reflecting a slight overestimation of the model, and F1 and F2 scores of 0.65 and 0.8, respectively, confirming that most of the measured positives are detected as such by the model (high F2 score) with a slight overestimation (lower F1 score).
Conclusions
We illustrates that metrics for evaluating model accuracy have to be selected according to the structure of the dataset and to the context of the assessment. We conclude that MERLIN-Expo accurately predicts positives (i.e. Pb concentration level above a threshold) and that it overestimates actual observed levels, but at a level that is compatible with accurate decision-making.

Th-PL-D1.5

Modeling Exposure to Traffic-Related Air Pollutants for the Residential Human Health Risk Assessment Study in Kyiv, Ukraine

Oksana Ananyeva, *State Institution "O.M. Marzeyev Institute for Public Health of the National academy of Medical Sciences of Ukraine", Kyiv, Ukraine*
Olena Turos, *State Institution "O.M. Marzeyev Institute for Public Health of the National Academy of Medical Sciences of Ukraine", Kyiv, Ukraine*
Arina Petrosian, *State Institution "O.M. Marzeyev Institute for Public Health of the National Academy of Medical Sciences of Ukraine", Kyiv, Ukraine*

Introduction: There is a strong body of evidence linking traffic-related air pollution exposure to adverse health effects. Although accurate air quality data is rarely available for the urban residential neighborhood. This provides the ground for the development of various modeling techniques for obtaining more accurate exposure estimates compared to monitoring data. Assessment of human health risks attributable to traffic related air pollution study was designed to evaluate the risks of development of respiratory health outcomes in population living near roadways in Kyiv, Ukraine.
Methods: A combination of modeling approaches was used to estimate exposures to traffic-related air pollutants for the residents living in close proximity to the roadways in two city districts. Field studies were conducted to provide measurement data for developing, evaluating and refining the models. CO, NO2, SO2, NMHC, PM2,5, PM10, PM1 concentrations, meteorological parameters, fleet intensity and structure data were obtained in two measurement campaigns (2012-2014). A land-use regression (LUR) models were developed to capture NO2, CO, PM10, PM2,5, PM1 near-road variability. ISC-AERMOD View air pollution dispersion model was used to model averaged 1-, 8-, 24-hour and annual concentrations of CO, NO2, SO2, NMHC and PM10 at 952 receptor points covering the area of the study. Agreement between ISC-AERMOD View modeled and measured concentrations was assessed based on the standard methods.
Results: Vehicles number, road distance, road length, land use and meteorological variables were the most important predictors of traffic-related air pollution variability for LUR models. Developed models explained up to 75% of the spatial variability for studied pollutants in near-roadway residential neighborhoods and demonstrated a good agreement between observed concentrations and predicted levels (average difference of 3-10%). Correlation analysis revealed strong association between ISC-AERMOD view modeled values and measurements of NO2, ($r=0,93$, $p<0,005$) and weaker associations for other pollutants ($p<0,005$): $r=0,88$ for CO, $r=0,90$ for SO2, $r=0,88$ for NMHC, and $r=0,74$ for PM10.
Conclusions: The results demonstrated an integrated measurement study design. Applied modeling techniques proved to be an effective instrument of human exposure assessment when limited monitoring data is available. Based on the models outputs traffic-related exposure maps were developed for the study area and corresponding human health risks formed by traffic- related air pollution were assessed.

Th-SY-E1: Methodologies in finding new and/or emerging risks of chemicals (NERCs) - I

Th-SY-E1.1

Development of priority EDCs list in integrated risk assessment and management for endocrine disrupting chemicals (IRAMe)

Kiyoung Lee, *Seoul National University, Seoul, Korea, South*
Ji Young Park, *Graduate School of Public Health Seoul National University, Seoul, Korea, South*
Miyoung Lim, *Graduate School of Public Health Seoul National University, Seoul, Korea, South*
Wonho Yang, *Graduate School of Public Health Seoul National University, Seoul, Korea, South*
Kyunghee Ji, *Graduate School of Public Health Seoul National University, Seoul, Korea, South*

Endocrine disrupting chemicals (EDCs) are emerging chemicals with possible adverse health effects from exposure to chemicals that can interfere with the endocrine system. The EDCs can be exposed through various exposure media and consumer products. To prevent health effects, we established Integrated Risk Assessment and Management for Endocrine Disrupting Chemicals (IRAMe) program in Korea. Because the EDCs were defined by toxicological effects, lists of EDCs were often different by organizations. The purpose of this study was to identify primary EDCs list that cause human health effects and determine priority EDCs from the list. The list of EDCs for the selection processes included EDIIS from Korea government, 68 EDCs from World Wildlife Fund, 60 EDCs from US EPA, 48 EDCs from US CDC, 67 EDCs from Japan EPA, 564 EDCs from EU Priority list. In addition WHO and UNEP established 182 EDCs. EPA endocrine Disruptor Screening Program (EDSP) added 159 EDCs. ChemSec established SIN(Substitute It Now!) of 80 EDCs. We initially selected all 179 EDCs from EDIIS from Korean government. We identified 117 chemicals which were listed both in EDIIS and at least one list. We also identified EDCs with at least two lists. Total of 260 EDCs were selected from the multiple list cross check. A primary list of 164 EDCs was established by excluding 90 pesticides including 2 prohibited chemicals, 4 heavy metals and chemicals. For prioritization, we developed chemical ranking and scoring system using three categories: exposure, toxicity, and social interest. The exposure category with 20 points included amount of use in nation, amount of commercial use in nation, degree of bioaccumulation, detection in consumer product. The toxicity category with 20 points included carcinogenicity and reproductive toxicity. The social interest category with 100 points included presence of regulation and media report. The priority score (maximum 500) was calculated by exposure (20) x toxicity (20) + social interest (100). For preliminary evaluation, 33 EDCs were analyzed. The highest priority score was DEHP, followed by DBP, BPA, DIDP, BBP, NP, NPE, DnOP, DIBP and DHP. After the 10th rank chemicals, the rest chemicals did not have distinctive scores. Methylparaben was the highest score among parabens.

Th-SY-E1.2

Identification, prioritization and evaluation of potential New Emerging Risk of Chemicals (NERCs) for Consumers

Elbert Hogendoorn, *RIVM, Bilthoven, Netherlands*
Gerlienke Schuur, *RIVM, Bilthoven, Netherlands*
Lya Soeteman-Hernandez, *RIVM, Bilthoven, Netherlands*

Background
People are exposed to hundreds of chemicals as a result of our living standard, via the environment or via consumer products, such as clothing, cosmetics and electronics, pharmaceuticals and detergents. Despite existing legislation and regulations new and/or emerging risks of chemicals (NERCs) continue to be reported, making an early warning system highly wanted to detect and evaluate chemical threats at an earliest stage
Aim
RIVM is developing an approach for early identification of NERCs combined with strategies to manage the identified risks to consumers.
Methods
The methodology under development for the identification of NERCs for consumers consists of data collection of chemicals in consumer products with reported adverse effects from various sources (i.e. literature search, network and consumer complaints. The information with regards to hazard, potency and exposure are systematically included in a database. The potential NERCs for consumers are evaluated, amplified and confirmed as outlined in Table 2 by an Expert Group resulting in the identification of NERCs in consumer products. The possible follow-up measures are identified taking into account the risk management options within the relevant (regulatory) frameworks of the identified NERC for consumers.
Results
With the method developed a number of substances were identified as possible NERC and evaluated with the proposed prioritization system providing a list of potential NERCs. Potential NERCs including CMR and/or SVHC chemicals were found in tattoo inks, children's products, cosmetics and textiles. Further evaluation by the network of experts is ongoing in order to indicate substances as NERCs requiring further action in order to reduce the risk of exposure.
Conclusions:
The proposed method is useful in the identification, prioritization and evaluation of NERCs for consumers.
Literature
Bakker J, Bruinen de Bruin Y, Hogendoorn E, et al. (2015) Progress report on New or Emerging Risks of Chemicals (NERCs), RIVM Report 2014-0040.

Th-SY-E1.3

Identification, prioritization and evaluation of potential New and Emerging Risk of Chemicals (NERCs) for Workers

Nicole Palmen, *RIVM, Bilthoven, Netherlands*

Background
New and emerging risks of chemicals (NERCs) continue to be reported despite existing laws and regulations put in place to limit the risks of dangerous substances at work. Quite often there is little or no knowledge of the harmful effects of substances that are used by workers. One of the reasons for this is the fact that the risk assessment is usually based on toxicological tests following oral exposure, while workers are exposed via the airways and the skin.

Aim
RIVM is arguing for a system that identifies NERCs as soon as possible to prevent workers from falling ill because of these NERCs.

Methods
An integrated approach was used for the identification of NERCs by the collection of case reports, periodic screening of literature and websites, data mining, health surveillance studies and analysis of secondary sources. Risk scores were identified by applying an Impact Analysis on potential NERCs based on the severity of the health effect and the likelihood of a causal relationship between exposure and health effect. Three risk priorities were identified by multiplying the risk scores with the manufacturing or use of the substance in the Netherlands. Subsequently, EU databases were consulted to check which measures were/are already being taken regarding registration, classification, Authorization and Restriction, etc..

Results
A list of 49 NERCs was published and subsequently prioritized to address those substances that deserve the most attention. Sixteen substances had a 'very high priority' suggesting that there is an urgent need to investigate a possible causal relationship between the exposure and the health effect. Nineteen substances were categorized with a 'high priority', meaning that action is necessary. Fourteen substances had a 'low priority', meaning that minimal action is needed. The inventory showing the extent to which the 49 substances already are being regulated by REACH or other legislation shows that most actions are being taken on substances with a very high priority, with exception of diacetyl and crystalline silica. Most substances with a high or low risk priority have no harmonized classification and are not regulated within REACH.

Conclusions:
The proposed method is useful in the identification, prioritization and evaluation of worker NERCs.

Literature

Palmen, NGM and Verbist KJM (2015) Prioritization of new and emerging chemical risks for workers and follow-up actions, RIVM report 2015-0091

Th-SY-E1.4

Identification, prioritization and evaluation of potential New Emerging Risk of Chemicals (NERCs) for the environment

Joost Bakker, *National Institute for Public Health and the Environment, Bilthoven, Utrecht, Netherlands*
Elbert Hogendoorn, *National Institute for Public Health and the Environment, Bilthoven, Utrecht, Netherlands*
Yuri Bruinen de Bruin, *National Institute for Public Health and the Environment, Bilthoven, Utrecht, Netherlands*

Background
New and emerging risks of chemicals (NERCs) continue to be reported despite existing laws and regulations to limit the risks of dangerous substances in the environment. Existing laws and regulations often address chemicals with extensive evidence on harmful effects. As a consequence there is a large time span between the first indication of the presence of a substance in the environment and actual measures taken to control the risk posed by these chemicals.
Aim
RIVM is developing a system that identifies NERCs as soon as possible and map out strategies or possibilities to manage the identified risks to the environment.
Methods
An integrated approach was used for the early identification of NERCs by the collection of signals from news reports, periodic screening of literature and websites, expert consultation and analytical environmental screening studies. The signals are valued on potential emerging risks by a tiered approach applying selection criteria and a prioritization scheme. Potential NERCs were identified and prioritized by applying an Impact Analysis based on the severity of the effect (hazard) and the likelihood of exposure. Based on the available information potential NERCs are categories into five risk priority classes. Information on hazardous properties and exposure is collected from public databases. If the required information is not available, the hazardous properties and exposure are estimated by applying for instance QSARs and exposure categories based on the characterization of the main type of use and the production volume. Before the data collection starts current measures that are already being taken regarding registration, classification, restriction, etc. are checked to see whether the identified risk is already being addressed.
Results
The methodology for selection and prioritization of NERCs has intensively been applied and tested. The prioritization scheme for instance has been applied to the results from two analytical screening studies providing a set of chemicals that were identified in fresh surface water samples and wastewater treatment effluent. This resulted in set of substances that deserve the most attention. The inventory also included the extent to which the substances are already being regulated by REACH or other legislation and shows that most actions are being taken on substances with a very high priority.
Conclusions:
The proposed method is useful in the identification, prioritization and evaluation of environmental NERCs.
Literature

Bakker J, Bruinen de Bruin Y, Hogendoorn E, et al. (2015) Progress report on New or Emerging Risks of Chemicals (NERCs), RIVM Rapport 2014-0040.

Th-SY-E1.5

General aspects in developing methods for identifying new and or emerging risks of chemicals threatening human health and/or environment

Yuri Bruinen de Bruin, *National Institute of Public Health and the Environment, Bilthoven, Netherlands*
Joost Bakker, *National Institute of Public Health and Environment (RIVM), Bilthoven, Netherlands*
Elbert Hogendoorn, *National Institute of Public Health and Environment (RIVM), Bilthoven, Netherlands*
Myrna Kooi, *National Institute of Public Health and Environment (RIVM), Bilthoven, Netherlands*
Nicole Palmen, *National Institute of Public Health and Environment (RIVM), Bilthoven, Netherlands*
Lya Hernandez, *National Institute of Public Health and Environment (RIVM, Bilthoven, Netherlands*

This presentation will demonstrate the operational version of the New and Emerging Risks of Chemicals (NERCs)-mechanism to identify new risks of chemicals following unknown exposure routes.
Aim
To identify new or emerging risks of chemicals for consumers, workers and the environment, the NERC-mechanism links causality between exposure and hazard in humans and the environment and provides regulators with information to optimize policy-making.
Methods
The NERCs-mechanism filters signals from media, literature and experts covering the environment, workers and consumers. Based on target-specific criteria additional exposure, hazard and policy data are discussed by experts. The data is translated into a risk score varying from 1 (lowest risk) to 25 (highest risk). The risk score prioritizes newly identified risks of chemicals requiring risk management options (RMO) like the derivation of a safety limit, enforcement or inspection, actions taken up by REACH or CLP (e.g. SVHC roadmap or harmonized classification and labelling) or making use or adaptation of other legislation.
Results
The filtering of a continuous stream of signals resulted into lists (environment, workers and consumers) containing new risks for which appropriate risk management options are being evaluated. This presentation will present 9 identified NERCs; 3 for each respective target and will present the proposed follow-up actions and its outcomes.
Conclusions
The NERCs mechanism is operational and identifies new or emerging risks of chemicals following unknown exposure routes and provides science-based options for risk management and policy improvement.
Acknowledgements
This initiative is financed by the Ministry of Infrastructure and the Environment, the Ministry of Social Affairs and Labor and the Ministry of Welfare and Sport. Special thanks to all involved experts.

Th-SY-A2: The exposome: a transdisciplinary paradigm for improved environment and health associations - II

Th-SY-A2.1

The Arylhydrocarbon Receptor in the context of the chemical exposome

Robert Barouki, *Université Paris Descartes, Paris, France*

The Arylhydrocarbon Receptor (AhR) has traditionally been considered as a sensor of a class of xenobiotics, primarily PolyAromatic Hydrocarbons (PAH). Its main function was thought to be the detection of those chemicals and the induction of a battery of genes including specific enzymes and transporters leading to their elimination. Paradoxically, the metabolic system activated by the AhR was also thought to mediate the toxicity of some of these contaminants either by leading to the generation of highly toxic intermediates or by being inactive towards some of its substrates, ie halogenated compounds.

Recently, the AhR was found to be more than a xenobiotic receptor. AhR knock out mice display a pathologic phenotype in the absence of any xenobiotic, suggesting that the receptor has endogenous functions. Additional studies have indicated that the AhR is the receptor of a number of endogenous compounds such as several metabolites of the amino acid tryptophan and that it is activated by dietary compounds such as polyphenols. Furthermore, the AhR was shown to bind several microbiome-derived chemicals, in particular virulence factors such as phenazines. It is likely that the activation of the AhR by those different classes of chemicals does not yield the same effects and this is possibly related to the structural plasticity of this receptor. In line with those studies, this receptor has been shown to have a number of different functions in the immune system, notably in barrier organs such as the gut, the skin and the lung as well as in the nervous system, the adipose tissue and hematopoietic tissues. Taken together, all those recent observations indicate that the AhR should be considered as a sensor of a fraction of the chemical exposome and not only of PAH and that it is involved in a variety of functions in vertebrates that go beyond the defence against polyaromatic contaminants.

La Merrill et al. Toxicological function of adipose tissue: focus on persistent organic pollutants. Environ Health Perspect. 2013 ;121:162-9

Guyot et al. The AhR twist: ligand-dependent AhR signaling and pharmaco-toxicological implications. Drug Discov Today. 2013;18:479-86

Barouki et al. The aryl hydrocarbon receptor system. Drug Metabol Drug Interact. 2012;27:3-8.

Barouki et al. The aryl hydrocarbon receptor, more than a xenobiotic-interacting protein. FEBS Lett. 2007; 581:3608-15

Th-SY-A2.2

Environmental origin of neurodevelopmental disorders: in vivo and human models to unraveling complex aetiologies

Gemma Calamandrei, *Istituto Superiore di Sanità. Rome, Italy, Rome, Italy*

Aim: Strong evidence exists that environmentally relevant exposure to chemical pollutants at critical developmental stages affects neural and behavioral development in children. Recent advances in research offer important clues into pathogenetic mechanisms of autism and other neurodevelopmental disorders (NDDs), indicating that environmental risk factors cannot be ruled out. Specifically, in NDDs, variations in several candidate genes may confer higher vulnerability to different kinds of adverse environmental stressors, including early exposure to chemicals.
Methods: There is an overall need to invest more in discovery research related to neurotoxicological hypothesis for NDDs. The exposomic approach aims at characterizing and quantifying the exogenous and endogenous exposures and modifiable risk factors that predispose to and predict NDDs. A critical issue is the identification and validation of peripheral biomarkers of effects that can inform on typical and atypical brain development, and help to establish biologically plausible links between chemical exposure and health effects. Behavioral toxicology studies might significantly contribute by modeling in "simpler" living organisms the complexity of the human exposure scenarios. Studies with laboratory rodents allow assessment of dose-response relationships, critical periods of susceptibility, and the relative contribution of genetic, epigenetic and environmental factors. In the neuropsychiatric disease field, the use of the in vivo models permits the selection of omic biomarkers anchored to the behavioral phenotype, which increases their translational value.
Results: As an example we present recent data on developmental neurotoxicity of the non-persistent organophosphorus insecticide chlorpyrifos (CPF), whose neurotoxic activity at low doses is currently a matter of concern for children's health. In mice exposed to CPF in utero and/or in early development several behavioural responses are altered in both sexes, in parallel with sex-dependent interference on neuroendocrine pathways regulating social behaviors (vasopressin, oxytocin, and steroid regulated systems). The route of exposure selected in our studies corresponds to relevant human exposure scenarios, supporting the view that neuroendocrine effects, especially in susceptible time windows, should deserve more attention in risk assessment of OP insecticides. Notably, in a mouse model of idiopathic autism, the BTBR strain, prenatal exposure to CPF induces more marked alteration of early behaviour than in wild type mice, associated to increased oxidative stress markers in both plasma and brain.
Conclusions: In synergy with mechanistic in vitro studies, PBPK models and human data, in vivo models may be pivotal to identify candidate biomarkers and pinpoint susceptible groups or lifestages to be translated to large prospective studies within the exposome context.

UTRECHT THE NETHERLANDS OCTOBER 9-13 ISES 2016

Th-SY-A2.3

Assessment of health risks for vulnerable population groups posed by exposure to mercury and its compounds

Milena Horvat, *Jozef Stefan Institute, Ljubljana, Slovenia*
Janja Snoj Tratnik, *Jozef Stefan Institute, Ljubljana, Slovenia*
Darja Mazej, *Jozef Stefan Institute, Ljubljana, Slovenia*
Marta Jagodic, *Jozef Stefan Institute, Ljubljana, Slovenia*
Ingrid Falnoga, *Jozef Stefan Institute, Ljubljana, Slovenia*
Anja Stajnko, *Jozef Stefan Institute, Ljubljana, Slovenia*
Alfred Kobal, *Jozef Stefan Institute, Ljubljana, Slovenia*

Aim: This presentation underlines investigation of mercury (Hg) exposure, effects, and susceptibility in Mediterranean population in early life and compares them with the outcomes of the population living in contaminated site due to historic mercury mining in the town of Idrija, Slovenia.
Methods: Maternal hair, maternal and cord blood samples were used to assess prenatal exposure to mercury and its compounds. Children (n=360) were genotyped for apolipoprotein E (Apoe) polymorphism and were assessed for neurodevelopment using Bayley Scales of Infant and Toddler Development, Third Edition (Bayley-III) at 18 months of age. In a subgroup of pregnant women, the potential interaction of Hg with selenium in plasma (P-Se) and antioxidative enzymes in erythrocytes was also assessed.
Results: The main predictor of Hg exposure in the Mediterranean population is fish consumption demonstrated by large proportion of methyl Hg in the collected samples. Population living in the contaminated site exhibited higher Hg concentrations, but with smaller proportion of Hg as methyl Hg. This indicates the importance of speciation as well as proper use of exposure biomarkers in human biomonitoring programmes in contaminated sites. The results of Bayley III assessment indicated that even low-to-median Hg exposure in children with normal neurodevelopmental outcome can result in lower cognitive and fine motor scores. The Hg-related decrease in cognitive sores was observed in children carrying at least one Apoe ε4 allele, while the decrease in fine motor scores was independent of the genotype. The number of examined mother-child pairs from the contaminated site was too small to show any significant effect. However, the internal doses of Hg received during pregnancy in the contaminated site did not decrease the bioavailability of Se, and the decrease in antioxidative capacity appeared to be mainly associated with pregnancy per se and not with an increased exposure to Hg.
Conclusion: The studies are on-going in the framework of the EU funded projects, HEALS (Health Environment associations in large population Studies) and Life+ CROME (Cross Mediterranean Heath Environment Network), which use the existing cohorts, and build on novel susceptibility markers and effect testing and integrate them within Health and Environment-wide Associations studies (EWAS) as part of the ‘exposome’ approach.

Th-SY-B2: Aspects to consider for Fungi and Mycotoxins occupational exposure and risk assessment

Th-SY-B2.1

Exposure and risk assessment in occupational exposure to fungi - Aspects to consider in highly contaminated settings

Carla Viegas, *ESTeSL-IPL, Lisbon, Portugal*

Although a clear correlation between levels of fungi in the air and health impacts has not been shown in epidemiological studies, fungi must be regarded as potential occupational health hazards. Fungi can have an impact on human health in four different ways: (1) they can infect humans, (2) they may act as allergens, (3) they can be toxigenic, or (4) they may cause inflammatory reactions. Fungi of concern in occupational hygiene are mostly non-pathogenic or facultative pathogenic (opportunistic) species, but are relevant as allergens and mycotoxins producers.
It is known that the exclusive use of conventional methods for fungal quantification (fungal culture) may underestimate the results due to different reasons. The incubation temperature chosen will not be the most suitable for every fungal species, resulting in the inhibition of some species and the favouring of others. Differences in fungi growth rates may also result in data underestimation, since the fungal species with higher growth rates may inhibit others species' growth. Finally, underestimated data can result from non-viable fungal particles that may have been collected or fungal species that do not grow in the culture media used, although these species may have clinical relevance in the context. Due to these constraints occupational exposure assessment, in setings with high fungal contamination levels, should follow these steps: Apply conventional methods to obtain fungal load information (air and surfaces) regarding the most critical scenario previously selected; Guideline comparation aplying or legal requirements or suggested limits by scientific and/or technical organizations. We should also compare our results with others from the same setting (if there is any); Select the most suitable indicators for each setting and apply conventional-culture methods and also molecular tools. These methodology will ensure a more real characterization of fungal burden in each setting and, consequently, permits to identify further measures regarding assessment of fungal metabolites, and also a more adequate workers health surveillance.
The methodology applied to characterize fungal burden in several occupational environments, focused in Aspergillus spp. prevalence, will be present and discussed.

Th-SY-B2.2

From lab to labor - detailed exposure characterization as basis for protective measures in rarely considered occupational settings

Stefan Mayer, *Statutory institution for accident insurance and prevention in the trade and goods logistics, Mannheim, Germany*

Exposure to fungi is a common situation in the private as well in the occupational settings. While the exposure in the private environment is usually limited, the occupational settings are of particular interest because workers are often exposed over a whole shift and longer periods of time. Some occupational environments like animal confinement, waste recycling or even moldy indoor environments are well recognized. However, there are still occupational settings which occur only seldom in the scientific and public discussion. Such settings are onion sorting, grape processing, processing of hemp, flax, hay, peas, reed etc. Common to all of these examples is that the products usually don't show any signs of fungal growth. Nevertheless, during processing fungi are released of these products to the ambient air in the range of 104 to 108 cfu/m^3.
An example for occupational fungal exposure with an increasing importance is the unloading of moldy items from freight containers. A large part of international and intercontinental transport is realized via freight containers. During sea transport, often condensation occurs within the freight containers what creates a sufficient environment for fungal growth especially on organic products.
Exposure measurements should not only characterize the level of airborne exposure, but also help the occupational health and safety consultants to improve workplace conditions. Publications about fungal concentrations at certain work places are helpful only if it becomes transparent if the conditions during the measurement are comparable to those at the workplaces which have to be evaluated. Therefore, we need an integrated approach from "lab to labor" covering all aspects of exposure. E.g., the information that workers are exposed to 105 cfu/m^3 in a grain elevator does not help unless it is stated at which task the measurement has been performed, which amount of grain has been handled, if there has been an aeration etc. Other important factors inter alia are the degree of microbial colonization, the tendency to release dust, the intensity of mechanical processing and the size of the products and their specific surface may have an influence. Additionally, exposure is more than the airborne fungal concentration. With respect to the question which protective measures are appropriate also the duration and the frequency of exposure are important factors.

Th-SY-B2.3

Occupational exposure to mycotoxins - Aspects to consider for the aggregate and cumulative risk assessment

Susana Viegas, *ESTeSL-IPL, Lisbon, Portugal*

Mycotoxins are an important group of naturally occurring substances known to contaminate a huge variety of agricultural products, feed and food commodities. The main concern is their widespread presence and toxic effects on humans and animals as they have been described as cytotoxic, nephrotoxic, hepatotoxic, teratogenic, immunosuppressive, mutagenic and/or carcinogenic. However, until now, risk assessments and regulations have usually been performed on individual mycotoxins despite humans and animals are being frequently exposed to a multitude of mycotoxins simultaneously. Moreover, even though some exposures through inhalation and dermal contact may potentially occur, only oral ingestion has been considered as the sole route of exposure in all the evaluations. However, more recent studies have also demonstrated airborne exposure to mycotoxins in different occupational settings with emphasis on agricultural professions. In these cases, skin contact with mold-infested substrates and inhalation of spore-borne toxins are the most important sources of exposure. Still, mycotoxins are not normally recongnize as na occupational hazard and exposure is different from the one ocurring by food intake. In this case, exposure is charaterized to be acute and simultaneous to other mycotoxins and also to fungi and dust. All these features increase the challenge implicated in the risk assessment process.
Some topics will be presented and discussed in detailed such as: What occupational settings should be consider in this case; possible exposure routes; exposure characterization; how to assess exposure; co-exposure; aggregate exposure and cumulative risk assessment.

Th-SY-C2: Development of personal sampling devices and chemical screening methods for large-scale epidemiology and human biomonitoring studies - II

Th-SY-C2.1

Gas Chromatography with Ultraviolet Photometric Detection for Elemental Mercury Analysis

Ronda Gras, *Dow Chemical Canada, Fort Saskatchewan, Alberta, Canada*
Jim Luong, *Dow Chemical Canada ULC, Fort Saskatchewan, Alberta, Canada*
Robert A. Shellie, *Trajan Scientific & Medical, Ringwood, Victoria, Australia*

Fossil fuel combustion such as the burning of coal contributes substantially to the release of elemental mercury into the environment. Even at the parts-per-billion level, exposure of elemental mercury can potentially have a negative impact on overall human health. We introduce a facile and practical analytical approach for the direct measurement of ultra-trace level of elemental mercury by combining the great separation power of gas chromatography with the highly sensitive and selective detection garnered from ultraviolet photometric detection approach. An inert sample flow path that utilizes surface deactivated tubing using the latest innovations in chemical vapor deposition chemistry was employed to prevent analyte adsorption for the highest degree of system inertness. Even without sample enrichment and a sample size of less than 1 mL, elemental mercury in various gas matrices can be directly measured with a detection limit of 1.7 μg per cubic metre. A total analysis can be rendered in less than 2 min [1]. A relative precision of less than 3% was attained with a concentration of 8.3 and 83 μg per cubic metre (n=20).
The analytical approach is useful and complementary to established methodologies such as cold-vapour atomic absorption or cold-vapour atomic fluorescence in the overall strategy of detecting mercury in various matrices.
In this lecture, method performance and examples to illustrate the utility of the technique described will be presented.
[1] R. Gras, J. Luong, R.A., Shellie, Anal. Chem, 2015, 87, 11429-11432

Th-SY-C2.2

Microsampling and Screening Technologies for Human Biomonitoring of Selected Persistent Organic Pollutants (POPs)

Chiara Calaprice, *University of Liege, Liege, Liege, Belgium*
Benjamin L'Homme, *University of Liege, Liege, Liege, Belgium*
Jef Focant, *University of Liege, Liege, Liege, Belgium*

Background. Human exposure to POPs (e.g. Dioxins, PCBs,...) is of concern. During biomonitoring campaign, individual bodyburden is estimated by measurements of serum levels. Depending on the analytes, up to 50 mL of serum are required from patients. This is negatively perceived and excludes young infants/elderlies from being sampled.
Objectives. In the quest for an easy to use and non-invasive sampling method, we have been investigating alternative microsampling approaches to reduce the level of invasiveness while maintaining the requested sensitivity. Several gas chromatographic (GC) and mass spectrometric (MS) methods have been studied to measure at sub picogram (pg) LOQ level. The aim was to keep sample volumes below 50 µL.
Methods. Dried-blood spots (DBS), processed using micro-extraction by packed sorbent (MEPS) were analyzed by cryogenic zone compression (CZC) coupled to negative chemical ionization (NCI), and high resolution time-of-flight MS (HRTOFMS). We also developed a very sensitive method based on the use of volumetric absorptive microsampling (VAMS) and GC coupled to triple quadrupole tandem in-space MS (GC-QQQMS/MS) for measurements. CZC measurements were also implemented using a sector MS instrument to take advantage of the most sensitive MS analyzer operating in selected ion monitoring (SIM) at high acquisition rates. Isotope dilution (ID) was used in all approaches.
Results. CZC applied to GC-NCI-IDHRTOFMS was used for the screening of markers of exposure (PCB-153, DDE) in 20 µL serum samples. The use of MEPS was automated with success and required only 500 µL of solvent for extraction. The non-scanning HRTOFMS analyzer makes analyses of other unknown and/or emerging compounds possible in the future. VAMS and GC-IDQQQMS/MS allowed to measure levels of 24 OCPs and 6 non dioxin-like PCBs (NDL-PCBs) in 40 µL whole blood. The sample preparation, involving micro-scale solid phase extraction (SPE), used 2 mL of solvent per sample. We reported analyte levels for a series of real human samples.
Conclusion. These minimally-invasive methods offer an alternative to conventional approaches in order to easily gather data from people in remote area, from young infants, or for purposes where blood volumes are restricted to a minimum.

Th-SY-C2.3

Diaper Use for Exposure Assessment of Infants and Toddlers

Michihiro Kamijima, *Nagoya City University Graduate School of Medical Sciences, Nagoya, Japan*
Yuki Ito, *Nagoya City University Graduate School of Medical Sciences, Nagoya, Japan*
Jun Ueyama, *Nagoya University Graduate School of Medicine, Nagoya, Japan*

Background: Exposure measurement is a critical step for risk assessment procedures. Of the various methods to characterize exposure, biological monitoring (or biomonitoring, BM) using urine, of which the collection is more feasible than other biological media such as blood, is a strong tool to directly assess overall exposure of individual study participants to environmental chemicals coming from all sources and via all pathways. However, urine sampling in infants and toddlers in epidemiological studies has been challenging since methods possibly bringing skin problems to non-toilet-trained children and/or emotional burden to parents apparently reduce the number of study participants, which has a negative impact on the validity of the study.
Objective: This presentation introduces our trial to develop a method of BM using disposable diapers to measure urinary concentrations of insecticide metabolites in infants and toddlers.
Our trial: Since commercial diapers are not developed for this purpose, the following issues needed to be investigated: urine extraction from the diapers, absorption of the target metabolites to the top sheet and urine absorber of the diaper, separation and recovery of the metabolites from the mixture of urine and extracting solvent, derivatization and analytical conditions, stability of the target metabolites, and measurement method of creatinine.
Conclusion: The established method is now ready for epidemiological studies. This method is apparently superior to other urine collecting techniques such as use of urine collection bags in terms of affordability and applicability to the population-based epidemiological studies focusing on early life exposure assessment.

Th-SY-C2.4

Panel discussion: Development of personal sampling devices and chemical screening methods for large-scale epidemiology and human biomonitoring studies

Shoji Nakayama, *National Institute for Environmental Studies, Tsukuba, Japan*
Andrew Gooley, *Trajan Scientific & Medical, Ringwood, Australia*

Panel discussion will be held after all talks. It will focus on the most recent development of personal sampling devices and chemical screening methods. Needs and requests for microsampling and screening will be heard from the audience.
Major discussion points about making meaningful measurements in populations are as follows:
- Microsampling strategies
- Fast but meaningful measurements
- Data analysis

Th-SY-D2: Environmental Exposure Monitoring in Birth & Early Life Cohort Studies

Th-SY-D2.1

Acceptability and Usability of Novel Technologies to Assess Environmental Exposure during Pregnancy

Artemis Doutsi, *King's College London, London, United Kingdom*
Evridiki Patelarou, *King's College London, London, United Kingdom*
Mireille Mireille B. Toledano, *Imperial College London, London, United Kingdom*
Vivien Bright, *University of Cambridge, Cambridge, United Kingdom*
Rod Jones, *University of Cambridge, Cambridge, Cambridge, United Kingdom*
Frank Kelly, *King's College London, London, United Kingdom*
Benjamin Barratt, *King's College London, London, United Kingdom*

Personalised exposure assessment utilising advancements in technology has the potential to strengthen associations between health outcomes and environmental quality in cohort studies. However, methods and tools require evaluation prior to large-scale deployment. Aim: To determine if novel techniques can better characterise a pregnant woman's environment within acceptable levels of convenience to the participant. Methods: 21 pregnant women of varying SES and ethnicity in London were recruited. Data were collected between March and June 2015. Real-time measurements for different air pollutants and noise were conducted over seven days using static units ('home platform'). Canister, dust sampling and rooms' spot checks for noise, moisture and electromagnetic radiation were also taken. Personal devices and smart phone apps were used to track mobility, activity and sleep patterns. Some metrics were duplicated to test different tools. A questionnaire survey collected information on home environment and participant views. Results: Of the 85 pregnant women that were eligible, 21 were enrolled (25% recruitment rate). No participant dropped out (100% completion rate). The home platforms were tolerated in all cases, likely due to the autonomous nature of the units. However, it reduced recruitment rates; 27 out of the 85 potential participants (32%) refused to participate due to accommodation of the home platform. The uptake for the portable devices and apps was 100% and 91%, respectively, reflecting the non-invasive method and minimal need for participant input. 30% of participants used their own smartphones. No inconvenience or privacy issues were recorded. The study also achieved high data collection rates across all selected metrics. However, the air pollution unit that integrated different high sensitivity and resolution sensors, for some pollutants exhibited variable reliability and required significant data processing resources. The noise device had poor performance due to battery life falling well below specification. With regard to personal units and apps, participants demonstrated persistent use and the completed data sets were from 84% - 95%. The feedback from the participants was overwhelmingly positive. Participants were eager to receive their results. Conclusions: Personalised monitoring technology was easily accepted and convenient. Mobile phone apps in combination with data algorithms, models and population datasets, present a scalable and low resource solution to the enhancement of environmental exposure assessment methods. Improved methods are required for feedback and communication to cohort participants of 'non-threshold' risks, such as noise and air pollution.

Th-SY-D2.2

The CHILD birth cohort: Ups and downs of exposure assessment to age five

Jeffrey Brook, *Environment and Climate Change Canada, Toronto, Canada*
Ryan Allan, *Simon Fraser University, Burnaby, Canada*
Michael Brauer, *University of British Columbia, VAncouver, Canada*
David Dia, *McMaster University, Hamilton, Canada*
Miriam Diamond, *University of Toronto, Toronto, Canada*
Greg Evans, *University of Toronto, Toronto, Canada*
Diana Lefebvre, *McMaster University, Hamilton, Canada*
Wendy Lou, *University of Toronto, Toronto, Canada*
Kathleen McLean, *Simon Fraser University, Burnaby, Canada*
James Scott, *University of Toronto, Toronto, Canada*
Malcolm Sears, *McMaster University, Hamilton, Canada*
Huan Shu*Karlstad University, Karlstad, Sweden*
Padmaja Subbarao, *Sickkids Hospital, Toronto, Canada*
Tim Takaro, *Simon Fraser University, Burnaby, Canada*
Amanda Wheeler, *Menzies Institute for Medical Research, University of Tasmania, Hobart, Tasmania, Tajikistan*

The Canadian Healthy Infant Longitudinal Development (CHILD) study involves an ambitious attempt to characterize environmental stressors from pregnancy onwards, including indoor and outdoor exposures. CHILD involves 3624 children and their parents from 4 Canadian cities in 4 provinces, with a subset in rural Manitoba. The first year of life was hypothesized to be a critical window in the development of asthma and allergy. Questionnaires were administered starting in pregnancy. We developed age-specific long and short questionnaires to assess the physical environment as validated tools with the desired detail did not exist. A key part of CHILD was a home visit at 3-4 months of age to assess a range of potential risk factors present in the indoor environment and to collect dust and biospecimens (urine, breastmilk, stool, nasal swab). Blood, stool, nasal swab and urine from additional time points were also collected leading to an extensive bank of samples for future analysis. To date urine specimens from multiple time points have been analyzed for phthalate metabolites and cotinine, while PAHs, phthalates, hopanes, 1-3 Beta-Glucan, endotoxin and allergens have been analyzed in subsets of dust samples. While results indicate relatively large exposure gradients among the children, optimizing the use of these multi-factorial data, including from the questionnaires and home assessment, remains a challenge. Parallel to CHILD approaches for refining exposure assessment through measurements have been explored. This has included concurrent measurement of semi-volatile organic compounds (SVOC) in dust and on window surfaces (wipes), and development of passive and active sampling methods for indoor and/or outdoor air. Comparison of traffic-related SVOCs in dust and wipes to land-use regression (LUR) estimates of NO_2 and NO indicated that accumulation of these SVOCs on the windows in the child's bedroom is most correlated with LUR-NO, indicating that they reflect fresh traffic emissions near the residence. Examples from these results and highlights from our experience to date will be presented. Balancing subject burden and cost versus the added value of additional data continues to pose a challenge for direct measurement in large cohorts. Furthermore, while additional measurements, looking at conditions external to the child, may provide unique information potentially indicative of exposure, none provide a complete picture. Exposure characterization through biospecimen analysis thus remains an attractive option, potentially helping to characterize individual exposomes. However, cost of analysis and the ability to link such data back to modifiable risk factors represents a significant hurdle to overcome.

Th-SY-D2.3

The SCAMP Study: Capturing Use of Mobile Phones, Wireless Technologies, and Electromagnetic Field Exposures in Today's Adolescents

Charlotte Fleming, *MRC-PHE Centre for Environment and Health, Imperial College London, London, United Kingdom*
Irene Chang, *MRC-PHE Centre for Environment and Health, Imperial College London, London, United Kingdom*
William Mueller, *MRC-PHE Centre for Environment and Health, Imperial College London, London, United Kingdom*
Paul Elliott, *MRC-PHE Centre for Environment and Health, Imperial College London, London, United Kingdom*
Martin Roosli, *Swiss Tropical and Public Health Institute, Basel, Switzerland*
Mireille Toledano, *MRC-PHE Centre for Environment and Health, Imperial College London, London, United Kingdom*

Aim: Scientists remain uncertain as to whether children's developing brains are more vulnerable than those of adults to radiofrequency electromagnetic fields (RF-EMF) emitted from mobile phones and other wireless devices. The Study of Cognition, Adolescents and Mobile Phones (SCAMP) is a three-year prospective cohort study of adolescents across London, UK which aims to investigate whether children's use of mobile phones and other wireless devices influences their neurocognitive/behavioural development.
Methods: Data on cognitive function, wireless device use and lifestyle are collected at baseline (study year 1) and follow-up (study year 3) via school-based computerised assessments and optional parent/pupil home-based questionnaires. Parents are invited to consent to linkage of their child's school assessment data with routine records (e.g. health and educational records, mobile traffic data), thus allowing for comparisons between self-reported and objective mobile phone use in terms of call frequency and duration, number of text messages and amount of data downloaded. Biological samples (e.g. urine, saliva) are also being collected to provide additional information about potential confounders such as puberty.
Additionally, a RF personal monitoring study is being conducted in a subset of the main cohort (n=200) to gain an in-depth understanding of personal RF exposure. Pupils are asked to carry a personal exposimeter with integrated GPS tracking for 48 hours, complete a smartphone activity diary and a paper questionnaire, and provide a urine and saliva sample. Study materials are distributed and collected from pupils at either their schools or their homes. These data will allow for differentiation of RF exposure from mobile phones and RF from other sources (near-field and far-field), and will be used to calibrate models estimating brain and whole body RF exposure metrics to better reflect 'real-life' exposure scenarios for epidemiological investigation in the SCAMP cohort study.
Results: Baseline data have been collected from 5,504 pupils (53.2% female, mean age 12.0 (SD 0.4); 42.3% White, 20.3% South Asian, 15.9% Black, 11.5% Mixed, 9.9% Other). Preliminary analysis shows that over 80% of participants own a mobile phone. To date, data for the personal monitoring sub-study have been collected from 40 pupils.
Conclusions: SCAMP will improve our understanding of children's RF exposures and will provide an evidence base to inform policy. In particular the personal monitoring sub-study will give a detailed assessment of children's personal RF exposure and the relative contribution of each RF source.

Th-SY-E2: Methodologies in finding new and/or emerging risks of chemicals (NERCs) - II

Th-SY-E2.1

Anses approach for the detection and investigation of emerging diseases in occupational health

Melina Le Barbier, *ANSES - French Agency for Food, Environmental and Occupational Health & Safety, Paris, France*
Gérard Lasfargues, *ANSES - French Agency for Food, Environmental and Occupational Health & Safety, Paris, France*
Natalie Vongmany, *ANSES - French Agency for Food, Environmental and Occupational Health & Safety, Paris, France*

Society has high expectations concerning the prevention of occupational risks. Identify emerging or re-emerging risks in occupational health is therefore an important project for the French Agency for food, environmental and occupational health & safety. This is done by means of the French National Occupational Diseases Surveillance and Prevention Network (RNV3P) and its working group on "Emerging occupational diseases". Missions of this working group are to establish a platform for sharing and expertise for the early detection of potentially new work related diseases, and to set up a process for reporting any occurrence of an emerging disease.
New work related diseases (WRD) are defined as new pairs {disease x exposure}, or as pairs {disease x exposure} that are already known but have been detected in a new occupation or a new industry sector (referred to as a "new triad").
The signals discussed by the group's experts involve not just clinical cases reported in the field but also information from statistics revealing emergence (data mining in the national RNV3P base), or even from proactive searches for cases in response to alerts on new diseases from other sources or organisations (literature, NIOSH, European Modernet Network). The second step is related to the "expertise" of each case report (checking of diagnosis, exposure and work-relatedness assessment rated). Finally actions (third step) are proposed according to a three dimensions decision-making tool in order to ensure transparency and reproductibility. This algorithm relies on the number of similar cases reported, severity of each of these cases and on work-attributability. This work-attributability is defined with four levels (impossible, not impossible, possible, very likely).
This three step approach and the related algorithm will be presented. Illustrations of new WRD detected and investigated will be given also. So far, 45 reports have been or are being assessed by the experts in this working group. Most of the reports have come from the clinical component. The work done at national level is coordinated with that at European level. This approach now offers a structured way for the capture and investigation of potentially new WRD.

Th-SY-E2.2

The International Health Regulations (2005): A Global Platform for Collaboration and Capacity Building to Identify, Prevent, Prepare for and Respond to Chemical Risks

Carolyn Vickers, *World Health Organization, Geneva, Switzerland*
Kersten Gutschmidt, *World Health Organization, Geneva, Switzerland*
Joanna Tempowski, *World Health Organization, Geneva, Switzerland*
Richard Brown, *World Health Organization, Geneva, Switzerland*

In 2005, the World Health Organization (WHO) Member States adopted the revised International Health Regulations (IHR) (2005). The Regulations provide a unique public health framework in the form of obligations and recommendations that enable countries to better prevent, prepare for and respond to public health events and emergencies of potential international concern, including chemical events.
The Regulations obligate States Parties to develop certain minimum core public health capacities (especially for early event detection and response) and to notify WHO of events that may constitute a public health emergency of international concern according to defined criteria. While the core capacities to control selected communicable diseases were already understood, international collaboration among the Regulations' 196 States Parties to detect diseases due to other hazards was new. In real life, disease outbreaks occur for which the cause is not immediately known, so many disciplines must work together.
WHO has specified a set of core capacities for chemical events and developed a system of national focal points and an on-line event information site for sharing information. Achievement of the core capacities is regularly assessed. As of 31 March 2015, 160 countries had reported on the implementation of the Regulations during 2014. Relatively low capacities for handling chemical events were reported. The global capacity score for chemical events was 56%, with large variations between WHO regions ranging from 28% in the African Region to 79% in the European Region. The scores for the other regions were Americas 54%, South-East Asia 50%, Eastern Mediterranean 53%, and Western Pacific 62%. Globally, the capacity indicator that most (76%) of countries had was a designated focal point for coordination during a chemical event. Only 40% of countries had an updated chemical event response plan, 56% had adequate laboratory capacity to confirm a chemical event and 63% had surveillance systems for chemical exposures. WHO is working with countries to strengthen chemicals capacities.
The establishment of surveillance systems for early event detection is a critical capacity. As well as acute outbreaks with known chemicals, surveillance can identify new threats, known threats with a changing pattern of occurrence, and emerging risks of a less acute onset. Sometimes the chemical cause of an outbreak is identified quickly, for example, methanol poisoning, but sometimes not. Examples will be presented.

Th-SY-E2.3

An integrated strategy for marine toxins of cell based bioassays and analytical tools to ensure safe seafood.

Toine Bovee, *RIKILT Wageningen UR, Wageningen, Netherlands*
Mirjam Klijnstra, *RIKILT Wageningen UR, Wageningen, Netherlands*
Marcia Bodero, *RIKILT Wageningen UR, Wageningen, Netherlands*
Jonathan Nicolas, *WUR, Wageningen, Netherlands Antilles*
Peter Hendriksen, *RIKILT Wageningen UR, Wageningen, Netherlands*
Arjen Gerssen, *RIKILT Wageningen UR, Wageningen, Netherlands*

Marine toxins are produced by algae that can accumulate in seafood. Consumption of contaminated products may lead to intoxication such as memory loss, paralysis, diarrhoea or in severe cases death. In order to protect consumers, methods have been developed and put into practice. Within Europe several analytical chemical alternatives for the widely applied mouse bioassay (MBA) are implemented within legislation. Current trends involve rapid LC-MS/MS or targeted screening with LC-hrMS for regulated and unregulated toxins. Despite the current alternatives, the unethical and unreliable test with mice is still being used, as it is capable to detect possible unknown toxins or new risks for consumers. Animal-free in vitro cell based effect assays offer the same opportunity as the mouse bioassay, i.e. to detect unknown toxins and new risks. A neuroblastoma cell assay was optimized at our laboratory for the detection of various toxins in seafood samples. This easy and relatively cheap in vitro cell assay is used as a first screening to differentiate between blank and suspect samples. Only suspect samples are further investigated with analytical chemical tools. For the regulated toxins such targeted LC-MS/MS methods have been established, and can be applied in monitoring programs (e.g. if countries miss laboratories with cell culture facilities). However, the combination of effect screening with LC-MS/MS confirmation is very powerful. If unexplained results are obtained, i.e. suspect samples in the cell assay that cannot be confirmed by LC-MS/MS, a second stage of cell based assays is applied which are more directed to the specific mode of action. For diarrheic marine toxins, these techniques are based on gene expression profiles in human Caco-2 cells, and a multielectrode array with neural cells was set-up for neurotoxic compounds. If suspect samples also show a response in these cell assays, and thus suggest presence of an unknown, analytical tools (LC-hrMS) in combination with library searching and/or statistical- and structure elucidation tools are used for identification. This strategy has been applied to over 100 mainly contaminated samples. The majority of the suspect samples could be confirmed by LC-MS/MS and contained clearly elevated levels of marine toxins. Till now only one false negative was observed. The assay also flagged some as suspect that could not be explained by the LC-MS/MS methods. These samples are currently under investigation. Results indicate that the complete proposed strategy, effect based assays combined with novel analytical tools, will be able to bypass the MBA.